About the Authors

Robert E. Reys is Curators' Professor of Mathematics Education at the University of Missouri-Columbia, where he teaches courses for elementary, middle and secondary mathematics teachers. He is a former mathematics teacher and district mathematics coordinator. His research interests are in the areas of calculators, mental computation, estimation, and number sense. Bob has authored over 220 articles in professional journals. He was General Editor for five yearbooks of the National Council of Teachers of Mathematics, and co-edited the 2010 NCTM Yearbook *Mathematics Curriculum: Issues, Trends, and Future Directions*. He was awarded the Lifetime Achievement Award for his service, leadership, and research by the Mathematics Education Trust of the National Council of Teachers of Mathematics.

Mary M. Lindquist is Fuller E. Callaway Professor of Mathematics Education, Emeritus at Columbus State University (Georgia). She taught undergraduate and graduate students in early childhood, middle grades, and secondary mathematics education. Mary was president of the National Council of Teachers of Mathematics and chair of the Commission on the Future of the Standards, the oversight committee for *Principles and Standards of School Mathematics* (NCTM, 2000). She has served or is serving on many national and international committees, including those involved with the National Assessment of Education Progress (NAEP) and the Trends in International Mathematics and Science Study (TIMSS). She was awarded the Lifetime Achievement Award for her service, leadership, and research by the Mathematics Education Trust.

Diana V. Lambdin is Armstrong Professor of Teacher Education and Professor of Mathematics Education at Indiana University in Bloomington, where she teaches courses for prospective elementary teachers, supervises students' field experience, works with masters and doctoral students in mathematics education, and previously served as Associate Dean for Teacher Education. Prior to entering the field of teacher education, she was a mathematics teacher in Massachusetts, Michigan, and Iowa. Diana has been active as an author, editor, project evaluator, and leader in the National Council of Teachers of Mathematics (NCTM). She was a member of the writing team for *Principles and Standards for School Mathematics* (NCTM, 2000) and currently serves as a member of NCTM's Board of Directors.

Nancy L. Smith has been an educator for over 30 years. She taught elementary school and middle school mathematics for ten years in Richmond, Missouri. She is currently Professor in the Department of Early Childhood/Elementary Teacher Education at Emporia State University in Emporia, Kansas, where she teaches elementary mathematics education courses for preservice and inservice teachers, teaches general elementary education courses, and supervises student teachers in the Olathe Professional Development Schools.

10TH EDITION

Helping Children Learn Mathematics

ROBERT REYS
University of Missouri

MARY M. LINDQUIST
Columbus State University

DIANA V. LAMBDIN
Indiana University

NANCY L. SMITH
Emporia State University

WILEY

John Wiley & Sons, Inc.

VICE PRESIDENT & EXECUTIVE PUBLISHER   Jay O'Callaghan
SENIOR ACQUISITIONS EDITOR   Robert Johnston
MARKETING MANAGER   Danielle Hagey
PHOTO EDITOR   Sheena Goldstein
DESIGN DIRECTOR   Harry Nolan
SENIOR DESIGNER   Wendy Lai
EDITORIAL ASSISTANT   Maura Gilligan
SENIOR PRODUCTION EDITOR   Patricia McFadden
SENIOR MEDIA EDITOR   Lynn Pearlman
PRODUCTION MANAGEMENT SERVICES   Ingrao Associates

Cover and Opener Photo Credits John Warden/Photolibrary

This book was set in 10/12 Electrica LH by Prepare and printed and bound by Courier/Kendallville. The cover was printed by Courier/Kendallville.

This book is printed on acid-free paper. ∞

Copyright © 2012 John Wiley & Sons, Inc. 9th edition: 2009, 8th edition: 2007, 7th edition: 2004, 6th edition: 2001, 5th edition: 1998, 4th edition: 1995. All rights reserved. No part of this publication may be reproduced, stored in a retrieval system or transmitted in any form or by any means, electronic, mechanical, photocopying, recording, scanning, or otherwise, except as permitted under Sections 107 or 108 of the 1976 United States Copyright Act, without either the prior written permission of the Publisher, or authorization through payment of the appropriate per-copy fee to the Copyright Clearance Center, Inc., 222 Rosewood Drive, Danvers, MA 01923, (978)750-8400, fax (978)750-4470 or on the web at www.copyright.com. Requests to the Publisher for permission should be addressed to the Permissions Department, John Wiley & Sons, Inc., 111 River Street, Hoboken, NJ 07030-5774, (201)748-6011, fax (201)748-6008, or online at http://www.wiley.com/go/permissions.

Evaluation copies are provided to qualified academics and professionals for review purposes only, for use in their courses during the next academic year. These copies are licensed and may not be sold or transferred to a third party. Upon completion of the review period, please return the evaluation copy to Wiley. Return instructions and a free of charge return shipping label are available at HYPERLINK "http://www.wiley.com/go/return" www.wiley.com/go/returnlabel. Outside of the United States, please contact your local representative.

To order books or for customer service, please call 1-800-CALL WILEY (225-5945).

ISBN-13 9781118001806

Printed in the United States of America

10 9 8 7 6 5 4 3 2 1

Preface

Welcome to the 10th edition of *Helping Children Learn Mathematics*. Earlier versions of *Helping Children Learn Mathematics* have helped several generations of elementary teachers. We are grateful to the many instructors and students who have written us with ideas, stories and suggestions about ways of helping children learn mathematics. This new edition of *Helping Children Learn Mathematics* reflects the ever changing world of learning and teaching elementary school mathematics. Change is everywhere. Teachers of mathematics in elementary school not only experience change but, more important, also have an opportunity to help lead the way.

This edition of *Helping Children Learn Mathematics* is built around three main themes:

- helping children make sense of mathematics
- incorporating practical experiences and research to guide teaching
- emphasizing major ideas from the 2010 *Common Core State Standards: Mathematics*.

This book is intended for those of you who are or who will be teachers of mathematics in elementary school. It is designed to help you help children learn mathematical concepts and skills, as well as important problem-solving strategies. In the process it will challenge your thinking and, we hope, will further stimulate your interest in learning and teaching mathematics.

Helping Children Learn Mathematics is a unique resource that consists of two main parts.

The first part (Chapters 1–6) provides a basis for understanding the changing mathematics curriculum and how children learn mathematics. These chapters offer guidelines for planning, instructing and evaluating. Attention is given to problem solving and assessment, both of which have profound implications for mathematics teaching. Their importance is reflected by the integration of problem solving and assessment throughout the book.

The second part (Chapters 7–18) discusses teaching strategies and techniques, as well as learning activities, related to specific mathematical topics. The emphasis is on making mathematics make sense for our diverse population of students.

FEATURES OF THIS TEXT

We have continued to revise this textbook in order to meet the changing needs of students preparing to become teachers, or teachers who are continuing to learn. This edition continues the rich tradition of this book—with contemporary new ideas interspersed. It has been updated to reflect current recommendations from the *Common Core State Standards: Mathematics*, and from the *National Council of Teachers of Mathematics* and other associations as well as recent research findings relevant to teaching mathematics. At the same time, we have maintained the characteristics and features that have made this book a popular choice of instructors for many years. A useable text for instructors, it is also readable and understandable by students who are being introduced to teaching elementary mathematics. Yet, its depth also makes it appropriate for teachers to use as they continue to learn about teaching.

UPDATED & EXPANDED! CLASSROOM APPLICATIONS

All "Snapshots of a Lesson" now include video clips, easily accessed online. Each chapter-opener Snapshot is an excerpt from an actual mathematics lesson or mathematical situation to show the vital role teachers play in helping students talk and learn about mathematics. Featuring student-teacher and student-student dialogue, they cover a wide range of grades (K to middle school) and mathematical topics. These videos demonstrate many effective classroom practices, and provide a smooth, practical segue into the body of the chapter. You'll see the Snapshots called out with this camera icon:

In addition to strategies woven into the text, the following two sections will help you practice translating theory into practice as you guide and teach children in their study of mathematics.

"In the Classroom" Activities: Found throughout the text in special boxes, these activities provide a wealth of ideas and strategies for helping children explore, learn and in some cases practice some mathematical topics of each chapter. We often include suggestions for using these activities and for reflecting on what happens as children do them. Additional suggestions for using these activities are given

v

in the 4th edition of *Teaching Elementary Mathematics: A Resource for Field Teachers* that is available on-line.

"Things to do" Sections: Found near the end of each chapter, these sections embody our active learning and teaching approach to mathematics. Divided into two parts, they are designed to engage you in inquiring and thinking about mathematics—to offer you experiences and introduce you to investigations that will help you achieve the understanding and insight you need to be a successful teacher. The first part, "From What You've Read," relates directly to this book. The second part, "Going Beyond This Book," offers four types of activities: *In the field, In your journal, With additional resources,* and *With technology*. These include some microteaching activities linked to *Teaching Elementary Mathematics: A Resource for Field Experiences*.

Updated & Expanded!
Topics in Mathematics Education

Technology & Computational Alternatives: Chapter 10 offers a balanced yet detailed discussion of calculators, and computational alternatives—mental computation, estimation, and written techniques, along with the importance of students' making wise choices among these alternatives. Calculators are powerful tools, and we have included calculator activities throughout the book. We have also made a conscious effort to integrate a wide variety of other technological tools, such as spreadsheets, simulation packages, and the Internet, to enhance mathematics learning.

The **Math Links** boxes within the chapters highlight additional technology resources. The Wiley web site (www.wiley.com/college/reys) connects to web resources, including virtual interactive manipulatives, **TeachScape** and, other video clips, lesson plans, and this text's instructor's manual.

Common Core State Standards-Mathematics (CCSS-M). The Common Core State Standard-Mathematics (CCSS-M) have been integrated into throughout *Helping Children Learn Mathematics*. In particular, the CCSS have been specifically integrated into each of the content chapters to highlight particular mathematical topics.

Equity. Equity means treating all students fairly and having equal expectations regarding all students' potential for learning and succeeding in mathematics, regardless of gender or of ethnic and cultural background. Equity is an important issue in mathematics classes and is highlighted in the NCTM's *Principles and Standards for School Mathematics*. Throughout the text, we discuss and illustrate many ways of providing for individual differences—from the wide range of student abilities portrayed in the "Snapshots of a Leson" to the "In the Classroom" activities.

Chapters 2 and 3 offer recommendations and practical suggestions for providing equitable instruction for a variety of learners. New information on Response to Intervention (RTI) has been added to chapter 3. Our book is intended to help you create a classroom environment that is positive and productive for everyone.

Diversity. Today's schools reflect our culturally diverse society, a diversity that is growing rapidly. A "Cultural Connections" section near the end of most chapters highlights differences in the mathematical experiences and mathematical performance of students from different cultures, and suggestions for learning activities. As mathematics teachers, the more we learn about different cultures the more prepared we will be to deal with diversity within our classrooms.

ADDITIONAL FEATURES

Other hallmark features of the text have been enhanced in this edition. These include:

Focus Questions. **Focus questions** are displayed early and point the way into each of the chapters. Think about these questions in advance, and use your readings and the activities in the chapter to help you form a response to each question.

Children's Literature. Although often underused in mathematics education, literature can be a powerful ally in helping children learn. We have made a concerted effort to identify books that you can use effectively to stimulate children's interest in mathematics and to promote mathematics learning. We cite and discuss specific books at various points within the text, and we provide an updated annotated list of useful books in the **Book Nook** section at the end of each chapter (except Chapter 1).

Updated and Streamlined Research. References to important research are found throughout the text. As a teacher, you are often called on to provide a rationale for curricular or instructional decisions, and we think you will find these references to relevant research useful; much of the research we cite has important implications for learning and teaching.

Manipulatives. In addition to the print manipulatives described above, some virtual manipulatives, available on the internet are identified in the **Math Links** found throughout the book.

Also Now Available Online for Instructors and Students: Teaching Elementary Mathematics: A Resource for Field Experiences, 4th Edition

Featuring a wealth of materials to use in observing, interviewing, teaching, and reflecting while students are in schools, this resource is designed for students to use when they participate in field experiences such as practicums, observations, and professional development school (PDS) experiences in K-8 classrooms. It is aligned with and designed to support *Helping Children Learn Mathematics* but may also be used alone. The activities engage preservice elementary teachers in collecting information about the school and school resources, observing and interviewing teachers and children, and doing mathematics with children in the form of games, technology activities, and mini-lessons. Pages from this book may be downloaded and printed for classroom use.

SUPPLEMENTS

In addition to links to videos and virtual interactive manipulatives as described above, the tenth edition of *Helping Children Learn Mathematics* is accompanied by the following instructor and student supplements:

THE WILEY RESOURCE KIT

The Wiley Resource Kit gives students access to premier, password-protected resources hosted by Wiley. Building upon what they learn in their courses, students can use interactive media, practice quizzes, videos and more at their own pace to further enhance mastery of key concepts. The Wiley Resource Kit also provides Respondus® Test Banks for many of Wiley's leading titles that instructors can assign and use for assessment through their campus learning management system. The Wiley Resource Kit and other resources can be accessed via the book companion site at www.wiley.com/college/reys.

ONLINE INSTRUCTOR'S MANUAL

The Instructor's Manual is a useful resource for both veteran and new elementary methods instructors. It is available on the Wiley web site at www.wiley.com/college/reys. Each chapter includes:

- Chapter overview, student objectives, and key vocabulary
- Supplemental lecture ideas, textbook extension ideas, and class and field activity suggestions
- Extensive resource list (both print and media)
- Extensive transparency masters highlighting key textbook ideas, NCTM Standards 2000 summaries, and children's work samples (available in Microsoft Word and PowerPoint formats)
- Provides additional connections to the Common Core State Standards-Mathematics

TEST BANK

Password protected in the instructor's section of www.wiley.com/college/reys, a test bank of over 500 items features both objective and open-ended questions with varying levels of complexity.

Teachscape Video Case Studies: This edition utilizes some videos from **Teachscape** that include (1) research-based best practices in action in the classroom; (2) commentaries by noted researchers that are designed to provide a research-based perspective on the practices illustrated; and (3) teacher reflections to promote better understanding of the featured teacher's instructional decisions. Some of these case studies are referenced, when appropriate, in the text; you will find others as you view these videos.

ABOUT THIS BOOK

Helping Children Learn Mathematics is an idea book. We believe that you will learn much from reading it and from talking about what you have read.

It is not possible—or desirable—to specify the exact steps to follow in teaching mathematics. Too much depends on what is being taught, to whom, and at what levels. In your classroom, it is you who will ultimately decide what to teach, to whom to teach it, how to teach it, and the amount of time to spend. This book will not answer all of these questions for you, but we think you will find it very helpful in making wise decisions as you guide elementary school students in their learning of mathematics. We believe this book will be a valuable teaching resource that you can use again and again in your classroom long after the course has been completed.

ACKNOWLEDGMENTS

We thank Marilyn Suydam for her hard work and insights during the early years of this text's development. Marilyn is now retired, but her legacy of significant contributions remains a vital part of this book. For this 10th edition, we are again grateful to Rick Callan—a veteran Indiana classroom teacher, expert and author on children's literature and mathematics, and presidential awardee-for providing additional suggestions for Book Nook updates. Also thanks to Robert Glasgow, Southwest Baptist University and Amanda Thomas for their contributions to this edition. Instructor's Test Bank Revision: John C. Yang, *Lakeland College*, Student Pre/Post Lecture Quizzes: William O. Lacefield, III, *Mercer University*, Instructor Manual, Sandi Cooper, *Baylor University*,Power Point Slide Revision; Yvelyne Germain-McCarthy, The University of New Orleans, Instructor Guide to Teachscape: Jane Strawhecker, The University of Nebraska, Kearney Student Guide to Teachscape Content: Barbara Ridener, *Florida Atlantic University*.

Our thanks to a number of people who provided thoughtful reviews and offered suggestions that have led to this edition. Specifically, we want to thank:

KIMBERLY ARP, Cabrini College
DELTA CAVNER, Southwest Baptist University
SANDI COOPER, Baylor University
MARTHA EGGERS, McKendree University
EDITH HAYS, Texas Woman's University
MICHELE KOOMEN, Gustavus Adolphus College
WILLIAM LACEFIELD, Mercer University
DAVID MARTIN, Florida Atlantic University

Sarah Murray, Centre College
Jason Silverman, Drexel University
Tina Sloan, Athens State University
Jane Strawhecker, University of Nebraska at Kearney
Elsa Villa, The University of Texas at El Paso
Beth Vinson, Athens State University
John Yang, Lakeland College

We also wish to acknowledge the many colleagues, friends, and students who have contributed in various ways to the development of this book over the years. In particular, we thank Barbara Reys, of the University of Missouri, Frank Lester, of Indiana University, and Paul Lindquist for their help and support.

We also wish to recognize the help of many reviewers and contributors of ideas and suggestions for prior editions, including:

Roda Amaria, Salem State College
Peter Appelbaum, Arcadia University
Tom Bassarear, Keene State College
Jennifer Bay-williams, Kansas State University
Martha Boedecker, Northwestern Oklahoma State University
Carol Bonilla Bowman, Ramapo College
Daniel Brahier, Bowling Green State University
Christine Browning, Western Michigan University
Lecretia Buckley, Purdue University
Grace Burton, University of North Carolina, Wilmington
Rick Callan, Franklin College
Richard Caulfield, Indiana University
Astrida Cirulis, National-Louis University
Bob Drake, University of Cincinnati
Dianne Erickson, Oregon State University
Skip Fennell, McDaniel College
Marvel Froemming, Moorhead State University
Jeff Frykholm, University of Colorado
K. Gaddis, Lewis & Clark College
Lowell Gadberry, Southwestern Oklahoma State University
Enrique Galindo, Indiana University
Madeleine Gregg, University of Alabama
Elsa L. Geskus, Kutztown
Claire Graham, Framingham State College
Andrea Guillaume, *California State University*
Janet Handler, Mount Mercy College
Kim Harris, University of North Carolina Charlotte
Kim Hartweg, Western Illinois University

Ruth M. Heaton, University of Nebraska, Lincoln
Karen Higgins, Oregon State University
Ellen Hines, Northern Illinois University
Robert Jackson, University of Minnesota
Gae Johnson, Northern Arizona University
Susan Johnson, Northwestern College
Todd Johnson, Eastern Washington University;
Mary Kabiri, Lincoln University
Henry S. Kepner, University of Wisconsin-Milwaukee
Diane H. Klein, Indiana University of Pennsylvania
Rick Kruschinsky, University of St. Thomas
Vena Long, University of Tennessee
Margie Mason, College of William and Mary
Robert Matulis, Millersville University
Sueanne Mckinney, Old Dominion University
William Merrill, Central Michigan University
Alice Mills, Quincy University
Jean Mitchell, California State University, Monterey Bay
Eula Ewing Monroe, Brigham Young University
Margaret (Maggie) Niess, Oregon State University
Jamar Pickreign, Rhode Island College
Don Ploger, Florida Atlantic University
Sara Powell, University of Charleston, (SC)
Frank Powers, University of Idaho
Jacelyn Marie Rees, McNeese State University
Gay Ragan, Southwest Missouri State University
Denise M. Reboli, King's College (PA)
Andy Reeves, University of South Florida-Tampa
Candice L. Ridlon, Towson University
Tom Romberg, University of Wisconsin
Thomas E. Rowan, University of Maryland, College Park
Mary Ellen Schmidt, Ohio State University-Mansfield
Linda Sheeran, Oklahoma State University
Marian Smith, Florida A&M University
Marilyn Soucie, University of Missouri
Frances Stern, New York University
David L. Stout, University of West Florida
Gertrude R. Toher, Hofstra University
Frederick L. Uy, California State University, Los Angeles
Juan Vazquez, Missouri Southern State College
Kay Wall, University of Central Oklahoma;
Pat Wall, Northern Arizona University
Judy Wells, Indiana State University
Tad Watanabe, Kennesaw State University
Dorothy Y. White, University of Georgia
Margaret Wyckoff, University of Maine-Farmington
Bernard Yvon, University of Maine

Contents

Preface, v
Acknowledgments, vii

CHAPTER 1
School Mathematics in a Changing World, 1
- SNAPSHOT OF A LESSON, 1
- FOCUS QUESTIONS, 2
- INTRODUCTION, 2

What Is Mathematics?, 2
What Determines the Mathematics Being Taught?, 3
- Historical Influences, 3
- Recent influences, 5

Where Can You Turn?, 7
- National Guidelines for School Mathematics, 7
- State and Local Guidelines, 8
- Research, 8
- Cultural and International Resources, 9
- Text Books and Other Materials, 9
- Electronic Materials, 9
- Professional Organizations, 10
- Professional Development, 10
- Other Teachers, 10

What Is Your Role Now?, 10
A Glance at Where We've Been, 10
- THINGS TO DO: FROM WHAT YOU'VE READ, 11
- THINGS TO DO: GOING BEYOND THIS BOOK, 11

CHAPTER 2
Helping All Children Learn Mathematics with Understanding, 12
- SNAPSHOT OF A LESSON, 12
- FOCUS QUESTIONS, 13
- INTRODUCTION, 13

What Do We Know About Learning Mathematics?, 13
How Can We Support the Diverse Learners in Our Classroom?, 14
- Creating a Positive Learning Environment, 14
- Avoiding Negative Experiences that Increase Anxiety, 15
- Establishing Clear Expectations, 16
- Treating All Students as Equally Likely to Have Aptitude for Mathematics, 17
- Helping Students Retain Mathematical Knowledge and Skills, 17

Helping Children Acquire Both Procedural and Conceptual Knowledge, 18
How Do Children Learn Mathematics?, 19
- Building Behavior, 20
- Constructing Understanding, 21

How Can We Help Children Make Sense of Mathematics?, 23
- Recommendation 1: Teach to the Developmental Characteristics of Students, 23
- Recommendation 2: Actively Involve Students, 25
- Recommendation 3: Move Learning from Concrete to Abstract, 27
- Recommendation 4: Use Communication to Encourage Understanding, 28

Cultural Connections, 29
A Glance at Where We've Been, 29
- THINGS TO DO: FROM WHAT YOU'VE READ, 30
- THINGS TO DO: GOING BEYOND THIS BOOK, 30
- BOOK NOOK FOR CHILDREN, 31

CHAPTER 3
Planning for and Teaching Diverse Learners, 32
- SNAPSHOT OF A LESSON, 32
- FOCUS QUESTIONS, 33
- INTRODUCTION, 33

Preparing to Teach: Questions to Ask, 34
- Do I Understand the Mathematics I am Teaching?, 34
- Where are My Students Developmentally?, 34
- What Do My Students Know?, 35
- What Kinds of Tasks Will I Give My Students?, 36
- How Will I Encourage My Students to Communicate?, 37
- What Materials Will We Use?, 39

Planning for Effective Teaching, 44
- Levels of Planning, 45
- Planning Different Types of Lessons, 46
- Meeting the Needs of All Students, 52
- Assessment and Analysis in Planning, 56

Cultural Connections, 56

A Glance at Where We've Been, 58
- THINGS TO DO: FROM WHAT YOU'VE READ, 59
- THINGS TO DO: GOING BEYOND THIS BOOK, 59
- BOOK NOOK FOR CHILDREN, 60

CHAPTER 4
Assessment: Enhancing Learning and Teaching, 61
- SNAPSHOT OF A LESSON, 61
- FOCUS QUESTIONS, 63
- INTRODUCTION, 63

Assessment *for* Learning: Formative Assessment, 63
- Phases of Assessment, 63
- Making Instructional Decisions, 65
- Monitoring Student Progress, 65
- Evaluating Student Achievement, 66

WAYS TO ASSESS STUDENTS' ABILITIES AND DISPOSITIONS, 66
- Observation, 68
- Questioning, 68
- Interviewing, 69
- Performance Tasks, 70
- Self-Assessment and Peer Assessment, 71
- Work Samples, 74
- Portfolios, 74
- Writing, 74
- Teacher-Designed Written Tests, 76
- Standardized Achievement Tests, 77

Keeping Records and Communicating About Assessments, 79
- Recording the Information, 79
- Communicating the Information, 81

Cultural Connections, 82

A Glance at Where We've Been, 83
- THINGS TO DO: FROM WHAT YOU'VE READ, 84
- THINGS TO DO: GOING BEYOND THIS BOOK, 84
- BOOK NOOK FOR CHILDREN, 85

CHAPTER 5
Mathematical Processes and Practices, 86
- SNAPSHOT OF A LESSON, 86
- FOCUS QUESTIONS, 87
- INTRODUCTION, 88

Problem Solving, 89

Reasoning and Proof, 92
- Reasoning Is About Making Generalizations, 94
- Reasoning Leads to a Web of Generalizations, 94
- Reasoning Leads to Mathematical Memory Built on Relationships, 94
- Learning Through Reasoning Requires Making Mistakes and Learning from Them, 95

Communication, 95

Connections, 97

Representations, 98
- Creating and Using Representations, 99
- Selecting, Applying, and Translating Among Representations, 99
- Using Representations to Model and Interpret Phenomena, 100

Cultural Connections, 101

A Glance at Where We've Been, 101
- THINGS TO DO: FROM WHAT YOU'VE READ, 103
- THINGS TO DO: GOING BEYOND THIS BOOK, 103
- BOOK NOOK FOR CHILDREN, 104

CHAPTER 6
Helping Children with Problem Solving, 105
- SNAPSHOT OF A LESSON, 105
- FOCUS QUESTIONS, 106
- INTRODUCTION, 106

What Is a Problem and What Is Problem Solving?, 107

Teaching Mathematics Through Problem Solving, 109
- Factors for Success in Problem Solving, 110
- Choosing Appropriate Problems, 111
- Finding Problems, 114
- Having Students Pose Problems, 115
- Using Calculators and Computers, 116

Strategies for Problem Solving, 117
- Act it Out, 118
- Make a Drawing or Diagram, 119
- Look for a Pattern, 119
- Construct a Table, 120
- Guess and Check, 121
- Work Backward, 122
- Solve a Similar But Simpler Problem, 122

The Importance of Looking Back, 124
- Looking Back at the Problem, 124
- Looking Back at the Answer, 124
- Looking Back at the Solution Process, 124
- Looking Back at One's Own Thinking, 125

Helping All Students with Problem Solving, 125
- Managing Time, 125
- Managing Classroom Routines, 125
- Managing Student Needs, 126

Cultural Connections, 127

A Glance at Where We've Been, 128
- THINGS TO DO: FROM WHAT YOU'VE READ, 128
- THINGS TO DO: GOING BEYOND THIS BOOK, 128
- BOOK NOOK FOR CHILDREN, 129

CHAPTER 7
Developing Counting and Number Sense in Early Grades, 130
- SNAPSHOT OF A LESSON, 130
- FOCUS QUESTIONS, 131

Number Sense, 131

Prenumber Concepts, 133
- Classification, 133
- Patterns, 136

Early Number Development, 137
- Conservation, 137
- Group Recognition, 137
- Comparisons and One-To-One Correspondence, 138

Counting, 139
- Counting Principles, 140
- Counting Stages, 141
- Counting Strategies, 142
- Counting Practice, 145
- Developing Number Benchmarks, 145
- Making Connections, 146

Cardinal, Ordinal, and Nominal Numbers, 148

Writing Numerals, 149

Cultural Connections, 150

A Glance at Where We've Been, 151
- THINGS TO DO: FROM WHAT YOU'VE READ, 152
- THINGS TO DO: GOING BEYOND THIS BOOK, 152
- BOOK NOOK FOR CHILDREN, 153

CHAPTER 8
Extending Number Sense: Place Value, 154
- SNAPSHOT OF A LESSON, 154
- FOCUS QUESTIONS, 155
- INTRODUCTION, 156

Our Numeration System, 157

Nature of Place Value, 157
- Modeling—Ungrouped and Pregrouped, 158
- Modeling—Proportional and Nonproportional, 158
- Grouping or Trading, 159

Beginning Place Value, 160
- A Place to Start, 160

Extending Place Value, 165
- Counting and Patterns, 166
- Composing and Decomposing, 167

Reading and Writing Numbers, 168

Rounding, 171

Cultural Connections, 173

A Glance at Where We've Been, 174
- THINGS TO DO: FROM WHAT YOU'VE READ, 175
- THINGS TO DO: GOING BEYOND THIS BOOK, 175
- BOOK NOOK FOR CHILDREN, 176

CHAPTER 9
Operations: Meanings and Basic Facts, 177
- SNAPSHOT OF A LESSON, 177
- FOCUS QUESTIONS, 178
- INTRODUCTION, 178

Helping Children Develop Number Sense and Computational Fluency, 180
- Facility with Counting, 180
- Experience with a Variety of Concrete Situations, 180
- Familiarity with Many Problem Contexts, 181
- Experience in Talking and Writing About Mathematical Ideas, 181

Developing Meanings for the Operations, 181
- Addition and Subtraction, 182
- Multiplication and Division, 183

Mathematical Properties, 185

Overview of Basic Fact Instruction, 185
- Start Where the Children Are, 187
- Build Understanding of the Basic Facts, 188
- Focus on How to Remember Facts, 189

Thinking Strategies For Basic Facts, 191
- Thinking Strategies for Addition Facts, 191
- Thinking Strategies for Subtraction Facts, 195
- Thinking Strategies for Multiplication Facts, 196
- Thinking Strategies for Division Facts, 199

Cultural Connections, 201

A Glance at Where We've Been, 202
- THINGS TO DO: FROM WHAT YOU'VE READ, 203
- THINGS TO DO: GOING BEYOND THIS BOOK, 203
- BOOK NOOK FOR CHILDREN, 204

CHAPTER 10
Computation Methods: Calculators, Mental Computation, and Estimation, 205
- SNAPSHOT OF A LESSON, 205
- FOCUS QUESTIONS, 207
- INTRODUCTION, 207

Balancing Your Instruction, 208

Calculators, 209
- Using Calculators Requires Thinking, 210
- Using Calculators Can Raise Student Achievement, 210
- Calculators are Not Always the Fastest Way of Doing Computations, 210
- Calculators are Useful for More than Just Doing Computations, 210

Mental Computation, 211
- Strategies Using Compatible Numbers and Decomposition, 211
- Encouraging Mental Computation, 213

Estimation, 216
- Background for Estimating, 216
- Front-End Estimation, 217
- Adjusting, 217

Compatible Numbers, 218
Flexible Rounding, 219
Clustering, 220
Choosing Estimation Strategies, 221

Cultural Connections, 222

A Glance at Where We've Been, 222
- THINGS TO DO: FROM WHAT YOU'VE READ, 223
- THINGS TO DO: GOING BEYOND THIS BOOK, 224
- BOOK NOOK FOR CHILDREN, 225

CHAPTER 11
Standard and Alternative Computational Algorithms, 226
- SNAPSHOT OF A LESSON, 226
- FOCUS QUESTIONS, 228
- INTRODUCTION, 228

Teaching Algorithms with Understanding, 231
Using Materials, 231
Using Place Value, 231

Addition, 232
Standard Addition Algorithm, 233
Partial-Sum Addition Algorithm, 235
Higher-Decade Addition, 236

Subtraction, 237
Standard Subtraction Algorithm, 237
Partial-Difference Subtraction Algorithm, 239

Multiplication, 240
Multiplication with One-Digit Multipliers, 240
Multiplication by 10 and Multiples of 10, 242
Multiplication with Zeros, 242
Multiplication with Two-Digit Multipliers, 242
Multiplication with Large Numbers, 243

Division, 243
Division with One-Digit Divisors, 244
Subtractive Algorithm, 245
Division with Two-Digit Divisors, 247
Making Sense of Division with Remainders, 248

Checking, 249

Choosing Appropriate Methods, 249

Building Computational Proficiency, 249

Cultural Connections, 249

A Glance at Where We've Been, 251
- THINGS TO DO: FROM WHAT YOU'VE READ, 251
- THINGS TO DO: GOING BEYOND THIS BOOK, 252
- BOOK NOOK FOR CHILDREN, 253

CHAPTER 12
Fractions and Decimals: Concepts and Operations, 254
- SNAPSHOT OF A LESSON, 254
- FOCUS QUESTIONS, 255
- INTRODUCTION, 255

CONCEPTUAL DEVELOPMENT OF FRACTIONS, 255
Three Meanings of Fractions, 256
Models of the Part–Whole Meaning, 257
Making Sense of Fractions, 258
Ordering Fractions and Equivalent Fractions, 263
Mixed Numbers and Improper Fractions, 267

Operations with Fractions, 267
Addition and Subtraction, 268
Multiplication and Division, 269

Conceptual Development of Decimals, 272
Relationship to Common Fractions, 273
Relationship to Place Value, 275
Ordering and Rounding Decimals, 275

Operations with Decimals, 276
Addition and Subtraction, 277
Multiplication and Division, 277

Cultural Connections, 278

A Glance at Where We've Been, 279
- THINGS TO DO: FROM WHAT YOU'VE READ, 280
- THINGS TO DO: GOING BEYOND THIS BOOK, 280
- BOOK NOOK FOR CHILDREN, 281

CHAPTER 13
Ratio, Proportion, and Percent: Meanings and Applications, 282
- SNAPSHOT OF A LESSON, 282
- FOCUS QUESTIONS, 283
- INTRODUCTION, 283

Ratios, 284

Proportions, 287

Percents, 291
Understanding Percents, 292
Applying Percents, 294

Cultural Connections, 296

A Glance At Where We've Been, 297
- THINGS TO DO: FROM WHAT YOU'VE READ, 297
- THINGS TO DO: GOING BEYOND THIS BOOK, 297
- BOOK NOOK FOR CHILDREN, 299

CHAPTER 14
Algebraic Thinking, 300
- SNAPSHOT OF A LESSON, 300
- FOCUS QUESTIONS, 301
- INTRODUCTION, 301

Problems, Patterns, and Relations, 302
Problems, 302
Patterns, 302
Relations, 305

Language and Symbols of Algebra, 307
Equality and Inequality, 308
Variables, 309
Expressions and Equations, 309

Representing, Generalizing, and Justifying, 310
  Routine Problems, 310
  Patterns, 312
  Nonroutine Problems, 314
  Relations: Functions, 316
  Relations: Properties of Numbers, 317
  Another Look at Representing, Generalizing, and Justifying, 318
Cultural Connections, 319
A Glance at Where We've Been, 320
  THINGS TO DO: FROM WHAT YOU'VE READ, 320
  THINGS TO DO: GOING BEYOND THIS BOOK, 320
  BOOK NOOK FOR CHILDREN, 321

CHAPTER 15
Geometry, 322
SNAPSHOT OF A LESSON, 322
FOCUS QUESTIONS, 323
INTRODUCTION, 323

Shapes, 324
  Three-Dimensional Shapes, 325
  Two-Dimensional Shapes, 329
Space, 338
Transformations, 340
Visualization and Spatial Reasoning, 342
  Using Geometric Physical and Pictorial Materials, 342
  Using Mental Images, 343
Cultural Connections, 344
A Glance at Where We've Been, 344
  THINGS TO DO: FROM WHAT YOU'VE READ, 345
  THINGS TO DO: GOING BEYOND THIS BOOK, 345
  BOOK NOOK FOR CHILDREN, 346

CHAPTER 16
Measurement, 347
SNAPSHOT OF A LESSON, 347
FOCUS QUESTIONS, 348
INTRODUCTION, 348

The Measurement Process, 349
  I. Identifying Attributes by Comparing, 350
  II. Choosing a Unit, 355
  III. and IV. Comparing an Object to a Unit and Finding the Number of Units, 357
  V. Reporting Measurements, 362
  Creating Objects Given the Measurement, 363
Comparing Measurements, 363
  Equivalences, 363
  Conversions, 364
Estimating Measurements, 365
Connecting Attributes, 367
  Area and Shape, 367
  Volume and Shape, 367
  Perimeter and Area, 367
  Volume and Surface Area, 368
Cultural Connections, 368
A Glance at Where We've Been, 369
  THINGS TO DO: FROM WHAT YOU'VE READ, 370
  THINGS TO DO: GOING BEYOND THIS BOOK, 370
  BOOK NOOK FOR CHILDREN, 371

CHAPTER 17
Data Analysis, Statistics, and Probability, 372
SNAPSHOT OF A LESSON, 372
FOCUS QUESTIONS, 373
INTRODUCTION, 373

Formulating Questions, 375
Collecting Data, 375
  Surveys, 376
  Experiments, 376
  Simulations, 376
Analyzing Data: Graphical Organization, 377
  Quick and Easy Graphing Methods, 377
  Plots, 378
  Picture Graphs, 380
  Bar Graphs and Histograms, 380
  Pie Graphs, 381
  Line Graphs, 382
  Graphical Roundup, 382
Analyzing Data: Descriptive Statistics, 383
  Measures of Central Tendency or Averages, 384
  Measures of Variation, 389
INTERPRETING RESULTS, 389
  Data Sense, 389
  Misleading Graphs, 391
  Communicating Results, 392
PROBABILITY, 392
  Probability of an Event, 392
Randomness, 395
  Independence of Events, 396
  Misconceptions About Probability, 397
Cultural Connections, 397
A Glance at Where We've Been, 398
  THINGS TO DO: FROM WHAT YOU'VE READ, 398
  THINGS TO DO: GOING BEYOND THIS BOOK, 398
  BOOK NOOK FOR CHILDREN, 400

CHAPTER 18
Number Theory, 401
SNAPSHOT OF A LESSON, 401
FOCUS QUESTIONS, 401
INTRODUCTION, 402
  Why Study Number Theory, 402

Number Theory in Elementary School Mathematics, 404
  Odds and Evens, 404

Factors and Multiples, 405
Primes and Composites, 407
Prime Factorization, 408

Divisibility, 410

Other Number Theory Topics, 412
Relatively Prime Pairs of Number, 412
Polygonal Numbers, 412
Modular Arithmetic, 413
Pascal's Triangle, 414
Pythagorean Triples, 414
Fibonacci Sequence, 415

Cultural Connections, 416

A Glance at Where We've Been, 417
THINGS TO DO: FROM WHAT YOU'VE READ, 417
THINGS TO DO: GOING BEYOND THE BOOK, 417
BOOK NOOK FOR CHILDREN, 418

References, 419

Appendix, 431

Index, I-1

CHAPTER 1

School Mathematics in a Changing World

> "THE SCHOOLS AIN'T WHAT THEY USED TO BE AND PROBABLY NEVER WERE."
> — Will Rogers

› SNAPSHOT OF A LESSON

KEY IDEAS

1. Has the teaching of mathematics really changed?
2. Mathematics should make sense.

BACKGROUND

This snapshot conveys a story of a classroom scene based on curriculum and classroom practices from the middle 1800s. Since there is no video to accompany it, try to picture the scene. Mr. Telford, the teacher, works with the fifth- and sixth-grade students in a one-room school. Luke was absent yesterday, so Mr. Telford has asked Esther to teach Luke about ten complements. John is finishing his lesson on adding and subtracting.

Mr. Telford: Luke, Esther, and John; we are ready to begin the arithmetic lesson for today. I am going to show you a way to multiply numbers using ten-complements. Luke, what is the ten complement?

Luke: Esther said the ten-complement of 8 is 2 and the ten-complement of 9 is 1. Together they make 10.

Mr. Telford: Copy these three steps [he shares these from a book] in your cipher book while I help Mary and Paul with their spelling.

Step 1. Write the ten-complement of the multiplicand and the multiplier to the right of each number.

Step 2. Multiply the two complements and write this product under the two complements.

Step 3. Subtract the multiplier's complement from the multiplicand. Write the difference to the left of the product of the two complements.

The three students write these steps in their cipher books with pen and ink.

Mr. Telford: Now let's use these steps to find 7×8.
He writes on a large slate and the three students copy in their cipher books.

$$\begin{array}{r} 8 \\ \times 7 \\ \hline \end{array}$$

Mr. Telford: Now lets use step 1. 8 is the multiplicand so I will write 2 to its right and 7 is the multiplier, so I will write 3 to its right.

Mr. Telford: Now for step 2. Multiply the two ten-complements. 2×3 is 6.

Mr. Telford: Read Step 3, John. So we subtract 3 from 8 and write 5 to the left of the 6.

```
  Step 1      Step 2      Step 3
    8 2         8 2         8 2
  r × 7 3     × 7 3       × 7 3
                  6         5 6      So 7 × 8 is 56.
```

Mr. Telford: Let's do 6×9.
He goes through this example:

```
    9 1
  × 6 4
    5 4      So 6 × 9 is 54.
```

He assigns the following problems: 7×6, 6×8, 8×8, 7×7, 9×9, and 6×6.

He clarifies the steps for Luke who was having trouble and leaves them to work with the other students.

1

FOCUS QUESTIONS

1. What is your view of mathematics?
2. What determines the mathematics currently taught?
3. What resources are available to help you continue developing your knowledge of mathematics and the learning and teaching of mathematics?

INTRODUCTION

As Will Rogers said, schools are not what they used to be, but our romantic view of the past is often flawed. Teaching mathematics often was similar to the way that Mr. Telford was showing how to multiply 7×8 in the Snapshot: "Do it this way because the teacher says this is the way." If you tried to use these steps without understanding, you would conclude that $6 \times 6 = 216$.

You have the opportunity to provide a positive and meaningful experience for those you help to learn mathematics. What is your vision of the mathematics you will be teaching? What is your vision of the classroom? Many of you will remember your experiences in elementary school. Some of you will remember memorizing multiplication tables, operating with fractions, or doing long division. Others may remember exploring patterns, doing geometry projects, or solving problems. Mathematics in elementary school may have been a positive experience for some of you, but for others it was filled with anxiety and frustration. Why were you learning mathematics and when would you use it? Teachers want students to learn mathematics and to learn that it is a useful subject. How can they make sure this happens?

This book is designed to expand your vision of mathematics teaching and learning and to help you help students learn mathematics. The book interweaves three main themes:

- *Theme 1. Best Practices and Research.* These best practices and research provide a basis for you to understand what mathematics children are expected to learn and how children learn mathematics.
- *Theme 2. Sense Making.* Mathematics must make sense to children. If children make sense of the mathematics they are learning, they can build on this understanding to learn more mathematics and use the mathematics to solve problems.
- *Theme 3. Practical Experiences.* Learning to teach mathematics requires experience. This theme is explicated by the many suggestions and ideas from teachers and our own experiences for you to use now and later in the classroom.

Learning to teach is a lifelong journey. During that journey, you will often ask questions such as these:

- What mathematical knowledge and understanding does each student bring to the class?
- What mathematics do students need to learn?
- How can I teach each unique child so that he or she will learn?
- How important is my own attitude toward mathematics?

Your answers to these questions will influence what you do when you are teaching. No matter what the age of the children you teach, we recommend three *general* goals:

- To help children make sense of specific mathematical content, including both procedures and concepts
- To help children learn how to apply mathematical ideas to solve problems
- To foster positive dispositions, such as persistence, flexibility, willingness to learn, and valuing mathematics

Developing ways to help you reach these three goals is considered in later chapters of this book. This first chapter focuses on what mathematics is and what determines the mathematics that is taught in schools. We also share where to turn for additional suggestions and help.

WHAT IS MATHEMATICS?

The view of mathematics in elementary school has changed from being mainly about numbers, especially computation, to a broader view. Figure 1-1 shows how the National Council of Teachers of Mathematics (NCTM) envisioned the distribution of mathematics content from

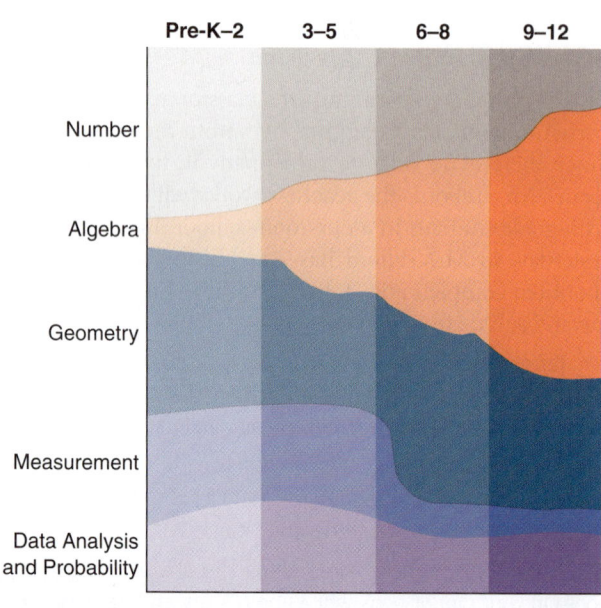

Figure 1-1 *Vision of the mathematics content distribution through the grade levels.* (Source: *Principles and Standards for School Mathematics*, p. 30, copyright 2000 by the National Council of Teachers of Mathematics. All rights reserved.)

prekindergarten through grade 12 in 2000. Numbers are an essential part of this vision, but algebra, geometry, measurement, and data analysis and probability are also important. As you can see, numbers, as well as geometry, are emphasized early and then receive progressively less instructional attention as algebra increases in importance. This relative emphasis on each of these five strands will continue to change, but together they constitute school mathematics.

Although we can consider mathematics as a collection of separate strands such as geometry and algebra, this may not be the best way of looking at it. It may be helpful to broaden your view of mathematics. Five views are presented to help you think of mathematics as being more than a collection of strands. Before reading these different views, take a minute to look at the quilt on the cover of this book. Do you see patterns and relations among the different parts of the quilt? Do you see the thinking underlying the quilt—the careful placement of the building blocks of various hues? Do you see the quilt as a piece of art? Without any language, what does the quilt communicate to you? Do you see the quilt as a tool—something that would keep you warm in the winter? Similarly, you can look at mathematics in these ways:

1. *Mathematics is a study of patterns and relationships.* Mathematics is filled with patterns and relationships providing threads that unify the curriculum. Children should come to see how one idea is like another. For example, children in first grade can see how one basic fact (say, 3 + 2 = 5) is related to another basic fact (say, 5 − 3 = 2). Older children can relate measuring to the nearest centimeter to rounding to the nearest hundred.

2. *Mathematics is a way of thinking.* Mathematics provides people with strategies for organizing, analyzing, and synthesizing information. Often symbolizing a real-life problem reduces it to a well-known mathematical procedure, making the problem easier to solve.

3. *Mathematics is an art, characterized by order and internal consistency.* Many children think of mathematics as a confusing set of discrete facts and skills that must be memorized. Children need guidance to recognize and appreciate the underlying orderliness and consistency to understand and use mathematics.

4. *Mathematics is a language that uses carefully defined terms and symbols.* Learning these terms and symbols enhances our ability to communicate about science, real-life situations, and mathematics itself. As with any language, you need to understand the meaning of these words and when it is appropriate to use them.

5. *Mathematics is a tool.* Mathematics has become an essential part of our world, both in everyday life and in the workplace. Children appreciate why they are learning mathematics if they know it is useful.

WHAT DETERMINES THE MATHEMATICS BEING TAUGHT?

Mathematics plays a prominent role in the elementary school program. It is second only to reading in the amount of time devoted to it and in the amount of money spent for curricular materials. Its importance is reflected in the degree of concern about school mathematics voiced by parents, politicians, and other social groups.

HISTORICAL INFLUENCES

Three factors—the needs of the subject, the child, and the society—have influenced what mathematics is to be taught in schools. Although many people think that "math is math" and never changes, a brief discussion of these factors paints a different picture of a subject that is ever changing.

NEEDS OF THE SUBJECT The nature of mathematics helps determine what is taught and when it is taught in elementary grades. For example, number work begins with whole numbers, then fractions and decimals. Length is studied before area. Such seemingly natural sequences are the result of long years of curricular evolution. This process has involved much analysis of what constitutes a progression from easy to difficult, based in part on what is deemed necessary at one level for the development of ideas at later levels. Once a curriculum is in place for a long time, however, people tend to consider it the only proper sequence. Thus, omitting a topic or changing the sequence of topics often involves a struggle for acceptance. However, research shows that all students do not always learn in the sequence that has been ingrained in our curriculum.

Sometimes the process of change is the result of an event, such as when the Soviet Union sent the first *Sputnik* into orbit. The shock of this evidence of another country's technological superiority sped curriculum change in the United States. The "new math" of the 1950s and 1960s was the result, and millions of dollars were channeled into mathematics and science education to strengthen school programs. Mathematicians became integrally involved. Because of their interests and the perceived weaknesses of previous curricula, they developed curricula based on the needs of the subject. The emphasis shifted from social usefulness to such unifying themes as the structure of mathematics, operations and their inverses, systems of notation, properties of numbers, and set language. New content was added at the elementary school level, and other topics were introduced at earlier grade levels.

Mathematics continues to change; new mathematics is created, and new uses of mathematics are discovered. As part of this change, technology has made some mathematics obsolete and has opened the door for other mathematics to be accessible to students. Think about all the mathematics you learned in elementary school. How much of this can be done on a simple calculator? What mathematics is now important because of the technology available today?

NEEDS OF THE CHILD The mathematics curriculum has been influenced by beliefs about how children learn and, ultimately, about how they should be taught. Before the early years of the twentieth century, mathematics was taught to train "mental faculties" or provide "mental discipline." Struggling with mathematical procedures was thought to exercise the mind (like muscles are exercised), helping children's brains work more effectively. Around the turn of the twentieth century, "mental discipline" was replaced by *connectionism*, the belief that learning established bonds, or connections, between a stimulus and responses. This led teachers to the endless use of drills aimed at establishing important mathematical connections.

In the 1920s, the Progressive movement advocated *incidental learning*, reflecting the belief that children would learn as much arithmetic as they needed and would learn it better if it was not systematically taught. The teacher's role was to take advantage of situations as they occurred to teach arithmetic, as well as to create situations in which arithmetic would arise.

During the late 1920s, the Committee of Seven, a committee of school superintendents and principals from midwestern cities, surveyed pupils to find out when they mastered various topics (Washburne, 1931). Based on that survey, the committee recommended teaching mathematics topics according to students' mental age. For example, subtraction facts under 10 were to be taught to children with a mental age of 6 years 7 months, and facts over 10 at 7 years 8 months; subtraction with borrowing or carrying was to be taught at 8 years 9 months. The recommendations of the Committee of Seven had a strong impact on the sequencing of the curriculum for years afterward.

Another change in thinking occurred in the mid-1930s, under the influence of *field theory*, or *Gestalt theory*. With William A. Brownell (1954, reprinted 2006) as a prominent spokesperson, this approach placed greater emphasis on a planned program to encourage the development of insight and the understanding of relationships, structures, patterns, interpretations, and principles. It contributed to an increased focus on learning as a process that led to *meaning and understanding*. The value of drill was acknowledged, but it was given less importance than understanding; drill was no longer the major means of providing instruction.

The relative importance of drill and understanding is still debated today. In this debate, people often treat understanding and learning skills as if they are opposites, but this is not the case. Clearly, drill is necessary to build speed and accuracy and to make skills automatic. But equally clearly, you need to know *why* as well as *how*. Both skills and understanding must be developed, and they can be developed together.

Changes in the field of psychology have continued to affect education. During the second half of the twentieth century, educators came to understand that the developmental level of the child is a major factor in determining the sequence of the curriculum. Topics cannot be taught until children are developmentally ready to learn them. Or, from another point of view, topics must be taught in such a way that children at a given developmental level are ready to learn them.

Research has provided increasing evidence that children construct their own knowledge. In so doing, they make sense of the mathematics and feel that they can tackle new problems. Thus, helping children learn mathematics means being aware of how children have constructed mathematics from their experiences both in and out of school.

Read more about the influence of psychological theories, in the chapter by Lambdin and Walcott (2007).

NEEDS OF SOCIETY The usefulness of mathematics in everyday life and in many vocations has also affected what is taught and when it is taught. In early America, mathematics was considered necessary primarily for clerks and bookkeepers. The curriculum was limited to counting; the simpler procedures for addition, subtraction, and multiplication; and some facts about measures and fractions. By the late nineteenth century, business and commerce had advanced to the point that mathematics was considered important for everyone. The arithmetic curriculum expanded to include such topics as percentage, ratio and proportion, powers, roots, and series.

This emphasis on *social utility*, on teaching what was needed for use in occupations, continued into the twentieth century. One of the most vocal advocates of social utility was Guy Wilson. He and his students conducted numerous surveys to determine what arithmetic was actually used by carpenters, shopkeepers, and other workers. He believed that the dominating aim of the school mathematics program should be to teach those skills and only those skills.

In the 1950s the outburst of public concern over the "space race" resulted in a wave of research and development in mathematics curricula. Much of this effort was focused on teaching the mathematically talented student. By the mid-1960s, however, concern was also being expressed for the disadvantaged student, as U.S. society renewed its commitment to equality of opportunity. With each of these changes, more and better mathematical achievement was promised.

In the 1970s, when it became apparent that the promise of greater achievement had not fully materialized, another swing in curriculum development occurred. Emphasis was again placed on the skills needed for success in the real world. The minimal competency movement stressed the basics. As embodied in sets of objectives and in tests, the basics were considered to be primarily addition, subtraction, multiplication, and division with whole numbers and fractions. Thus, the skills needed in colonial times were again being considered by many to be the sole necessities, even though children were now living in a world with calculators, computers, and other features of a much more technological society.

By the 1980s, it was acknowledged that no one knew exactly what skills were needed for the future but that everyone needed to be able to solve problems. The emphasis on problem solving matured through the last 20 years of the century to the point where problem solving was not seen as a separate topic but as a way to learn and to use mathematics (see Chapter 6).

RECENT INFLUENCES

The three influences of what mathematics is taught are still shaping mathematics teaching and learning. In this section, we will look at more recent documents and policies that are changing what mathematics is taught and learned today.

NCTM STANDARDS At various times, different national professional organizations have made recommendations about curricula. The NCTM, the largest professional organization of teachers of mathematics, developed standards for curriculum and for evaluation, teaching, and assessment (NCTM, 1989, 1991, 1995). Because states and localities in the United States have the right to determine their own school policies, these standards were not prescriptive, but they have provided vision and direction for schools.

In 2000, the NCTM published an update of the standards in a document titled *Principles and Standards for School Mathematics* (NCTM, 2000). The principles represent fundamental beliefs about the characteristics of a high-quality, equitable mathematics program. The standards, discussed in later chapters of this book, describe the mathematical content and mathematical processes that should be taught in school mathematics. Combined, the principles and standards present a vision for mathematics education programs in a changing world.

In 2006, NCTM published *Curriculum Focal Points for Prekindergarten through Grade 8 Mathematics* as one possible response to the question of how to organize curriculum standards within a coherent, focused curriculum, by showing how to build on important mathematical content and connections identified for each grade level (NCTM, 2006, p. 3).

These documents have been influential both in current state standards and in the current thrust in the United States for higher and more rigorous statements described in the section, Common Core State Standards for Mathematics, which follows on page 6.

NO CHILD LEFT BEHIND LEGISLATION A serious effort to hold schools accountable for student learning began with the new century. Accountability is one of the pillars of the No Child Left Behind Act (NCLB) of 2001, which requires states to show that each school is making yearly progress in academic subjects and that all students are becoming mathematically proficient. However, it was left to the states to decide what is meant by mathematically proficient. The No Child Left Behind legislation, which calls for annual testing in reading and mathematics each year in grades 3–8 and once in grades 10–12, provides "an example of federal funding that is linked, at least in part, to the results of standardized achievement tests at the state level. Every state has now instituted some sort of statewide testing program, partly in response to this legislation" (Wilson, 2007, p. 1099).

Along with its focus on accountability, the NCLB called for supporting supplementary services and professional development for teachers. Unfortunately, the majority of the available funds have gone into testing for accountability. Poorly performing schools are still struggling and are still in need of well-prepared teachers to meet the challenge of helping all students become mathematically proficient.

One of the strengths of NCLB is its focus on closing the achievement gap—that is, on bringing all students, including the disadvantaged, to a proficient level. There is evidence that there has been some progress in closing this gap.

HIGH-STAKES ASSESSMENTS Today's society is focused on assessments in the form of tests that are used to compare students' performance across schools, states, and nations. These summative assessments are primarily designed to document what students know and are able to do. Tests are sometimes given to make decisions about students—which class to place students, what grade to record on their report card, or whether to promote them to the next grade. When assessments have serious consequences such as these, we call them "high-stakes assessments." In the situations just described, high-stakes decisions about the students themselves are being made as a result of the assessments.

However, more frequently these days, it is not just the individual students who are affected by the consequences of high-stakes assessments. Tests may be administered in order to document the achievement of a group of students or to compare one group of students with another. Individual teachers, or schools, or districts may be held accountable for their students' test scores, taken as a group. Teachers and schools may be rated or ranked according to the results of such tests. Many schools, districts, and states now use student test results to make funding decisions, to help determine teacher's pay or even decide who will keep their jobs.

Teachers feel pressure to "teach to the test." This is not all bad if the tests actually measure what is important and allow students access to show what they really know and can do. However, many of the tests focus only on lower level skills. While mathematics proficiency requires a level of skill, say, in computation, it requires much more. If children have the opportunity to learn the content on the tests in a manner that makes sense to them, they will do well on such tests. Teach to the standards that are set for your school district and community.

Society needs a citizenry and a workforce that can solve problems, reason mathematically, process and interpret data, and communicate in a technological world. Will high-stake tests measure these things? Will these tests help schools focus on improving mathematics programs in order to prepare students for life in today's society? The answer to all such questions will be negative if these tests are not aligned with school and community goals.

COMMON CORE STATE STANDARDS FOR MATHEMATICS (CCSSM) The most recent effort, the Common Core State Standards Initiative (CCSSI) to set standards and associated assessments was led by the National Governors Association for Best Practices and the Council of Chief State School Officers. These standards for reading and for mathematics were informed by present state standards and standards from around the world. The *Common Core State Standards for Mathematics* (CCSSI, 2010) define the mathematical knowledge and skills students should obtain during their K–12 education so they are prepared for college or workforce training programs.

This is an attempt to have a common understanding across the United States of what students are expected to learn at each grade level. The states that choose to adopt these standards (and most have) will have a period of time to align their standards with these. Common assessments, both summative and formative, are being developed by centers with consortia of the states that will use these assessments.

The emphasis on common core should bring more standardization to the mathematics education of our youth. It is expected that the level of rigor of the standards will also raise the performance of students. As active participants in the world's economy, we need to ensure that our students are given the opportunity to compete. The standards and associated assessments will also free individual states from having to produce documents and assessments. It is hoped that they can turn their attention and support to their teachers who may need professional development and to students who may need extra or special help in reaching the standards.

In this text, we have included the relevant standards in the content-focused chapters. It is important that you understand these standards and begin to use them in your preparation for teaching. The full text of CCSSM is available at the Web site: http://www.corestandards.org/.

NCTM PRINCIPLES In today's world of change, we can often lose focus on what is really important. The six NCTM Principles (NCTM, 2000) represent fundamental beliefs about the characteristics of a high-quality, equitable mathematics program. We close this section by examining these statements.

The Equity Principle Excellence in mathematics education requires equity—high expectations and strong support for all students.

The Equity Principle states clearly that excellence in mathematics education means ensuring that all students learn mathematics. This vision can be realized only if each person involved in education firmly believes that all children can learn mathematics and that each child should be expected to do so. Every child must be given the opportunity to learn worthwhile mathematics. This means designing instructional programs that can encompass all the different interests, strengths, needs, cultures, and mathematical backgrounds of students. Plenty of evidence supports the idea that all students can learn mathematics. High-quality instructional programs are needed that let well-prepared teachers and other school personnel respond to students' varied strengths and needs.

Our schools are characterized by diversity—students from many different cultures and languages; from many different economic and home backgrounds; with different strengths, ways of learning mathematics, and past experiences with mathematics. Equity *does* mean that all children must learn worthwhile mathematics, but it *does not* mean that all should have the same instruction. In fact, it means that children can reach the high expectations set for them only if we meet the individual needs of each child. Your repertoire of ways to reach children will grow as you teach and learn. At this point, you can begin by challenging the popular belief that only some children can learn mathematics. This is an important first step in becoming a teacher who can help every child learn mathematics.

The Curriculum Principle A curriculum must be coherent, focused on important mathematics, and well articulated across the grades.

- *Coherent.* A curriculum that fits mathematical ideas together in a meaningful way
- *Focused.* A curriculum that focuses on the important mathematics topics and ideas at each grade, not on every possible topic
- *Well articulated.* A curriculum that builds on previous learning and grows across the grades

No one knows exactly what mathematics will be needed as the twenty-first century progresses, but it is clear that students will need to know how to reason mathematically and how to apply mathematical thinking to a wide range of situations. How you view mathematics will determine how you view teaching mathematics. If you view mathematics as a collection of facts to learn and procedures to practice, then you will teach that to your students. If you view mathematics as a logical body of knowledge, you will design your program to guide children in making sense of mathematics. Chapters 7–18 look at specific content and ways to help you help children.

The Learning Principle Students must learn mathematics with understanding, actively building new knowledge from experience and prior knowledge.

What it means to learn mathematics has changed a great deal over the past century. Currently, the phrase mathematical proficiency is used to describe what it means to learn mathematics successfully. Ideas about developing mathematical proficiency are considered in more depth in Chapter 2.

In a changing world, learning mathematics with understanding is essential in order to meet this goal of mathematical proficiency. Research has shown that if children are able to make sense of the mathematics they are learning, they can build on this understanding to learn more mathematics and use that mathematics to solve problems in order to become mathematically proficient.

The Teaching Principle Effective mathematics teaching requires understanding what students know and need to learn and then challenging and supporting them to learn it well.

To teach mathematics effectively, teachers must know more than just mathematics. They need to know their students as learners, and they must adjust their pedagogical strategies in response to students' varying experiences. Teachers must design lessons that reveal to them what students already know, that reveal students' misunderstandings, and that guide students to construct more complex understandings of mathematics. Teachers must create challenging and supportive classroom learning environments that help children make sense of mathematics. Teachers must also encourage students to think, question, solve problems, and discuss their ideas. Chapter 3 initiates the discussion of teaching, and succeeding chapters focus on ways to teach and on useful types of activities.

The Assessment Principle Assessment should support the learning of important mathematics and furnish useful information to both teachers and students.

People often think of assessment as testing to see what students have learned. The Assessment Principle presents a much broader view of assessment. Helping all students learn mathematics requires that assessment be an integral part of the instructional program. But assessment should not be something that is done to students; rather, a mathematics program must include assessments that are done for students, to guide and enhance their learning. The Assessment Principle is considered in more detail in Chapter 4.

The Technology Principle Technology is essential in teaching and learning mathematics; it influences the mathematics that is taught and enhances students' learning.

You will teach at a time when technology dominates activities both in and out of school. The Technology Principle acknowledges that technology will continue to be important in teaching and learning mathematics, as long as it enhances what is being learned and how it is being taught. As you teach your classes, you should keep asking three questions:

1. How can I help children use technology appropriately?
2. What mathematics do children need in order to use technology wisely?
3. What mathematics is no longer necessary because of technology?

Some parents continue to be concerned about the use of calculators in learning mathematics in elementary schools. A meta-analysis of 54 research studies on the use of and attitudes toward calculators (Ellington, 2003) suggests that using calculators does not hinder the development of mathematical skills and that students who used calculators had better attitudes toward mathematics than those who did not. Of course, children need to learn to use calculators appropriately, as they do any other tool (see Chapter 10).

WHERE CAN YOU TURN?

There are many places you can turn to develop your knowledge of mathematics and of mathematics learning and teaching. In this section, we discuss a few of the resources that we reference throughout this book and that you can use now and in teaching.

NATIONAL GUIDELINES FOR SCHOOL MATHEMATICS

We have discussed the role of *The Common Core State Standards for School Mathematics* (CCSSI, 2010) previously in this chapter. The relevant standards are discussed in the remaining Chapters, but you may access the full document as shown in Math Links 1.1.

> **Math Links 1.1**
>
> You can access *The Common Core State Standards for Mathematics* from http://www.corestandards.org/ or from this book's Web site.
>
> www.wiley.com/college/reys

We also frequently refer to the *Principles and Standards for School Mathematics* (NCTM, 2000). In this chapter, we have briefly discussed the principles. The five process standards form the framework for Chapter 5 and the background for a more extensive look at problem solving in Chapter 6. The five content standards give structure to Chapters 7–18. The Wiley site provides a complete version of the NCTM's expectations for preK through grade 8, giving a quick overview of what children at different grade bands are expected to know and be able to do. The document, *Principles and Standards*, more fully explains the standards at each of the grade bands (preK–2, 3–5, 6–8, and 9–12).

> ## Math Links 1.2
>
> The entire NCTM *Standards* document is available, both in print and electronically, from NCTM. The electronic version also provides interactive activities to help you better understand the intent of each standard. You can access this from http://www.nctm.org/standards/default.aspx?id=58 (either as a member of NCTM or via the 120-day free access) or from this book's Web site.
>
>
> www.wiley.com/college/reys

As mentioned earlier in this chapter, NCTM has also published *Focal Points*. To familiarize yourself with the purpose of this document, use Math Links 1.3 to read answers to the questions frequently asked about these recommendations. The relevant parts of Focal Points can be found on the Wiley site.

Other professional organizations also have recommendations for school mathematics that complement the NCTM's view. In particular, the recommendations from the National Association for the Education of Young Children (www.naeyc.org) and the American Statistical Association (www.amstat.org/education/) are referenced in later chapters.

> ## Math Links 1.3
>
> Read the Frequently Asked Questions about the *Focal Points*. You can access this from http://www.nctm.org/standards/faq.aspx?id=270 or from this book's Web site.
>
>
> www.wiley.com/college/reys

STATE AND LOCAL GUIDELINES

Almost every state has a document with standards, guidelines, or frameworks for school mathematics. Although there is much commonality among states, a study (Reys, 2006) of the grade-level expectations of states found a wide variety in the specificity and in the grade in which a skill was targeted. For example, expectation of mastery of basic addition facts ranges from grade 1 to grade 3. Of the 38 states that specify the grade level, 21% indicate grade 1, 74% indicate grade 2, and 5% indicate grade 3 (Reys, 2006). Consequently, textbooks often contain material not appropriate for your state's guidelines. One reason for developing CCSSM is the vast difference among states.

You need to become familiar with your state document and use it to plan when you are teaching. Links to state mathematics curriculum frameworks are available from the Center for the Study of Mathematics Curriculum (http://www.mathcurriculumcenter.org/). Many localities have their own versions of the state's document. They often expect more than the state documents and connect the expectations to their mathematics program.

RESEARCH

Research is referenced throughout this book, not only to acquaint you with research in mathematics education, but also to illustrate or support discussions in the text. There is a substantial body of research in mathematics education, both about children's learning and about teaching. We often use the following two sources: *Second Handbook for Research on Mathematics Learning and Teaching* (Lester, 2007), and *Teaching and Learning Mathematics: Translating Research for Elementary School Teachers* (Lambdin, 2010). The second reference is written especially for easy and practical access to research.

The National Assessment of Educational Progress (NAEP, pronounced "nape") is the nation's measure of students' achievement and trends in achievement in the academic subjects. With the passage of the NCLB in 2001, NAEP has become more prominent. The present framework for NAEP elementary mathematics is closely aligned with the NCTM's standards. The NAEP assessment contains a variety of types of items (multiple choice, short response, and open-ended response). The mathematics assessment is given to a sample of students at grades 4 and 8 in every state. Results are reported by states as well as by race, gender, and socioeconomic status. Figure 1-2 shows the overall national results for the years 1990 through 2009. Note that both grades showed significant improvement over this period.

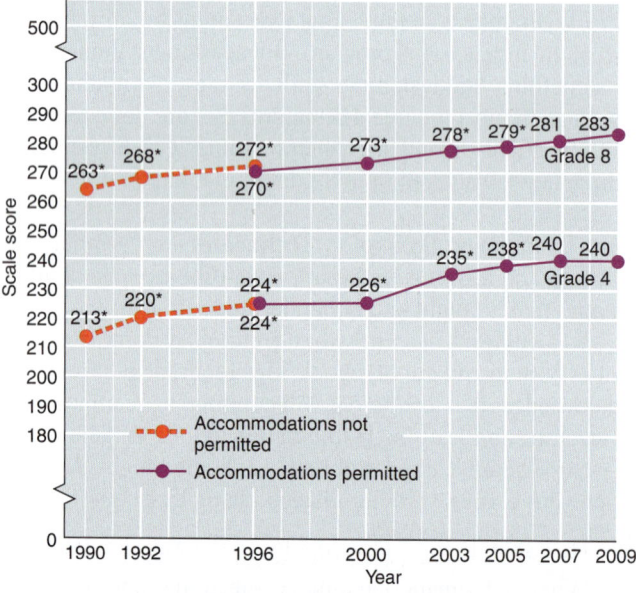

Figure 1-2 *Trends in mathematics performance*. The Nation's Report Card: Mathematics 2009. National Assessment of Educational Progress at Grades 4 and 8. U. S. Department of Education: NCES 2010-451, p. 1.

Research reports appear in many journals. We often use articles from the research journal of the NCTM, Journal for Research in Mathematics Education (JRME), to guide our recommendations in this book. JRME articles often lead directly to the classroom ideas and recommendations found in other NCTM journals, such as Teaching Children Mathematics. The Research Clips and Briefs found on NCTM's Web site give snippets of recent and relevant research.

Math Links 1.4

More information about NAEP, including sample items, may be found at http://nces.ed.gov/nationsreportcard/mathematics or from this book's Web site.

www.wiley.com/college/reys

CULTURAL AND INTERNATIONAL RESOURCES

The Trends in Mathematics and Science Study (TIMSS), an international study also given at grades 4 and 8, clearly shows that U.S. students could do better in mathematics. The results of a recent TIMSS study, in 2007, showed that U.S. fourth-grade students were above the international average but had improved little since the previous assessment, in 2003. Eighth-grade U.S. students were also above the international average and did show significant improvement since 2003. Performance varies greatly within the United States and is closely linked to economic status.

The TIMSS study also collects information about curriculum, teaching, and teachers. Video studies of classrooms suggest that Japanese teaching in grade 8 more closely resembles the recommendations of the NCTM's standards than does teaching in the United States. These results suggest that U.S. educators can learn much from analyzing how other countries teach mathematics.

Many of the other chapters in this book suggest resources that will help you understand how mathematics is taught and learned in other cultures and that show you ways to use culturally oriented activities as you strive to reach each child.

Math Links 1.5

You can find more information about TIMSS, including sample items, at http://www.nces.ed.gov/timss, watch videos of math classes at http://www.timssvideo.com, or from this book's Web site.

www.wiley.com/college/reys

TEXT BOOKS AND OTHER MATERIALS

There are many different types of textbooks. Some provide a lot of drill and practice but offer little help in developing understanding and using mathematics. Others may do the opposite, providing a great deal of help in developing students' understanding but falling short on practice of necessary skills. As you gain experience, you will be better able to judge the quality of a textbook and better able to depart from it as appropriate. If you have not had much experience, you may want to stick with the textbook until you become more comfortable with teaching. But be sure always to ask why you need to teach a given lesson. Does it help children develop the mathematics they need? Does it help children make sense of the mathematics?

Today's textbooks provide supplementary materials such as assessments, problems to solve, and extra practice. Teachers' manuals also provide a wealth of materials and teaching suggestions.

ELECTRONIC MATERIALS

The kind and amount of materials readily available are increasing every day. The Web provides immediate access to lesson plans, help with the mathematics itself, assessment items, and information that can be used in teaching mathematics. Additionally, many sites have videos of mathematics classes. Each chapter of this book begins with a Snapshot of a Lesson. Most of these snapshots are excerpts from videos that illustrate teaching related to the topic of the chapter. Watching the entire video will give you a better feel for interactions of the teacher and students.

The Math Links included in each chapter recommend sites that are worth investigating. A simple search of the Internet for information about mathematics for elementary students teachers will lead you to many other Web sites. Availability on the Web is no guarantee of quality. A prospective teacher cited an activity from a Web site in which young children found the capacity of their mouths by filling them with marshmallows! The mathematics is questionable because the unit of measurement could change as the marshmallows are squashed. More importantly, children could easily choke on a mouthful of marshmallows. Teaching involves making good judgments on many fronts.

Math Links 1.6

The Math Forum Web site includes a mathematics library, a discussion board for teachers, lesson plans, activities created and submitted by teachers, Problems of the Week for students, and answers to math problems from Ask Dr. Math. You can access this Web site at http://www.mathforum.org/ or from this book's Web site.

www.wiley.com/college/reys

PROFESSIONAL ORGANIZATIONS

Professional association with others and the support you can find from being a part of a professional organization will enhance your teaching career. Journals, conferences, and other materials of a professional organization are often available through schools.

The NCTM (www.nctm.org) offers many publications, including a journal for elementary teachers of mathematics (*Teaching Children Mathematics*) and one for middle school teachers (*Mathematics Teaching in the Middle School*). You will find many references in this book to these journals. The NCTM also sponsors conferences, e-workshops, and other support activities. There are many affiliated state and local groups of NCTM that offer publications and programs.

PROFESSIONAL DEVELOPMENT

Start taking advantage of professional development opportunities. Some of these will be formal, such as workshops, college courses, and conferences; others will include informal study groups. Your school, district, or state will provide some opportunities for you. Others will be commercially sponsored or sponsored by a professional organization. An increasing number of opportunities on the Web are designed so you can participate as your schedule permits. For example, *Teachscape* is a professional development program that we have referenced in several chapters. Several modules are available if you have purchased this book. They have many videos of students along with commentaries from the teacher and other experts.

Often, schools or districts have funds set aside for professional development that are available on request. Some districts have teachers design their own professional development plans and support them in carrying out those plans.

OTHER TEACHERS

Teachers learn from each other. You will learn from your school experiences, but do not let learning stop there. Look for schools where sharing ideas about helping students learn mathematics and sharing teaching tips and materials are the norm. Look for teachers in other schools, either near you or far away but connected electronically, who are willing to discuss and to share. A good teacher who is willing to work with you is an invaluable resource.

> ### Math Links 1.7
> Listen to what teachers say about teaching—the challenges and the rewards. A video of various teachers is available at http://teachers.nework.org/Videos/TeachersonTeaching.htm or from this book's Web site.
>
>
> www.wiley.com/college/reys

WHAT IS YOUR ROLE NOW?

As you prepare for teaching mathematics, be sure to think about the broader context of your work and carefully consider these three challenges:

- *Examine your own disposition toward mathematics and your beliefs about who can learn mathematics.* Be ready to question your beliefs, to evaluate proposed changes, and to make a difference in helping children learn mathematics.

- *Take seriously the title of this book. Teaching means helping students learn, not merely giving out information.* As you begin working with children, stop and listen to them, individually and collectively. Reflect on what you are hearing, and learn with and from the children.

- *Realize that doing mathematics and teaching mathematics are different.* Teaching mathematics requires a depth of understanding about mathematics (Ma, 1999), about students, about schools, about curriculum, and about pedagogy. If you come to this realization and actively seek knowledge and experiences that integrate these areas, you are well on your way to becoming a good teacher.

A GLANCE AT WHERE WE'VE BEEN

Teaching mathematics in a changing world means that the curriculum and instruction must change to reflect the needs of the subject, the child, and society. In this chapter you have been challenged to consider your view of mathematics as a subject. You have seen a glimpse of the changes through the past as well as recommendations for teaching mathematics in the twenty-first century. The six principles put forth by the NCTM underpin many of the recommendations for today. Resources have been identified to support your study throughout the rest of this book and, more important, as you teach. The challenge is to keep an open mind and continue your own learning about teaching children mathematics. Prepare to help your students make sense of mathematics.

Things to Do: From What You've Read

1. What are the three general goals mentioned in the introduction? Which do you think is the most important? Explain why.
2. Give an illustration (different from those in this chapter) of how mathematics is a study of patterns and relationships, a way of thinking, an art, and a language.
3. Which of the resources in Where Can You Turn have you already used? Which ones do you think will be most helpful to you? Why?
4. Explain in your own words the six principles that underpin *Principles and Standards for School Mathematics*.

Things to Do: Going Beyond This Book

In the Field

1. [1]*Mathematics in the School.* As you observe in a school, look for signs of the role that mathematics plays in that school. Does it differ from class to class?
2. [1]*Equity: Interview the Teacher.* What does the statement "all children can learn mathematics" mean to you? Interview a teacher and compare his or her answer to yours.

In Your Journal

3. The Technology Principle recommends that technology should support effective mathematics teaching. Write a statement of your experience using calculators and describe your philosophy regarding calculators in learning elementary mathematics.
4. For what numbers do you think the procedure in the Snapshot would be helpful? Were the steps in the snapshot clear to you? Write about how Mr. Telford's teaching was similar and how it was different from how you were taught math in elementary school? If you were teaching this procedure, what would you differently?
5. Modify the method used in the snapshot to multiply 18×19 using 20-complements. What changes did you need to make in the steps?

With Additional Resources

6. Read Chapter 2 of *Principles and Standards for School Mathematics*. Choose the principle that you think is the most important; critique the discussion of that principle.
7. Find a recent issue of *Teaching Children Mathematics* or *Teaching Mathematics in Middle Grades*. Select an article that describes a classroom application. Describe the principle(s) and/or the standard(s) that are considered in the recommendations for this classroom application.

With Technology

8. NCTM has a Web site that illuminates *Principles and Standards*. Visit the site (illuminations.nctm.org) and select the button called "imath investigations." You need to select a particular grade level, preK–2, 3–5, 6–8, or 9–12. Select one of the elementary levels. Explore at least one of the ready-to-use, online, interactive, multimedia math investigations Write about how you could use the investigations when you teach.
9. Navigate the Web site in Math Links 1.0 "Ask Dr. Math" Web site. Review the questions that elementary teachers and students send for Dr. Math to answer. Send a question to Dr. Math. How long is it before your question is answered? Was the answer helpful?
10. Watch the video suggested in Math Links 1.7. What reason for remaining in teaching that teachers described is the most important to you? Why?

---

[1]Additional activities, suggestions, and questions are available in the field experience manual on the Student Companion site at www.wiley.com/college/reys.

Note to Instructors: You can find additional resources, learning activities, and blackline masters in this text's accompanying Instructor's Manual, at www.wiley.com/college/reys.

CHAPTER 2

Helping All Children Learn Mathematics with Understanding

> "I NEVER TEACH MY PUPILS; I ONLY ATTEMPT TO PROVIDE THE CONDITIONS IN WHICH THEY CAN LEARN."
> — Albert Einstein, Physicist and Nobel Prize winner, 1879–1955

>> SNAPSHOT OF A LESSON

KEY IDEAS

1. Use models to divide a set into equal parts.
2. Represent equivalent parts using whole numbers and fractions.
3. Represent fractions using symbols and models.

BACKGROUND

This snapshot is from the video *Cookies to Share*, from Teaching Math: A Video Library, K–4, from Annenberg Media. Produced by WGBH Educational Foundation. © 1995 WGBH. All rights reserved. To view the video, go to http://www.learner.org/resources/series32.html (video 36).

Ms. Kincaid is teaching a fraction lesson to her fourth graders. She begins by reviewing a previous discussion.

Ms. Kincaid: When we make something even, when we cut something in half, what are we doing? What are we looking for?

Richard: We're looking for an even part, like in a piece of pie. You cut that so with two people one gets an even part and the other does, too.

Ms. Kincaid: So what do we call even parts?

Carly: Equal

Ms. Kincaid: So we're going to be looking at equal parts today. Before we do, we're going to read a story by Pat Hutchens, *The Doorbell Rang*. [Reads story.] What was the problem presented in the story? [*Two children describe the problem in the story.*]

Ms. Kincaid: Let's go back to the one part where Joy and Simon came. Let's say Joy and Simon came without their cousins. How many children would there be? [*8*] How many cookies did they start with? It's another name for a dozen. [*12*]

So here is the problem I want you to work on today. Imagine that you have one dozen, or 12, cookies and you have to divide that equally among 8 people. Who knows what I mean, when I say equal parts, when we're talking about equally?

Girl: It means when you divide all the cookies to each people, they would each have enough and there wouldn't be any left.

Ms. Kincaid: Will I have more cookies than you? [*no*] Will you have more cookies than me? [*no*] Would anybody have more cookies? [*no*] That's what we mean by equal parts.

12

This is your task for right now. I'm going to give each group a piece of paper and a Baggie. Inside are cookies. You'll have to cut out the cookies so you have 12 cookies. You're going to show a way, within your groups, to divide these 12 cookies. You may cut. You may use your pencil to draw equal parts. It is up to you, but I want you to divide the 12 cookies into 8 equal parts. Then I want you to tell me how you know you are right. How can you prove to me that you have divided equally among those 8 people? I'm going to see words and I'm going to see pictures. [*A student summarizes the directions, and they get started. Ms. Kincaid circulates among several groups, observing and asking probing questions.*]

Ms. Kincaid: Okay, you have 8 people, right? OK. What did you discover? [*Each person gets one and a half.*] How do you know you are right?

Girl: I got 4 and they got 2 apiece, so that is 8. Then we cut 4 cookies in halves. We gave every person what they needed. We ended up with 8 cookies and 8 halves.

Ms. Kincaid: So how much is that for each person? [*one cookie and one half, one and a half*] Does everybody have the same amount? Does anybody have more? [*no*]. Okay, it sounds like you are on to something. Why don't you show me so you can prove it to the rest of the group? Find a way to organize your information. Then what you just told me, I want you to write that down. [*Ms. Kincaid circulates as students continue to work in groups. Some need more trial and error with the paper cookies before they figure it out. Once the problem has been solved, they write their explanation.*]

Ms. Kincaid: OK, ladies and gentlemen. Let's come back together and look at the processes you used. We're going to take a look at how you know you are right. [*One person from each group presents their solution and explanation.*]

Alicia: We gave each person 1 cookie, there were 4 left, so we gave each one half.

Boy: We got confused. We tried many different things like splitting 2 cookies with each other but it only made it for 4 people. We didn't really understand that you could cut them with scissors. Then we figured out cutting in half, and we gave each person a half, each got a cookie and a half.

Ms. Kincaid: You showed me how you can take 12 cookies and divide them equally among 8 people. We can make other problems from the story. Could we figure out how to split those 12 cookies with 7 people? [*yes*] Could we find a way to split them with 10 people? [*yes*] Maybe so. That is something we might explore later on.

FOCUS QUESTIONS

1. What does it mean to learn or know mathematics?
2. How do children learn mathematics?
3. How do behaviorist approaches to learning differ from constructivist approaches to learning?
4. What are four recommendations for helping children make sense of mathematics based on what is known about how children learn mathematics?

INTRODUCTION

How do children learn mathematics? This important question has no simple answer, but the safest response is "Children learn mathematics in very different ways." The Snapshot shows how Ms. Kincaid used modeling and materials to help the students understand the problem and move toward solutions. She got her students actively engaged in the problem, and challenged them to describe their solutions. Often children have a difficult time describing the processes used, which makes our jobs as teachers more like a detective as we look for clues to better understanding what they did and how they did it.

Each elementary classroom is filled with children who have different backgrounds, interests, strengths, and needs. It is important for teachers not only to know the children in their classroom well but also to answer the question, "How will my students best learn mathematics?" Teachers provide their answers through classroom practices. In fact, every instructional activity within the classroom expresses the teacher's view of learning. The way in which you plan lessons, present topics, and handle questions reflects how you perceive learning and influences what happens in the classroom. The purpose of this chapter is to build on your previous knowledge from educational psychology and stimulate your thinking about how to help all children learn mathematics.

WHAT DO WE KNOW ABOUT LEARNING MATHEMATICS?

One of the clearest models for describing the complexity of mathematics learning was proposed by the National Research Council in *Adding in Up: Helping Children Learn Mathematics* (Kilpatrick, Swafford & Findell, 2001), and more recently used as the foundation for Mathematical Practices put fourth in the *Common Core State Standards-Mathematics*. This model shown in Figure 2-1 focuses on mathematical proficiency, and shows that mathematical proficiency is not one dimensional, but has multiple elements that are intertwined together.

Figure 2-1 briefly characterizes each of the five strands, which together highlight the practices, skills, and habits of mind that lead to mathematical proficiency. Figure 2-1 illustrates how these proficiencies are intertwined and, together, compose what we think of as learning and understanding mathematics. Each of them has implications for our discussion of how children learn mathematics.

• *Conceptual Understanding*	Comprehension of mathematical concepts, operations, and relations
• *Procedural Fluency*	Skill in carrying out procedures flexibly, accurately, efficiently, and appropriately
• *Strategic Competence*	Ability to formulate, represent, and solve mathematical problems
• *Adaptive Reasoning*	Capacity for logical thought, reflection, explanation, and justification
• *Productive Disposition*	Habitual inclination to see mathematics as sensible, useful, and worthwhile, coupled with a belief in diligence and one's own efficacy

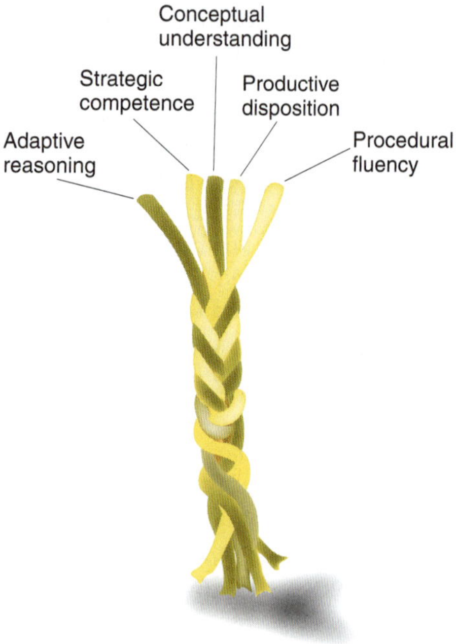

Figure 2-1 *Five Intertwined Strands of Mathematical Proficiency.*

HOW CAN WE SUPPORT THE DIVERSE LEARNERS IN OUR CLASSROOM?

Each school year brings a new mix of children with varying personalities, strengths, and needs. Elementary classrooms are becoming increasingly more diverse (Knowles, 2006). Nationwide, the last decade has seen large increases in the number of students who are members of ethnic minorities, are English-language learners, and receive services for disabilities. Data from the National Center for Educational Statistics (http://nces.ed.gov) indicate that a typical classroom of 25 students includes approximately

- 11 students who are members of a minority group
- 4 students who speak a language other than English at home
- 10 students who qualify for free or reduced-cost lunches
- 4 students who live in poverty
- 4 students with disabilities who receive special services, a little more than half of whom spend more than 80% of their school day in a regular classroom

The students in our classrooms also vary in their cognitive, physical, and social development and abilities. They come from different cultures and family structures. They have different background experiences, interests, levels of motivation, and styles of learning. In the end, every child in every class is unique. As teachers, we are charged with helping each and every one of these children learn mathematics and develop to their maximum potential. The current national focus on accountability in education reinforces that responsibility. In the words of Felix Frankfurter in the chapter opener, treating all students the same will not help us reach this goal.

NCTM's vision, described in the Equity Principle, is that "All students, regardless of their personal characteristics, backgrounds, or physical challenges, must have opportunities to study—and support to learn—mathematics. Equity does not mean that every student should receive identical instruction; instead, it demands that reasonable and appropriate accommodations be made as needed to promote access and attainment for all students" (NCTM, 2000, p. 12). Providing appropriate learning opportunities for all the children in your mathematics classroom is a challenge.

In Chapter 3 we will examine how you can adapt lessons to meet the needs of individual students. There are many other strategies you can use to support all students without having to individualize instruction for each one. In the following section, we describe five of these strategies.

CREATING A POSITIVE LEARNING ENVIRONMENT

Creating a positive learning environment means being concerned both with the physical setting and with other factors: "If we want students to learn to make conjectures, experiment with alternative approaches to solving problems, and construct and respond to others' mathematical arguments, then creating an environment that fosters these kinds of activities is essential" (NCTM, 1991, p. 56). The teacher is largely responsible for creating an appropriate environment. Here are some of the things you should do.

- Make sure the classroom arrangement is safe and comfortable and that it supports the lesson's learning activities. When you do a demonstration, be sure all students can see you, the board, or the screen. When students are working in small groups, arrange their desks in clusters before the lesson begins. If they need to use manipulatives or other supplies, have them packaged and available for easy access or distribution. Consider a room arrangement that lets both you and the students move around the room easily.

- Make sure the classroom atmosphere is intellectually stimulating for learning mathematics. Encourage intellectual risk-taking, and help children feel safe about taking risks. Help children understand that confusion, partial understanding, incorrect answers, conceptual errors, and some frustration are natural as they construct their mathematical knowledge. Children who do not inhibit their intuitive responses and are not overly concerned about giving wrong answers are more likely to search for patterns, feel free to make conjectures, engage in discussions, and take risks when doing mathematics.

- Make sure students understand that they will not all learn the same things at the same time and that they will not all be equally proficient, but that everyone can indeed become proficient. Learning mathematics is a long-term process. Sometimes progress will be slow; sometimes learning will jump ahead in moments of aha and insight, when they will say, "I've got it!" or "Now I understand!"

- Reward students for critical thinking and creative problem solving so that students learn to value and respect those approaches. Students who experience a problem-solving approach to mathematics consistently perform higher than those who experience a focus on skills and procedures (Sutton and Krueger, 2002). Table 2-1 shows some additional recommendations from NCTM for shifting the classroom learning environment so it better supports all students.

AVOIDING NEGATIVE EXPERIENCES THAT INCREASE ANXIETY

Many children experience some degree of mathematics anxiety, or "mathophobia"—a fear of mathematics or other negative attitudes toward mathematics. Mathematics anxiety can be expressed as poor performance, more than the usual number of misunderstandings, lack of confidence about doing mathematics, and so on (see Figure 2-2). Data from the National Assessment of Educational Progress (NAEP) suggest that primary-grade children generally express positive attitudes about mathematics, but attitudes toward mathematics tend to become progressively more negative as children move into middle school and high school (Lubienski, McGraw, and Strutchens, 2004). Students who experience mathematics anxiety tend to take less mathematics during secondary school, thereby blocking their access to many careers.

What can be done about mathematics anxiety? Here are some suggestions for ways you can help students cope with this problem:

- Emphasize meaning and understanding rather than memorization. Children attempting to memorize mathematics without understanding are likely to fall into the "anxiety gorge" in Figure 2-2. Helping students make connections between the concrete (e.g., models and manipulatives) and the abstract (e.g., generalizations and symbolic representations) facilitates understanding, promotes success at learning, and helps relieve mathematics anxiety.

- Model problem-solving strategies rather than presenting finished solutions. Help students realize that using incorrect strategies and taking unnecessary steps are a natural part of developing problem-solving skills. Focusing on the process rather than the answer helps reduce the anxiety associated with getting "wrong" answers.

- Show a positive attitude toward mathematics. Students' attitudes are greatly influenced by the attitudes

TABLE 2-1 • Five Shifts in Classroom Environment

Mathematics Instruction Should Shift:	
Toward	**Away from**
• Classrooms as mathematics communities	• Classrooms as collections of individuals
• Logic and mathematical evidence as verification	• The teacher as the sole authority for right answers
• Mathematical reasoning	• Mere memorizing of procedures
• Conjecturing, inventing, and problem solving	• Finding answers mechanistically
• Connecting mathematics, its ideas, and its applications	• Treating mathematics as a body of isolated concepts and procedures

Adapted from *Professional Standards for Teaching Mathematics* (NCTM, 1991).

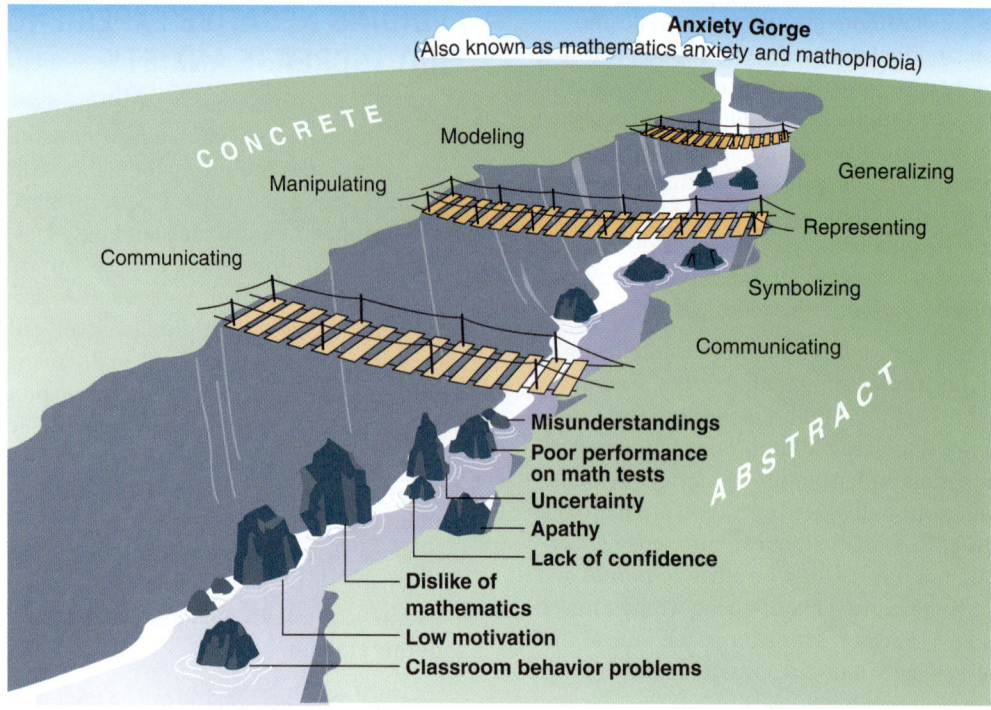

Figure 2-2 *Bridges linking meaning to mathematics.*

of their teachers. Research has shown that teachers who enjoy teaching mathematics and who share their interest and enthusiasm for the subject tend to produce students who like mathematics (Bay-Williams & Karp, 2011).

- Give students mathematical experiences they will enjoy and that will interest and challenge them while allowing them to be successful. Successful experiences in learning mathematics result in self-confidence, which greatly influences the persistence that students will exhibit when confronted with challenging problems.
- Encourage students to tell you how they feel about mathematics. What do they like, and why do they like it? This self-reflective (or metacognitive) diagnosis can help you detect symptoms of mathematics anxiety.
- Don't overemphasize speed tests or drills in your classroom. Some children may enjoy the challenge of competition, but others are uncomfortable with it—for these students, timed races breed apprehension and fear of mathematics. Use diagnostic techniques to identify students who are experiencing particular difficulty or need special help, and provide this help quickly to get them back on track.

ESTABLISHING CLEAR EXPECTATIONS

Students want to meet teacher expectations, so teachers must make sure their expectations are understood. To make their expectations clear, "teachers must think through what they really expect from their students and then ensure that their own behavior is consistent with those expectations" (Good and Brophy, 2010, p. 127). Ways to establish clear expectations include the following:

- Make it clear that you respect and value student ideas and ways of thinking; also make it clear that you expect your students to respect and value each other.
- Establish a mathematics class motto for your students: "Do only what makes sense to you." This motto encourages students to question, reflect, and seek explanations that make sense to them. It also paves the way for constructing knowledge that students find meaningful and that they understand.

  Another class motto might be: "Maybe one answer—certainly many paths." This motto will help your students realize that the process is just as important as the answer. Ask students to explain their thinking so you can see what path they took to get to their answers.
- Encourage children to reflect on their learning—not just on what they have learned, but also on how they learned it, their thought process. Metacognition is

an important part of learning. Individual reflection or interaction with others (both teachers and peers) encourages students to communicate and explain their thinking. Ms. Kincaid in the lesson snapshot had each group share their thought processes.

TREATING ALL STUDENTS AS EQUALLY LIKELY TO HAVE APTITUDE FOR MATHEMATICS

Do not let your words or actions suggest that some students—for example, boys or students of Asian background—are more likely than others to excel in mathematics. Doing so sends a message to those other students that you expect them not to succeed. Make it clear that you expect students to succeed in mathematics regardless of their gender, economic, racial, or cultural and ethnic background. Expecting high achievement from everyone sends a powerful message that mathematics is for everyone. Research consistently confirms that teacher expectations greatly affect student performance.

A complex assortment of social forces produces or influences inequities related to mathematics. For example, parents of young children may indicate that they expect their sons to be better at mathematics than their daughters. School counselors may subtly discourage students from particular ethnic or socioeconomic backgrounds from studying mathematics or pursuing careers where mathematics is important.

Research has shown that teachers can help minority students succeed in mathematics. To be an effective teacher, you must have high expectations for all students, and you must challenge all students equally. You must also consider minority students' languages, cultures, and community backgrounds as assets, not as liabilities. If you develop flexible assignments and assessments that help you identify and use student strengths, you can increase the cognitive level of interactions of minority students. You should also provide immediate and effective remediation when needed (Holloway, 2004).

Research suggests that teachers may actually treat girls and boys differently in the mathematics classroom. For example, teachers may call on boys more and may be less likely to praise girls for correct responses and less willing to prompt girls who give wrong answers. Teachers also tend to attribute boys' failure at mathematics to a lack of motivation, whereas they tend to attribute girls' failure to a lack of talent. Girls may take such criticism to heart and think that it truly indicates their mathematical abilities.

Girls are less impulsive than boys and are better able to sit still and read. In general, girls are better at literacy. Boys are usually more competitive (King and Gurian, 2006). The willingness of students to take risks—to take a chance on answering a question when they are not certain of the answer—may be a factor behind gender differences on tests. Gender differences also arise with respect to *learned helplessness*—the belief that the individual cannot control outcomes and is destined to fail without the existence of a strong safety net. Learned helplessness includes feelings of incompetence, lack of motivation, and low self-esteem. It usually develops from what is perceived as failure or lack of success in learning, and it is often associated with mathematics. Students feel there is little sense in trying because the opportunity for success is beyond their control. Both boys and girls experience *learned helplessness*, but girls are particularly susceptible to this syndrome.

Within the NCTM's *Principles and Standards*, the Equity Principle (Chapter 1) should lead teachers to make strong efforts to confront and eliminate biases. Actions you can take to address inequities include the following:

- Dispelling myths (such as "mathematicians work in complete isolation" or "only white males do mathematics") that discourage women and some minorities from pursuing careers in mathematics (Mewborn and Cross, 2007).
- Having equally high expectations for all students, and clearly communicating those expectations to both students and their parents.
- Communicating to parents the importance of encouraging all their children—both girls and boys—to aspire to success in mathematics (Moschkovich, 2011).
- Engaging both boys and girls in solving difficult problems, raising questions, and communicating their mathematical thinking—that is, making sure boys and girls participate equally in the class discussion and receive equal shares of your time and attention.
- Making relevant connections between mathematics and students' lives (Coates, 2007).
- Calling attention to role models of both genders and the widest range of racial, cultural, and ethnic backgrounds, in both mathematics and science; also, helping students increase their awareness of career opportunities for people with strong mathematics backgrounds. (The Resources and Book Nook at the end of this chapter includes books that will help you find biographies and stories about mathematicians and scientists of varied backgrounds.)
- Discussing learned helplessness with students having problems and developing ways to prevent learned helplessness or to remedy it.
- Using a variety of ways to assess student performance (e.g., a range of test formats, interviews, and portfolios).

HELPING STUDENTS RETAIN MATHEMATICAL KNOWLEDGE AND SKILLS

Retention reflects the degree to which students can hold onto and use what they have learned. For example, if students can read a clock in class but have forgotten how to do so by the time they get home, we would say that their retention of this skill is very limited. Clearly, retention of knowledge and skills is an important aspect of learning.

Forgetting is a problem in all disciplines, but the cumulative nature of mathematics increases its importance. Forgetting occurs over a summer, a spring vacation, a weekend, a day, or even shorter periods. It can make the retention of skills and specific knowledge decrease dramatically from the peak during instruction. For instance, the knowledge that lets students answer questions like the following is quickly forgotten if it is not used regularly:

What is a prime number?

What is the transitive property?

How many pints are in a gallon?

Skills, too, are quickly lost if not used regularly. For example, students may have trouble doing exercises like these:

What is the quotient of 2/3 and 1/5?

Use the formula to find the area of a trapezoid.

Thus, classroom and achievement tests of mathematical skills and knowledge often report very changeable levels of performance.

Performance at problem solving, in contrast, is more stable over time and less susceptible to big declines. One reason is that problem solving is a complex behavior supported by several higher-level thinking processes. Such processes take time to develop, but once established, they are retained longer than many skills, and problem solving often improves as time goes by.

Retention is an important goal and instruction should be geared to maximize it. Research suggests several ways you can help children improve their retention:

- *Meaningful learning is the best way to increase retention.* All phases of mathematics (knowledge, skills, and problem solving) that have been developed with meaning and learned with understanding are retained longer.

- *The manner in which a concept was learned can aid long-term retention.* (National Research Council, 1999). For example, physically measuring the diameter and circumference of many different circles, observing patterns, and recording them helps students remember that the ratio of the diameter and circumference is constant more than simply being told by the teacher. You can view an example of this lesson in the Chapter 3 Snapshot of a Lesson.

- *Establishing connections aids long-term retention.* Connections help children see how mathematical ideas are related to each other and to the real world. Mathematical topics must not be taught in isolation, but in conjunction with problem solving and with applications in meaningful, real-world contexts. Research documents the value of establishing connections, not only to gain better understanding but also to promote retention.

- *Periodically reviewing key ideas helps anchor knowledge and can contribute substantially to retention.* The spiral development of high-quality mathematics programs reflects the importance of periodic reviews of mathematical topics for children at every age. These reviews may be explicit or implicit. Some teachers spend the first few minutes of each math lesson explicitly reviewing previously learned concepts; others implicitly incorporate reviews of previous topics in discussions of new topics. In either case, reviews help remove rustiness, reinforce and refresh knowledge in ways that improve immediate performance, and contribute to higher achievement and greater retention.

HELPING CHILDREN ACQUIRE BOTH PROCEDURAL AND CONCEPTUAL KNOWLEDGE

The overriding goal for mathematics education is for students to become mathematically proficient. The intertwining strands associated with being mathematically proficient were highlighted earlier in Figure 2-1. In this section, we will elaborate on procedural fluency and conceptual understanding. In mathematics education, the relative importance of teaching students procedural knowledge or fluency versus conceptual knowledge or understanding has long been debated. But this debate involves a false dichotomy. Clearly, not only are both procedures and concepts necessary for expertise in mathematics but they too are intertwined. As teachers, we need to understand what constitutes procedural knowledge and conceptual knowledge and the importance of helping students make meaningful connections between them. Both types of knowledge are essential in order for students to be able to learn mathematics with understanding.

Procedural knowledge is reflected in skillful use of mathematical rules or algorithms. A student with procedural knowledge can successfully and efficiently use a rule or complete a process, a sequence of actions. For example, a student who has procedural knowledge of two-digit division can perform the steps in the long-division algorithm quickly and accurately. *Conceptual knowledge* involves understanding what mathematical concepts mean. For example, one aspect of conceptual knowledge for division is that one meaning of division means forming equal groups. Students with conceptual knowledge can link ideas in networks of connected meanings, incorporate new information into those networks, and see relationships among different pieces of information (Hiebert and Grouws, 2007). In the Snapshot of a Lesson, the students are developing conceptual knowledge of fractional parts by using models in a way that makes sense to them.

Procedural knowledge alone helps students answer specific questions, but it may lack important connections.

Conceptual knowledge requires the learner actively think about relationships and make connections, while also making adjustments to fit the new learning into existing mental structures. Research shows that conceptual knowledge not only doesn't diminish skills but may even help students recall and use skills (Hiebert and Grouws, 2007). For example, in response to the question "What is a square?" procedural knowledge might lead a student to respond, "A square is a figure with four congruent sides and four right angles." Procedural knowledge would not, however, help the student understand other relationships—for example, that any square is also a rectangle, a parallelogram, a regular polygon, and an equilateral quadrilateral. In the Classroom 2-1 requires that students demonstrate conceptual knowledge—that is, a deeper understanding of four-sided figures.

There is a negative consequence of teaching procedural knowledge without conceptual knowledge. If teachers indicate that the only important thing is to get the right answer, then students will have no motivation to learn why algorithms work. The addition algorithm for 23 + 49 could be memorized as: "Add the 3 and the 9 to get 12. Bring down the 2 and carry the 1…" Rote learning (memorization without meaning) has no place in school mathematics, but it highlights one of the ever-present dangers associated with teaching algorithms. Research suggests that students with highly developed rules or procedures for manipulating symbols are reluctant to make the effort to connect these rules to other representations that might help them understand the mathematical meaning of the algorithms. In addition, a student who learns a procedure without meaning will have difficulty knowing when to use it, remembering how to do it, applying it in new situations, and judging if the results are reasonable.

As teachers of mathematics, we must help students establish connections and see relationships between conceptual and procedural knowledge. Students may not do it automatically. Research suggests that instruction that focuses on developing conceptual understanding can also yield efficient skills (Hiebert and Grouws, 2007). One way to focus on conceptual understanding is to focus on students' explanations (Lehrer, 2003). Our challenge as teachers is to construct learning experiences that help students build connections between mathematical ideas. A closer look at what we know about how children learn mathematics will help you meet this challenge.

HOW DO CHILDREN LEARN MATHEMATICS?

The vision for mathematics education promoted by the National Council of Teachers of Mathematics (NCTM, 2000) is for all children to learn mathematics with understanding. More specifically, the Learning Principle says: "Learning with understanding is essential to enable students to solve the new kinds of problems they will inevitably face in the future" (NCTM, 2000, p. 21). The National Research Council concurs: "All young Americans must learn to think mathematically, and they must think mathematically to learn" (Kilpatrick, Swafford & Findell, 2001, p. 16).

The premise that mathematics must be learned with understanding is based on research that has been accumulating for many years and reflects advances in mathematics education, psychology, and other areas of science. Mathematics builds on itself, becoming more abstract and symbol oriented as the ideas build (e.g., from arithmetic to algebra). Evidence from science suggests that learning changes the physical structure of the brain, and different parts of the brain may be ready to learn different things at different times (National Research Council, 1999). Ultimately, to learn more abstract mathematical concepts, children need to have developed enough both physically and psychologically to handle the abstraction.

Early in the twentieth century, John Dewey asserted that learning comes from experience and active involvement by the learner. Much has been discovered since then about how children learn mathematics, but the importance

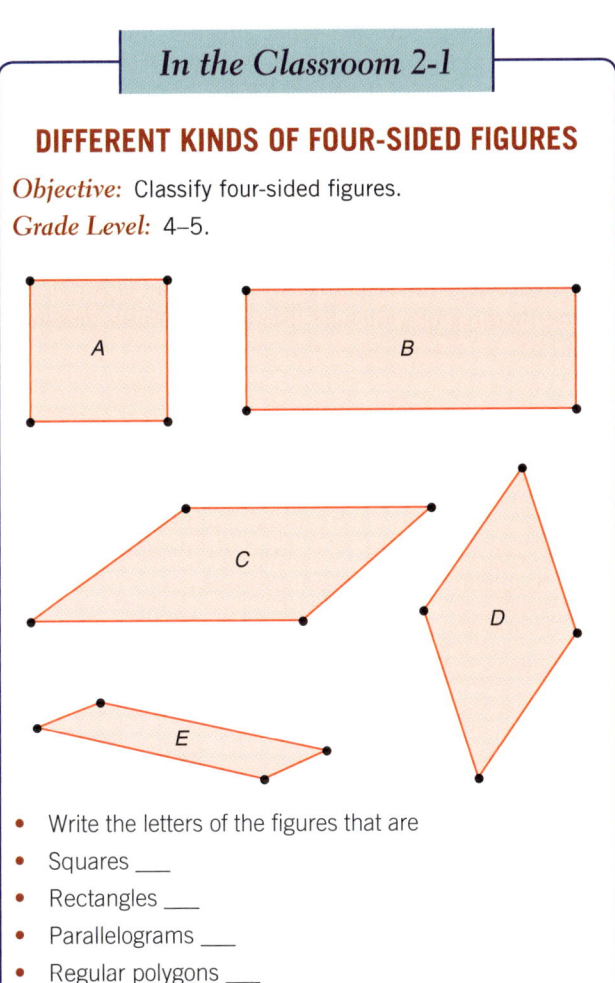

In the Classroom 2-1

DIFFERENT KINDS OF FOUR-SIDED FIGURES

Objective: Classify four-sided figures.
Grade Level: 4–5.

- Write the letters of the figures that are
- Squares ___
- Rectangles ___
- Parallelograms ___
- Regular polygons ___
- Equilateral quadrilaterals ___

of meaningful experience remains unchallenged. Later, Jean Piaget argued that learners actively construct their own knowledge. This view of learning, known as *constructivism*, suggests that rather than simply accepting new information, students interpret what they see, hear, or do in relation to what they already know. As the Learning Principle indicates, students learn mathematics with understanding by actively building new knowledge from their personal experiences and prior knowledge. In this chapter's Snapshot of a Lesson, students used everyday objects, from their own experiences, to explore equal parts.

Teachers develop their lessons based on how they believe children learn. There are currently two prevailing theories of learning, behaviorism and constructivism. *Behaviorism* focuses on observable behaviors and is based on the idea that learning means producing a particular response (behavior) to a particular stimulus (something in the external world). From this perspective, students learn specific skills (behaviors) by observing teachers demonstrating those skills in relation to specific stimuli (e.g., a mathematics problem). For example, a teacher demonstrates how to find the mean of a set of numbers, and students learn to produce that type of response (finding the mean) when given that type of stimulus (a second set of numbers). Behaviorism's focus on behaviors excludes consideration of any thought processes students may use to arrive at their answers.

Constructivism has a different focus: It concentrates on what happens between the stimulus and the response. That is, the focus is on the thinking students do. From this perspective, learning depends not only on what the teacher does but also on the students themselves—how they integrate new ideas with their experiences and with what they already know.

Both views of learning hold implications for teaching mathematics. Both are key to a further understanding of how children learn mathematics.

BUILDING BEHAVIOR

Behaviorism focuses on external actions and observable behaviors—on stimuli and responses. The main advocates of the behaviorist perspective over the years include Edward L. Thorndike, B. F. Skinner, and Robert Gagné; today, few learning theorists argue for an exclusively behaviorist approach to mathematics learning. Behaviorism has had a significant impact on mathematics programs, and teachers must consider behaviorist psychology when developing instructional goals and planning lessons. Nevertheless, strict adherence to a behaviorist approach to mathematics learning in elementary school is inappropriate because of its lack of consideration of the child's own thinking. Thus, as teachers, we can draw on behaviorist ideas, but we must do so wisely and with an awareness of their limitations. Keeping this point in mind, we will examine behaviorism a bit more closely.

A major tenet of behaviorism is that behavior can be shaped through reinforcement—that is, through rewards and punishments. Teachers can use reinforcement (e.g., feedback on a student's performance, praise or criticism in class) to get students to practice desired behaviors (e.g., written algorithms). It is well documented that meaningful practice has great value and power in mathematics learning, but excessive practice, premature practice, or practice without understanding is associated with negative effects. Such practice often leads to a fear or dislike of mathematics and an attitude that mathematics does not need to make sense, despite the fact that making sense of mathematics should be a major goal of mathematics learning.

A behaviorist approach can be useful in helping children learn a fixed set of skills in a fixed order—that is, to help children acquire procedural knowledge. Many behaviorists think of mathematics as being sequenced in a linear fashion, where one idea builds on another and where proficiency in one skill is used to develop proficiency in the next skill. For example, the concept of multiplication as repeated addition suggests that students need to master addition before moving to the concept of multiplication. This hierarchical view of mathematics learning makes it a popular candidate for a behaviorist approach.

The first consideration in planning a lesson from a behaviorist perspective is to state precisely the objectives, or goals, of instruction. Such statements give the teacher direction in planning lessons and give the students clear expectations—both valuable outcomes that are consistently supported by research. Then, once an objective has been clearly stated, the teacher should identify the prerequisites for achieving that goal and use those prerequisites as building blocks in planning instruction. For example, suppose you have stated this objective: Use the formula $A = 1/2\ ba$ to find the area of a triangle. Some prerequisite questions would be

What is a triangle?

What is a triangle's base?

What is a triangle's altitude?

Clear answers to these prerequisite questions are necessary in order to reach the objective; however, the following questions could also be considered prerequisites:

What is area?

How do you multiply by a fraction?

How do you multiply two whole numbers?

Divide two numbers?

This indicates the difficulty in constructing a complete set of prerequisites for any objective in mathematics, no matter how simple the objective may seem. Of course, you have to consider prerequisites when preparing lessons, but you must be guided by common sense, not by zeal to state every prerequisite.

From the behaviorist perspective, clarifying the goals for instruction focuses the instruction on the desired learning outcomes. What behavior do you want the students to exhibit at the end of instruction? That outcome provides the focus of the lesson. This process of forming behaviorally oriented objectives may be useful, but it ignores a more important goal—learning mathematics with meaning. Students may be able to demonstrate a desired behavior without understanding what that behavior means. If the objective of the lesson is to use the formula correctly to find the area of a triangle, the lesson's focus is on lower-level cognitive outcomes, where the students are expected only to correctly identify and use the base and altitude of the triangle to calculate a numbers the area. If, on the other hand, the objective of the lesson is to explain how the area of the triangle is half the area of a specific rectangle (see Figure 2-3), the students are directed toward a higher cognitive level of understanding of the area of the triangle. From this higher-level perspective, the prerequisite questions for the lesson must include the question of understanding the area of a rectangle. The lesson must then focus the students on identifying the rectangle that is formed by the base and altitude of the triangle (rectangle BEFC in Figure 2-3). Look at Figure 2-3 and decide for yourself *why* the area of the shaded triangle (ABC) must be one-half the area of the rectangle (BEFC).

This example shows how emphasizing behavioral outcomes for a lesson may result in lower-level cognitive procedures. Unfortunately, many mathematics programs include a hefty proportion of such outcomes, probably because they are easily measured. The result is programs where students are "shown" algorithms and mathematical relationships are "illustrated" on the textbook pages, but where learning with understanding is sadly underemphasized.

The attractive features of the behaviorist approach are that it provides instructional guidelines, allows for short-term progress, and lends itself well to the current focus on accountability. Knowing what outcomes will be assessed in state or standardized tests gives teachers specific direction for designing their lessons; however, a real and constant danger in using a behaviorist approach is that it leads to a focus on simple, short-term objectives that are easily measured. These simple objectives, in turn, lead to mastery of specific skills, but may not lead to higher-level understandings and the development of connections that make knowledge meaningful and useful. The emphasis on short-term objectives often results in a de-emphasis on long-term goals and higher-level cognitive processes such as problem solving.

Nevertheless, careful use of a behaviorist approach can result in more learner involvement and can even promote higher-level thinking in mathematics. Lesson outcomes can be identified that encourage children to develop concepts and use critical thinking. This possibility is illustrated by the use of behaviorally oriented verbs in stating outcomes, verbs such as *explore, justify, represent, solve, construct, discuss, use, investigate, describe, develop,* and *predict*. In sum, the behaviorist approach does lead to some useful ideas in teaching mathematics:

- Behavior can be shaped by reinforcement of drill and practice.
- Students can be helped to learn specific skills in a fixed order.
- Clear statements of objectives help teachers design lessons directed at specific learning outcomes.
- Clear statements of objectives and learning outcomes give students a clear idea of expectations.

Although identifying specific outcomes is an important part of mathematics instruction, the constructivist perspective helps teachers focus more directly on helping students understand and make sense of mathematics.

CONSTRUCTING UNDERSTANDING

During the first half of the twentieth century, William Brownell advanced a notion of meaningful learning that was a forerunner of constructivism. Brownell conceived of mathematics as a closely knit system of ideas, principles, and processes—a structure that should be the cornerstone for learning mathematics. Connections among concepts should be established so that "arithmetic is less a challenge to the pupil's memory and more a challenge to his [or her] intelligence" (Brownell, 1935, p. 32).

In recent years, research has consistently confirmed that isolated "learnings" are not retained (Hiebert, 2003a). Mathematics can and should make sense to learners. If it does, it will have meaning to learners and will be understood as a discipline with order, structure, and numerous relationships—and will be likely to be called on in a variety of problem-solving situations. Meaningful learning provides the basis for mathematical connections and is an integral part of the constructivist perspective.

In addition to Brownell, Jean Piaget, Jerome Bruner, and Zoltan Dienes have each contributed to the growth of constructivism. Figure 2-4 summarizes their frameworks of the learning process. Many of the major recommendations for

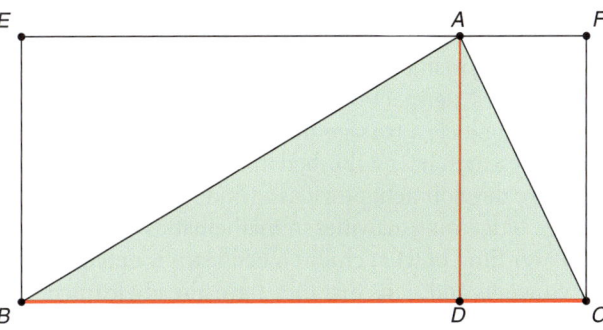

Figure 2-3 *The area of triangle ABC is half the area of rectangle BEFC. The length of the rectangle equals the base of the triangle and the width of the rectangle equals the altitude of the triangle.*

Levels of Thinking by Elementary School Children as Characterized by Piaget	Levels of Developmental Learning as Characterized by Bruner	Levels of Mathematical Learning as Characterized by Dienes
Formal Operational: Considers the possible rather than being restricted to concrete reality. Capable of logical thinking that allows children to reflect on their own thought processes.	**Symbolic:** Manipulation of symbols. Child manipulates and/or uses symbols irrespective of their enactive or iconic counterparts.	**Formalization:** Provides an ordering of the mathematics. Fundamental rules and properties are recognized as structure of the system evolves.
		Symbolization: Describes the representation in language and/or mathematical symbols.
Concrete Operational: Thinking may be logical but is perceptually oriented and limited to physical reality.	**Iconic (semi-concrete):** Representational thinking based on pictures, images, or other representations. Child is involved with pictorial and/or verbal information based on the real world.	**Representation:** Provides a peg on which to hang what has been abstracted. Images and pictures are used to provide a representation.
		Generalization: Patterns, regularities, and commonalities are observed and abstracted across different models. These structural relationships are independent of the embodiments.
Preoperational: Represents action through thought and language but is prelogical in development.	**Enactive:** Firsthand manipulating, constructing, or arranging of real-world objects. Child is interacting directly with the physical world.	**Free Play:** Interacts directly with physical materials within the environment. Different embodiments provide exposure to the same basic concepts, but at this stage few commonalities are observed.

(Left axis: Early → Advanced; Right axis: Introductory → Abstractions)

Figure 2-4 *Frameworks of the learning process.*

teaching mathematics advocated by the *Professional Standards for Teaching Mathematics* (NCTM, 1991) and more recently in the Mathematical Practices in the CCSS-M are based on their theories of learning mathematics. Both of these documents provide strong support for changing from the traditional behaviorist approach to a constructivist approach. In fact, research has shown that students learn mathematics well only when they *construct* their own mathematical understanding.

What, then, does it really mean for students to construct their own mathematical understanding? It means different things to different people, but three basic tenets of constructivism help us answer this question:

1. *Knowledge is not passively received; rather, knowledge is actively created or invented (constructed) by students.* Piaget (1972) suggested that mathematics understanding is made (constructed) by children, not found like a rock or received from others as a gift.

2. *Students create (construct) new mathematical knowledge by reflecting on their physical and mental actions.* They observe relationships, recognize patterns, and make generalizations and abstractions as they integrate new knowledge into their existing mental structure (Dienes, 1960).

3. *Learning reflects a social process in which children engage in dialogue and discussion with themselves as well as with others (including teachers) as they develop intellectually* (Bruner, 1986). This tenet suggests that students are involved not only in manipulating materials, discovering patterns, inventing their own algorithms, and generating different solutions, but also in sharing their observations, describing their relationships, explaining their procedures, and defending the processes they followed.

These tenets have significant implications for learning and teaching mathematics. They also suggest that, from the constructivist perspective, learning is a process that takes time and reflects a passage through several developmental stages. Research has established that each stage of children's cognitive development provides a window of opportunity for a range of learning activities in mathematics. At each stage, the lower limit of what children can learn is determined by the concepts and skills that they have already learned. The upper limit is determined by tasks that they can successfully complete only with scaffolding or support from someone more skilled or knowledgeable. The Russian psychologist Lev Vygotsky referred to the child's *zone of proximal development* to describe this range of learning activities and experiences (Bay-Williams and Herrera, 2007).

Research suggests that learning activities that fall within a child's zone of proximal development have a high probability of success, whereas activities outside the zone have much less likelihood of success. While we want to provide children with tasks that will help them move to a higher level of thinking or to mastery of a higher-level skill, we do not want the tasks to be so far beyond their reach that they cannot succeed and give up. Vygotsky's notion of a child's zone of proximal development challenges you to know your students well and to have a reasonably good understanding of the limits of their zones.

Learning is active and internally monitored; it is a process of acquiring, discovering, and constructing meaning from experience. In this context, teaching mathematics with the use of concrete models and making connections with the children's experiences help them make sense of mathematics. This chapter's Snapshot of a Lesson provides a good example. The process results in learning that is filtered through the student's unique knowledge base and thoughts, thereby impacting their thinking.

Piaget, Bruner, and Dienes characterize children's levels of development somewhat differently (see Figure 2-4), but overall, their proposed frameworks are remarkably similar. A careful examination of these frameworks reveals four important observations about how children learn:

- Several characteristic and identifiable stages of thinking exist, and children progress through these stages as they grow and mature.
- Learners are actively involved in the learning process.
- Learning proceeds from the concrete to the abstract.
- Learners need opportunities for talking about or otherwise communicating their ideas with others.

HOW CAN WE HELP CHILDREN MAKE SENSE OF MATHEMATICS?

Teaching occurs only to the extent that learning occurs. Therefore, effective teaching of mathematics rests heavily on considerations about how children learn. The process of building bridges from the concrete to the abstract and helping children cross them is at the heart of good teaching—and it is a continual challenge.

A blend of learning theory, research, teaching experience, and thinking about how children learn mathematics leads to practical recommendations for teaching mathematics in a manner consistent with the NCTM's Learning Principle—the principle that emphasizes the importance of personal experiences and prior knowledge in actively building new mathematical knowledge. What follows are four recommendations for helping children make sense of mathematics, based on the four observations listed above, and derived from the frameworks of Piaget, Bruner, and Dienes. These recommendations provide a strong foundation for mathematics instruction, and we extend and apply them throughout this book. No priority of importance is suggested by the order in which they are listed.

RECOMMENDATION 1: TEACH TO THE DEVELOPMENTAL CHARACTERISTICS OF STUDENTS

As children grow, they progress through identifiable stages of development (see Figure 2-4), and children learn best when mathematical topics are appropriate for their developmental level. Topics should be presented in enjoyable and interesting ways that challenge children's thinking and allow them to add new concepts and skills to existing ones. Effective and efficient learning of mathematics depends on this type of teaching—it doesn't just happen.

The learners in our classrooms have many diverse characteristics—each child is different from every other child. Nevertheless, there are common characteristics that are typical of children in a particular age group, and that is useful for teachers to know about. In this section, we focus on three types of development that influence mathematics instruction: cognitive development, physical development, and social development. *Cognitive development* has to do with how a child thinks and reasons. It also influences how the child learns new information. As educators, we probably pay more attention to our students' cognitive development than to any other type. The knowledge and skills we want children to learn are mainly cognitive in nature. *Physical development* has to do with the child's muscles and motor skills. If we want children to participate actively in mathematics, then children's physical development must also be considered when we plan lessons. *Social development* has to do with how children interact with others. It also helps describe their self-concept, how they feel about themselves. If lessons require children to interact with you and other students, you must consider their social development.

Children aged 4–7 are in the primary elementary grades (preschool through grade 2). Children in this age range are rapidly developing cognitively, physically, and socially. They are active learners. With each year of growth, their cognitive ability to understand concrete and abstract ideas increases. Their physical abilities to focus, control muscle movements, and demonstrate motor skills also increase during these years. And socially, they are learning to understand and express themselves and to understand and interact with others. Table 2-2 summarizes some key ideas to keep in mind when teaching mathematics to children in the primary grades (Kostelnik, Soderman, and Whiren, 2004).

Children aged 8–11 are in the intermediate elementary grades (grades 3–6). Children in this age range are becoming more independent learners. Their thinking is becoming increasingly sophisticated, as is their physical and social development. These children are also active learners who enjoy the challenges of working independently and with others. Table 2-3 summarizes some key ideas to keep in mind when teaching mathematics to children in the intermediate grades (Kostelnik et al., 2004).

It is useful to consider the general developmental characteristics of children in the primary and intermediate grades that may impact your mathematics instruction. Keep in mind, however, that there is a big difference in development between children at the beginning of these age ranges and those at the end. Note that Tables 2-2 and 2-3 include some suggestions to get you started thinking about how you can plan lessons that support learners at particular developmental levels.

TABLE 2-2 • Characteristics of Learners in the Primary Grades (Ages 4–7)

Cognitive Characteristics (thinking and ways of learning)	Suggestions for Teachers
• Piaget's preoperational stage • Concentration—able to focus on only one idea or stimulus at a time • Irreversibility—unable to recognize the reversibility of changes to actions or objects • Begins to understand ideas beyond firsthand concrete experiences	• Use everyday experiences to connect mathematics concepts to children's life experiences • Break tasks into manageable parts • Encourage informal exploration, choice, and experimentation with concrete materials • Use questions to stimulate thinking and challenge misconceptions • Begin using pictures and symbols to represent concrete actions
Physical Characteristics (muscle and motor skills)	**Suggestions for Teachers**
• Developing control of large and small muscles, fine-motor skills • Short attention span	• Provide short, hands-on activities with choice and opportunity for movement • Focus more on process than product • Provide materials that can be easily handled or manipulated
Social Characteristics (self-concept and interpersonal skills)	**Suggestions for Teachers**
• Egocentric—focuses more on self than others, talks at rather than with others • Developing sense of self and abilities • May have a best friend and exclude others • Learning to express oneself and make decisions	• Encourage friendship skills, sharing, taking turns, and working with another person • Provide positive support, challenge, and feedback • Allow children to make their own plans for manipulating things in the environment and solving problems

TABLE 2-3 • Characteristics of Learners in the Intermediate Grades (Ages 8–11)

Cognitive Characteristics (thinking and ways of learning)	Suggestions for Teachers
• Piaget's concrete operational stage • Able to decenter, can focus on part/whole • Understands reversibility of actions or objects • Able to classify and sequence by attribute • Can use logic and concrete objects to solve problems	• Use concrete materials and link to symbols • Provide opportunities to manipulate, classify, and sequence both objects and numbers • Explore multiple representations and strategies • Have students regularly explain thinking
Physical Characteristics (muscle and motor skills)	**Suggestions for Teachers**
• Able to complete more complex physical skills	• Provide experiences that promote active, physical involvement • Provide opportunities for children to use real tools and materials, make models and diagrams, and conduct experiments
Social Characteristics (self-concept and interpersonal skills)	**Suggestions for Teachers**
• Growing more independent • Small groups of friends have major influence • Hard to accept failure or criticism	• Provide opportunities for problem solving and projects • Provide opportunities for group work • Provide positive support, challenge, and feedback

How Can We Help Children Make Sense of Mathematics? 25

Math Links 2.1

To see more examples of children in various stages of cognitive development, you may view the following short video clips. At www.davidsonfilms.com, click on film clips and view excerpts from, "Piaget's Developmental Theory: An Overview," "Growing Minds: Cognitive Development in Early Childhood," or "Concrete Operations." From http://teachertube.com, enter "preoperational conservation child" into the search section and view the video clip.

www.wiley.com/college/reys

RECOMMENDATION 2: ACTIVELY INVOLVE STUDENTS

Actively involving students in the learning process encourages them to make sense out of what they are doing and thereby develop greater understanding of mathematics. There is an ancient Chinese proverb that emphasizes the importance of active involvement for students as they construct their own mathematical meaning:

I hear and I forget;
I see and I remember;
I do and I understand.

Active involvement sometimes requires physical activity but always demands mental activity. One of the daily challenges of teaching is to do the things that encourage and reward active involvement. These include directly interacting with the children, offering them hands-on experience with manipulatives, having them interact with learning materials such as textbooks and technology-based materials, and having them reflect on what they have done, thought, and experienced in class. Broadly speaking, we can classify such methods of actively involving students into two categories: exploration and sense making, and reflection and use of metacognition.

ENCOURAGING STUDENT EXPLORATION AND SENSE MAKING
One way to actively involve students is to encourage physical involvement. Suppose, for example, you were developing a lesson on volume. A traditional approach might include giving students the formula for finding the volume of a right prism: V = bhl. You could then show students drawings of various right prisms and ask them to use the formula to compute the volume of each.

A more active, developmentally appropriate approach would help children understand volume through an activity such as In the Classroom 2-2, where children are given blocks and asked to build different boxes (right prisms).

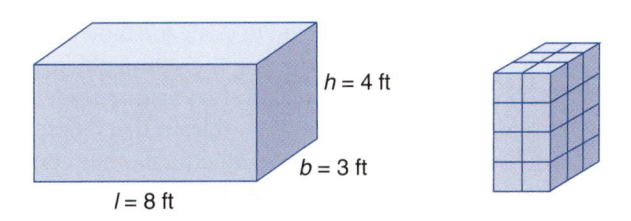

$h = 4$ ft
$b = 3$ ft
$l = 8$ ft

In the Classroom 2-2

PATTERNS WITH BLOCKS

Objective: Find the volume of rectangular prisms and explore how a prism's volume relates to its dimensions.

Grade Level: 3–4.

Each of these boxes holds 16 blocks. Record the length, width, height, and volume (number of blocks) in the table.

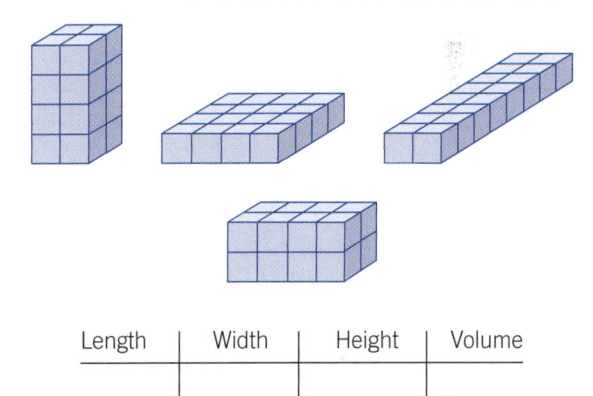

Length	Width	Height	Volume

- Take 24 blocks. Estimate how many different boxes you can make ___
- Make as many different boxes as you can. Record the length, width, height, and volume in the table. How many different boxes did you make? ___

Length	Width	Height	Volume

- You have four piles of blocks. You must use all the blocks in a pile to make boxes.
- Pile A has 16 blocks   Pile B has 25 blocks
- Pile C has 36 blocks   Pile D has 50 blocks
    — Which pile of blocks would let you build the most different boxes? Why?
    — Which pile would let you build the fewest different boxes? Why?
- Build the boxes, fill in the table, and see if you are right. What did you find out?
- What patterns do you see in your tables?

As they build the boxes, they begin to relate the dimensions of the boxes to their volumes. Finding the volume of boxes is a primary objective for this activity; however, as students find the volumes of the boxes, they also learn that different boxes can have the same volume. Furthermore, they come to understand why they need to multiply the dimensions of a box to find the volume, so the formula for the volume of a box evolves naturally from this active involvement.

An activity like this involves children in using models, making decisions, and thinking about mathematics, rather than just methodically applying a formula. It also lets them physically experience important mathematical properties. For example, using 16 blocks, they build a $4 \times (2 \times 2)$ box and they build a $(4 \times 2) \times 2$ box. Because both boxes have 16 blocks and the same dimensions, they are demonstrating that $(4 \times 2) \times 2 = 4 \times (2 \times 2)$. The boxes are the same, just oriented differently. Similarly, as they continue to build boxes with 16 blocks, they construct models that show $(2 \times 2) \times 4 = (2 \times 8 \times 1)$ and, more important, they practice arithmetic and develop algebraic thinking. If you compare the "teacher telling" approach with the "children discovering" approach, you can see that the "children discovering" approach, which actively involves children in discovering the formula, will probably result in greater understanding and retention. Children are more likely to remember something they have figured out for themselves than something they have just been told.

ENCOURAGING STUDENT REFLECTION AND THE USE OF METACOGNITION You can also actively involve students mentally by helping them develop their metacognitive skills. *Metacognition* refers to thinking about one's own thinking (i.e., cognition about cognition). It includes what one knows or believes about oneself as a learner and how one regulates and adjusts one's own behavior. Children need to become aware of their own strengths, weaknesses, and typical behaviors and of the repertoire of procedures and strategies they use to learn and do mathematics and, more specifically, solve problems.

In a sense, metacognition means looking over your own shoulder—observing what you do as you work and thinking about what you are thinking. Students who monitor their mathematical thinking search for understanding and strive to make sense of the mathematics being learned. Competent problem solvers are efficient at keeping track of what they know and of how well or poorly their attempt to solve a problem is proceeding. They continuously ask:

> What am I doing?
>
> Why am I doing it?
>
> How will it help me?

More and more research suggests that what students know or believe about themselves as mathematics learners not only greatly affects their performance but also influences their behavior as they do mathematics (Kilpatrick, Swafford & Findell, 2001).

Children who are skillful at metacognition can learn, for example, that practice improves their performance on particular tasks, that drawing a picture often helps them understand a problem, and that "I get scared when I see a word problem." This type of metacognitive knowledge often helps students control and adjust their behavior. For instance, if Sheena knows that she frequently makes keystroking errors with her calculator, she is more likely to work slowly and to check on the reasonableness of her calculator's answers.

The development of metacognitive skills requires that children observe what they know and what they do and that they reflect on what they observe. Encouraging students to think about their thinking is an important aspect of mathematics teaching. Here are a few things to help students develop metacognitive awareness:

1. Help students be explicit about how they work when solving problems. For example, consider asking questions like these:

   - Why did you do that?
   - How did you know not to use that information?

Teachers routinely consider such questions themselves when solving problems, but if we then go on to present complete, polished solutions to students, it may inhibit students from discussing their own decisions. There is value in helping children be aware of your behind-the-scenes decision making. When we share aloud our own thinking, it provides a model for students.

2. Prepare students to accept and be aware of common problem-solving experiences by pointing out various aspects of problem solving:

   - Some problems take a long time to solve.
   - Some problems can be solved in several different ways.

3. Encourage students to become more aware of and to think about their own mathematical thinking. For example, you might ask students to discuss the following questions:

   - What mathematics problems do you like best? Why?
   - What mathematics problems are most difficult?

Some educators recommend that teachers have students with special needs brainstorm how they learn best and then help them to use those strengths (Diezmann, Thornton, and Watters, 2003). This may include helping them use specific learning approaches or styles, helping them use specific methods for communicating their thinking and their approaches to problem solving, and encouraging them to rephrase tasks or problems in their own words.

RECOMMENDATION 3: MOVE LEARNING FROM CONCRETE TO ABSTRACT

When you think about how learning should move from the concrete to the abstract, it is important to keep in mind that *concrete* is a relative term. To one child, joining two blocks and four blocks is concrete, but 2 + 4 is not; another child may view 2 + 4 as concrete and *x + y* as abstract. Symbols and formal representations of mathematical ideas follow naturally from the concrete level, but only after conceptualization and meaningful understanding have been established. Without such understanding, children do not feel comfortable working with mathematical symbols, and the mathematics does not make sense to them. As Bruner noted in his framework, children need opportunities to work with objects in the physical world before they will be ready to work with pictures and other representations. Then, after working with pictures and other representations, they will be ready to work with symbols.

USING MANIPULATIVES AND MODELS Manipulative materials and models have a critical role throughout elementary school in helping students learn mathematics. Mathematical ideas are abstract by their very nature, so any model that embodies them is imperfect and has limitations. Even though the model is not the mathematics, models provide a context for the mathematical concept under consideration. Research shows that learning occurs best when students have a meaningful context for the mathematical knowledge. The children in the Snapshot of a Lesson used everyday objects to make the concept of dividing with equal fractional parts meaningful to them. Helping children establish meaningful connections between the model (context) and the mathematics is a challenge, but it is a rewarding struggle.

Suppose you were developing the concept of a circle. You might use a plate to illustrate this concept, but the plate would also illustrate many other mathematical concepts: area, boundary, circumference, and diameter, to name but a few. Moreover, when a concept is first being formed, the learner has no way of knowing which attributes characterize it. Thus, the learner may think that irrelevant attributes matter (e.g., the design on the plate, its attractive finish, or a chip or crack) and may even fail to recognize the relevant attributes. The learner may end up not being at all clear about what characterizes a circle.

You could introduce a model such as a coin, but students might still focus on the interior rather than the boundary. In contrast, models such as a bike tire and a ring would reinforce the roundness associated with a circle but would also make it clear that a circle is associated with the outer edge or boundary of the models (see Figure 2-5). Research has shown that lightly directing children's attention to the important attributes enhances learning, so you might take a piece of chalk or a water-soluble pen and trace around the outer part of the coin or plate to highlight the circle.

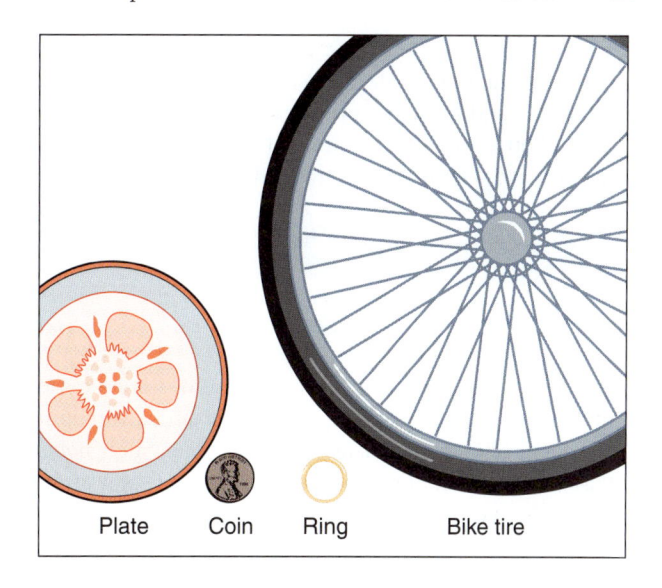

Figure 2-5 *Some models of a circle.*

The use of perceptually different models, such as those shown in Figure 2-5, is called *multiple embodiment*, or *multiembodiment*. Research has shown that transfer across contexts is difficult if children have experienced the concept in only a single context rather than in multiple contexts. The more different the models or contexts, the more likely the students are to focus only on the common characteristics and make the correct abstractions. You cannot expect children to abstract correctly from a single model in mathematics. Multiembodiment helps students abstract or generalize appropriately. It also decreases the likelihood that children will associate a mathematical concept with a particular model, which is a danger whenever a single model is used to illustrate the concept (Jordon, 2010).

Mathematical learning depends heavily on abstraction and generalization. Multiembodiment rests on the value of experiencing a mathematical concept in a variety of different physical contexts. Each embodiment has many different attributes. Children need to experience many different embodiments (along with models that do not embody the concept) before they can recognize which attributes all embodiments share, allowing them to make the correct generalizations.

Students also need to recognize that even embodiments of the same kind have variable characteristics. For instance, consider again the plate shown in Figure 2-5 as an embodiment used to develop the concept of a circle. To help children realize that they should ignore such features as size, design, and chips, you could show them several different kinds of plates. By pointing out how these features vary, you can encourage children to focus their attention on a feature that does not vary—the plate's roundness. By using varied examples of a single type of model, you can increase the likelihood that learners will focus on only the mathematically significant attributes.

Similarly, you could use rings of different designs, styles, and materials to show children that these attributes don't matter and help them recognize that the rings model a circle. You should also point out that some rings don't model a circle—for example, this "adjustable ring":

This ring does not model a circle because it is not a closed figure. If you were using plates to model a circle, you could use an oval serving platter as a nonexample. Such *nonexamples* (models that don't embody the concept) play an important role in concept formulation. Research confirms that students learn more when presented with a combination of examples and nonexamples of a mathematical concept than with examples alone.

MAKING FORMAL REPRESENTATIONS FOLLOWS CONCEPTUALIZATION Children need many experiences with concrete models before they can work meaningfully with abstract symbols. How long should a model be used? That depends on both the student and the mathematical content. In general, too little time is spent with models. That is, students are rushed (or dragged) too quickly through firsthand experiences with models and then confronted with symbolic representations. Students need to feel comfortable with a model and have opportunities to both observe and talk about the key mathematical features it embodies. It is a good idea to keep models available for students. Students know when they no longer need a model and when symbolic representations have been established. Even at this stage, leaving the model doesn't mean that they will never use it again. The same model may be used at various levels throughout elementary school to help children develop new or more sophisticated concepts.

RECOMMENDATION 4: USE COMMUNICATION TO ENCOURAGE UNDERSTANDING

The use of models, manipulatives, and real-world examples provides many opportunities for thinking, talking, and listening. The importance of communication in mathematics learning is demonstrated by the fact that communication is one of the key process standards in the NCTM's *Principles and Standards* (2000). You will learn more about these important standards in Chapter 5. When children explain their approaches, talk about mathematics, make conjectures, and defend their thinking orally, as well as in writing, their understanding deepens. Talking and writing about mathematics are essential in learning mathematics.

USING ORAL COMMUNICATION TO ENCOURAGE SENSE MAKING Lev Vygotsky, who originated the concept of the zone of proximal development, is also well known for his ideas about the role of language in learning. Vygotsky believed that learning is a social experience—that is, interactions with others challenge learners to make sense of new ideas (Kostelnik et al., 2004). This theory supports the idea that we should provide children with many opportunities to talk about their thinking. Often, communicating our thoughts to others can clarify them for ourselves as well.

> **Math Links 2.2**
>
> To see more about Piaget or Vygotsky, you may view the following short video clip. At www.davidson-films.com, click on Giants of Psychology and view an excerpt from "Piaget's Developmental Theory: An Overview" or "Vygotsky's Developmental Theory: An Introduction."
>
> www.wiley.com/college/reys

Students at all levels should talk about mathematics before being asked to communicate about mathematics symbolically. Just as speaking precedes writing for children, so should talking about mathematics precede writing about it using symbolic representations. Both student-to-student communication and student-to-teacher communication are important in the learning process.

Also, in talking about mathematics, students are likely to give you valuable insights into their thinking and understanding. This talk may take different forms. For example, by listening to the discussion in the Snapshot of a Lesson, we know that different groups of children approached and solved the problem in a variety of ways. This kind of talking between students is natural and gives them many opportunities to explain what they did, justify their thinking, and share their methods. You should do all you can to stimulate and encourage this kind of student-to-student communication, even if much of it takes place outside your hearing.

Now consider an example of student–teacher interaction as a fourth-grade class is exploring primes and composite numbers by building rectangles with tiles.

BOB: So every even number is composite.

TEACHER: Every even number? What about 2?

This question stimulates additional thinking and encourages Bob to try justifying his overgeneralization or to modify it. Questions are a vital element in the learning process. Students should ask each other questions and ask teachers questions, and teachers should ask students questions. Such interactions give students opportunities to talk about their ideas, get feedback on their thinking, and hear other points of view. Thus, students learn from one another as well as from teachers.

USING WRITTEN COMMUNICATION TO CONVEY THINKING

In writing about mathematics, students provide insight into what they are thinking and what they understand. Just like talking, putting our thoughts in writing forces us to think more deeply or helps us clarify our thoughts. Young children who are still developing their writing skills can communicate in writing by drawing pictures or dictating their thoughts to a more skilled writer. You can write prompts to help students get started. For example,

> I think the answer is… because…
> Another way to do it is…
> The thing I liked best was … because…
> I still don't understand.…

In addition, after students have had experience with a concept through the use of models and oral language, they should begin to learn the mathematical notation to communicate those same ideas. Using conventional symbols to represent mathematical concepts is a valuable part of being able to communicate mathematically.

Talking and writing about mathematics makes it more alive and more personal, and it keeps student interest high. Reading students' writing carefully and listening carefully to what students say, as well as noticing what they do *not* say, help you tailor your teaching to all the unique individuals in your class.

CULTURAL CONNECTIONS

For approximately 20 years, standards published by NCTM have included recommendations for effective mathematics instruction that reflect the recommendations about what we know about how children learn described in this chapter. They include making mathematics more meaningful and teaching mathematics through problem solving. (Teaching through problem solving is addressed in more detail in Chapter 6.) As new curricula and instructional strategies are introduced that reflect these recommendations, we must also consider how the diverse learners in our classrooms will need support so they can be successful learners of mathematics. Mathematics sometimes serves as a gatekeeper to advanced educational opportunities and careers.

As mentioned at the beginning of this chapter, there are many ways in which the students in our classrooms are diverse. When we consider cultural diversity among our students, we usually think about race or ethnicity. However, another cultural characteristic of students, which has been shown to have a large impact on learning, is socioeconomic status (SES). Many students who are struggling to learn the material in their grade level are those of low SES.

Lubienski (2007) researched how low- and high-SES students responded to a mathematics curriculum that was problem-centered. She found that students of low SES were more resistant to learning mathematics through problem solving and discussion. They preferred to have the teacher "explain how to do it" or "tell the answer." They were sometimes confused by class discussions because they weren't sure which ideas were right or wrong. They also focused more on the real-world aspects of a problem and missed the mathematical idea being presented.

Lubienski found previous research that showed "that whereas working-class jobs tend to require conformity to externally imposed routines, middle-class occupations allow for more autonomy and intellectual work" (p. 55). She also found that working-class parents tend to be more directive with their children. She cautions us to be careful not to stereotype low-SES students but to also consider that how children have been raised may impact their approaches to learning and their response to curriculum reforms.

She suggests that while mathematics achievement has steadily increased since the standards were published, we still need to provide additional interventions to narrow the achievement gap between SES groups. She provides four recommendations for teachers. First, we need to push for meaningful learning rather than rote memorization. Students are less likely to forget content they have learned meaningfully. Second, teachers need to be sure students are learning what was intended from the problems presented in the curriculum. Teachers need to provide scaffolding so students do not rely only on explicit teacher instruction. Third, teachers need to analyze achievement data and identify important topics that are causing difficulty for struggling students so that remediation may be provided. Finally, she suggests that teachers of low-SES and minority students need to advocate that those students have access to the highest-quality teachers and curriculum.

Just as children have diverse backgrounds, so do teachers. It is our responsibility as teachers to learn all we can about the culture of socioeconomic groups to which we do not belong. Such information will serve us well in helping children learn mathematics.

A GLANCE AT WHERE WE'VE BEEN

The children in classrooms today are increasingly more diverse. Teachers must create a classroom climate to support their diverse learners. This includes setting a positive learning environment, avoiding negative experiences, establishing clear expectations, showing positive attitudes toward mathematics, treating aptitudes as equal, and working to improve student retention.

Mathematics learning can and must have meaning. This statement is the cornerstone of all instructional planning and teaching. Conceptual and procedural knowledge are essential elements of learning mathematics. Mathematics learning is a slow process that requires years of development. Many individual differences exist, and the rate of learning varies greatly among children. Given these variables, the essential role of teachers is to help children construct mathematical knowledge that is meaningful to them. In performing this role, teachers use behaviorist and constructivist learning theory to make countless decisions to plan appropriate learning activities; establish an inviting classroom environment; and organize the classroom to ensure that all children are actively participating in experiencing, learning, abstracting, and constructing mathematics that is meaningful to them.

Research and learning theory suggest that children move through distinct stages of development, that there is significant value in children being actively involved constructing their own knowledge, that learning moves from concrete to abstract, and that communication plays an important role in facilitating construction of new knowledge. In addition to hands-on experiences, children learn from telling, explaining, clarifying, making conjectures, and reflecting on what they have done. They also learn from watching, listening, reading, following directions, imitating, and practicing. While all of these experiences contribute to learning mathematics, teachers have the responsibility of deciding the proper balance of these experiences.

Things to Do: From What You've Read

1. Discuss the Albert Einstein quotation at the beginning of this chapter. What does it mean to you? What are some of the conditions that support learning?

2. Provide some mathematics examples to help distinguish between procedural and conceptual knowledge.

3. Describe how you would help a child understand that in Figure 2-3 triangle ABC has an area that is half the area of rectangle BEFC using the different regions in the diagram.

4. Examine the learning frameworks in Figure 2-4. Tell how the models proposed by Piaget, Bruner, and Dienes are alike. How are they different?

5. Select Table 2-2 or 2-3. Then describe how this information will influence your teaching as you help children learn mathematics.

6. What is metacognition? Examine the activity described in In the Classroom 2-1. Describe how metacognition might be used to think about your thinking during that activity.

7. Of the four recommendations discussed for helping children make sense of mathematics, which one do you think will be most challenging for you to implement? Tell why.

Things to Do: Going Beyond This Book

In the Field

1. [1]*Patterns with Blocks.* Use In the Classroom 2 with a child. Did the child realize that two of the pictured boxes are the same box, just in a different position? Would it make sense to try this activity with children if blocks were not available? Why?

2. [1]*Classroom Manipulatives.* Identify models and manipulatives that are available to use in a mathematics classroom that you visit. Did you see them being used? If so, what models and manipulatives were being used and how?

3. [1]*Equity in Whole-Class Lessons.* Observe a mathematics class and be sensitive to the verbal and physical clues that suggest boys are treated differently from girls. If you see evidence that boys are treated differently from girls, discuss some examples.

4. [1]*Student Attitudes.* Ask a child some questions to learn about his or her attitudes and perceptions about mathematics. For example:

   If I say, "Let's do some mathematics," what would you do?
   What school subject do you like most? Are you good at it?
   Do you think knowing mathematics will help you when you grow up? Tell why.
   Do you think your teacher likes to teach mathematics? Tell why.
   Summarize what you learned about this child's feelings about mathematics.

In your Journal

5. Reflect on your own experiences learning mathematics. Describe how you "learned" mathematics.

6. Consider the two learning theories, behaviorism and constructivism. Describe personal experiences that suggest the influence of these theories on your learning of mathematics.

---

[1]Additional activities, suggestions, and questions for this activity are available in the field experience manual on the Student Companion site at www.wiley.com/college/reys.

With Additional Resources

7. Google *A Framework for Understanding Poverty* by Ruby Payne and examine the PDF. Tell how understanding poverty will help you be a better teacher of mathematics?

8. Examine the Learning Principle in the *Principles and Standards for School Mathematics* (NCTM, 2000). Describe some specific ideas that would be useful in planning instruction and helping children learn mathematics.

9. Examine the NCTM Yearbook *Multicultural and Gender Equity in the Mathematics Classroom* (Trentacosta and Kenney, 1997). Read one of the articles, and then identify and discuss several specific implications for mathematics teaching.

10. Examine the book *Mathematicians are People Too: Stories from the Lives of Great Mathematicians* by Reimer and Reimer in Book Nook for Children and report how you might use this resource. Do you think this personalization helps promote more interest in learning mathematics? Do you think helping children learn more about women's contributions in mathematics is an important instructional goal? Tell why.

With Technology

11. Check the Women's Educational Equity Act (WEEA) at the Equity Resource Center Web site (www.edu.org/womensequity/) for a history of Title IX, a list of publications, current articles, and other Web links. Report on something related to equity and mathematics that you find at this site.

12. Conduct a Web search for the Ask Dr. Math Web site. Review the questions that elementary teachers and students send for Dr. Math to answer. Do you find any evidence of mathematics anxiety in the questions that you see? Do you see a difference in the questions submitted by girls and by boys?

Note to Instructors: You can find additional resources, learning activities, and blackline masters in this text's accompanying Instructor's Manual, at www.wiley.com/college/reys.

Book Nook for Children

Hopkins, L. B., and Barbour, K. *Marvelous Math: A Book of Poems.* New York: Aladdin, 1998.

Sixteen poems cover different mathematical topics that include division, multiplication, fractions, counting, and measurement. All of these poems will relate to everyday living in our world. The book will let children see how mathematics plays an important role in our everyday living. This book will be a good springboard for introducing different topics in mathematics.

A Day with No Math. Orlando, FL: Harcourt Brace Jovanovich, 1992.

This book will let children see how we use mathematics in our everyday living. Children will see many topics in mathematics in the story line. Addition, measuring, estimating, geometry, statistics, and probability are just some of the topics presented in this story about a boy who does not like to do math. As he falls asleep, he dreams what it would be like with no math in our daily living, and then realizes through his dream just how important mathematics is in our daily living.

Reimer, L., and Reimer, W. *Mathematicians Are People, Too: Stories from the Lives of Great Mathematicians* (2 volumes). New York: Pearson, 1997.

Share moments of mathematical discovery experienced by famous mathematicians with your students. Use these stories to demonstrate that both males and females are mathematicians, that mathematicians come from a variety of cultures, and that the work of mathematicians is applied to many diverse fields. Use this book with children in grades 3–7 to experience the stories of great mathematics, including Pythagoras, Galileo, Pascal, Albert Einstein, and Ada Lovelace.

W. Kieran, *Math and My World.* Vero Beach, FL: Rourke Publishing, (2004).

This six-book series includes real-life examples and photographs of mathematics in our world. Topics include math in the kitchen, on maps, in medicine, with money, and with travel and weather. Upper elementary students could use these books as resources and make connections to the topics being studied in their mathematics class.

Scieszka, J., and Smith, L. *Math Curse.* New York: Viking, 1995.

This book will let children see that if you let math become an anxiety, it will become a curse. Many core topics are presented in this book, such as fractions, counting, dividing, Fibonacci numbers, physical education, problem solving, and measurement, to name just a few topics. This book will let children appreciate mathematics in a new way. This book can be used at all levels of mathematical learning.

CHAPTER 3

Planning for and Teaching Diverse Learners

> "ONE GOOD TEACHER OUTWEIGHS A TON OF BOOKS."
> — Chinese proverb

>> SNAPSHOT OF A LESSON

KEY IDEAS

1. Measure the circumference and diameter of various circles to the nearest centimeter.
2. Collect and analyze data.
3. Discover the relationship between the circumference and diameter of a circle.

BACKGROUND

This snapshot is from the video *Circumference/Diameter*, from Teaching Math: A Video Library, K–4, from Annenberg Media. Produced by WGBH Educational Foundation. © 1995 WGBH. All rights reserved. To view the video, go to http://www.learner.org/resources/series32.html (video 22).

Ms. Scrivner is teaching a geometry lesson to her fourth graders. They have been learning about circumference and diameter for the last two days. Today she begins by reviewing the vocabulary.

Ms. Scrivner: What did we do yesterday to learn about circles?

Ian: We drew a circle and drew a diameter and a, what was that word again?

Ms. Scrivner: The *R* word? Look over here, we have our vocabulary words written on the board over here. Let's read them. [*Class reads* circumference, radius, center, compass, radii.] Daniel, how did you tell us we could remember radii.

Daniel: *Radii* has two *i* and and so it means 2 or more.

Ms. Scrivner: I'm going to pick the octagon team. Can you show us on the floor what we did yesterday? [*Children kneel around a circle diagram on the floor, touching various parts.*]

Child: We took a pencil and put it in the center. Then we put a string on it. Then we went out as far as the string could go.

Ms. Scrivner: So you have the pencil as the center of the circle and you are using the string as a what? [*A radius of the circle.*] Can you remember where we learned about circumference in math before?

Chris: We used a pumpkin to learn about the circumference.

Ms. Scrivner: OK. What was the circumference of the pumpkin? Can you show me with your arms? Close your eyes and imagine you have a big pumpkin in your arms. [*Students make a circle with their arms in front of them.*] Now open your eyes and look at your arms and your body. That is the circumference of the pumpkin. Now we know all about that. Let's look at the word *circumference*. [*Teacher writes* circumference *and* circle *on the overhead projector sheet.*] What do you see is the same about these two words?

James: circ

Ms. Scrivner: There are some other words that have *circ* in them. [*Writes* circus.] Do you know why a circus might be named a circus? What does it have to do with something round?

32

Girl: They have different rings that they do stuff in. They call the circus master the ringmaster.

Ms. Scrivner: Excellent! That's exactly right. So all of these things have to do with something round. So today, what we're going to do is see if we can discover a relationship or something that is special about the circumference of a circle. Do this [*has the students draw a ring in the air with their finger*] and the diameter, now do this [*has the students draw a horizontal line in the air*]. Alright, I would like you to get your math kits out of your desk, please. [*Students remove a large zipper bag containing manipulatives from their desks.*] I want to make sure that you have a centimeter tape. Pull out your tape and take a look. We're going to be measuring in centimeters today. Say that word. [*Class says* centimeters.] When you are measuring, I want you to round off to the nearest centimeter. I have a chart for each team. We are going to use this chart to collect some measuring data. James, can you tell me what *data* means?

James: Collecting information to find something out.

Ms. Scrivner: Good. We are going to discover something. We are going to find something out. It looks like you are going to find some items around the classroom and measure the what? [*Circumference*] So can it be a pencil box? Will it be something round or square? [*Round*] You are going to be searching around the room for round things. You are going to measure the circumference [*hand motions*] and the diameter [*hand motions*]. Next, I want you to come to a team consensus and come back to your table and record that data. Alright? Let's get busy. [*Students get out of their seats, measuring various objects such as a circle on a Venn diagram poster, a baseball hat, the trashcan, a pencil can, etc. Then they have a whole-class discussion.*]

Ms. Scrivner: Right now we are going to share some of our data. Look at your team chart and figure out what you know you measured well. [*Students discuss choices with team. Ms. Scrivner notices an error on one team's paper.*]

Ms. Scrivner: Show me the diameter with your hands. [*horizontal line in the air*] Show me the diameter of that trashcan. [*oops!*] Put your three heads together, and measure the diameter of that trashcan. [*Ms. Scrivner has teams share data and records it on the overhead projector chart.*]

Ms. Scrivner: I want you to look and see if you can come to a quick decision about what you found in this data. You may not be able to see it. This one [*6 cm and 2 cm*] is the easiest to see. Think about your multiplication facts. Is there a relationship that these numbers share? Does anybody think they see it? James thinks he does. Put your heads together and talk about it for a minute. [*Students discuss in groups.*]

Girl: It's times 3.

Ms. Scrivner: Two times 3 is 6. Six divided by 2 is 3. Let's take the circumference divided by diameter and see what we get.

[*Students use calculators to check other pairs of numbers, since they had not had experience yet with the division algorithm.*]

FOCUS QUESTIONS

1. What questions must an elementary mathematics teacher answer when planning for teaching?
2. Why does the teacher plan mathematics lessons so carefully?
3. What levels of plans does the teacher create?
4. What are three types of lessons used to teach mathematics, and what is the purpose of each?
5. How can the teacher meet the needs of all students?
6. How does the teacher integrate planning with assessing and analysis?

INTRODUCTION

Teachers play many different roles in the classroom, such as counselor, gardener, police officer, parent, actor, friend, guide, cheerleader, and others that you can probably think of. That is why, as the opening quote says, one teacher is really valuable. Which roles will you play? To some extent, that depends on your personality, your school, the students in your class, and many other factors, but all teachers must play the role of instructor.

The instructor role involves four basic functions: planning, teaching, assessing, and analyzing. Consider how the teacher in the Snapshot of a Lesson performed these functions: First, she made plans—deciding on the focus of the lesson (the objective) and then figuring out how to carry out the lesson from the textbook or curriculum. Then she taught the lesson (as illustrated in the Snapshot discussion)—presenting the problem, giving directions, asking questions, as well as listening to, observing, and providing feedback to the students. During and after the lesson, she assessed the students' understanding. Finally, she reflected on the lesson and analyzed what went on.

In this chapter, we focus on the planning and teaching functions. First, we discuss the kinds of general questions a teacher must ask and answer before beginning to make specific plans—questions that address such topics as what the teacher and students already know and how to get students talking and questioning. We then discuss levels of planning (for the year, for units, for lessons), planning for different types of lessons, and planning to meet the needs of all students. This chapter also briefly touches on assessment and analysis (covered in detail in Chapter 4). In later chapters, we explain how to help children learn specific mathematics concepts and skills, and we elaborate on each of these instructional functions. Thus, this chapter serves as an organizer for much of what you will learn throughout the rest of this text.

And, like every other chapter in this book, this chapter reflects our acceptance of the idea that children actively construct their own new understandings as they build on what they already know.

PREPARING TO TEACH: QUESTIONS TO ASK

Teaching children mathematics is a complex task. As a teacher, you will have to make many on-the-spot decisions during instruction. But there is much you can do beforehand to increase the chance that the year, each unit, and each individual lesson will be a success for all the diverse learners in your classroom. Making this happen means asking and answering several important questions before lessons begin, even before you begin writing lesson plans. We discuss these questions and answers in the following sections.

DO I UNDERSTAND THE MATHEMATICS I AM TEACHING?

Teachers today are being asked not only to teach mathematics in ways different from how they were taught in elementary school, but also to teach topics and concepts they did not learn in school. When this happens to you—when you have to teach something you don't know much about—your first job will be to learn all you can about it. This point is made in NCTM's *Principles and Standards* (2000), which states, "Students' understanding of mathematics, their ability to use it to solve problems, and their confidence in, and disposition toward, mathematics are all shaped by the teaching they encounter in school .... *Teachers must know and understand deeply the mathematics they are teaching and be able to draw on that knowledge with flexibility in their teaching tasks.*" (pp. 16–17; emphasis added). Similarly, the National Research Council (2002) notes that teachers must be proficient at a much deeper level than their students in order to teach mathematics well. We know from a variety of research that teachers with strong content knowledge are better able to develop their students' conceptual understanding of mathematics by providing multiple representations of the same concept. Teachers with strong content knowledge are better at posing questions and interpreting and responding to students' answers. They can effectively respond to students' questions about why a procedure works. These teachers possess mathematical knowledge for teaching (MKT) (Thames and Ball, 2010). Teaching in the manner described above can lead to higher student achievement in mathematics. You won't be able to teach for understanding if you don't have a strong understanding yourself (Burns, 2007).

> ### Virtual Classroom Observation
>
> Go to **www.wiley.com/college/reys**,
> Access the Wiley Resource Kit
> Click on the Virtual Classroom Observations Section
>
> Module 1: Foundations of Effective Mathematics Teaching, Part B, Focus on Teacher Content Knowledge, Section 2, Learn About It
>
> 1. Read about the importance of teacher content knowledge and view the video.
> 2. List some examples of how deep mathematical understanding informs a teacher's practice.

There are many ways you can improve your subject-matter knowledge, and doing so will not only make you more skilled but will also increase your confidence! Educators have emphasized how important it is for you to work through lesson problems and activities before presenting them to students, because by working through the lesson and taking the role of the learner, you deepen your understanding of the mathematics and you experience firsthand what your students will be doing (Brodesky, Gross, McTigue, and Tierney, 2004). These same educators suggest that teachers identify the mathematical goals of each lesson, and what is most important for all students to learn, in order to avoid teaching in ways that do not meet those goals or that even conflict with them. And this process, too, requires that teachers understand the mathematical content at a deep level. Even if the lesson topic is familiar to you, you need to review the mathematics you will be teaching in order to experience it in the same way the students will and so that you can focus on students' conceptual understanding.

In other words, *be sure you understand the material well*. Chapter 1 gives several suggestions about where you can turn for help in case you don't; one of the best of these resources is the teachers' guide for your school's textbook. Many of these guides not only help you understand the mathematics being taught; they also alert you to common student misconceptions and describe effective methods and materials you can use to teach the topic.

WHERE ARE MY STUDENTS DEVELOPMENTALLY?

Before planning mathematics lessons, you should review the general developmental characteristics of the children in your classroom, as well as the strengths, needs, and interests of individual learners. Understanding your students' characteristics, especially their levels of development, can help you plan lessons that are appropriate for everyone. For example, a teacher planning a first-grade lesson should realize that his students need to manipulate concrete objects to solve addition problems and are just beginning to use symbols to represent those objects. Also, remembering that first graders are still egocentric in their thinking could give the teacher the idea of making up story situations, with the students themselves as characters.

In Chapter 2, we discussed some of the cognitive, physical, and social developmental characteristics of students in elementary school, and these overall trends provide a good starting point as you begin planning mathematics experiences for your students. Remember, however, that children do not all progress through developmental stages at exactly the same time and that they may, therefore, vary greatly in their readiness for learning. Thus, some first graders will understand addition and develop fluency with basic facts before some third graders. Likewise, some intermediate-grade students may have difficulty visualizing the blocks shown in In the Classroom 2.2; they may need to "build" these boxes before the task becomes meaningful. Children at the concrete operations stage must root their ideas in concrete objects; they have little ability to manipulate abstract relationships. These children would find it difficult (if not impossible) to learn how to find the volume of a cube by looking at a two-dimensional drawing.

You should also focus on your students' strengths as well as on their weaknesses. This will lead you to ways of using areas of strength to help students compensate for their weaknesses. It will also help you identify parts of a lesson that may give students problems and think of ways to make those parts more accessible. For example, a student who has trouble focusing but loves to talk and read might be paired with a student who focuses well but has reading difficulties. What is a strength for one student may be a weakness for another (Brodesky et al., 2004).

Finally, considering your students' social and cultural experiences when planning can make your lessons more relevant and meaningful. Children may find it easier to learn and be more motivated to learn if they can see connections to their lives. For example, when planning a lesson on patterns, one teacher had the children bring in small, countable food items from home to create the patterns. Students' learning styles and preferences also are a reflection of their social and cultural experiences, so if you can be aware of and teach with these things in mind, the lesson will be more likely be successful.

WHAT DO MY STUDENTS KNOW?

As a teacher, you must identify and build on what your students already know. Instructional time is precious, not to be wasted on lessons that are too easy or difficult for your students. Of course, if students in your classroom are at different academic levels or if they have learning problems or disabilities, providing equitable treatment may mean adapting instruction to meet their particular needs. To meet the needs of all of the children in your class, you first need to assess them. Chapter 4 offers a variety of assessments you might use. Paper-and-pencil assessments can be useful tools in learning what your students know and can do, especially if you go beyond looking at final scores and analyze how children reached their answers. Also, you can learn a lot about your students by using other assessment strategies such as observing and talking with them; this can reveal much that would not be noticeable from tests. You should use the results of assessments to help select instructional materials and approaches, group students for instruction, adapt materials, and decide what needs to be taught or re-taught to individuals and to the class as a whole.

After assessing your students, you must organize the mathematics so you teach it in a way that is appropriate and understandable to them. Mathematics includes both conceptual and procedural knowledge; thus, you face two challenges—helping students develop both types of knowledge and helping them understand the relations between them. In no other discipline is previous knowledge more critical—for example, it would clearly be pointless to ask students to estimate a distance in kilometers if they didn't already have some sense of the length of a kilometer.

The importance of children's previous knowledge is clear from the way that quality mathematics programs are organized, both to provide continuous development and to help students understand the basic structure of mathematics. Scope-and-sequence charts give an overview of how particular topics are developed across grade levels. A careful examination of such a chart reveals how the sequence of activities related to a particular topic is organized in a *spiral approach*, giving students repeated opportunities to develop and broaden concepts. The spiral approach incorporates and builds on earlier learning to help guide the child through the increasingly intricate study of a topic. For example, angle measurement is informally introduced in primary grades and returned to many times in later grades, each time at a higher level of sophistication. Figure 3-1 shows how the spiral approach uses previous experience to develop the concept of "angle."

The description we just gave of the spiral approach represents an ideal; the reality, however, may be somewhat different. In theory, the spiral mathematics curriculum provides for continuous growth and development, but in practice many topics are revisited without appreciable change in the level of treatment from one year to the next. Consequently, much valuable time may be spent each year treading water, merely reviewing topics rather than building on prior learning or introducing new topics. Of course, a certain amount of review is probably necessary, for reasons like those discussed in the next paragraph. However, unnecessary repetition is detrimental because it turns students off. It robs both teachers and students of the excitement inherent in exploring new topics. It also uses up time that would otherwise be available for new learning, which means that fewer new ideas can be experienced each year.

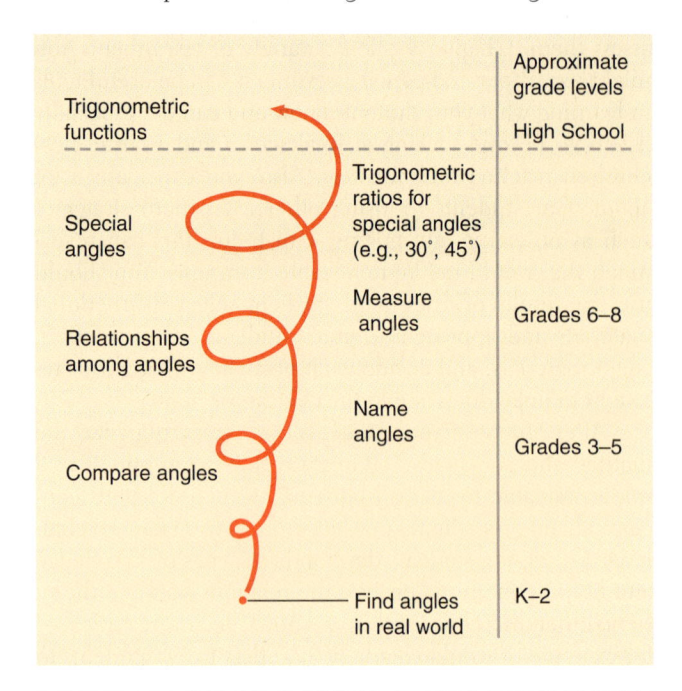

	Approximate grade levels
Trigonometric functions	High School
Trigonometric ratios for special angles (e.g., 30°, 45°)	
Special angles	
Measure angles	Grades 6–8
Relationships among angles	
Name angles	Grades 3–5
Compare angles	
Find angles in real world	K–2

Figure 3-1 *Spiral approach toward developing the "angle" concept.*

The spiral approach holds profound implications for learning and teaching. For example, when planning a lesson, you must consider what prerequisites students must have to succeed in the lesson, and you must check to see if students have them. Often, you will find students who have never learned, incorrectly learned, or simply forgotten such prerequisites. Detecting these weaknesses early and quickly will let you plan the right kind and amount of review, so you can properly prepare students for later lessons without wasting time.

Besides knowing what students learned in earlier grades, teachers must look ahead to see what their students will need to know tomorrow, next month, and next year. Third-grade teachers must know what their students learned in kindergarten, first grade, and second grade, as well as what their students will have to know in fourth grade and beyond. This broad perspective helps teachers appreciate the importance of their role. It also helps them detect gaps in their students' mathematics learning, so they can fill in any gaps as quickly and effectively as possible.

Of course, a gap for one student may not be a gap for another. If an individual student has a gap in understanding that most students do not have, you will need to work with that student to fill in the gap. It is not appropriate to force the whole class to review something that only one or two students do not understand. As a teacher, you will need to make many instructional decisions directed at helping individual students while continuing to challenge *all* your students, extending their experiences and their levels of understanding, and pass this information on to teachers in the following grade. This will help those teachers continue to extend the learning of individual students.

WHAT KINDS OF TASKS WILL I GIVE MY STUDENTS?

What are the appropriate kinds of tasks for your students? This is another key question you must answer before you begin planning lessons. Appropriate tasks might be "projects, questions, problems, constructions, applications, and exercises" (NCTM, 1991, p. 20). Tasks must encourage students "to reason about mathematical ideas, to make connections, and to formulate, grapple with, and solve problems" (p. 32). One educator states:

> No other decision that teachers make has a greater impact on students' opportunity to learn and on their perceptions about what mathematics is than the selection or creation of the tasks with which the teacher engages the students in studying mathematics. Here the teacher is the architect, the designer of the curriculum.
>
> (Lappan, 1993, p. 524)

ENCOURAGE HIGHER-LEVEL THINKING Your students will define what it means to "do mathematics" based on their experiences completing the classroom tasks you provide. If you provide lower-level tasks requiring only memorization or learning procedures without conceptual understanding, your students will believe that learning mathematics does not require understanding or justification, just recall and repetition. If you provide higher-level tasks such as learning procedures by connecting conceptual understanding and multiple representations or using their acquired knowledge to solve problems in a variety of ways and justify their solutions, your students will believe that learning mathematics involves critical thinking, exploration, and sense making (Wagener, 2009). Researchers have classified mathematical tasks from lowest to highest cognitive demand.

Lower-level demands

- Memorization (i.e., reproducing previously learned facts, rules, formulas)
- Procedures without connections (i.e., focusing on producing correct answers)

Higher-level demands

- Procedures with connections (i.e., using procedures to develop deeper understanding or concepts)
- Doing mathematics (i.e., exploring concepts, applying relevant knowledge to solve complex tasks)

In addition, it was found that the highest student learning gains on a mathematics performance assessment resulted from engaging students in those higher-level tasks that required elevated levels of cognitive thinking and reasoning (Smith and Stein, 2008). Examine Figure 3-2, found later in this chapter. Compare the cognitive levels of the two sample problems. Which one would encourage higher-level thinking?

FOCUS ON THE MATHEMATICS   It is extremely important that mathematics lessons maintain a focus on the important mathematical topics. Thus, one of your top priorities when you start planning lessons will be to decide on the mathematics objectives of each lesson. Certainly, you can improve your lessons and make them more fun for students by using children's literature as a springboard, by involving students in hands-on explorations, and by setting problems in real-world contexts. But your first question when planning a lesson should be, "What mathematics do I want the children to learn from this?" It is not appropriate to use manipulative materials just as toys, to engage students in talking and writing only about their feelings and not about the mathematics, or to do lessons that are just fun for fun's sake.

As a teacher, you will be responsible and held accountable for helping children meet district and state standards, and you can do this only if your mathematics lessons involve students in tasks that are mathematically rich. When children are genuinely engaged in solving mathematical problems that make sense to them, the learning they take away from that experience is likely to be deep and lasting. The Snapshot of a Lesson at the beginning of this chapter shows children engaged in important mathematical processes such as those recommended by the *Principles and Standards for School Mathematics* (NCTM, 2000) and discussed in Chapter 5. The objectives you teach are often provided in your district or state curriculum guides. Insight can also be gained by consulting national documents such as NCTM's *Principles and Standards for School Mathematics* (2000) or the *Common Core State Standards for Mathematics* (CCSSI, 2010).

PROVIDE NECESSARY PRACTICE   Part of selecting appropriate tasks is considering how to provide necessary practice. Clearly, students need practice in order to acquire fluency with a mathematical idea or procedure. Just as we must practice while learning to walk or drive a car, we must practice basic addition facts or how to make patterns. The choice is not if, but when. Research has long indicated that drill and practice should follow, not precede, the development of meaning (Brownell and Chazal, 1935). In addition, drill and practice should not consume as much time as developmental work that helps children to understand the idea or procedure. Whenever possible, practice should be meaningful—it should be much more focused on reinforcing conceptual understandings than on going over rote procedures. Finally, practice should help children apply their knowledge in new contexts.

Many textbook series provide practice on the content presented, lesson by lesson. Your decision whether to assign this practice material (in addition to all the other practice that students get—for instance, in the course of problem solving) must be based on your students' needs. You have to strike a balance between two conditions: One is that children must understand the mathematics before practicing it; the other is that practicing a skill that has already been mastered is boring and thus can be harmful.

Therefore, you should include practice on previously learned topics as a small but regular component of each lesson. Keep in mind that completing a meaningful task will often provide just as much practice as completing textbook exercises, if not more.

HOW WILL I ENCOURAGE MY STUDENTS TO COMMUNICATE?

You can help your students learn mathematics by encouraging them to communicate their thoughts to you and to each other, by asking them the right questions at the right times, and by grouping students in ways that are appropriate to their strengths and needs and to the mathematics being taught.

TALKING   You can use classroom talk, or discourse, to make certain that the classroom discussion helps students make sense of the mathematics you are teaching. Classroom discourse also helps children, who are learning English, develop and practice their language skills. Just listening to others discuss concepts can help clarify those ideas for some students. You will be the organizer, cheerleader, and conductor of the entire process of discourse, making appropriate tools available and finding ways to help children assume responsibility for their own learning. One teacher describes her efforts to improve and increase her students' discourse:

> After some experimentation, I realized that if I rephrased questions to start with "How," "What," or "Why," students needed to give a longer response than just "yes" or "no." I also discovered that I could give a request such as "Tell me about…" rather than ask a specific question. Frequent use of the think-pair-share strategy also encouraged discussion by allowing time for students to compose their thoughts and by helping shy students become more confident in responding in front of the class. The length and depth of student responses gradually increased. In addition, I began to insist that students explain their reasoning and justify their responses for even factual textbook questions. At first, many students seemed to feel that their response was incorrect whenever I asked for an explanation. With reassurance that I was just curious about their thinking, they began to share more often and soon began to justify their responses without being prompted.
>
> (Cady, 2006, p. 461.)

A flow of ideas—not only from teacher to student but also from student to teacher and from student to student helps every child learn mathematics (NCTM, 1991). Here are some ways you can help create effective classroom talk.

- Step back and allow your students to do most of the talking. In our enthusiasm to clarify an idea, or affirm correct answers, we may be communicating to our students that their ideas aren't important or clearly expressed. You should be the one who facilitates, not dominates, classroom discussion (Sims, 2008).

- Show your students that you expect them to communicate their thinking, to you and to other students. In addition, model ways in which your students can question and be questioned by their peers, and show them that you expect them to do this. Students must explain and justify their thinking, (CCSSI, 2010).
- Make sure your mathematics class is a safe place for students to conjecture and to make mistakes. Students often learn a great deal by listening to others and analyzing mistakes. Be certain to give students the opportunity to change their minds about conjectures and possible solutions.
- Pose problems and assign tasks that encourage children to talk about their mathematical thinking as well as to listen and respond to other students' ideas. Interactions among children give them opportunities to talk about their ideas, get feedback, and see things from other points of view. Students can learn from one another as well as from the teacher.
- Explicitly teach mathematics vocabulary. Rubenstein (2007) notes that some vocabulary may be confusing for students. For example, words like *round* or *square* have multiple meanings in mathematics. Words like *product* or *reflection* have different meanings in mathematics than when used in everyday language. Learning to understand and use correct terminology will enhance students' abilities to communicate mathematical ideas effectively. It is important to develop the concept or mathematical idea before introducing the vocabulary. Then students already have the knowledge and the focus is on communicating it effectively (Burns, 2007).

Virtual Classroom Observation

Go to www.wiley.com/college/reys,
Access the Wiley Resource Kit
Click on the Virtual Classroom Observations Section

Module 1: Foundations of Effective Mathematics Teaching, Part C, Listening To and Interpreting Student Thinking, Section 2, Learn About It

1. Read about the importance of listening to students' thinking and making inferences and view the video.
2. List some questions teachers can ask to draw out student thinking.

Teachers also benefit from listening to their students' thinking. Notice how the teacher in the Snapshot of a Lesson at the beginning of this chapter helped the children make meaningful connections and find ways to remember important terms.

QUESTIONING Good questions are vital in facilitating learning. Questions that are aimed at checking children's knowledge of a fact or their ability to perform a skill are relatively low level and are the easiest to create during a lesson. Higher-level questions require children to analyze, synthesize, or evaluate information, such as when we ask students to explain why a procedure works or how they decided an answer is correct. Research has shown that in mathematics lessons, teachers tend to use lower-level questions much more frequently than higher-level questions. Children's ability to answer lower-level questions may correlate with high scores on some achievement tests, but such questions give children the wrong picture of mathematics by implying that mathematics involves only simple answers and that there is only one correct answer per question. So plan on being a teacher who asks plenty of higher-level questions that encourage children to do more talking about mathematics—talking that explores why a procedure works, what would happen if something were changed in the problem, and how mathematics could be applied in real-life situations.

To be this type of teacher, you will need to consider the questions you want to ask and write them in your lesson plans. Try to prepare a range of questions, but focus on questions that require children to think rather than merely supply a fact from memory or perform a learned procedure. If you ask only lower-level questions, students may come to believe that the main goal in mathematics is to memorize facts rather than to understand and make sense of mathematical ideas. Remember, however, that creating higher-level questions can be challenging—you'll have much more success if you develop such questions before the lesson begins.

The NCTM *Professional Standards* (NCTM, 1991, pp. 3–4) includes five categories of higher-level questions teachers should ask:

1. *Questions that help students work together to make sense of mathematics:*
"What do others think about what Janine said?" and "Can you convince the rest of us that that makes sense?"

2. *Questions that help students rely more on themselves to determine whether something is mathematically correct:*
"Why is that true?" and "How did you reach that conclusion?"

3. *Questions that seek to help students learn to reason mathematically:*
"How could you prove that?" and "What assumptions are you making?"

4. *Questions that help students learn to conjecture, invent, and solve problems:*
"What would happen if …" and "Do you see a pattern?"

5. *Questions that relate to helping students connect mathematics, its ideas, and its applications:*
"How does this relate to …" and "Have we ever solved a problem like this one before?"

Math Links 3.1

To see a teacher and students communicating, view the video "Developing Estimation Strategies by Making Connections Among Number, Geometry, Measurement, and Data Concepts: Estimating Scoops," at the NCTM Standards Web site: http://standards.NCTM.org/document/eexamples/chap4/4.6/index.htm or at this book's Web site.

www.wiley.com/college/reys

GROUPING   One way to increase student communication in your mathematics class is to use a variety of grouping patterns. There are three basic patterns of grouping for mathematical instruction:

- Whole class, with teacher guidance
- Small group, either with teacher guidance or with student leaders
- Individuals working independently

Many teachers use whole-class instruction too extensively. You should keep in mind that small-group work can help children develop their reasoning abilities through active involvement, can ensure that children work on content focused on their particular needs, and can at the same time help them learn to work together to solve problems. Having individuals work alone can also serve important purposes. Here are some guidelines to help you decide when to use which grouping pattern:

1. Use large-group instruction:
   - If the topic can be presented to all students at approximately the same point in time (i.e., if all students have the prerequisites for understanding the initial presentation)
   - If students need continuous guidance or modeling from the teacher to master the topic
2. Use small-group instruction:
   - If students can profit from student-to-student interaction with less teacher guidance
   - If you are trying to encourage exploration and communication about mathematics
   - If you are trying to help students benefit from the acquisition of cooperative learning skills
3. Use individual instruction:
   - If students can follow a sequence or conduct an activity on their own
   - If the focus is individual practice to attain mastery

Most effective teachers use a combination of grouping patterns. For example, a teacher might develop new material with the whole class and then have children work individually or in small groups for one or more activities. Communication with one another is especially valuable because the students help each other master mathematical ideas. The teacher moves from group to group, providing assistance by asking thought-provoking questions as needed. The whole class is then brought together again for a teacher-mediated discussion so that children can hear ideas from students in other groups. A variation on small-group learning involves just two children working together on a task. This is particularly appropriate for young children who may not be developmentally ready to work in larger groups. The process of working together promotes the learning of both children.

Educators have pointed out that placing students in small groups will not guarantee a successful learning experience. You will need to teach students skills in communication, team building, small-group socialization, and helping so that all group members benefit from the experience. Before you place children in groups, carefully consider their individual abilities and personalities. For example, you might think that placing low, middle, and high achievers in separate groups would better meet their needs, but research has found that homogeneous (similar ability) groups do not help increase overall student achievement and may promote inequity—that is, this type of grouping has not been found to narrow achievement gaps. In fact, low-achieving students actually perform worse when placed in homogeneous groups than they do in heterogeneous (mixed ability) groups. Middle- and high-achieving students do show some gain when placed in homogeneous groups; this may be helpful at times, but grouping them heterogeneously holds even more promise (Marzano, Pickering, and Pollock, 2001).

Small-group learning may require more planning and class time than whole-class instruction, but this is more than outweighed by the positive results: increased mathematical communication, social support for learning mathematics, opportunities for all students to experience success in mathematics, increased likelihood that children will see more than one approach to solving a problem, and opportunities to deal with mathematics through discussion of meaningful problems.

WHAT MATERIALS WILL WE USE?

Planning for instruction also involves selecting materials to enhance or support the text or curriculum. Teachers have varying levels of freedom when selecting materials to support their lesson objectives. Some can start with the textbook lessons and supplement them creatively, finding ways to adapt the textbook to the learners in their classroom. Others may be required by the school to use particular curriculum materials or textbooks and not to make more than the minimum adaptations needed for the students. Either way, you will need to make the best use of the available materials, given your goals for the lesson and the children in your class. Whatever curriculum materials you use, sometimes you will want to locate additional resources for yourself or your students. Educational publishers and the Internet provide lots of options, and you can also consider resources such as children's literature, manipulatives, and technology.

TEXTBOOK LESSONS Beginning teachers often wonder how much to depend on a textbook for lesson planning. To what extent can you use textbook lessons "as is"? How and why should you adapt them for your individual use? Obviously, answers to these questions depend on the textbook, on how closely its goals and methods align with your district and state standards, and on your own teaching preferences. It certainly does not make sense to develop your own original lesson plans if you have a textbook that provides an appropriate outline for each lesson. If you determine that the lessons are developmentally appropriate for your students, that they engage the students in genuine problem solving, and that they promote conceptual development, there is no reason not to use them, adapting them as seems necessary. Teachers' guides to textbooks can give you a variety of ideas for teaching lessons using the textbook. Guides also often provide suggestions for follow-up activities, such as written work and games. And many guides offer suggestions for remediation, enrichment, and alternative tasks for children with learning disabilities and other special needs.

There are several effective ways you can supplement a textbook lesson. If you are required to teach from a textbook with a more traditional approach, you can considerably enhance the lessons by embedding them in a motivating theme or problem-solving context. Such a context will give you and your students an organizing framework for teaching and learning and will help tie the lesson (or several lessons) together. Contexts for lessons may be taken from real-life situations (e.g., from current events or from special days such as holidays noted on a calendar) or may be developed from integrative themes tied to other subjects such as science, social studies, or children's literature. To get a sense of what a context is, look at Table 3-1, which shows some examples of how you can use contexts to enhance mathematics lessons.

Make sure you give students open-ended tasks that allow for multiple solution paths or multiple solutions. This allows students at different levels to participate and learn where they are. It also lets students select the approach that best fits their learning preferences. For example, one student may prefer to draw a picture to solve a problem while another needs to use blocks. It is relatively easy to start with traditional textbook tasks and make them more open ended (Kabiri and Smith, 2003), as illustrated in Figure 3-2. Notice that not only does the open-ended problem in Figure 3-2 encourage critical thinking more than the traditional problem does, but it also gives students the opportunity to practice using the formula for volume many times rather than just once.

You may also need to supplement lessons in order to increase the focus on problem solving and conceptual development. For example, if your text does not provide enough meaningful experiences or hands-on practice with manipulatives for the students in your class, you may need to find other ways of providing such experiences and practice. Ultimately, you will be the one who best knows your students and their needs. Treat your textbook as one of many resources available to help you teach mathematics to your students. Adapting textbook lessons can be challenging

---

Traditional Problem: A rectangular aquarium is 12 in. wide by 14 in. long by 12 in. high. What is the volume of water needed to fill the aquarium? (Houghton Mifflin 2002, p. 475).

Open-ended Problem: You have been asked to design an aquarium in the shape of a rectangular prism for the school visitor's lounge. Because of the type of fish being purchased, the pet store recommends that the aquarium should hold 24 cubic feet of water. Find as many different dimensions for the aquarium as possible. Then decide which aquarium you would recommend for the lounge and explain why you made that choice.

Figure 3-2 *A measurement problem shown in both a traditional and an open-ended form.* (Source: Reprinted with permission from *Mathematics Teaching in the Middle School*, 2003 by the National Council of Teachers of Mathematics. All rights reserved.)

TABLE 3-1 • Using Contexts in Mathematics Lessons

This is a context	This is not a context
• Sorting M&Ms for party favors gives a reason for using this manipulative. • Planning a schedule for a class field trip involves telling time. • Ordering carpet for classrooms in the school is a context. We need to know the area of the room to find out how much to order. • Measuring classroom objects to decide how to fit them into storage containers gives purpose to a lesson	• M&Ms (alone) are a manipulative, not a context. • Time is a math topic, not a context. • Area of rectangles is a math concept, not a context. • Measuring classroom objects is a mathematics procedure, not a context.

but also very rewarding. If you have the freedom to do so, another option is to select curriculum materials that already incorporate the kind of instruction recommended by NCTM and CCSSM.

STANDARDS-BASED CURRICULA Research tells us that implementing NCTM's standards recommendations for instruction can improve student achievement on both traditional objectives like computation and reform objectives like problem solving (Ross, McDougall, Hogaboam-Gray, and LeSage, 2003). Research has also shown that curriculum materials (textbooks) have a powerful impact on what is taught and how it is taught.

In the 1990s, the National Science Foundation began funding several mathematics curriculum development projects aimed at designing and implementing curriculum materials that more closely aligned with NCTM's standards. These materials, called standards-based texts or reform texts, tend to use an inquiry approach to learning mathematics. They provide more opportunities for students to use reasoning and problem solving and fewer opportunities to practice basic skills. Conventional texts provide more practice of basic skills and less opportunities to apply the skills in problem-solving situations. Research studies comparing achievement of students taught using standards-based texts or conventional texts have revealed several consistent and positive findings.

Students taught using standards-based texts have greater conceptual understanding and problem solving skills than those using conventional texts. Students using standards-based texts also perform at about the same level on tests of mathematical skills and procedures as those using conventional texts (Stein and Smith, 2010). Many conventional programs report that they have incorporated NCTM's standards in their texts, but this often means they have incorporated the content standards without fully implementing the instructional approaches recommended by NCTM. In this chapter, we provide sample lessons from three of the standards-based elementary programs: *Everyday Mathematics* (for grades pre K–6), *Investigations in Number, Data, and Space* (for grades K–5), and *Math Trailblazers* (for grades K–5). These programs challenge students to understand the mathematics, to use higher-level thinking, and to learn mathematics through problem solving. Teachers may find it challenging to teach mathematics in the ways recommended in the standards. Often, they must significantly change their approach to mathematics instruction. At first they may need to spend more time preparing and planning for lessons. Sometimes teachers seeking to give their students additional support in these challenging standards-based curricula end up inadvertently eliminating essential components of the program. If you use these curricula, you must be careful to include classroom discourse, allow students to learn the mathematics concepts through problem solving and inquiry, and apply the mathematics in real situations (Sutton and Krueger, 2002).

You should also realize that new ways of teaching and learning mathematics are not only challenging for teachers, but may also be confusing for parents. So it is important to keep communication open with parents. Some schools invite parents for math nights where teachers can explain the rationale for new instructional methods. Math nights can also give parents opportunities to directly experience these new types of math lessons, while giving teachers opportunities to suggest ways for parents to help their children at home. Some teachers send letters home for parents at the start of each unit so that parents know what is happening in math class. Remember that parents and teachers are partners in educating children. Anything you can do to help parents understand your goals should help you reach those goals.

CHILDREN'S LITERATURE Teachers often use children's trade books to enhance lessons in language arts, social studies, science, music, art, dance, and drama, but it is also easy, effective, and appropriate to use children's literature in mathematics lessons. "Children's books can spark students' imaginations in ways that exercises in textbooks or workbooks often don't" (Burns, 2007 p. 45). Stories can provide children with a common starting point from which to share and discuss mathematics. Mathematics and language skills develop hand in hand as children talk about problems and read and write about mathematical ideas. In addition, you can incorporate multicultural children's literature in your mathematics lessons as a way of making relevant connections for your students. An example of how this may be done is included in the Cultural Connections section at the end of this chapter.

Use of children's literature in mathematics lessons can enhance learning in many ways. Children's literature:

- Provides a context or story line that can launch or develop mathematical concepts
- Provides illustrations that clearly represent mathematical concepts
- Provides quality illustrations that are motivating to the reader
- Is a source of problems or a basis for generating problems
- Provides styles and formats that can motivate the class to write problems or design your own book (Thiessen, 2004)

Many reference books for teachers can help you identify titles that work well for integrating children's literature with lessons on particular mathematics topics. The Book Nook for Children at the end of each chapter in this book includes numerous such suggestions.

MANIPULATIVES As indicated by Bruner in Figure 2-4, learning moves from enactive or concrete objects to iconic or semi-concrete pictures to symbolic or abstract symbols. Effective teachers use manipulatives when appropriate to provide concrete experiences that help children make sense of mathematics and build their mathematical thinking. The use of manipulatives is also an effective way to meet your students' diverse learning needs. Manipulatives let teachers illustrate mathematical ideas in multiple ways using a variety of different models and representations. Manipulatives support problem solving and also increase the active engagement of the children. They can test and verify their ideas, eventually forming mental models from the concrete models. The time children take to become familiar with the materials is time well spent—during this process, children construct knowledge as they make connections between the models and the mathematical ideas.

All through this book, we stress the importance of having children use manipulatives to model mathematical ideas. Commonly used materials in elementary school include chips or tiles, interlocking cubes, pattern blocks, attribute blocks, tangrams, base-ten blocks, fraction models, geoboards, measuring instruments, spinners and dice, and play money. (Patterns for many of the better-known manipulatives are available on this textbook's Web site.)

> **Math Links 3.2**
>
> Go to the National Library of VirtualManipulatives for Interactive Mathematics Web site to try out a variety of electronic manipulatives that could be used with children. This site includes several common manipulatives, such as base-ten blocks and fraction pies, as well as graph tools, puzzles, and games. You can access this from http://matti.usu.edu/nlvm/nav/vlibrary.html or from this book's Web site.
>
> www.wiley.com/college/reys

While some teachers or parents may be concerned that children will become overly reliant on manipulatives, research has shown that children will progress from random guessing, to using of tools such as manipulatives, to using their own mental strategies. It is also suggested that children need opportunities to explore different tools and strategies and discuss how and why they are useful (Jacobs and Kusiak, 2006).

Roberts (2007) reminds us that having students use manipulatives and models does not guarantee they will be able to construct the desired mathematical understanding. The teacher will need to provide careful planning and delivery of the task and encourage thoughtful reflection by the students so they can make connections between the manipulatives and the mathematics symbols. When selecting a manipulative, the teacher should ask the following questions:

- Does the manipulative or model clearly and accurately represent the concept?
- Do the materials lend themselves to an efficient guided-discovery activity?
- Does the activity include a structure or recording procedure that will facilitate students' ability to construct the desired knowledge?

Practical suggestions for incorporating manipulatives into your instruction are provided by Burns (2007):

- Talk with children about how manipulatives will help them learn. These conversations are essential from time to time.
- Set ground rules. Manipulatives are not toys or games; the children should be expected to stay on task, completing assigned problems or activities.
- Set up a system for storage of the manipulatives. Children should be able to reach and remove things easily themselves without your assistance.
- Provide time for free exploration. Students will be more willing to complete the assigned tasks once they have satisfied their own curiosity.
- Post charts listing the names of manipulatives the children may use. This communicates that you value manipulatives and helps students learn the names and spellings.
- Send home a letter explaining why the children will be using manipulatives. You might also have children take home materials to use with their families.

Remember that young children, whose fine-motor skills aren't fully developed, may have difficulty if manipulatives are too small or too difficult to connect. To help children who have special needs, you might use materials with built-in organization (such as an abacus) or materials that fit together (such as connecting cubes), rather than separate objects (such as base-ten blocks). Or you can use containers that keep counters separated, such as egg cartons or mats with clearly separated sections. Finally, package manipulatives into sets for one student, two students, or a small group for easy distribution and collection. Remember to have the children pass out and collect the materials as much as possible. You don't have to do it all!

TECHNOLOGY The increased availability of technology in today's homes and classrooms is having a profound effect on teaching and learning. There are several compelling reasons for incorporating technology in our mathematics instruction. One is to prepare students to function in our technological world. Another is given by NCTM's Technology Principle: "Electronic technologies ... furnish visual images of mathematical ideas, they facilitate organizing and analyzing data, and they compute efficiently and accurately. They can support investigation by students in every area of mathematics ... When technological tools are available, students can focus on decision making, reflection, reasoning, and problem solving" (NCTM, 2000, p. 24). Third, applied effectively, technology has the potential to increase student learning, understanding, achievement, and motivation. Students using technology can also learn collaboration, critical thinking, and problem solving (Pitler, Hubbell, Kuhn, and Malenoski, 2007). In The Common Core State Standards for Mathematics (CCSSI, 2010), it is recommended that students use appropriate tools strategically. To be able to make good decisions about tool use, students need sufficient opportunities to use a variety of mathematical tools, including technology.

When selecting technology to use in mathematics lessons, it important to identify the purpose it will serve. Soucie, Radovic, and Svedrec (2010) list several worthwhile reasons to use technology:

- Engage students in discovery and exploration;
- Promote higher-level thinking;
- Improve students' visualization;
- Enable students to engage in real-life applications of mathematics;
- Prepare students for the demands of this century; and
- Make mathematics more engaging and fun (p. 468).

Computers have become essential tools in our larger society and are one of the most common forms of technology used in elementary classrooms. Many students today have access to desktop, laptop, and handheld computers. According to a recent report, in 2009, 97% of teachers had one or more computers located in the classroom every day. Internet access was available on 93% of the computers in the classroom. The ratio of students to computers was 5.3 to 1. Teachers reported computers were used during instructional time often (40%) or sometimes (29%) (Gray, Thomas, and Lewis, 2010). The way we choose to use technology also can impact student achievement. For example, the 1996 NAEP mathematics assessment examined how computers were used in grades 4 and 8 and found that the use of computers to develop lower-level skills was negatively related to academic achievement. In contrast, for both fourth- and eighth-grade students, using computers to develop higher-level thinking skills was associated with higher student achievement (North Central Regional Laboratory, 2005).

Research shows that even preschool children benefit from using computers, especially from programs that are open ended, encourage discussion and solving problems, and support the development of conceptual knowledge. Unfortunately, many teachers are unsure how to use computers to do these things and often opt for lower-level uses. For example, of the teachers in the NAEP questionnaire who reported using computers, the primary use in grade 4 was for mathematics games, followed by drill-and-practice activities.

There are basically six types of computer software available. Students use each type in a slightly different way, and the different types also offer different benefits to children. One educator describes each software type in terms of its potential for developing problem-solving and higher-level thinking skills (Crown, 2003).

- *Drill-and-practice software* provides practice for a skill that has been previously taught. The computer presents a problem to the student, waits for an answer, and indicates whether the student is right or wrong. Some drill-and-practice programs keep records and provide a score, or they may even adjust the difficulty level of problems based on student responses (e.g., number and type of errors). The primary focus of drill-and-practice software is on mastery of lower-level skills and basic facts.

- *Tutorial software* provides instruction on new skills. It may introduce information, present examples, and provide practice. Depending on whether the student grasps the concept easily or has difficulty, the program will move to a more difficult lesson or provide additional instruction. Some tutorials are quite effective in presenting new information, but they give students little opportunity to think creatively. Also, tutorial software might encourage children to focus on procedures or rules prematurely, before they have solidified their conceptual knowledge of the topic.

- *Simulation software* allows students to experience events and explore environments that might be too expensive, messy, dangerous, or time-consuming to experience and explore in reality in the classroom. For example, through computer simulation, students can run a business or go on an expedition, or they can conduct a probability experiment (such as flipping coins or drawing randomly from a collection) with many more trials than would be feasible to conduct by hand. Simulation software has the potential to support problem solving and critical thinking, but teachers have to carefully plan related discussions and activities to help students gain these benefits.

- *Educational game software* engages students in fun activities that address specific educational skills or may aid in the development of logical thinking or problem-solving skills. Games usually include randomized events, offer an opportunity to "win," and present some obstacles to winning that the child must overcome. Often these games provide drill-and-practice opportunities. In addition, they can help children improve their higher-level thinking skills by analyzing game situations and developing strategies. However, games do not typically develop key mathematical content knowledge.

- *Problem-solving software* is designed to aid in the development of higher-order problem-solving strategies. Problem-solving programs are similar to simulation programs, where students are placed in a situation in which they can manipulate variables and receive feedback, but problem-solving programs do not necessarily model real-life situations. One type of problem-solving software involves programming. Students write programs in a computer language such as Logo to instruct the computer to operate in a particular way; in the process, they develop logical thinking and problem-solving skills. Problem-solving software can improve students' performance on problem-solving tasks but may not develop their mathematics skills or concepts unless the teacher can help them look beyond the task itself and focus on the mathematics.

- *Tool software* can enhance both teaching and learning and may be the most effective type of software for developing mathematics through problem solving. Tools help efficiently and effortlessly with graphing, visualizing, and computing. Useful classroom tool software includes word processors, databases, spreadsheets, graphing programs, hypermedia, dynamic geometry software, computer microworlds, and applets (electronic versions of manipulatives, also called *virtual manipulatives*).

Today, many software programs of the types described above are available on the Internet, which gives teachers and students access to people, information, and resources that have never before been so accessible. Throughout this text, look for Math Links that will direct you to a wide variety of additional Internet resources that can enhance the teaching and learning of mathematics.

Math Links 3.3

You can not only locate resources for students on the Web, but you can also find resources to support your teaching. For example, Thinkfinity is a teacher site that provides standards-based Internet materials for K–12 teachers. You can access this from http://www.marcopoloeducation.org or from this book's Web site.

www.wiley.com/college/reys

Calculators are another important technology tool in elementary mathematics classrooms. According to NCTM (2000):

Calculators should be available at appropriate times as computational tools, particularly when many or cumbersome computations are needed to solve problems. However, when teachers are working with students on developing computational algorithms, the calculator should be set aside to allow this focus. Today the calculator is a commonly used computational tool outside the classroom, and the environment inside the classroom should reflect this reality. (pp. 32–33).

In NCTM's Number and Operation standard, students in grades preK–2 are expected to "use a variety of methods and tools to compute, including objects, mental computation, estimation, paper and pencil, and calculators" (NCTM, 2000, p. 78). Students in grades 3–5 should be able to "select appropriate methods and tools for computing with whole numbers from mental computation, estimation, calculators, and paper and pencil." (NCTM, 2000, p. 148).

Calculators not only let students compute quickly and accurately, but they also let them examine patterns, solve problems, develop mathematical concepts, and perform other higher-level tasks. Research has shown that the use of calculators can help develop increased problem-solving skills. They can also be used as pedagogical tools to help students learn mathematics (Battista, 2010). See Chapter 10 for more about calculators and how they can be used to develop higher-level thinking.)

Today, more and more students have access in school to technology that goes beyond calculators and desktop computers, including handheld computers or data collection devices; classroom networks; interactive whiteboards; classroom response systems; digital cameras; video cameras; calculator-based laboratory systems; and CDs, videos, MP3s, and DVDs. In and out of school, students are connected to multimedia digital content, cell phones, online resources, blogs, wikis, and social media. As a teacher, you must not only learn to be comfortable using these technologies but must also consider how they might be integrated into your mathematics lessons. Do not bring in technology simply to impress or motivate your students. Remember that technology will improve student achievement only if it is used to develop the higher-level thinking skills involved in collecting and analyzing data, investigating patterns, and solving problems.

Finally, teachers must ensure equitable access to technology. Make sure not to use technology only as a reward for early finishers; rather, plan opportunities for all children to use technology during the lesson. If there is not enough technology available for all to use it at the same time, design a schedule that gives each student an opportunity to use it on a regular basis.

PLANNING FOR EFFECTIVE TEACHING

Planning lies at the heart of good teaching. Children learn best from lessons that are interesting and carefully organized, directed by thoughtful questions, and enriched by activities and materials that give them the opportunity to develop ideas about mathematics. Careful development of ideas, based on clear explanations, careful questioning, and effective use of manipulatives and technology, is particularly important in helping children learn mathematics. Teachers make many of the decisions related to the questions just discussed. Once those decisions have been made, it is time to organize them into plans for the year as a whole, for units, and for individual lessons.

Teachers plan in a variety of ways. Some just list the objectives they want children to attain or at least take a step toward attaining. Some jot down key questions they want to ask. Some lay out materials for children to use or run off worksheets for them. And some read the comments in the teachers' guide of the textbook. All of these approaches require teachers to think through what they plan to do and how they will do it.

Few experienced teachers take the time to write out a complete, detailed plan for every mathematics lesson they teach. Some do it occasionally, when they know that the idea they want to teach must be developed especially carefully or when the lesson is one they haven't taught before. Experienced teachers often have detailed mental plans, although they may not write more than a page of sketchy notes. They may have taught a similar lesson many times before and therefore have a good sense of appropriate sequencing, timing, questioning, and potential pitfalls. Beginning teachers, however, lack that experience, and so it is particularly important for them to write detailed lesson plans. Careful planning helps make initial teaching experiences good ones, for both the children and the teacher. In fact, research shows that both experienced and preservice teachers benefit from thinking deeply about the lesson tasks, teacher activity, students' responses to them, and possible interventions before teaching the lesson (O'Donnell and Taylor, 2007).

Good plans give you the security of knowing what you will do and say, of having interesting activities and materials ready for the children's use, and of anticipating what the children might do. A good, detailed lesson plan gives you a way to judge how well the lesson went. Even though you might not be able to follow your plan precisely, it will help you evaluate your teaching and assess your students' learning. A written plan can also be shared with other teachers, who can provide helpful comments and suggestions. Moreover, writing detailed plans helps you learn how to plan in your head.

LEVELS OF PLANNING

Planning for your mathematics class is done at three levels. You start at the broadest level with a plan that sets out your goals and objectives for the year. Then, moving down a level, you plan how to achieve those goals and objectives by organizing the mathematics content into units. Finally, at the narrowest level, you plan daily lessons for teaching specific parts of that content. Each level of planning is important in ensuring that you meet all your grade-level requirements in a logical fashion.

PLANNING FOR THE YEAR Before the school year begins, you need to consider what you want to have the children in your class accomplish during that year. Of course, you do not need to develop these goals on your own. Most schools prepare scope-and-sequence charts or curriculum guides, or they rely on those provided with the textbook series they use. Your district or state may provide a curriculum framework for your grade level, and the NCTM's *Principles and Standards for School Mathematics* (2000) indicates, in broad terms, the mathematical content and processes appropriate for students at each of the grade bands (preK–2, 3–5, 6–8, and 9–12). You should be familiar with the NCTM standards, your state education requirements (most states have adopted the Common Core State Standards), and your school's scope-and-sequence guide or curriculum before you do any planning. Such materials are designed to ensure that children are taught the desired range of content across grade levels. You should also check with your principal and other teachers to determine whether any changes have been made to meet the needs of the children in your school and to learn how much flexibility you have in making changes for your class.

After determining the goals of mathematics instruction for the year and the order in which topics are to be taught, you need to consider the approximate amount of time you want to spend on each phase of the curriculum, based on the relative importance of each. Doing this helps you be sure that you can fit in all the mathematics content you want to include. The goals for the year can be organized by textbook chapters or units. Your school's curriculum guide or the textbook's planning chart can help you with this and with the decision about how much time to assign to teaching each topic. Most mathematics textbooks contain 130 to 150 lessons. Given the typical 180-day school year, you may have the flexibility to spend extra time on some topics or to teach topics that are not in the textbook. Prepare a brief outline listing each unit, its objectives, and the estimated number of days and dates you will teach it.

PLANNING FOR UNITS Once you have outlined your year, you can begin planning your units. Begin by outlining, in sequence, the topics that are to be taught and how much time you will spend on each. Then decide on supplemental materials, instructional strategies, and assessments. Next, outline what you want to accomplish each week—that will make the process of developing daily lesson plans much easier. Some schools require teachers to maintain a lesson plan book in which they note the objectives (and sometimes other details as well) for each day's lessons for a week. Even if this is not required, it is a good idea because it keeps you aware of progress toward meeting your goals for the year and it gives you a guide to follow each day. You cannot expect to follow this guide exactly, but as you plan each week or unit, you can review the progress of the children in your class and then vary or pace the content to be taught to meet their individual needs. Or you can plan to teach different content to small groups or individuals.

PLANNING FOR DAILY LESSONS Finally, once you have outlined the units, you are ready to write daily lesson plans. Putting your lesson plan into writing will help you clarify many of your ideas, give you a "road map" to follow while teaching, and give you a record that you can use to evaluate each lesson and to plan subsequent lessons. Consider the questions discussed at the beginning of this chapter and use the process of writing the plan to help you think deeply about the lesson, the steps involved in teaching it, and the time you will need for each step. Then, if you diverge from the plan, you will be able to pinpoint where you diverged and return later to pick up where you left off. Your plan should be concise yet complete. Since it is just for your use, you may reach the point where you can write just short sentences or phrases to cue your memory. Use a format that lets you refer easily to the plan while you are teaching.

PLANNING DIFFERENT TYPES OF LESSONS

When writing daily lesson plans, you must also consider the format of the lesson. The most familiar format for mathematics lessons, review-teach-practice, may also be one of the least effective, especially when it is used to introduce new content. In this lesson format, the teacher begins by reviewing homework or problems worked during a previous lesson; then the teacher briefly explains a new concept or demonstrates a new skill, using sample problems; finally, the teacher assigns exercises (of the same type as the sample problems) for student practice. Typically, for the remainder of the lesson, students practice solving similar problems on their own while the teacher assists individual students.

According to the Trends in International Mathematics and Science Study (TIMSS), this mathematics lesson format is commonly used in the United States and Germany (*Pursuing Excellence*, 1996), and students from both countries have not performed particularly well on tests of mathematics achievement. Many educators attribute much of this poor performance to the ineffectiveness of the review-teach-practice lesson format. By contrast, the typical lesson format in Japan (whose students were top TIMSS scorers) is similar to the investigative lesson (described below): The teacher poses a challenging problem that students then think about; individual students present ideas for solutions to the class; they then discuss these ideas. The teacher uses the discussion to highlight the key mathematical concepts, and then the students practice similar problems.

Review-teach-practice lessons in the United States and Germany focus on skill acquisition, on teaching students to do something; in Japan, the typical lesson focus is on understanding mathematical concepts. In U.S. and German classrooms, students spend more than twice as much time practicing routine procedures as students in Japanese classrooms (although many Japanese students practice skills in paid tutoring sessions after school). Japanese students spend most of their class time inventing new solutions and thinking about mathematical concepts. Analysis of TIMSS data suggests that what occurs in lessons is critical to students' learning. The specific topics taught and how these topics are presented and developed shape what students learn and are able to do. In the United States, lessons often consist of episodic encounters between students and curricular content, rather than ensuring that students stay focused on the content, as is done in other countries. Topics and concepts are presented in a fragmented and disjointed manner, and underlying themes and principles are either not identified or merely stated but not developed.

Instead of the traditional review-teach-practice format, we suggest using three different lesson formats: investigative lessons, direct instruction lessons, and explorations or learning centers. These formats can provide structure for most of the situations you will encounter during mathematics lessons. You will probably choose to use different types of lessons on different days, depending on your goals and your students' previous experiences with the content. But no matter which format you are using, be sure to keep in mind the overarching goals of ensuring student understanding and maintaining active student involvement. You will probably design many lessons that use aspects of more than one of these formats. In other words, the three lesson formats should not be seen as entirely distinct. Nevertheless, it is useful to outline the general characteristics of each form, as we do in the sections below.

No matter what lesson format you select, the beginning of a lesson plan usually shows basic information, such as the lesson objective(s), how you will assess students, what adaptations you might make (i.e., gearing up and gearing down), and what materials you will need. The lesson itself generally consists of three phases: (1) launch (how you will begin the lesson); (2) investigate, instruct, or explore (depending on the lesson format); and (3) summarize (how you will bring the lesson to a close). Figure 3-3 shows a generalized lesson plan covering all three formats; in this figure, you can see how the lesson outline differs among the formats. The descriptions below should help you learn how to write lesson plans using each format; we also provide a sample lesson plan for each, which you could try out with children.

The kind of teaching envisioned by the NCTM *Standards* (NCTM, 1991, 2000) is probably very different from the kind of teaching you experienced when you were in elementary school. The emphasis is shifting from a teacher-directed classroom to one in which students are actively involved in learning. Though still in control, of course, the teacher involves the students through discussions and activities, usually by incorporating small-group work. Direct instruction and discussions with the whole class have not disappeared, but children learn much more effectively when they work with materials to solve problems and talk about their results with each other.

Generalized Lesson Plan for All Three Formats

Introductory Information
Grade Level _____ Date _____
Topic/Title _____
Objective(s) Determine what you want your students to know or do. Align with district and state standards.
Assessment What types of information will you collect to help you determine how well individual children achieved the objective(s) of your lesson and where they had trouble? Assessment information may come from a variety of sources—for example, classroom observations, written class work (either individual or group), homework. Your dual goals in planning for assessment are to be able to make appropriate follow-up instructional decisions and to be able to report on students' progress (e.g., by assigning grades or writing a narrative report). Be sure to describe the criteria you will use to decide who "got it" and how you might translate your assessment to a grade.
Gearing Down What will you do if your lesson, as planned, turns out to be too advanced for some or all students? How can the lesson be modified (on the spot) for slower students?
Gearing Up What will you do if your lesson, as planned, turns out to be too easy for some or all students? How can the lesson be modified (on the spot) for faster students?
Materials List all materials needed:
- math manipulatives, technology (and how many of each)
- books, including page numbers (and how many of each)
- other supplies (scissors, markers, posters, etc., and how many of each)
- handouts (attach a copy of each)
- references (include bibliographic information for books or resources used in planning the lesson)

Lesson Outline (include approximate times for each segment of your lesson)

LAUNCH (setting the stage)
Describe briefly how you will begin the lesson. Your aim is to motivate students to get involved in the lesson activities and to ensure that all the children understand what they are to do. Include questions you might ask. Introduce the context you will use (e.g., real-life examples, manipulatives, problems, or puzzles).

INVESTIGATE (for investigative format)
This part of the lesson could consist of one continuous problem-solving activity (whole class, small group, or individual), or it could be divided into several segments, each involving different activities or groupings. For each segment, give a step-by-step description of what you (the teacher) will be doing and what the students will be doing. In particular, provide examples of the types of questions you might ask and of any products that the students will create. If your lesson is divided into segments, give an approximate time for each segment; also, describe how you plan to make transitions from each segment to the next.
OR
INSTRUCT (for direct instruction format)
Whole-Class Teacher Instruction and Modeling What concepts, definitions, formulas, algorithms, etc., will you explain and model? List all the examples you will use, and order them appropriately. Include directions for the activities that will follow.
Activities Activities may be done by individuals or small groups. Students work with you on problems or exercises that follow rather closely from what they learned during your whole-class instruction and modeling. Then, when they are ready to practice independently from you, circulate, assisting individual students, answering questions, and, for students who finish early, providing tasks that let them extend their learning. Make a list of the questions you will ask as you circulate. Include a copy of any written materials that students will work on, such as a worksheet. Include several simpler problems that you could use in helping students who have trouble. Include several more challenging problems that you could assign to early finishers.
OR
EXPLORE (for learning center format)
For each exploration, give a step-by-step description of what you will be doing and what the students will be doing. In particular, give examples of the types of questions you might ask and of any products that the students will create. If you plan to rotate students through several activities, include the approximate time for each and describe how you plan to make transitions from each segment to the next.

SUMMARIZE (closure)
Describe how you will bring the lesson to a close. Find a way that motivates the students to reflect on their activities and what they learned. Include questions you might ask yourself and/or your students to help you determine whether the lesson was a success.

Figure 3.3 *Generalized lesson plan.*

Regardless of the lesson format, the teacher orchestrates the classroom discourse by "deciding when to provide information, when to clarify an issue, when to model, when to lead, and when to let students struggle with a difficulty" (NCTM, 1991, p. 35). You learn to make these decisions through planning and through experience. No one mode of instruction can be considered best. You should learn many instructional modes and use them as appropriate. The NCTM *Standards* (2000) urges teachers to consider the goal, the task, and the students (i.e., to ask what will best help them learn).

INVESTIGATIVE LESSONS Investigative, or problem-based, lessons are most appropriate for helping children develop problem-solving skills, learn new concepts, or apply and deepen their understanding of previously learned topics. In an investigative lesson, students solve a problem or conduct an investigation on their own. The problem or investigation may have been brought up by the teacher or by one or more of the students, but the lesson itself revolves around ideas that the students generate through their own investigations. The teacher is responsible for guiding the lesson, but the students are expected to come up with their own approaches, strategies, and solutions.

In the *launch phase* of an investigative lesson, you provide a motivating introduction for the lesson and explain the problem at hand. The launch might come from a student presenting a problem that he or she has been struggling with and asking classmates to work on, too. Or you might provide motivation by reading a children's story or a newspaper clipping, playing a game, presenting a puzzle, or using some other context that involves the problem to be worked on.

Once all the students understand the challenge, the *investigate phase* of the lesson begins. In this phase, the students work on the problem. This might involve learning stations, individual work, paired problem solving, or small groups, and students might use models, manipulatives, computers, calculators, or other tools. In an investigative lesson, the problem should truly challenge the students, so it is important that you do not explain up front how to approach the problem. The students should not simply mimic strategies or skills that you have just shown them. Instead, they should decide for themselves how to get started and what to try. Your role as teacher is to let them go. You should circulate around the room, listening in on conversations, observing what individuals or groups are doing, and occasionally interjecting questions or comments to help students recognize where they may be going wrong or to suggest a different approach.

During the *summarize phase* of the lesson, the class talks together about its findings. Using information you gathered from observing the children at work, you orchestrate a discussion in which various groups or individuals report what they tried and what they discovered. It is important to be neither too negative nor too positive about student ideas and responses. Encourage students to challenge each other's ideas. They need practice in learning how to judge when mathematical ideas are valid and when they are not. Also be sure not to finalize the solution to the problem prematurely.

Students learn from listening to each other and from hearing about alternative approaches. Your job is to encourage the students to share their ideas while maintaining control of the discussion and trying to guide it in ways that advance your curricular agenda. During the final portion of the summarize phase, it may be important to help students be explicit in stating what they have learned from the investigation. This is the time for stating generalizations or rules that have been formulated. It may be tempting to cut short this phase, but do not do so. Talking together and sharing ideas are critical to meaningful learning of mathematics.

Figure 3-4 shows an investigative-format lesson plan for second grade created from a lesson from the *Math Trailblazers* textbook series. In the lesson outline portion of this plan, the teacher details how he or she wants to conduct the lesson. Read the outline to get a sense of what the teacher intends to do. Notice that the launch phase of the lesson includes background information and discussion that students need before they start their investigation. Now look at the outline of the investigate phase and note that the teacher has considered how the students will be organized, moving the children back and forth between small-group and whole-group discussion as needed. Finally, the outline of the summarize phase includes additional questions to be asked at the end of the lesson.

DIRECT INSTRUCTION LESSONS In direct instruction lesson, the teacher plays a more central role in directing the instruction than in investigative lessons. Direct instruction lessons are appropriate when the teacher wants to communicate specific knowledge, to introduce new vocabulary, or to teach specific procedures. In a direct instruction lesson, the teacher exercises more control than in an investigative lesson, and the lesson generally has a tighter focus. The three phases of a plan for a direct instruction lesson are usually (1) launch, (2) instruct, and (3) summarize (see Figure 3-3); each phase may be subdivided into smaller segments of various types, depending on the content of the lesson.

As with investigative lessons, in the *launch phase* of the lesson, the teacher provides a motivating introduction and introduces the students to the problem; led by the teacher, the class talks together about how to approach the problem and what previously learned concepts and skills might be useful. In the *instruct phase*, the teacher models for the students how to proceed but also elicits ideas from the students while helping them move toward the concept or skill that is the goal of the lesson. Guided practice is included during the instruct phase—students actively practice using the concept or skill, with guidance and feedback from the teacher. Afterward, the students have a chance to work with the ideas presented earlier in the lesson and to make them their own, working independently from

Sample Investigative Lesson Plan

Introductory Information

Grade Level 2 **Date** March 15, 16

Topic/Title Measurement/When Close Is Good Enough

Objective(s) The students will: 1. Estimate length and distance with nonstandard units.
2. Measure length and distance with nonstandard units.
State standards: The student estimates and measures using standard and nonstandard units of measure with concrete objects in a variety of situations.

Assessment Objective 1: Answers to questions 1 and 4 on p. 40 and questions 1–6 on p. 41 should be reasonable estimates. 6/8 correct shows mastery.
Objective 2: Observe while measuring, jot down notes. Questions 2, 5, and 6a,b on p. 40—3/4 correct shows mastery.

Gearing Down Do the first hand span and cubit measure together. Assign partners if needed.

Gearing Up Encourage students to measure additional items with hand spans and cubits, or have students estimate and then use a different nonstandard unit of their choosing.

Materials Overhead projector; Discovery book p. 41, Hand Spans and Cubits activity pages; pp. 39–41 for each student.
Reference: *Math Trailblazers: 2nd Grade*, Unit 4, Lesson 4, by Wagreich et al., 2008, Kendall/Hunt Publishing Company.

Lesson Outline

LAUNCH—Day 1 (15 minutes)

- Who remembers measuring length? What tools have you used? Today we're going to use two nonstandard tools.
- Hand span—With your fingers spread wide, the distance from tip of thumb to tip of little finger. Everyone show the length of their hand span. Compare with your neighbor. Are they all the same length? That is why we call it a nonstandard unit.
- Cubit—The distance between your elbow and the tip of your longest finger. Everyone show your cubit. Compare. Are they all the same? Also a nonstandard unit.
- Hand spans and cubits let us make approximate, not exact, measurements. Think of some examples when an approximate measurement with nonstandard units would be OK.
- Demonstrate how to measure the length of the dry erase board with hand spans, then cubits. Show how to use free hand to mark place before moving.

INVESTIGATE—Day 1 (30 minutes), Day 2 (30 minutes to finish charts and do pp. 40, 41)

- Assign students to groups of 4 and send to one of the locations on p. 39. Each is to measure with hand spans. If time permits, measure other objects and record on the sheet. Students write their measures in the table on the overhead transparency. (15 minutes)
- Whole-class discussion: Discuss why there are differences. Discuss how to handle partial hand spans or cubits. Decide on a rule for the future. (10 minutes)
- Do you think it will take more or fewer cubits to measure the dry erase board? About how many?
- Groups finish p. 39 with cubits. Record results on transparency. (15 minutes)
- Now let's estimate how many hand spans and cubits for the width and height of the door. Explain how you estimated. Did you use the information on our chart to help you?
- Individually, students complete pp. 40, 41. Send p. 41 in Discovery book as homework.

SUMMARIZE—Day 2 (15 minutes)

- Why did we get so many different measurements on our chart? Why are these approximate measurements?
- What is the difference between a standard unit and a nonstandard unit?
- What other tools could be used as nonstandard measures? Does your family use any nonstandard tools?

Figure 3-4 *Sample investigative lesson plan.*

the teacher. The task during this phase should be closely related to the original task but not so similar that the students are simply copying what the teacher has just shown them. Finally, during the *summarize phase*, the teacher and students summarize what was learned.

Figure 3-5 shows a direct instruction lesson for third grade created from a lesson from the *Everyday Mathematics* textbook series. Notice the role of the teacher during the instruct phase of the lesson. The teacher provides important information, models the procedure, guides the students to practice the procedure, and gives them opportunities to practice. During the summarize phase, students not only have an opportunity to reflect on their learning but also share the strategies they used.

EXPLORATIONS OR LEARNING CENTERS Sometimes teachers want to provide a series of learning activities, or *explorations*, for students to simultaneously work on either independently or in small groups around the room. These activities are often called learning centers or learning stations; they are commonly used in early childhood classrooms with very young children who are not ready to participate in more formal lessons, but they can also be appropriate for older elementary and middle school students. Explorations should be structured so the children can complete them without teacher assistance and allow for more individual choice, while working with assigned materials, than in an investigative lesson. Explorations are not just free time. They provide opportunities for students to work independently on important mathematical tasks. Students can work at their own pace and make decisions about their learning. The learning activities can all be related to a particular topic and provide various levels of difficulty or complexity. They can also provide a variety of opportunities to practice a skill and may encourage the development of conceptual knowledge or of problem-solving skills. In addition, explorations can provide remedial or enrichment activities for students who need them. A set of explorations can be made available on particular days of the week, over the course of a unit, or over a period of days in place of whole-class lessons.

Explorations don't necessarily resemble a typical whole-class lesson, but you still need to identify the student objectives and carefully plan the activities. You need to consider how the activities will be managed and think about how you can package the necessary materials for easy access and cleanup.

Explorations let you plan to meet the needs of students with varied learning styles and interests. Some students may prefer to work with technology, while others may want to use manipulatives. Planning a variety of activities will give students choices. Also, by considering how you can make particular activities less difficult or more challenging, you can plan to meet individual needs.

Consider your class size and how many children you want to work with each activity at the same time. You might plan for five or six stations and have groups of children rotate through all of them, spending a predetermined amount of time at each location. Or you could provide several sets of materials for each activity so several groups or individuals can be working on the same activity at the same time. One educator calls these types of experiences *menus* and suggests providing a menu of activities related to a particular unit or topic (Burns, 2007). You can require that all students complete certain explorations on the menu, while others could be optional. Once again, this is also an effective way for you to meet students' individual learning needs. You could assign particular activities to students who are struggling, while assigning more challenging activities to those who are more advanced. Alternatively the students could decide in what order they would like the activities and how much time they will spend on them. Once students are familiar with the routine of completing these types of activities and are able to work fairly independently, explorations can also give you time to assess individual students or small groups.

An exploration generally involves three phases: (1) launch, (2) explore, and (3) summarize (see Figure 3-3). The *launch phase* of an exploration will probably involve telling the students about the available activities, what you expect from them for each activity, how they should manage materials, how they will move from one activity to another, and what products they will create and you will check. You can give older students written instructions for each activity, but young children will need you to model the activities and provide oral instructions. Often, it is best to gradually introduce the activities involved in an exploration to the whole class over several class periods. Then, when students are sent to work independently, they have previous experience with the activities and are more likely to be successful when working on their own.

During the *exploration phase*, students work independently for a specified amount of time. Be sure you have planned the step-by-step procedures for each activity. Be sure students understand what to do when they have completed one activity and are ready for another. Also consider your role. Will you be circulating? What questions will you ask about particular activities? You might carry a clipboard and make notes about student learning as you observe or keep a list so you can check products that particular groups have created and can mark groups off whether the products look satisfactory.

In the *summarize phase*, find time to talk with the students, letting them share strategies or products they created. You may want to instruct everyone to complete one particular activity and make that the focus of the discussion for that day.

Figure 3-6 shows a kindergarten exploration plan from *Investigations in Number, Data, and Space* textbook materials. Notice that in the launch phase, the teacher provides instructions so the children can work independently while completing their activities. In the explore phase, the teacher lists organizational details. In the summarize phase, the teacher uses questions to get the children to talk about their experiences.

Sample Direct Instruction Lesson Plan

Introductory Information

Grade Level 3   **Date** November 15

Topic/Title   Unit 4: Multiplication and Division   Lesson 4: Division Ties to Multiplication

Objective(s)   The students will model division number stories with arrays, multiplication/division diagrams, and number models.
State: The student identifies multiplication and division fact families.

Assessment   Check Home Link Master 92 for accuracy. Need to get 2/3 correct to show mastery.

Gearing Down   If students are having trouble drawing the array, they can draw hops on a number line.

Gearing Up   Challenge students to think of their own division number stories. They can trade and solve them.

Materials   pennies, dice for Division Array game
Math Journal p. 86, Home Link Master p. 92
References: *Everyday Mathematics, Grade* 3, 3rd edition, Unit 4, Lesson 4, by the University of Chicago School Mathematics Project, Copyright © 2007 by Wright Group/McGraw-Hill.

Lesson Outline
LAUNCH (15 minutes)

- Begin by having pairs use pennies to play a few rounds of the game, *Division Arrays*. (Introduced yesterday.)
- Have students solve the following Math Message problem: 12 pennies are shared equally by 4 children. How many pennies per child is that? If you wish, you may use your tool-kit pennies to act out the story.
- Discuss strategies used. If any used arrays, have them draw them on the board.

INSTRUCT (35 minutes)

- Next to the 4 × 3 array from the math message, write the following multiplication/division diagram.

Children	Pennies per child	Pennies in all
4	?	12

- How many pennies per child? Write 3 different number models to much the story.

    $12 \div 4 = 3$ $12/4 = 3$ $4\overline{)12}^{\,3}$

- Label the following vocabulary.
- The <u>quotient</u> is the answer in each number model.
- The <u>dividend</u> is the total before sharing.
- The <u>divisor</u> is the number of equal parts or the number in each equal part.
- Pose the following problem: There are 15 pennies. Each child receives four pennies. How many children are there?
- Model making the array and filling in the multiplication/division diagram. Have volunteers write the number models to match the story.
- The <u>remainder</u> is the quantity left over when a set of objects is shared equally.
- Have students try a couple more problems, using the same array, diagram, and number models. Discuss solutions.
- 21 puppies are placed equally in three pens. How many puppies are there per pen?
- 17 markers are shared equally among six children. How many markers does each child get?
- Turn to page 86 in math journal. Students work in small groups to solve the four number stories. Draw a picture of the array made with the pennies, fill in the diagram, and write the number models for each problem.

SUMMARIZE (10 minutes)

- Give students opportunities to discuss their strategies and solutions for each of the four problems. Using the diagrams, discuss the relationship between multiplication and division.
- Your assignment tonight is to complete the Home Link Master p. 92. Use counters or pictures to show someone at home how you can use division to solve number stories. Fill in the diagrams.

Figure 3-5   *Sample direct instruction lesson plan.*

52 Chapter 3 • Planning for and Teaching Diverse Learners

MEETING THE NEEDS OF ALL STUDENTS

Once you have created a lesson plan, one more step remains before you are ready to teach it. Be sure to review the special needs of your students and make adaptations to your plan so all students can have the opportunity for success and challenge. Equity of educational opportunity is one of the NCTM's fundamental principles. Similarly, the *Professional Standards for Teaching Mathematics* call for every child to have equal opportunities to learn mathematics. According to these standards, "every child" includes

---

Sample Exploration Lesson Plan

Introductory Information

Grade Level K   **Date** October 15

Topic/Title Algebra/Making Patterns

Objective(s) The students will: 1. Use concrete objects to create linear, repeating patterns.
2. Differentiate between a pattern and a nonpattern.
State standard: The student uses objects to generate repeating patterns, and identifies and continues patterns.

Assessment Objective 1: Check place mats for patterns. Should be 100% accurate.
Objective 2: Show each child 4 premade place mats and ask the child to tell which are patterns. Should get 3 out of 4 correct to show mastery.

Gearing Down Just use 2 different shapes/objects to create patterns. (Station 1)

Gearing Up Use 3 or more objects. Allow students to label their patterns orally—AB, ABC, etc.

Materials 12″ × 18″ construction paper with 2-inch squares marked along each 18-inch side.
Station 1: orange and yellow leaf stickers.
Station 2: fall rubber stamps and stamp pad.
Station 3: markers and fruit stencils.
Reference: *Kindergarten, Investigations in Number, Data, and Space*, Investigation 1: Exploring Patterns, by Susan Jo Russell, 2004, Pearson Education, Inc.

Lesson Outline
LAUNCH (10 minutes)

- What have we been doing at our math centers this week? (Making patterns with pattern blocks, connecting cubes, and buttons.) What are we having at school next week? (School Council Fall Harvest Party)
- Today we're going to decorate placemats for the party. (Show blank mat with squares marked.) We're going to make patterns by putting one object in each square. We want to be sure the objects make a pattern. How do we know if it is a pattern? (Repeating element)
- Demonstrate using each material, discussing how to apply to placemats. Show one pattern and one nonpattern. Ask children to tell if an arrangement is a pattern or not with thumbs up or down. Review how patterns can be described with AB, ABC, etc.
- When you get to your center, you may make your placemat. Show it to me when you are finished. Then you can make more patterns with manipulatives in the tubs.

INVESTIGATE (20 minutes)

- Children move to their preassigned center groups. When they finish their placemat, they'll tell me about it and I'll check it off. They may make more patterns on blank paper or using manipulatives such as pattern blocks.
- While circulating, I'll show 4 premade placemats and ask each child to tell me which are patterns and which are not.
- Other notes I can record: Who is making patterns and who is making random arrangements? What kinds of patterns are they making? How are they describing their pattern?

SUMMARIZE

- Children move back together as a whole class and show a partner their placemat. They should also tell why it is a pattern.
- Select 5 placemats. Final questions: How can we tell this is a pattern? How could we describe this pattern?

Figure 3-6 *Sample exploration lesson plan.*

Planning for Effective Teaching 53

- Students who have been denied access in any way to educational opportunities as well as those who have not.
- Students who are African American, Hispanic, Native American, and other minorities as well as those who are considered to be a part of the majority.
- Students who are female as well as those who are male.
- Students who have not been successful in school and in mathematics as well as those who have been successful (NCTM, 1991, p. 4).

Throughout this book, we discuss ways of promoting equity. It is important to note that providing equal opportunities to learn does not necessarily mean providing exactly the same instruction to every child. *Equity* implies fairness and justice. What is fair and just for one student may be different from what is fair and just for another. Providing equitable instruction means adapting instruction appropriately for the needs of each student.

Recent federal laws, including the No child Left Behind Act of 2001 (NCLB) and the Individuals with Disabilities Education Improvement Act of 2004(IDEA), direct schools to focus on helping all children learn. As part of IDEA, Response to Intervention (RTI) is a tiered process for providing instruction, monitoring progress, and providing support for student learning. Grade level teachers frequently work together in a professional learning community (PLC) to plan and deliver appropriate tiered instruction for their students. Figure 3-7 shows the levels of support.

In Tier 1, all students are provided with focused, effective instruction of their grade level curricula. Student learning is assessed on a regular basis and differentiated instruction and support is provided for those who need it. In Tier 2, additional instructional interventions are provided for those students who are not successful after receiving Tier 1 instruction. Their progress is continually monitored and support may be increased or decreased, based on their success. If assessment shows a student is still not successful after receiving Tier 2 instruction for a designated period of time, they may be referred to Tier 3 for specialized evaluation to determine the cause of their difficulty and more intensive instructional interventions are provided.

Research has shown that RTI can improve academic achievement for students at risk and reduce the total number of students referred for special education placements. It is a collaborative, data-driven process that supports equitable education for all students (Brown-Chidsey, 2007).

Teachers are often concerned that planning to meet the needs of each student in their classroom will be an overwhelming task. However, rather than create a different lesson plan for each student, you should create a single lesson plan and then consider how you can adapt it to make it more accessible for the different learners in your class. To help in planning, think of three students from your class: a student who is ready for the lesson, a student who will require help with the lesson, and a student who is ready to work beyond the lesson. Keep the different needs of these three students in mind while you plan. In addition, if you have any students who are learning English, keep their needs in mind, too. Next, think of some support or enrichment strategies that you can use to meet students' individual needs while helping all the children meet the mathematical goal of the lesson. In this text, we refer to adaptations for students who need help with a lesson as strategies for *gearing down* the lesson and adaptations for students who are ready to work beyond the lesson as strategies for *gearing up* the lesson. Be careful not to go too far and "lose the integrity of the mathematics content and pedagogy or set expectations too low for students" (Brodesky et al, 2004, p. 150).

Some educators believe that a teacher must ask three questions while planning to individualize instruction for students who need extra help (Karp and Howell, 2004):

1. What organizational, behavioral, and cognitive skills are necessary for students with special needs to derive meaning from this activity?
2. Which students have important weaknesses in any of these skills?
3. How can I provide additional support in these areas of weakness so that students with special needs can focus on the conceptual task in the activity? (p. 119)

It is useful to begin collecting ideas and approaches for adapting lessons. Currently, much is being written about how teachers can provide equitable instruction for students from different cultures, students who are English-language learners, and students with identified special needs. Knowing some of these recommendations will give you a head start at becoming an effective teacher.

TEACHING STUDENTS FROM OTHER CULTURES   As the United States becomes more culturally diverse, the task of making mathematics relevant for all students both becomes

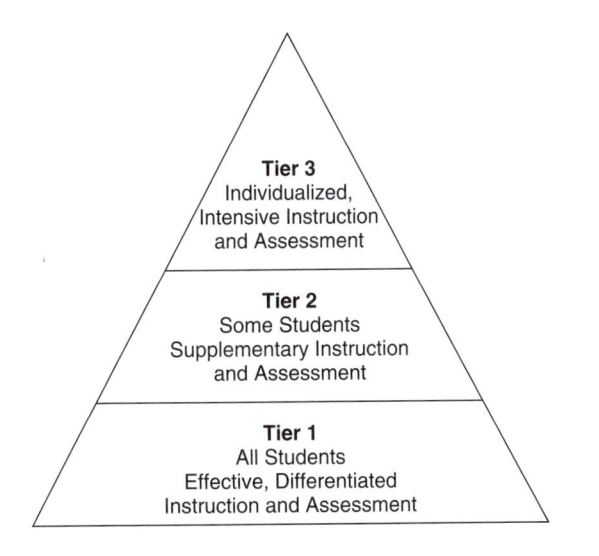

Figure 3-7   *Response to Intervention Pyramid.*

Tier 3
Individualized,
Intensive Instruction
and Assessment

Tier 2
Some Students
Supplementary Instruction
and Assessment

Tier 1
All Students
Effective, Differentiated
Instruction and Assessment

more challenging and offers more opportunities. You can increase the engagement of students, regardless of their background, by integrating the mathematics of world cultures into mathematics instruction. This allows students to

- See mathematical connections to their cultural roots
- Gain respect for the cultures of others
- Learn how mathematics was developed by people from many countries
- Connect mathematics to other aspects of life such as art, literature, games

It is important to plan lessons and activities that take into account the realities of students' daily lives as well as to ensure that students' cultural backgrounds are reflected in the mathematics curriculum and materials. Doing this allows teachers to model interest in learning about cultures different from their own. The Cultural Connections sections throughout this book can help you learn more about mathematics in various cultures and make connections in your instruction. Even if you do not have students from those cultures in your classes, they are interesting topics for all children to explore. The Cultural Connection for this chapter shows how you can connect a high quality, multicultural children's book to mathematics activities.

Students' cultures may also affect their learning preferences. While being careful not to stereotype students from a particular culture, you may benefit from learning more about the general learning preferences of children from that culture. Keep in mind, too, that ethnicity is not the only cultural factor to be considered. Cultural differences can also be recognized among students from families of low, middle, and high socioeconomic status (SES) (Payne, 2005). For example, students from low-SES families may lack important background knowledge or may not have mastered certain cognitive strategies. Once such deficits are identified, you can find ways to support these students so they may be successful.

You should also understand cultural norms and how they will show themselves in your classroom. For example, you should be aware that some students from other countries may need help adapting to new teaching and testing styles (Abrams and Ferguson, 2005). Students whose previous instruction focused on memorization and repetition may become frustrated in a classroom where the teacher uses a more open-ended approach. In addition, some behaviors considered acceptable in the students' home culture may not be as well received in their new location. For example, in some cultures students do not look their teachers in the eye when speaking with them. That behavior may be viewed as disrespectful if we do not understand the reason it occurs. With support, students can maintain their own cultural identity and still learn to function successfully in class in the new culture. Teachers must communicate instructions and expectations clearly and help students adjust to the new culture. Be aware that some of your students may feel and act in ways that could make it difficult for them to succeed in class without your help and understanding: "In many countries, students are not to speak unless the teacher asks them a question directly. To volunteer answers might be considered boastful or conceited. Many students will not question what the teacher says, even if they know it to be wrong" (Miller and Endo, 2004, p. 789).

TEACHING ENGLISH-LANGUAGE LEARNERS The U.S. Department of Education reports that more than 10% of all students in grades K-12 are English language learners (ELL). This is an increase of more than 53% since 1997 and the numbers continue to grow. For these statistics and resources for teaching ELLs, Math Links 3.4 connects you to the National Clearinghouse for English Language Acquisition (NCELA).

Students who are learning English will need particular supports in our classrooms. Some educators assume that mathematics should not cause difficulty for ELLs because the focus is on numbers and symbols rather than language. However, that is not the case. The ways various countries read and express symbols and numbers have enough differences to cause confusion. Some examples are included in Table 3-2 (Brown, Cady, and Taylor, 2009). In addition, the context in which mathematics is found in word problems may also cause difficulty. Cummins identified two levels of

TABLE 3-2 • Systems That Can Cause Confusion Among ELLs

Examples	U.S. Notation	Foreign Country Notation
• Dates	• 5/26/50 (Month/day/year)	• 26/5/50 (i.e., Mexico Day/month/year)
• Times	• 9:35 a.m. 3:45 p.m. 12-hour clock	• 09:35 15:45 24-hour clock (i.e., France)
• Place-value notations, Or marking place value	• 9,427,813	• 9.427.813
• Marking decimal places	• 15.6 (fifteen and six-tenths)	• 15,6 (fifteen and six-tenths)

(Source: Reprinted with permission from Mathematics Teaching in the Middle School, 2009 by the National Council of Teachers of Mathematics. All rights reserved.)

language proficiency; Basic Interpersonal Communication Skills (BICS) and Cognitive Academic Language Proficiency (CALP). Those who may seem fluent in spoken English may not be proficient in the academic language of English (Murrey, 2008).

According to one expert, there are five basic principles of instruction that teachers should incorporate in order to meet the needs of English-language learners (Herrell, 2000, p. xiv).

- *Give students comprehensible input (i.e., language they will understand).* This may involve using clear and concise language and avoiding slang. Emphasize and repeat key vocabulary and concepts and accompany these key points with explanatory visuals, models, or physical actions.

- *Give students opportunities to increase verbal interaction during class activities (i.e., to talk more with you and with the other children).* This will help English-language learners practice their language skills. Be aware, however, that some children may need to listen to and make sense of a new language for an extended period of time before they will try to speak it in class. Be sure to respect this silent period. Also, match questions to the child's proficiency level. For children who are not yet using the language, allow them to respond with actions (e.g., by demonstrating a skill). For children who are able to speak only in single-word sentences, phrase your questions so one-word answers will suffice, or provide choices of answers from which they may select. Respond to the students' intent, not to their pronunciation, grammar, or choice of words. In addition, you may need to wait longer for English-language learners to respond to questions, giving them time to translate the question, think about it, and then translate the answer. Finally, remember that not all cultures value verbal communication as much as ours does.

- *Teach in a way that contextualizes language.* A manipulative or graphic organizer may help students connect old and new knowledge and make your instruction more meaningful. Approach new ideas by beginning with students' current knowledge and by connecting to their lives to provide context. If you can, use both English words and words from students' native languages.

- *Use teaching strategies and groupings that reduce the anxiety of the students.* Look for ways to give English-language learners additional support. Involve as many of their senses as possible. Give students ways to respond besides writing or speaking in English. Allow them to interact with a partner or in a small group, as this is less stressful than having to speak in front of the entire class. Often, working with a peer who speaks the same language can make a student more comfortable.

- *Assign activities in the classroom that offer students opportunities for active involvement.* Using manipulatives and everyday objects can bring meaning to a concept. Some students will be more proficient demonstrating with objects than explaining with words.

Math Links 3.4

For more information on and resources for teaching English-language learners, look under Resources for Teachers or Practice on the home page of the Web site for the National Clearinghouse for English Language Acquisition. You can access this from http://www.ncela.gwu.edu or from this book's Web site.

www.wiley.com/college/reys

TEACHING STUDENTS WITH IDENTIFIED SPECIAL NEEDS
Students with identified special needs are those students who have been tested, have been admitted into special education services, and have an individual education plan. These students can include gifted children, children with physical handicaps, and children with learning disabilities. There are two common myths about teaching students with learning disabilities (Karp and Howell, 2004, p. 119):

- Myth 1: "*Students with special needs are vastly different from the regular school population and must be spoon-fed information or they will not be able to learn it.*" If you teach based on this belief, the result may be passive students who rely on others to tell them how to approach a problem. Since the goal is to develop independent learners, you must find ways to provide support to these students while encouraging them to develop self-reliance and confidence.

- Myth 2: "*Students with special needs are just like all other children in the class, and 'good teaching' is good teaching for all students.*" In order to be successful, students with identified special needs require different learning conditions and depend on different learning strategies than the other students in the class. They often possess a significant deficit in one or more areas, and these deficits can be barriers to their learning (see Figure 3-8).

To teach children with special needs of any kind (gifted, physically handicapped, or learning disabled), you will need to make adaptations in the curriculum and in your instruction. Figure 3-9 shows nine ways to make such adaptations (Ebeling, Deschenes, and Sprague, 1994). To decide which adaptations are appropriate for which students, you must be familiar with three things: (1) individual students and their special needs, (2) the activity and its subject-content demands

> **Memory:** visual memory, verbal/auditory memory, working memory
>
> **Self-regulation:** excitement/relaxation, attention, inhibition of impulses
>
> **Visual processing:** visual memory, visual discrimination, visual/spatial organization, visual-motor coordination
>
> **Language processing:** expressive language, vocabulary development, receptive language, auditory processing
>
> **Related academic skills:** reading, writing, study skills
>
> **Motor skills:** writing legibly, aligning columns, working with small manipulatives, using one-to-one correspondence, writing numerals
>
> Reprinted with permission from *Teaching Children Mathematics*, 2004 by the National Council of Teachers of Mathematics. All rights reserved.

Figure 3-8 *Potential barriers for students with special needs.*

(what is required to successfully complete the assignment), and (3) the options available for making adaptations. The adaptation categories in Figure 3-9 can be useful when planning for all kinds of learners in your classroom.

ASSESSMENT AND ANALYSIS IN PLANNING

Assessment should be an integral aspect of mathematics instruction. You need to ascertain whether you taught what you think you have taught and whether each child has learned what you think he or she has learned.

All the assessment information teachers collect is useful as they plan lessons. They will know more about the achievement and progress of each child as well as what his or her individual needs are.

Many teachers also find it helpful to keep an evaluative record of the effectiveness of their lessons. They jot down notes in their plan books or their teachers' guides about the things that went well and the things that didn't during each lesson (see Figure 3-10). They keep records of activities tried, articles read, ideas they want to try, and other anecdotal records. These notes help them to plan the following year. You may think you will remember what happened, but you will probably forget.

Finally, the *Professional Teaching Standards* (NCTM, 1991) considers *analysis*: "the systematic reflection in which teachers engage. It entails the ongoing monitoring of classroom life—how well the tasks, discourse, and environment foster the development of every student's mathematical literacy and power" (p. 20). Such analyses are a primary source of information for planning and improving instruction during the course of a lesson and as lessons build during the year.

CULTURAL CONNECTIONS

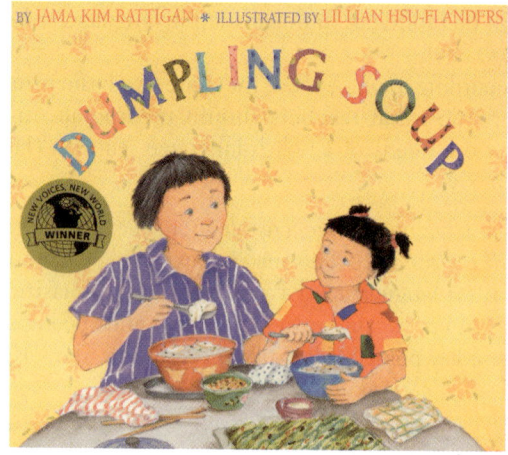

(Source: Cover of DUMPLING SOUP by Jama Kim Rattigan and Illustrated by Lillian Hsu-Flanders - Little Brown and Company, a division of Hachette Book Group, Inc.)

One way to connect culture to mathematics lessons is to integrate multicultural children's literature. High-quality multicultural books provide insight into a culture or lifestyle. While many children's books were written with mathematics concepts in mind, you can also find mathematics in many ordinary children's books. *Dumpling Soup*, by Jama Kim Rattigan, is a compelling story about Marissa, an Asian-American girl, who celebrates the Chinese New Year with her family in Hawaii. Marissa is excited because she is finally old enough to help make dumplings for Grandma's soup. The story shows the connection between creating and sharing food to expressing love within a large extended family. This book was not specifically designed for teaching math, but contains a variety of events in the story that could be connected to mathematics activities. As you read children's books, be on the lookout for math connections. Below are a few ideas for connecting mathematics to *Dumpling Soup*. An article, by Smith, Babione, and Vick, elaborates on these lesson ideas and may be found in one of this chapter's Book Nook resources, *Exploring Mathematics through Literature*, edited by Thiessen, 2004.

- **Algebra: Marissa hears sound patterns in the kitchen.**
  Patterns: Have children bring kitchen utensils from home. Use them to make sound patterns.
- **Number and operations: The family shares dumplings.**
  Fractions: Give students paper squares or circles. Have them figure out how to share them with various numbers of people. For example, two people share 4 dumplings or four people share 6 dumplings. Add and subtract dumplings.
- **Geometry: The dumplings cover rectangular trays.**
  Area: Cut irregular shapes out of graph paper. Have students estimate which "pan" would hold the most

Cultural Connections 57

Nine Types of Adaptation

Size
Adapt the number of items that the student is expected to learn or complete.

For example:
Reduce the number of division problems a student must complete.

Input
Adapt the way instruction is delivered to the learner.

For example:
Use different visual aids; plan more concrete examples; provide hands-on activities; place students in cooperative groups.

Participation
Adapt the extent to which a learner is actively involved in the task.

For example:
In geometry, have a student hold a shape while others point out attributes.

Time
Adapt the time allowed for learning, task completion, or testing.

For example:
Individualize a timeline for completing a task; pace learning differently (increase or decrease) for some learners.

Difficulty
Adapt the skill level, the problem type, or the rules on how the learner may approach the work.

For example:
Allow the use of a calculator to figure out math problems; simplify task directions; change rules to accommodate learner needs.

Alternate Goals
Adapt the goals or outcome expectations while using the same materials.

For example:
For fractions, expect one student to be able to add fractions while others learn to simplify the sum as well.

Level of Support
Increase the amount of personal assistance with a specific learner.

For example:
Assign peer buddies, teaching assistants, peer tutors, or cross-age tutors.

Output
Adapt how the learner can respond to instruction.

For example:
Instead of answering questions in writing, allow a verbal response; use a journal for some students; allow students to show knowledge with hands-on materials.

Substitute Curriculum
Provide different instruction and materials to meet a learner's individual goals.

For example:
During a class session, one student is working on a computer tutorial while the teacher provides a whole-class lesson on a different topic.

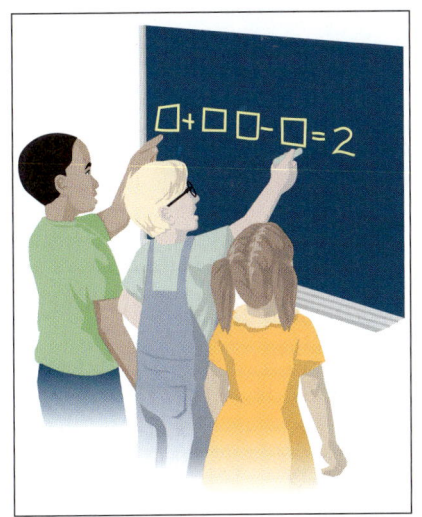

Figure 3-9 *Adopted from Nine types of curriculum and instruction adaptations.* (Source: Reproduced with permission from Ebeling, D., Deschenes, C., and Sprague, J. Adapting Curriculum and Instruction in Inclusive Classrooms: Staff Development Kit. Bloomington, IN: Institute for the Study of Developmental Disabilities, 1994.)

Figure 3-10 *A sampling of stick-on notes from a fifth-grade teacher's textbook.*

dumplings (color tiles). After estimating, students cover the shape with color tiles. Change colors after each 10 tiles used, then students can count the number of tens and ones and record in a place value chart.

- **Measurement:** The family makes the dumplings together.
  Recipes: Use measuring cups and spoons to follow a simple recipe. Discuss how some cooks do not use instruments, but instead estimate, such as a pinch.
- **Data:** Children try on the family's shoes, sitting on the front porch.
  Shoe Graphs: Have students develop estimates for their shoes such as "How many shoes in our class?", "How many tennis shoes?" etc. Then make real graphs with their shoes on the floor to answer the questions.
- **Probability:** Children play shoe store with the family's shoes.
  Make a match: Make pairs of paper shoes (different colors) and place in a box or sack. Have students predict how many draws they will need to make before they have a matched pair. Record estimates on sticky notes. Place on the board from smallest to largest. Have students do the experiment.

A GLANCE AT WHERE WE'VE BEEN

Effective elementary mathematics teachers must understand the mathematics being taught; be aware of their students' developmental characteristics; consider what their students know; think about the kinds of tasks they will give their students in order to maintain a focus on the mathematics, provide necessary practice; think about how they will encourage students to talk, what kinds of questions they will ask, and how they will group students; and consider the kinds of materials they will use in class, including curriculum materials, children's literature, manipulatives, and technology.

Planning lies at the heart of good teaching. Planning helps ensure that all essential content is included, permits scheduling the work in feasible periods of time and in a sensible sequence, helps control the pace of a lesson, aids in holding children's attention, helps avoid unnecessary repetition while ensuring necessary review and practice, and helps the teacher feel confident. Planning must be done for the year, the unit, and the lesson. Lesson plans should include clearly stated objectives, procedures, time allotments, and assessment. Different types of lessons (investigative, direct instruction, and exploration) are appropriate for the most common teaching goals. When planning, teachers must also consider equity—how to meet the needs of all their students, including students from other cultures, English-language learners, and students with identified special needs Finally, teachers' plans must provide for assessment and analysis. Now that you have read Chapter 3, go back to the Focus Questions at the beginning of the chapter and review your new learning.

Things to Do: From What You've Read

1. From the questions to ask before planning, select one and discuss why it is important to you.
2. Discuss one way you plan to supplement your math textbook materials when you teach.
3. Name a specific special need that students in your class may have. How can you adapt lessons for those students?
4. Briefly describe the most important differences among the three different types of lessons described in this chapter. When and why would you choose to use each type?

Things to Do: Going Beyond This Book

In the Field

1. [1]*Analyzing Classroom Discourse.* Observe in a classroom, keeping a record that lets you answer the following questions: What proportion of time does the teacher talk? What types of questions does the teacher ask? What evidence do you see that children have learned some mathematics? How many of the children have an opportunity to talk? Make a list of what they say (either to the class or to each other). In your opinion, is the classroom discourse about significant mathematics? Why?
2. [1]*Equity for Students with Special Needs.* Observe in a mathematics class where students identified for special education services are included in the room. What types of adaptations are made for them? If you were this teacher, what else would you do for these children?
3. [1]*Teaching a Mathematics Lesson* Watch a teacher teach a mathematics lesson. Use an appropriate lesson plan template and fill in the components of the lesson you observed. What was effective about the lesson? What would you do differently?

In Your Journal

4. Reflect on your skill and confidence related to lesson planning. What areas are strengths? What areas do you hope to improve? What do you plan to do to increase your skill and confidence?
5. Describe a lesson you think would lend itself to working with children in cooperative learning groups. How would cooperative learning groups support children in this lesson?

With Additional Resources

6. Select a topic to teach. Locate resources such as literature, manipulatives, or technology to enhance your teaching of the topic. Write a lesson plan you might teach.
7. Select a content topic. Compare the way it is taught in a standards-based text versus a more conventional text.
8. Examine a textbook and the supplemental materials that accompany it to see what assistance is provided for adapting instruction for various levels of students. Are suggestions for modifying lessons included? Are there extra-challenge ideas? Are there review activities for students who need extra practice? How would you use these materials if your class included students with widely varying levels of understanding and skill?

---

[1]Additional activities, suggestions, and questions are provided in *Teaching Elementary Mathematics: A Resource for Field Experiences*, found on this text's accompanying Web site, at www.wiley.com/college/reys.

With Technology

9. Explore NCTM's Electronic Examples at http://standards.NCTM.org/document/eexamples/. Select one of the examples and develop a lesson plan that integrates the example in the children's activities. Describe any special preparation needed in order for the lesson to be effective.

10. Search the Internet for teacher resource Web sites. Find a lesson plan that you could use with your grade level of interest. Modify the lesson plan to fit a specific classroom situation with which you have had experience.

Note to Instructors: You can find additional resources, learning activities, and black-line masters in this text's accompanying Instructor's Manual, at www.wiley.com/college/reys.

Book Nook for Children

Bresser, R. *Math and Literature (Grades 4–6)*. 2nd ed. White Plains, NY: Cuisenaire, 2004.

This resource gives intermediate-grade teachers 20 lessons based on popular children's books, including "Counting on Frank," "The Giraffe That Walked to Paris," and Jumanji."

Evans, C.W, Leija, A. J., and Falkner, T.R. *Math Links: Teaching the NCTM 2000 Standards Through Children's Literature*. Englewood, CO: Teacher Ideas Press, 2001.

Children's Literature titles with accompanying math activities are organized under the five content standards and five process standards.

Simmons, K., and Guinn, C. *Math, Manipulatives & Magic Wands: Manipulatives, Literature Ideas, and Hands-On Math Activities for the K–5 Classroom*. Gainesville, FL: Maupin House, 2001.

Students learn national math standard skills as part of literature-based hands-on projects. Projects, ideas, literature links, and blackline masters make this a good resource that enriches the language-arts block and helps you teach the language of math to all your students.

Thiessen D. *Exploring Mathematics through Literature*. Reston, VA: NCTM, 2003.

This book contains a collection of articles and lesson from NCTM journal that demonstrate how to use children's literature to teach mathematics in grades preK–8. Blackline masters are included for most lessons.

Whitin, D., and Whitin, P. *New Visions for Linking Literature and Mathematics*. Reston, VA: NCTM, 2004.

This book provide ideas for integrating mathematics and literature in grades K–6. Resources include criteria for evaluating mathematics-related books, strategies for using the books, and an annotated list of recommended books.

CHAPTER 4

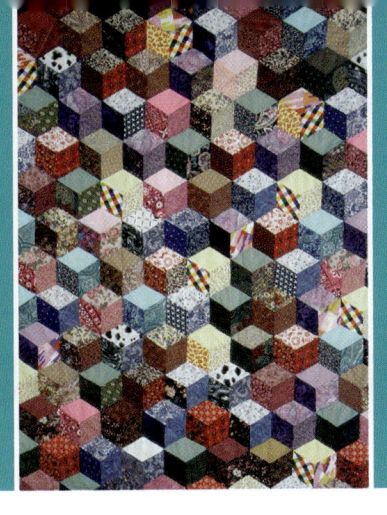

Assessment: Enhancing Learning and Teaching

> "HIGH EXPECTATIONS ARE THE KEY TO EVERYTHING."
> — Sam Walton, Walmart founder

> **SNAPSHOT OF A LESSON**

KEY IDEAS

1. Understand that documenting student understanding is important and complex.
2. Realize that students' classroom responses provide incomplete evidence of their understanding.
3. Understand that multiple assessment methods are needed to get a more complete picture of students' understanding.

BACKGROUND

Portions of the classroom and individual interview episodes below are available on the NCTM Illuminations Web site (see Math Links 4.4 for details). The entire interviews are captured on the video in *Mathematics: Assessing Understanding, Grades K–6*, by Marilyn Burns, available from ETA/Cuisenaire ® Company. In the excerpts below, we first see two second graders, Cena and Jonathan, counting and answering questions about place value during a whole-class discussion. We also gain additional insights into their thinking by hearing how each of them responds to questions posed during an individual, task-based interview. Their teacher will, no doubt, also examine their written work to assess their understanding. Teachers gain deeper insights into their students' understanding when they use multiple methods of assessment.

The camera zooms in on Cena, a 7-year-old student, waving her hand, eager to respond, during a whole-class lesson. The teacher has drawn many stars on the chalkboard. In order to count the stars, the children have instructed the teacher to count one by one, pausing to put a circle around each group of 10 stars. After the teacher has circled four groups of 10 stars, and has counted the 9 stars that are left, she calls on Cena to tell how many stars there are altogether.

49 stars

Cena: [Pointing at the board as she explains] Look, you got, um, 1, 2, 3, 4 tens and you, like, put 4 ... 4 right there. And put ... yeah ... put 4. [*The teacher writes a 4.*] And then if you have 9, um, stars leftover, then you put 9 right there. [*The teacher writes a 9 next to the 4.*]

Cena: Yeah, you have 49.

Cena seems confident in counting using tens and ones, and she also appears to understand how to write the numeral associated with the total number of stars.

Next we see Cena on her own, on a different day, during an individual task-based interview conducted by mathematics educator Marilyn Burns. Cena's first task is to count a pile of tiles into groups of ten. Cena counts aloud as she separates the tiles into two piles of ten with 4 tiles left over. Ms. Burns then asks Cena how many tiles she has altogether. Although Cena has successfully grouped the tiles by tens, she looks confused. She hesitates. She is unable to say how many tiles there are in all. She chooses to count all of the tiles, one by one, and then gives the answer of 24. Ms. Burns poses several other tile-counting tasks. At one point, Cena counts out 18 tiles, one by one. Ms. Burns asks Cena to write the number 18 on paper and she does this correctly, with ease.

Ms. Burns: Great! That's how you write 18. Cena, can you tell me ... can you show me with the tiles what the 8 means?

Cena: Um, 8. um.

Ms. Burns: Put them [*waving her hand over some of the tiles*] right up here next to the 8. [*There's a pause. Cena is not sure what to do.*]

Ms. Burns: So, I just want to see just the 8. You can count out loud.

Cena: 1, 2, 3, 4, 5, 6, 7, 8 [*as she pulls 8 of the 18 tiles aside and puts them near the numeral 8 on her paper*].

Ms. Burns: Eight. So, this is just 8 tiles. When you wrote the number 18, you wrote a 1 and an 8. Can you show me what the 1 means? [*Cena moves 1 tile forward.*]

Ms. Burns: And when you counted 18, there were all these tiles together [*motioning toward the entire group of 18*]. So if this is the 8 [*motioning toward the 8 tiles now pushed near the numeral 8*] and *this* is the 1 [*motioning toward the 1 tile now pushed near the numeral 1*], where do those fit [*motioning toward the rest of the 18 tiles*]?

Cena: Um, over here? [*Cena discards the remaining tiles and does not seem bothered that, although she was asked to represent 18 with the tiles, only 9 tiles remain on her paper.*]

In another video clip of the same whole-class star-counting lesson, we see Jonathan, one of Cena's 7½-year-old classmates. Jonathan appears to demonstrate a firm understanding of place value by counting a large number of stars drawn on the chalkboard by first grouping them in tens. The teacher calls on Jonathan to count the stars.

Ms. Burns: What shall we do?

Jonathan: Um, count by tens!

Ms. Burns: OK, and how will we do that?

Jonathan: Just circle [*each*] 10. And then say 10, 20, 30, 40, 50, 60. [*Then Jonathan indicates how to find the total number of stars by counting on: reciting 61, 62, 63, 64, and so on.*]

Later, in an individual interview, after Jonathan has counted out a group of 16 tiles and has successfully written the numeral "16," Ms. Burns asks him to show her, using the tiles, what the 6 means. At first, he has difficulty with understanding the task.

Jonathan: OK. OK, what do you mean?

Ms. Burns: Well, I am interested in how you knew how to write a 1 and a 6 to mean 16. So I thought maybe you would start by telling me what does the 6 mean and what does the 1 mean and how together that means 16.

Jonathan: Well, sometimes, in my reason, in like that star thing—

Ms. Burns: Uh-huh.

Jonathan: Um, my reason is, you count and put it in tens. And then, and then you count the tens. And then there is some left over. So, like, um, there probably would be ... The right side of the thing means the ones that are left over, and the other side means how many tens.

Ms. Burns: Let's try it and see. So how many are left over in there [*motioning toward the numeral 16*]?

Jonathan: Left over? Here? or ... ? [*He waves his hand around the table, confused.*]

Ms. Burns: No, left over in the number, on the right side. If you put this into tens [*motioning toward the pile of 16 tiles*], do you have any leftovers with 16?

Jonathan: Probably. [*Said with a questioning tone. He seems quite skeptical.*]

Ms. Burns: What do you think? How many leftovers do you think you'd have?

Jonathan: Um. Well. Um. Maybe 3 or so.

Ms. Burns: Try it.

Jonathan: Around 3.

Ms. Burns: Try it and see.

Jonathan: [*Reaches over to the pile of 16 tiles and counts out*] 1, 2, 3, 4, 5, 6, 7, 8, 9, 10.

Ms. Burns: And how many leftovers?

Jonathan: [*counts the leftovers silently*] 6. [*He seems rather surprised.*]

In the whole-class lesson, Jonathan appeared confident in counting the stars by forming circles of 10 and counting the extras to state how many altogether. However, during the individual interview, the fragility of his understanding was revealed. Amazingly, Jonathan was unable to predict how many groups of 10s and how many extra tiles there would be in a pile of 16 tiles, even immediately after counting out the 16 tiles (one by one) and writing the numeral 16. Even though he can say and write that there are 16, he can only guess (by inspection) that there will be just one pile of 10, and he has to rely on counting to find the number of leftovers (6).

FOCUS QUESTIONS

1. How are assessments *of* learning (summative assessment) and assessments *for* learning (formative assessment) alike and different?
2. How do the four phases of classroom assessment help teachers inform their instruction?
3. What different methods can teachers use to gather information about their students' abilities, dispositions, and interests, and what do each of these methods communicate to students about what is valued in teaching and learning mathematics?

INTRODUCTION

Assessment should support the learning of important mathematics and provide useful information to teachers and students.

(NCTM, 2000, p. 22)

What comes to mind when you think of assessment? Grading? Exams? Quizzes? Homework? Projects? Individual teacher–student interviews? Standardized tests? According to NCTM's *Assessment Standards for School Mathematics* (NCTM, 1995), assessment is the "process of gathering evidence about a student's knowledge of, ability to use, and disposition toward mathematics and of making inferences from that evidence for a variety of purposes" (p. 3).

It's important to recognize the distinction between two rather different types of assessment: assessment *of* learning and assessment *for* learning (Stiggins, 2002). As a classroom teacher, you will need to be familiar with both. In brief, assessments *of* learning (sometimes called summative assessments) primarily provide evidence of student achievement for purposes of public reporting and accountability. Common examples are end-of-year exams or standardized tests. When summative assessments have serious consequences (such as deciding whether to retain or promote a student from one grade-level to the next, or determining if a student graduates or not, or influencing teacher salaries or school funding), then the assessments are called "high-stakes" assessments.

In Chapter 1, you read about the recent history in the United States of the development of standards and summative assessments—including those relating to the federal government's No Child Left Behind (NCLB) legislation, and the Common Core State Standards Initiative, developed by the National Governor's Association and the Council for Chief State School Officers (CCSSI, 2010). We do not discuss summative assessments further in this chapter.

Assessments *for* learning (sometimes called formative assessments) are the focus of this chapter—because they are more under the control of the classroom teacher.

Formative assessments should serve not only to document students' achievement but also to help students learn more (Petit and Zawojewski, 2010). Common examples include homework, in-class assignments, performance assessments, teacher observations, and classroom tests. Of course, as a classroom teacher, you will be involved with both sorts of assessments (and in many cases the distinction between them may become blurred). You should always ensure—no matter what sort of assessment—that students understand what you expect them to know and be able to do, and that you provide the support necessary for every student to do his or her personal best. Just as Sam Walton, founder of Walmart (who became a billionaire), attributed his success to "high expectations," NCTM's *Principles and Standards for School Mathematics* (2000) calls for "high expectations and strong support for all students" (p. 11).

Math Links 4.1

You can read more about these ideas in a full-text view of NCTM's Assessment Standards on the NCTM Web site. It may be accessed from http://www.standards.nctm.org (requires NCTM membership) or from this book's Web site.

www.wiley.com/college/reys

ASSESSMENT *FOR* LEARNING: FORMATIVE ASSESSMENT

As a classroom teacher, much of your effort and time will need to be focused on classroom assessment—"assessment *for* learning." Infrequent, high-stakes tests may be important, but they cannot provide you with the information about student achievement that you will need, moment by moment, day by day, in order to make crucial instructional decisions (Leahy, Lyon, Thompson, and Wiliam, 2005).

PHASES OF ASSESSMENT

NCTM's *Assessment Standards* identify four phases for the classroom assessment process: planning, gathering, interpreting, and using. Let's consider an actual classroom example (from Moskal, 2000) as we explore these phases.

While planning for an upcoming unit on decimals with her sixth-grade class, Ms. Lee decided it would be important to gather information about what her students already understood about place value, ordering, and density of decimals. So she posed the following question in class: "Write all the numbers between 3.4 and 3.5 on the board."

One of the students, Dana, wrote: 3.41 3.42 3.43 3.44 3.45 3.46 3.47 3.48 3.49.

Nakisha disagreed, asserting there were more numbers between 3.4 and 3.5. Ms. Lee asked the class, "Who is correct? Dana or Nakisha?" She assigned them to use paper and pencil to take a stance and explain their reasoning.

Figure 4-1 shows the response of three students from Ms. Lee's class. Jim thought Nakisha was right. To support his claim, he offered examples of more numbers between 3.4 and 3.5. Afwandi also agreed; but instead of offering specific examples of other numbers, she simply claimed, "There are infinitely many numbers between 3.4 and 3.5." Juan offered an example of how to find additional numbers between any pair of decimals.

Jim's response

> Because these numbers are in between
> 3.410, 3.412, 3.413, 3.414, 3.415, 3.416, 3.417, etc.

Afwandi's response

> There are infinitely many numbers between 3.4 and 3.5.

Juan's response

> 3.415 is between 3.41 and 3.42
> 3.4155 is between 3.415 and 3.42
> 3.41555 is between 3.455 and 3.42
> and it just keeps going

Figure 4-1 *Responses from three of Ms. Lee's students.* (Moskal, 2000, p. 192).

Ms. Lee reviewed all of her students' papers. Juan appeared to have some rudimentary knowledge of the concept of infinity. Jim's response concerned her because he seemed to be (incorrectly) claiming that 3.410 was missing from Dana's list. Dana had listed 3.41, but maybe Jim didn't realize that the trailing zero on 3.410 didn't change the value of the number. It wasn't clear from Afwandi's response what she understood about the density of decimal numbers. It seemed possible that her answer about "infinitely many numbers" was merely a phrase she had heard before and echoed without really understanding.

Ms. Lee decided more instruction was needed, particularly regarding zeros in decimal numerals. She decided to try—by posing another task—to help her students think about zeros in decimals at the same time that she would gather more information about their thinking. So she assigned this task:

> "Identify the numbers below that are equal:
> 02.3 20.3 20.03 20.030 2.3.
> Explain how you know which numbers are equal"
> (Moskal, 2000, p. 193).

Ms. Lee's approach to teaching about decimals provides an example of the planning, gathering, interpreting, and using phases of the assessment process. She began by *planning* a task (write all the numbers between 3.4 and 3.5). After her students had completed the task, she *gathered* their responses and *interpreted* them. Finally, she *used* the information she had gathered to plan another assignment for her students. Of course, the events in each phase of this model may vary, depending on the type of assessment you use, the mathematical topic, and your students' responses. However, making instructional decisions is one of the primary ways that teachers use information gained through the classroom assessment process. For example, Ms. Lee might also decide to talk with Afwandi, or to engage her in a task-based interview, to gain more insights into her understanding of "infinitely many." Note that it was Jim's response, and similar responses from other students, that convinced Ms. Lee to assign another task to help students reconsider the role that zero plays in decimals.

Now that we have examined a specific example, let's think, more generally, about the four phases of classroom assessment.

When you are *planning an assessment*, it is important to think about the purpose of the assessment, the methods you will use to collect data, how you will interpret the evidence you collect (e.g., how you will award points for students' written problem-solving work), and how you will summarize and use your findings. Will the outcome of your assessment change your lesson plans for the next week? Or will you use the information when explaining to parents what their children's strengths and weaknesses are? How you answer these questions may affect the sorts of data you decide to collect.

When you are *gathering evidence*, think about the variety of tasks or activities you might use and try to choose appropriately. For example, if you want to find out what your students understand about decimals, you could have them complete some worksheets on addition and subtraction of decimals. But that task might not provide you with as much information as having them show you how they could use base-ten blocks to represent an addition or subtraction problem with decimals or having them write about using base-ten blocks and draw pictures to accompany their writing.

When you are *interpreting evidence*, it is important to think ahead of time about the criteria you will use to judge the adequacy of responses. You should think about how you *hope* your students respond to a task or question and what sorts of difficulties or misconceptions they *might* have. What sorts of responses will you consider exemplary, adequate, or flawed? What will you consider adequate evidence of understanding?

Finally, when *using results* from assessments, think about how you will communicate your judgments and how they will affect future instructional decisions. For example, how well will a numeric or letter grade help your students understand where they went wrong and what they need to do differently? Can you supplement the grade with written comments? Can you find time for

MAKING INSTRUCTIONAL DECISIONS

As a teacher, you will be making many instructional decisions. As with any decision, the better informed you are, the wiser are the choices you usually make. The NCTM Assessment Standards suggest ways that you may use assessment information differently in making instructional decisions than teachers did years ago (see Table 4-1).

One shift in assessment practices is a move toward gathering assessment data continuously rather than periodically or at the end of a chapter, when it is too late to modify your instruction. Another shift is toward using a wide variety of sources of information (rather than primarily tests). The third shift is related to keeping the needs and progress of students in mind when making both short-range and long-range teaching plans (rather than following a set plan of study regardless of how well students are doing). Indeed, teachers use these three shifts in assessment when implementing Response to Intervention (RTI)—a contemporary assessment/instruction process designed to ensure that every child's progress is individually monitored and that children receive appropriate just-in-time interventions in the form of alternative or supplemental instruction. (See Chapter 3 for a more detailed description of RTI.)

MONITORING STUDENT PROGRESS

Several shifts in monitoring student progress are also called for by NCTM's assessment *Standards* (Table 4-1). These include assessing a wide range of student abilities rather than focusing just on testing students' factual knowledge and skills; providing more extensive and elaborated feedback to students on their progress; using a wider variety of methods for gathering information (rather than relying primarily on tests and quizzes); and involving students more in the assessment process.

TABLE 4-1 • Major Shifts in Assessment Practices

Shifts in Assessing to Make Instructional Decisions

Toward	Away From
Integrating assessment with instruction	Depending primarily on scheduled testing
Using evidence from a variety of assessment formats and contexts	Relying on any one source of information
Using evidence of every student's progress toward long-range planning goals	Planning primarily for content coverage

Shifts in Assessing to Monitor Students' Progress

Toward	Away From
Assessing progress toward mathematical power	Assessing knowledge of specific facts and isolated skills
Communicating with students about performance in a continuous, comprehensive manner	Simply indicating right or wrong answers
Using multiple and complex assessment tools	Primary reliance on answers to brief questions of quizzes and tests
Students learning to assess their own progress	Teachers and external agencies as the sole judges of progress

Shifts in Assessing to Evaluate Students' Achievement

Toward	Away From
Comparing students' performance with performance criteria	Comparing student with student
Assessing progress toward mathematical power	Assessing knowledge of specific facts and isolated skills
Certification based on balanced, multiple sources of information	Relying on only a few narrowly conceived sources of evidence
Profiles of achievement based on public criteria	Single letter grades based on variable or nonpublic criteria

Adapted with permission from *Assessment Standards for School Mathematics*, Copyright 1995 by the National Council of Teachers of Mathematics. All rights reserved.

EVALUATING STUDENT ACHIEVEMENT

Teachers today also do several things differently in terms of grading and certifying students' achievement (Table 4-1). One change is a move toward judging students' performance against stated criteria (standards or expectations) rather than by comparing students with one another. The second and third shifts were mentioned earlier: assessing a broad range of components of mathematical power rather than isolated skills and knowledge, and using a variety of sources for gathering assessment information. Finally, there are trends toward making achievement criteria more open and public.

The number of shifts listed in this section is an indication of the need for change in many of our assessment practices. This change will come through time and open minds. This chapter is an opportunity for you to begin asking yourself questions about how you will assess students in order to make decisions, monitor progress, and evaluate achievement.

Math Links 4.2

As you can see, there is much to consider when you assess your students. However, there are many helpful Web sites that can assist you, such as the following.

1. Balanced Assessment in Mathematics program—offers a free on-line library of over 300 mathematics assessment tasks:
   http://www.balancedassessment.concord.org/
2. Math Forum: Internet Mathematics Library on Assessment/Testing:
   http://www.mathforum.org/library/ed_topics/assessment

Each site may be accessed from its Web address listed above or from this book's Web site.

www.wiley.com/college/reys

WAYS TO ASSESS STUDENTS' ABILITIES AND DISPOSITIONS

There are many different ways to gather information about the abilities, dispositions, and interests of students. What type of information to collect and how to obtain it depend on the purpose for which it will be used. You first need to ask yourself what you want to assess.

For example, if you want to assess the principle discussed in Chapter 2—that mathematics learning should make sense to students—then you want to observe students and ask them questions that elicit their level of understanding, to find out whether they have attached meaning to what they are doing. If you want your students to be persistent and willing to approach problems in a variety of ways, then you need to assess these characteristics through problems of a challenging nature. If you want your students to be able to communicate well and to work well with others, then you should make those characteristics part of your assessment.

Once you have engaged your students in activities likely to elicit the behaviors that you want to judge, you need some sort of guidelines or framework to use in looking at the data you collect. For example, if you want to see how well your students can communicate about the notions of area and perimeter, you might ask them to write a letter to a younger child comparing and contrasting these two ideas and providing an example of a real-world problem using each measure. But how will you judge which student letters are better than others? You may choose to use a rubric, or scoring guide.

A *rubric*, or *scoring guide*, is a rating scale that can be designed or adapted for use with a certain class of students or a particular task. Generally, a scoring guide is used to assign anywhere from 0 to 10 points to student work to provide a rating of performance. This is different from awarding points when grading a quiz or test, where you generally count the number of correct answers to arrive at a score. Instead, with a rubric you look at performance on the task and rate that performance along a continuum. As Stenmark and Bush (2001) explain, "*Scoring* is comparing students' work to criteria or rubrics that describe what we expect the work to be. *Grading* is the result of accumulating scores and other information about a student's work for the purpose of summarizing and communicating to others" (p. 118).

Rubrics, or scoring guides, can be either holistic or analytic. Holistic criteria assign a single score based on the overall quality of the student's work. Holistic scoring guides may have just two or three levels, or more, and should include general descriptors of the achievement necessary to attain them. For example, you might award points for a student's problem-solving work, as shown in the scoring guide in Figure 4-2. On the other hand, when it seems important to score various dimensions or traits of student work separately, analytic scoring rather than holistic scoring is appropriate. (See Figure 4-10 later in the chapter for an example of an analytic scoring guide for assessing open-ended questions.)

One very straightforward way of constructing a scoring guide is to think about sorting students' work into three piles:

1. Does not understand the outcome (concept or process)
2. Developing understanding (but not quite there)
3. Understands the idea and can apply and communicate about it

Ways to Assess Students' Abilities and Dispositions   67

4	Exemplary work	Task solved correctly and efficiently, or student may even have gone beyond expectations of the task (e.g., problem solving in more than one way, or student has extended the problem and completed a more difficult version). Communication of process and thinking is clear and effective.
3	Good work	Task was completed correctly but possibly inefficiently, or not communicated clearly, or with some errors so minor that the teacher is confident of the student's understanding. Any errors or communication difficulties should be able to be corrected with minimal feedback from teacher.
2	Marginal work	Task partly completed, or completed with major errors. There may be either a lack of evidence of understanding or evidence of some misunderstandings. Some further teaching will be required.
1	Needs more instruction	No progress or only minimal progress. Work may include evidence of faulty reasoning, or that the problem was not understood, or that inappropriate methods were used. Teacher intervention is essential.

Figure 4-2 *General holistic scoring guide for math problem solving.*

For any specific task, you can develop *performance indicators* to help you judge which student work to put in which categories. Ultimately, these performance indicators provide descriptions of how well the students are making sense of the mathematics.

For example, you might design a task for students to show what they know about the meaning of decimals by using base-ten blocks to build specified decimal numbers (e.g., 3 tenths, or 1 and 5 hundredths) and to demonstrate decimal relationships (e.g., build 2 tenths and build 22 hundredths and explain which one is larger). Each flat block could represent a whole number, each rod a tenth, and each little cube a hundredth. In this example the student has displayed 2.46.

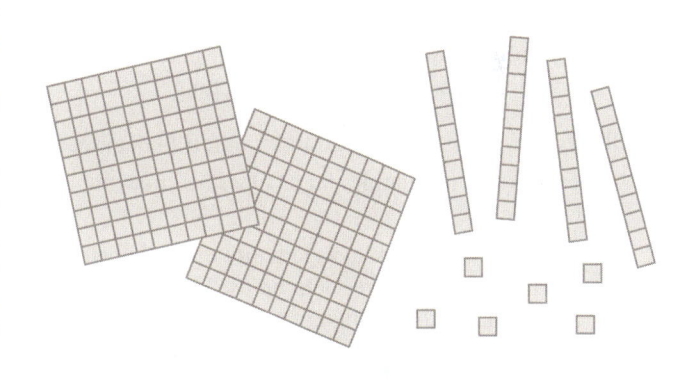

SAMPLE PERFORMANCE INDICATORS FOR A STUDENT WHO:

Does not understand	Is developing understanding	Understands
• When told to use one flat to represent a whole, cannot use blocks to show how many tenths would be equivalent to that one whole	• When told to use flat to represent a whole, can state which block would represent one tenth (rod) and which block would represent one hundredth (little cube)	• Understands that the same blocks can be used to represent different powers of 10 in different problems
• When told to use one flat to represent a whole, cannot use blocks to show how many hundredths would be equivalent to that one whole	• Can use blocks to show why 10 tenths and 100 hundredths are each equivalent to one whole	• When told which block represents one, can state what number is represented by any collection of blocks
• Has difficulty understanding that the same blocks can be used to represent different powers of 10 in different problems (may think a flat must represent 100 if earlier experiences assigned a value of 1 to each small cube)	• May interchange value of blocks when building decimals	• Can build, order, and compare decimals using blocks to illustrate
	• May have difficulty building decimals with blocks, especially when the decimal includes zeros (e.g., 3.04)	• Can build alternative models for the same decimal (e.g., can represent 3.45 either with 3 flats, 4 rods, and 5 little cubes or with 3 flats and 45 little cubes)
	• May have difficulty stating what number is represented by a given collection of blocks (e.g., 5 flats, 2 rods, and 3 little cubes)	

Figure 4-3 *Scoring guide with performance indicators for base-ten block task on decimals.*

Figure 4-3 shows a three-category rubric, or scoring guide, with performance indicators for this task. This type of scoring guide is often called an annotated holistic rubric.

OBSERVATION

Observing students as they work and making judgments about their performance from observation is probably one of the most commonly used assessment methods in the classroom. Yet many teachers never go one important step further: to make notes about their observations or think carefully about what they want to look for and why. Watching and listening may seem easy and commonplace, but it takes practice and planning to hone your observation skills. For example, it is helpful to plan what you will observe on a given day.

Suppose you are teaching first grade and your students are solving problems involving addition facts. You may want to observe and record which students are using physical materials, which are doing most of the problems mentally, which are using thinking strategies to determine basic facts (which will be discussed in Chapter 9), and which are relying on memorized facts. At times, you may plan to observe only one student in a cooperative group setting:

- Does José jump right in or wait for others to begin?
- Does he know his doubles facts? Does he know the facts that make 10?
- Is he able to use known facts to figure out unknown facts? If so, what strategies does he use? What sense does he make of the mathematics? Does he accept or challenge the ideas of others?

From such observations, you can gain insight into a student's attitude and disposition toward mathematics. This knowledge, in turn, can help you plan ways to encourage strengths and work on weaknesses.

Your notes about what you have observed will certainly be useful for anecdotal records that you can use for individual assessment, for future planning, and for reporting to parents. Your observations can also help you decide what to do immediately in the classroom, while you are leading a discussion or presenting a new concept.

Some practical tips are helpful in organizing yourself for classroom observations. Some teachers find it helpful to set aside a page in a notebook for each student or to use individual notecards in a file. But it can be clumsy to page through the notebook or fumble through the cards when you are walking around the room. To make observation notes easier to handle, some teachers like to tape an index card for each student in the class on a single page of heavy cardboard, with the cards overlapping as shown in Figure 4-4.

When you want to write notes about a particular individual, you just flip to his or her card. When a card is full, you can file it away in that student's file and replace it with a fresh card. Another handy idea is to carry a page of computer labels or a packet of stick-on notes on your clipboard as you circulate around the class. Make notes about individual students

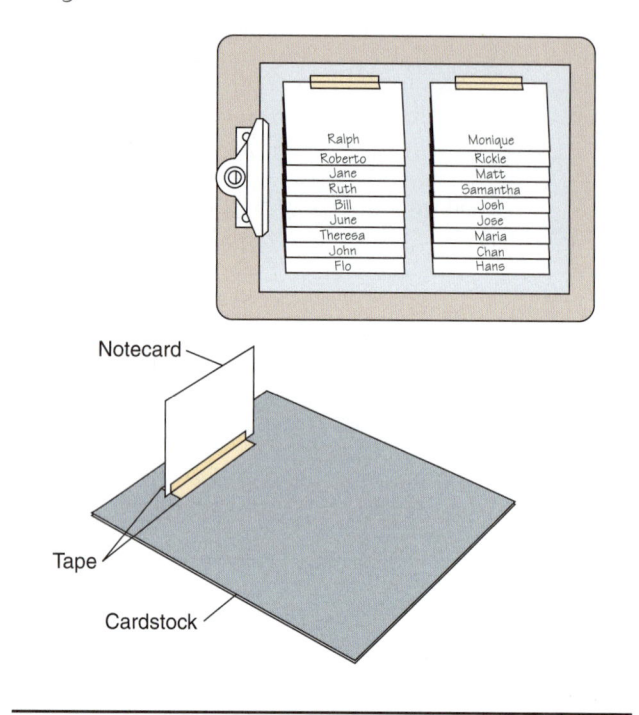

Figure 4-4 *Flip cards for recording classroom observations. Cards can be arranged alphabetically or by classroom seat assignments, whatever will help you find the right card quickly.*

on separate labels or notes (putting name and date on each). Later, you can paste these notes into the individual students' folders or pages in your observation notebook. If you want to remember to collect data on certain students on a given day, you can write their names on labels or stick-on notes in advance as a reminder. Whichever form or process you use, it is extremely important to remember to date your observations.

You can read more about assessing via classroom observation in Albert, Mayotte, and Sohn (2002); Bush (2001); Glanfield, Stenmark, and Bush (2001); and Stenmark and Bush (2001).

QUESTIONING

In Chapters 2 and 3, we discussed the value of asking questions of students to help direct their mathematical thinking. Good questioning techniques can also complement and enlighten your assessment observations. Asking good questions is an art that needs to be developed and practiced. When you teach through questioning, you actively involve students and know more about what they are thinking. In planning your lessons, you should think of questions that will help you gauge whether students are doing the following:

- Making sense of the mathematics
- Approaching a problem in different ways
- Clearly explaining their thinking
- Showing evidence of generalizing

It can be very effective to "engineer" your formative assessment questions to draw out evidence of your students' abilities by designing the questions based on what educational research tells us about students' typical learning trajectories of the mathematics under consideration (Petit and Zawojewski, 2010). For example, when students are learning to add and subtract fractions, they sometimes inappropriately bring their whole-number reasoning to working with fractions, treating the numerator and denominator of the fraction as two unrelated whole numbers, rather than as parts of a single quantity. If you are aware of this tendency, and use it to design some questions to see whether this is a problem for your students, you are likely to gain better insights into how well your students understand about fractions.

Limit the number of questions that can be answered yes or no or with one-word responses. For example, instead of asking "Should we add, subtract, multiply, or divide to solve this problem?" ask, "What could we do to understand this problem better?" or "What sort of picture can we draw to help decide what operation to use?" As a teacher, you should also work to develop more high-level and open-response questioning, asking students to explain or defend their answers or to describe another way to solve the problem. Remember that students need time to think about their answers to such questions. Practice waiting silently.

Finally, recognize that you can involve many more of your students in thinking deeply about mathematics if you think of ways to get everyone involved in asking and answering questions (Reinhart, 2000). For example, you can pose a question to the class and ask everyone to "think-pair-share." The students should first think about their own answers, then pair up with a partner to talk about their ideas, and finally be ready to offer suggestions to the entire class. (For even more opportunities for small-group discussion, you can have the pairs join up for discussion in groups of four or six before you bring the whole class together to talk about the question.) With this questioning model, everyone is encouraged to think and talk about the question. The conversation in the classroom is no longer "ping-pong" from teacher to individual student and back.

Some classic and recent references on questioning include Burns (1985); Hufferd-Ackles, Fuson, and Sherin (2004); Manouchehri and Lapp (2003); Mewborn and Huberty (1999); Parks (2009); and Vacc (1993).

Math Links 4.3

To view some online lesson videos, with accompanying guidelines for thinking about teacher questioning visit NCTM's Reflections Web site at http://www.nctm.org/resources/content.aspx?menu_id=598&id=6372 or link from this book's Web site. Choose "Reflect by Topic" (on left), and then "Discourse".

www.wiley.com/college/reys

INTERVIEWING

As illustrated in the Snapshot of a Lesson that opened this chapter, interviewing is a combination of questioning and observing, usually done with one student in a quiet place. It is a powerful way to learn about a student's thinking and to give her or him some special attention. Key factors in a successful interview are establishing rapport with the child, accepting responses without judging, and encouraging the child to talk and explain.

For example, at the beginning of the school year, Ms. Morihara decided to conduct short interviews with her new fourth graders to assess their understanding of odd and even numbers. Here is what Tonya said during that interview:

Ms. Morihara: Tonya, is 9 an odd or even number?

Tonya: It's odd ... unless you go by 3s, then it's even.

Ms. Morihara: Oh, that's interesting. Can you show me what you mean with tiles? [*Ms. Morihara gave Tonya 9 tiles to explain her thinking. Tonya stacked the tiles into 4 groups of 2 and showed her the 1 that was left over.*]

Tonya: See, if we divide evenly by 2s, there is 1 left over so it is odd.

[*Tonya then divided the 9 tiles into three groups of 3.*]

Tonya: But, if we divide it into groups of 3, there aren't any left over, so it is even. That is why 9 is both odd and even.

Ms. Morihara: Thank you very much, Tonya, for explaining your thinking to me.

This interview with Tonya shows how just a short interview can provide a wealth of information regarding a student's understanding of mathematics. Tonya is doing some interesting thinking about number relationships, particularly what it means to separate various numbers of tiles into even-sized piles. She seems to recognize that there may be different numbers of tiles left over depending on what sized piles the tiles are sorted into, and she is interested in focusing not just on the number of piles but also on the number of leftovers. Though she is confused about the definitions of "odd" and "even," she actually seems on the verge of understanding the more complex notions of "factor" and "multiple." If you had a chance to probe Tonya's thinking even further, what questions would you ask her? How would you attempt to change her misconceptions?

Although you will certainly not have time to ask in-depth questions of each of your students each day or even every week, it can be enlightening to choose a few children to interview each week until you have had a chance to talk with each one individually. Teachers are often surprised at how students value private time with them and how much they learn in a short time about the individual student. You may be able to squeeze in an interview or two each day during whole-class silent reading time, workstation time (when other students are circulating from station to station), or seatwork time.

Before interviewing a student, you need a basic plan of what concepts or skills you want to assess, what materials you will need, what questions you will ask, and when and how you will record the information. You may want to have alternative paths to take if the interview proceeds in different ways.

The interview itself generally has three parts: initiation, questioning/hypothesis formulation, and questioning/hypothesis testing (Long and Ben-Hur, 1991). During the initiation phase, the interviewer puts the student at ease by chatting informally, asking nonthreatening personal interest questions, showing the child the materials that will be used and ensuring that he or she is familiar with them, and explaining the purpose of the interview.

Next, the interviewer begins posing tasks, always making certain to rephrase questions in the student's own language when necessary and to encourage the student to explain, elaborate, and show what he or she knows. Students often understand more than it appears at first glance.

It is the interviewer's responsibility to ask questions in a variety of ways in order to formulate hypotheses about how the child is thinking about the task and what the child does and does not understand. It is critical that the interviewer remain nonjudgmental about student responses, even though students often look for subtle clues about the correctness of their answers. Some noncommittal phrases that are helpful in prompting children to explain their thinking further (without indicating whether previous responses have been right or wrong) include the following:

- I am interested in knowing more about your thinking. Talk to me about it.
- Pretend you are the teacher and I am your student. Please help me understand.
- Can you explain that in a different way?
- I like it when you take the time to explain your thinking.
- I think I understand now, but what if … ?

As the second phase of the interview proceeds, the interviewer begins to formulate hypotheses about what the child knows and where the child has trouble.

The third phase of the interview is the time to ask questions specifically designed to test hypotheses, to see if the difficulties really are what they seem to be. A big temptation for many teachers is to slide into teaching during an interview. It is important to resist the temptation to fix errors or misconceptions on the spot. If there is a lot of time, teaching may be appropriate, but re-teaching or tutoring is often better reserved for another time or another day. The focus of a classroom interview should be to figure out what a student knows, not to help him or her on the spot. In the Classroom 4-1 is a plan for interviewing third-grade students about their understanding of place value.

A classic and very useful chapter about classroom interviews is "Listening to Children: The Interview Method" (Labinowicz, 1985). Additional sources of information on interviews include Buschman (2001), Bush (2001), Ellemor-Collins and Wright (2008), Glanfield and colleagues (2001), Moyer and Milewicz (2002), and Stenmark and Bush (2001). To read about using game-based individualized interviews to assess number sense in young children, including those at risk because of socioeconomic level, disability, or the necessity of learning a second language, see Moomaw, Carr, Boat, and Barnett (2010). For an example of using interviews with English Language Learners (to assess their understanding of measurement ideas), see Fernandes, Anhalt, and Civil (2009).

> **Math Links 4.4**
>
> It is interesting to see the kind of insight into a child's thinking a teacher can gain from an interview. You can do this by viewing video clips at the NCTM Illuminations Web site. They may be accessed from http://www.illuminations.nctm.org/index.asp or from this book's Web site. Choose "Standards," then scroll down to "Video Reflections for Grades PreK–2" and click on "Understanding a Child's Development of Number Sense."
>
> www.wiley.com/college/reys

PERFORMANCE TASKS

In a class that is alive with problem solving and investigations, many valuable opportunities arise to observe students working on performance tasks. In fact, the only way to assess some skills is through performance tasks. For example, if one objective is for children to know how to measure with a ruler, then they need to be assessed doing such measuring. Questions on written tests that present a picture of a ruler perfectly lined up next to a picture of some other object to be measured do not reveal much about the actual skill of measuring. From this task, you can judge whether the child can read measurements from the ruler, but you have no idea whether he or she would actually be able to use a ruler effectively.

Performance tasks generally mirror the real world, are open-ended, and require time for grappling with a problem. It is often helpful to pair children when observing performance of such tasks so that you can hear their conversation as they work. In planning, you should list some of the areas you want to observe as the pairs work on appropriate tasks.

As an example of a performance task, look back at In the Classroom 2-2 (Chapter 2). In that activity, children are asked to build different boxes (right prisms) with some blocks. As they build boxes of various dimensions and various volumes, the formula for the volume of a box evolves naturally. The children are involved in using models, making decisions, and thinking mathematically, rather than just applying a formula. Figure 4-5 lists three areas you might plan to observe and examples of the notes you might make

In the Classroom 4–1

PRIMARY INTERVIEW ON PLACE VALUE

Objective: To determine students' understanding of place value.
Grade Level: 1-3.

Tasks:

1. *Can you show me 24 Unifix cubes?* Observe if child counts accurately (says numbers in proper order, uses one-to-one correspondence, has a method for keeping track of which cubes have been counted).

2. *Now let's suppose we want 34 cubes. Can you show me that?* Does child count out 10 more or count over from the beginning?

3. *Can you write the number 34 here?*

4. *Let's put the 34 cubes into groups of ten. How many tens do you think we can make? Do you think you'll have any cubes left over? Shall we count out the 34 cubes into piles to make sure?*

5. *Let's look at 34, the number you wrote. Can you show me with the cubes what the 3 means? Can you show me what the 4 means?*

6. *Now let's make four piles of 10 cubes each. And let's have 5 left over. Do you know how many cubes we have here all together? How can you tell?* Does student know to count the piles by 10 and add the leftovers? Or does he or she still count all?

7a. If the child has trouble with the questions above, check further by asking him or her again to count out several different numbers of cubes (e.g., 17, 26, or 30) and to predict how many tens and how many left over. Ask the reverse question: *For example, If we have 2 tens and 7 left over, how many cubes would that be?*

7b. If the child has no trouble with the questions above, probe for understanding of place value of larger numbers, such as 123 or 347, using base-ten blocks and asking how many piles of 100 could be made and how many piles of 10 and how many ones. Also ask the reverse question: *If we have three flats and six sticks and two units, how many cubes would that be altogether?* Extension: Check to see how many different ways the child can show 136.

Interview Reminders: Listen and watch carefully! Let them do it! Don't teach!

Be flexible. Ask for more examples, if needed.

Ask: How would you show a friend? How would you explain it to a little kid?

---

while observing the performance task from In the Classroom 2-2. Similarly, to read about a teacher who used a series of self-designed number-sense performance tasks with her 7- and 8-year-old students, see Diezman (2008).

As indicated in Chapter 3, *Professional Standards for Teaching Mathematics* (NCTM, 1991) emphasize worthwhile mathematical tasks as a core dimension. You may want to begin with simple tasks and then build up to longer, more complicated ones. Often, you will be surprised at the tenacity of very young children on a task that is engaging them, so you should not rule out richer tasks for this age. Badger (1992), Moon (1993), and Sanford (1993) are classics that provide interesting reading about performance assessment in the elementary math classroom, as do the more recent Bush (2001), Dubon and Shafer, 2010), Glanfield and colleagues (2001), Kitchen et al. (2002), and Stenmark and Bush (2001).

SELF-ASSESSMENT AND PEER ASSESSMENT

Self-assessment is an activity that engages many people. For example, actors, athletes, and musicians often study videotapes of their performances in order to figure out how to improve. In a similar way, when math students engage in activities that promote self-awareness and self-evaluation, they can eliminate weaknesses and become better problem solvers. Students are often the best assessors of their own work and feelings. When students evaluate their own work, the responsibility for learning is theirs.

You can begin the self-assessment process by having students validate their own thinking or their answers to selected exercises. For example, in a nonthreatening way, you could ask Dwayne to show you how he arrived at the answers to each of the problems below:

$$\begin{array}{r} 28 \\ \times\, 46 \\ \hline 168 \\ 112 \\ \hline 280 \end{array} \qquad \begin{array}{r} 24 \\ \times\, 39 \\ \hline 216 \\ 72 \\ \hline 936 \end{array}$$

It is important to ask questions about correct answers as well as incorrect ones, and to ask about standard approaches as well as nonstandard, so that children do not

72 Chapter 4 • Assessment: Enhancing Learning and Teaching

	Adam	Amy	Lawande	Tyler	Zaria	Kristin	Becca	Andrew
Strategy								
Used trial and error								
Organized by length (or height or width)								
Found and used patterns								
Made other comparisons								
Used a combination of strategies								
Result								
Found few or all combinations								
Found the connection to factors of the number								
Found/did not find the formula								
Able/Unable to describe the procedure								
Attitude								
Worked well with others								
Curious; explored other shapes, other numbers								
Enthusiastic								
Remained engaged								

Figure 4-5 *Observation guide for block-building activity (In the Classroom 2-2).*

think they are questioned only when something is wrong. Asking only about incorrect answers does not help build their self-esteem or establish the feeling that they have control of the mathematics.

It is helpful for students to learn to ask themselves a variety of mental self-assessment questions while they are engaged in problem solving. You can help students develop the habit of self-assessment by prompting them with questions such as "What are you doing now?" "Why are you doing that?" "How will it help you find the solution?" Eventually, these questions will become second nature to the students, and you will no longer need to prompt them. Being able to think about one's own thinking and to monitor one's own problem-solving efforts (*metacognition*) is one of the goals of problem-solving instruction. To read the story of a fifth-grade teacher who helps her students to self-assess both before and after more formal classroom assessments, see Weiser (2008). Similarly, a fourth-grade teacher reports that his students' standardized test scores soared after he taught them to peer review and recap their daily mathematics lessons with an easy-to-implement index card format (Quinn, Kavanaugh, Boakes, and Caro, 2008).

Engaging students in self-assessment of their problem-solving efforts after they have finished solving problems is another useful way to improve their monitoring abilities. For example, you might ask students to write a sentence or two in response to prompts like those in Figure 4-6 after they have finished a problem-solving activity.

Students can also analyze each others' strategies for solving problems. As they listen to and discuss how another student or group of students solved a problem, they begin to see different ways to proceed and to make judgments about which way makes the most sense to them, which seems easier or different, and which leads to stumbling blocks. Not only will they learn from such discussions and self-assessments, but you will also learn a lot about the students.

Sometimes knowing about students' attitudes, beliefs, and feelings about mathematics and mathematical tasks helps you, as a teacher, know how to design lessons more effectively. No one is better at assessing how a student feels about a given task than the student who is doing it. You may want to give a simple attitude inventory (see Figure 4-7). How would you plan a lesson that involves problem solving if you find from the survey that most of your students have a negative attitude toward problem solving? How could this information help you in grouping students in the lesson?

> Think about the problem you just worked on. Then answer these question by circling what you think:
>
> 1. How sure are you that your answer is right?
>
> ABSOLUTELY SURE   PRETTY SURE   SORT OF SURE   NOT SO SURE   I KNOW I GOT IT WRONG
>
> Why are you sure (or not sure) about your answer?
>
> 2. How hard was this problem for you?
>
> VERY, VERY HARD   PRETTY HARD   SORT OF HARD   NOT SO HARD   REALLY EASY
>
> Why was this problem at this difficulty level for you?
>
> 3. Have you ever solved a problem like this one before?
>
> YES, MANY TIMES   YES, ONCE OR TWICE   NOT SURE   DON'T THINK SO   NO, DEFINITELY NOT
>
> If so, describe the other problems and tell why they were like this one. If not, what was different about this problem from the others you've seen?

Figure 4-6 *Self-assessment questions for problem solving.*

Attitude Inventory Items

Pretend your class has been given some math story problems to solve. Mark true or false depending on how the statement describes you. There are no right or wrong answers for this part.

_____ 1. I will put down any answer just to finish a problem.
2. It is no fun to try to solve problems.
3. I will try almost any problem.
_____ 4. When I do not get the right answer right away, I give up.
5. I like to try hard problems.
6. My ideas about how to solve problems are not as good as other students' ideas.
_____ 7. I can only do problems everyone else can do.
8. I will not stop working on a problem until I get an answer.
_____ 9. I am sure I can solve most problems.
10. I will work a long time on a problem.

_____ 11. I am better than many students at solving problems.
12. I need someone to help me work on problems.
_____ 13. I can solve most hard problems.
_____ 14. There are some problems I will just not try.
15. I do not like to try problems that are hard to understand.
_____ 16. I will keep working on a problem until I get it right.
17. I like to try to solve problems.
18. I give up on problems right away.
_____ 19. Most problems are too hard for me to solve.
20. I am a good problem solver.

Figure 4-7 *A problem-solving attitude survey for elementary students.* (Source: Charles, Lester, and O'Daffer, 1987. Reprinted with permission from *How to Evaluate Progress in Problem Solving,* copyright 1987 by the National Council of Teachers of Mathematics. All rights reserved.)

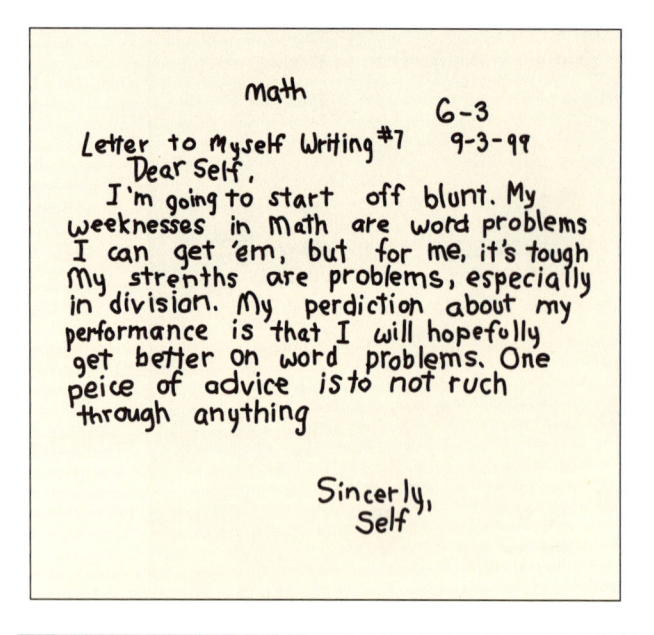

> **Math** 6-3
> Letter to myself Writing #7 9-3-99
>   Dear Self,
>   I'm going to start off blunt. My weeknesses in math are word problems I can get 'em, but for me, it's tough. My strenths are problems, especially in division. My perdiction about my performance is that I will hopefully get better on word problems. One peice of advice is to not ruch through anything
>
>   Sincerly,
>   Self

Figure 4-8 *Sample student self-assessment "letter to myself" written at the beginning of the year. Students can compare this letter with a letter written at the end of the year to show their growth over the school year.*

You might assign students to write math journals or letters to themselves or to others. One teacher had her students write letters to themselves about their strengths and weaknesses at the beginning of each grading period. Months later, they read back over their letters to see if they had made progress in any of the areas they had previously self-identified as weaknesses (see Figure 4-8). This information can be shared with parents during conference time.

You may learn more from children by asking them questions such as the following:

What are you really good at in math? What things do you need to improve on?

If you were telling a friend about the favorite thing you do in mathematics class, what would you tell him or her? What other things about math class might you tell your friend? Why?

If you and a friend got different answers to the same problem, what would you do?

How do you know when you have solved a problem correctly?

WORK SAMPLES

Work samples can include written assignments, projects, and other student work that you collect and evaluate. Figure 4-9 reproduces the work of four children on the following problem:

A carpenter makes only 3-legged and 4-legged tables. At the end of one day, he had used 31 legs. How many stools and tables did he make?

Figure 4-10 provides a typical analytic scoring scale for assessing children's work in problem solving on three measures (understanding the problem, planning a solution, and getting an answer).

Table 4-2 shows one teacher's scores for each of the four children on the three measures in Figure 4-10. There are, in fact, many different schemes for scoring problem solving, and each involves judgment. For example, try scoring the papers in Figure 4-9 using the same analytic scoring scale to see whether you agree with the scorer in Table 4-2. (You may well disagree, but you should be able to justify your position.)

Perhaps more important than scoring children's work is analyzing it to see what you can learn about the students. Looking again at the work of the four students in Figure 4-9, for example, you might conclude that only Suzy searched for more than one answer to the problem.

PORTFOLIOS

Portfolios have long been used to evaluate works of art, but in recent years this technique has also been used for assessment in mathematics education and for other school subjects. Generally, portfolios are purposeful collections of work. The items chosen for inclusion in a portfolio display the student's effort, progress over time, and level of accomplishment. Usually the student plays a key role in deciding what examples will be included. He or she may also help determine criteria for judging merit and evidence of student self-reflection. A student's mathematics portfolio might include such things as special problem-solving tasks, writings, investigations, projects, and reports. These samples could be presented through a variety of media, including paper and pencil, audio- or videotapes, and computer disks. Some teachers keep both classroom working portfolios (in which students keep the majority of their work) and assessment portfolios (which contain selected samples of their work for purposes of making judgments about students' attitudes, abilities, and dispositions in mathematics). See Figure 4-11 for a list of the items one teacher chose to require in student portfolios.

A particular benefit of portfolios is their value as a self-assessment tool for students. It is important for students to date the entries so that they (and others) can see their growth over time. It is also helpful if they describe each task and reflect on it. Figure 4-12 provides some advice for getting started with math learning portfolios.

WRITING

The NCTM *Standards* call for more emphasis on communication in mathematics. As stated in Figure 4-13, numeracy and literacy go hand in hand. Student writings as

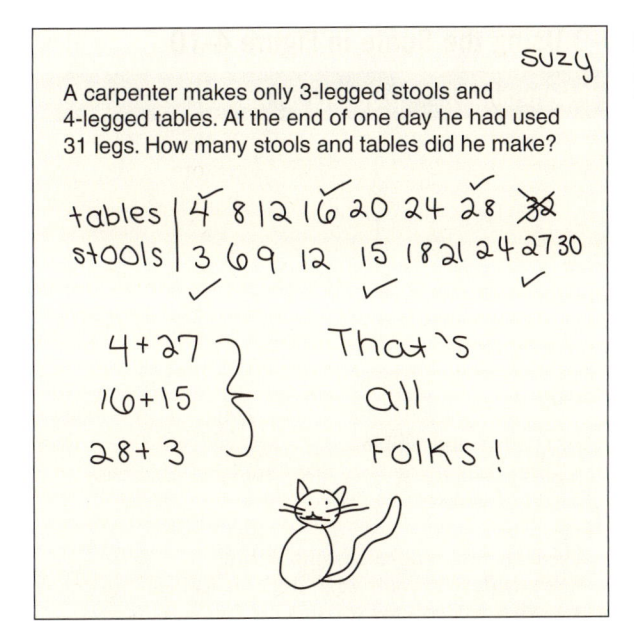

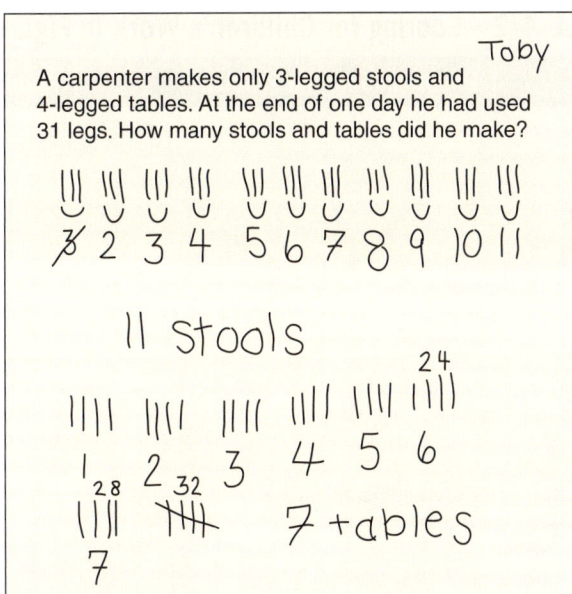

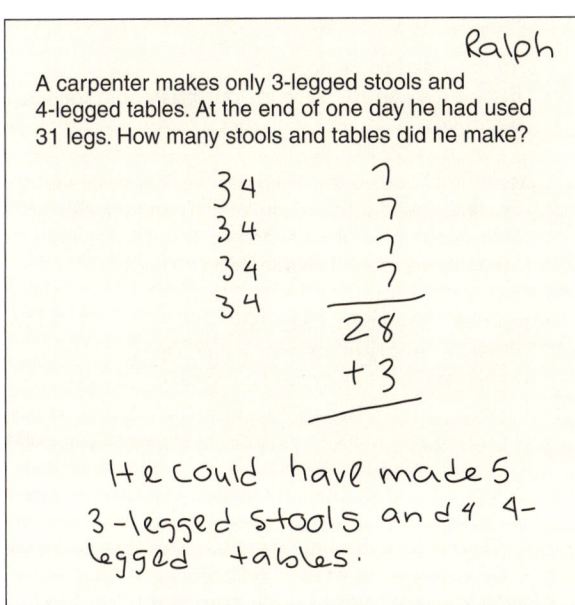

Ways to Assess Students' Abilities and Dispositions   75

Figure 4-9 *Samples of children's problem-solving work.*

Analytic Scoring Scale

Understanding the Problem
0: Complete misunderstanding of the problem
1: Part of the problem misunderstood or misinterpreted
2: Complete understanding of the problem

Planning a Solution
0: No attempt or totally inappropriate plan
1: Partially correct plan based on part of the problem being interpreted correctly
2: Plan could have led to a correct solution if implemented properly

Getting an Answer
0: No answer, or wrong answer based on an inappropriate plan
1: Copying error, computational error, partial answer for a problem with multiple answers
2: Correct answer and correct label for the answer

Figure 4-10 *A scale for scoring problem solving.* (Source: Charles, Lester, and O'Daffer, 1987. Reprinted with permission from *How to Evaluate Progress in Problem Solving*, copyright 1987 by the National Council of Teachers of Mathematics. All rights reserved.)

76   Chapter 4 • Assessment: Enhancing Learning and Teaching

TABLE 4-2 • Scoring for Children's Work in Figure 4-9 Using the Scale in Figure 4-10

	Understanding the Problem	Planning a Solution	Getting an Answer	Total
Suzy	2	2	1	5
Toby	1	1	0	2
Katie	2	2	1	5
Ralph	2	2	1	5

Minimum requirements for inclusion in your portfolio at the end of each grading period:

1. 3 homework assignments
2. 3 writing samples
3. 2 quizzes
4. 2 examples of class notes
5. 1 example of special project work
6. 1 example of cooperative group work

Your portfolio is not limited to the above. At the end of each grading period, you will choose five pieces of work to remain in your portfolio to represent the grading period's work.

Figure 4-11   *One teacher's specifications for portfolio contents.*

---

a form of communication can provide another source for assessment. You may want students to keep a journal or to add writing to other assignments.

Virtual Classroom Observation

Go to www.wiley.com/college/reys,
Access the Wiley Resource Kit
Click on the Virtual Classroom Observations Section
Module 5: Geometry-Calculating the Area of a Triangle.

1. Watch video: Teaching Example: Posing a Challenge to Assess Understanding
2. Read: Perspectives—Journals, and Perspectives—Teacher Reflection—Journaling in Mathematics
3. Do: From the Classroom—Assessing Student Work. (Look at 4 student journal entries and compare your assessment of the journals to Ms. Wittingham's)

A simple way to begin is to ask students to write about what they did or did not understand for one assignment, how they felt about an activity, what they learned today in class, or what they like about mathematics. Your creativity in providing suggestions will help spark theirs. For example, you might ask them to write a letter to a friend about math class

---

Getting Started with Portfolios

Start small and simple.
- Especially if you have many students, portfolios are a lot of work.

Developing a rationale is important.
- Why do you want your students to keep portfolios?
- What will you say to students and parents about your goals?

Portfolios need to be accessible.
- Establish a central place in your classroom for keeping portfolios.

Class time allocated for portfolio work is time well spent.
- Students need guidance in labeling and choosing.
- Sharing samples of students work in class is important.
- Students need guidance in being reflective.

A table of contents is essential.
- Provides an organizational structure
- Facilitates locating portfolio contents

The criteria you identify for judging students' portfolios reflect your goals and values.
- What categories of mathematical tasks will you require be included?
- Will you require examples of both draft and revised work?
- Will you require written reflections?
- Will you put grades on the portfolios? If so, how (what criteria)?

* How often will you give feedback? (Monthly, quarterly, at the end of each grading period?)

Figure 4-12   *Tips for getting started with portfolios.*

---

or to write a poem about triangles. As writing becomes a part of math class, you can use it to assess children's knowledge of and attitudes toward mathematics. Figure 4-14 provides some sample ideas for writing prompts.

TEACHER-DESIGNED WRITTEN TESTS

Tests can inform and guide your instruction, rather than simply determine grades. Students learn from tests, but too often the lessons they learn may not be those intended.

Ways to Assess Students' Abilities and Dispositions   77

When too much emphasis is placed on right/wrong answers from written tests, students may believe that it is not important to be able to show why procedures work, or to explain how they solved a problem, or to be able to solve word problems (because a good grade may be obtained without ever doing these things). As some educators have observed, students often decide what is valued by what gets tested—"or what you test is what you get" (Wilson and Kenney, 2003).

Carefully constructed and correctly analyzed tests can tell us a lot about students. For example, the children's papers in Figure 4-15 are for a simple test on subtraction of two-digit numbers, but they reveal a lot about the students. Notice how many of them missed the problems with zeros. If the first or seventh item had not been on the test, the teacher might not have realized that Jim regroups—and then ignores—even when regrouping is not necessary. (Try analyzing the kinds of errors the other students made and what characterizes each student.)

Most paper-and-pencil tests do not give you the opportunity to learn how a student arrived at an answer. One way to gain more insight is to ask the students to explain in writing what they did. If children are accustomed to explaining orally in class, then this task will be easier for them, and you will receive more meaningful explanations. Ask for explanations on a few items, and accept them in the children's language. Otherwise, they will soon learn to parrot explanations that are meaningless to them.

There is no reason why paper-and-pencil tests cannot include estimation items and require manipulatives and calculators. Tests can also permit the use of textbooks or notebooks, ask thought-provoking questions, and require students to connect new learning to previous learning. They need not be races against time in which the students regurgitate all they remember from the week or about the procedure last taught.

Thoughtful, well-constructed tests are one way—often a very efficient way—to gather information. Alone, they do not give a complete assessment of students' knowledge, but they can add one more piece to the puzzle.

STANDARDIZED ACHIEVEMENT TESTS

At some point you will undoubtedly be asked to administer standardized or statewide tests. The results of these tests may be used to take a look at your students' achievements in general. For example, if you see that your students did poorly in measurement, then you need to rethink what you are doing in that area. If student test results come back late in the year at your school, you might ask to see the results from the previous year for students in your class. Sometimes this information can provide valuable insight into what areas of study need extra work this year.

You should be aware that there are two types of standardized tests: *criterion referenced tests* and *norm-referenced tests*. Criterion referenced tests (such as most state accountability tests) are generally designed to test students on pre-published expectations. Furthermore, students' scores on

Figure 4-13 *An idea to engage students in writing about numbers.* (Source: From *The Phantom Tollbooth* by Norton Juster. Text copyright 1961 by Norton Juster. Text copyright renewed 1989 by Norton Juster. Reprinted by permission of Random House Children's Books, a division of Random House, Inc.)

Describe an activity our class might do to help us understand how big one million really is.

Let's play teacher! You want to teach some first graders how to subtract two-digit numbers. Give examples of the problems and materials you will use. Tell what you will say and what you will have the children do.

Audrey claims that her teacher could improve the average test score in the class by 10 points if she just added 10 points to everyone's score. Kelly doesn't believe that would work. What do you think? Explain your thinking.

Which decimal is closer to one-half, 0.307 or 0.32? How do you know? Explain how you could show or prove it.

What connections are there between fractions and decimals? Explain and use examples in your explanation.

Figure 4-14 *Sample writing prompts for having children explain their understanding of mathematical concepts.*

Figure 4-15 *Samples of children's papers for a written test of subtraction.*

these tests are usually compared against pre-established benchmark scores. For example, if multiplying fractions is an expectation for grade five in your state, then you would expect to see some questions about multiplying fractions on your state-level test. To identify the mathematics expectations for your grade level, you would refer to your school, district, or state curriculum framework or to the Common Core State Standards for Mathematics (CCSSI, 2010) if your state has adopted those standards. By contrast, norm-referenced tests (such Iowa Test of Basic Skills or CAT5) are designed to report how well a given student performed in comparison with all other students who took that same test, rather than comparing that student's performance against a fixed benchmark score. As a result, norm-referenced tests are designed to spread children across the whole normal distribution. On a norm-referenced test, there will always be 50% of the students below average and 50% above average. Thus, some items on norm-referenced tests are not intended for everyone to answer correctly. The fact that an item on multiplying fraction appears on a norm-referenced third-grade test would not necessarily mean that this is an expected skill for all students at this level.

In recent years, standardized tests have been changing. Some allow the use of calculators; some ask questions in an open format rather than multiple-choice format; some ask for explanations; and some include problem-solving situations. Be sure your students are familiar with the format of whatever standardized tests they are required to take. In addition, you should strive to help your students learn mathematics in ways that are meaningful to them and that help build their confidence and the attitude that they can do mathematics. These are some of the best ways to ensure that your students will be able to show what they know on standardized achievement tests.

KEEPING RECORDS AND COMMUNICATING ABOUT ASSESSMENTS

It is important to keep both informal and formal records of students' learning and their disposition toward mathematics. Certain types of record keeping may be required by your school system, but you can always keep additional records of your own.

A word of caution: Do not become burdened with a multitude of records, but keep enough that you can reflect on your students' progress and can justify any major decisions about them or what to teach. The type of records you keep will also depend on the ways you report information to students, parents, and the school administration.

RECORDING THE INFORMATION

Teachers are aware of many things about their students and keep many informal records about them. For example, you may know from daily interactions that Treena is always willing to answer in math class, that Joshua does not work well with Katrina, that Yong knows how to compute but seems to lack understanding about when and why certain operations are used, and that Cary is often absent on math test days. But maybe you need to jog your memory about Rachel before her parents come for the next parent conference. Or maybe you are thinking about changing the composition of cooperative groups and wondering if Jerry and Richard worked well together in the past.

Because of the volume of information about students that may be useful, it is often helpful as well as necessary to record some of it. Several techniques for recording information are described here. You may find others or modify these to suit your needs.

CHECKLISTS Checklists may be used in a variety of ways to record individualized information about students' understandings, attitudes, or content achievement. They may be simply lists that are marked in different ways, rating scales, or annotated checklists (see Figures 4-16 through 4-19). The beauty of checklists is that they can be adapted to your situation and, in fact, may help you think about your goals and the needs of each child you are teaching.

You cannot spend all your time keeping records, so select a few significant aspects of students' learning and attitudes and target a few children each day. For example, you may want to observe and keep records about children working in cooperative groups one week and the next week focus on individual children's understanding of a new topic.

MATH CLASS PARTICIPATION									
	Center A	Center B	Center C	Group Leader	Group Recorder	Group Reporter	Bulletin Board	Materials	Other
NAME	DATE: March								
Atkins, Willie	2	4	6	5	13	10	✓	✓	mCounts
Bero, Chuck	9	10	11	10	5			March	
Connel, Brenda	2	3	5			10		✓	
Coroi, Troy	11	10	9	5		13	✓	March	mCounts

Figure 4-16 *Checklist of student participation: useful for tracking which children need to be encouraged to participate and in which activities or roles. For example, Brenda has worked at math centers and has been responsible for materials. She may need encouragement to become a group leader or group recorder.*

CLASS OBSERVATION			
Class 5th period		Week Feb. 6-10/13-17	
		ACTION	
NAMES	COMMENTS	NEEDED	TAKEN
Atkins, Willie			
Bero, Chuck	Needs a challenge	*	Yes
Connel, Brenda			
Coroi, Troy			
Cosby, Kim	No homework for 3 days	*	Called

Figure 4-17 *Annotated class observation checklist: provides teacher with reminders of points to discuss when communicating with parents and students.*

Chapter 4 • Assessment: Enhancing Learning and Teaching

STUDENT DISPOSITIONS CLASS: 3rd Grade Math MONTH: October	Helen	Art	Whit	Beverly	Anita	Jim
Confidence Is sure of answer Knows how to proceed	✓ ✓			✓ ✓		
Flexibility Will change direction Tries several ways	If prodded		✓ ✓		✓	
Perseverance Stays with task Enjoys involved problems	✓	✓		✓		
Curiosity Wants to find out why Challenges			✓ ✓	✓		✓
Sharing Works well with others Shows leadership			✓	✓		✓

Figure 4-18 *Checklist of students' dispositions. The checklist is helpful in deciding which students might need special activities or encouragement, or as information to inform conferences with parents or students.*

Student Profile in Mathematics K–4

Date **Feb.**
Student **Jeremy Gelder**   Teacher **Matt Bell**   Grade **2**

Content Areas
 Number Sense & Numeration *Still sees numbers as ones - needs to develop place value*
 Estimation *Good feel for numbers - size of quantities*
 Concepts of Operations *Good problem solver*
 Computation
 Geometry & Spatial Sense *Geometry he loves*
 Measurement
 Statistics & Probability
 Fractions & Decimals
 Patterns & Relationships *Better at spatial patterns*

Math Power
 Problem Solving *Likes to think in creative ways*
 Reasoning
 Communication *Explains well*
 Connections
 Concepts *Developing*
 Procedures *Doesn't like to follow procedures - invents his own*

Dispositions
 Confidence *Growing*
 Flexibility
 Perseverance *On many tasks*
 Curiosity *Great*
 Reflection

Other Comments

Figure 4-19 *Sample student profile.*

If you use a checklist, you will want to keep it handy so that you can make quick entries as children are engaged in tasks. If you wait until the end of the day, you may forget some of the day's gems or be too preoccupied with other tasks to jot them down.

If you are recording information that is meant only for you, keep the checklist away from the eyes of your students. Be especially sensitive to children's feelings. A public checklist that shows the progress of each class member's skill attainment may be a great boost for those at the top but detrimental to those at the bottom, who often need the most encouragement.

STUDENT FILES   Many teachers keep a record of students' learning in the form of a file of work samples from each child. If children are keeping their own portfolios, those may suffice, and you will not need to keep additional samples of their work in your own student files. In either case, you may want to keep a summary profile for each child, like the one shown in Figure 4-19. Sometimes the school's format for reporting to parents requires that you keep such a record.

CLASS RECORDS   Often, the only class record is the grade book. Although it may be necessary to keep such a record, you must realize the limitations of a grade book. Alternatively, you could modify it with shorthand entries of your choosing to tell you more than attendance; grades on assignments, quizzes, and tests; and cumulative grades.

You could also supplement the grade book with a cumulative checklist made from your daily or weekly checklists. You can determine which items to include by reviewing the usefulness and frequency of your entries in these earlier checklists. The cumulative checklist will then give you a picture of the class as a whole and help you plan.

COMMUNICATING THE INFORMATION

Most teachers have three main audiences to whom the information they have gathered will be communicated: students, parents or guardians, and administration. Each group will receive different amounts of information in different ways.

TO STUDENTS   Much of your communication of information to students will be done orally or through actions, but you will also be writing comments on work samples, portfolios, tests, journals, or other forms of assessment materials. You may give a letter or number grade. All of these communications influence children's feelings about the value of different aspects of mathematics, their expectations of what they can accomplish, and their sense of their own worth.

Be positive and fair. Use information from your checklists or from student portfolios when letting each student know whether he or she is meeting expectations, and look for ways to demonstrate to each that he or she can do what is expected. Finally, remember that it is important that children grow in their ability to self-assess. One of your goals should be to help children become independent learners. If they always have to rely on you to validate their work and their thinking, they will not reach this goal.

TO PARENTS OR GUARDIANS   You will be reporting to parents or guardians both in written form and orally. The format of these reports is often determined by the school system, but you are responsible for the quality. You will grow in your ability to communicate with parents, but it is important from the beginning that you keep records and use them to illustrate or justify your oral and written comments. Phone calls can be useful for immediate feedback to parents not only when there is a problem but also to applaud good behavior or outstanding achievement. Parents and children are often surprised and gratified when a teacher calls home to report something good. Some teachers try to make good calls home to all parents early in the year. This practice can set up good relationships with parents and establish the expectation that a teacher phone call need not necessarily mean bad news. Some teachers use regular newsletters to keep parents informed about classroom activities and goals or letters to parents to tell them about special events or request their help. With today's broad access to e-mail, many teachers use that form of communication to maintain contact with parents. (Figure 4-20 shows a letter one teacher sent to parents to enlist their assistance in developing and maintaining their children's math portfolios.)

At parent conferences, you have an opportunity to explain goals for your instruction and to enlist parents' aid in helping their child meet those goals. It is helpful if you have collected a wide variety of samples of student work to share with the parents as you talk about their child's progress. Parent conferences are also times for gathering information. Parents often provide insights about their children that may not be evident to you in the classroom. The information you glean from them should be included in notes in your student files.

TO THE SCHOOL ADMINISTRATION   When you begin teaching, you will need to find out what types of records and grades are required in your school and how they are used:

- What sort of report card format is used? A/B/C letter grades? S/I/N (satisfactory, improving, needs improvement) progress indicators? Numeric grades? Narrative reports?
- Does the administration expect written reports on each child?
- Is there an official checklist?
- Is the class grade record sufficient?
- Is information on students passed to the next teacher, kept in a permanent file, or used to make tracking decisions?
- Is the information used in teacher evaluations?

> Dear Parents,
>
> Your child will be developing a "portfolio" in math class this year. One of the primary purposes of this project is to allow students the opportunity to reflect on their work, including exemplary work, work that is considered a "personal best," or work that they feel is in some way memorable. Through this project, students can begin to develop critical self-evaluative tools that will be extremely valuable in the years ahead.
>
> A second purpose of the portfolio project is to allow students to share some of their work with you. Please take some time this week to discuss your child's portfolio with him or her. Your child has been given written guidelines for the portfolio project. Please review the guidelines together.
>
> Please help your child this week as he or she finalizes selections for the portfolio. There is no set number of pieces required but the selection of 5 or 6 pieces is suggested. Help your child select pieces that best meet the guidelines and that are *most* representative of the major math content areas that we have covered during this grading period. The attached "News Bulletin" from this grading period may provide you with a better idea of the topics your child has studied and the activities he or she has participated in. Students have been asked to provide written accounts about *why* they chose particular pieces for their portfolios, so be sure to talk with your child about why he or she is choosing each piece for inclusion in the portfolio. (If you wish, you too may include some written comments about the pieces selected or the selection process.)
>
> After you have read the "News Bulletin" and have helped your child with his or her portfolio selections, please sign and detach the form below. Your child is to return the signed form, along with his or her portfolio, by Friday, November 15.
>
> Thank you for helping to make your child's mathematical experience a memorable one!
>
> Sincerely,
>
> _____
>
> I have read and discussed the "News Bulletin" with my child and have helped him or her select choice pieces for his or her portfolio.

Figure 4-20 *Letter to parents about math portfolios.*

The type and use of records vary from school to school, so it is necessary to find out about your own situation. When you know how the records are to be used, you will be better able to provide information that is suitable for particular uses.

CULTURAL CONNECTIONS

Ever since the early 1970s, when the U.S. government began documenting the mathematics achievement of American children through the National Assessment of Educational Progress (NAEP), certain groups of students have consistently been shown to lag behind other groups. In particular, black and Hispanic students generally have lower NAEP scores than white students. Socioeconomic status (SES) has generally been a reasonable predictor of achievement as well (Strutchens et al., 2004). It is well known that, on average, students in poverty tend to have lower test scores than students who come from affluent backgrounds.

The good news is that, in general, both fourth- and eighth-grade students' performances in mathematics showed consistent improvement from 1990 to 2007. In 2009, eighth graders continued the trend, with even higher scores than in 2007. Fourth-graders performance stayed as high in 2009 as in 2007—in other words, the scores were not significantly different. Breaking results down by ethnicity, white, black, Hispanic, and Asian/Pacific Islander, students from fourth and eighth grade all showed higher mathematics scores in 2007 than in any of the previous assessments (and eighth graders' scores in 2009 were higher than in 2007). On the other hand, the bad news is that the scores of minority students have not increased enough to result in a significant closing of the persistent performance gap between minority students and white students (www.nationsreportcard.gov/math_2009).

One way that each student's performance on the NAEP mathematics assessment is reported is by assigning it an achievement level (either basic, proficient, or advanced). In 2009, nearly half of all fourth graders scored at the basic level (and another 18% scored below basic). Put another way, only 39% of all students at the fourth-grade level were judged either proficient or advanced. However, the performance of black and Hispanic students was markedly weaker. While 51% of the fourth-grade whites were judged proficient or higher, only 16% of the fourth-grade blacks and only 22% of the fourth-grade Hispanics scored at those levels (www.nationsreportcard.gov/math_2009).

Both students who take the NAEP tests and their teachers complete questionnaires about classroom practices that offer us glimpses into what goes on in our nation's schools. In many cases, the NAEP questionnaire data indicate that students of all racial, cultural, and socioeconomic status have similar school experiences across the nation. However, the data also reveal certain interesting (and rather disturbing) differences.

In particular, NAEP questionnaire data, combined with NAEP assessment data, can offer insights into how mathematics assessment and instruction often differ in American classrooms depending on the race, ethnicity, or social class of that classroom's students. Different questions are asked on the NAEP questionnaires in different years. From analyzing NAEP data from 1992–2000, researchers discovered that multiple-choice testing increased in American classrooms during the 1990s and that it increased more for students of color and students in poverty. (Unfortunately, data for this same question are not available for years after 2000.)

> The percentage of students whose teachers reported using multiple-choice assessments once or twice a month increased between 1992 and 2000 from 49 to 60 for 4th graders and 34 to 44 for 8th graders. This increase could be a result of teachers preparing their students to take state-mandated multiple-choice tests. However, the percentages vary widely by race. In 2000, more teachers of black (70%) and Hispanic (69%) 4th graders than white (56%) 4th graders reported assessing their students with multiple-choice tests at least monthly. The disparities were even larger at the 8th-grade level, where 63% of black students, 53% of Hispanic students, and 38% of white students were assessed with multiple-choice tests at least once monthly. Students who qualified for free or reduced-price lunch were more likely to be assessed with multiple-choice tests than their higher SES counterparts within each race/ethnicity group. However, an analysis of race/ethnicity and SES together reveals that even high-SES black and Hispanic students were assessed with multiple-choice tests more than average white students.... Hence SES differences between black and white students do not account for the black–white disparities in multiple-choice testing.
>
> *(Strutchens et al., 2004, pp. 286–287).*

If certain portions of the school population are being tested with multiple-choice assessments more frequently than others, what effect does this have on their ability to think, reason, and perform on higher-level tasks (often more difficult to assess with multiple choice items)? Data from the 2000 NAEP have helped researchers answer this question, because NAEP includes not only multiple-choice tasks but also short constructed-response tasks (where students must write in their own answers) and extended constructed-response tasks (where students must not only write answers but also must provide an explanation of their reasoning). In 2000, white students performed considerably better than black or Hispanic students on each of the three NAEP item types; however, "score differences related to the extended constructed-response items were most dramatic" (Strutchens et al., 2004, p. 300). Similar score differences were found when comparing students by economic condition. These findings have led some researchers to hypothesize that schools enrolling large numbers of students of color or students in poverty may "be plagued by a 'pedagogy of poverty,' which is characterized by instruction that is very directive, controlling, and debilitating for students (Haberman, 1991). The pedagogy of poverty often does not lead to students' developing critical-thinking or problem-solving skills" (Strutchens et al., 2004, p. 301).

Your goal, as a teacher, must be to help ALL students develop high levels of problem-solving ability and conceptual understanding, as well as robust procedural skills. The findings cited above from the National Assessment of Educational Progress help us gain appreciation for the inequities that can result when teachers treat students differently according to their racial, ethnic, cultural, or socioeconomic backgrounds.

A GLANCE AT WHERE WE'VE BEEN

Assessment is an integral part of teaching. It should "support the learning of important mathematics and provide useful information to teachers and students" (NCTM, 2000, p. 22). Assessments *of* learning (summative assessments) are primarily designed to document what students know and are able to do. Final exams, as well as state and national standardized tests, are familiar examples of assessments *of* learning.

Assessments *for* learning (formative assessments) help teachers make informed decisions about the varied aspects of their day-to-day work: planning instruction, making instructional decisions, monitoring student progress, and evaluating student achievement. The assessment process itself has four phases: planning the assessment, gathering data, interpreting the data, and using the results to plan further instruction.

There are many methods used to gather, analyze, and present the information from assessments, whether the assessments are focused on student learning or on students' dispositions (attitudes and beliefs) toward mathematics. In the classroom, these methods include informal observations, interviewing, peer assessments, portfolios, writing, testing, and many other methods. For many forms of assessment, rubrics, or scoring guidelines, provide a framework for examining the information collected.

Assessment information needs to be recorded in ways that make it easy to analyze and use. There are numerous techniques for recording and communicating information. Portfolios, in particular, can provide a broad and useful picture of student progress and show growth over time.

Assessment can make a difference in how you help your students learn mathematics. You can use it in a positive way to encourage the children to become independent learners, to modify your instruction, and to communicate with parents and others with an interest in students' achievement.

Things to Do: From What You've Read

1. Explain the difference between assessment of learning (summative assessment) and assessment for learning (formative assessment), and give two examples of each.
2. Identify, and briefly describe, each of the four phases of assessment.
3. A narrow view is that assessment in mathematics focuses on how well students can perform during a timed test. Describe a broader view of mathematics assessment.
4. Try scoring the papers in Figure 4-9 using the analytic scoring scheme shown in Figure 4-10. Compare your scores with those shown in Table 4-2. Be prepared to justify the scores you give.
5. List three ideas for keeping records of your observations of students as they work during class. Do you have to observe every child during each class session? If not, how will you be sure to give each student adequate attention?
6. What sorts of assessment records would be good to keep to assist with a parent–teacher conference? What other ways can you communicate with parents about their children's progress?

Things to Do: Going Beyond This Book

In the Field

1. [1]*Analytic/Holistic Problem-Solving Scoring.* Collect a set of problem-solving papers from students. Analyze and score them using the analytic scale in Figure 4-10.
2. [1]*Observing Students in Class.* Observe one or two students during a mathematics class. Describe what you observed. If you were to work with these students in mathematics, what course of action would you take?
3. [1]*Performance Task.* Try a performance task with a couple of students. Make a list of observations and use it as the basis for your description of the children's performance. You may find samples in Mathematics Assessment: A Practical Handbook for Grades 3–5 (Stenmark and Bush (2001)).
4. Talk to one or more classroom teachers about how they grade their students in math. Ask what sorts of tasks are figured into the grade and how they are weighted. Ask to see the teacher's grade book so you can understand what sorts of records it includes. Find out what other records the teacher keeps (e.g., folders, portfolios, notes).

In Your Journal

5. Which of the recommendations for shifts in assessment (Table 4-1) have you experienced in your own years as a student? Briefly describe when and how.
6. Write a "letter to self" similar to that shown in Figure 4-8. In your letter you should reflect on your current strengths and weaknesses, as well as your future goals as a teacher of elementary school mathematics.

With Additional Resources

7. Use a Web-based library search to identify a teacher-authored journal article about assessment or grading. (Good journal options include *Teaching Children Mathematics, Phi Delta Kappan, Educational Leadership,* or *Elementary School Journal.*) Write a brief summary of the article, followed by your own reflection about two good points in the article and one point that you disagree with or that you think could have been improved.
8. Analyze a test in an elementary mathematics textbook. Would it make a difference to results on this test if the students had access to calculators? To get valid results, what items might need to be changed or put into a "no-calculator" section? Explain.
9. Examine several end-of-chapter tests from elementary school textbooks. How well do they align with the content of the chapter? To what extent do the tests emphasize vocabulary, concepts, procedures, and processes (such as problem solving, connections, communication, reasoning, and representation). Do any of the tests require writing about mathematics. If not, what changes could you make so that they would do so? What other changes could you make to the tests so that they would be more authentic?

With Technology

10. Use Math Links 4.4 to view the video for this chapter's opening Snapshot of a Lesson.
    a. In the first video clip, Cena explains how she would group objects in tens and ones during a whole-class discussion. What can you say about her understanding of numbers after listening to her explanation? Why might Cena be able to write two-digit numerals correctly without being able to explain the meaning of digits? What reasons might there be for the differences in Cena's performance during the whole-class lesson as compared to the individual interview?

b. In the second video clip, notice Jonathan's confidence as he answers in class, as contrasted with his hesitancy in responding to the interview's questions. How would you describe Jonathan's understanding of place value as compared to Cena's understanding?

c. Consider the additional questions listed under "Conclusion, Reflection, and Discussion" on the Web site.

11. Visit the Web site of the Nation's Report Card (www.nces.ed.gov/nationsreportcard/), which offers public access to test items from the National Assessment for Educational Progress as well as state and national performance data from years of administration of this test. Search for fourth-grade mathematics test items in a content area of your choice (e.g., number or algebra or geometry). Identify an item that was easy for most students, and another item that was difficult for many students. Write a paragraph about each item, offering your reasoning about why the difficulty level of these items was different.

---

[1] Additional activities, suggestions, and questions are available in the field experience manual on the Student Companion site at www.wiley.com/college/reys

Note to Instructors: You can find additional resources, learning activities, and blackline masters in this text's accompanying Instructor's Manual, at www.wiley.com/college/reys.

Book Nook for Children

Cohen, M. *First Grade Takes a Test.* New York: Star Bright Books, 2006.

Sammy, Jim, and George can't find the correct answers to some of the questions. Classmate Anna Marie has no trouble finding the answers, and she is eventually moved to a special class. The first-grade class is sad without Anna Marie. Eventually, Anna Marie returns and tells the class she has missed them. But is getting the highest score that important? This book prompts teachers to consider how children think when given a test.

Finchler, J. *Testing Miss Malarkey.* New York: Scholastic, 2000.

This is the story of a class and how upset everyone is—including the teacher and principal—over The Instructional Performance Through Understanding test. See how things change as teacher and students prepare for the test.

Leedy, Loreen. *Missing Math: A Number Mystery.* Tarrytown, NY: Marshall Cavendish Corporation, 2008. Everyday activities come to a halt when a town loses its numbers.

Murphy, S. *Pepper's Journal.* New York: HarperCollins, 2000.

Joey and Lisa keep a journal on their new cat, Pepper, for one year. This book will help provide students with insights into how a journal is kept and may offer them a good introduction to journal writing. There are activities in the back of the book for teachers and parents.

Prelutsky, J., and Smith, L. *Dr. Seuss Hooray for Diffendoofer Day!* New York: Random House, 1998.

This is the story of a school with a very diverse faculty who teach in a wide variety of ways. Then, one day, the principal brings in a test for everyone to take. The students are worried, but the teacher reminds them that they know everything that is on the test, plus more. The principal tells them that all students from miles and miles around will be taking this test to see who is learning and what school is the best. If the school doesn't do well, the school will be torn down and they will have to attend a school that teaches and behaves in only one way for all students. At the end, the school does very well and everyone passes the test. Readers see that everyone learns differently and that people can learn by doing things differently.

CHAPTER 5

Mathematical Processes and Practices

> "DO NOT CONFINE YOUR CHILDREN TO YOUR OWN LEARNING, FOR THEY WERE BORN IN ANOTHER TIME."
>
> —Hebrew proverb

>> SNAPSHOT OF A LESSON

KEY IDEAS

Understand that doing mathematics involves

1. Posing and solving problems, both purely mathematical and from the real world.
2. Talking about mathematical ideas, justifying reasoning, and using a variety of representations.

BACKGROUND

This lesson is based on the video *What's the Price?*, from Teaching Math: A Video Library, K–4, from Annenberg Media. Produced by WGBH Educational Foundation ©1995 WGBH. All rights reserved. To view the video, go to http://www.learner.org/resources/series32.html (video #11).

Lorraine Barr is a third-grade teacher at Wilson Elementary School in Wisconsin. Her class began working with the concept of division just last week, so the children understand that division involves fair sharing, but they have not yet learned any methods for computing the answers to division problems. In this lesson, they use problem-solving approaches—such as role playing and drawing pictures—to investigate and understand real-world problems involving division. They make connections to everyday life and use calculators as they determine unit costs for two different boxes of cereal and for a variety of other consumer products.

The lesson unfolds in three parts: (1) open-ended problem solving in small groups to figure out unit prices by drawing pictures and using estimation, (2) whole-class discussion, and (3) using calculators to find unit prices (which requires that the children interpret lengthy decimal results appropriately).

Ms. Barr launches the lesson by holding up a box of her favorite breakfast cereal, pointing out that the package claims 1 ounce is a serving. She asks the class what information she would need to find out how much one serving of the cereal would cost. The children recognize that they would need to know the price of the box of cereal and how many ounces are in the package.

Ms. Barr organizes the class into small groups. She gives each group some paper, a marker, and a package brought from home that contains multiple items. Each group in the class has a different package (for example, an $8.99 package of 3 pairs of socks, a $2.05 package of 20 plastic storage bags, etc.). Ms. Barr instructs the groups to figure out how much *one* of their items would cost and to illustrate their thinking by drawing a picture.

The group with the package of socks is working on figuring how much each pair of socks costs. One child suggests dealing out the cost, like dealing cards, one penny at a time for each of the pairs of socks. Ms. Barr is watching and helping guide the students.

Ms Barr: David said, "You could give each sock 1 cent, each sock another cent, each sock another cent."

A girl [enthusiastically]: "Or you could use dollars!" [*She apparently realizes that sharing 1 cent at a time is inefficient.*]

Figure 5-1 How much is one pair of socks if they are 3 for $8.99?

The children talk among themselves as they write numbers under their drawing of 3 pairs of socks (see Figure 5-1).

A girl: What if we're 1¢ over?

Ms. Barr: Well, you have to divide it. You can't be 1¢ over.

A boy: So, then that would be $3.

Another boy: Yeah, $9.

Ms. Barr: But they only cost $8.99.

Another group is working on the problem of finding out the unit cost for a box of 20 plastic storage bags. Although they initially suggest starting by assigning a 2¢ cost to each bag, they, too, quickly realize that starting so low will be inefficient. So they decide to start by estimating that each bag might cost 5¢. Pointing to the 20 bags in their picture, and counting by 5s, they find that the total doesn't reach $2.05. They try again, counting by 11¢ each (which turns out to be too much). Then they try again using 9¢ each, laboriously counting aloud by 9s to figure the total (and making some mental computational errors along the way). They reach a sum that is less than $2.05. So they decide their answer must be between 9¢ each (which was too little) and 11¢ each (which was too much).

When most of the groups have come close to finding their unit costs, Ms. Barr brings the class back together and asks them to share their approaches with the rest of the class. She narrates the video with this comment: "I knew that we would end up with some difficulty with figuring out the price, but I also wanted it to be a real-life situation. If they're going to use this in the grocery store, the items aren't going to come out exact." Ms. Barr asks what could help with the difficulty (how could they could determine the unit prices more efficiently and deal with prices that don't come out evenly)? The students suggest using a calculator.

The students now use calculators to perform their divisions. Ms. Barr asks the groups to raise their hands when they have figured out the price of their item.

A boy: We had the taco shells. We put in $2.09. ... There's 18 taco shells in here ...

Ms. Barr: What did you get for your answer?

Boy: It was high.

Ms. Barr: Tell me what you got on your calculator.

Boy: I got one-one-six.

Ms. Barr [interrupting]: Does there happen to be a point in there?

Boy: Yes ... We got point one-one-six-one-one-one.

Ms. Barr: What does that tell us?

Another boy: It's worth 11¢ with a remainder.

Ms. Barr discusses with the students how to interpret the digits after the hundredths place, and then challenges the whole class to use their calculators to help her find out which of two different sizes of boxes of cereal is the better buy. The first computation comes out 0.159375 and the children seem ready to agree that means 15¢ per ounce. Then Kellin raises his hand.

Kellin: With this first number as 9, that's almost close to making it be another penny. So when I mark it on here, I'm going to mark it as costing 16¢ per ounce.

Nicole observes that the other box of cereal comes out to $0.145 per ounce. She correctly interprets that as 14.5¢ per ounce.

FOCUS QUESTIONS

1. What five processes are identified in *Principles and Standards for School Mathematics* (NCTM, 2000) as key to an active vision of learning and doing mathematics?

2. What eight mathematical practices are highlighted in the Common Core State Standards for Mathematics (CCSSI, 2010)?

3. How is teaching mathematics through problem solving different from simply teaching students to solve problems?

4. For young children, what does mathematical reasoning involve and how does it help them make sense of mathematical knowledge and relationships? How can elementary children be encouraged to communicate their mathematical thinking?

5. What connections are important to aid elementary children in learning mathematics? What are three major goals for representation as a process in elementary school mathematics?

INTRODUCTION

As we have seen in earlier chapters, school mathematics is moving in new directions—and for good reason, as suggested by the Hebrew proverb "Do not confine your children to your own learning, for they were born in another time." In many mathematics classrooms these days, it is rare for a lesson to be taught to rows of children studiously bent over worksheets, practicing computations, rules, and formulas. Instead, it is more common to observe children behaving as in our Snapshot of Lesson—first by working in small groups—while talking and using drawings or tiles or blocks to model a problem—and by sharing their ideas, observations, and problem-solving processes. They listen carefully to each other, and they challenge and question each other. The lesson is often consolidated through whole-class discussion.

In doing mathematics, students are actively involved in a wide variety of physical and mental actions—actions that can be described by verbs such as *exploring*, *investigating*, *patterning*, *experimenting*, *modeling*, *conjecturing*, and *verifying*. Students learn mathematics by doing mathematics.

> **Math Links 5.1**
>
> Read more about the five process standards by choosing from the left-hand menu of NCTM's index to *Principles and Standards for School Mathematics* on the Web at http://www.standards.nctm.org/document/chapter3/index.htm or link from this book's Web site.
>
> www.wiley.com/college/reys

The *Principles and Standards for School Mathematics* (NCTM, 2000) highlight this active vision of learning and doing mathematics by identifying the five process standards listed in Table 5-1 (problem solving, reasoning and proof, communication, connections, and representations). These process standards share equal billing with the five content standards (i.e., number, algebra, geometry, measurement, and data).

TABLE 5-1 • Five Process Standards for School Mathematics (NCTM, 2000)

	NCTM's Process Standards
Problem Solving	Instructional programs from pre-K through grade 12 should enable all students to • Build new mathematical knowledge through problem solving • Solve problems that arise in mathematics and in other contexts • Apply and adapt a variety of appropriate strategies to solve problems • Monitor and reflect on the process of mathematical problem solving
Reasoning and Proof	Instructional programs from pre-K through grade 12 should enable all students to • Recognize reasoning and proof as fundamental aspects of mathematics • Make and investigate mathematical conjectures • Develop and evaluate mathematical arguments and proofs • Select and use various types of reasoning and methods of proof
Communication	Instructional programs from pre-K through grade 12 should enable all students to • Organize and consolidate their mathematical thinking through communication • Communicate their mathematical thinking coherently and clearly to peers, teachers, and others • Analyze and evaluate the mathematical thinking and strategies of others • Use the language of mathematics to express mathematical ideas precisely
Connections	Instructional programs from pre-K through grade 12 should enable all students to • Recognize and use connections among mathematical ideas • Understand how mathematical ideas interconnect and build on one another to produce a coherent whole • Recognize and apply mathematics in contexts outside of mathematics
Representation	Instructional programs from pre-K through grade 12 should enable all students to • Create and use representation to organize, record, and communicate mathematical ideas • Select, apply, and translate among mathematical representations to solve problems • Use representations to model and interpret physical, social, and mathematical phenomena

TABLE 5-2 • Eight Standards for Mathematical Practice (CCSSI, 2010, pp. 6–8)

Mathematical Practice	Standard Elaboration
1. Problem Solving	Make sense of problems and persevere in solving them
2. Reasoning	Reason abstractly and quantitatively
3. Argumentation	Construct viable arguments and critique the reasoning of others
4. Modeling	Model with mathematics
5. Using tools	Use appropriate tools strategically
6. Precision	Attend to precision
7. Structure	Look for and make use of structure
8. Regularity	Look for and express regularity in repeated reasoning

This summary was made by the authors of *Helping Children Learn Mathematics*. The full document (CCSSI, 2010) may be obtained at www.corestandards.org/the-standards/mathematics.

The NCTM *Standards* make it clear that doing mathematics means engaging in these fundamental processes.

In 2010, the *Common Core State Standards* for Mathematics (CCSSM) drew attention to eight standards for "mathematical practices." The CCSSM mathematical practices are closely related both in spirit and in substance to NCTM's five "process standards." The eight mathematical practices from CCSSM are listed, and briefly described, in Table 5-2.

This chapter provides an introduction to NCTM's five *processes* for doing and making sense of mathematics. Additionally, we have included attention to how CCSSM's eight mathematical *practices* are related to NCTM's *processes*. (In Chapter 6, we give additional attention to problem solving as a foundation on which the entire mathematics curriculum can be organized.)

PROBLEM SOLVING

What comes to mind when you think of problem solving? Some people think of challenging situations they may have encountered in real life, such as when their car got stuck in a snow bank, and they were unable to get it out until they solved the problem by putting floor mats under the wheels to provide greater traction. Similarly, when you think of problem solving in the mathematics classroom, you may think of challenging situations involving numbers or shapes or patterns. For example, a state official may need to figure out how many distinct license plates can be produced if each must be printed with a unique identifier consisting of exactly six characters. Perhaps the characters may be chosen only from among the 10 numeric digits (0–9) and 25 of the 26 letters of the alphabet (with the letter O excluded because it can be confused with the numeral zero). How many different license plates are possible? That's a problem! In this chapter's Snapshot of a Lesson, third graders who had not yet learned how to divide drew pictures to figure out unit prices. For them, that was truly a problem, too!

Alternately, when *you* think of mathematical problem solving, you may remember the story problems or word problems that often came at the end of each chapter of the mathematics text when you were in elementary school. After you had learned to perform certain computations (say, multiplication of fractions), your text may have provided problems in context that used those same skills. In many cases, these may not have been genuine problems for you because the techniques to use had been clearly outlined in the preceding pages.

Problem solving, as envisioned by the NCTM *Standards* and by the CCSSM, is much more than just finding answers to lists of exercises. Problem solving is a "major means of developing mathematical knowledge" (NCTM, 2000, p. 116). By general agreement, a problem is a situation in which a person wants something and does not know immediately what to do to get it. Problem solving is the foundation of all mathematical activity. As such, problem solving should play a prominent role in the elementary school mathematics curriculum. The children in the Snapshot of a Lesson were confronted with a problem in which they were interested—how to figure out unit prices for packages of socks or plastic storage bags, or for servings of a box of cereal. Although they didn't yet know how to perform long division, they demonstrated that they understood the concept of division by drawing pictures of the items in their packages and by sharing the entire cost of the package as evenly as possible among all the items in the package. The children connected their real-world shopping experiences with their understanding of division as fair sharing. They helped one another by talking and by developing a pictorial representation of the problem. Through these approaches to the task at hand, these children were certainly "doing mathematics."

Consider a different type of problem, actually a game that can be played by children in upper elementary grades. To investigate this particular problem, you will need to play the game with one or more opponents. You'll need two dice, 12 small chips or tiles or markers each, and a piece of paper on which to make a game board for each person (see In the Classroom 5-1).

Place your chips on the game board, putting as many or as few as you like on each of the 12 numbered spaces. (For example, you might choose to put all your chips on your favorite number, 4. Or you could put one chip on each space. Or you might put two chips on each of the even spaces and

In the Classroom 5–1

ROLLING THE DICE

Objective: Using a dice game to investigate experimental and theoretical probabilities and to develop analytical reasoning skills.

Grade Level: 3–6.

Materials: Twelve counters or chips for each player; one pair of dice for each four players.

Rules of the Game: Up to four players can play together with one pair of dice. Each player has his or her own game board. You begin by placing all 12 of your counters on your game board. The game board has 12 spaces numbered 1–12. You may place as many counters as you choose (0–12) on each space. You may leave spaces blank. And you may put one or more than one counter on any space.

Players take turns rolling the dice. The first player rolls the two dice and finds their sum. (For example, if 2 and 3 are rolled, the sum is 5.) Each player may remove one counter from his or her 5 space. Even if there is more than one counter on that space, only one may be removed. If there are no counters on that space, no counters may be removed from any space. The next player rolls the two dice and finds their sum (e.g., 4+4=8). Each player now removes one counter from his or her 8 space, and so on. The goal of the game is to empty your board. The first player with no counters left on his or her board is the winner.

Analyzing the Game: When you have played several times, talk with each other about these questions:
- Which sums were rolled most often?
- Which sums were rolled, or not rolled, very often?
- Why do you think some sums came up more often than others?
- Can you prove which sums are most liked to occur?
- What do you think is a good strategy for placing your counters on the game board? Why?

Writing about the Game: Write advice to a friend who is new to the game. Tell him or her your favorite strategy for placing the counters on the game board, and explain why you believe this strategy is a good one.

Rolling the Dice Game Board

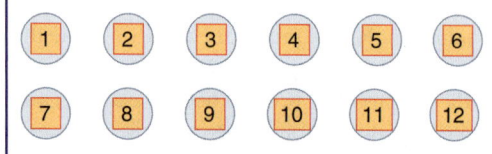

① ② ③ ④ ⑤ ⑥
⑦ ⑧ ⑨ ⑩ ⑪ ⑫

---

also be removing a chip from the 8 space on their boards, if they can. Roll the dice again, sum the results, and remove another chip. The goal of the game is to be the first person to remove all your chips from your board. Try the entire game a couple of times before reading on.

Here's the problem: Find a good strategy for placing your chips at the start of this game so that you are more likely to be able to clear your board quickly. Which spaces are good to avoid? Are there certain sums that rarely (or never) came up when you rolled the pair of dice? Why? Which spaces are good to put chips on? Are there certain sums that came up rather often? Why? Can you explain why some sums are more likely than others?

A fourth-grade class played this game several times, worked on the problem of finding a useful game strategy, talked about what they discovered, and then wrote about their findings. See Figure 5-2 for what some of the children wrote.

In a classroom where mathematics is taught through problem solving, students given the challenge of the dice game might already have had previous experiences with making organized lists or making tables. Nevertheless, their teacher would *not* have provided them with advice about

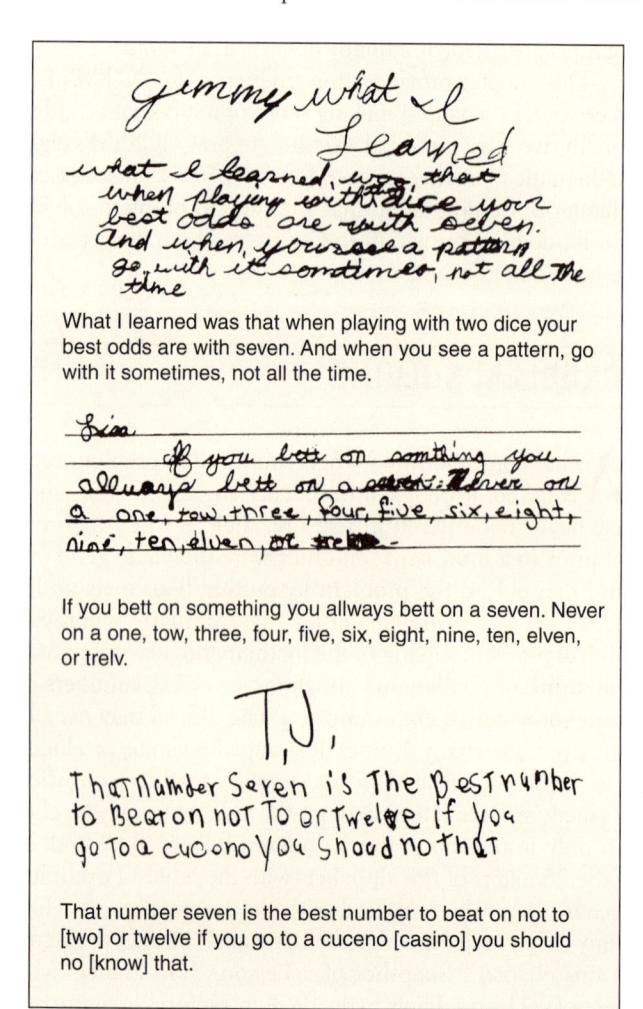

What I learned was that when playing with two dice your best odds are with seven. And when you see a pattern, go with it sometimes, not all the time.

If you bett on something you allways bett on a seven. Never on a one, tow, three, four, five, six, eight, nine, ten, elven, or trelv.

That number seven is the best number to beat on not to [two] or twelve if you go to a cuceno [casino] you should no [know] that.

Figure 5-2 *Fourth-grade students' writing about playing the dice game.*

---

leave all the odd spaces empty. It's your decision.) Roll the dice. Say you get 3 and 5. Find the sum (3+5=8). If you have any chips on the 8 space, you may remove one of them. When anyone rolls, everyone plays, so your opponents should

exactly how to solve the problem if he or she wanted this to be a true problem-solving challenge. A reasonable approach is to begin experimenting by playing a few games. It is helpful to keep a list of the sums obtained when the dice are rolled. If you don't keep a list, it may be hard to be sure which sums come up more often than others. Haphazard experimentation is not likely to produce a good, well-justified solution. Solving this problem requires students to think logically and to make some important decisions, particularly about how to keep track in a systematic way of the sums that come up. (Perhaps even better is keeping track of the pairs of dice that lead to those sums; writing down 5+2=7 is more informative than just writing 7, since 7 could be obtained in several different ways.) A skill such as making an organized list or a table can be useful in a wide variety of problem situations.

When you roll two dice, you are much more likely to roll a sum of 7 than a sum of 3, since there are only two ways to get 3 (1+2 and 2+1), but there are many more ways to get 7. As this problem helps develop general problem-solving expertise, it also deepens students' understanding of probability (helping them to recognize that we can compare the likelihood of rolling various sums by seeing how many ways those sums can be formed). The probability of rolling a 7 is actually 6/36 whereas the probability of rolling a 3 is only 2/36 (refer to Figure 5-3 to see two ways of representing the 36 different possible sums: 1+1, 1+2, ..., all the way up to 6+6). Students who work on this problem are learning about probability *through* problem solving. They are also developing and using a strategy for playing a game that can be used to solve other problems in mathematics as well as in other contexts.

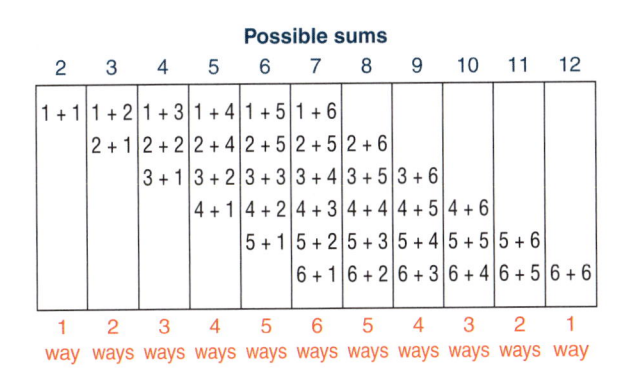

What kinds of problems are appropriate for elementary school students? In the early grades, most school mathematics problems are related in some way to the children's own experiences, as in the Snapshot of a Lesson, because their world is relatively circumscribed and children relate best to concrete situations. By upper elementary school, however, the universe of problem contexts should diversify. Increasingly, problems can grow out of situations in the world at large or from the investigation of mathematical ideas. Mathematics instruction in the upper grades can take advantage of the increasing sophistication of students and their growing knowledge of such topics as probability, statistics, and geometry.

Upper elementary students can deal with messier and more complex problems than primary children can, not only because they are more capable and confident in working with mathematical ideas than younger children but also because they can use technology to alleviate much of the drudgery that—until recently—often constrained school mathematics to considering problems with "nice numbers." Computers, calculators, and electronic data-gathering devices such as calculator-based laboratories (CBLs) and calculator-based rangers (CBRs) provide simpler methods for gathering and analyzing data that in years past might have been considered too much trouble. Similarly, classroom Internet connections make it possible for students to look up facts and figures quickly and easily for use in posing and solving a wide variety of real-world problems. Graphing calculators and easy-to-use computer software enable students to move effortlessly between different representations of problem data. They can also compute with large quantities of data and with "messy" numbers, both large and small, with relative ease. As a result, problems in the elementary school can and should be responsive to student questions and interests.

This discussion about problem solving as one of the five fundamental processes for school mathematics has described the general nature of problem solving and briefly considered the types of problems that may be appropriate for elementary school students. However, the most important point is that problem solving can serve as a foundation for all mathematics teaching because problem solving also involves students in work with all the other fundamental processes of doing mathematics (reasoning, communicating, connecting, and representing). Problem solving is a way of teaching and learning. This means it involves more than the presentation of word problems; it involves the way you encourage children to approach mathematical learning. A situation is posed, as in a word problem, and then there is a search for a resolution. But the situation that is posed often has a mathematical basis beyond the application of some procedures. Students or the teacher may pose the problem. In either case, using problems as a jumping-off point for mathematics instruction involves the teacher in posing questions that provoke student thought and also in encouraging students to pose their own questions. Using

Possible sums

2	3	4	5	6	7	8	9	10	11	12
1+1	1+2	1+3	1+4	1+5	1+6					
	2+1	2+2	2+3	2+4	2+5	2+6				
		3+1	3+2	3+3	3+4	3+5	3+6			
			4+1	4+2	4+3	4+4	4+5	4+6		
				5+1	5+2	5+3	5+4	5+5	5+6	
					6+1	6+2	6+3	6+4	6+5	6+6
1 way	2 ways	3 ways	4 ways	5 ways	6 ways	5 ways	4 ways	3 ways	2 ways	1 way

Ways to make each sum

+	1	2	3	4	5	6
1	2	3	4	5	6	7
2	3	4	5	6	7	8
3	4	5	6	7	8	9
4	5	6	7	8	9	10
5	6	7	8	9	10	11
6	7	8	9	10	11	12

Figure 5-3 *Two tables show the number of ways to get each possible sum when rolling two dice.*

problem solving as a foundation for mathematics instruction requires students to engage in a search for a reasonable solution or solutions.

Researchers have documented changes in children's perceptions about mathematics when they are taught in a problem-rich environment reflecting the spirit of the NCTM *Standards*. One example is students who were involved in the Cognitively Guided Instruction (CGI) program (Franke and Carey, 1997). In the CGI model of instruction, the teacher poses a rich mathematical task. Students take time to work individually or in small groups to solve the problem and then share their approaches with each other. Students are encouraged to listen carefully to each other and to question each other about processes and strategies. The teacher's role is to choose appropriate tasks and to orchestrate the classroom discourse, using what he or she knows about the students' developmental level. A key aspect of CGI is the teacher's ability to analyze the children's thinking and to guide classroom problem solving accordingly. The CGI first graders "perceived of mathematics as a problem-solving endeavor in which many different strategies are considered viable and communicating mathematical thinking is an integral part of the task" (Franke and Carey, 1997, p. 8). In a study of another problem-centered mathematics program, after two years in the program, third graders scored significantly higher on standardized measures of computational proficiency and conceptual understanding and held stronger beliefs about the importance of finding their own or different ways to solve problems than those in "textbook classes" (Wood and Sellers, 1996).

Problem solving has been the focus of numerous articles, books, collections of materials, and research studies in recent years (e.g., Buschman, 2003a, 2003b; Cai, 2010; Leatham, Lawrence, and Mewborn, 2005; Lester and Charles, 2003). Because it is so fundamental to teaching and learning mathematics, we have devoted another entire chapter to it in this textbook. Chapter 6 focuses on helping you probe deeper into questions such as the following:

What is involved in teaching mathematics through problem solving?

What strategies for problem solving are helpful for elementary students?

How can problem solving be evaluated?

REASONING AND PROOF

From their earliest experiences with mathematical challenges and problems, children should understand that they are expected to supply reasons for their arguments. The question "Why do you think so?" should be commonplace in the classroom. Teachers should not be the only ones asking "Why?" Asking why comes naturally

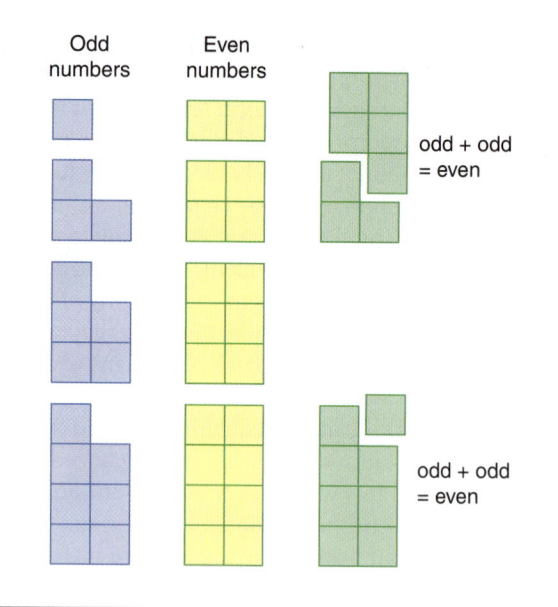

Figure 5-4 *Pictures of odds and evens can help students justify why the sum of two odd numbers is always even.*

to small children. They should be encouraged to sustain their natural curiosity for justification as they share their mathematical ideas with each other. When children observe a pattern (e.g., whenever you add two odd numbers, the answer is even), they should be encouraged to ask why. One second grader used square tiles to represent and sort various numbers (see Figure 5-4) and then explained his reasoning as follows:

All the even numbers are just rectangles because you can make them go two by two. But all the odd numbers can't. Odd numbers make rectangles with chimneys. If you put two rectangles with chimneys together, the new one doesn't have a chimney. So two odds makes an even.

Three interesting articles offer examples of students' thinking about odds and evens as they learn to formulate conjectures and justify their reasoning (Ball and Bass, 2003; Knuth, Choppin, and Bieda, 2009; Martin and Kasmer, 2009/10). If students are consistently expected to explore, question, conjecture, and justify their ideas, they learn that mathematics should make sense, rather than believing that mathematics is a set of arbitrary rules and formulas. This notion underlies both CCSSM and NCTM's Reasoning standards.

Being able to reason mathematically is essential to making sense of mathematics and, ultimately, to justifying mathematical conjectures. Proof is often interpreted as a formal process, reserved for students in advanced mathematics. On the contrary, in the elementary classroom, clear articulation of one's reasoning is the goal because this serves as an important precursor to formal proof.

CCSSM highlights three mathematical practices that relate to reasoning and proof: (1) reason abstractly and quantitatively, (2) construct viable arguments and critique the reasoning of others, and (3) look for and express regularity in repeated reasoning.

Young children have limited experience with what constitutes a reasonable argument. To help them develop their reasoning powers, it is helpful to pose problems that encourage children to examine, explain, and justify their own thinking—and the thinking of their peers. Children learn to reason by being encouraged to detect fallacies and to critique their own thinking and the thinking of others (Fillingim and Barrow, 2010; NCTM, 2000; Martin and Kasmer, 2009/10; Shifter, Russell, and Bastable, 2009; Yackel and Hanna, 2003).

Figure 5-5 shows two related problems used with a third-grade class to engage the children in logical thinking and in critiquing one another's reasoning (McGivney-Burelle, 2004/2005).

As soon as their teacher introduced the Playground Problem: Version A, children immediately raised their hands and began guessing how many boys and girls were on the playground. A variety of answers were proposed, but the most common suggestion was that there were probably an equal number of boys and girls on the playground. As one student explained, "Well, it has to be fair!" (McGivney-Burelle, 2004/2005, p. 273). After further discussion, however, some children recognized that there was not enough information given in Version A to determine how many boys and girls would be on the playground. Eventually the children convinced each other that "it could be anything that adds to 10. There could be 10 girls and no boys or 10 boys and no girls and anything in between" (McGivney-Burelle, 2004/2005, p. 273).

Their teacher then posed Playground Problem: Version B, which includes additional information. She helped them understand the new situation by examining a simpler problem first—asking two children to come forward to act as brother and sister. Marcus pointed to Jessica (pretending she was his sister), while Jessica left her arms by her sides. When Rosie joined them, representing another sister, the students figured out that Marcus should point to both girls and the girls should point to each other. The children then broke into groups to discuss what could be concluded from the Playground Problem network with red arrows showing sister relationships. After some thought, they noted the following:

- Arrows from both A and C to B mean B must be a sister to both A and C. Because there are no arrows pointing to A or C, they must be brothers. So B must be a girl, while A and C must be boys.

- Arrows from E to D and from E to F mean D and F must be sisters of E, so D and F are girls. No arrows pointing toward E mean that E must be a boy.

- Mutual arrows between G and H mean they are sisters to each other, therefore both girls.

What about I and J? The children suggested it was impossible to determine their gender and thought they must be "only children."

Playground Problem, Version A

There are 10 children on a playground. Some of them are brothers and sisters. The oval represents the playground and the dots represent the children. The letters next to each dot represent the children's names.

Which of the children are brothers and sisters?

How many of the children are boys and how many of them are girls?

How many families are on the playground?

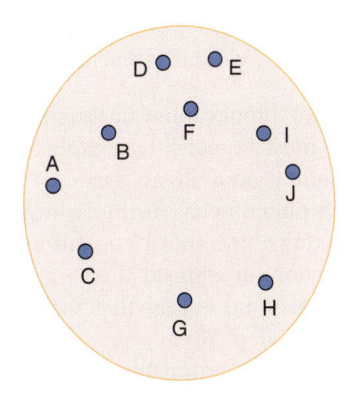

Playground Problem, Version B

There are 10 children on a playground. Some of them are brothers and sisters. The oval represents the playground and the dots represent the children. The letters next to each dot represent the children's names.

To help us out, each child on the playground will point to his or her sister. On paper, we will draw a red arrow to represent this. For example, if there is a red arrow leaving from C and pointing to B, then we know that B is a sister of C.

Which of the children are brothers and sisters?

How many of the children are boys and how many of them are girls?

How many families are on the playground?

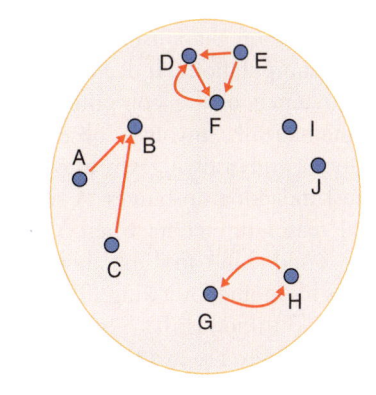

Figure 5-5 *Two logical reasoning problems.* (Source: Adapted from Jean M. McGivney-Burelle's "Connecting the Dots: Network Problems That Foster Mathematical Reasoning," *Teaching Children Mathematics* (December 2004/January 2005), pp. 272–227. Reprinted with permission from *Teaching Children Mathematics*, copyright 2004 by the National Council of Teachers of Mathematics.)

"But why couldn't I and J be brothers?" a student wondered aloud. "I mean, since kids are only pointing to their sisters, couldn't it be that I and J are brothers and so there are no arrows between them?" After some discussion along these lines, the students concluded that it was impossible to infer the relationship between I and J. ... The most definitive answers that the class could come to were that there were "*at least* five girls (B, D, F, G, H), *at least* three boys (A, C, E), and *at least* four families (ABC, DEF, GH, with I and J from either the same family or different families). Beyond that it was purely speculation.

(McGivney-Burelle, 2004/2005, p. 275).

Reasoning and proof cannot be taught in a single unit or lesson; they must be consistent expectations throughout all units and lessons in all grades. "By developing ideas, exploring phenomena, justifying results, and using mathematical conjectures in all content areas and—with different expectations of sophistication—at all grade levels, students should see and expect that mathematics makes sense" (NCTM, 2000, p. 56).

But what must elementary teachers consider in order to promote an environment where children are nurtured and encouraged to make sense of mathematical ideas and thus develop their abilities to reason mathematically, providing proofs of mathematical conjectures? Russell (1999) identifies four important points about active mathematical reasoning in elementary school classrooms: (1) Reasoning is about making generalizations. (2) Reasoning leads to a web of generalizations. (3) Reasoning leads to mathematical memory built on relationships. (4) Learning through reasoning requires making mistakes and learning from them.

REASONING IS ABOUT MAKING GENERALIZATIONS

Mathematics is much more than just finding answers to specific problems or computations. Indeed, in today's world, machines can do much of the computational drudgery. Reasoning mathematically involves observing patterns, thinking about them, and justifying why they should be true in more than just individual instances. A simple example of mathematical reasoning occurs when a kindergartner, proud of his newfound ability to think about big numbers, eagerly challenges his mother by asking, "Do you know how much 2 trillion plus 3 trillion is? It's 5 trillion!" There is no doubt that this boy has never seen a trillion of anything, nor is he able to write numbers in the trillions. But he is familiar with the pattern in smaller cases (2 apples + 3 apples = 5 apples; 2 hundred + 3 hundred = 5 hundred), so he is able to generalize to adding trillions.

Here is another example of making generalizations—the story of Katie, a third grader who was working with a partner to find factor pairs for 120 (Russell, 1999). Katie and her partner believed at first that 3×42 might be such a factor pair (a result obtained by incorrectly counting squares on a rectangular array on graph paper); however, Katie reasoned that answer must be incorrect because she remembered that 6×20 was a valid factor pair for 120, and from previous experience in finding factor pairs, she had figured out that if you halve one factor, you should double the other factor to keep the product the same. This is a powerful and useful generalization. Using this line of reasoning, Katie reported that the correct factor pair must be 3×40, not 3×42 because—thinking about 6×20—if 6 is halved to get 3, then 20 should be doubled to get 40 to keep the product the same: $6 \times 20 = 3 \times 40$. When children go beyond specific instances of mathematical ideas to consider general cases, they are reasoning mathematically.

REASONING LEADS TO A WEB OF GENERALIZATIONS

A second point about mathematical reasoning is that it "leads to an interconnected web of mathematical knowledge within a mathematical domain" (Russell, 1999, p. 1). Students should expect newly encountered mathematical ideas to fit with ideas they have already learned. Students have much more mathematical power if they have many ways to think about a number or fact or assertion. Students might understand in the first grade that "three-quarters" is between "one-half dollar" and "one dollar"; however, they will have a much more extensive understanding of the fact that three-quarters is between one-half and one when they are older, and have come to understand this relationship in terms of fractions $\frac{1}{2}, \frac{3}{4}, 1$, in terms of equivalent fractions $\frac{2}{4}, \frac{3}{4}, \frac{4}{4}$, in terms of decimals (0.5, 0.75, and 1.00), and by visualizing pieces of a pie or portions of a collection of trading cards. When students incorrectly claim that 0.25 must be larger than 0.5 because 25 is more than 5, they apparently have not developed this robust web of connections for these ideas. They are unable to reason about the other meanings they may know for these two numbers and are thereby unable to recognize the contradiction in their thinking.

REASONING LEADS TO MATHEMATICAL MEMORY BUILT ON RELATIONSHIPS

A third point about mathematical reasoning is that the development of a web of mathematical understandings is the foundation of what Russell calls *mathematical memory* (or mathematical sense), a capability that provides the basis for insight into mathematical problems. For example, consider the problem of finding the sum of the first 100 counting numbers: $1 + 2 + 3 + \ldots + 100$. Some time during your years of studying mathematics, you probably encountered a formula that would allow you to calculate the sum of any arithmetic series (a list of numbers where the difference between consecutive numbers is always the same—here, the difference is always just one). If you are like most people, you have long since forgotten that formula because it is not something that you use every day. On the other hand, if you have ever seen a geometric illustration of that formula, your web of mathematical understanding may be strong enough to help you reconstruct the formula with little trouble.

Picture the sum from 1 to 100 as a set of stairs. Picture another identical set of stairs. Turn the second set of stairs upside down on the first, and you have a rectangle (100 wide, 101 high). (See Figure 5-6 for a picture of a related, simpler problem: a 10-by-11 rectangle built from two staircases from 1 to 10.) It's easy to find the area of the 100-by-101 rectangle (100×101=10,100). The sum you want is half of this (100×101 divided by 2, or 5,050) because the rectangle is made of two staircases instead of just the one original staircase. So 1+2+3+…+100=100(101)/2.

Using the same idea, you can figure out how to find the sum, S, from 1 to any number, n. It's just S=n(n+1)/2. A similar line of reasoning can be used to find the sums for various other arithmetic series (such as 1+3+5+…+175 or 12+15+18+…+639). In other words, if you can connect the idea of summing numbers in a series with the geometric illustration of the dual staircases, you'll never again need to worry about forgetting the formula for the sum of an arithmetic series.

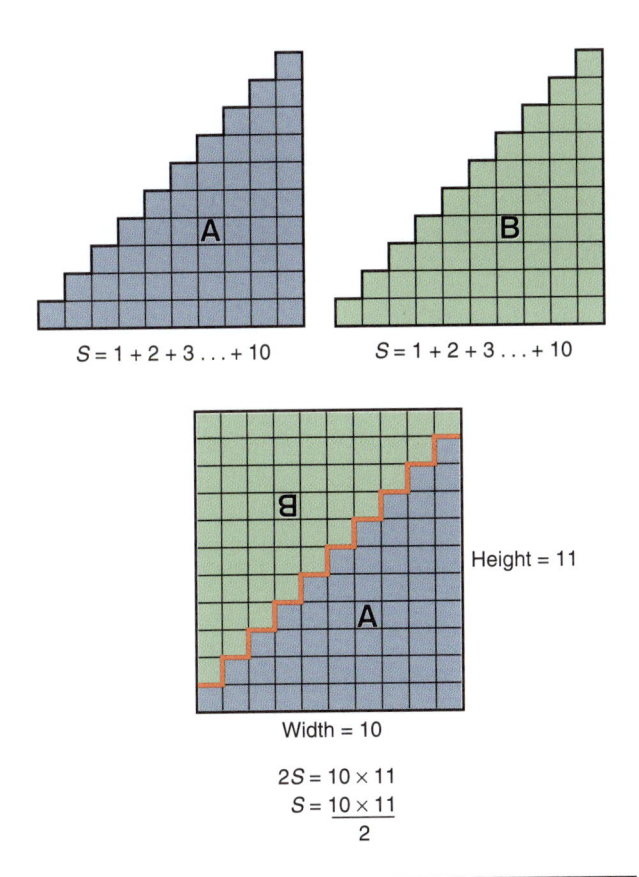

Figure 5-6 *A 10-by-11 rectangle built with two staircases from 1 to 10 can help you remember the formula for the sum of a series of numbers.*

LEARNING THROUGH REASONING REQUIRES MAKING MISTAKES AND LEARNING FROM THEM

A final, important point about mathematical reasoning and proof is that one of the best ways to develop stronger reasoning and proof abilities is to study flawed or incorrect reasoning. There will be many times when your students think they've figured something out, but their reasoning just isn't quite right. That's just human nature. For example, a student might observe that you can make the problem 29+95 easier to do mentally by adding 29 and 100, then subtracting 5 (obtaining the correct sum, 124). Can this same shortcut be used to make the problem 29×95 easier to do mentally (i.e., can you multiply 29 times 100, and then subtract 5)? It turns out that just isn't correct! Why? Rather than just tell your students that this doesn't work, it would be much better to help them investigate why the shortcut works for addition and why it does not work for multiplication. Some possible approaches would be to try adding and multiplying a wide variety of examples to try to figure out what is going on. Or it might help to represent the problems geometrically (using a 29-by-95 array to represent the multiplication and comparing it with a 29-by-100 array may offer some insights; see Figure 5-7).

Your role as a teacher is to encourage students to examine their own thinking and the thinking of others as well as to help them uncover and understand flawed reasoning when it occurs.

COMMUNICATION

Because language is a powerful tool for organizing thinking about mathematical ideas, it is extremely important for students to have many experiences with talking and writing about mathematics, describing and explaining their ideas. Conversely, it is also important that students often be on the receiving end of communications: hearing about, reading about, and listening to the descriptions and explanations of others. Two-way communication about mathematical ideas helps students identify, clarify, organize, articulate, and extend their thinking (Amos, 2007; Chapin and O'Connor, 2007; Lampert and Cobb, 2003; Tyminski, Richardson, and Winarski, 2010). Have you ever noticed how struggling to explain an idea helped you figure out what you were really thinking? Reflection and communication are intertwined. Sharing your ideas with others through talking or writing forces you to think more deeply about those ideas. Also, in thinking about your ideas, you often deepen your understanding and, consequently, are able to communicate even more clearly. Communication is obviously a process rather than an end in itself.

The CCSSM's mathematical practices call for clear communication. Students are expected to "construct viable arguments and critique the reasoning of others," as well as to

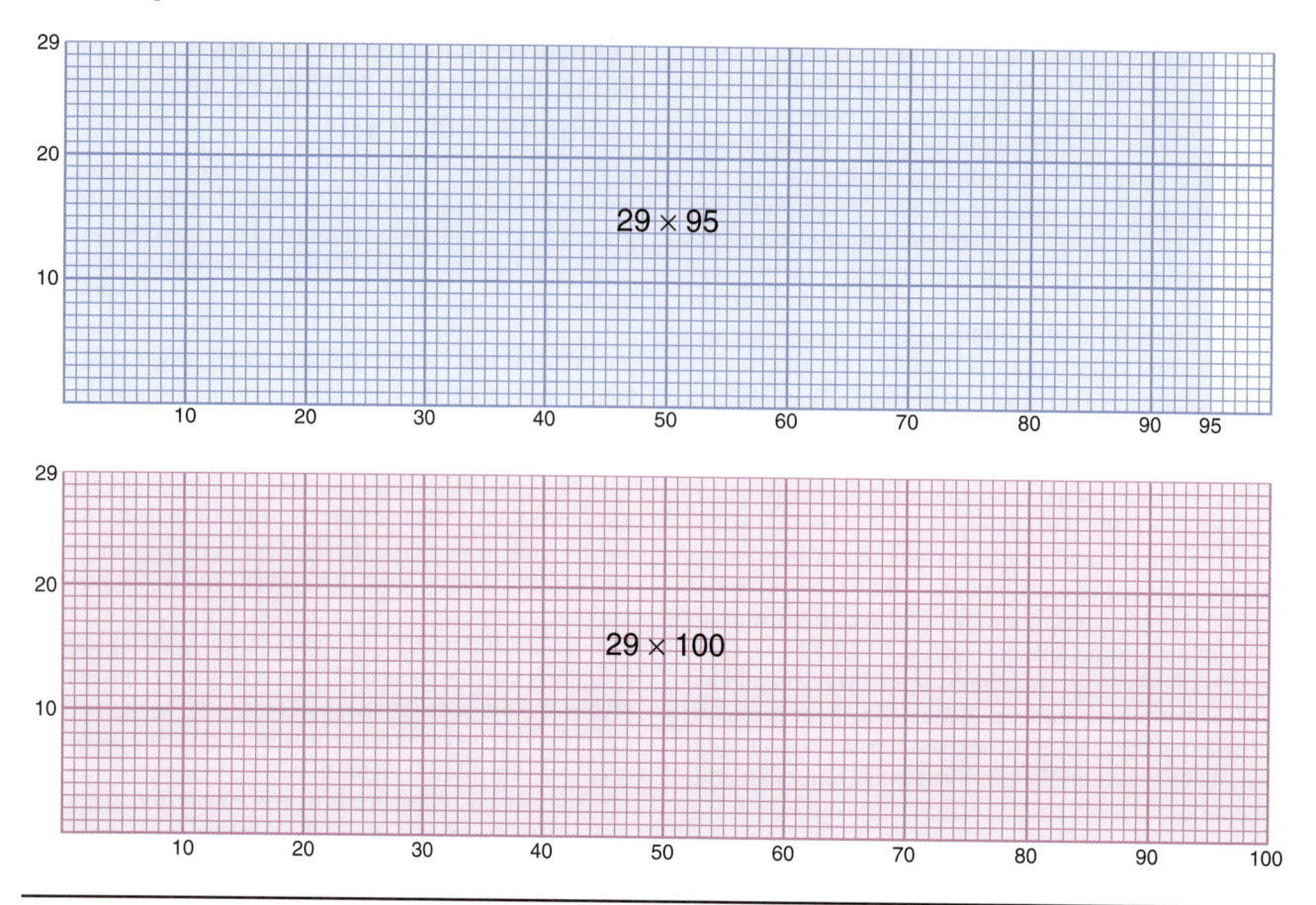

Figure 5-7 Visualizing 29×95 and 29×100 helps students think about why $29 \times 95 \neq (29 \times 100) - 5$.

"attend to precision"—which includes "communicat[ing] precisely to others" and using "clear definitions with others and in their own reasoning" (CCSSI, 2010, pp. 6-7).

Students should be encouraged to communicate their mathematical thinking in a variety of modes: through pictures, gestures, graphs, charts, and symbols, as well as through words (both spoken and written). Figures 5-4, 5-5, 5-6, and 5-7 clearly illustrate the power of visuals in communicating about mathematics. Such nonverbal communication is often useful in promoting learning.

Especially at first, students' efforts at communicating about mathematical ideas may be idiosyncratic (they may use symbols, expressions, or notations that they make up on their own). Angela Andrews, a teacher who has written a fascinating book about the problem-solving experiences of her kindergartners, shares several examples of their naive uses of language and notation. For example, she reports that a student used the notation shown in Figure 5-8 to represent the number of scoops of rice needed to fill a certain bottle at the rice table. When she questioned him about his writing, he confidently reported, "Oh, that jar held $10\frac{1}{2}$ scoops, Teacher." His symbol for $10\frac{1}{2}$ was certainly not the standard representation, but it was effective in communicating his idea" (Andrews and Trafton, 2002, p. 94).

Math Links 5.2

If you would like the opportunity to see a teacher facilitating children's communication, you can view the video clips Teaching, Learning, and Communicating about Fractions at the NCTM Illuminations Web site. At this site, reflection questions are also provided to help you focus your viewing. The videos may be accessed from http://www.illuminations.nctm.org/info/standards.asp or from this book's Web site, by scrolling down to grades 3–5 and selecting "Teaching, Learning, and Communicating about Fractions."

www.wiley.com/college/reys

Figure 5-8 Child's representation of $10\frac{1}{2}$, from Andrews and Trafton, 2002, p. 94.

Over time, with more experience and practice, students learn to use conventional and more precise language to express their ideas. Indeed, mathematics as a language has a vocabulary, syntax, and symbolism all its own (cf. Perry and Adkins, 2002). Sometimes words or phrases used in everyday conversation may be used in mathematics with different, more precise, meanings. The symbolism of mathematics (particularly equations and graphs) often helps clarify concepts and promote understanding. Throughout their elementary school years, students should have daily opportunities for communicating about mathematical ideas. Gradually, they should be expected to incorporate more precise mathematical terms in their explanations.

Mathematics journals can be used to prompt students to write about mathematics. The regularity of writing in a journal can help students monitor their own understanding of mathematical concepts. Students can be encouraged to write about such questions as "What am I puzzled about?" and "What mistakes do I make and why?" When teachers take time to respond to student journals on a regular basis, however briefly, the journals can become a regular chain of communication between student and teacher. Also, teachers can use students' writing diagnostically. When the writings of many students in a group or class reveal similar confusions or misconceptions, the teacher can more appropriately plan future lessons.

Other forms of writing in math class include open-ended writing as follow-up to a lesson, having students write their own word problems, having them describe their solutions to problem-solving activities, having them describe a procedure or process, and writing about connections between and among ideas. Young children can be encouraged to use their own invented spelling, and all students can benefit by accompanying words with pictures and symbols so they can express themselves as fully and completely as possible.

CONNECTIONS

Although mathematics is often represented as a list of topics or a collection of skills, this is a shallow view. Mathematics is actually a well-integrated domain of study. The ideas of school mathematics are richly connected. It is important for the elementary school curriculum to provide children with ongoing opportunities to experience and appreciate the connectedness of the subject. Using slightly different words, the CCSSM recognizes the importance of connections by indicating that "look for and make use of structure" is a key mathematical practice (CCSSI, 2010, p. 8).

At least three types of connections are important in learning mathematics. First, ideas within mathematics itself are richly connected with one another. Students who learn about fractions, decimals, and percentages in isolation from one another miss an important opportunity to see the connections among these ideas. For example, in Figure 5-9, we

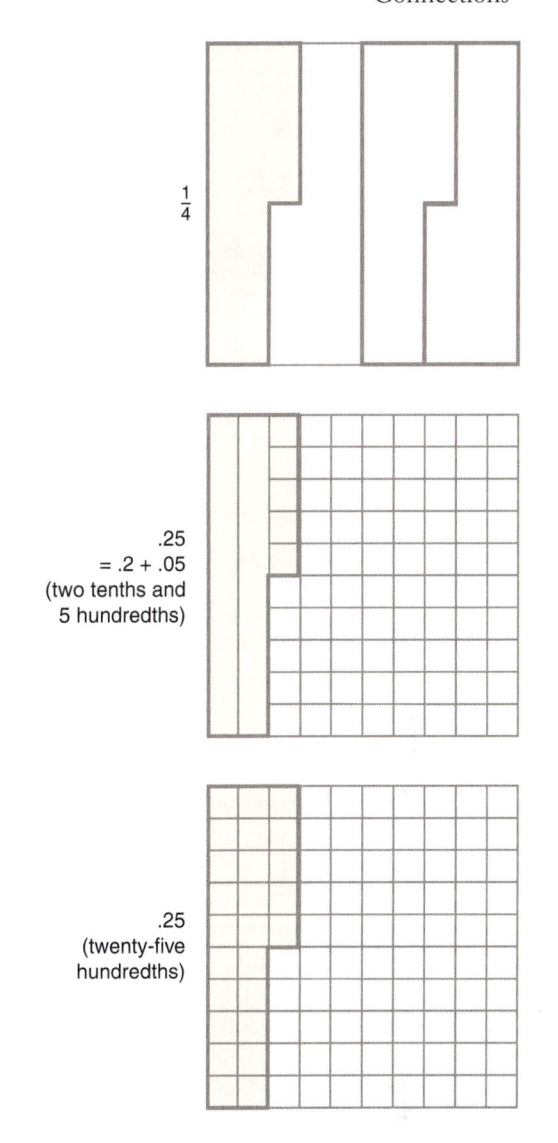

Figure 5-9 Some representations for one fourth ($\frac{1}{4}$), 2 tenths plus 5 hundredths (0.25=0.2+0.05), and 25 hundredths (0.25) show that they are equivalent.

can see why $\frac{1}{4}$, 2 tenths + 5 hundredths, and 0.25 are actually all names for the same quantity.

A second important type of connection is between the symbols and procedures of mathematics and the conceptual ideas that the symbolism represents. For example, why do we refer to 3^2 as "3 squared"? 3^2 is 3×3, or 9. A drawing of 9 dots, arranged in a 3-by-3 array, forms a square. Similarly, any array of x-by-x dots would form a square; thus students can recognize why x^2 is read as "x squared." The area of any geometric figure is generally reported in square units (e.g., square feet, square centimeters, square miles). Why? Because measuring area is actually just measuring how many "squares" it would take to cover a surface. If the squares are 1 inch on each side, then you are measuring in square inches. In fact, you can write "in.^{2}" instead of "square inches" for the same reason. So here we can *see* connections between

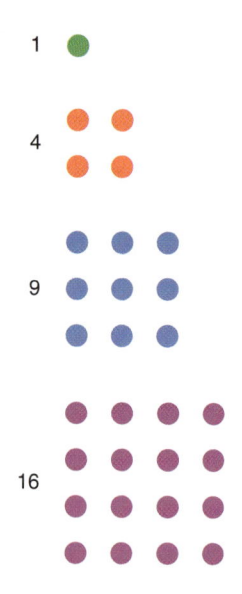

Figure 5-10 *Arranging dots in square patterns connects the numbers 1, 4, 9, and 16 to their reference as square numbers.*

number theory (the "square numbers" 4, 9, 16, etc., as shown in Figure 5-10), algebraic language (*x* squared), and measurement (square inches).

A third type of connection is between mathematics and the real world or between mathematics and other school subjects. For example, engaging students in problems as they occur naturally in the classroom or in the lives of the children often provides a totally natural connection to mathematics. For example: *Our class has 24 students. Today 4 desks are empty. Tony, Melissa, and Gaby brought their lunches from home. Everyone else who is here today wants to buy the school lunch. How many orders for lunch should be sent to the office?* As children encounter problems from real-world contexts where mathematics is a significant part of the solution, they come to recognize and value the utility and relevance of the subject. Classroom instruction should also provide many opportunities for children to experience mathematics as found in domains such as science, business, home economics, social studies, literature, and art. Mathematical connections can be highlighted through integrated or thematic curricula as well as through mathematics lessons motivated by situations in children's literature. The famous poet Carl Sandburg (1960) wrote a poem entitled "Arithmetic" that included images such as "arithmetic is where numbers fly like pigeons in and out of your head" and "arithmetic is numbers you squeeze from your head to your hand to your pencil to your paper till you get the answer." After reading and discussing Sandburg's poem, children might write about their reactions or brainstorm other aspects of mathematics (with examples); they could even be prompted to write their own poems about mathematics after the style of Sandburg (focusing on their own views of arithmetic or their own thoughts about other aspects of mathematics).

Math Links 5.3

To find many resources for connecting mathematics and other school subjects, visit the Education-World site, which can be accessed from http://www.educationworld.com/a_curr/curr146.shtml or from this book's Web site.

www.wiley.com/college/reys

REPRESENTATIONS

When most people think of mathematics, they may think of numbers such as 2, 29, or 5280, or of numeric or symbolic expressions such as 5×2, $(a+b)(a-b)$, or $5798 \div 13$, or of equations such as $x^2 + y^2 = r^2$ or $2x + 7 = 13$. Alternatively, they may think of tables of numbers or graphs or geometric figures. All are commonly used representations for mathematical ideas. Interestingly, it is often possible to use a variety of these different representations to illustrate or model the same mathematical ideas. Different representations for an idea can lead us to different ways of understanding and using that idea. This is the power of representation. CCSSM's mathematical practice "model with mathematics" can be viewed as closely related to NCTM's process standard on "representations." Figure 5-11 identifies five ways that many mathematical ideas can be modeled or represented. Helping students

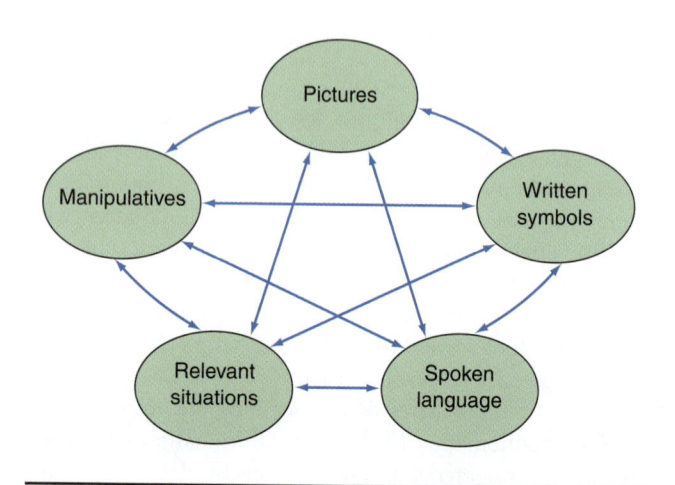

Figure 5-11 *Five representations for mathematical ideas.* (Source: Adapted from Lesh and Landau, 1983, p. 271.)

become comfortable moving among these representations is an important goal of school mathematics.

The school mathematics curriculum has traditionally involved children in learning about a variety of representations for mathematical ideas; unfortunately, however, different representations have sometimes been learned in isolation from one another. Consider the following scenario (Clement, 2004, p. 97).

TEACHER: I will tell you a number, and you write it. Please write for me the number *one-half*.

STUDENT: [Writes $1\frac{1}{2}$.]

TEACHER: Please write for me the number *one and one-half*.

STUDENT: [Writes $1\frac{1}{2}$.] It is the same thing.

TEACHER: [Writes $\frac{1}{2}$ and points to it] What would you call this?

STUDENT: Half.

Even children who have a reasonable understanding of fractions as encountered in their everyday lives may not have made connections between those ideas and the language or symbolism traditionally used to represent them. "The connections children make between language and written symbols may be different from the connections adults make" (Clement, 2004, p. 97). Children often use the word *half*, not *one-half*, when they share in real life, for example, asking "Can I have half your brownie?" As a result, when the teacher asks them to write "one-half," they may write $1\frac{1}{2}$—writing "1" to represent the "one" in "one-half" and writing "$\frac{1}{2}$" to represent the "half."

Similarly, students might learn about fractions in one chapter and decimals in another, but they also should be provided with opportunities to connect these two different representations for the same numbers. It is important to challenge students to connect, compare, and contrast the utility and power of different representations.

The NCTM *Standards* (2000) discuss three major goals for representation as a process in school mathematics: (1) creating and using representations to organize, record, and communicate mathematical ideas; (2) selecting, applying, and translating among representations to solve problems; and (3) using representations to model and interpret physical, social, and mathematical phenomena.

CREATING AND USING REPRESENTATIONS

In this chapter's Snapshot of a Lesson, children were developing their own individual representations for keeping track of how to divide money fairly among a number of objects. Although finding the answer by drawing pictures and making lists was certainly less efficient than using the traditional long-division algorithm, the pictures and lists were more meaningful to the children because they were just beginning to solidify understanding of the concept of division.

It is important for young children to have repeated opportunities both to invent their own ways of recording and communicating mathematical ideas and to work with conventional representations. The mathematical symbols and representations that are used every day (e.g., base-ten notation, equations, graphs of various types, and traditional computational algorithms) have been polished and refined over many centuries. When students come to understand them in deep ways, they have a set of tools that expands their capacity to think mathematically.

SELECTING, APPLYING, AND TRANSLATING AMONG REPRESENTATIONS

As mentioned earlier, mathematical ideas can often be represented in different ways. Each of these representations may be appropriate for different purposes. For example, a student who can think flexibly about numbers is probably able to think about the number 24 in many different ways: 2 tens and 4 ones, 1 ten and 14 ones, a little less than 25, double 12, the perimeter of a square with side 6, the area of rectangles with sides 2×12 or 4×6, and so on. Depending on the problem at hand, some of these representations may be more useful than others. Technology now offers students many opportunities for experiences with translating among representations. Data analysis software such as Tinkerplots can help students easily compare and contrast various graphical representations. It is important that students consider the kinds of data and questions for which the graphical representation is appropriate.

In fact, CCSSM identifies "use appropriate tools strategically" as one of their eight key mathematical practices. Tools might include "pencil and paper, concrete models, a ruler, a protractor, a calculator, a spreadsheet,... a statistical package, or dynamic geometry software. Proficient students are sufficiently familiar with tools appropriate for their grade or course to make sound decisions about when each of these tools might be helpful, recognizing both the insight to be gained and their limitations" (CCSSI, 2010, p. 7).

For example, Figure 5-12 shows the results when a class was asked to describe the heights of the students in the class. The children recorded their heights in the appropriate rows in a spreadsheet and prepared three graphs: a bar graph, a circle graph, and a line graph. Which graph best represents the data collected? Children are used to being lined up by height. However, the circle graph does not order the heights as clearly as either the bar or line graph. The line graph incorrectly gives the impression that there are children of heights between the measurement points. On the other hand, the bar graph shows that the most common height of the children is 4 feet 7 inches and that the second most common height is 4 feet 5 inches.

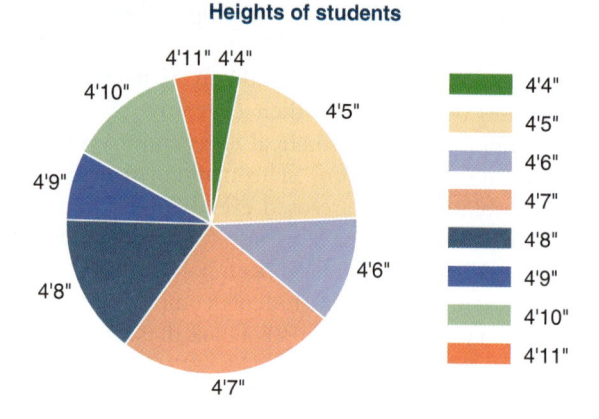

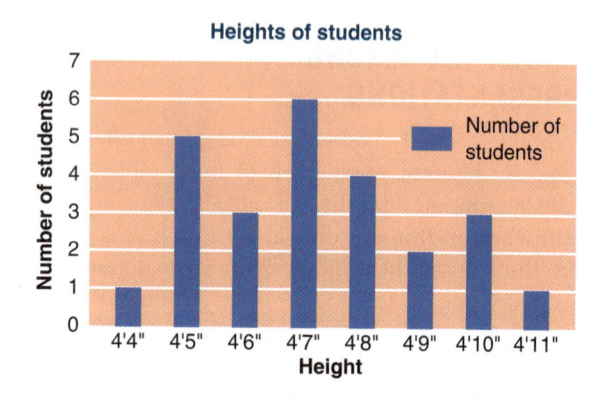

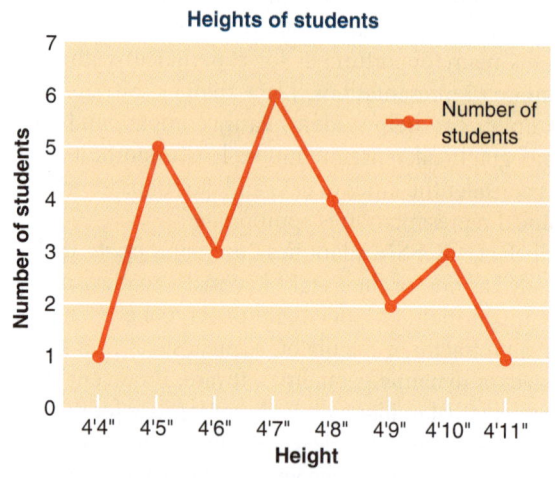

Figure 5-12 *Which graph best represents the height of students in the class?*

USING REPRESENTATIONS TO MODEL AND INTERPRET PHENOMENA

Much of mathematics involves simplifying problems—stripping away context and excess information to reduce the problem to symbols or representations that are easier to work with. This is mathematical modeling. For example, to solve the following word problem, you might reduce it to a picture or to a table of numbers:

> Alice is stacking soup cans for display on a shelf at the end of a supermarket aisle. She wants the display to look like a pyramid. On the top row she wants to put just 1 can, on the next-to-top row 3 cans, on the next row down 5 cans, and so on. Alice decides the display can be 6 rows high. How many cans should she start with on the bottom row?

Once you've made a picture or a table of numbers, the fact that the problem is about stacking soup cans is no longer really important. You've modeled the problem and used the power of mathematics to solve it. Similarly, when students ask, when solving word problems, "Do I add or do I subtract?" they are asking for advice about modeling the situation at hand. It is important to encourage students not to move too quickly and unthinkingly from real-world situations to abstract models. The best answer to the question about adding or subtracting is to ask, "What's going on in the problem?" or "Can you draw a picture or can you act it out to help you decide which operation to use?" Note that this is exactly what the children were doing in the opening Snapshot of a Lesson while finding unit prices.

It is also important to check back at the end of solving a problem to ensure that the solution obtained from the mathematical model fits the original situation. A classic example is a problem where students are asked to determine how many buses are needed to take a group of children on a field trip. The problem tells how many students need to be transported (e.g., 130) and how many can fit in each bus (perhaps 40). Many students correctly pick out the numbers in the problem and divide (130÷40), but when the division doesn't come out evenly, they may offer decontextualized answers such as $3\frac{1}{4}$ or 3.25. They have forgotten to check back with the context of the problem to see that a realistic answer would be 4 buses or perhaps 3 buses and a large van.

Math Links 5.4

Children can use electronic "virtual manipulatives" to model and solve problems. Visit the Web site of the National Library of Virtual Manipulatives (NLVM) to see a wide variety of such tools. For example, click on "Virtual Library," then "Algebra," and then "Towers of Hanoi" to experiment with using a virtual manipulative to solve that problem. The NLVM is at http://www.nlvm.usu.edu/en/nav/index.html or can be accessed from this book's Web site.

www.wiley.com/college/reys

In sum, representations are ways of thinking about ideas. Individuals develop their own idiosyncratic ways of thinking, but mathematics offers a broad repertoire of conventional representations that are helpful in problem solving and in communicating about mathematical ideas (Goldin, 2003; Lubinski and Otto, 2002; Monk, 2003; NCTM, 2001;

Smith, 2003; Wu, An, King, Ramirez, and Evans, 2009). One of the most important goals of mathematics instruction should be to help students build bridges from their own ways of thinking to the conventional, so that they come to understand, value, and use these powerful mathematical tools. For a collection of readings all focused on the roles of representation in school mathematics, see Cuoco (2001).

> ### Virtual Classroom Observation
>
> Go to www.wiley.com/college/reys,
> Access the Wiley Resource Kit
> Click on the Virtual Classroom Observations Section
> Module 6: Data Analysis and Probability: Measures of Center
> 1. Watch 2 videos: Teaching Examples: #3. Navigating Through the Landmarks and #4. Seeing the Landmarks
> 2. Do: From the Classroom—Assessing Student Work (See how students' representations of data help the teacher assess their understanding.)

CULTURAL CONNECTIONS

As we have seen in this chapter, NCTM's *Principles and Standards for School Mathematics* (2000) recommends classroom instruction that focuses on engaging students in learning mathematics through problem solving, communicating, reasoning, connecting, and using a variety of representations. While there can be no doubt that all students benefit from these sorts of experiences, special attention may be necessary to ensure that students whose first language is not English are genuinely included in classroom conversations and problem-solving activities. These students are sometimes referred to as learners of English as a second language (ESL) or learners of English as a new language (ENL).

Teachers involved in Project IMPACT (Increasing the Mathematical Power of All Children and Teachers)—an effort that focused on schoolwide reform in elementary schools in predominantly minority urban schools outside Washington, D.C.—focused on establishing a mathematics culture in each classroom that encouraged "language, communication, mathematical content, mathematical connections, decision making, and equity" (Campbell and Rowan, 1997, p. 60). In the Project IMPACT schools, an extremely high proportion of the student population either spoke some language other than English as their native tongue or came from a different cultural heritage. Prior to the start of Project IMPACT, teachers' response to this language and cultural diversity had been to strip away as much of the language and context from the mathematics curriculum as possible, in an effort to make math lessons more accessible to the language-minority students. This is a common first reaction to the challenge of teaching ENL or ESL learners. However, while this approach may help these learners acquire proficiency with computation and symbol manipulation, it unfortunately narrows their mathematics education to arithmetic only, stripping much of the challenge, beauty, and relevance from the mathematics curriculum and removing any focus on the process standards that we have discussed in this chapter.

Instead, the Project IMPACT teachers supported their students in using problem solving, reasoning, connections, and representations in their mathematics lessons by implementing a variety of helping strategies. They used mixed-ability groups so that children could help each other with language challenges. They ensured that everyone understood problem contexts by encouraging students to talk through the word problems during whole-class sharing before going off to work individually, in pairs, or in small groups. They assigned buddies to help each other with language translation or with using concrete diagrams or materials to show their thinking. Figure 5-13 provides several additional teacher tips that are useful for all students, but particularly helpful when your class includes students with different language or cultural backgrounds. (For additional research-based advice on equitable teaching practices for English-language learners, see Moschkovich, 2010.)

Additionally, the Project IMPACT teachers worked together in grade-level teams to help and support each other with helping their ESL learners. The result of Project IMPACT was instruction that took account of "diversity in the children's racial-ethnic heritage and primary languages" at the same time that it supported mathematics instruction aligned with NCTM's process standards (Campbell and Rowan, 1997, p. 60).

A GLANCE AT WHERE WE'VE BEEN

Learning and doing mathematics involves engaging in five key processes: problem solving, reasoning and proof, communication, connections, and representations. These processes are inextricably linked with each other and with the mathematics content that students learn (number, algebra, geometry, data, and measurement). Through these processes, students engage actively in making sense of the mathematics they are learning.

Problem solving requires that students be engaged in questions where the solution is not known and cannot be easily identified. Good problems give students an opportunity to extend what they know, building their understanding of mathematical ideas. An important role for teachers is to select challenging problems, mathematics tasks, or opportunities that engage their students in the process of problem solving. Careful attention to establishing this environment will support students in enhancing their mathematical understanding.

> 1. Use tools to help you communicate with students.
>   a. Use manipulatives, pictures, contexts, stories, and drama.
>   b. Write directions, key questions, terms, or ideas on chart paper, the chalkboard, or the overhead projector.
>   c. Demonstrate (show, don't just tell) what you want students to do.
>   d. If you use handouts, have students read the directions. Have them repeat them in their own words before following them. Check for understanding.
>
> 2. Listen to students. Don't assume you know what they will say.
>
> 3. Watch for clues that students are understanding.
>   a. Use eye contact.
>   b. Watch their facial expressions and other body language.
>   c. Notice if they are asking one another to explain what you are saying.
>   d. Ask them to say back to you what you have explained (directions, new ideas, etc.).
>
> 4. When students respond and you don't understand, try one of the following:
>   a. Ask them to speak slower or use different words.
>   b. Ask another student to explain what the first student said.
>   c. Ask them to show what they mean by coming to the chalkboard, drawing a diagram, or using concrete materials.
>   d. If appropriate, promise to speak personally with them later, then do so.
>   e. Identify what you think you *do* understand and say it back to the student: "What I heard you say was that you have another method you'd like to use. But I'm not following your method. Please explain it again slower or show us at the board."
>
> 5. Use lesson openings and closings to your advantage (and your students'!).
>   a. Begin by asking them where you left off yesterday or ask key questions to review what they should have learned. Ask more than one student to say what was learned in a different way. Listen to how they phrase things. Ask other students for corrections or modifications.
>   b. Include a summary at the end of your lesson. Have students tell what they have learned. Listen to how they say it. As above, ask for other ways to express the new ideas.
>
> 6. Use guided practice in your lessons, as appropriate, for example, for skill development. "We'll do number 1 together." Then develop it with them. "Now, you do number 2." Give them an item just like the first, monitor them, and see how they do. In this way you get evidence of their learning, and you see which students understand and can be resources for explaining to others. Continue alternating between examples done together and those done by students individually or in pairs.

Figure 5-13 *Teacher tips for interactions in diverse classrooms.* Source: Rubenstein, 1997, p. 220. Reprinted with permission from *Multicultural and Gender Equity in the Mathematics Classroom: The Gift of Diversity*, copyright 1997 by The National Council of Teachers of Mathematics.

Being able to reason mathematically is essential in order to make sense of mathematics and, ultimately, to justify or prove mathematical conjectures. To meet this challenge, children must be expected to supply reasons for their arguments from their earliest experiences with challenges and problems. Reasoning is about making generalizations and connecting new ideas with old ones to develop a web of mathematical understandings. Beginners' reasoning is often faulty, but that's OK. One of the best ways to help students develop stronger reasoning abilities is to encourage them to examine flawed arguments and to figure out where the thinking went wrong.

Communication provides a way for children to share their ideas with others. Their knowledge deepens as they help others understand their mathematical ideas and conjectures. Listening to the explanations of other children is equally important in the communication process. Two-way communication about mathematical ideas helps children identify, clarify, organize, articulate, and extend their thinking. Children need to be encouraged to present their ideas in a variety of modes (e.g., pictures, gestures, graphs, charts, and symbols) in addition to spoken and written words.

Connecting mathematical ideas helps children expand their understanding. At least three types of connections are important in learning mathematics: mathematics connected with other mathematical ideas, symbols connected to mathematical procedures, and mathematics extended and connected to contexts outside of mathematics. Teachers need to emphasize mathematical connections to encourage their students to routinely seek connections and to enjoy that search.

A variety of representations can be used to express the same mathematical idea: pictures, manipulatives, written symbols, relevant situations, and spoken language. The different representations lead to different ways of understanding and using that idea. Children should be encouraged to (1) create and use a variety of representations; (2) select, apply, and translate among representations to solve problems; and (3) use representations to model and interpret physical, social, and mathematical phenomena.

Things to Do: From What You've Read

1. Name the five key mathematical processes described in this chapter. For each process, briefly describe how the Snapshot of a Lesson that opens this chapter shows children engaging in that process.
2. Name the eight mathematical practices recommended by the Common Core State Standards. For each of these practices, consider whether the Snapshot of a Lesson that opens this chapter shows children engaging in that practice. If so, explain when and how. If not, explain why you think not.
3. Make a list of as many different representations for the number 75 as you can think of. Compare your list with the lists of others. You might use equivalent expressions, models, words, pictures, and so on. It is a sign of flexibility in thinking to be able to represent numbers and other mathematical ideas in many different ways.
4. What are the three distinct types of connections that are important in school mathematics? Give two examples of each type of connection.

Things to Do: Going Beyond This Book

In the Field

1. [1]*Mathematical Processes.* Observe an elementary school classroom while children are engaged in a math lesson or investigation. Make a list of instances in which the children show evidence of using one or more of the five mathematical processes (NCTM) or the eight mathematical practices (CCSSM). Tell what the children were doing or saying and what tasks were involved. What role did the teacher play?
2. Check a textbook for evidence that students are encouraged to make connections. Describe three examples.
3. [1]*"Doing Math"?* Talk to one or more elementary teachers about their vision of what it means to "do mathematics." Ask whether they connect math to other areas of their curriculum and, if so, how. Write a reflection about how teachers' views of mathematics fit (or do not fit) with a view involving the five mathematical processes and eight mathematical practices discussed in this chapter.

In Your Journal

4. Which of NCTM's five process standards are familiar from your own study of mathematics? Give some examples. Which of these have not been a part of your experiences in learning math? Explain.
5. Which of CCSSM's mathematical practices are familiar from your own study of mathematics? Give some examples. Which of these have not been a part of your experiences in learning math? Explain.

With Additional Resources

6. Identify a journal article or book chapter about writing and communication in mathematics classes (perhaps from one of the NCTM's teacher journals, *Teaching Children Mathematics* or *Mathematics Teaching in the Middle School*). Summarize the main ideas for supporting communication in the classroom.
7. Begin a resource folder for yourself that contains sections for each of the five mathematical processes and/or eight mathematical practices. Collect and review recent articles in journals, such as *Teaching Children Mathematics*, to locate ideas that support these processes and/or practices.

---

[1]Additional activities, suggestions, and questions are available in the field experience manual on the Student Companion site at www.wiley.com/college/reys.

Note to Instructors: You can find additional resources, learning activities, and blackline masters in this text's accompanying Instructors Manual, at www.wiley.com/college/reys.

Book Nook for Children

Harness, C. *Ghosts of the White House.* New York: Scholastic, 1998.

This is the story of a class that takes a field trip to the White House. One girl, Sara, wishes that she could meet many of the presidents from the past. As Sara continues walking through the White House, many ghosts of the presidents come out and talk to Sara, giving her information about the White House and themselves. This book is good for connecting graphs and time lines of historical data.

Hulme, J. N. "Wild Fibonacci: Nature's Secret Code". *Illustrations by Carol Schwaratz.* Berkeley, CA: Tricycle Press, 2005.

The book helps readers see how Fibonacci numbers can show different patterns by showing recurring sequences using numbers, plants, and different animals. The author gives a brief history of Fibonacci numbers and how the reader can find these sequences in nature.

Nagda, A.W., and Bickel, C. *Tiger Math: Learning to Graph From a Baby Tiger.* New York: Scholastic, 2000.

This is the story of TJ, a Siberian tiger cub, who was born in the Denver Zoo. The book shows readers how TJ grew in inches and weight using four different types of graphs: picture graphs, circle graphs, bar graphs, and line graphs. This is a good book about representation. Readers will learn about the Siberian tigers, too.

Schmandt-Besserat, D. *The History of Counting.* New York: Morrow Junior Books, 1999.

This resource book for students and teachers informs readers about how our number system was invented and how other cultures and societies invented their own number systems so they could be precise in counting. The author shows many different cultures in this book and also focuses on our modern decimal system and tells how it, too, evolved. The book was written by an archaeologist who took 20 years in authoring it.

Sciedzka, J. *Math Curse.* New York: Viking, 1995.

Mrs. Fibonacci has told her class, "You know, you can think of almost everything as a math problem." The Math Curse has begun. Suddenly the book's hero cannot look anywhere without thinking of math problem after math problem—in the closet, at the dinner table, on the bus. He is agitated until he realizes he can solve all the problems he sees. Encourage your students to solve the problems and then challenge them to find more math problems in their world.

Turner, P. *Among the Odds and Evens. A Tale of Adventure.* New York: Scholastic, 1999.

The Kingdom of Wontoo is visited by X and Y, two Adventurers that soon learn that there are two types of numbers: the Odds, which are eccentric, and the Evens, which are orderly and predictable. As the story unfolds, the reader will see how odd and even numbers are represented in a variety of pictures. The reader will also learn that adding two Evens or two Odds always result in an Even number. This book is a great start for algebraic reasoning for young students.

CHAPTER 6

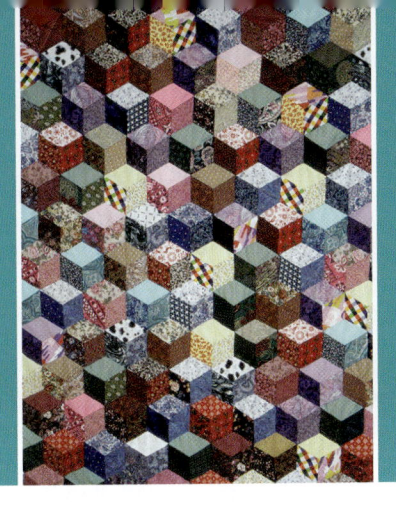

Helping Children with Problem Solving

> "IF THE ONLY TOOL YOU HAVE IS A HAMMER, YOU TEND TO SEE EVERY PROBLEM AS A NAIL."
> —Abraham Maslow (father of humanistic psychology)

> SNAPSHOT OF A LESSON

KEY IDEAS

1. Have students work collaboratively to solve an unfamiliar problem.
2. Have students look for patterns and use representations in problem solving.

BACKGROUND

This lesson is based on the video *Valentine Exchange*, from Teaching Math: A Video Library, K–4, from Annenberg Media. Produced by WGBH Educational Foundation © 1995 WGBH. All rights reserved. To view the video, go to http://www.learner.org/resources/series32.html (video #42).

Lilia Olivas is a fourth-grade bilingual teacher at an elementary school in Tucson, Arizona. The day after students exchange Valentine's Day cards in her classroom, Ms. Olivas poses a problem involving patterns and functions. Ms. Olivas models a valentine exchange between herself and a student, Gina, at the front of the class. She asks the students how many valentines are exchanged if she gives Gina 1 valentine and, in turn, Gina gives Ms. Olivas 1 valentine. Another student, Chelsea, offers an explanation about why the exchange involved 2 valentines.

Ms. Olivas: What if I had 3 people exchange valentines?

Student: You'll have 6 exchanges. [actually 6 valentines exchanged, from 3 exchanges]

Ms. Olivas: We'll have 6 exchanges? Will you show us how you came to that number, 6?

In both English and Spanish, the class discusses whether or not there would indeed be a total of 6 valentines exchanged. One student asks why the number of valentines exchanged has to be 6. To help students understand, Ms. Olivas models the exchange of valentines with two other students at the front of the class and finds there are 6 valentines exchanged. Ms. Olivas calls a third student to the front of the class and asks how many valentines are involved when four people exchange valentines with each other. They find that there are 12. In Spanish, Ms. Olivas poses a problem to the class: Figure out how many valentines would be exchanged in their class of 24 students if everyone gave everyone else a valentine. She asks the students to investigate this problem by working on their own, with a partner, or in small groups.

105

Ms. Olivas encourages students to use manipulative materials in solving this problem. One pair of students chooses to use cubes of different sizes to represent people and valentines. Another group uses stacks of pattern blocks. Students draw figures and tables on paper to keep track of the exchanges.

Ms. Olivas: [*to camera*] The benefit is that you don't have a controlled situation. I'm not controlling their thinking. I'm allowing them to explore freely, to develop new ideas, to share, to reason, and that's what I want out of my students. I want them to be confident mathematical learners.

Ms. Olivas circulates around the classroom, listening to students, observing their work, and occasionally asking questions.

Ms. Olivas: I'm looking at what Winston has here. What does this 24 mean?

Winston: [*working in a group of three students*] 24 valentines.

Ms. Olivas: And you each had a valentine? Twenty-four people? If you started making exchanges, would there be 24 exchanges?

Second Student: [*in the same group of three*] This is how I did it. Two, four, six, eight, … twenty-four. [*The student counts by 2s the number of cubes he has arranged to represent the valentines.*] Like, if you and Winston exchange, that's 2. [*He points to the first two cubes and continues to explain that the two students in each pair exchange with each other. He then explains that you add them up to get 48 valentines exchanged. It seems he is thinking that the number of valentines exchanged will always be double the number of people involved, which is actually not true.*]

Third Student: [*in the same group, questioning his teammate's solution*] Forty-eight is probably not the answer. I think it's much larger. [*He describes valentine exchanges that his teammate has left out.*]

Ms. Olivas: [*to camera*] The questions that I concentrate on are questions that help them to think about what they are doing. I would ask something like, "Can you explain what these mean?" tell me your thinking." "Tell me how you're doing these things."

Ms. Olivas speaks with groups of students to discuss their thinking and the meaning of the manipulatives they are using. Students talk within their group and with the teacher about the mathematical meaning of the strategies they choose.

One student in the class develops his own approach to the problem. Instead of finding the number of valentines exchanged by 24 students, he works on a simpler problem first, focusing on 12 people making exchanges.

Ms. Olivas: [*to camera*] There was a student that understood what was expected, then decided that he was going to do half of the problem. And even though the answer was not right, he knew where he was going. He knew there was a strategy involved. It was making sense to him.

Ms. Olivas calls the class back together to discuss their experiences as they worked on the valentine exchange problem. Many students offer their thoughts about the day's activity, including issues that confused or frustrated them and strategies they employed for solving the problem.

Ms. Olivas: [*to camera*] The whole focus is, "How were you thinking about the problem? How are you making sense of the information you have on the table? How was it making a connection in your mind? What are you looking at as far as strategy building and problem solving?"

FOCUS QUESTIONS

1. What is the difference between solving problems and practicing exercises?
2. What does it mean to teach math through problem solving? What "signposts" for teaching guide this approach?
3. What types of problems can be used in teaching through problem solving?
4. What strategies for problem solving are helpful for elementary students?
5. Why is looking back such an important phase in problem solving? What questions should students learn to ask themselves when they are solving problems and reflecting on their solutions?

INTRODUCTION

Mathematical problem solving is a skill people need throughout their lives. In school, students must solve problems in order to understand mathematical concepts, discover new mathematical relationships, and make sense of connections between mathematics and other subjects. Both children and adults confront mathematical problems in their daily lives—as consumers, citizens, and workers. Thus, it makes sense that the *Common Core State Standards for Mathematics* include "Make sense of problems and persevere in solving them" as the first of their eight standards for mathematical practice (CCSSI, 2010, p. 6).

A problem-solving approach can pervade the mathematics curriculum. Teachers can use problems to introduce new topics, to form threads that connect topics throughout instruction, and to ascertain whether children can apply what they have learned to new situations (Cai, 2003, 2010; Lester and Charles, 2003; Midgett and Trafton, 2001). *Principles and Standards for School Mathematics* (NCTM, 2000) recommends that "instructional programs from prekindergarten through grade 12 should enable all students to

- build new mathematical knowledge through problem solving;
- solve problems that arise in mathematics and in other contexts;

- apply and adapt a variety of appropriate strategies to solve problems;
- monitor and reflect on the process of mathematical problem solving. (p. 52)

This chapter describes how elementary school teachers can support children's mathematical growth through problem solving. For example, children can build new mathematical knowledge by having concepts introduced through problems set in familiar contexts (e.g., sports, games, and everyday activities) or through problems involving a variety of representations (e.g., hundreds charts, fraction bars, and counting cubes). A strong mathematics program builds on the natural, informal problem-solving strategies that the child has encountered before entering school. As the opening quote by Abraham Maslow suggests, the more problem-solving strategies we become familiar with, the more appropriately we can handle unfamiliar problem situations. Many of the best problems for elementary children involve everyday situations—for example, first graders will relate to problems such as "How many more chairs will we need if we're having five visitors and two children are absent?" and "How many cookies will we need if everyone is to have two?" while fifth graders might be more interested in problems such as "Who has the higher batting average, Benny or Marianne?" or "Which of these dice games is a fair game?"

No matter what type of problem is involved, children who are effective problem solvers plan ahead when given the problem, ask themselves if what they are doing makes sense, adjust their problem-solving strategies when necessary, and look back afterwards to reflect on the reasonableness of their solution and their approach.

WHAT IS A PROBLEM AND WHAT IS PROBLEM SOLVING?

A *problem* is something a person needs to figure out, something where the solution is not immediately obvious. Solving problems requires creative effort and higher-level thinking. If a child immediately sees how to get the answer to a problem, then it is not really a problem for that child.

Skill in solving problems comes through experiences with solving many problems of many different kinds. Children who have worked on many problems score higher on problem-solving tests than children who have worked on few. This chapter offers many suggestions for helping children become more proficient in mathematics through problem-solving experiences in the mathematics classroom.

As a teacher, you must be sure not to shield children from problem-solving challenges by assigning "problems" that really are just exercises. For example, a page in a traditional children's textbook (after children have been taught how to add large numbers) might begin with exercises such as

```
  3194        5479
  5346        3477
 +8877       +6399
```

Next, there might be a story problem such as this one:[1]

(A) 7809 people watched television on Monday.
9060 people watched on Tuesday.
9924 people watched on Wednesday.
How many people watched in the 3 days?

But this story problem is merely an exercise with words around it, not a true problem. The only challenge is doing the computation. Sometimes tasks like this—which can be solved by applying a mathematical procedure in much the same way as it was learned—are called *routine problems* or exercises. Looked at this way, many traditional textbook word problems are routine problems. If the past week's work has been on addition, a textbook's routine word problems are typically solved by adding; if the topic has been division, then students can simply look for two numbers in the routine problems and divide. It is little wonder that children taught in this way flounder on tests or in the real world, where problems are not conveniently grouped by operation. By contrast, *nonroutine problems*—that is, true problems—generally require thinking because the mathematical procedures that children must use to solve them are not obvious. In the remainder of this chapter, when we refer to problems, we usually mean nonroutine problems.

Try the following nonroutine problem:

(B) Begin with the digits

1 2 3 4 5 6 7 8 9

Use each digit at least once, and form three four-digit numbers with the sum of 9636.

While solving this problem, children will get considerable practice in addition, but they will also have to try many possibilities. If they are thoughtful about applying previously learned mathematical ideas, they may be able to take a more efficient approach. For instance, the desired sum 9636 has a 6 in the ones place, so students who know something about sums of odd numbers and sums of even numbers might realize that the digits in the ones place must either be all even or one even and two odd, since only these choices would result in an even sum. Any child who can add four-digit numbers has the knowledge required to solve this problem, but it is still a genuine problem because the solution is not immediately apparent.

Whether a problem is truly a problem or merely an exercise certainly depends on the person trying to solve it. For example, figuring out how to share 75¢ equally among three people is probably not a problem for you, but it probably would be a good problem for a second grader. What is a problem for

---

[1]Solutions for the problems presented in this chapter may be found in the *Helping Children Learn Mathematics Instructor's Manual* (available at the Wiley Book Companion Web site, www.wiley.com/college/reys).

Ann now may not be a problem for her in three weeks, and it may not be a problem now for Armando. The problems that you select for children must truly be problems for them in order to give them appropriate experiences in problem solving. Children who think they should always be able to solve problems immediately and easily are likely to view as impossible any problem where the solution is not immediately apparent, and they are unlikely to persist in working toward a solution. Finding the right level of challenge for students is not easy, but you can do it by trying out a range of problems, providing time, and encouraging students to explore many ways around obstacles.

> **Math Links 6.1**
>
> The Internet provides access to a wealth of problems. For example, take a look at the University of Mississippi's Math Contest site. It offers an Elementary Brain Teaser Contest for elementary and middle students. Archives of previous brain teasers are also available. See http://www.mathcontest.olemiss.edu/ or link from this book's Web site.
>
> www.wiley.com/college/reys

You shouldn't underestimate children's abilities, as is illustrated by the following anecdote. Ms. Chewning, a teacher in Virginia, had been using a problem-solving approach to mathematics instruction with her second graders. She read in a newspaper about a problem that 40% of the nation's eighth graders had not been able to answer at even a basic level on a recent national mathematics test. She wondered how well her second graders could do with that national test problem (quoted here):

> Jill needs to earn $45 for a class trip. She earns $2 each day on Mondays, Tuesdays, and Wednesdays. She earns $3 each day on Thursdays, Fridays, and Saturdays. She does not work on Sundays. How many weeks will it take her to earn $45?

Ms. Chewning was delighted to find that

> every single student attempted the problem, which was presented as optional. Such risk-takers they have become. Two students, using mental math only, presented me with the correct answer by the time I had completed writing the number story on the board. A total of 82% of the students, using a variety of strategies, successfully solved the problem in less than five minutes. Of the three students who struggled, two were right on track, making only minor computational errors, and the third achieved success after extensive trial and error. Needless to say, I was astounded. While I had expected them to be successful to some extent, I had not anticipated the speed and comfort with which they approached the task. *(The University of Chicago School Mathematics Project, 2001, pp. 204–205).*

> **Math Links 6.2**
>
> You can view hundreds of released items from the National Assessment of Educational Progress (NAEP), along with data on student performance, by visiting the Nation's Report Card Web site maintained by the National Center for Education Statistics (NCES). It may be accessed from www.nces.ed.gov/nationsreportcard/ or from this book's Web site. Click on SUBJECT AREAS, Mathematics, Sample Questions.
>
> www.wiley.com/college/reys

Results from national assessments have shown that many American students have difficulty with problems that require analysis or thinking, though scores have been rising in recent years. In fact, findings from the 2009 National Assessment of Educational Progress (NAEP)—which was administered at grades 4 and 8—showed that the percentage of students performing at or above the proficient level in both fourth and eighth grade increased consistently from 1990 to 2007 for both 4th and 8th grade (and was not significantly different from 2007 to 2009 for 4th grade) (Kloosterman et al., 2004; Kloosterman and Lester, 2007; www.nationsreportcard.gov/math_2009/). Fourth-grade students who perform at the proficient level are expected to consistently apply integrated procedural knowledge and conceptual understanding to problem solving in the five NAEP content strands.

American students have generally been found to be successful in solving routine one-step problems like those found in most textbooks but to have more difficulty in solving multistep or nonroutine problems, particularly those that involve application of more than one arithmetic operation. The NAEP test includes some "extended-constructed-response" (ECR) items for which students must construct their own answers and provide an explanation of their solution. Students at all grade levels tend to have more difficulty with these ECR items than with multiple-choice items or "short-constructed-response" (SCR) items (where students must provide their own answers but are not required to explain their solutions). Although no analyses of student performance on ECR and SCR items are available for the most recent administrations of the NAEP test, results from previous years are quite informative. "With respect to SCR items administered in both 1992 and 2000, the percentages of 4th-grade students responding correctly increased significantly for 20 of the 25 common items, with no significant differences for the other 5 common items" (Arbaugh, Brown, Lynch, & McGraw, 2004). The same researchers reported that American fourth graders' skill in solving routine word problems in subtraction improved during the 1990s, according to findings from the 2000 NAEP. Fourth graders' performance on word problems in multiplication was stable over the decade, but their performance on division-with-remainder word problems improved (Kloosterman et al., 2004).

TEACHING MATHEMATICS THROUGH PROBLEM SOLVING

The primary goal of school mathematics instruction should be to ensure that students make sense of the mathematics they are learning, and we emphasize this goal throughout this text. It is certainly important for students to learn—for example—how to subtract three-digit numbers or how to add fractions or how to compute the area of a circle, but this is not enough. Students must also learn to make sense of mathematical concepts and procedures so that they will be able to use them flexibly and appropriately in unfamiliar situations both during their school years and afterward. For example, figuring out what fraction is colored (not white) in each of the quilt patterns shown in Figure 6-1 is a motivating problem that could lead to class discussion not only about the fraction answers but also about how students figured them out. Students could also be challenged to develop their own quilt patterns to illustrate designated fractions or to exchange as fraction challenges with one another. Using dot paper (Appendix A) can help with block design. (See Westegaard, 2008, for an entire article about using quilt blocks in helping children construct understanding in mathematics.)

It may seem efficient to teach by telling students exactly how to do things and then to have them practice until they are able to perform confidently, but this approach is actually not very effective. When children are told things, rather than figuring things out for themselves, they are less likely to remember them or to be able to apply the ideas in the future. Mathematical sense making is best supported by a teaching approach in which students are confronted with problems, supported in their efforts to solve those problems, and helped to discuss and consolidate the learning that results (O'Donnell, 2009; Wood, Williams, and McNeal, 2006).

> **Virtual Classroom Observation**
>
> Go to www.wiley.com/college/reys,
> Access the Wiley Resource Kit
> Click on the Virtual Classroom Observations Section
> Module 3: Number and Operations: The Magnitude of Fractions
>
> 1. Watch video: Teaching Examples: Introduction (and, if time, watch videos #1–#5)
> 2. Do: From the Classroom—Assessing Student Work (Think about the different problem-solving strategies used, and what you—as a teacher—might do next)

Here are some "signposts" to guide you in teaching mathematics through problem solving (Hiebert, 2003a):

- *Signpost 1: Allow mathematics to be problematic for students.* This signpost means it is important to give students problems that challenge them, to allow them to struggle, and to help them examine and make sense of the approaches that they use. All students can benefit from problem-solving experiences. This teaching approach is probably unfamiliar to many teachers, and many may even find it questionable. They may think that good teachers should explain things clearly and completely so that students do not need to struggle at all. But it is precisely through the experience of solving challenging problems that students grow in their mathematical understanding.

  Allowing mathematics to be problematic does not mean you must search for lots of extra problems to assign. Instead, it means allowing students to grapple with the everyday mathematical challenges that the school curriculum includes, rather than immediately stepping in to tell students what to do. Imagine, for example, that a group of primary students had no experience with subtracting multidigit numbers, though they understand the concept of subtraction. A problem-based approach to teaching multidigit subtraction would not have the teacher begin by showing children how to subtract when no regrouping is required (as in 579–342) and later explaining how to regroup (as in 523–279); rather, it would have the teacher encourage the students themselves to think about how subtraction is the same and different in the two cases and to figure out how to solve the second type of problem in a way that makes sense.

- *Signpost 2: Focus on the methods used to solve problems.* When students are challenged to solve an unfamiliar problem, they should be encouraged to

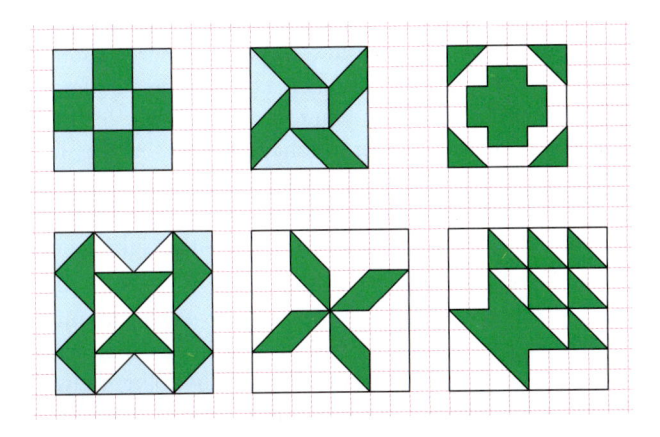

Figure 6-1 *What fraction is colored green in each quilt block?*

talk with one another about their methods, to compare the methods, and to think about their advantages and disadvantages. Given the problem 523−279, some students might think about adding up to get from 279 to 523 (279+1=280, 280+20=300, 300+223=523, so the difference between 279 and 523 is 1+20+223, or 244). Another student might think, "523−279 is hard to find, but I know it's the same as 524−280 (adding 1 to each number), and that's the same as 544−300 (adding 20 to each number). Aha! 544−300 is easy to do! It's 244. So I know that 523−279 is 244, too!" Still other students might choose to represent 523 with base-ten blocks and to think about trading in certain larger blocks for smaller blocks to make it possible to take away 279 blocks. This method most closely resembles the traditional paper-and-pencil approach.

Learning begins when the students are challenged to make sense of a problem. Learning continues when they search for ways to solve it. And learning is extended when they share their approaches with one another. Students deepen their mathematical understanding by thinking about which approaches are easier and harder and by questioning each other about why various approaches make sense.

- *Signpost 3: Tell the right things at the right time.* Teachers who use a problem-solving approach to mathematics instruction must decide when to share information and when to let children figure things out for themselves. Speaking up too soon can eliminate the challenge and much of the learning, but it is also important not to leave students floundering when you can see they aren't making progress.

Here are some suggestions for telling the right things at the right time. First, it's OK (indeed, important) to show students the written symbols of mathematics (e.g., notation for fractions, decimals, percents, operations, and equations) and to define technical language for them (e.g., the meaning of words such as *difference, quotient, quadrilateral,* and *median*). Words and symbols are all just social conventions, so there is no way that students will "discover" these things for themselves. However, the best time to tell students about these things is when they arise naturally during the course of students' mathematical endeavors.

Second, it's also OK to tell students about alternative strategies for solving problems if these strategies have not been suggested by class members. The important thing is to encourage students to compare their own methods with those of their peers and with those that you might suggest.

Finally, you should highlight the big mathematical ideas that come up during discussions of problem solving. In the course of solving a problem, children will uncover many important mathematical ideas. It is your job to guide the class in recognizing and talking about these discoveries.

FACTORS FOR SUCCESS IN PROBLEM SOLVING

Problem solving can be difficult to teach and to learn. Research on the characteristics of children who are successful or unsuccessful at solving problems, on the characteristics of problems, and on teaching strategies and classroom conditions that may help children be more successful at problem solving has led to some broad generalizations (McClain and Cobb, 2001):

- Young children enter school able to solve many problems. Instruction should build on what children already know.
- Children can begin solving problems in the earliest grades. They do *not* need to become skillful at computation before engaging in problem solving.
- Children's problem-solving abilities are related to their developmental level. They need to be given problems at appropriate levels of difficulty.
- Children should be taught a variety of problem-solving strategies to draw from as they meet a variety of problems. They should be encouraged to try solving a range of problems using the very same strategy and to try solving a single problem with more than one strategy. These efforts will help them recognize why some strategies are more appropriate for certain problems than for others. They can also be encouraged to adapt strategies to suit new types of problems.

Major factors that impact students' problem-solving skills are knowledge, beliefs and affects, control, and sociocultural factors. At all levels, teachers should be aware of the importance of all these factors when they teach problem solving:

- *Knowledge.* Students must learn to make connections between new problems and problems they have solved in the past. They must learn to recognize underlying structural similarities among problems and to choose the appropriate approach for solving each type of problem. (Do you see how this relates to the opening quote by Abraham Maslow?) That is, children should choose an approach based on a clear understanding of the problem rather than relying on surface features such as key words—for instance, the fact that the statement of a problem includes the words *in all* doesn't always mean that addition is the right approach, and *how many left* doesn't always indicate that subtraction is required.
- *Beliefs and affects.* Students' problem-solving abilities often correlate strongly with their attitudes, their level of self-confidence, and their beliefs about themselves as problem solvers. Teachers must show students that they believe *all* students can be good problem solvers. Teachers must also encourage students to develop

their own strategies for and approaches to problem solving. Teachers who believe there is only one way to solve a problem prevent students from truly experiencing what it means to be a problem solver and to do mathematics.

- *Control.* It is extremely important for students to learn to monitor and control their own thinking about problem solving. Research indicates that good problem solvers often spend a considerable amount of time up front, making sure they understand a problem, and at the end, looking back to see what they did, analyze how their solution might be modified or improved, and think about how the problem is similar to and different from other problems. By contrast, weaker problem solvers tend to be impulsive, often jumping right in and crunching numbers with little regard for what they mean, without stopping to think about what approach might be most productive. Teachers must give students tasks and activities that encourage them to monitor, reflect on, and control their own thinking.

- *Sociocultural factors.* The atmosphere of the classroom should encourage students to use and further develop the problem-solving strategies that they have already developed naturally through experiences outside the classroom. Furthermore, the classroom climate itself (with its opportunities for discussion, collaboration, sharing, and mutual encouragement among students) plays an important role in helping students become more skillful problem solvers. To use problems effectively, teachers need to consider the time involved, planning aids, needed resources, the role of technology, and how to manage the class.

CHOOSING APPROPRIATE PROBLEMS

Teaching through problem solving requires planning and coordinating the problems that you assign so that students have the chance to use a variety of problem-solving strategies and to analyze, write about, and discuss their solutions. You will probably be expected to teach from a mathematics textbook, so you need to consider how to use it most effectively. Begin by thinking about the big mathematical ideas that are involved in each chapter or unit. Then look at the problems provided in the text and think about how you can use them to prompt your students to get actively involved with the big mathematical ideas. Try to identify additional problems you could assign to accomplish this goal, and remember the *signposts* we discussed above as you select problems to assign. The problems you assign do not need to be extremely clever or original. Most good problems are quite simple. Sometimes an effective approach is to assign a problem that involves exactly the mathematical ideas you want to emphasize *before* students get to those ideas in the textbook. For example, if your students will soon be learning about multiplication, consider assigning a problem like this:

How many colored pencils would be needed if there are 18 children and each child needs 3 pencils? Their approaches will give you insight into what they already know and the strategies they already have, and trying to solve the problem will help them begin to think about the important idea of equal-sized groups.

The following list describes a wide variety of problem types, with examples of each type:

- Problems that ask students to represent a mathematical idea in various ways:

  (C) Think about the number 10 separated into two parts. Draw a picture to show ways that 10 things could be put in two parts. Make up a story to go along with your picture. Write a number sentence to go along with your picture.

  (D) Make up a story and draw a picture about marbles for this number sentence: 18÷6=3. (Figure 6-2 shows the responses of two third graders to this problem.)

- Problems that ask students to investigate a numeric or geometric concept:

  (E) Begin with a square that measures 6 cm by 6 cm. The area of the square is 36 cm². How many different shapes can you think of that have this same area? For each shape, explain your thinking. For example, how many different rectangles can you find with an area of 36 cm²? How many different triangles? Can you think of trapezoids or parallelograms with an area of 36 cm²?

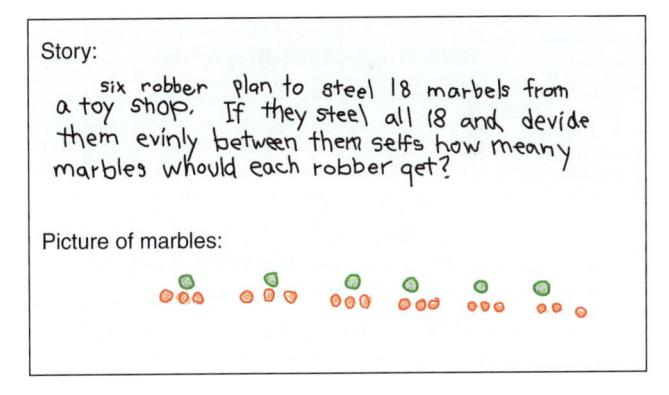

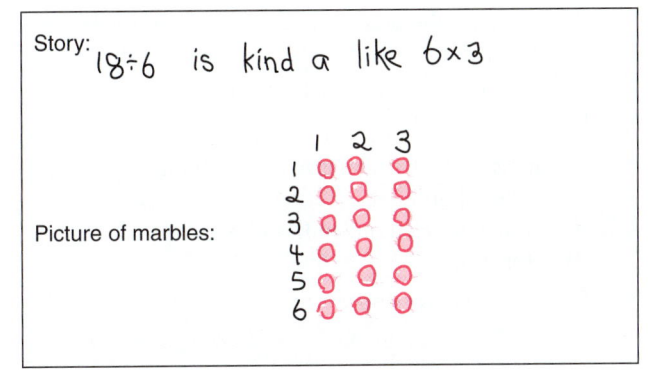

Story: six robber plan to steel 18 marbels from a toy shop. If they steel all 18 and devide them evinly between them selfs how meany marbles whould each robber get?

Picture of marbles:

Story: 18÷6 is kind a like 6×3

Picture of marbles:

Figure 6-2 *Stories about 18 ÷ 6 = 3 written by two third graders.*

What other shapes can you think of with that area? (Tayeh and Britton, 2005).

- Problems that ask students to estimate or to decide on the degree of accuracy required or to apply mathematics to practical situations (e.g., buying or measuring or building something):

(F) Eric wondered, "how many Cheerios are in an entire cereal box?" His younger sister, Lori, predicted there were a million Cheerios, but his older sister, Myrah, said, there is no way a million Cheerios could fit in that box." How can Eric estimate about how many Cheerios are in a 15-oz box of Cheerios? If Myrah is correct, about how many boxes would you need to have a million Cheerios? (See Tayeh, 2006/2007, for a description of how first and fourth graders from Texas tackled this problem together.)

Math Links 6.3

Visit the National Library of Virtual Manipulatives on the Web to see what's available for help with problem solving. For example, you can use a virtual balance scale and a collection of virtual coins to figure out which coin weighs less than the others in the Counterfeit Coin problem. See if, with practice, you can use logic and reasoning to solve the problem in fewer moves. Visit http://www.matti.usu.edu (or connect from this book's Web site), then select "Virtual Library," "Algebra," "Coin Problem" (listed under grades 6–8).

www.wiley.com/college/reys

(G) Erica is helping her father build a pen for her rabbit. She finds four pieces of lumber in the garage that they can use for corner stakes, and 36 feet of chicken-wire fencing. She recognizes that she can use these materials to make a variety of rectangular-shaped pens. Name three different sizes of rectangular pens they could build. What are the dimensions of the rectangular pen with the largest area? How do you know?

(H) Which is the better buy, a 6-ounce jar of jelly for $1.79 or a 9-ounce jar for $2.79? (Note that this problem is very similar to those posed in the Snapshot of a Lesson in Chapter 5.)

- Problems that ask students to conceptualize very large or very small numbers:
  (I) Have you lived one million hours?
  (J) How thick is a piece of paper?

- Problems that ask students to use logic, to reason, to strategize, to test conjectures, or to gauge the reasonableness of information:
  (K) Three children guessed how many jelly beans were in a jar. Their guesses were 80, 75, and 76. One child missed by 1. Another missed by 4. The other child guessed right. How many jelly beans were in the jar?
  (L) You and a friend have two dice to play a game. You can choose whether you will add or multiply the two numbers on the dice after you throw them. If you choose to multiply, you will multiply on all your turns, and your friend will add on all turns. If you choose to add, your friend will multiply. You get to go first. The winner will be the first person whose dice give an answer of 12. Would you choose to add or multiply? Explain your thinking (Stenmark and Bush, 2001).

- Problems that ask students to perform multiple steps or use more than one strategy:
  (M) Ellie has $10.00 in her pocket. She spends $5.50 for a movie ticket. The theater offers a popcorn and drink special for $3.79. If Ellie buys the popcorn and drink, does she have enough money for a $1.45 candy bar, too?
  (N) How many days have you come to school so far this school year?

In addition to giving students problems like these, you should sometimes give them open-ended problems—that is, problems that can have more than one correct answer. With open-ended problems, the answer depends on the approach taken (of course, the answer must be reasonable). Different students approach open-ended problems in very different ways, so such problems are ideal for ensuring that students at all levels can experience some measure of success. Open-ended problems are especially appropriate for cooperative group work, in which case they should be followed with a class discussion in which the mathematical ideas and planning skills involved are explored and students get a chance to clarify their thinking and validate their decisions. Traditional textbook problems are rarely open ended, but you can often make them open ended by modifying them in minor ways. Figure 6-3 shows a problem in traditional form and its transformation into a problem in open-ended form (O).

Math Links 6.4

To see an entire unit of study (8 lessons) built around realistic, open-ended problems, examine *Planning a Trip*, found at the NCTM Illuminations Web site. Students use a variety of resources, including the Internet, to plan short and long trips. It may be accessed from http://www.Illuminations.nctm.org/index_d.aspx?id=359 or from this book's Web site.

www.wiley.com/college/reys

An open-ended problem, is much more effective when constraints ensure that students must do mathematics at a high level to solve the problem and that the focus stays on the mathematics. Of course, it is also important that the

Teaching Mathematics through Problem Solving   113

MENU

Side Dishes

Taco Salad	$3.50
Quesadillas	$2.25
Cheese Nachos	$1.50
Baked Potato	$1.75
Cole Slaw	$1.00

Main Dishes

Chicken and Rice	$6.95
Beef Fajitas	$8.95
Chicken Fingers	$4.50
Chimichangas	$5.95
Beef Burritos	$4.95

Desserts

Flan	$3.25
Baked Apples	$2.45
Empanadas	$3.15

Traditional form: Randy, Becky, CJ, Lauren, and Ty go to eat dinner at their favorite restaurant. Ty orders quesadillas, beef fajitas, and flan. CJ has chicken fingers and does not order a side dish or dessert. How much do these two meals cost? (Houghton Mifflin 2002, p. 54).

Open-ended form: Randy has $13.00 to spend at his favorite restaurant. He wants to order one main dish, two side dishes, and one dessert. He also knows he will spend $1.50 on video games while he waits for his order. Find three different meals that Randy could choose. Show your calculations and explain how you thought about the problem.

Figure 6-3 *A menu problem in two forms: traditional and open ended.* (Source: Kabiri and Smith, 2003, p. 187. Reprinted with permission from *Mathematics Teaching in the Middle School*, copyright 2003 by the National Council of Teachers of Mathematics.)

constraints seem reasonable or realistic, not arbitrary. In open-ended problem (O), the constraints are that Randy is limited to $13, must spend $1.50 on video games, and must order a main dish, two sides, and a dessert. In a problem that involves buying things from a catalog, the constraints might be that the students have to buy between 5 and 10 items, that no item can cost less than $1 or more than $50, and that the total cost of the items must be between $175 and $200. If the teacher does not impose appropriate constraints on the problem, students are likely to spend most of their time simply debating what to buy, with very little time spent estimating or computing how to spend their money. This can be fun for the children, but it involves little or no mathematics.

In the Classroom 6-1 shows an open-ended problem (P) adapted from a problem-oriented elementary curriculum, *Investigations in Number, Data, and Space*. The children cut out seven rectangles and are instructed to arrange them in order from biggest to smallest. A third-grade teacher reports having used this problem with her class of bilingual third graders, with good success in encouraging them to think about length and area. The problem allowed for multiple entry points and a variety of approaches for that diverse group of students. Her account of her students' conversations about their thinking is fascinating (Dwyer, 2003).

Problems (Q) and (R), shown in Figures 6-4 and 6-5, represent a different sort of open-ended problem. In both problems, the numeric answer is not at all difficult to determine (and everyone should get the same answer). The challenge is to see how many different and interesting ways there are to find that answer.

(Q) Figure 6-4a shows that a pentomino is an arrangement of five square tiles where each tile has at least one side adjoining a side of another tile. Using graph paper and scissors, how many different pentominos can you make? How can you know that you have identified all the possible pentominos? Figure 6-4b shows one pentomino that will fold into a box and one that will not. How many of the possible pentominos will fold into an open box?

a. Pentomino   Not a pentomino

b. Box pentomino   Not a box pentomino

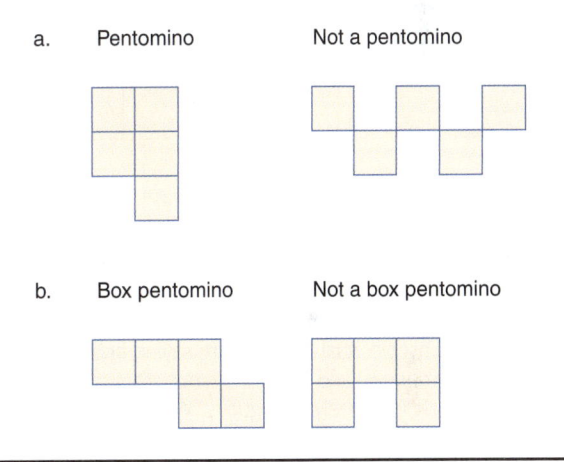

Figure 6-4 *How many box pentominos can you make?*

(R) Figure 6-5 shows marbles arranged in a pattern. How many marbles are there? Find the answer by counting or computing in as many different ways as you can.

Figure 6-5 *How many different ways can you find the number of marbles in the picture?*

In the Classroom 6–1

WHICH RECTANGLE IS BIGGEST?

Objective: Ordering rectangles, defining "biggest" in different ways, finding the number of tiles that cover - rectangles.

Grade Level: 2–3.

Materials: Color tiles (1 tub per 6–8 students). For each student: scissors, glue, some black paper, and a worksheet that pictures 7–10 rectangles of various sizes and shapes (skinny, fat, a few with the same length or width, etc). Choose rectangle sizes so that color tiles will fit on top perfectly, since Part III of the activity involves covering the rectangles with color tiles to find areas. Do not arrange the rectangles in any particular order by size. Label rectangles with letters, A,B,C, etc.

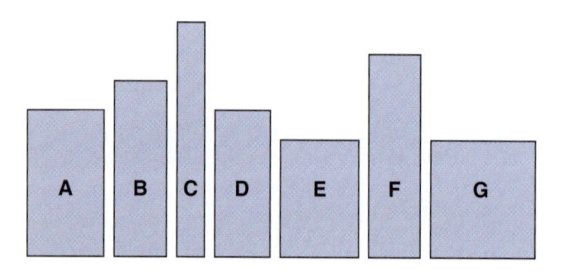

Part I: Ordering Rectangles

A. Paired Work

Have the students work with partners to arrange the rectangles from smallest to biggest on their paper, and to glue them in place after they have decided. (The teacher should observe what students are saying and doing. Different orderings and disagreements are expected. The question is purposely ambiguous. What's important is to observe how students justify their reasoning.)

B. Class Discussion

Bring the group together, with their papers. Ask one pair to show how they ordered the rectangles. (For example, a common ordering is to arrange the rectangles "standing up"—with their longer sides vertical—and to order them from tallest to shortest, with the wider one first whenever two have the same height.) Ask the students to explain why they chose their own particular order. Ask pairs who ordered the rectangles differently to show their papers and explain their thinking. For example, a pair may have chosen to arrange the rectangles "lying down," or they may have tried to order by area rather than by height. Encourage students to talk about the different ways they decided what was "big" and what was "small."

Part II: Covering Rectangles

While all the rectangles are still visible, tell the children to imagine they are chocolate bars. Which would have the most chocolate? Which would have the least? How can you figure it out? Students may suggest different approaches. One possibility is covering them with color tiles. Each color tile can represent one square of chocolate.

A. Paired Work

Distribute a new sheet of paper showing all the original rectangles. Have students work in pairs to find how many color tiles it will take to cover each rectangle. They should write the corresponding number inside each rectangle.

B. Class Discussion

Bring students together to compare findings. Were they surprised about any results? Were any rectangles covered by the same number of tiles? Encourage students to discuss whether they think these rectangles are the same size or different sizes. (If they think different, which is "bigger" and why?) Were there any rectangles that were considered "big" in the first activity (ordering rectangles) but not as "big" in the second activity (covering rectangles)? [Note: Students may need numerous experiences before understanding that what we consider "big" may change, depending on the attribute we are considering, e.g., ordering by height rather than by area.]

Activity adapted from: Akers; J., Battista, M., Godrow, A., Clements, H.D., & Sarama, J. "Shapes, Halves and Symmetry," from *Investigations in Number, Data, and Space: Grade 2*. Menlo Park, CA: Dale Seymour Publications, 1996, pp. 52–54.

FINDING PROBLEMS

Where can you find problems to challenge your students? Most textbooks include a wide range of problems, and we have seen that textbook problems in traditional form can often be converted easily to open-ended problems. In addition, there are many resources for problems that will stimulate and challenge your students. To find problems, a teacher might

- Investigate articles and books referenced in this chapter.
- Investigate a variety of Web sites.
- Write problems yourself (possibly using ideas from newspapers or from events in your community). See, for example, a description of a broad assortment of problems drawn from a single newspaper circular

describing activities related to Washington, D.C.'s annual cherry-blossom festival (Silbey, 1999). Barrow (2010) offers a framework to guide teachers in designing problems tailored to meet their students' needs and experiences.

- Use situations that arise spontaneously, particularly children's questions or conjectures. For example, after reading a story about a robbery, children might wonder, "If a bank robber stole a million dollars, how heavy and bulky would the money be? Could he run down the street with it in his pocket? In a shopping bag?" For another example, read Phyllis Whitin's account of her fourth-graders' surprise when, while working on the pentomino problem (Q), they noticed that some of the pentominos had perimeter 10, while others had perimeter 12, even though all the pentomino shapes were made from exactly 5 square tiles. This led Whitin's students to make and explore a variety of conjectures as they wrestled with this unexpected observation (P. Whitin, 2004).
- Attend problem-solving sessions at professional meetings. Share problems with other teachers.
- Have children write problems to share with each other (Hildebrand, Ludeman, and Mullin, 1999; Wu, An, King, Ramirez, and Evans, 2009).
- Use children's literature as a context for solving mathematical problems embedded in or related to the story (Ameis, 2002; Bay-Williams and Martinie, 2004; Bresser, 2004; Burns and Sheffield, 2004a, 2004b; Whitin and Whitin, 2004).

Math Links 6.5

Collect problems from professional journals such as *Teaching Children Mathematics*, newspapers, magazines, resource books such as the Navigations Series from the National Council of Teachers of Mathematics, and Web sites such as http://www.figurethis.org/challenges/toc.htm or this book's Web site.

www.wiley.com/college/reys

It is never too soon to start a problem file, with problems categorized so you can locate them readily. Categorize them by mathematical content, by strategies, or by how you are going to use the idea. Laminating the cards permits them to be used repeatedly by students for individual or small-group problem solving.

Math Links 6.6

On the Web, teachers and students can find a wealth of challenging and interesting mathematics problems or search for data for real-world problem-solving investigations. Some examples of sites to consider include http://www.mathforum.org/pows/ or http://www.pbs.org/teachers (select math under any grade span). You can also link to these sites from this book's Web site.

www.wiley.com/college/reys

HAVING STUDENTS POSE PROBLEMS

Encouraging students to write, share, and solve their own problems is a good way to help them develop their problem-solving skills. By posing problems, students learn how problems are structured, they develop critical thinking and reasoning abilities, and they learn to express their ideas clearly.

It is often helpful to begin by having students modify familiar problems. For example, third-grade students may have read the classic story *The Doorbell Rang* (Hutchins, 1986) and considered problems like this one: Suppose Mama baked 12 cookies. If she has 2 children (or 4 or 6 or 12) and each child gets the same number of cookies, how many cookies does each child get? This problem can be rewritten in various ways. A simple modification would be to change the numbers. Some changes in number don't really affect the difficulty of the problem (e.g., 18 cookies and 6 children is hardly more difficult than 12 cookies and 2 children). But other changes in number might make a big difference (e.g., 12 cookies shared by 8 children or 12 cookies shared by 5 children are harder because they involve remainders or fractions). This could be turned into a more open-ended problem. By leaving the number of children unspecified, there can be more than one answer: Mama baked 18 cookies. She can share them equally among her children without breaking any cookies. How many different numbers of children can she have? Another interesting way to reformulate a problem is to exchange the known and unknown information—here, for example, making the number of cookies unknown: Mama baked a lot of cookies. If she has 2 children (or 4 or 6 or 12) and each child got 3, how many cookies did Mama bake?"

Here are four principles for helping students as they learn to pose problems (Moses, Bjork, and Goldenberg, 1990):

- Focus students' attention on the various kinds of information in problems: the information a problem gives them (the *known*), the information they are supposed to find (the *unknown*), and the *restrictions* that are placed on the answer. Encourage students to ask "what if" questions. For example, what if you make the known information different? What if you switch what is known in the problem and what is unknown? What if you change the restrictions?

- Begin with mathematical topics or concepts that are familiar.
- Encourage students to use *ambiguity* (what they are not sure about or what they want to know) as they work toward composing new questions and problems.
- Teach students about the idea of *domain* (the numbers they are allowed to use in a particular problem). Extending or restricting the domain of a problem is an interesting way to change it. For example, the problem "Name three numbers whose product is 24" is very different depending on whether you consider a domain of all whole numbers, only even whole numbers, all integers (both positive and negative), or perhaps even fractions and decimals.

The teacher plays a key role in establishing a classroom environment where students are encouraged to think deeply about how problems can be changed or rewritten. The teacher can model an inquiring mind by frequently asking "What if?" when discussing problems and by encouraging students to make conjectures and to reformulate problems.

Having students write their own problems (rather than modify problems already at hand) is also useful. Generally, this is best done after students have had considerable experience with modifying familiar problems. When you ask students to compose problems on their own, it is often important that you specify certain goals or constraints; otherwise, the assignment may become nothing more than an exercise in creativity. (Students may write fantastic stories with no discernible mathematics content or pose problems that are so complicated that no solution is possible.) You could specify goals or constraints, for example, by assigning students to write a word problem that matches a given mathematical expression or a given figure:

- Write a word problem for $250 \div 5 = 50$.
- Write a comparison word problem for $12 - 8 = 4$. (If you are not familiar with "comparison" subtraction problems, see Chapter 9, where they are defined, along with "separation" subtraction problems and "part-whole" subtraction problems.)
- Write a multiplication word problem for the tree diagram shown below. (For example, two kinds of ice cream and three possible toppings for each gives six different types of sundaes.)

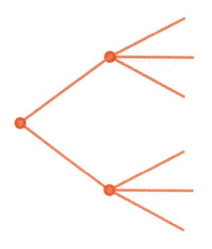

Situations or information in magazine advertisements, newspaper articles, books of world records, sales flyers, and so on can also be effective in prompting students to pose problems. An approach used by a sixth-grade teacher was to challenge her students to write and illustrate problems that involved multiplication or division and that included extraneous information. Students responded with creative problems such as this:

> At midnight, the wind over Jamaica started to increase as Hurricane Lucas came closer. Every 10 minutes from then on, the wind doubled, and five trees were pulled from the ground. At 12:30, the wind speed was 120 miles per hour. At what speed was the wind blowing at 12:00 midnight?

Students need help learning to write problems, and it is easy to integrate such lessons with language arts instruction. You could use a writing workshop approach, consisting of stages such as brainstorming and prewriting, writing, several rounds of peer critiquing followed by rewriting, and, finally, editing and publication. Student problems may be published on bulletin boards, on cards (to be made available for other students to solve), or in a class book of problems. The sixth-grade teacher mentioned above obtained a small classroom grant to produce enough copies of her class's illustrated book of problems to distribute one to each sixth-grade class in the school district. Her students proudly went on a field trip that took them from school to school delivering their books to sixth-grade classrooms, where they shared problem-solving experiences with other students their own age. In other states, a third-grade teacher and a second-grade teacher engaged their students in taking "the mathematician's chair" as they challenged their peers with problems they had written themselves. The third-grade teacher assessed the children's writing by considering three aspects of each problem on a scale of 1–4: the attributes of the problem, the structure of the problem, and the student's use of language conventions (Hildebrand, Ludeman, and Mullin, 1999). The second-grade teacher encouraged her students to use a graphic organizer to help focus their thinking when they compose problems (Wu, An, King, Ramirez, and Evans, 2009).

USING CALCULATORS AND COMPUTERS

The calculator's potential for helping children become more proficient at problem solving has been recognized since handheld calculators became widely available more than 35 years ago. Of course, if the focus of a lesson is on practicing paper-and-pencil computation or mental computation, students should not be permitted to use calculators. But when the focus is on problem solving or concept development, then calculators can often be an important aid, as we discuss in more detail in Chapter 10. You should consider having students use calculators whenever

- They let children solve more complex problems or problems with realistic, rather than contrived, data.
- They eliminate tedious and time-consuming computations and help reduce children's anxiety about being able to do computations correctly.
- Their special functions can help children explore mathematical objects, concepts, and operations.

Because calculators present answers in decimal form, what comes up when students use a calculator while solving problems may surprise and challenge them. For example, consider this problem: "If one bus can transport 34 students, how many buses will be required to take 489 students?" A student working on this problem keyed in 489÷34 and was surprised when her calculator showed 3.456138. How should this answer be interpreted? In this case, the calculator's answer presents an excellent opportunity for talking about division, remainders, and fraction-decimal equivalents. (See the Snapshot of a Lesson in Chapter 5 for a firsthand look at how a teacher handled this sort of situation in her classroom.)

One educator reports the story of a fifth grader named Jonathan who enjoyed using calculators to explore mathematical problems (Battista, 2003). Jonathan overheard a third grader working on finding pairs of whole numbers whose product was 9. Jonathan knew quite well that 9×1, 1×9, and 3×3 were the answers to the question. But it prompted him to think about a related problem: Are there more numbers that multiply to make 9, if you can use decimals or fractions? Jonathan was able to come up with quite a few number pairs mentally ($4\frac{1}{2} \times 2$, 6×1.5, $8 \times 1\frac{1}{8}$, $4 \times 2\frac{1}{4}$, etc.). He spent the next 30 minutes brainstorming a long list of other possibilities and checking them with a calculator. Eventually, an adult asked Jonathan, "How many pairs do you think there are?"

> Jonathan: "Billions." (Short pause.) "There's one for every number. Like a googolplex times 9 googolplexths [sic]" (To Jonathan, a googolplex was an unimaginably large number; he had encountered the term in his reading.)
>
> (Battista, 2003, p. 273)

Jonathan uses his calculator as a tool that helps him explore challenging mathematical problems—problems that he certainly could not easily explore without a calculator.

A popular type of problem-solving activity involving calculators is the "broken key" activity. The children are assigned a series of problems and are told they may use a calculator but must pretend that certain keys on the calculator are broken. (The specified "broken keys" can be number keys, operation keys, or a combination of the two. It may be helpful to give the children a small piece of masking tape to stick on the "broken keys" as a reminder.) Suppose one of the problems is: Compute 5×39 when the "3" key is broken. The children may suggest various ways to work around the broken key—for example, 40−1+40−1+40−1+40−1+40−1, or (5×29)+(5×10), or (5×40)−5. The teacher should ensure that students talk about and compare their approaches and the various operations they used. (You can read about how fourth and fifth graders tackled the problem of multiplying 88 × 8 without using the 8 key or the × key in Ellis, Yeh, and Stump, 2007.)

Computers can also be important problem-solving tools. Like calculators, computers allow students to work on problems with realistic data. But computers also let students experience problems of different types, such as problems involving graphics and graphing. For example, students could collect data from classmates on their favorite types of music, enter the data into a spreadsheet, and then generate circle graphs or histograms to represent the data. Students could also try to determine whether musical preferences differ among children of different grade levels or between boys and girls—they could separate the data by grade level or gender and draw a histogram that compares the preferences of one group with those of the other.

Many fine software programs provide a variety of problem-solving experiences. Some, such as *Numbers Undercover*, involve computation. Others, such as *The Factory Deluxe*, address spatial visualization. Still others, such as *Math Shop*, provide direct experiences with problem solving. *The Cruncher* teaches spreadsheet skills for solving such real-life problems as how many weeks of allowance equal a new CD player. *Geometer's Sketchpad*, lets students create geometric figures, make conjectures, and explore relationships. For example, students can explore the sum of the angles of triangles, quadrilaterals, and other polygons and try to predict the sum for 10-sided polygons.

In recent years, the World Wide Web has become an extremely rich resource for mathematics education. On the Web, teachers and students can find a wealth of challenging and interesting mathematics problems, search for data for real-world problem solving, and seek answers to mathematical queries and conundrums (see the Math Links throughout this book).

STRATEGIES FOR PROBLEM SOLVING

The Hungarian-born mathematician George Polya proposed a now-classic four-stage model of problem solving (Polya, 1973):

1. *Understand* the problem.
2. *Devise* a plan for solving it.
3. *Carry out* your plan.
4. *Look back* to examine your solution.

This model forms the basis for the problem-solving approach used in problem-solving lessons in most elementary school mathematics textbooks. However, Polya's model can be less than helpful if taken too literally. Except for simple problems, it is rarely possible to go through the steps in lockstep sequence. Moreover, the steps are not discrete, and it is not always necessary to perform every step. For instance, while trying to understand a problem, students may move into the planning stage without realizing they have done so. Or simply understanding a problem may enable students to see a solution without any planning. In addition, going through the steps does not always help students find a solution. Many children become trapped in an endless process of reading, thinking, rereading—and rereading and rereading—until they give up.

Children need specific strategies that will help them move through the steps in a productive way. Polya himself delineated many of these strategies (or heuristics), and many textbooks provide lists of the strategies geared to various grade levels. It is important, however, to distinguish between Polya's model itself and these strategies. Polya's four-stage model provides a general picture of how to move through the process of solving a problem, whereas strategies are tools that may be useful for helping students interpret and describe mathematical situations or for helping them move forward at various points in the problem-solving process (Lesh and Zawojewski, 2007).

In this section, we discuss seven strategies that can help children solve a wide variety of problems (see Table 6-1). Many textbooks introduce problem-solving strategies in a systematic way. If you are teaching from a textbook that does not cover some of the strategies we discuss here, you can probably devise a plan for fitting them in by referring to your textbook's scope and sequence.

For additional guidance in helping children with problem-solving strategies, you can look at supplemental texts (e.g., Charles, Lester, and Lambdin, 2005). Of course, you should not limit students to using only the strategies that you have discussed in class. Rather, you should always encourage students to generate their own ideas about how to approach a new situation. If you see some students successfully using a strategy that you haven't discussed, you should encourage them to share their ideas with the rest of the class. You might also help the children think of a label for that new strategy so it can be referenced easily in future class discussions. You could list problem-solving strategies on a bulletin board for quick reference by students. For example, Jessica might discover that a good way to get started on a problem involving large numbers is to solve a similar but simpler problem with smaller numbers. When she shares her idea during whole-class discussion, you might suggest naming this the "make-it-simpler strategy"; this would give you the option of beginning future problem-solving discussions by asking whether the make-it-simpler strategy might be useful in attacking the problem at hand.

The following discussions of specific problem-solving strategies include illustrative problems, covering a range of mathematical topics and grade levels, that could be used to develop each strategy. Most problems can be solved with any of a variety of strategies. But even so, for a given problem it is often the case that some strategies are more effective than others. For this reason, children should have a repertoire of strategies that they feel comfortable using. And sometimes students will have to use more than one strategy to solve a problem. Being able to draw on a repertoire that includes a wide variety of strategies allows students to attack many different types of problems. Moreover, when one strategy fails, the children have other strategies to turn to, and this can help them develop confidence in their ability to find a path to a solution.

As you read, do stop and try to solve the problems!

ACT IT OUT

Acting out a problem helps children visualize what is involved in the problem. In using this strategy, either the children themselves perform the actions described in the problem or they manipulate objects. When teaching the strategy, you should stress that the objects used do not have to be the real thing—for example, real money is obviously not needed to act out a problem involving coins, only something labeled "5¢" or "25¢." Children are adept at pretending, so they will probably suggest substitute objects themselves, but make sure they focus their attention on the actions, not on the objects.

In the early grades, you can develop the act-it-out strategy using simple real-life problems:

1. Six children are standing at the teacher's desk. Five children join them. Now how many children are at the teacher's desk?

In later grades, when the problems are more challenging, it is probably unrealistic for students actually to act them out. But they might find it valuable to act out simpler variations of the problem to see if they can identify patterns of actions. Or they simply *think* about the actions involved, and keep notes of what would happen at each step of acting out the problem if the problem situation involves multiple steps.

2. There are 24 children in a class. Each child gives a valentine to each of the other children in the class. How many valentines are exchanged? (This is the problem discussed in the Snapshot of a Lesson at the beginning of this chapter. Thinking about acting this out requires thinking about how many valentines each child will need to bring to class. We can envision the first child giving a valentine to each other child—that's 23, and the second child doing the same—that's 23 more, and so on.)

TABLE 6-1 • A List of Useful Problem-Solving Strategies

Act it out	Guess and check
Make a drawing or diagram	Work backward
Look for a pattern	Solve a similar but simpler problem
	Construct a table

Strategies for Problem Solving   119

3. A man buys a horse for $60, sells it for $70, buys it back for $80, and sells it for $90. How much does the man make or lose in the horse-trading business?

MAKE A DRAWING OR DIAGRAM

Within the past week or so you have probably used the drawing or diagramming strategy to help solve a real-life problem. Perhaps you had to give directions to your house, so you drew (or referred to) a sketch of the route. Or maybe you wanted to rearrange a room and drew diagrams to see how the furniture could be placed. This strategy of making a drawing or diagram lets you depict the relationships among the different pieces of information in a problem in a way that makes those more apparent.

When teaching this strategy, stress to the children that there is no need to draw detailed pictures. Rather, encourage the children to draw only what is essential to represent the problem. For example, if the problem is about children in a classroom, stick figures (or even just circles) can represent the children and a square can represent the classroom—it is not necessary to draw the children's clothing or the walls of the classroom.

Here are two problems where students could apply the drawing or diagramming strategy:

4. Aunt Katrina wants to hang six decorative plates on her dining room wall, in a straight line and spaced evenly apart (from each other and from the edges of the wall). Each plate is 8 inches in diameter, and the wall is 104 inches long. How far apart should the plates be hung? (Hartweg, 2004/2005). See Figure 6-6 for two fourth-graders' attempts at this problem.

5. A snail is at the bottom of a jar that is 15 cm high. Each day the snail crawls up 5 cm, but each night it slides back down 3 cm. How many days will it take the snail to reach the top of the jar? (*Note:* The answer is not $7\frac{1}{2}$ days. Draw a picture to see why not!)

At times you can use this strategy as the basis for an activity, by presenting a picture or diagram for which the children have to make up a problem. For instance, you could show children the following picture, which could prompt them to pose the problem of how far each car goes before they crash or how long it takes them to crash:

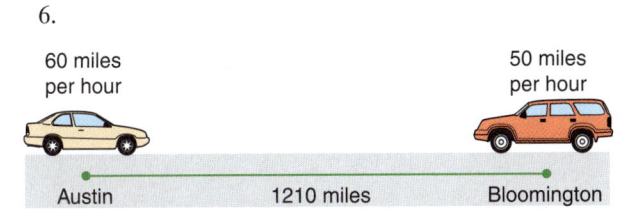

6.

60 miles per hour — Austin — 1210 miles — Bloomington — 50 miles per hour

LOOK FOR A PATTERN

Understanding "patterns, relations, and functions" is one of the major goals of the Algebra Standard (across all grades K–12) in NCTM'S *Principles and Standards for School Mathematics*. Recognizing, describing, extending, and generalizing patterns are important components of algebraic

---

Aunt Katrina wants to hang six decorative plates on her dining room wall, in a straight line and evenly spaced apart. Each plate is 8 inches in diameter and the wall is 104 inches wide. How far apart should the plates be hung? (Hartweg, 2004/2005).

A 4th grade student named Kelly attempted to solve this problem without drawing a picture. She understood that she needed to divide up the wall distance to figure out how far apart to space the plates. Her work shows that she tried dividing 104 by 6 and also by 8.

A.

Kelly explained her thinking as follows: "I knew dividing by 6 wouldn't work because there was some left over, so I divided by 8 and got 13. Then I checked my work by multiplying 13 by 8 and got 104." When pressed about why she chose 6 and 8 as divisors, she replied that there were 6 plates with a diameter of 8. She believed that her answer was 13, but was unable to explain what the 13 represented. Without a drawing to help her see relationships, she was unable to solve the problem.

Lizzie was a classmate of Kelly's. Her solution is shown below.

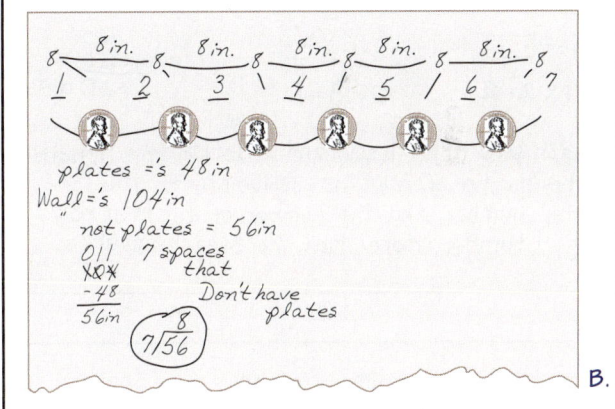

B.

Lizzie taped pennies onto a piece of paper to represent the 6 plates and determined that the plates took up 48 inches because they were each 8 inches across ("plates = 's 48 in."). She subtracted this amount from 104 to determine that the empty spaces totaled 56 in. ("not plates=56 in."). She numbered the "7 spaces that don't have plates" on the wall (before, between, and after the 6 plates), and then divided the 56 by 7 to arrive at 8 inches for each of these spaces.

Figure 6-6 *Two fourth graders attempt the plate-hanging problem.* (Source: From Hartweg, 2004/2005, pp. 280 and 282. Reprinted with permission from *Teaching Children Mathematics*, copyright 2004/5 by The National Council of Teachers of Mathematics.)

thinking. In many early learning activities, children have to identify a pattern in pictures or numbers. In problem solving, children look for patterns in more active ways—for example, by constructing a table that might help them see a pattern.

Here is a problem that appeared on the 1992 National Assessment of Educational Progress (NAEP) for mathematics:

7. A pattern of dots is shown below. At each step, more dots are added to the pattern. The number of dots added at each step is more than the number added in the previous step. The pattern continues indefinitely.

Marcy has to determine the number of dots in the 20th step, but she does not want to draw all 20 pictures and then count the dots. Explain or show how she could do this and give the answer that Marcy should get for the number of dots.

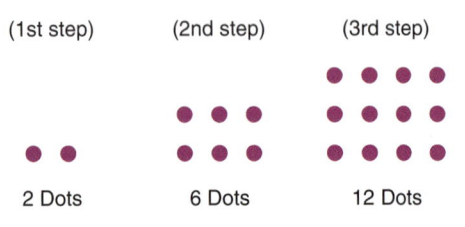

(1st step)   (2nd step)   (3rd step)

2 Dots   6 Dots   12 Dots

Figure 6-7 shows how four different children used the strategy of looking for a pattern to find the correct number of dots.

Here is another problem where finding a pattern is useful.

8. In a town of 90,000 people, one person starts a rumor by telling it to 3 other people. If each person who hears the rumor tells it to 3 new people every 15 minutes, how long would it take to spread the rumor to every person in town?

CONSTRUCT A TABLE

Organizing information into a table often helps children discover a pattern and identify missing information. Constructing a table is an efficient way to classify and order large amounts of information; also, it provides a record of what's been tried so that children need not retrace nonproductive paths or repeatedly do the same computations. The following two problems lend themselves to the strategy of constructing a table:

9. Can you make change for a quarter using exactly 9 coins? 17 coins? 8 coins? How many different ways can you make change for a quarter?

10. Your teacher agrees to let you have 1 minute of recess on the first day of school, 2 minutes on the second day, 4 minutes on the third day, and so on. How long will your recess be at the end of 2 weeks?

Figure 6-7 *Examples of strategies used for finding the correct number of Marcy's dots.* (Source: From Stylianou et al., 2000, p. 138. Reprinted with permission from *Mathematics Teaching in the Middle School*, copyright 2000 by the National Council of Teachers of Mathematics.)

The mathematical idea involved in Problem (10) can be stated in terms of other situations, and such reformulations can alter the difficulty level of the problem. Reformulation can also give children practice in recognizing similarities in the structure of different problems—an ability that appears to be closely allied to good problem-solving skills. Here is a reformulation of Problem (10):

11. Suppose someone offers you a job for 15 days. They offer you your choice of how you will be paid. You can start at 1¢ a day, get 2¢ the next day, 4¢ the next day, and continue doubling the amount every day. Or you can start at $1 the first day, get $2 the next day, $3 the next day, and continue adding $1 to the amount every day. Which would you choose? Why?

Textbooks frequently teach part of the table-construction strategy by having students read a table or complete a table that is already structured. However, it is vital that children also learn how to construct a table from scratch. They need to determine for themselves what form the table should have (e.g., how many rows and columns are needed), what the columns or rows should be labeled, and so on. To teach children these skills, you can present problems that require children to collect information and then organize it into a table in order to report it. A spreadsheet can be helpful with this task, as it would be with Problems (10) and (11). Table 6-2 shows the beginning of a spreadsheet that children could use to solve Problem (11).

GUESS AND CHECK

For years, children have been discouraged from guessing. Of course, random guessing is not good problem solving, but guessing can be a useful strategy if students incorporate what they know into their guesses—that is, if their guesses are educated guesses rather than wild guesses. Educated guesses are based on careful attention to pertinent aspects of the problem and on knowledge gained from previous work on similar problems. *The guess-and-check strategy involves making repeated educated guesses, using what has been learned from earlier guesses to make subsequent guesses better and better.* Too often, children just check a guess and, on finding that it is wrong, make another guess that may be even more off the mark. But when this happens, instead of saying you're only guessing" in a derisive tone, you must help your students learn how to refine their guesses efficiently. Consider the following problem:

12. Suppose it costs 30¢ to mail a postcard and 46¢ for a letter. Bill wrote to 12 friends and spent $4.40 for postage. How many letters and how many postcards did he send?

Suppose Roberto begins by making a guess of 6 letters and 6 postcards—total postage $4.56 (too much). You should encourage him not to just make another guess randomly; instead, you should help him understand that his next guess should be fewer letters than 6. (Why?) Maybe he guesses 4 letters, so there must be 8 postcards—total postage $4.24 (too little). Now you must make sure he sees that there have to be more letters than 4 but still fewer than 6. That is, the only possibility is 5, and when he tries 5 letters and 7 postcards, he experiences the reward of having discovered the solution after only two wrong guesses—much better than he is likely to have done with random guessing.

Here are three more problems that you could use to teach children the guess-and-check strategy:

13. A restaurant advertises twin-flavor milkshakes—that's 2 flavors of milkshake poured side-by-side into a glass. They brag that they offer 28 different twin-flavor combinations. How many different single flavors of milkshake must be available?

TABLE 6-2 • A Spreadsheet Beginning to Solve Problem (11)

Day Number	Double the Amount Each Day		Add $1 Each Day	
	Day's Pay	Total Paid	Day's Pay	Total Paid
1	.01	.01	1.00	1.00
2	.02	.03	2.00	3.00
3	.04	.07	3.00	6.00
4	.08	.15	4.00	10.00
5	.16	.31	5.00	15.00

122 Chapter 6 • Helping Children with Problem Solving

14. Maggie hit the dartboard with 4 darts. Each dart hit a different number. Her total score was 25. Which numbers might she have hit to make that score?

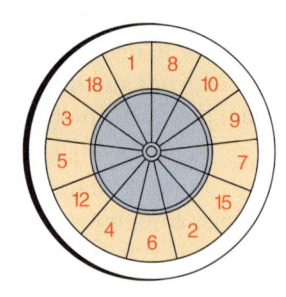

15. Use the numbers 1 through 6 to fill the 6 circles. Use each number only once. Each side of the triangle must add up to 9.

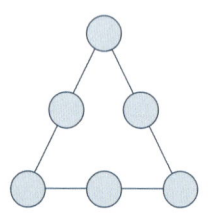

WORK BACKWARD

Children must work backward to solve a problem if the problem states a result or an endpoint and the children have to figure out the initial conditions or the beginning (many mazes are solved by working backward from the end to the beginning). Here are two problems that require children to work backward from given results:

16. Complete the following addition table:

+		3			
	12		11	15	
6	6			7	
2			5	9	
					13
5					14

17. Sue baked some cookies. She put half of them away for the next day. Then she divided the remaining cookies evenly among her two sisters and herself, so each got 4 cookies. How many cookies did she bake?

SOLVE A SIMILAR BUT SIMPLER PROBLEM

Children who know how to solve a given problem can usually solve a second problem that is somewhat similar, even if the second problem is also somewhat more difficult. The insight and understanding they gain from solving easier problems, where relationships are more apparent, carry through and let them solve harder problems. When given a problem that seems too hard, children can apply this strategy by setting the problem aside for a moment and solving a similar but simpler problem. Then they can try using the same method to solve the original problem.

Some problems are difficult just because they involve large numbers or complicated patterns, which make it hard for children to see how to solve them. Solving a similar but simpler problem first may help children figure out how to approach the original problem. In the Classroom 6–2 shows how to use this similar-but-simpler strategy to solve the following problem, which is difficult because of the large number 64:

18. Sixty-four students play in pairs in a checkers tournament. Losers are out of the tournament. Winners play until only one winner is left. How many games must be played before there is one winner left? (See In the Classroom 6–2, where this problem is used to introduce an entire lesson on solving a simpler problem.)

In the next problem, the difficulty arises from the "complicated" number 32.7. Students who are unsure how to solve this problem could try substituting simpler numbers, such as 30 miles per gallon and a 15-gallon tank; doing this might help them recognize that it makes sense to multiply the two numbers.

19. We get 32.7 miles per gallon of gas in our car. If the tank holds 14 gallons and we fill it up, about how far can we go without filling up again?

When solving a problem requires a series of steps, children may need help in recognizing that they have to answer an intermediate, "hidden" question before they can answer the final question. Answering a hidden question is another form of solving a simpler problem first, as shown in the following problem:

20. Fred bought 6 of the fish-flavored cat nibbles on sale. How much did he pay for all 6 nibbles? (Hidden question: What is the sale price of a fish-flavored cat nibble?)

SALE 3¢ Off Regular Price
CAT NIBBLES

Flavor	Regular Price for 1
Beef	8¢
Liver	5¢
Chicken	15¢
Fish	12¢

Students might find the following problem difficult because it involves so many numbers:

21. Place the numbers 1 to 19 into the 19 circles so that the center number plus any two numbers on opposite sides of the center always make the same sum.

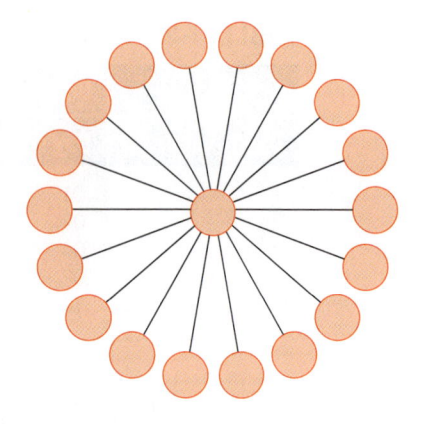

Students might tackle problem 21 by first trying simpler problems such as placing the numbers 1 to 5 or 1 to 7 or 1 to 9 in a similar pattern of circles. For example, Jeon and Bishop (2008) describe how a class of fifth graders explored the related problem of arranging sets of "9 jumping numbers" in a 9-number Ferris wheel. Each student chose his or her own "starting number" (for example, 10) and "jumping number" (for example, 4) to create a set of 9 jumping

---

In the Classroom 6–2

SOLVE A SIMILAR BUT SIMPLER PROBLEM

Objective: Solve difficult problems by first identifying and solving a related, simpler problem.

Grade Level: 4–5.

Directions to Teacher
Work with the entire class to solve the first problem together, as an example of using the "solve a similar, but simpler" problem-solving strategy. Then have children work on the later problems alone, in pairs, or in small groups.

Problem
Sixty-four people enter a checkers tournament. They play individually against each other. Losers are out of the tournament. Winners play another game against a different player, until just one winner is left. That person wins the tournament. How many games must be played until there is a winner?

Understand
How many games will be played in the first round of the tournament? (32 games, 64 people in pairs) What do you need to figure out? (the total number of games before there is a winner)

Plan
Is there a simpler problem you could solve first? (How many games would be played if there were only 2 players? 3 players? 4 players?) Can you find a pattern in the simpler problems that can be extended to the original problem?

Solve

Number of Players	Games to Play	Total Number of Games
2 (A and B)	A vs. B	1
3 (A, B, and C)	A vs. B, C vs. winner of AB	2
4 (A, B, C, and D)	A vs. B, C vs. D, winners of AB and CD play each other	3
5 (A, B, C, D, and E)	A vs. B, C vs. D, E plays winner of AB, winner of that game plays winner of CD	4
Any number		The number of games to play is always one less than the number of players

Note to teacher: Students may find it helpful to draw stick pictures of players, labeling them A, B (or even with names) and connecting with lines to show who wins and who plays whom next. Making a table is not necessary but helps in displaying the information discovered.

Look Back
How did solving a simpler problem help with solving the original problem? (The smaller numbers were easier to work with, and made it possible to see the pattern.)

Try These!
1. Jason and his 5 friends each gave one sticker to the 5 others. How many stickers were given out altogether?
2. Boxes of colored markers are labeled with their color on the top and on every side of the box except the bottom. Ten boxes are lined up against the wall on the shelf. How many labels can you see without moving any of the boxes?
3. After school at Clown Club, there were 8 kids. Every afternoon, each kid made a new funny face for each of the other clowns. They took pictures of all the faces for their clown scrapbook. How many pictures did the clowns add to their scrapbook each afternoon?

Adapted from pp. 62–63 from *Silver Burdett Ginn Mathematics, Grade* 4 by Francis Fennell et al. (1999).

numbers (in this case, 10, 14, 18, 22, 26, 30, 34, 38, 42). Then they were challenged to arrange those numbers in a 9-number Ferris wheel. The children found that exploring with a simpler set of 9 jumping numbers first (for example, 1, 2, 3, … 9—with starting number 1 and jumping number 1) helped them think about how their own jumping numbers might need to be placed in the Ferris wheel.

When children do not understand a problem, you can try asking them to restate the problem in their own words. Hearing how they restate the problem can sometimes help you identify what it is that the children do not understand. At other times, restating the problem helps the children themselves figure out what the problem is asking. As a teacher, you yourself might try restating a problem as a way of eliminating unimportant words or using words that are more easily understood. Try rewording each of the following problems so that children will understand them:

22. Find three different integers such that the sum of their reciprocals is an integer.

23. I bought some items at the store. All were the same price. I bought as many items as the number of cents in the cost of each item. My bill was $2.25. How many items did I buy?

THE IMPORTANCE OF LOOKING BACK

Some of the most important learning that results from problem solving occurs after the problem has been solved, when students look back at the problem, at the solution, and at how they found the solution. Even more important may be the learning that results from looking back at their own thought processes and at the strategies they used as they worked on the problem (Lesh and Zawojewski, 2007). In fact, research indicates that helping children discuss their own thinking with you and with each other may be one of the best ways to help them become better problem solvers. In Chapter 2, you read about the importance of metacognition—thinking about one's own thinking. You should regularly schedule looking-back time after classroom problem solving to help children develop their metacognitive abilities and thereby improve their mathematical competencies.

LOOKING BACK AT THE PROBLEM

It is important to help students generalize after solving a problem. Generalizing involves relating the problem to other problems. Every problem solver should get in the habit of asking himself or herself this important question: How is this problem similar to and different from other problems I've seen or solved before? Problems that are very different in context or detail can be very similar in their structure. Focusing on how a problem is structurally similar to other problems often results in more significant insights than does focusing on the details of the problem or on the answer.

LOOKING BACK AT THE ANSWER

When teachers tell their students "Be sure to check your work!" they usually mean that the students should go through the problem-solving process again to be sure they did not make any careless errors. But a more important type of checking is to look back at the answer to make sure it is reasonable. Problem solvers must learn to step back and ask themselves whether their answers make sense. Estimating the answer before solving the problem can help, as can thinking about whether the problem could have more than one answer. Healthy skepticism about answers is important for good problem solving.

LOOKING BACK AT THE SOLUTION PROCESS

Too often, children are given problems with only one right way to solve. Many textbook exercises are like that. However, in real-life situations, two or more approaches are often quite possible, and the same is true of most interesting mathematics problems. Moreover, approaching a problem in different ways helps students understand the problem better. For example, think about the problems we have presented in this chapter. Maybe you do not agree with how we classified a particular problem under a particular strategy—you might have used a different strategy to solve that problem. But even for the problems where you felt the classification was satisfactory, there are probably other strategies that could be used to solve them. (Try and see!)

Beyond thinking about different ways of approaching problems, it's also important for students to look back and put the solution process into perspective. They should consider what they did at each stage of the process—what facts they uncovered, what strategies they used, and what was productive and unproductive. Sometimes, you can help children look back in this way by giving them a similar problem without numbers. This lets them focus on the relationships in the problem rather than on the specific details. As they think about how they would find the answer, no matter what numbers were involved, they must focus more clearly on the solution process. You can ask different students to talk about how they proceeded, so everyone can see that there are different ways of reaching the same answer.

Other techniques you can use to help children increase their understanding of the problem-solving process include having them write about how they solved a problem and encouraging them to evaluate their solution process, to see what was successful and what was not.

LOOKING BACK AT ONE'S OWN THINKING

People who are good problem solvers are generally good at thinking about their own thinking—that is, they are good at metacognition. They monitor the skills they have and the things they already know, they think about how they can use their skills and knowledge to solve new problems, and they make judgments about what they are doing. Older children are better at these things than younger children, but by regularly teaching this type of looking back and thereby giving children experience with it, you can help all your students develop their metacognitive abilities.

An effective teaching technique is to model the sorts of questions students should ask themselves when thinking about their own thinking, both during problem solving and while looking back afterwards. You could even post a set of questions so students can refer to them—for example:

Was this problem like any I had ever seen before? If so, how were they similar and how were they different?

Was this problem easier (or harder) than I expected it to be? Why?

Did it take me more time (or less time) to solve this problem than I expected? If so, why?

What stumbling blocks did I come up against? How did I handle them? Are there other ways I could have done it?

How confident am I about my solution? Why?

HELPING ALL STUDENTS WITH PROBLEM SOLVING

As we have seen, giving students problems that are just within their reach, that challenge them to reach solutions, helps them make sense of mathematics. You must think carefully about how you organize and manage classroom instruction to ensure that all students have worthwhile problem-solving experiences. This includes paying careful attention to helping students with special needs—those with learning difficulties as well as those who are gifted in mathematics. Figuring out how to structure problem-solving sessions for a full class of diverse students will always be a challenge. Important considerations are how to manage time, classroom routines, and the needs of individual students.

MANAGING TIME

Teaching mathematics effectively through problem solving requires time (Cai, 2010). Students need time to think about the problems, to mull over the relationships, and to explore methods of finding solutions. They need to be encouraged to keep working and not to give up prematurely. It takes more time to tackle a problem that you do not know how to solve than to complete an exercise where you know how to proceed. Consider tasks (A) and (B) (p. 107) in this chapter. Didn't (B) take you a lot more time than (A)? This is because (B) was probably an actual problem for you, whereas (A) was merely an exercise. You also need time for helping students to share and compare their solutions and to learn from one another. You can gain time for problem solving by reorganizing instructional activities so that some of the time previously allotted for practicing computational and other skills is redirected toward problem solving. This makes sense because students will be using and practicing such skills as they solve problems.

MANAGING CLASSROOM ROUTINES

When you teach problem solving, you will sometimes find it useful to teach the whole class, sometimes to divide the class into small groups, and sometimes to have children work in pairs or individually.

Large-group instruction (teaching the whole class) is effective for presenting and developing a new problem-solving strategy and for examining how different strategies can be used for solving the same problem. You can focus children's attention on a problem's structure, pose questions to help them use one strategy or find one solution, lead them to use other strategies or find other solutions, and encourage them to generalize from one problem to other problems. However, if you try to have the whole class actually solve a problem together, some students will come up with an answer before others have finished considering the problem carefully; in that case, the slower children will not benefit much. Moreover, what may be an appropriately challenging problem to some students may be a trivial exercise or an impossibly difficult task for others. Discussions about problem solving are feasible with large groups, but the actual process of solving problems should be practiced in small groups as well as individually and in pairs.

Small-group instruction makes it possible to group students by problem-solving ability and interests. Students in small groups have the opportunity to work cooperatively and at an appropriate level of difficulty for everyone in the group.

Children feel less anxiety when everyone in a group is working together, discussing problems, sharing ideas, debating alternatives, and verifying solutions. Also, small groups of students can generally solve a wider range of problems than students working alone, although the groups may take longer on each problem. In addition, research indicates that when groups discuss what a problem means and how it might be solved, they achieve better results than when they are simply told how to solve it. Groups are clearly a means of promoting communication about mathematics.

Having children work in pairs is useful because they can teach each other, and this can happen no matter if you pair children of comparable abilities or of slightly different abilities. Peer teaching—with each child learning from the other—can occur in both situations.

Having children solve problems individually is also necessary, so children can progress at their own pace and use the strategies they find most comfortable. You may want to have problems available in the classroom that individual children can work on in their free time; a bulletin board, a problem corner, or a file of problems can be useful for this.

MANAGING STUDENT NEEDS

In any given classroom, students are likely to vary widely in their ability to learn mathematics through problem solving and in the amount and kind of help they need to increase that ability. There are a variety of ways in which you can allow for different ability levels by challenging different students differently during problem-solving sessions (Diezmann et al., 2003).

Compensatory strategies are specific strategies that individual students can use to deal with their own specific learning needs. These strategies can be as different as the students' needs, but certain strategies can benefit almost all students. These strategies, which are particularly important when students are expected to tackle challenging problems, include

- Using an approach that fits their way of learning. For example, students who tend to learn visually can try representing a problem and a solution method by making a drawing; other students can try talking through problems or using manipulatives.
- Using their strengths to communicate their thinking and to keep a record of how they solved a problem—for example, using actions, symbols, words, or manipulatives. Sometimes a learning buddy can help bridge the gap between what a student is able to do or say while solving a problem and what needs to be in a written record of the solution.
- Rephrasing a problem in their own words in order to understand the problem better. (As we discussed above, this is a variant of the strategy of solving a similar but simpler problem.)

For students with specific needs, it may be important to:

- Help them get started by reading the problem aloud or having another student explain the problem. Also consider posing a simpler problem to get a student started or having hint cards prepared ahead of time for students to consult if they wish.
- Help them create a journal or card file of "types of math problems I can solve."
- Allow extra time for students with abstract reasoning or reading difficulties to break down problems by making notes summarizing "what I know" and "what I need to find out." Alternatively, you or another student could help them do this. Students with these types of difficulties are more likely to be successful at understanding and solving a problem if they can break it down in this way. (It is important to recognize that helping students understand a problem more clearly should not involve telling them how to do it; rather, you should help them figure out what is expected and what approaches they can use.)

There are a variety ways you can modify your instruction to help all children with problem solving, but these modifications are especially important for students with special learning difficulties.

- Assign problems that different students can approach in different ways.
- Routinely call on individual students to rephrase problems; this will help all the students understand the problem better.
- Ask pairs or small groups of students to share their solutions with the rest of the class; then ask one student from the group (sometimes a student with special needs) to talk about what was discussed in the group as they worked on the problem.
- Allow students to write on the chalkboard or on a lap-size whiteboard, rather than on paper (especially students with vision or spatial–motor difficulties).
- Allow students to work on problems in quiet places (especially students with attention deficits).

Many students with learning disabilities are accustomed to instruction that focuses on learning through imitation, rather than learning through solving problems, so they may need time to adjust to this new approach. Your goal is to supply just the right amount of support and help, while still ensuring that each child is challenged by problems at the appropriate level of difficulty. For gifted students, you should prepare problem extensions ahead of time, so they can move on to thinking about a problem extension if they successfully complete the problem before the rest of the class. The extension might be a more general problem, a problem with larger numbers, or a problem with different constraints. For example:

24. Andrew has 20 cookies. How many will he have left over if he shares them equally (without breaking any) with himself and 1 friend? Himself and 2 friends? Himself and 3 friends? Himself and 4 friends?

25. Andrew has some cookies, but we're not sure how many. He can share his cookies equally without any left over (and without breaking any) with 1 or 2 or 3 or 4 or 5 friends. What is the smallest number of cookies that Andrew could have?

Problem (24) might be appropriate for a third-grade class just being introduced to division, and Problem (25) could be an extension for students who need more challenge.

CULTURAL CONNECTIONS

In both the United States and other countries, the pendulum of curricular change in mathematics education has swung back and forth over the years between more attention to basic skills and more attention to problem solving. Early in the twenty-first century, researchers noted that problem solving seemed to be gaining more attention and interest worldwide, in response to an anticipated demand for more workers with higher-order skills (Lesh and Zawojewski, 2007; Maclean, 2001).

Problem solving is incorporated into school mathematics instruction differently in different countries around the world. In the Trends in International Mathematics and Science Study (TIMSS), Japanese students attained some of the highest overall test scores, while American students' scores were quite a bit lower. Interestingly, the TIMSS study also gathered videotapes of actual classroom instruction in different countries, and these give us a glimpse of how Japanese and American lessons tend to be structured differently (Shimizu, 2003; Stigler and Hiebert, 1999).

American lessons videotaped for the TIMSS study generally had two phases and involved students doing many "problems" (but the problems were often really just exercises, rather than true problems):

1. The teacher demonstrated or explained how to do sample problems.
2. The students worked on numerous problems (really exercises) on their own while the teacher helped individual students who were having trouble.

By contrast, a typical Japanese mathematics lesson—which lasts 45 minutes in elementary school—consisted of four (or sometimes five) phases but typically focused on just one (or at most two) genuine problems:

1. The teacher posed a complex problem (something challenging that the students had not seen before).
2. The students worked on the problem on their own or in pairs.
3. The whole class discussed different students' approaches to the problem, orchestrated by the teacher. The teacher summed up.

Optionally, depending on the time available and on students' facility at solving the original problem, students worked on extensions of the original problem. A key difference between American and Japanese teaching is the role of the teacher while children are working on the initial problem (Shimizu, 2003). The Japanese teacher moves around the classroom not only to give guidance and assistance but also to take note of the different ways in which students are approaching the problem. The Japanese teacher is planning which students to call on during the upcoming class discussion. In contrast, the American teacher is likely to call on whichever students raise their hands, or on the students who seem likely to give a correct response, or on students whose minds are wandering. The American teacher is less likely to think about what approach the student used to solve the problem and which approaches to discuss in which order during the class discussion.

Before Japanese teachers even assign a problem, they think about what approaches students are likely to take to the problem (and what errors are likely), and they decide on the order for talking about these approaches during the class discussion. As they circulate around the room, they look for particular approaches, and they plan to call on particular students to present their thinking. For example, a teacher might first call on a student who has solved the problem using a correct but very laborious method. Then the teacher might call on students with various incorrect solutions so that the class can debate whether the answers are correct and discuss and understand common errors. Later the teacher might call on students with elegant but non-obvious solutions.

As one researcher (Shimizu, 2003) notes:

> For Japanese teachers, a lesson is regarded as a drama, which leads up to at least one climax, or *yamaba*. In fact, a central characteristic of Japanese teachers' lesson planning is their deliberate structuring of their lessons around the yamaba. Thus, when a whole-class discussion begins, students listen carefully to the solutions proposed by their classmates and present their own ideas, because during this discussion, the lesson highlights, or yamaba, appear. (p. 207)

This same researcher (Shimizu, 2003) offers five practical suggestions from Japanese teachers:

- *Suggestion 1: Label students' methods* ... during the whole-class discussion to ensure student ownership of the presented method and to make the discussion more interesting for the class.

- *Suggestion 2: Use the chalkboard effectively.* Whenever possible, record everything discussed during the lesson on the board in an organized fashion, and do not erase. This helps everyone compare different solution methods, and—at the end—offers a written bird's-eye view of the entire lesson.

- *Suggestion 3: Use the whole-class discussion to polish the students' ideas.* The Japanese word *neriage*—which translates as "polishing up"—describes what should happen during the collaborative class discussion, as everyone involved participates in pulling all the various students' ideas together into one coherent mathematical idea.
- *Suggestion 4: Choose the numbers in, and the context of, the problem carefully.* The teacher must think about the variety of responses that he/she hopes the students will present and choose the problem accordingly.
- *Suggestion 5: Consider how to encourage a variety of solution methods.* If the teacher anticipates four different approaches, but her class comes up with only two of those four, she must be ready with preplanned questions that can be asked to lead students to thinking about the problem in other ways, if these alternative approaches are important to the goal of the lesson. (p. 213)

Japanese teachers understand the importance of thinking deeply about the relationship between the mathematical content to be taught and the problems they assign, and they realize that anticipating student responses is a crucial aspect of lesson planning when teaching through problem solving.

A GLANCE AT WHERE WE'VE BEEN

Problem solving is one of the most important skills in mathematics; it should pervade the mathematics curriculum. Children need many experiences with genuine problems, not just with exercises. American students tend to have difficulty with nonroutine problems but have shown improving performance on routine problems. Problem solving helps students make sense of mathematics.

Signposts for teaching mathematics through problem solving include (1) allowing mathematics to be problematic for students, (2) focusing on the methods used to solve problems, and (3) telling students the right things at the right time. Factors that impact children's problem-solving skills include knowledge (making connections between what they know and new problems), beliefs and affects (self-confidence and a good attitude), control (the ability to monitor and direct their own thinking), and sociocultural factors (using nonschool experience to solve problems; also, the classroom atmosphere). Effective teaching through problems means choosing appropriate types of problems (problems that test the right skills and that are at the right difficulty level), finding good problems, having students pose problems, and using calculators and computers appropriately.

Children should be encouraged to use a variety of problem-solving strategies. Important strategies that can help children develop their problem-solving skills include acting out the problem, making a drawing or diagram, looking for a pattern, constructing a table, guessing and checking, working backward, and solving a similar but simpler problem. It is extremely important that children learn to look back after engaging in problem solving. They should look back at the problem to see how it is the same as and different from other problems, look back at the answer to make sure it is reasonable, look back at the solution process to assess whether they used an appropriate strategy, and, most important, look back at their own thinking, at how they thought about the problem and why.

Teachers can ensure that they help all children with problem solving, including children with special needs, by managing their time appropriately, managing the classroom routines (using large-group instruction and small-group instruction as appropriate, as well as having students work in pairs and individually), and managing student needs by using compensatory strategies to adjust instruction to the needs of individual students.

Things to Do: From What You've Read

1. Solving problems is considered a key route to learning with understanding. What is the difference between a genuine problem and a mere exercise?
2. What does teaching mathematics through problem solving mean? What signposts for teaching guide this approach?
3. Give an example of a traditional mathematical story problem, and show how you might change the problem to make it open ended.
4. Choose a content topic for a particular grade level (e.g., number, geometry, data). Make up at least one interesting problem for that topic that can be solved by each of these strategies: look for a pattern, make a drawing or diagram, and construct a table.
5. Why isn't finding the answer the final step in solving a problem? Explain what good problem solvers do after they have obtained an answer.

Things to Do: Going Beyond This Book

In the Field

1. Choose a problem from this chapter and pose it to two children. Identify the strategies each child uses. Based on what you learn about the children from

their solutions, identify a problem that you might use next with them and explain why you chose it.

2. [2]*Mathematical Processes*. Observe an elementary school classroom while children are engaged in a math lesson or investigation. Make a list of instances in which the children show evidence of using problem-solving strategies. Tell what the children were doing or saying and what tasks were involved. What role did the teacher play?

With Additional Resources

3. Start a file with problems from this chapter, and then add other problems, especially nonroutine and open-ended problems. Many math education Web sites offer long lists of interesting problems. Solve your problems and identify the strategies that you found useful. Categorize the problems in the way you find most useful.

4. Plan an interactive bulletin board focused on problem solving.

5. Read one of the chapters from Lester and Charles (2003), an edited book for teachers about teaching mathematics through problem solving. Write a brief summary and a critique.

6. Read the classic, and very humorous, children's book *Counting on Frank* (Clement, 1991). Write a lesson plan in which this book could be read aloud to grade 4–6 students as a motivator for a problem-solving lesson.

7. Read Dwyer (2003), where a teacher reports about teaching problem solving to her class of bilingual third graders. Solve the problem the children solved and write about the features of this problem that allow for multiple entry points and a variety of approaches for that diverse group of students.

8. Read Westegaard (2008), where a teacher suggests many ways that quilt blocks may be used as a context for mathematical problem solving. Identify a math concept that can be approached through problems involving quilt blocks, and develop a problem-based math lesson to engage children in investigating that concept.

With Technology

9. Identify problems in this chapter for which calculators or spreadsheets would be particularly useful. Are there problems for which using a calculator or a spreadsheet would take away all of the challenge? (These are important considerations when you are choosing problems for your class and deciding when students may and may not use calculators or spreadsheets.)

10. *Spreadsheet*. Use a spreadsheet to help solve Problems (10) and (11) of this chapter.

---

[2]Additional activities, suggestions, and questions are available in the field experience manual on the Student Companion site at www.wiley.com/college/reys

Note to Instructors: You can find additional resources, learning activities, and blackline masters in this text's accompanying Instructor's Manual, at www.wiley.com/college/reys.

Book Nook for Children

Murphy, S. J. *The Penny Pot.* New York: HarperCollins, 1998.

Jesse decides she wants her face painted at the school fair but realizes that she doesn't have enough money. At the face-painting booth, there is a penny pot. People place their extra pennies in the pot after having their face painted. Soon there are enough pennies in the pot for Jessie to have her face painted. Different combinations of money are used to let the reader see how much money Jessie needs to have her face painted. There are also activities and games in the back of the book for teachers and parents.

Zaccaro, E. *Challenge Math for the Elementary and Middle School Student.* Bellevue, IA: Hickory Grove Press, 2000.

Challenge children in grades 4–8 using some of more than 1000 problems (presented with three levels of difficulty) in areas such as algebra, astronomy, trigonometry, probability, and more.

CHAPTER 7

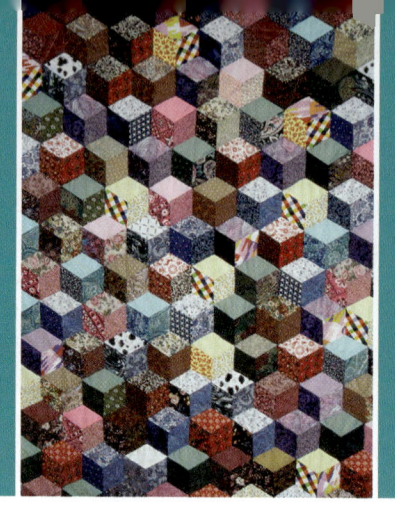

Developing Counting and Number Sense in Early Grades

> "WHENEVER YOU CAN, COUNT."
> —Sir Francis Galton, Statistician, Geneticist, and Psychologist

>> SNAPSHOT OF A LESSON

KEY IDEAS

1. Engage students in oral counting.
2. Provide a number of different contexts for counting practice.

BACKGROUND

This lesson is based on the video *Math Buddies*, from Teaching Math: A Video Library, K–4, from Annenberg Media. Produced by WGBH Educational Foundation. © 1995 WGBH. All rights reserved. To view the video, go to http://www.learner.org/resources/series32.html (video #3).

Mr. Chuck Walker teaches kindergarten at the Eissler Elementary School in Bakersfield, California. In a lesson he calls Math Buddies, Mr. Walker pairs kindergartners with sixth graders in an effort both to support the kindergartners' mathematical understanding of number and to orient them to the school environment. The lesson is composed of 20 stations, both inside and outside the classroom, where students investigate the numbers 1–50 through hands-on manipulative-based counting activities.

Mr. Walker: Now, the little buddies would like to welcome all of their big buddies to their classroom. Today, boys and girls, we're going to be exploring numbers. Numbers from 1 to 50.

Mr. Walker: [*To camera*] I use buddies to teach math to reinforce a lot of the skills that we teach in the classroom. The buddies know that they are to try to pull information from them, not to give them answers. That they are to try to help them discover and that's the whole idea.

Mr. Walker's intention is to show students how numbers occur naturally in our lives and environments. By exposing students to a variety of contexts, Mr. Walker underscores the power of numbers.

The video shows kindergartners and sixth graders working at a variety of counting stations. At one station, students glue objects onto a number grid. At another station, students count portions of snack food and place them into plastic storage bags. Outside, students count the number of times they swing back and forth on the swing set and the number of steps they take on a balance beam. In some of the instances pictured in the video, the older students intercede in the kindergartners' counting process. This is in conflict with Mr. Walker's instructions to the older children, in which they were told not to supply answers to the kindergartners, but to help pull information from the younger buddies.

Toward the end of the class, Mr. Walker calls the younger and older buddies back together to discuss the work done at the stations.

Mr. Walker: [*To camera*] I like to pull all of the children together and talk about what we have accomplished throughout the day and try to put a little bit of closure to what has happened. And I want them also to have that warmth of the buddies to make contact with one another as a group, to sing a few songs together, to bring us even closer together. And then to send them off with good feelings and knowing that they're going to return the following week and continue to build upon the foundations we've already established.

As Mr. Walker and the children sit in a circle, the teacher plays the guitar and asks students to sing along using the rules of a mathematics game.

Mr. Walker: You tell me what comes after ... 0.

Mr. Walker continues to play the guitar and asks students to sing the counting numbers that follow the numbers he sings. The students respond enthusiastically to his request, singing the succeeding counting numbers.

Once Mr. Walker has worked his way through the counting numbers to 10, he changes the rules of the game.

Mr. Walker: Alright, I'm going to get a little more difficult for you. You ready? Tell me what comes before ... 1.

Just as they had done before, the students sing counting numbers back to Mr. Walker. This time, they sing the counting number preceding Mr. Walker's number.

Mr. Walker plays the guitar while he and his class sing a song to the big buddies as they leave the classroom.

Mr. Walker: I saw today exactly what I hoped would happen. I watched both the big buddy and the little buddy working together, interacting with one another. I saw them discovering. And you look for that little spark to see if something has happened, if something has clicked.

FOCUS QUESTIONS

1. Why is classification a prerequisite for counting?
2. What is subitizing? Why is it an important early step in number sense?
3. What characteristics are associated with the different counting stages?
4. How can counting with a calculator help children develop number sense?

NUMBER SENSE

This chapter begins the second part of *Helping Children Learn Mathematics*, in that it discusses teaching strategies, techniques, and learning activities related to specific mathematical topics—in this case, beginning number sense. Number sense, like common sense, is difficult to define, but here are some characteristics. Number sense includes

- An understanding of number concepts and operations on these numbers
- The development of useful strategies for handling numbers and operations
- The facility to compute accurately and efficiently, to detect errors, and to recognize results as reasonable

- The ability and inclination to use this understanding in flexible ways to make mathematical judgments
- An expectation that numbers are useful and that work with numbers is meaningful and makes sense

These characteristics suggest that people with number sense are able to understand numbers and use them effectively in everyday living. Elementary school mathematics programs play a critical role in children's development of number sense.

These ideas regarding number sense are reminiscent of Brownell's ideas about meaningful learning (see Chapter 2). Students need to develop concepts meaningfully so that they use numbers effectively both in and out of school. Helping students develop number sense requires appropriate modeling, posing process questions, encouraging thinking about numbers, and, in general, creating a classroom environment that nurtures number sense.

Number sense is not a finite entity that a student either has or does not have. Its development is a lifelong process, and in early childhood and elementary school, number sense development involves several stages:

1. Prenumber development
   - Classification
   - Patterns
2. Early number development
   - Conservation
   - Group recognition
   - One-to-one correspondence
   - Comparisons
3. Number development
   - Connecting groups with number names, including oral and written cardinal and ordinal numbers
   - Counting forward and backward
   - Skip counting
   - Establishing benchmarks of quantities, such as 5 or 10
   - Place value

Math Links 7.1

You will find detailed information about the *Common Core State Standards for Mathematics* at http://www.corestandards.org/. It shows a learning trajectory for these key mathematical topics related to counting and place value that begin in early childhood and evolve throughout the elementary grades.

www.wiley.com/college/reys

132 Chapter 7 • Developing Counting and Number Sense in Early Grades

TABLE 7-1 • Some Early Classification and Number Expectations from the CCSSI. This summary of the CCSSM standards was made by the author of *Helping Children Learn Mathematics.* The full document (CCSSI, 2010) may be obtained at www.corestandards/the-standards/mathematics.

Classify objects and count the number of objects in each category
- Classify objects into given categories.
- Count the numbers of objects in each category.
- Sort the categories by count.

Know number names and the count sequence
- Count to 100 by ones and by tens.
- Count forward beginning from a given number within the known sequence.
- Write numbers from 0-20.
- Represent a number of objects with a written numeral 0-20.

Count to tell the number of objects
- Understand the relationship between numbers and quantities: connect counting to cardinality.
  - When counting objects, say the number names in the standard order, pairing each object with one and only one number name and each number name with one and only one object.
  - Understand that the last name said tells the number of objects counted. The number of objects is the same regardless of their arrangement or the order in which they were counted.
  - Understand that each successive name refers to a quantity that is one larger.
- Count to answer 'how many?' questions about as many as 20 things arranged in a line, a rectangular array, a circle, or as many as 10 things in a scattered configuration'.
- Given a number from 1-20, count out that many objects.

Compare numbers
- Identify whether the number of objects in one group is greater than, less than, or equal to the number of objects in another group.
- Compare two numbers between 1 and 10 presented as written numerals.

These stages form the basis of whole-number development and provide the underpinnings for developing basic facts as well as mental and written computation involving addition, subtraction, multiplication, and division of whole numbers. Computations arising from meaningful problem-solving situations will promote children's development of number sense and encourage them to reflect on the reasonableness of their results to make sure their answers make sense. Facility with whole numbers provides the foundation for work with fractions and decimals, topics that reflect the continuing development of number sense.

The *Common Core State Standards for Mathematics* (CCSSM) discussed earlier in Chapter 1 offer specific expectations for developing early number sense for young children. Table 7-1 highlights some specific expectations as young children progress in their number development.

Let's explore several examples of number sense in action. What does the number 5 mean to young children? It can mean many things. It might be their current age or their age next year. It might be how old they were when they started kindergarten. Figure 7-1 illustrates a few uses and interpretations of the number 5 suggested by young children.

Children begin to develop some sense about numbers long before they begin to count. For example, young children can answer these kinds of questions:

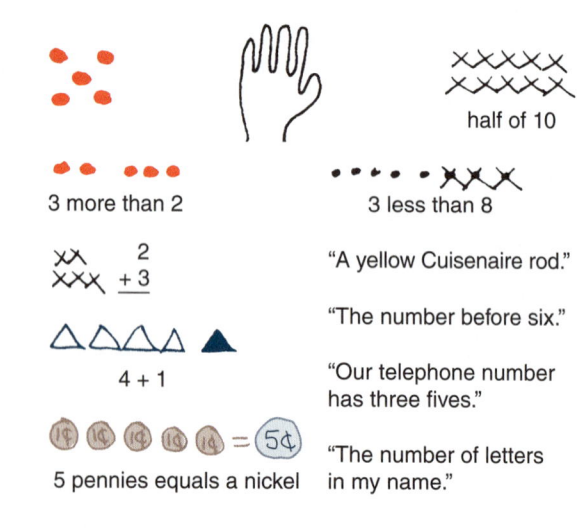

Figure 7-1 *The meanings of the number 5 suggested by young children.*

How old are you? [2]

What channel should we watch? [13]

On what floor is your doctor's office? [4]

How many sisters do you have? [1]

Such early experiences introduce the number names as well as their symbols—13 on the channel indicator or 4 on the elevator. These names and symbols are memorized through sound and sight recognition and provide an important beginning, but a child's knowledge of these concepts alone does not indicate the child's grasp of number.

For one thing, these experiences underscore a very important characteristic of number. It is an abstraction. It can't be adequately illustrated in just one situation. The multiple meanings of 5 illustrated in Figure 7-1 demonstrate how quickly the concept of *five* becomes associated with different situations. Research into how children develop number sense makes it clear that the more varied and different their experiences, the more likely it is that they will abstract number concepts from their experiences (Fuson, 2003a, 2003b; Kilpatrick et al., 2001). Helping children further their development of number and number sense has a high instructional priority. The goal of this chapter is to stimulate your thinking about number sense and its development during the early years.

PRENUMBER CONCEPTS

Numbers are everywhere, and thus even young children have a vast amount of early number experience, as shown in Figure 7-1. Many of these experiences do not rely on numbers per se but provide the basis for building early number concepts and the foundation for later skills. Such experiences are called *prenumber experiences*. As a teacher, you need to help children take advantage of them. Different steps are involved in developing prenumber concepts that lead eventually to meaningful counting skills and number sense. Prenumber work is not all done before children do anything with numbers in school; rather, it typically occurs simultaneously with activities involving number. Although the learning paths that children take are bound to differ greatly, they all begin with classifying whatever is to be counted.

CLASSIFICATION

Classification is fundamental to learning about the real world, and it can be done with or without numbers. For example, children can be separated into groups of boys and girls (which is classification) without considering number. Yet classification skills are prerequisite to any meaningful number work. If children want to know how many girls are in the class, they must be able to recognize (i.e., classify) the girls. Thus, before children can count, they must know what to count, and classification helps identify what is to be counted.

Classification is naturally integrated in other subjects, such as reading, science, and social studies (Gallenstein, 2004). For example, recognizing consonant and vowel letters from the alphabet requires classification, as does separating flowers from weeds or distinguishing between a leaf and a stem in plants. In social studies, listing the states whose names begin with the letter *M* or naming the presidents in the twenty-first century are classification activities. Young children learn to distinguish between dogs and cats, reptiles and mammals, and toys they enjoy and those they never use. These are all examples of classification in action. Classification not only helps children make some sense of things around them but also helps them become flexible thinkers. Classifying objects in different ways fosters the development of reasoning and thinking skills.

As children classify, or sort, materials, they must decide whether each object has the given characteristic. If children disagree on how an object should be classified, it forces them to defend their answers and clarify how the classification process was done. This type of argumentation is the beginning of helping children see the need for explaining their thinking. At this point, there may be no counting as materials are classified, although words such as *more*, *few*, *many*, *most*, and *none* will likely be used in describing the resulting collections.

Classification allows people to reach general agreement on what is to be counted. For example, consider a pile of buttons and the following question: How many plastic buttons have two holes? The answer is a number that tells *how many*. When a number is used in this way, it is called a *cardinal number*. Before finding the specific cardinal number, however, you must first decide which buttons are plastic and how many of them have exactly two holes. Once this classification is done, the members are well defined and ready to be counted.

Stories provide opportunities for classification. Books such as *Dave's Down to Earth Rock Shop* (Murphy, 2000) provide practice in visual discrimination as well as in classification.

Attribute blocks, sometimes called *logic blocks*, provide an excellent model for classification activities and help develop logical thinking. These attribute blocks can be made from cardboard (see Blackline Master in Appendix), but they are also commercially available in wood or plastic. They differ in several attributes, including color, shape, and size. Consider, for example, the 24 pieces shown in In the Classroom 7–1. These pieces illustrate three attributes:

Size: large, small (L, S)

Color: blue, red, green (B, R, G)

Shape: square, triangle, pentagon, circle (S, T, P, C)

The first block can be described in words as the "large blue square." Later in primary and upper elementary school, symbols, such as LBS, can be meaningfully attached to the attribute pieces, but at this stage a clear verbal description of the pieces by young children is the goal. As children manipulate the blocks and describe them, they begin to make natural connections between the concrete model and different ways of representation. Many of the 24 pieces are

134 Chapter 7 • Developing Counting and Number Sense in Early Grades

---

In the Classroom 7–1

WHO AM I?

Objective: Using attribute pieces to develop classification and reasoning.

Grade Level: K-2.

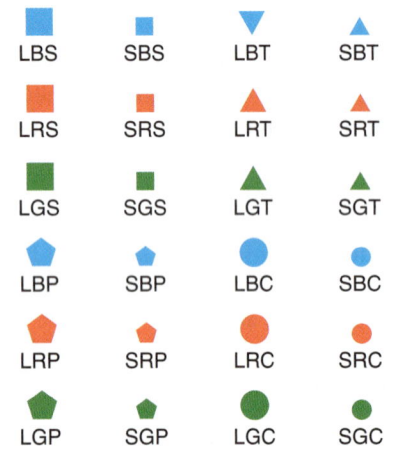

LBS	SBS	LBT	SBT
LRS	SRS	LRT	SRT
LGS	SGS	LGT	SGT
LBP	SBP	LBC	SBC
LRP	SRP	LRC	SRC
LGP	SGP	LGC	SGC

Use these attribute blocks:

A. Pick up a piece and ask:
What is its color? What is its shape? Is it large?

B. Pick up another piece.
Are there any pieces exactly like this piece?

C. Show a piece that is "almost" like this piece. Tell why.

Play who am I?

D. I have three sides. I am blue. I am large. Who am I?

E. I am green. I am a square. Who might I be?

F. I am not blue. I am not green. What is my color?

G. I am not large? I have more than 4 sides. I am green. Who am I?

Your Turn: Play "Who am I?" with a partner.

---

alike in some attributes, but no two pieces are alike in all attributes, which provides opportunities for Who Am I?" games, as presented In the Classroom 7–1. Such activities encourage children to think logically and develop communication skills. In the process, children informally explore fundamental notions, including matching, comparison, shape, sets, subsets, and disjoint sets as well as set operations.

Communication and language can be further developed as the set operations of union and intersection are encountered. The combining, or union, of *disjoint sets* (sets with no members in common) is a natural model for addition. The logical connection *or* can be used to develop the union of two or more sets. For example, the union of triangles and squares produces a set that contains all attribute blocks that are either triangles *or* squares, as shown below:

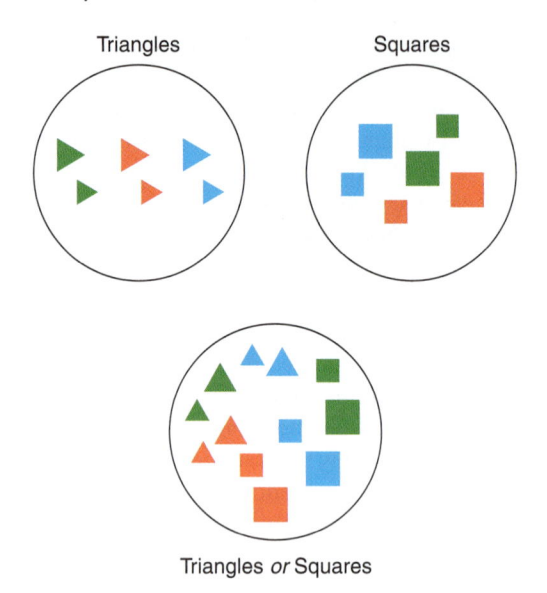

Triangles

Squares

Triangles *or* Squares

The intersection of sets can be used to explore the logical connective *and*. Using the attribute blocks, you could examine the pieces that are pentagons *and* blue. Children might place these in yarn loops as shown below:

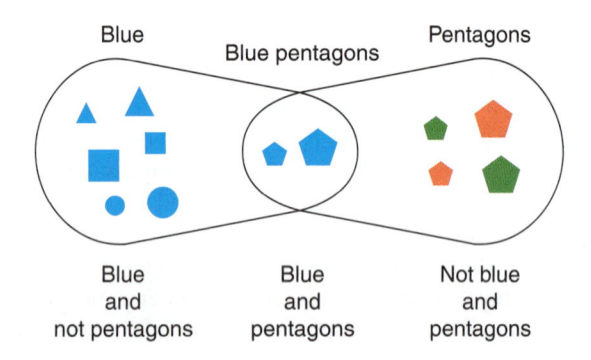

Blue | Blue pentagons | Pentagons

Blue and not pentagons | Blue and pentagons | Not blue and pentagons

This arrangement also allows children to identify other subsets, such as pieces that are blue and not pentagons. Using *not* to describe a relationship is an important step in development.

The logical connectives *and*, *or*, and *not* can be used to help children classify pieces according to their attributes. For example, the "alike-and-difference trains" shown In the Classroom 7–2 provide opportunities for students to use attribute blocks to classify and to search for patterns and use logical thinking as well. Children at all grade levels can benefit from structured activities with these materials.

Attribute blocks almost guarantee student involvement, but they also require teachers to assume an active role. When children are engaged in activities with attribute blocks, directed questions and probes can provide clues about their thinking processes. Observing children's actions reveals much about their maturity. For example, when asked to choose a piece that is blue and a triangle, one child might choose a blue piece but not a triangle. Another might select a triangle that is not blue. These responses may only reflect poor listening skills; however, additional questioning may show that the two children don't understand what the word

Prenumber Concepts 135

In the Classroom 7–2

ALIKE-AND-DIFFERENCE TRAINS

Objective: Using attribute pieces to recognize patterns and identify relationships.

Grade Level: 2–4

Each car in a train is like the car it follows in one or two ways, or it is different from the car it follows in one or two ways.

- Find the alike-and-difference pattern in each train, and describe the missing car:

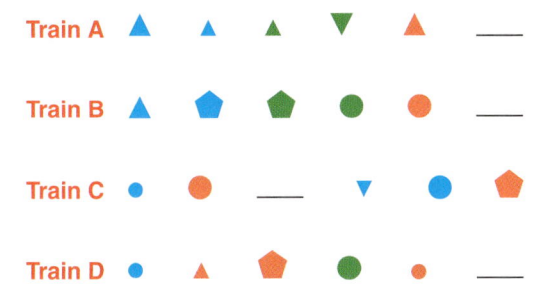

    Train A ▲ ▲ ▲ ▽ ▲ ___

    Train B ▲ ⬟ ⬟ ● ● ___

    Train C ● ● ___ ▽ ● ⬟ ___

    Train D ● ▲ ⬟ ● ● ___

Which of these cars are one-difference trains? ___
Which of these cars are two-difference trains? ___

- Your Turn:

    Begin with ■. Make a train (with at least six cars) in which each car has exactly one attribute different from the car it follows. Compare your train with someone's train. Are they the same?

    Begin with ■. Make a train (with at least six cars) in which each car has exactly one attribute the same as the car it follows. Compare your train with a classmate's train. Tell how they are alike.

---

and means or are unable to keep two different attributes in mind simultaneously. Carefully observing and questioning children as they are using these materials will help you better understand what they are thinking, which, in turn, will help you design more appropriate learning activities.

Many different experiences are needed to sharpen children's observation skills and provide them with the basis on which to build the notion of numbers. Consider another example, in which children are asked to count money:

1¢   1¢   5¢

How much money is this?
Three, if coins are counted.
Seven, if cents are counted.

This example provides a reminder that a number name alone, such as three or seven, is rarely reported. In this case, "three coins" describes both *cardinality* (i.e., how many) and what was actually counted. This example also provides another reminder that what is to be counted must be well defined or clearly understood. If there is any confusion about what is being counted, then counting discrepancies are certain to happen.

Such discrepancies occur in many different forms but are particularly troublesome with a number line. Two children standing on a number line that has been made from a roll of adding machine tape and fastened to the floor provide an example:

Barb   Scott

You can ask,

How far is it from Barb to Scott?

Is it 4? or 3? or 5?

The solution depends on what is to be counted:

Should the intervals between the dots be counted?
Should the dots be counted?
All of the dots?

The distance between Barb and Scott is determined by length, or in this case counting the intervals.

Research confirms that confusion between dots and intervals often contributes to later misunderstanding with a number line (Kloosterman and Lester, 2004). For example, about one-third of fourth graders were able to correctly answer the item on the Third International Mathematics and Science Study shown in Figure 7-2. Confusion over what should be counted is a classification problem that must be solved before counting can be meaningful. Thus classification is a very important step in developing number sense and early counting skills.

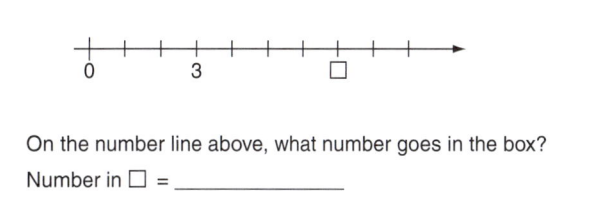

0   3   □

On the number line above, what number goes in the box?
Number in □ = _____

Figure 7-2 *Fourth-grade item from the Third International Mathematics and Science Study.*

PATTERNS

Mathematics is the study of patterns. Creating, constructing, and describing patterns require problem-solving skills and constitute an important part of mathematics learning. Patterns can be based on geometric attributes (shape, symmetry), relational attributes (sequence, function), physical attributes (color, length, number), or affective attributes (like, happiness). Sometimes patterns combine several attributes. For example, a child's list of favorite colors provides a pattern involving physical attributes (color) and affective attributes (like).

Paper, cubes, attribute blocks, pattern blocks, and other manipulatives (objects) provide opportunities for children to stack, arrange, and order objects in various ways. Number sense and mathematical exploration grow from such patterning. In the early grades, patterns help children develop number sense, ordering, counting, and sequencing. Later, patterns are helpful in developing thinking strategies for basic facts, discussed in Chapter 9, and in developing algebraic thinking in Chapter 14.

Exploring patterns requires active mental involvement and often physical involvement. The opportunities to do patterning are limitless. Here are four different ways that patterns might be used in developing mathematical ideas.

COPYING A PATTERN Children are shown a pattern and then asked to make one "just like it." The original pattern might take many different forms. For example, children might be given a string with beads and asked to make the same pattern:

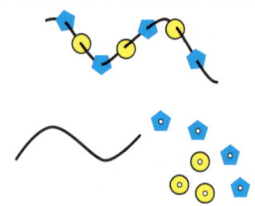

Or pattern blocks (see Blackline Master in Appendix) could be laid out for children to copy:

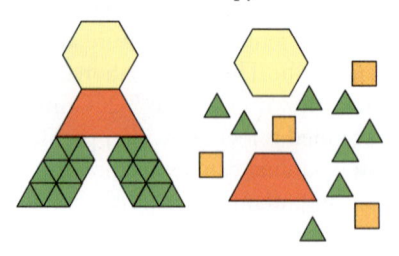

This experience requires students to choose the same pieces and arrange them in the same order. Or one could model a figure on a geoboard (see Blackline Master in Appendix) and ask children to copy the figure on an empty geoboard.

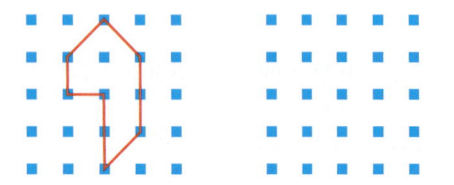

FINDING THE NEXT ONE The trains In the Classroom 7–2 illustrate problems where children "find the next one." In that case, "the next one" is the next car in the train. Consider a somewhat easier pattern suggested by "stairs" of Cuisenaire rods:

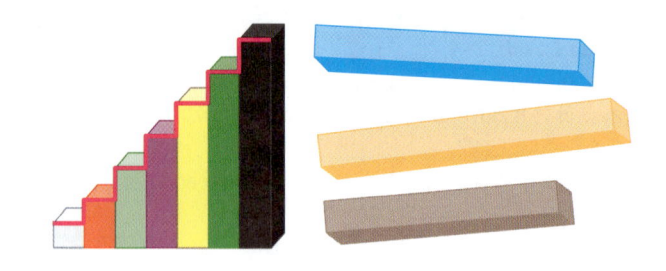

Children might be asked to find the next rod for the staircase. This find-the-next activity naturally leads to continuing or extending the pattern.

EXTENDING A PATTERN Children are shown a pattern and asked to continue it. For example, an initial pattern can be made with blocks or sticky notes, and children can be asked to continue the pattern:

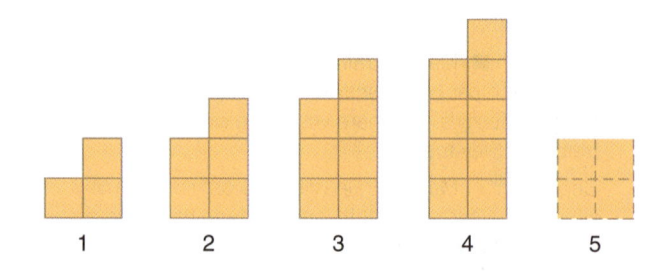

Notice how this visual pattern might serve as the foundation for exploring several important mathematical ideas. It could lead to classifying odd numbers. It could be used to observe something common about the representations—that they are all a rectangle plus one. This latter observation might lead to the algebraic generalization $2N + 1$ to describe odd numbers (See Chapters 14 & 18).

MAKING THEIR OWN PATTERNS Children need opportunities to create their own patterns and are eager to do so. Sometimes the patterns they make are highly creative and reveal insight into their mathematical thinking.

Language and communication are important elements of patterning activities. Encourage children to "think out loud" as they search for patterns. Ask them to explain why they selected a certain piece or why they did what they did. Sometimes children "see" different patterns than you anticipate. As teachers, we must try to learn and understand children's patterns and encourage them to share their thinking.

EARLY NUMBER DEVELOPMENT

Making comparisons and recognizing more and less of quantities may be done without counting, and may also provide opportunities for counting. Visual comparisons help develop the recognition of small quantities, and yet research by Piaget and others shows that for young children this number recognition may be unstable (Kilpatrick et al., 2001). This leads to important stages of number development.

CONSERVATION

The phenomenon of *conservation of number*—that a given number does not vary—reflects how children think. You need to be aware of the symptoms of the lack of conservation of number in children and its implications for early number development and counting. This idea occurs in different forms, but let's look at an example involving counting and numbers.

Two rows of blocks are arranged side by side, and a teacher and a 5- or 6-year-old child look at them together. The teacher asks the child to make a comparison to decide whether there are more orange blocks or more purple blocks, and initiates the following dialogue:

TEACHER: How many purple blocks?
STUDENT: [*counting them*] Nine.
TEACHER: How many orange blocks?
STUDENT: [*counting again*] Nine.
TEACHER: Are there more purple blocks or orange blocks?
STUDENT: They are the same.

Now the teacher spreads out the orange blocks as follows:

TEACHER: How many purple blocks now?
STUDENT: Nine.
TEACHER: How many orange blocks now?
STUDENT: Nine.
TEACHER: Are there more purple blocks or orange blocks?
STUDENT: More orange blocks.
TEACHER: I thought you said there were nine purple and nine orange.
STUDENT: I did, but this nine [*pointing to row of orange blocks*] is bigger.

This example illustrates a typical case where a young child thinks a number varies and depends on its arrangement or configuration. Here the child believes that stretching out the row of blocks changes the number of blocks and verbalizes this by saying "one nine is bigger than another nine" and sees no flaw in his or her reasoning. Thus, at this stage children believe the number in a quantity may change depending on its arrangement.

Look at the following grouping of marbles.

Some children count six marbles in each of these groups but report one group has more. For adults it seems inconceivable that "this six" could be more than "that six," but as this and other examples in this book show, children's logic and adults' logic can be very different.

Conservation was described by Jean Piaget and has been the subject of much research. Rarely do children conserve number before 5 or 6 years of age. Children up to this age don't realize that moving the objects in a set has no effect on the number of the objects. Thus, many children in kindergarten and some in first and second grade are nonconservers. A child can be adept at counting and remain naive about conservation. Whenever this happens, instructional activities should be used to increase the child's awareness of the invariance of number.

GROUP RECOGNITION

The patterns encountered in classifying and making comparisons provide many number-sense experiences. In fact, before actually counting, children are aware of small numbers of things: one nose, two hands, three wheels on a tricycle. Research shows that most children entering school can identify quantities of three things or less by inspection alone without the use of counting techniques (Clements, 1999).

The skill to "instantly see how many" in a group is called *subitizing*, from a Latin word meaning "suddenly." It is an important skill to develop. In fact, one instructional goal for first-grade students is to develop immediate recognition of small groups. Sight recognition of quantities up to five or six is important for several reasons:

1. *It saves time.* Recognizing the number in a small group is much faster than counting each individual member of that group.
2. *It is the forerunner of some powerful number ideas.* Children who can name small groups give evidence of knowing early order relations, such as 3 is more than 2 and 3 is 1 less than 4. Some may also realize that 5 contains a group of 2 and a group of 3.
3. *It helps develop more sophisticated counting skills.* Children who recognize the number in a small group will more quickly begin counting from that point.
4. *It accelerates the development of addition and subtraction.* Being able to recognize the quantity in a small group frees children of the burden of counting small quantities to be joined or removed and allows them to concentrate on the action of the operation.

Sight recognition is evidenced by children's skills in reading the number of dots on the face of a die or on a domino. In fact, both of these materials provide natural as well as interesting models for developing and practicing this skill. As children grow older, their ability to recognize quantities continues to improve, but it is still limited (Glanfield, 2010). Certain arrangements are more easily recognized or subitized. For example, look at these arrangements:

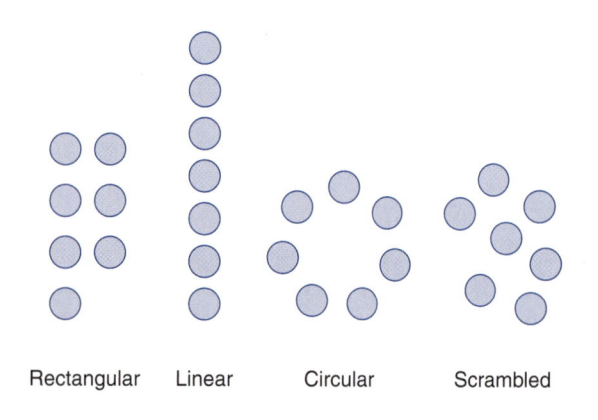

Rectangular   Linear   Circular   Scrambled

Children usually find rectangular arrangements easiest, followed by linear, and then circular, whereas scrambled arrangements are usually the most difficult. If the arrangement does not lend itself to some grouping, people of any age have more difficulty with larger quantities. Few adults can recognize by inspection groups of more than 6 or 8, and even these groups must be in common patterns such as those found on playing cards or dominoes. For example, look at these pictures of birds:

How many birds do you see? Each picture shows 12 birds, but you probably used different processes to count them. The picture at the left provides no clear groups, so you could either count every bird or perhaps immediately identify the numbers in some subset of the birds and then count the rest. In the other two pictures, some natural groupings are suggested: four groups of 3 and two groups of 6. Small-group recognition or subitizing is a powerful ally in counting larger groups. Research supports the notion that subitizing is a prerequisite for counting (Clements & Sarama, 2007).

COMPARISONS AND ONE-TO-ONE CORRESPONDENCE

Comparison of quantities is another important part of learning to count and is also essential in developing number awareness. Comparisons are plentiful in classrooms as children use materials. Teacher-led activities frequently provide opportunities for comparisons, with questions such as:

Does everyone have a piece of paper?

Are there more pencils or desks?

These questions either directly or indirectly involve comparisons that may lead to the important and powerful mathematical notion of one-to-one correspondence. Look at Figure 7-3a and consider this question: Are there more hearts or gingerbread cookies? Counting provides a solution, particularly with the cookies scattered on a plate; however, if the cookies are arranged in an orderly fashion (Figure 7-3b), you can make direct comparisons and answer the question without counting. Sometimes placing connectors (laying string or yarn, drawing lines or arrows) provides a visual reminder of the one-to-one correspondence that underlies many comparisons, as in Figure 7-3c. In this case, the one-to-one correspondence confirms that there is one more heart than cookies. The process of recording tallies to keep track of the number of objects (see Chapter 17) rests on the notion of one-to-one correspondence.

When making comparisons, students must be able to discriminate between important and irrelevant attributes. In Figure 7-4a, for example, who has more leaves—Bonnie or Sammy? The leaves are very different; their sizes, shapes, and colors vary. Still, the procedure for setting up a correspondence is the same.

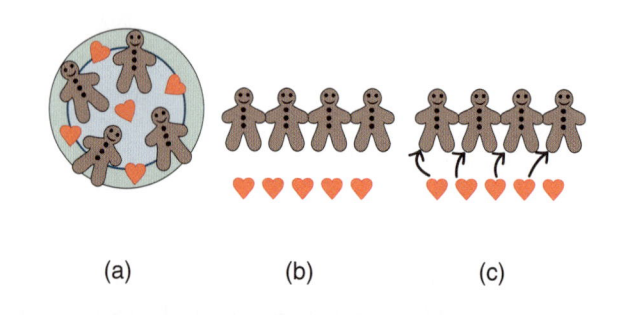

(a)   (b)   (c)

Figure 7-3 *Models for making comparisons by (a) counting, (b) physically comparing without counting, and (c) one-to-one correspondence.*

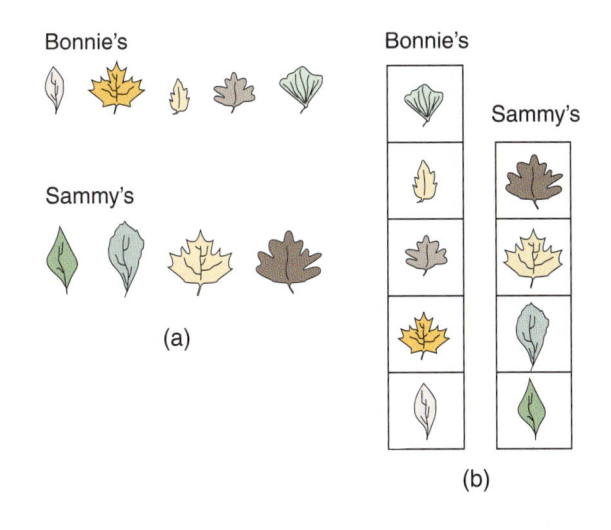

Figure 7-4 *A framework for comparison that facilitates a one-to-one correspondence.*

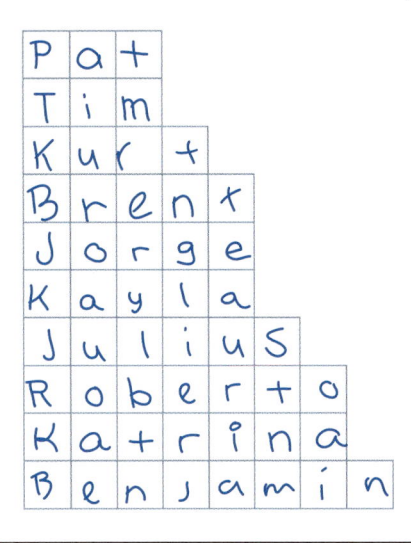

Figure 7-5 *Classification of children's names on a graph for comparison.*

To ensure that members of two sets are arranged in an orderly fashion for comparison, a method that is sometimes helpful involves using pieces of square paper or index cards. In this case, placing each leaf on a card and then stacking the cards on a common base, as in Figure 7-4b, provides a helpful framework. This method provides a graphical representation of the information, allowing quick and accurate visual comparisons.

Several different but equally valid verbal descriptions may be used for the example given in Figure 7-4:

Bonnie has more leaves than Sammy.

Sammy has fewer leaves than Bonnie.

Children need to become familiar with descriptions of relationships such as *more than, less (fewer) than,* and *as many as.* A grasp of these terms can be followed by more explicit characterization:

Bonnie has one more leaf than Sammy.

Sammy has one less leaf than Bonnie.

In these cases, the notion of order and succession are being developed. Children must come to realize that 4 is the number between 5 and 3 as well as 1 more than 3 and 1 less than 5. Understanding such relationships evolves naturally to comparisons, such as $5 > 4$ (five is greater than four) or $4 < 5$ (four is less than five). Representing and describing relationships among numbers in different ways is an important step in sense making.

When comparisons are made among several different things, ordering is involved. For example, children can print their first names on some grid paper:

Then they can physically compare their names with others' names to answer questions such as these:

Who has the longest name?

Who has the shortest name?

Can you find someone with a name the same length as yours?

Can you find someone whose name has one more (fewer) letter than your name?

Ordering often requires several comparisons, and graphs help organize information. The graph in Figure 7-5 was constructed by classifying children's names according to length. It summarizes much information and presents it in an organized form. The graph could be used to answer the previous questions and additional questions such as these:

Which length name is most popular?

Greg wasn't here today. Where should his name go on our graph?

Can you think of anyone you know who has a shorter name than Tim?

As more things are ordered, the ordering process becomes more complicated, and most children need some guidance to be able to order things efficiently. That's why organizational techniques, such as graphing (see Chapter 17), are particularly helpful and will contribute to the early development of numbers.

COUNTING

Patterns facilitate the counting process; however, there are no sound patterns within the first 12 number names. Children learn these number names by imitating adults and older children. As young children practice counting, they often say nonconventional sequences of number names.

140   Chapter 7 • Developing Counting and Number Sense in Early Grades

It is not unusual to hear a young child count "one, two, five, eight, fifteen, twenty, six, hundred." This counting may sound strange, but it is perfectly natural. It reflects the child's struggle to remember both the number names and their order, both of which are necessary in order to connect counting to cardinality as shown in Table 7-1.

Eventually children may count apples, blocks, cards, rocks, stones, twigs, even petals on a flower. Try counting the petals on the flowers shown in Figure 7-6. They provide a very interesting setting for practicing counting and a reminder that numbers are everywhere in nature.

Items such as blocks or petals are *discrete objects* (i.e., materials that lend themselves well to handling and counting). *Continuous quantities*, such as the amount of water in a glass or the weight of a person, are measured rather than counted.

What is counting? It is a surprisingly intricate process by which children call number values by name. A close look at the counting process shows that finding how many objects are present involves two distinct actions. A child must say the number-name series, beginning with one, and point to a different object as each number name is spoken. Children exhibit several different but distinct stages of counting, which we now discuss.

COUNTING PRINCIPLES

How do children count? Look at an actual counting situation. Suppose seven shells are to be counted. A child who is what is called a "rational counter" says each number name as the shells are counted, as indicated in Figure 7-7. A rational counter also realizes that the last number named, "seven," reports the total, or the cardinality of the set of shells being counted.

As adults, you probably cannot recall your own struggle with counting. Yet, observing young children can remind you how counting strategies vary and are developed sequentially over a period of years (Fuson, 2003b, 2003a; Schwerdtfeger & Chan, 2007). Here are four important principles on which the counting process rests:

1. *Each object to be counted must be assigned one and only one number name.* As shown in Figure 7-8, a *one-to-one* correspondence between each shell and the number name was established.

2. *The number-name list must be used in a fixed order every time a group of objects is counted.* The child in the figure started with "one" and counted "two, three, ..., seven," in a specific order. This is known as the *stable-order rule*.

3. *The order in which the objects are counted doesn't matter.* This is known as the *order-irrelevance rule*. Thus, the child can start with any object and count them in any order.

Figure 7-6 *Models from nature for counting practice.*
(Source: Top photo, PhotoDisc/Getty Images; bottom photo, Digital Vision/Getty Images.)

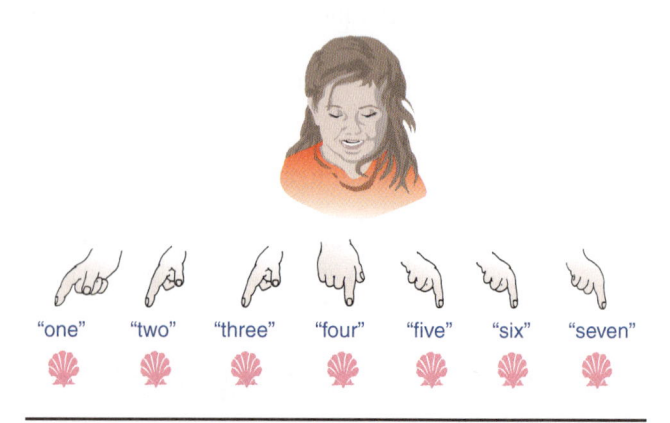

Figure 7-7 *Rational counting: Correct sequence, with correct correspondence.*

(a) **Incorrect sequence, correct correspondence**

"one two three five nine ten seven twenty eight"

(b) **Correct sequence, incorrect correspondence (counts too fast)**

"one two three four five six seven eight nine ten eleven twelve thirteen ..."

(c) **Correct sequence, incorrect correspondence (points too fast)**

"one two three four five ..."

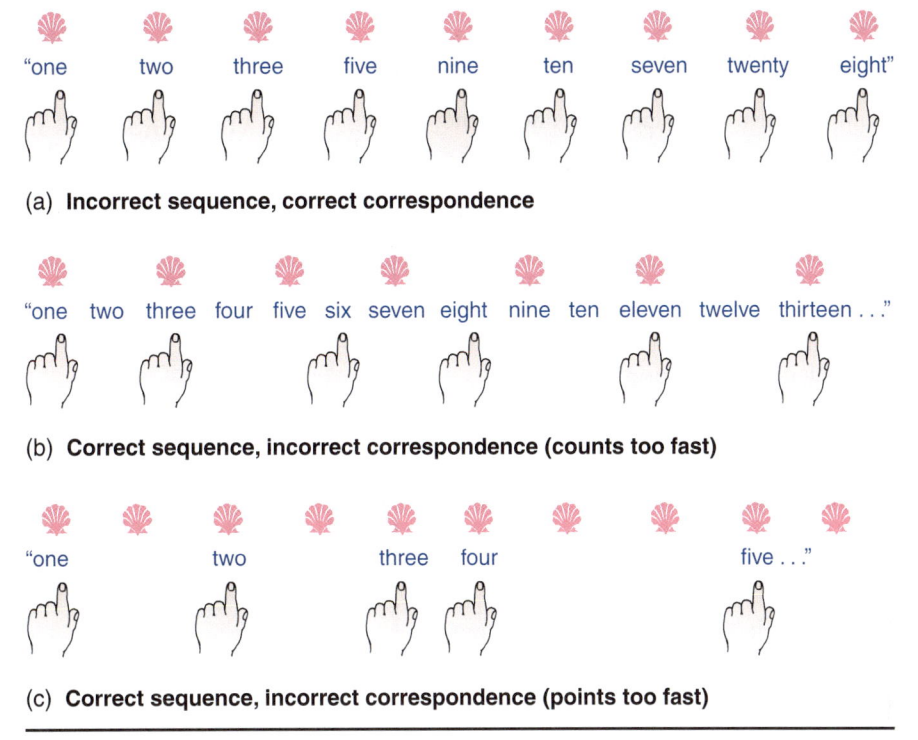

Figure 7-8 *Rote counting errors.*

4. *The last number name used gives the number of objects.* This principle is a statement of the *cardinality rule*, which connects counting with *how many*. Regardless of which block is counted first or the order in which they are counted, the last block named always tells the cardinal number of the objects being counted.

Children reflecting these counting principles may show confusion or uncertainty when counting, yet research shows that with encouragement and opportunities to count, young children will develop efficient counting strategies without direct instruction (Clements & Sarama, 2007). Knowing these principles will help you recognize the levels of children's counting skills. Careful observation of children, coupled with a good understanding of these principles, can pinpoint counting errors. Once the trouble is diagnosed, instruction can focus on the specific problem.

COUNTING STAGES

There are several identifiable counting stages, and each reflects one or more of the counting principles. For example, some children may count the objects correctly and still not know how many objects have been counted. In response to the question "How many shells are on the table?" a child might correctly count "one, two, three, four, five, six, seven," as shown in Figure 7-7, and yet be unable to answer the question. This child does not realize that the last number named tells how many, which is one of the expectations for the CCSSM shown in Table 7-1.

The shells in Figure 7-8 are shown in a linear arrangement because this configuration is easier to count than objects in a rectangular or scattered formulation. Sometimes it is helpful to encourage children to organize their objects prior to counting to decrease the likelihood of counting errors that result in double counting or omitting objects.

ROTE COUNTING A child using rote counting knows some number names but not necessarily the proper sequence, as shown in Figure 7-8a. In this case, the child provides number names, but these names are not in correct counting sequence. These children may "count" the same objects several times and use a different counting sequence each time. Children who exhibit this error need to spend more time on the *stable-order rule*.

Rote counters may know the proper counting sequence, but they may not always be able to maintain a correct correspondence between the objects being counted and the number names. Figure 7-8b shows an example in which the rote counter is saying the number names faster than pointing, so that number names are not coordinated with the shells being counted. Rote counters may say the number names until they perceive all the objects as being counted. It is also possible that the rote counter points faster than saying the words, as illustrated in Figure 7-8c. This rote counter is pointing to the objects but is not providing a name for each of them. Asking children to slow down their counting and stressing the importance of one-to-one correspondence helps children with these types of rote counting errors.

It is important that children demonstrate all four counting principles. Otherwise, children may not have their number names in the proper sequence, or they may not consistently provide a number name for each object being

142 Chapter 7 • Developing Counting and Number Sense in Early Grades

counted. A one-to-one correspondence may not be shown, which is a critical distinction between rote and rational counting. Using a one-to-one correspondence in counting represents significant progress and establishes one of the prerequisites to rational counting.

RATIONAL COUNTING   In rational counting, the child gives a correct number name as objects are counted in succession; however, in rational counting the child not only uses one-to-one correspondence but also is able to answer the question about the number of objects being counted. Rational counters exhibit all four counting principles.

Rational counting is an important skill for every primary-grade child. Children notice their own progress in developing this skill and become proud of their accomplishments. Early in kindergarten some children will count to 10, others to 20, some to 50, and a few to over 100. The *Common Core State Standards* in Table 7-1 sets a goal of rational counting to 100 by ones and tens for children by the end of kindergarten. Instruction should provide regular practice and encourage each child to count as far as he or she can.

COUNTING STRATEGIES

Once mastery of rational counting to 10 or 20 has been reached, more efficient and sophisticated counting strategies should be encouraged.

COUNTING ON   In counting on, the child gives correct number names as counting proceeds and can start at any number and begin counting. For example, a child may count a pile of 8 pennies and when 3 more pennies are added to the pile, count "nine, ten, eleven"; or begin with 78 pennies and count "seventy-nine, eighty, eighty-one"; or begin with 98 pennies and count "ninety-nine, one hundred, one hundred one." Counting on is preceded by counting all, where a child, when the 3 pennies are added to the 8 pennies that had already been counted, would count all of them again by beginning to count from one and then end with eleven. Counting-on practice leads children to discover valuable patterns. Counting on is also an essential strategy for developing addition.

Being able to count on or back requires children to recognize the starting number and the nested inclusion of previous numbers. Thus, if we start counting on from 8, a child must know the numbers that come immediately before and after 8 as well as the sequence of numbers preceding or "nested" within the number 8. Research suggests that even children reflecting all the counting principles show confusion or uncertainty about the nested inclusion of previous numbers (Fuson, 2003a, 2003b).

COUNTING BACK   When children count back, they give correct number names as they count backward from a particular point. For example, to count back to solve the problem, "Bobbie had 22 rabbits and 3 were lost," a child might

---

In the Classroom 7–3

NUMBERS ON A CALENDAR

Objective: Using a calendar to count forward and backward.

Grade Level: 1–2

• Look at the calendar pictured below.

SUNDAY	MONDAY	TUESDAY	WEDNESDAY	THURSDAY	FRIDAY	SATURDAY
		1	2	3	4	5
6	7	8	9	10	11	12
13	14	15	16	17	18	19
20	21	22	23	24	25	26
27	28	29	30	31		

1. A calendar has many patterns. Tell a pattern you see.
2. What day and date is 7 days after the 7th?
3. What day and date is 1 week after the 5th?
4. If you start on Monday and count on 7 days, what is the day?
5. If you start on Sunday and count back 4 days, what is the day?
6. What day and date is 5 days before the 30th?
7. What day and date is 5 days after the 21st?
8. Count on 7 days after the 11th. What is the date? The day?
9. Count back 7 days from the 27th. What is the date? The day?
10. Make up some counting problems on your own that use the calendar. Give them to a friend to solve.

---

count "twenty-one, twenty, nineteen" and conclude there were 19 left. At an early stage, counting back can be related to rockets blasting off (counting down—five, four, three, two, one, blast off); later, it becomes helpful in developing subtraction.

Instruction in counting should include practice counting backward as well as forward. Counting backward—"Five, four, three, two, one"—helps children establish sequences and relate each number to another in a different way. In the Classroom 7–3 provides an activity that uses a calendar to practice counting forward and backward.

As shown earlier in Figure 7-2, children need to understand what is to be counted on the number line. Once the number line is clearly understood, the number line is a powerful model for counting on from any starting point A and counting to the right to B

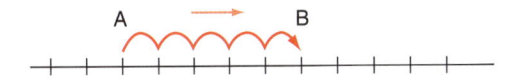

and counting back to the left starting from B and counting to A.

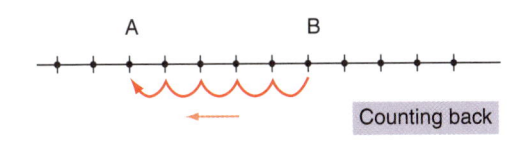

Just as counting on models addition, counting back models subtraction. For example, use the counting line below to start at B on the number line and count 5 steps back. This action models $7 - 5 = 2$.

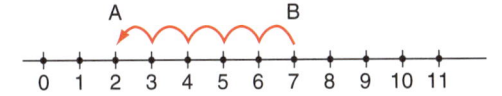

If the counting line is replaced with the number line to the right, then counting back 5 from B would model $3 - 5 = -2$ and illustrates how the number line provides a natural entry into integers.

In the Classroom 7–4

COUNTING ON ... AND ON ... AND ON

Objective: Using calculators to count and develop number sense.

Grade Level: 1–3

- Use your [calculator] to count.

  Enter $1 + 1$.

  Now press $= = = = \ldots$ and count as long as you want.

- Time yourself:

  How long did it take to count from 1 to 100? ___

  Guess how long it will take to count from 100 to 200 by 1s? ___

  How long did it take to count from 100 to 200 by 1s? ___

  Guess how long it will take to count from 1 to 1000 by 1s? ___

In the Classroom 7–5

COUNTING ON ... AND BACK

Objective: Using calculators to count and develop number sense.

Grade Level: 1–3

- Enter $1 + 1$ in your [calculator], Then press $= = = = \ldots$

Stop when you get to 8. Predict the next number. Count to 15. Cover your display. Press the equal button 4 times. What number do you think is hidden? Check your display to see if your are right.

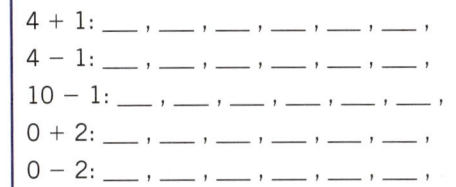

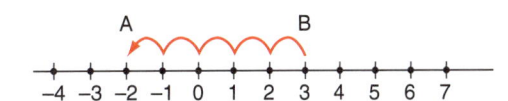

$4 + 1$: ___ , ___ , ___ , ___ , ___ , ___ ,

$4 - 1$: ___ , ___ , ___ , ___ , ___ , ___ ,

$10 - 1$: ___ , ___ , ___ , ___ , ___ , ___ ,

$0 + 2$: ___ , ___ , ___ , ___ , ___ , ___ ,

$0 - 2$: ___ , ___ , ___ , ___ , ___ , ___ ,

A number line never ends, so if a different part of the number line is shown, then counting back 5 from B would model a different computation. For example, in the number line below B denotes -1, so $-1 - 5$ is -6.

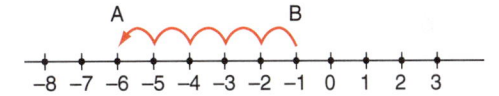

This demonstrates the importance of labeling a number line accurately and reminding children that since we can't show all the numbers, we are only looking at a piece of a number line. The piece of the number line shown is usually chosen to focus on certain numbers.

Many children find it difficult to count backward, just as many adults find it difficult to recite the alphabet backward. The calculator provides a valuable instructional tool to help children improve their ability to count backward. Children are often surprised to learn that it is as easy to count backward on a calculator as it is to count forward.

In the Classroom 7–4 uses the calculator to count forward and back, and In the Classroom 7–5 shows how counting backward with a calculator introduces negative integers.

You might ask: Why introduce zero and negative numbers when counting? Since both counting on and back are important skills, children will encounter zero and negative numbers, so it makes sense to introduce them. Furthermore, most young children already have an intuitive knowledge of nonpositive numbers, and counting back on a calculator

In the Classroom 7–6

HUNTING FOR NUMBERS

Objective: Using a hundred chart to identify patterns and develop number sense.

Grade Level: 1–2

- Look at this chart:

1	2	3	4	5	6	7	8	9	10
11	12	13	14	15	16	17	18	19	20
21	22	23	24	25	26	27	28	29	30
31	32	33	34	35	36	37	38	39	40
41	42	■	44	45	46	47	48	49	50
51	52	53	54	55	56	57	58	59	60
61	62	63	64	65	66	67	68	69	70
71	72	73	74	75	76	77	78	79	80
81	82	83	84	85	86	87	88	89	90
91	92	93	94	95	96	97	98	99	100

- What number is hidden by the ■?
- What number is after the ■?
- What number is before the ■?
- Put a ● on any number.
    - Begin at ●: Count forward five.
    - Begin at ● again: Count backward five.
- Put a ▲ on a different number.
    - Begin at ▲: Count forward ten.
    - Begin at ▲ again: Count backward ten.
- Tell about any patterns you see.

In the Classroom 7–7

SKIP COUNTING

Objective: Using a hundred chart to skip count and identify resulting patterns.

Grade Level: 2–3

- Use your calculator and a hundred chart to count.

1	2	3	4	5	6	7	8	9	10
11	12	13	14	15	16	17	18	19	20
21	22	23	24	25	26	27	28	29	30
31	32	33	34	35	36	37	38	39	40
41	42	43	44	45	46	47	48	49	50
51	52	53	54	55	56	57	58	59	60
61	62	63	64	65	66	67	68	69	70
71	72	73	74	75	76	77	78	79	80
81	82	83	84	85	86	87	88	89	90
91	92	93	94	95	96	97	98	99	100

- Start at 3 and count by 3s.
- Circle every number you counted.
- Describe a pattern.

- Which of these numbers would be counted?
  13, 51, 61, 62, 63, 100, 113

Math Links 7.2

Here are two excellent Web sites that provide opportunities for children to make patterns, practice counting, and develop a better sense about numbers:

Go to the Math Forum Web site at http://www.mathforum.org/mathtools and then go to Math Topics. Under Kindergarten, Math 1 and 2, you will find a number of K – 2 virtual resources related to early number development, including those designed to help children practice counting, such as Sort and Classify Counting, Patterns, and Writing Numbers.

Go to the NCTM Illuminations Web site at http://www.illuminations.nctm.org. Click on Activities and you will find a wide range of tools, several of which are related to early number development. Among those tools are Five-Frame and Bobbie Bear-Counting Strategies.

You can access both of these sites from this book's Web site.

www.wiley.com/college/reys

---

provides one tool for representing them on a number line. Other models include thermometers, losses and gains, pebbles in a bag, and elevators. Research suggests that young students are capable of understanding negative numbers far earlier than was once thought (Kilpatrick et al., 2001).

SKIP COUNTING In skip counting, the child gives correct names, but instead of counting by 1s, counts by 2s, 5s, 10s, or other values. The starting point and direction are optional. In addition to providing many patterns, skip counting on the hundred chart as shown In the Classroom 7–6 and 7–7 provides counting practice and provides readiness for multiplication and division.

Different sports provide opportunities for skip counting, such as by 2s and 3s in basketball, and by 2s, 3s, 6s, 7s, and 8s in football. Skip counting, coupled with counting on and counting back, provides excellent preparation for counting change in a monetary transaction. Thus, given these coins

(Source: © Daniel R. Burch/iStockphoto)

children would be encouraged to choose the largest-valued coin and then begin counting on—"twenty-five, thirty, thirty-five, forty." Such counting experiences help children make change (Chandler & Kamii, 2009).

Counting change is a very important skill whose usefulness children recognize. It holds great appeal for them. It should be introduced and extended as far as possible in the primary grades. As teachers, we should take advantage of every opportunity to encourage accurate and rapid skip counting.

COUNTING PRACTICE

Counting practice should include counting on and counting back. Books such as *Anno's Counting Book* by Mitsumasa Anno and *I Can Count the Petals of a Flower* by John and Stacy Wahl provide a variety of rich and stimulating contexts for counting. These books engage children in counting and provide insight into how children count. Other books, such as *How Many Snails? A Counting Book* by Paul Giganti and *Count on Your Fingers African Style* by Claudia Zaslavsky, focus on numerical relationships. Such books are useful for counting-based discussions between children and adults.

Research has shown predictable trouble spots for children when counting. For example, children often slow down, hesitate, or stop when they reach certain numbers, such as 29; however, as soon as they establish the next number as 30, their counting pace quickens, until they are ready to enter the next decade (set of 10 numbers). Bridging the next century (set of 100 numbers) poses a similar challenge. As children count ". . . one hundred ninety-eight, one hundred ninety-nine," they may pause and be uncertain how to name the next number. Bridging to the next 10 or 100 is among the common transitional points of counting difficulty:

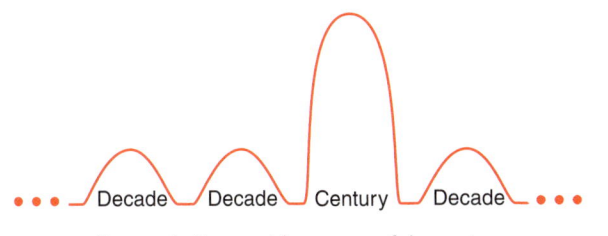

Bumps in the road for successful counters

Calculators are beneficial tools to illustrate how numbers move in a repetitive pattern and how transitions are made across the decades, 39–40, 69–70, and beyond 99–100, 109–110, and 199–200. Not only is the calculator a valuable instructional tool that helps improve children's ability to count, but it is also a powerful counting tool that they love to explore. Early counting with the calculator should emphasize the physical link between pressing the keys and watching the display. Because the display changes constantly, the students begin to recognize patterns. Calculator counting involves a physical activity (pressing a key each time a number is counted) through which students can relate the size of a number to the amount of time needed to count it.

Children are usually surprised to learn it is as easy to count by any number on the calculator as to count by 1s. A calculator can start at zero and count by 1s (In the Classroom 7–4). Calculators can also begin at any starting point and skip count either forward or backward by any number (In the Classroom 7–5).

Calculator counting opens exciting mathematical explorations and promotes both critical thinking and problem solving. According to Huinker (2002), "using calculators as learning tools can empower young children with the capacity to investigate number ideas in ways that were previously inaccessible to them" (p. 316).

DEVELOPING NUMBER BENCHMARKS

Number benchmarks are perceptual anchors that become internalized from many concrete experiences, often accumulated over many years. For example, the numbers 5 and 10 (the number of fingers on one and two hands) provide two early number benchmarks. Children recognize four fingers as being one less than five and eight as being three more than five or two less than ten:

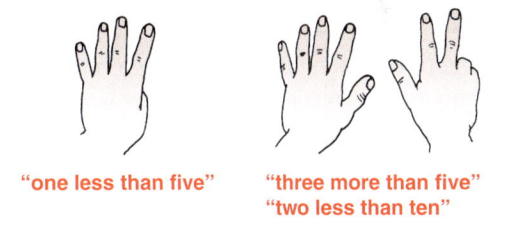

"one less than five"   "three more than five"
                        "two less than ten"

The five-frame (5 × 1 array) and the ten-frame (5 × 2 array) (Blackline Master in Appendix) use these early benchmarks:

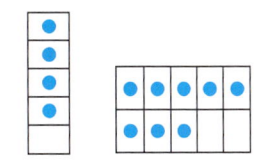

In Japan, the early benchmarks of 5 and 10 are later used with the ten-frame and the Japanese *soroban* (which is similar to an abacus) to promote counting, quick recognition of quantities, and mental computation (Shigematsu, Iwasaki, and Koyama, 1994).

Figure 7-9 shows some of the connections that might be constructed as children examine different representations on the ten-frame. These relationships encourage children to think flexibly about numbers, thereby promoting

Four
- Double 2
- Half of 8
- 1 less than 5

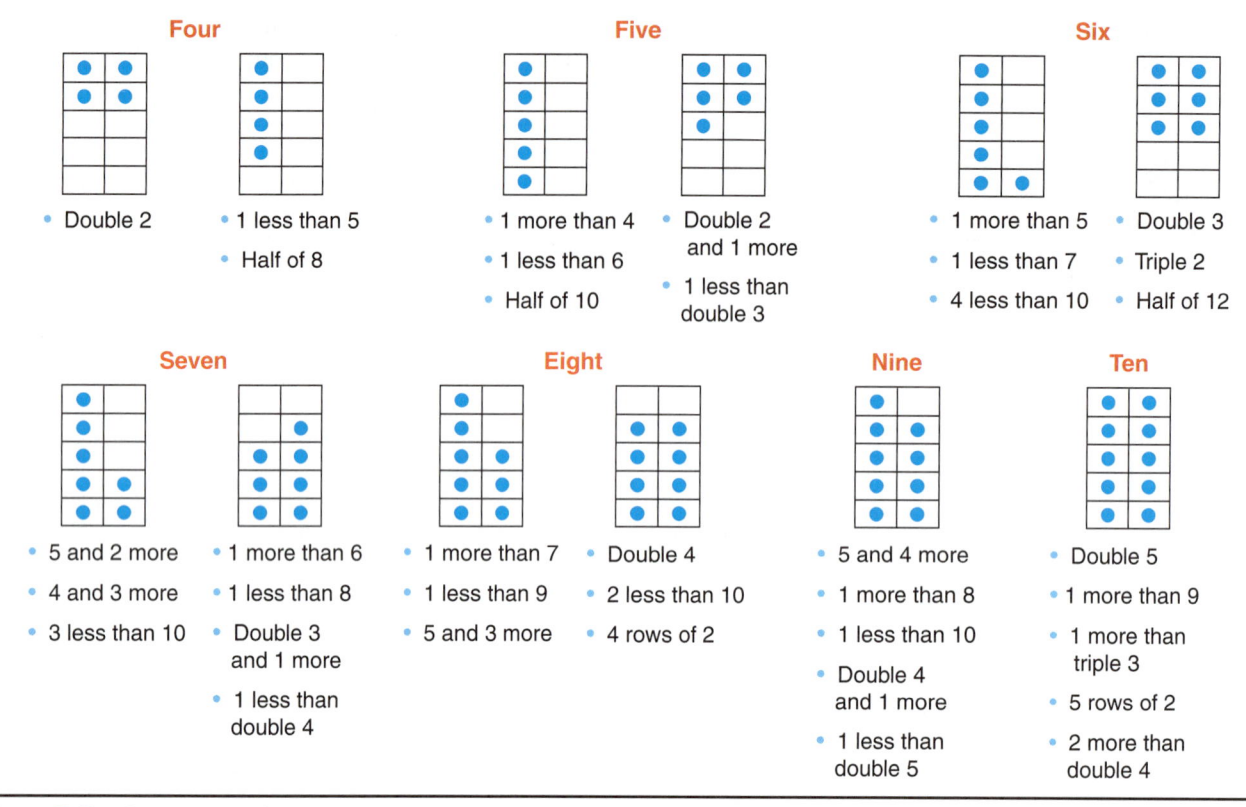

Five
- 1 more than 4
- 1 less than 6
- Half of 10
- Double 2 and 1 more
- 1 less than double 3

Six
- 1 more than 5
- 1 less than 7
- 4 less than 10
- Double 3
- Triple 2
- Half of 12

Seven
- 5 and 2 more
- 4 and 3 more
- 3 less than 10
- 1 more than 6
- 1 less than 8
- Double 3 and 1 more
- 1 less than double 4

Eight
- 1 more than 7
- 1 less than 9
- 5 and 3 more
- Double 4
- 2 less than 10
- 4 rows of 2

Nine
- 5 and 4 more
- 1 more than 8
- 1 less than 10
- Double 4 and 1 more
- 1 less than double 5

Ten
- Double 5
- 1 more than 9
- 1 more than triple 3
- 5 rows of 2
- 2 more than double 4

Figure 7-9 *Connections form representations on the ten-frame.*

greater number sense. Experiences with the ten-frame also facilitate the development of addition, subtraction, multiplication, and division, as well as place value.

MAKING CONNECTIONS

Development of the numbers 1 through 5 is principally done through sight recognition of patterns, coupled with immediate association with the oral name and then the written symbol. For example, here is a picture of a tricycle and the question, "How many wheels?"

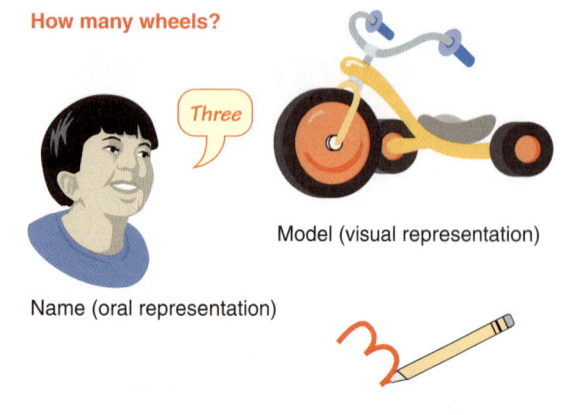

How many wheels?

Three

Name (oral representation)

Model (visual representation)

Symbol (written representation)

It is important that the number of wheels be linked to both oral name and written symbol. It is also important to provide different configurations of dots, blocks, and other objects, as well as different forms of the numerals, such as 3 and 3, to broaden their experiences.

Many valuable relationships are established as the numbers 1 through 5 are developed, but none are more useful than the notions of *one more* and *one less*. These connections are fundamental in early counting and also in learning place value with larger numbers. The notions *one more* and *one less* evolve from many different real-world experiences, such as these:

David has one less cookie than Jean-Paul.

Mira has one more apple than Beth.

They have one less player on their team.

Their group has one more girl than our group.

The concepts of *one more* and *one less* can be modeled in different ways. Figure 7-10 shows three such models. These arrangements provide a basis for developing the concept of 5 as well as discussing *one more* or *one less*. Using these models for discussion can help children establish important connections that link numbers such as 4 and 5. For example, children might say

The yellow rod is one step longer than the purple rod.

The five card is just like the four card except it has an extra heart in the middle.

Having many experiences with such models and patterns helps children abstract numbers and establish useful connections between them.

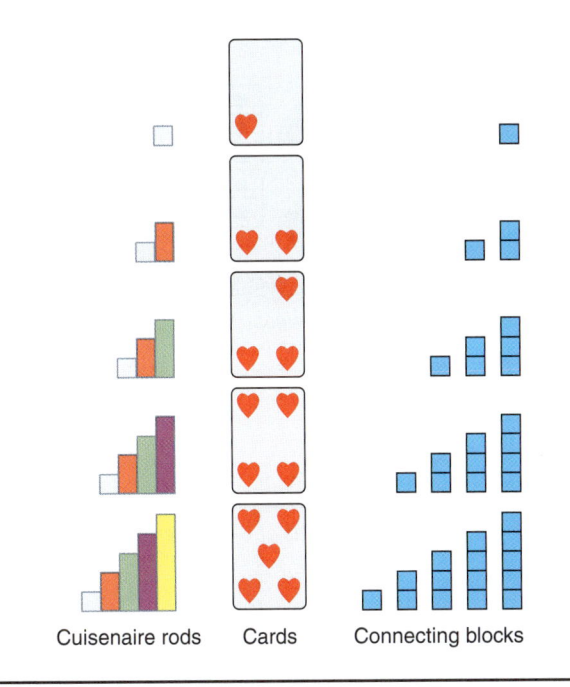

Counting   **147**

Figure 7-10   *Three models for one more or one less.*

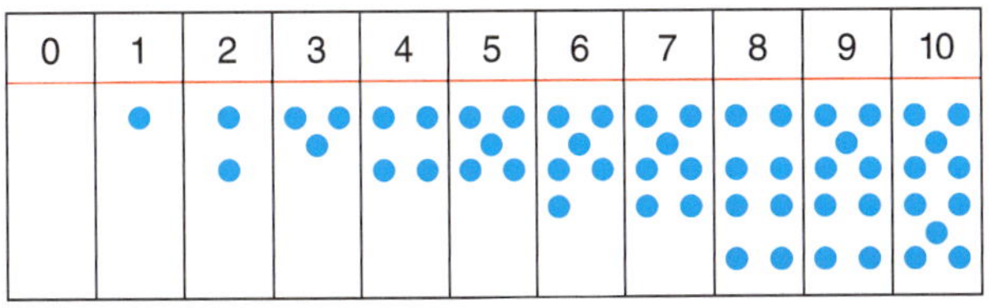

> ### *In the Classroom 7–8*
>
> ***Objective:*** Recognize different visual representations of the same number.
>
> ***Grade Level:*** K–1
>
> Give children blank index cards.
>
> Have them make 3 different sets of dot cards 1-6.
>
> Each child decides how to arrange the dots on their cards.
>
> The children mix their cards.
> You select a card.
> Ask students to find a card with the same number of dots.
> Repeat several times.
>
> Gearing up.
> Select a card.
> Ask students to find a card with one more dot. One less.
> Have children made additional dot cards for 7-10.
> Use dot cards 1-10 for similar activities.

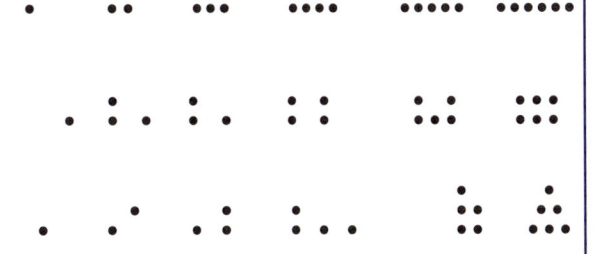

The hearts on the cards and the trains of blocks in Figure 7-10 provide clear reminders of the numbers represented and the notion of *one more* or *one less*. Both the staircase of rods and the trains of blocks vividly illustrate not only these concepts but also an important, yet subtle, difference between the models. The staircase of rods illustrates the concept of *more*, but it is not absolutely clear how much more until the length of the rods has been made clear. If you used a single rod without identifying a unit rod, it would not be possible to associate the rod with a unique number. Thus the rods are a different model for developing numbers than the cards or trains of blocks.

Some models illustrate zero more clearly than others. For example, a rod of length zero is more difficult to grasp than a card with zero hearts. Care should be taken to introduce zero as soon as it becomes natural to do so, using models appropriate for the purpose, including the number line. Help children distinguish between zero and nothing by encouraging the use of zero to report the absence of something. For example, when reporting the score of a game, it is better to say "Cardinals three, Bears zero" than to say "three to nothing."

As the numbers through 10 are developed, it is important that various patterns among them be discovered, recognized, used, and discussed. Many patterns suggesting many different relationships are shown in this number chart:

For example, the number 7 is shown by 5 dots and 2 more dots. The number 10 is composed of two groups of 5 dots. On the number chart, 6 is shown as 5 dots and 1 dot, but other representations are possible, as shown in Figure 7-11, and they should be explored and discussed with students. Figure 7-11b, for example, shows that 6 can be represented as one group of 3, one group of 2, and one group of 1. It can also be shown as two groups of 3, as in parts (a) and (d), or three groups of 2, as in part (c). No mention is made of addition or multiplication in this context, but such observations provide helpful connections when these operations are developed.

Similar illustrations and applications of the numbers 7 through 10 should be presented. For example, 7 days in a

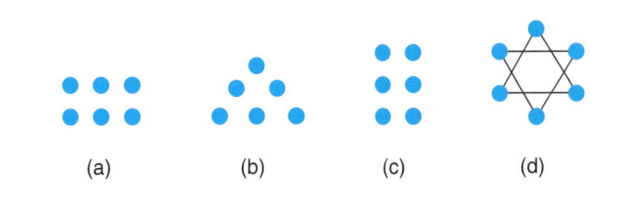

Figure 7-11   *Some representations of 6.*

Math Links 7.3

You will find Learning About Number Relationships and Properties of Numbers Using Calculators and Hundreds Boards at the NCTM Standards Web site at http://www.nctm.org/standards/content.aspx?id=24600 or from this book's Web site. These virtual tools provide an opportunity to count with a calculator that is synchronized with a hundred board. Together they provide connections between calculator counting and a wide range of patterns that can be explored on the hundred board.

www.wiley.com/college/reys

Two more make ten, so it's eight.

Eight — five and three more.

Eight — two groups of four.

Figure 7-12 *Representations of 8 on a ten frame.*

week may suggest a natural grouping; 8 vertices (corners) of a cube suggest two groups of 4; and the number of boxes in a tic-tac-toe grid suggest three groups of 3. These different representations of the same number provide opportunities to check on students' conservation of number. In the Classroom 7–8 illustrates an approach with numbers 1-6.

Children should realize very early that 10 is a special number. At the early stage of number development, the most unusual thing about 10 is that it is the first number represented by two digits, 1 and 0. In addition to having 10 fingers and toes, children encounter the number 10 in many situations, such as in playing games and changing money. These experiences can be extended to include discussion about different representations of 10:

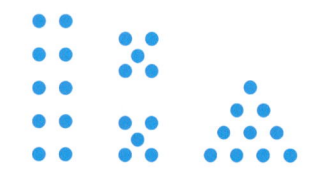

Can you find two groups of 5?
Can you find five groups of 2?
Does the group of 4, 3, 2, and 1 remind you of bowling?

The number 10 provides the cornerstone for our number system, and its significance is developed further in Chapter 8.

A ten-frame is one of the most effective models for facilitating patterns, developing group recognition of numbers, and building an understanding of place value. The ten-frame can be made from an egg carton shortened to contain 10 pouches, or it may simply be outlined on paper or tagboard (see Blackline Master in Appendix). This frame is a powerful organizer and helps provide the base for many thinking strategies and mental computations. Initially, children might use counters to make different representations of the same number in the ten-frame, as illustrated in Figure 7-12. Encountering a variety of groupings on the ten-frame should stimulate discussion about different patterns.

CARDINAL, ORDINAL, AND NOMINAL NUMBERS

Thus far we have discussed some important considerations in number development. A major goal has been counting and then finding a correct number name for a given group. This aspect of number provides a *cardinal number*, which answers the question "How many?" Another important aspect of number emphasizes arranging things in an order and is known as *ordinal number*; it answers the question "Which one?"

An emphasis on ordering or arranging things in a given sequence leads to ordinal numbers. The order may be based on any criterion, such as size, time of day, age, or position in a race. Once an order is established, however, the counting process not only produces a set of number names but also names each object according to its position. Thus, in counting the rungs on this ladder, the number 1 is first, 2 is second, 3 is third, and so on.

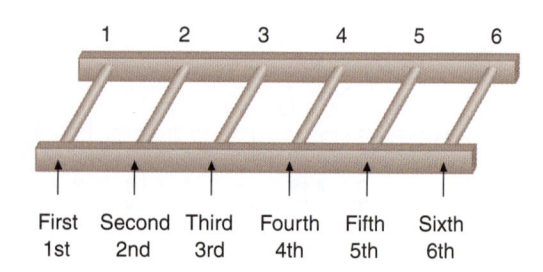

| 1 | 2 | 3 | 4 | 5 | 6 |

First Second Third Fourth Fifth Sixth
1st   2nd   3rd   4th   5th   6th

Many children know some ordinal numbers such as first, second, and third before they begin school. Encounters with statements such as the following provide early and valuable experience with ordinal numbers:

The first letter of the alphabet is A.

Bob is second in line.

Cary was third in the race.

It is important that the development of early number concepts provide children with opportunities to learn both ordinal and cardinal numbers. Don't worry about which to teach first; just be sure both are given attention.

A knowledge of ordinal relationships, along with logical thinking, leads to more challenging experiences, such as those suggested in Figure 7-13. These questions are guaranteed to generate much discussion as they help children further clarify notions of ordinal numbers.

Another aspect of number provides a label or classification and is known as *nominal*. Examples are the number on a player's uniform, the license plate of a car, a postal zip code, and a telephone number. Nominal numbers provide essential information for identification but do not necessarily use the ordinal or cardinal aspects of the number.

When using cardinal, ordinal, and nominal numbers, children do not need to distinguish between the terms. Distinctions can be made informally by asking questions within problem situations such as:

How many pieces are on the chessboard?

Which runner is third?

What is your phone number?

These questions not only help children think about numbers but also illustrate that numbers have different uses.

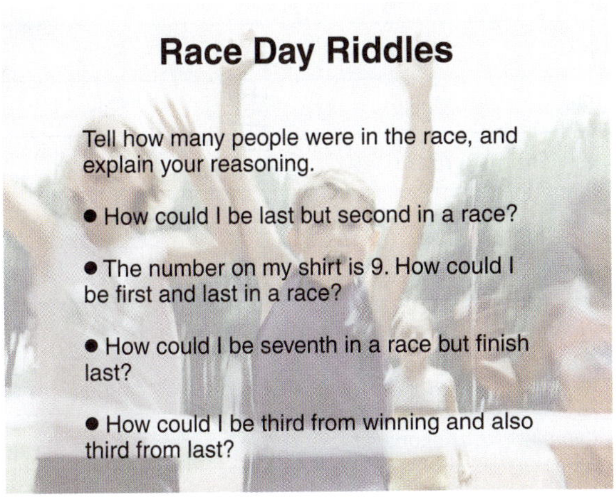

Race Day Riddles

Tell how many people were in the race, and explain your reasoning.

- How could I be last but second in a race?
- The number on my shirt is 9. How could I be first and last in a race?
- How could I be seventh in a race but finish last?
- How could I be third from winning and also third from last?

Figure 7-13 *Thinking about ordinal relationships.* (Source: Julia Fishkin/The Image Bank/Getty Images).

WRITING NUMERALS

Young children typically have difficulty writing numerals as well as letters, they should spend less time writing numerals. The lack of development of the small muscles needed to write presents one problem, and the limited eye–hand coordination of many young children constitutes another difficulty. Both of these make it difficult or impossible for some young children to write numerals. If children are pressured into premature symbolization, it can create unnecessary frustrations and anxieties. Therefore the focus should be on number development and relationships among numbers and not on writing.

Children can usually recognize a number symbol and say it correctly long before they write it. Many young children initiate early writing of numerals on their own, and they get a feeling of great accomplishment from it; however, children's writing skills develop much more quickly as second graders than as kindergartners.

Children should begin by tracing the digits; here is a recommended stroke sequence:

0 1 2 3 4 5 6 7 8 9

Textbooks provide guidance to children in different forms. Usually a starting point is indicated as well as the direction:

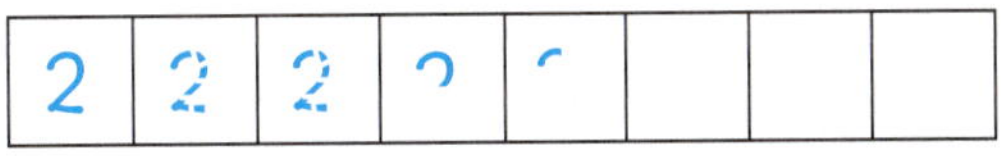

| 2 | 2 | 2 | 2 | 2 | | | |

Encourage children to draw the appropriate number of objects beside the numeral being written to help them connect the number concept with its symbolic representation. When children are learning to write numerals, have them work on this writing skill aside from "mathematics." Thus, children can master this skill and use it freely before applying it in written computational situations; otherwise, the writing task consumes all of their concentration, and they forget about the mathematics being done.

Many children develop the necessary writing skills on their own. Even then, these youngsters need monitoring and maybe some occasional guidance. Others, however, need systematic step-by-step procedures to help them. Although there is no one best way to form a numeral, some patterns may help. Guiding a child's hand until the child takes the initiative in writing helps him or her get started. Later, outlines of numerals for children to trace are helpful. Here are some additional suggestions; these will not be necessary in all cases, but some children will profit from them:

- *Cut out shapes of numerals and have class members trace them in the air.* As the tracing is being done, describe it verbally such as "go to the right and then down." This activity can be extended to using only dots to form the outline or a pattern for the children to follow.

- *Have a child who can make numerals stand behind someone who cannot.* Ask the skilled child to use a finger and gently "write" a numeral on the other child's back. The child in front should identify the numeral and write it on the whiteboard, trace it on a poster, or write it in the air. This approach calls on the tactile sense and helps some children better develop their writing skills.

- *Use numerals that have been cut from sandpaper and pasted to cards, or take some cord and glue it in the shape of numerals.* Place a mark on each numeral to show the child where to begin tracing it with his or her finger. This approach is particularly helpful with children who persist in reversing numerals.

- *Cover numerals to be traced with a transparency.* Then give the child a water-soluble pen and have him or her practice tracing the numerals.

- *For children having difficulty writing numerals, a calculator is helpful.* The calculator display provides a visual reminder of a number's symbol and removes the burden of writing complicated numerals.

The fact that numerals take different forms also should be mentioned but not belabored.

4 4 4 4

Some familiarity with these forms will help avoid confusion when a 4 appears as a printed number or as a digital display on a clock or calculator. The wide use of digital numbers in everyday living demands planned instruction to make children familiar with them.

CULTURAL CONNECTIONS

Children enjoy learning words and symbols from other countries. This expands their worldview and may cast a new light on their native system. Good teachers show an interest in the cultures of their students, and the diversity of cultures in classrooms provides many opportunities for sharing and learning.

Counting and early number development occur in every culture. While in some countries the symbolism is similar to that used in the United States (Mexico and Germany), in others countries, such as Asia, it is different. Figure 7-14 shows how the numbers 0–10 are represented and named in several different countries. A look at this figure shows some connections, such as *quatro* in Spanish meaning "four" in

Number	English	Spanish	German	Japanese Symbol	Japanese	Japanese	Chinese Symbol	Chinese
0	Zero	Cero	Null	〇	Rei		零	Ling
1	One	Uno	Eins	一	Ichi	Hitotsu	一	Yi
2	Two	Dos	Zwei	二	Ni	Futatsu	二	Er
3	Three	Tres	Drei	三	San	Mittsu	三	San
4	Four	Quatro	Vier	四	Yon, Shi	Yottsu	四	Si
5	Five	Cinco	Funf	五	Go	Itsutsu	五	Wu
6	Six	Seis	Sechs	六	Roku	Muttsu	六	Liu
7	Seven	Siete	Sieben	七	Nana, Shichi	Nanatsu	七	Gi
8	Eight	Ocho	Acht	八	Hachi	Yattsu	八	Ba
9	Nine	Nueve	Neun	九	Kyuu	Kokonotsu	九	Jiu
10	Ten	Diez	Zehn	十	Jun or Ju	To	十	Shi

Figure 7-14 *Symbols and number names for 0 to 10 in several different countries.*

English and connected with a quadrilateral, meaning a polygon with four sides. Likewise, *null* in German is used in English to reflect the cardinal number of an empty set. Although three in Japan and China both use the number name, *san*, the symbols are different and, in general, connections between terms in Asian countries are not possible because of the dramatic differences in the languages.

Yet children in all of these countries learn to count, applying the same counting principles that were discussed earlier. In Asian countries, such as China, Japan, and Taiwan, students learn their own system for counting and communicating their results. In addition, the international availability of technology, including calculators, means that students in these countries must also become familiar with Hindu-Arabic notation. Thus Asian students must learn to recognize and use both Hindu-Arabic numerals and those used in their own country as well as the proper vocabulary with each of them.

The Japanese counting system is complex to us. Figure 7-14 shows the number symbols and two different number names used in Japan for the numbers four and seven. These numbers have alternate number names (e.g., 4 may be *shi* or *yon*) and although either is fine, there are preferred uses. For example, July or the seventh month is Shichigatsu, whereas 7 yen is *nana* yen. Figure 7-14 also shows two different Japanese columns with number names up to ten, which means Japan has two counting systems for the first 10 numbers. Figure 7-14 has a blank for the name of zero in Japan for the counting numbers. Counting begins with one, so there is no Japanese name for zero when counting. After 10, only one system is used, and it will be explored in the next chapter.

Ordinal numbers in English require students to memorize new words, such as *first*, *third*, and *sixth*. In China, making ordinal numbers is easier. It is accomplished by simply adding *di* to the front of a number. So to make "one" become "the first," *yi* becomes *di yi*, and to make "six" become "the sixth," *liu* becomes *di liu*. These examples of how culture influences counting provide a vivid reminder of the challenges of helping children from different backgrounds learn to count.

Math Links 7.4

Here are two Web sites that provide opportunities to count in other languages:

1–10 in over 5000 languages at http://www.zompist.com/numbers.shtml; 1–100 in 18 languages at http://www.marijn.org/everything-is-4/counting-0-to-100. Both sides can be accessed from this book's Web site.

www.wiley.com/college/reys

A GLANCE AT WHERE WE'VE BEEN

Good number sense is a prerequisite for all later computational development. Young children need to recognize small groups of objects (up to 5 or 6) by sight (*subitizing*) and name them properly. Subitizing saves time, helps develop more sophisticated counting skills, and accelerates the later development of addition and subtraction. Activities involving sight recognition of the numbers of objects in small groups provide many opportunities to introduce and use key terms such as *more*, *less*, *after*, *before*, *one more*, and *one less*. To foster a better number sense, instruction on the numbers through 5 should focus on patterns and develop recognition skills. Tools such as the five- and ten-frame, number line, hundred chart, and calculators provide powerful resources for helping children explore and construct relationships for numbers to 10 and beyond.

Counting skills usually start before children begin school but must be developed by careful and systematic instruction before written work is appropriate. Rational counting is a goal for all young children and is characterized by several principles, including using a fixed number-name list, assigning one and only one name to each object counted, and realizing that the last object counted represents the number in the group.

Counting processes reflect various levels of sophistication, beginning with rote counting and eventually leading to rapid skip counting forward and backward. Counting skills are extended in the intermediate grades and often are further refined throughout our lives. Oral counting leads to ways of writing and representing cardinal, ordinal, and nominal numbers.

Competence in counting as highlighted by the *Common Core State Standards* (Table 17-1) is essential for meaningful later development of larger numbers. Counting, connecting number names with written symbols, and recognizing relationships among numbers, such as, more, one less, and greater than, must be well understood. This knowledge is the basis for the successful study of elementary mathematics, and it prepares children for the necessary understanding of large numbers and place value.

Things to Do: From What You've Read

1. Suppose you send a note home to parents encouraging them to help their children improve sight-recognition skills. One parent responds, "Why should my child learn to recognize a group of 5? After all, you can just count them." How would you respond?
2. Four fundamental principles of counting were identified in this chapter.
   a. Describe in your own words what each of the principles means.
   b. Tell why that principle is essential to becoming a rational counter.
   c. Choose one of the principles and describe an activity that would help a child progress toward developing that principle.
3. Distinguish between rote and rational counting.
4. What is meant by conservation of number? Why is its development an important part of number sense?
5. Describe how counting could be used to answer these questions:
   a. How many floors are between the seventh and fifteenth floors?
   b. If Alfinio has read to the bottom of page 16, how many pages must he read to reach the top of page 21?
6. What is subitizing? Tell why it is an important skill to help children develop.

Things to Do: Going Beyond This Book

In the Field

1. [1]*Observing Counting.* Observe a young child counting in your classroom. Identify any counting errors and where they took place, such as bridging to the next decade. Also identify any counting principles that were exhibited.
2. [1]*Counting Resources.* Find the instructor's guidebook or teacher's edition being used in a K–2 mathematics classroom in your school. Describe how resources, such as calculators and children's books, are used to support counting.
3. *Using Attribute Pieces.* In the Classroom 7–1 and 7–2 provide several patterning and logical reasoning activities using attribute pieces with children. Explore these trains with several children. How many different correct solutions did children find for Train A? Did your children find one-difference trains to be the same difficulty as two-difference trains? Did students feel comfortable knowing that more than one correct answer exists?

With Additional Resources

4. Examine *Number Sense and Operations* (Burton, Mills, Lennon, and Parker, 1993), *Developing Number Sense in Middle Grades* (Reys et al., 1991), *Navigating Through Number and Operations in Prekindergarten–Grade 2* (Cavanagh et al., 2004), or the *SENSE* series (McIntosh, Reys, Reys, and Hope, 1997). How do these books characterize number sense? Select and report on an instructional activity designed to foster number sense.
5. Read the article about zero by Anthony and Walsh (2004). Describe some of the misconceptions children have and share some of the suggestions made to help children gain a better understanding of zero.
6. Read the article by Kamii & Rummelsburg (2010). Describe how the 'bowling game' could be used to help children practice subitizing by providing different arrangements, and knocking over some pins helps develop addition and subtraction.
7. Compare two current textbook series:
   a. Find how far each series expects children to be able to count when they begin first grade and when they complete first grade.
   b. Find examples of activities designed to develop number sense. Describe the activities and identify the grade level. Do you think these activities would be effective? Tell why.
   c. Find an example of a visual pattern that connects numbers to geometry.

With Technology

8. *Counting with Calculators.* In the Classroom 7–4 and 7–5 provide structured calculator counting activities. Try one or more of them with some children in a classroom that has calculators available. Were the children experienced at using the calculators or were they viewed as a novelty? Were the children comfortable making estimates for the counts? Describe some patterns that emerged.

---

[1]Additional activities, suggestions, and questions are provided in the field experience manual on the Student Companion site at www.wiley.com/college/reys.

Note to Instructors: You can find additional resources, learning activities, and blackline masters in this text's accompanying Instructor's Manual, at www.wiley.com/college/reys.

9. Virtual Calculators. Investigate the NCTM Web site at http://www.nctm.org/standards/content.aspx?id=24600 and try the Calculator and Hundreds Boards activity. Explore this resource by entering 5 or 10 and then repeatedly enter the symbol on the calculator. How does the hundred board change? Describe how this virtual tool can be used as an instructional tool. Identify advantages of using this tool over the handheld calculator and the paper hundred board.

10. Go to the Center for the Study of Mathematics Curriculum at http://www.mathcurriculumcenter.org and click on Curriculum Databases. Then click on State Mathematics Standards Database. Check to see if your state has mathematics standards available. If so, identify the grade-level expectations for counting that are idenfied for your state. Compare the expectations for your state with another state or the Common Core State Standards-Mathematics.

Book Nook for Children

Anno, M. *Anno's Counting Book.* New York: Thomas Y. Crowell, 1977.

How would you describe a picture that refers to zero? This book begins with a barren winter landscape—a hazy, blue sky above a hazy, white hill. On the next page the scene brightens: one tree, one bird, one house. Turn the page again and the snow has started to melt and you find two buildings, two trucks, two trees, two children, two dogs, and two adults. Suddenly there is almost more than you can count on each page! And the numbers continue to increase.

Giganti, P. *How Many Snails? A Counting Book.* New York: Greenwillow Books (Pearson K–12), 1998.

Use these many opportunities to help children practice their counting. A series of simple questions directs them to determine the differences between seemingly similar objects, encouraging them to develop powers of observation, discrimination, and visual analysis.

Murphy, S. *Dave's Down-to-Earth Rock Shop.* New York: HarperCollins, 2000.

Josh and Amy have a rock collection. They learn that they can sort their collections by many different ways such as color, size, type, and hardness. Josh and Amy also learn that the same objects can be organized in many different ways. This is a great book for students to learn about attributes. There are activities and games in the back of this book for parents and teachers to use.

Murphy, S. *Just Enough Carrots.* New York: HarperCollins, 1997.

This is the story of a bunny and his mother and how they shop in a grocery store for the lunch guests that are coming to their house. As they get the items they need, the bunny learns about counting and comparing different amounts from their cart of items to the other shoppers and their carts. This is a great introduction to comparing different amounts. There are activities and games in the back of this book for teachers and parents to use.

Schlein, M. *More Than One.* New York: Greenwillow Books, 1996.

This book will let the reader see how the number 1 can be more than 1. This will depend on how the number 1 is attached to another number. For example: 1 basketball team, 1 month, 1 week, 1 year, etc. This is a great book to introduce number sense.

Wahl, J., and Wahl, S. *I Can Count the Petals of a Flower.* Reston, VA: NCTM, 1976.

This book provides a variety of rich and stimulating context teaching children to count using the petals of a flower.

Zaslavsky, C. *Count On Your Fingers African Style.* New York: Crowell, 1980.

The focus is on numerical relationships in the African marketplace where people buy and trade using many different languages, including various methods of finger counting. This book explores the practicality of math within the context of African culture and helps children see that math can be fun and creative.

CHAPTER 8

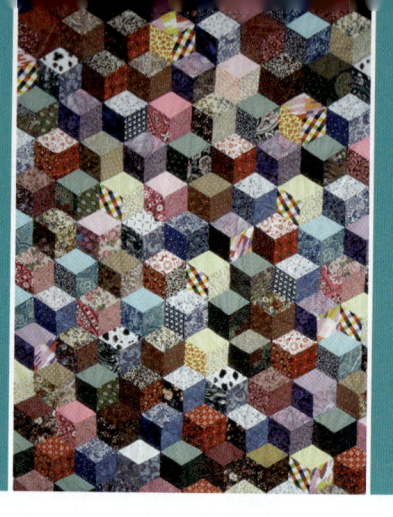

Extending Number Sense: Place Value

> "WITHOUT PLACE VALUE, WE WOULD GET NO PLACE WITH NUMBERS."
> — Carl Friedrich Gauss, Mathematician

>> SNAPSHOT OF A LESSON

KEY IDEAS

1. Illustrate the power and importance of grouping by ten.
2. Demonstrate the place value concept with different models.
3. Provide concrete representations of two-digit numbers and their corresponding symbolization.

BACKGROUND

This lesson is based on the video *Place Value Centers*, from Teaching Math: A Video Library, K–4, from Annenberg Media. Produced by WGBH Educational Foundation. ©1995 WGBH. All rights reserved. To view the video, go to http://www.learner.org/resources/series32.html (video #4).

Shaylene Vickstrom is a first-grade teacher. Today she introduces the concept of place value through a familiar daily task, updating the counting center that keeps track of the number of days they have been in school so far this year. Ms. Vickstrom has placed pockets on the bulletin board to hold straws. At the beginning of the school day, she adds a straw to the ones pocket. Once they have 10 ones, they bundle them and place bundles of 10 in the tens pocket.

Ms. Vickstrom asks her students to explain what happens when they find 10 ones in the ones pocket.

Ms. Vickstrom: Amanda, what do we do to them?

Amanda: Put them in a bundle of ten.

Ms. Vickstrom: Put them in a bundle of ten. That's right. And then, do they get to stay on the ones side?

Amanda: No.

Ms. Vickstrom: Where do we have to move them to?

Amanda: In the tens side.

Ms. Vickstrom: On to the tens side. That's right. So all of these over here are a bundle of ten. Aren't they? That's why they're on the tens side. But we have the single ones over here on the ones side.

Ms. Vickstrom transitions the students' attention from the calendar to the place-value mat by pointing out their similarities.

Ms. Vickstrom: I have something up here for you guys. This is my giant place-value mat. And guess what? This is an awful lot like the thing that's over there on the calendar that holds our ones and our tens. This is my ones side and it says "ones" right here. This is my tens side.

Ms. Vickstrom displays a large place-value mat and shares to the camera that the overall goal of today's lesson is to allow students to become familiar with place value. In particular, she wants to concentrate on the exchange of ones and tens so that students are prepared to move into addition and subtraction with double-digit numbers.

Ms. Vickstrom asks the age of two students in her class. She uses the place-value mat to model the age with unit tiles. She places six unit tiles on the ones area of the mat to represent the first student's age and then places seven additional unit tiles in the ones area to represent the second student's age. The class works together to find the sum of the students' ages. They bundle ten unit tiles into one tens rod and transfer the rod to the tens area of the mat.

Ms. Vickstrom: There's my bundle of ten. Now, if we wanted to add Racillia's age to Javier's age, do you think you know how much it is altogether if we are adding 6 plus 7? Kayla, what do you think?

Kayla: 10, 11, 12, 13.

Ms. Vickstrom: Why did you start off on 10?

Kayla: Because there's a bundle of ten.

Ms. Vickstrom: Because there's a bundle of 10 right there [*as she points to tens rod*]. So you counted, you said I already know there's 10 there, right? So you went 10 …

Amanda: 11, 12, 13.

Ms. Vickstrom explains how to write the numeral 13 by counting the number of tens rods and unit tiles from the place-value mat. She writes the numeral 1 under the tens area and the numeral 3 under the ones area.

Ms. Vickstrom: So, one special rule that we need to remember today is that on your ones side, once you get up to 10, you have to transfer them over here to your tens side.

Ms. Vickstrom explains the four activity centers. The first is a measuring center where children measure several items by linking together Unifix cubes to form a train. Children then break down the trains into ones and tens on their place-value mats. Children record their results on a recording sheet.

At the second center, using cups and place-value mats, students take inventory of piles of beans, buttons, and stones.

At the third center, Mr. Montez, a student teacher in Ms. Vickstrom's class, works with students to model numbers from a hundred chart.

In the last center, students compete to see who can make the longest train of Unifix cubes. Once all of the cubes in a pile are used, students break down the train into tens and ones using their place-value mats.

After students visit the centers, Ms. Vickstrom pulls the class back together to discuss what they have learned. As a class, they discuss the measuring center and the hundred board center.

Ms. Vickstrom: Now if Mr. Montez has 95 in his little mystery box, can somebody tell me what you would show on your place-value boards for 95 [*as she displays a place-value mat to the class*]? What would you show? Matthew?

Matthew: 95.

Ms. Vickstrom: 95. Ok, where would all of the cubes be? Would you have 95 cubes right here [*as she motions to the ones area of the place-value mat*]? Just all in a pile? What would you have?

Matthew: You'd have nine tens there.

Ms. Vickstrom: You'd have nine tens here [*as she motions to tens area of place value mat*]? Ok, nine bundles of 10.

Matthew: And you'd have five ones there.

Ms. Vickstrom: Five ones. So what does … when we're looking at a number, Matthew, what does the number on the right-hand side stand for? This number [*points to the five*]. Right now the 5 is right there.

Matthew: The ones.

Ms. Vickstrom: That stands for the ones. What does the number on the left-hand side stand for?

Matthew: The tens.

Ms. Vickstrom: The tens.

In Ms. Vickstrom's reflection on the day's lesson, she shares the idea of providing multiple problem-solving activities in order to reach different types of learners. She also stresses the importance of hands-on activities and providing a culminating class discussion at the end of free exploration activities to share the group findings.

FOCUS QUESTIONS

1. Why is place value important?
2. What role do models play in developing an understanding of place value?
3. Why is composing and decomposing numbers important in developing place value?
4. How can calculators help children develop place-value concepts?

Math Links 8.1

You will find the full text version of the *Common Core State Standards for Mathematics* at http://www.corestandards.org/. It highlights various aspects of place value across the elementary grades.

www.wiley.com/college/reys

INTRODUCTION

Children must make sense of numbers and the ways in which numbers are used in and out of school. Place value is critical to this understanding or sense making and is one of the cornerstones of our number system. The famous mathematician Gauss underscored the importance of place value when he said, "Without place value, we would get no place with numbers." Place value is a central and continuing theme in the *Common Core State Standards for Mathematics* (Figures 7-1 and 8-1). An examination of the grade level development of place value in Figure 8-1 shows how understanding place value evolves over several grades, helping children to compare and operate with numbers. Place value is a visible component that begins in the early primary grades and is extended to adding, subtracting, multiplying, or dividing larger numbers. Place value patterns with whole numbers provide a natural connection when working with decimals in the upper grades.

The Snapshot of a Lesson shows children counting and how place value provides an organizational structure for counting. The same principles are involved as larger numbers are counted. Representations of larger numbers are based on learning operations with numbers in a sense-making or meaningful way. So in counting from 9 to 10, or 99 to 100, or 999 to 1000, place value is an important part

Understand place value (Grade 1)

- Understand that the two digits of a two-digit number represent amounts of tens and ones. Understand the following as special cases.
  - 10 can be thought of as a bundle of ten ones—called a "ten".
  - The numbers from 11 to 19 are composed of a ten and one, two, three, four, five, six, seven, eight, or nine ones.
  - The number of 10, 20, 30, 40, 50, 60, 70, 80, 90 refer to one, two three, four, five, six, seven, eight, or nine tens (and 0 ones).
- Compare two two-digit numbers based on meanings of the tens and ones digits, recording the results of comparisons with the symbols >, =, <.

Understand place value (Grade 2)

- Understand that the three digits of a three-digit number represent amounts of hundreds, tens, and ones: e.g., 706 equals 7 hundreds, 0 tens and 6 ones. Understand the following special cases:
  - 100 can be thought of as a bundle of ten tens—called a "hundred".
  - The numbers 100, 200, 300, 400, 500, 600, 700, 800, 900 refer to one, two, three, four, five, six, seven eight, or nine hundreds (and 0 tens and 0 ones).
- Count within 1000; skip-count by 5s, 10s, and 100s.
- Read and write numbers to 1000 using base-ten numerals, number names, and expanded form.
- Compare two three-digit number based on meanings of hundreds, tens, and ones digits, using >, =, < symbols to record the results of comparisons.

Use place value understanding and properties of operations to perform multidigit arithmetic (Grade 3)

- Use place value understanding to round whole numbers to the nearest 10 or 100.
- Fluently add and subtract within 1000 using strategies and algorithms based on place value properties of operations, and/or the relationship between addition and subtraction.
- Multiply one-digit whole numbers by multiples of 10 in the range 10–90 (e.g., 9 × 80, 5 × 60) using strategies based on place value and properties of operations.

Generalize place value understanding for multi-digit whole numbers (Grade 4)

- Recognize that in a multi-digit whole number, a digit in one place represents ten times what it represents in the place to its right.
- Read and write multi-digit whole numbers using base-ten numerals, number names, and expanded form. Compare two multi-digit numbers based on meanings of the digits in each place, using >, =, and < symbols to record the results of comparisons.
- Use place value understanding to round multi-digit whole numbers to any place.

Further understand the place value system (Grade 5)

- Recognize that in a multi-digit number, a digit in one place represents 10 times as much as it represents in the place to its right and 1/10 of what it represents in the place to its left.

Figure 8-1 Selected place-value standards from CCSSM. This summary of the CCSSM standards was made by the authors of *Helping Children Learn Mathematics*. The full document (CCSSI, 2010) may be obtained at www.corestandards.org/the-standards/mathematics.

of the symbols and language. In a similar manner, in adding 1 to 9 to get 10, or 1 to 99 to get 100, or 1 to 999 to get 1000, changes in place value are again experienced. These transitions from one-to-two-, two-to-three-, or three-to-four-digit numbers are fundamental to place-value work.

OUR NUMERATION SYSTEM

Although we say "our" numeration system, it is multicultural, and it is "ours" only to the extent that it is a part of our cultural heritage. History tells us that "our" numeration system is really the result of continuous development and refinement over many centuries (Zaslavsky, 2003). The number system we use, called the Hindu-Arabic system, was probably invented in India by the Hindus and transmitted to Europe by the Arabs, but many different countries and cultures contributed to its development.

The Hindu-Arabic numeration system has four important characteristics:

1. *Place value*: The position of a digit represents its value; for example, the 2 in $23 names "two tens" or "twenty" and has a different mathematical meaning from the 2 in $32, which names "two ones."
2. *Base of ten*: The term *base* simply means a collection. Thus, in our system, ten is the value that determines a new collection and is represented by 10. The system has 10 digits, 0 through 9.
3. *Use of zero*: A symbol for zero exists and allows us to represent symbolically the absence of something. For example, 309 shows the absence of tens in a number containing hundreds and ones.
4. *Additive property*: Numbers can be written in expanded notation and summed with respect to place value. For example, 123 names the number that is the sum of 100+20+3.

These properties make the system efficient and contribute to the development of number sense. That is, once children understand these characteristics, the formation and interpretation of numbers—either large or small—is a natural development.

Roman numerals are still in use today in our society, but not for computation. Roman numerals are different from the Hindu-Arabic system in that they lack place value, have no symbol for zero, and no base. As a result, computation with Roman numerals is a difficult and cumbersome process.

NATURE OF PLACE VALUE

A thorough understanding of place value is necessary if computational algorithms for addition, subtraction, multiplication, and division are to be learned and used in a meaningful way. Place value develops from many various experiences, such as counting and mental computation (e.g., $1 plus $0.25 to get $1.25). Development of place value promotes number sense and rests on two key ideas:

- *Explicit grouping or trading rules are defined and consistently followed.* These ideas are illustrated in the opening Snapshot of a Lesson as different models are used and are implicit in the bulletin board display shown in Figure 8-2. Such a display provides a constant reminder of the importance of grouping by tens to place value. Our base-ten system is characterized by trading 10 ones for 1 ten (or 1 ten for 10 ones), 10 tens for 1 hundred, 10 hundreds for 1 thousand, and so on. The two-way direction of these trades (e.g., 10 tens for 1 hundred or 1 hundred for 110 tens) should be stressed because there are times when each type of trade must be used. These equivalent trades illustrate the composing and decomposing of numbers mentioned in Figure 8-2. It should also be noted that similar trades are followed with numbers less than

Figure 8-2 *A bulletin board display for place value.*

1—decimals. Thus, 1 can be traded for 10 tenths (or 10 tenths for 1), 10 hundredths for 1 tenth (or 1 tenth for 10 hundredths), and so on.

- *The position of a digit determines the number being represented.* For example, the 2 in 3042 and the 2 in 2403 represent completely different quantities: 2 ones in 3042 and 2 thousands in 2403. Furthermore, the zero plays a similar yet different role in each of these numbers. It has positional value in each case, and reports the lack of a quantity for that place. Although the notion of zero will continue to be expanded and developed throughout elementary school mathematics, children should experience the role of zero in place value early and often.

In the Hindu-Arabic number system, place value means that any number can be represented using only 10 digits (0–9). Think about the problems of representing numbers without place value! Each number would require a separate and unique symbol. Your memory storage would quickly be exceeded, and you would probably have to use only the few numbers with symbols you could remember. Nevertheless, place value is difficult for some children to grasp. Oral counting or rote recitation of numbers by young children is often interpreted as understanding place value. Yet many children who can count correctly have absolutely no concept of place value. In most cases, the confusion or misunderstanding can be traced to a lack of counting and trading experiences with appropriate materials and the subsequent recording of these results. Early and frequent hands-on counting activities, similar to those described in Chapter 7, are essential prerequisites to meaningful understanding of place value.

Place-value concepts are encountered before starting school. For example, many children distinguish between the one- and two-digit numbers on a channel indicator of a television, a timer for a microwave oven, and house or apartment numbers. Children learn early that channels 43 and 34 are different, but they may not make a connection to place value.

MODELING—UNGROUPED AND PREGROUPED

Two types of materials help young children develop place value: ungrouped materials and pregrouped materials. Ungrouped materials include beans, cubes, or straws that children form into groups, as shown in Figure 8-3. If there are enough beans to make a group of 10, then those beans can be placed in a cup or bag or glued on a stick. Once these beans are glued on a stick, then they can be considered pregrouped materials.

Pregrouped materials are formed into groups before the child uses them, and Figure 8-3 also shows three examples of pregrouped materials. Hands-on experience with manipulatives is essential in establishing and developing the concept of place value. Research suggests that instruction should focus on concrete models that are simultaneously connected to oral descriptions and symbolic representations of the models (Thompson, 2000; Wearne and Hiebert, 1994).

MODELING—PROPORTIONAL AND NONPROPORTIONAL

Place-value models may be either proportional or not proportional. Base-ten blocks, beans glued in groups of 10 on a stick, and tongue depressors bundled together are proportional models (Figure 8-4). In proportional models, the material for 10 is 10 times the size of the material for 1, 100 is 10 times the size of 10, and so on. Measurement provides another proportional model, as was shown in this chapter's Snapshot of a Lesson, where children use interlocking cubes to measure their height and then form groups of 10. Meter sticks, decimeter rods, and centimeter cubes could also be used to model any three-digit number.

Nonproportional models do not maintain any size relationships. Money is a real-world example of a nonproportional model where size relationships are not maintained. For example, 10 pennies are bigger than a dime but are a fair trade in our monetary system. Ten dimes are the same value as a dollar, but they are not proportional in size.

Figure 8-4 illustrates some proportional and nonproportional models that are effective in helping students understand not only place value but larger numbers as well.

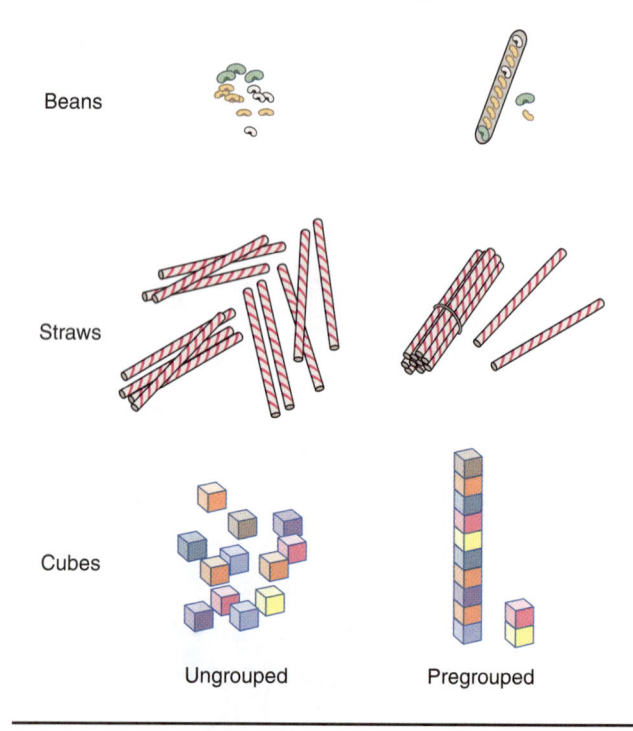

Figure 8-3 *Ungrouped and pregrouped models for developing place value*

Proportional Models

Base-ten blocks
- Flats
- Longs
- Units

Bean sticks
- Bundles
- Sticks
- Beans

Tongue depressors
- Sacks
- Bands
- Depressors

Nonproportional Models

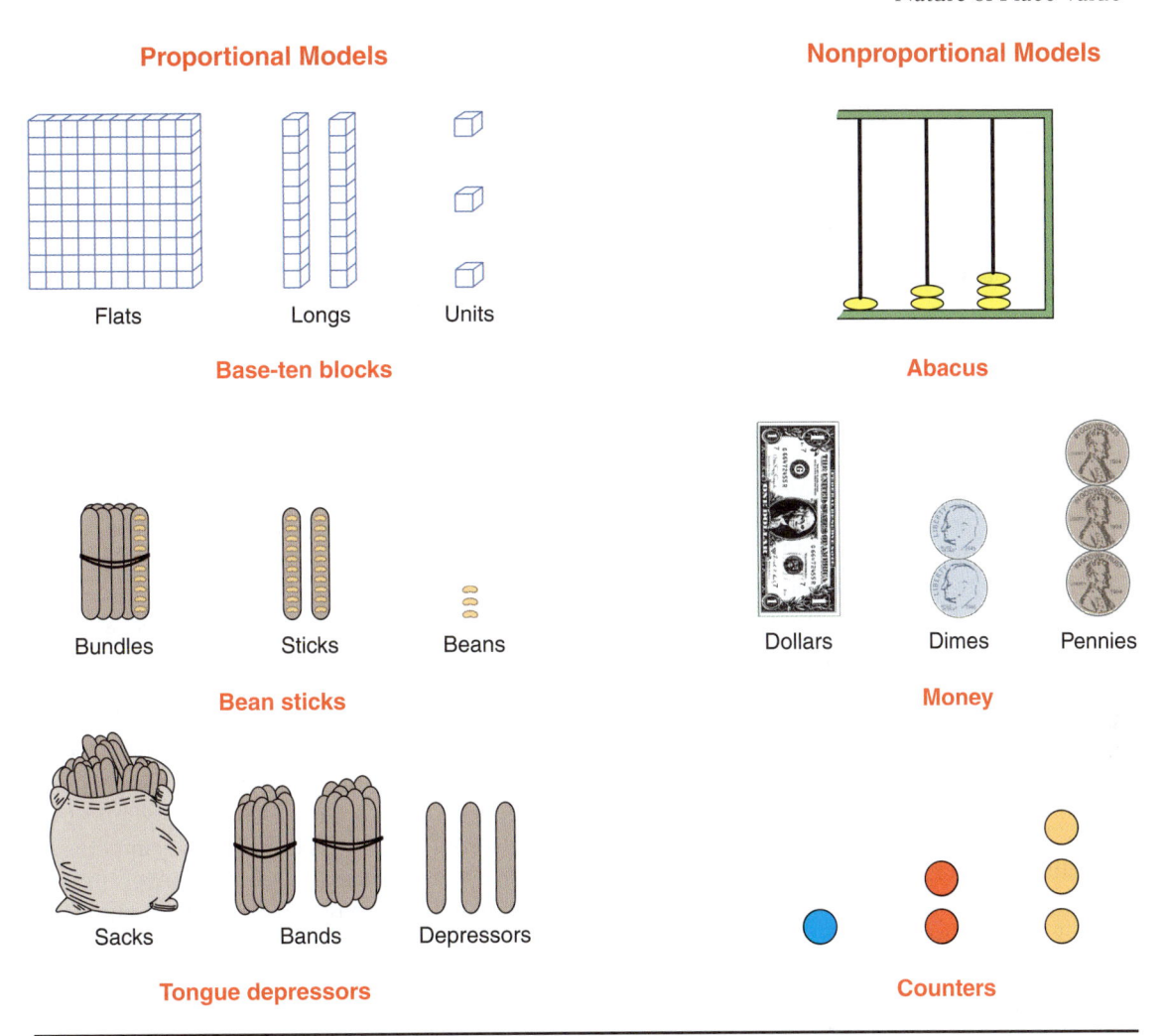

Abacus

Money
- Dollars
- Dimes
- Pennies

Counters

Figure 8-4 *Place value models: Proportional models and nonproportional models.*

All of the models shown in Figure 8-4 represent the same three-digit number, 123. The value of using different embodiments is that a child is less likely to associate place value with a particular model. In fact, a key instructional goal is to develop concepts to a level that does not depend on any one physical model, instead providing for abstraction of the commonality among all models (Chapter 2).

Since nonproportional materials are not related by size, the child must understand the exchange relationship among the pieces. Thus, the child must already understand base-ten relationships, and these relationships take time to develop. These ideas introduce children to a level of abstraction based entirely on trading rules. Young students often focus on size proportionality, which is why they are often willing to trade a dime for one or two pennies or why they prefer to have a few dimes rather than a dollar bill.

Although both types of embodiments are important and should be represented, proportional models are more concrete, and children need to use and clearly understand them before moving on to nonproportional models.

Of the nonproportional models shown in Figure 8-4, the abacus and the counters are similar. In each model, different-colored beads or counters provide the basis for trading. For example, 10 orange beads (or counters) might be traded for 1 red, 10 red for 1 blue, and so on. Use of a trading mat can help keep the counters in order. The beads on an abacus are arranged in a fixed order. The color distinction is important for the early establishment of proper trades, but it should be dropped as soon as possible so that attention shifts from color of the beads or counters to their position. Only the position of the counter has long-range significance.

GROUPING OR TRADING

Children need experience in counting piles of objects; trading for grouped tens, hundreds, and thousands; and talking about the results. The bean sticks and ten-frame provide two early models for counting and grouping. As children work with these models, they use these materials (e.g., beans, buttons, cubes, or other counters) to practice counting and grouping. These ungrouped materials provide valuable experience and prepare children for using pregrouped materials. Notice this transition from an ungrouped pile of beans to grouping by tens to representing that same quantity with pregrouped beans sticks to recording symbols.

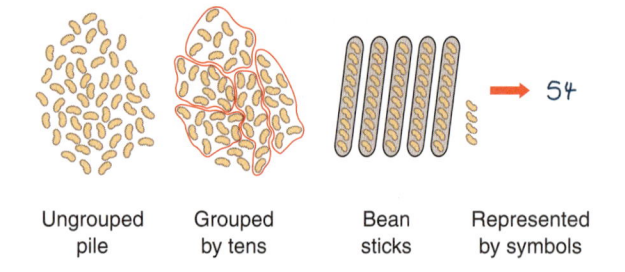

160 Chapter 8 • Extending Number Sense: Place Value

| Ungrouped pile | Grouped by tens | Bean sticks | Represented by symbols |

The transition from 10 beans to 1 stick with 10 beans illustrates how the number 10 is composed of 10 ones.

While ungrouped materials provide practice in counting and grouping, it is tedious and time consuming to count beans and to build models as quantities increase. For example, a large pile of beans is impressive and useful for demonstration but often not practical for use at individual desks. Pregrouped materials make it easy to build larger numbers.

Trading and grouping by tens provide problem-solving activities that contribute to number sense and provide opportunities for developing mental computation. Asking children to group by tens as they count the larger piles serves several valuable purposes. First, if a child loses count, correction is easier if these smaller groups have been formed. It is also easier to check for errors by inspecting groups of 10 than to recount the entire pile. The most important purpose of this practice, however, is that it shows children how an unknown quantity can be organized into a form that can be interpreted by inspection. This process of grouping by tens is the framework for place value.

BEGINNING PLACE VALUE

A PLACE TO START

You may have seen some children's errors, such as those shown in Figure 8-5. These examples illustrate confusion involving different aspects of place value. Later errors in computation can often be traced back to a lack of understanding about place value (Whitenack et al., 2004). The importance of place value makes getting a good start essential.

In developing place value and establishing number names, it is far better to skip beyond the teens and start with the larger numbers. Figure 8-1 says "the numbers from 11 to 19 are composed of a ten and one, two, …or nine ones." However, Figure 8-5 illustrates that the names for the numbers 11–19 are not consistent with the names for other numbers, even though the symbolization or visual pattern is wholly consistent. This lack of consistency between the oral and written form of the teen numbers is confusing, and children need to explore the vocabulary of larger numbers before the place-value counting patterns are evident. To exhibit oral patterns consistent with place-value characteristics, the numbers 11-19 would have to be renamed onety-one (1 ten and 1 one), onety-two, …, onety-nine, which would make them consistent with larger numbers, such as forty-one, forty-two, …, forty-nine. In many other countries, such as Japan, the naming pattern for the numbers 11–19 is consistent with the naming of larger numbers (Yoshikawa, 1994).

What does a number such as 25 mean? The number 25 can be thought of as 25 ones, but it is also composed of 2 tens and 5 ones. It can be decomposed into 1 ten and 15 ones, as well as 5 groups of 5 ones. It is important that children have experiences thinking of numbers in various ways. The ability to compose and decompose numbers in different ways reflects good number sense. For example, when thinking of money, several combinations of coins—such as 25 pennies, 2 dimes and 5 pennies, 5 nickels, or 1 quarter—may be imagined. With certain items, such as eggs, 1 more than 2 dozen and 4 six-packs plus 1 are two different but equivalent representations of the quantity 25 and thus different ways of recomposing or decomposing 25 into other meaningful representations. With metric measures, 25 centimeters might be thought of as 2 decimeters and 5 centimeters; however, it could also be thought of as 0.25 meter. Children with good number sense know when a particular form is useful.

Of course, if place value is to be called on, the tens and ones model is needed. It might use money (2 dimes and 5 pennies) or another model, such as bean sticks. With the bean sticks, 25 could be represented several different ways:

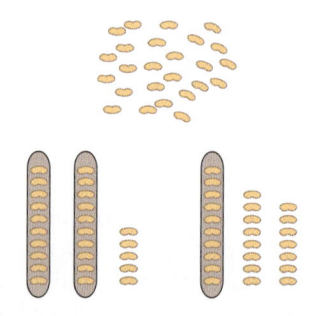

Counting	Becoming confused when counting the teen numbers because of the lack of pattern in the numbers from eleven to nineteen
Bridging the decade or hundred	Making the transition when counting; for example, counting aloud "thirty-eight, thirty-nine, thirty-ten" or writing "38, 39, 3010"
Reversing digits when writing number	For example, writing 52 and 25 and not recognizing any difference in these numbers
Writing numbers that were read aloud	For example, writing "one hundred sixty-four" as 100 604

Figure 8-5 *Some common errors related to place value.*

Which way is better? It depends on what is to be done with the 25 beans because there are times when each form may be useful. The grouping at the top, for example, would be easier to divide among several people. Either of the groupings on the bottom would be easier to count.

The notion of representing a quantity with the least number of pieces for a particular model is critical in place value. Establishing its importance at an early stage can eliminate some later errors such as:

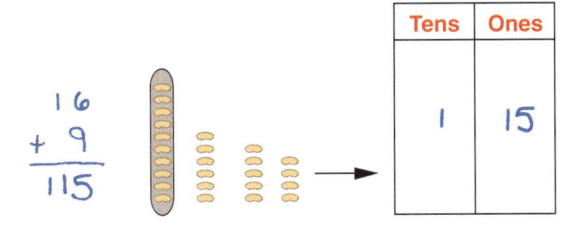

Because 10 or more of something (namely, ones) exist on the place-value mat, a trade must be made. Making the ten-for-one trade results in the least number of pieces and thus 25 becomes the only representation that is meaningful:

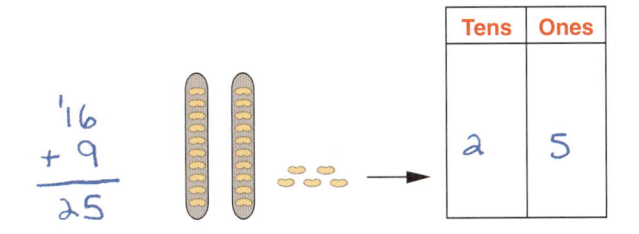

Figure 8-6 highlights the advancement from a concrete model to a symbolic representation. The bridges from the physical models to the symbolic representation must be crossed back and forth many times if meaningful learning is to occur. Careful attention must be given to linking modeling with the language. As children become fluent in talking about their models, it will become natural for them to describe 25 in different ways.

Many children reverse the digits of numbers. Although this error is generally caused by carelessness, it may be symptomatic of a disability known as *dyslexia*. In either case, it is important that children understand the consequences of such reversals. Consider, for example, the numbers 25 and 52. The ten-frame (Blackline Master in Appendix) provides a convenient model for representing these two-digit numbers.

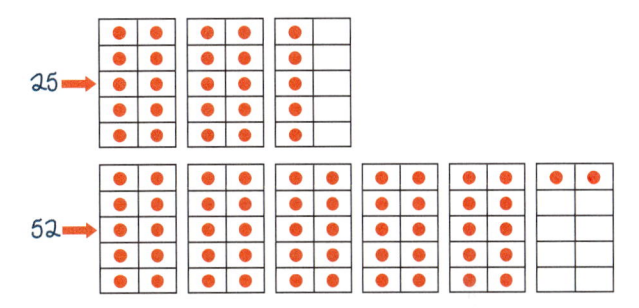

This provides a visual reminder of the physical differences between 25 and 52, which are formed with the same digits but are different because of place value. Children should compare the modeled numbers and talk about them in an effort to better appreciate the magnitude of the differences.

When quantities are grouped by ten, it not only illustrates place value, but it facilitates counting. For example,

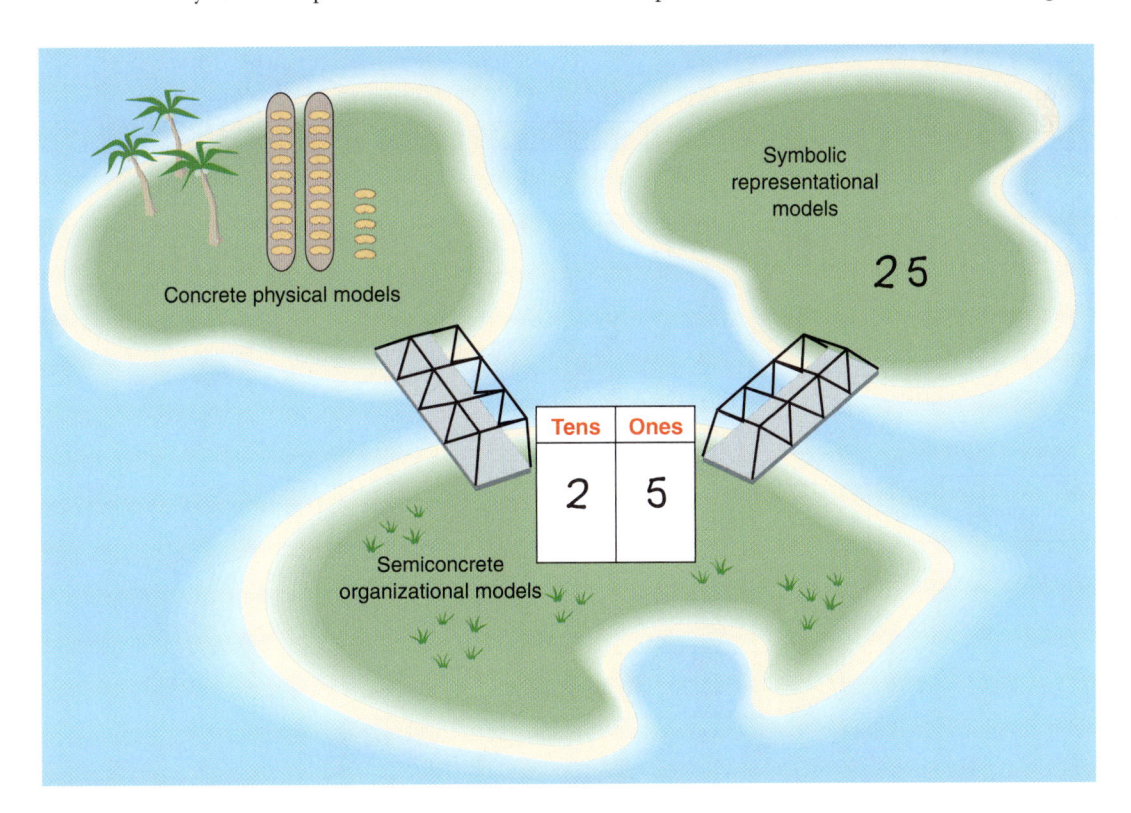

Figure 8-6 *The idea of place value can help children cross from concrete models of numbers to their symbolic representations.*

162 Chapter 8 • Extending Number Sense: Place Value

these two piles show the same number of buttons. One has all of the buttons in a string. The other has recomposed the buttons into groups of tens and ones. Decide how many buttons are in each pile.

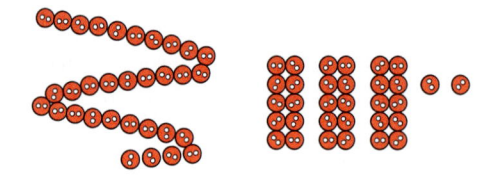

If you wanted to know how many buttons are in each pile, which arrangement would you use? Explaining which pile and telling why that arrangement was chosen leads to a discussion of how grouping by tens facilitates counting and organizing larger quantities.

The base-ten blocks, together with the place-value mat, as shown in Figure 8-7, can be used to model the additive property and illustrate expanded notation for 123. A variety of strips, such as those in Figure 8-7, which connect the base-ten blocks to number symbols in an expanded form, can be used effectively to review key ideas. While 123 is composed of 1 hundred, 2 tens, and 3 ones, if you were dividing the number by 3, it would be easier to think of 123 as 12 tens and 3 ones. Encouraging children to name the same number in different ways promotes number sense.

In the Classroom 8–1 shows two slightly different hundreds charts. One chart goes from 1 to 100, the other from 0 to 99 (Blackline Master in Appendix). One advantage of the latter chart is that the tens digit in each row is constant, whereas in the 1-to-100 chart the tens digit always changes in the last column. The 1–100 chart eliminates a common problem of students starting on 0 and becoming confused with their result. Either hundred chart can be used to explore place value, patterns, mental computation, and algebraic thinking. This chapter's Snapshot of a Lesson shows children modeling the tens and ones for numbers that are identified on a hundred chart.

A hundred chart also provides many opportunities for counting and doing mental computation. For example, 46+30 can be determined by counting mentally 46, 56, 66, 76 and is a natural by-product of counting by tens on the hundred chart. Furthermore, 46+29 can be found by counting 46, 56, 66, 76 and then dropping back one to 75. Counting by tens and then dropping back or bumping up illustrates how to adjust numbers and be flexible when using and thinking about numbers.

Consider the following diagram and how patterns and generalizations lead to algebraic thinking:

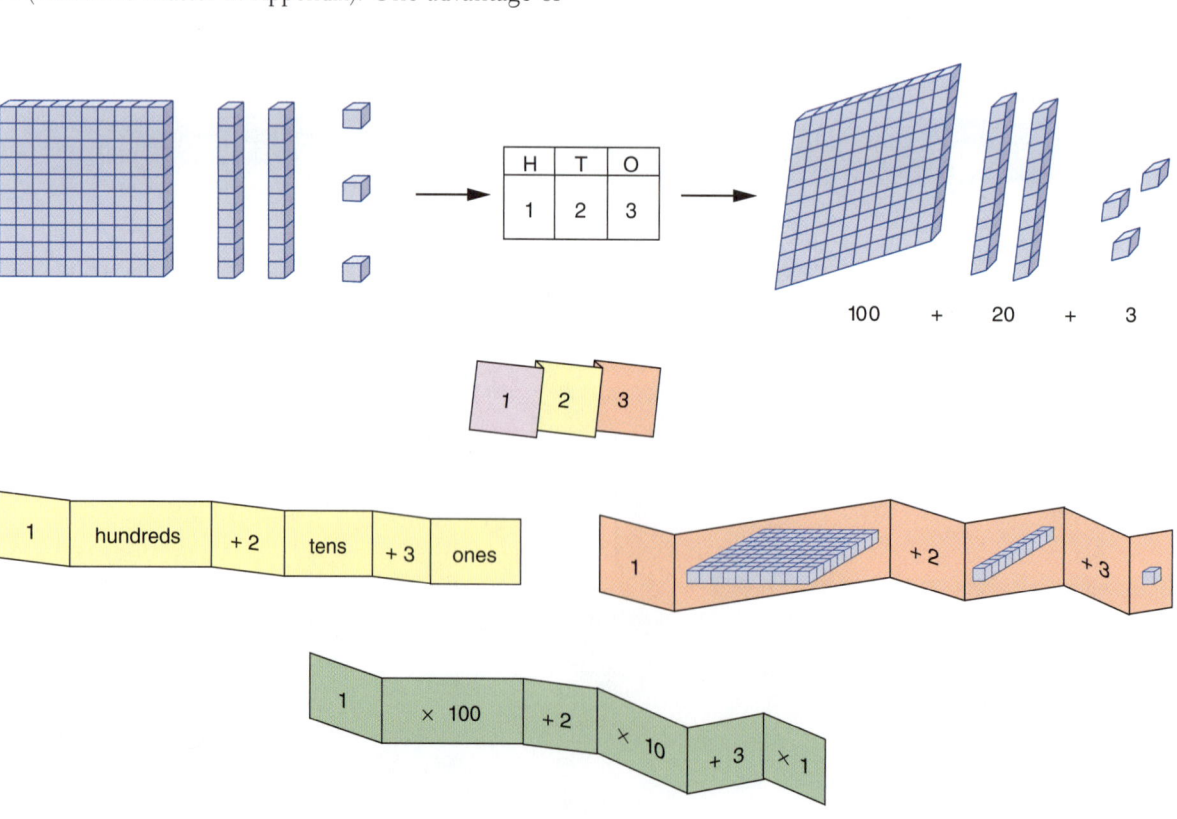

Figure 8-7 *Connecting models and symbols that reinforce place value.*

Beginning Place Value   163

In the Classroom 8–1

MOVING ABOUT ON A HUNDRED CHART

Objective: Using a hundred chart to explore place value and pattern.

Grade Level: 2–3.

1	2	3	4	5	6	7	8	9	10
11	12	13	14	15	16	17	18	19	20
21	22	23	24	25	26	27	28	29	30
31	32	33	34	35	36	37	38	39	40
41	42	43	44	45	46	47	48	49	50
51	52	53	54	55	56	57	58	59	60
61	62	63	64	65	66	67	68	69	70
71	72	73	74	75	76	77	78	79	80
81	82	83	84	85	86	87	88	89	90
91	92	93	94	95	96	97	98	99	100

↑ Fifth column

0	1	2	3	4	5	6	7	8	9
10	11	12	13	14	15	16	17	18	19
20	21	22	23	24	25	26	27	28	29
30	31	32	33	34	35	36	37	38	39
40	41	42	43	44	45	46	47	48	49
50	51	52	53	54	55	56	57	58	59
60	61	62	63	64	65	66	67	68	69
70	71	72	73	74	75	76	77	78	79
80	81	82	83	84	85	86	87	88	89
90	91	92	93	94	95	96	97	98	99

▼ Use the hundreds charts to answer these questions:
- What is alike for all the numbers in the fifth column?
- How are the numbers in the fourth and sixth columns alike? Different?
- Tell where you stop if you start on any square in the first three rows and count forward 10 more squares.
- Start on a different square in the first three rows and count forward 10 more squares. Where did you stop?
- After you have done this several times, tell about a pattern that you found.

▼ Cut out a piece like this and lay it on the chart.
- What numbers are covered? Tell how you found them.
- Move the piece to a different place, and tell what numbers are covered.
- Do it again with different shapes like and

▼ Here is only a part of a hundred chart:
- Use what you know about a hundred chart to
- Find A _____ B _____ C _____ D _____
- Tell two different ways to find C.
- Suppose 46 is replaced by N.
- Find A _____ B _____ C _____ D _____

(positions shown: A above-right, 46 center, B to left, C D to right)

Patterns from the hundred chart suggest these solutions directly from the hundred chart:

A = 46 − 10
B = 46 + 10
C = 46 + 20 + 1
D = 46 + 30

If the diagram is shifted on the hundred chart, then 46 changes to some number N. Although the starting number N has changed, the solutions are similar, as

A = N − 10
B = N + 10
C = N + 20 + 1
D = N + 30

Such activities illustrate a variable and the power of patterns in promoting algebraic thinking.

Modeling three-digit numbers provides opportunities to show how the rearrangement of digits impacts a number. Notice the physical difference between 134, 314, and 413.

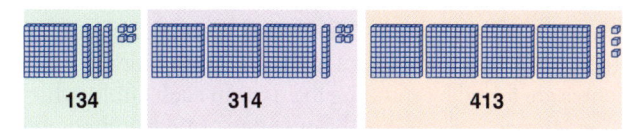

134   314   413

Looking at the different arrangements shows that 413 is about 100 more than 314 and that 134 is more than 100 less than 314. This ability to focus on the lead, or "front-end," digits is an important part of number sense.

In the Classroom 8–2

WHAT'S IN A SCORE?

Objective: Counting by multiples of ones, tens, and hundreds to develop place value.

Grade Level: K–2.

Here is how darts landed for this score.

- Determine the score for these boards:

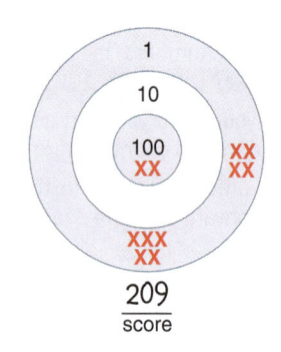

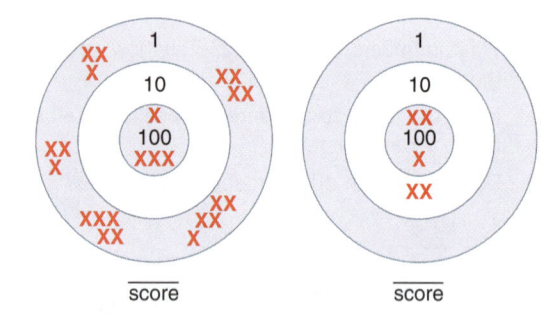

(Board 1: 1 ring – empty; 10 ring – XXX XX; 100 ring – XX XX; score: 209)

(Board 2: 1 ring – XX X; 10 ring – XX XX; 100 ring – X XXX; outer – XXX XX, XX X; score: ___)

(Board 3: 1 ring – (empty); 10 ring – XX; 100 ring – XX X; score: ___)

- Place darts on these boards to show how these scores might have been made.

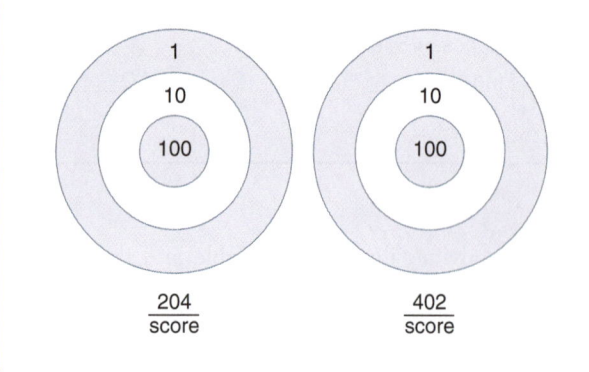

(Board: score 204)

(Board: score 402)

In the Classroom 8–2 reinforces place value and provides practice in important mental computational skills. Scores are found by counting the darts in each circle, which provides practice counting by ones, tens, and hundreds and then computing the totals.

A calculator provides many opportunities to practice and develop important place-value concepts. Wipe Out is a place-value game that involves either addition or subtraction using a calculator. The goal is to change (wipe out) a predetermined digit by subtracting or adding a number. This activity can be made into a competitive game for two people. The players take turns entering a number and naming a specific digit the other player must change to 0.

For example, Kelly enters 431, naming the 3 to be wiped out:

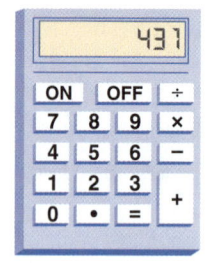

(Calculator display: 431)   (Calculator display: 401)

Tanya wipes out the 3 by subtracting 30, which also leaves the other digits unchanged.

A player scores a point for changing the digit to 0 on the first try. A record of the game in table form reinforces the identification of the correct place value:

WIPE OUT RECORD   Name: Tanya

Entered	Wiped Out	Keys Pressed	Display	Score
431	3	−30	401	1
24	4	+6	30	1
849	8	−800	49	1
206	2	−200	6	1

Although Wipe Out or variations of it can be played without a calculator, it is much more exciting with one. Children don't get bogged down with computation. The focus remains on place value. Furthermore, they are often surprised by what happens when they make a place-value error, which increases their place-value understanding.

Math Links 8.2

There are a number of excellent Web sites that provide opportunities for children to make patterns, practice counting, and develop a better sense about numbers. Go to the National Library of Virtual Manipulatives at http://nlum.usu.edu. There you will find Base Blocks, an Abacus, and Chip Abacus in the K–2 resources all of which provide opportunities to construct and model numbers. You can use the Base Blocks to provide explorations in different bases.

These sites can be accessed from this book's Web site.

www.wiley.com/college/reys

EXTENDING PLACE VALUE

Research reports that many children lack an understanding of the relative sizes of numbers greater than 100 (Fuson, 2003a, 2003b). This results from many factors, one of which may be the lack of opportunities to model, which helps children develop a visual awareness of the relative sizes of numbers. Figure 8-7 showed a physical mode for 123. Here are several different equivalent representations:

One hundred twenty-three

12 tens and 3 ones

$100+20+3$

$1(100)+2(10)+3$

$1(10^2)+2(10)^1+3$

For large numbers, children can use variations of the models shown in Figure 8-7. For the base-ten blocks, lay another place-value mat to the left of the mat holding hundreds, tens, and ones. Figure 8-8 shows the two mats, where the thousands mat holds thousands, ten thousands, and hundred thousands. In addition to these physical models, stories such as those in the Book Nook for Children provide real-world connections to larger numbers.

Figure 8-8 provides another demonstration of how two numbers can have the same digits but be different. How are 2130 and 1032 alike and different? As children engage in discussion to answer this question, their knowledge of place value and their sense of numbers will grow. Using the same digits to represent different numbers helps children appreciate the importance of representing the place values accurately. Although the numbers 2130 and 1032 use the same digits, the models that represent these numbers are dramatically different. Furthermore, the symbolic representation of them is different. For example:

Two thousand one hundred thirty is 2130 or in expanded form is $2(1000)+1(100)+3(10)$ or $2(10^3)+1(10^2)+3(10)^1$.

One thousand thirty-two is 1032 or in expanded form is $1(1000)+3(10)+2$ or $1(10^3)+3(10)^1+2(10)^0$.

Notice how the language used to describe these expanded notations, namely, *ten cubed* and *ten squared*, can be linked directly to the base-ten model in Figure 8-8. This type of experience helps develop number sense and alerts children to the importance of the front-end, or lead, digits, which in this case denote thousands.

As the number of digits increases, children should be encouraged to focus on the front-end digits. The front-end digits are used when comparing and ordering numbers as well as when computing mentally and estimating:

3000 + 4000

3 + 4 = 7. That's seven thousand.

Mental computation

5286

That's more than five thousand, but less than six thousand.

Estimating

Thousands

H	T	O		H	T	O
		2		1	3	0

How many thousands? [2]
How many hundreds? [1]
How many tens? [3]
How many ones? [0]
Can you say the number?
[two thousand one hundred thirty]

Thousands

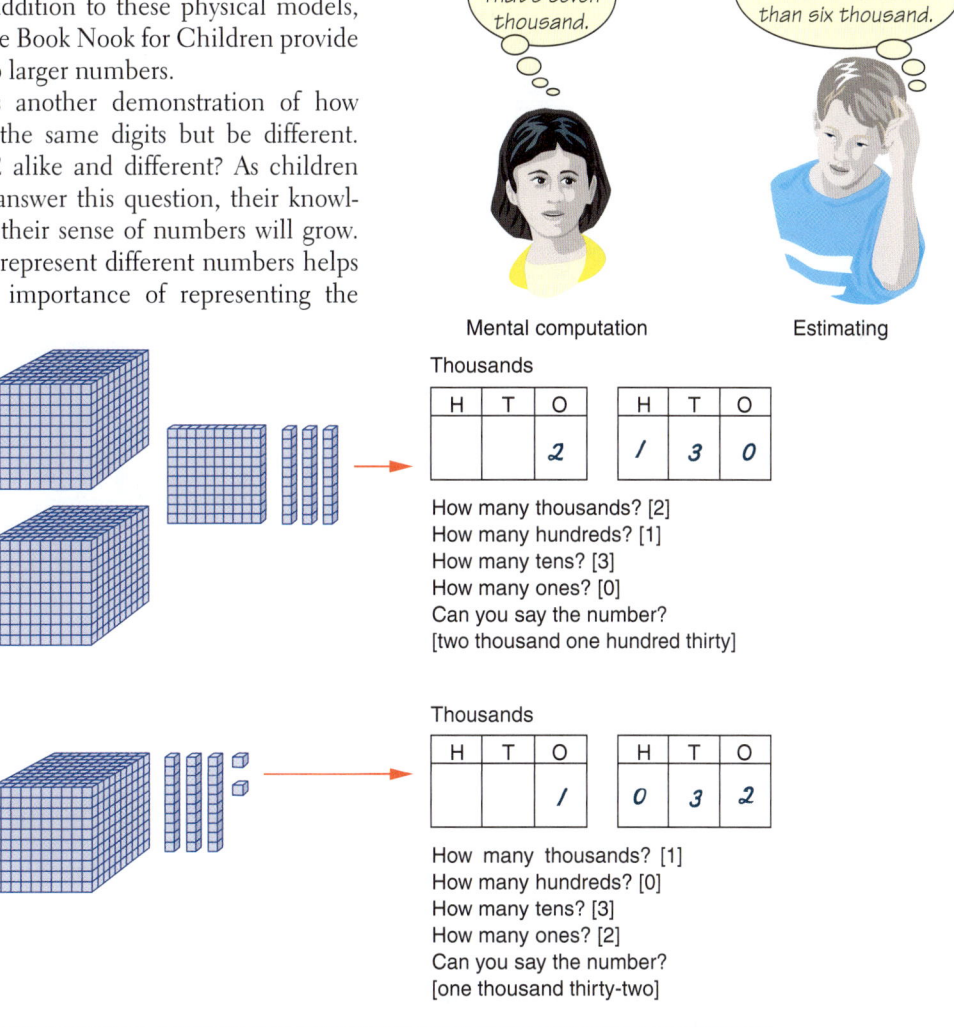

H	T	O		H	T	O
		1		0	3	2

How many thousands? [1]
How many hundreds? [0]
How many tens? [3]
How many ones? [2]
Can you say the number?
[one thousand thirty-two]

Figure 8-8 *Using a thousands mat to demonstrate how two numbers can have the same digits but be different.*

166   Chapter 8 • Extending Number Sense: Place Value

The front-end approach can be naturally extended and applied to larger numbers. For example, students could be asked to decide which number is larger when the front-end digits are the same (a). And then asked the same question when the backend digits are the same (b).

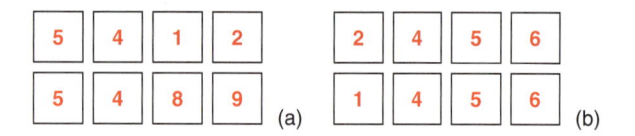

5	4	1	2		2	4	5	6
5	4	8	9		1	4	5	6

(a)   (b)

Using front-end digits helps students compare and order larger numbers. When the front-end digits are the same as in (a), it may be necessary to compare additional digits. However, the front-end approach is used only when the whole numbers have the same number of digits. For example, the front-end digits are not needed to compare 2456 with 897 because one whole number has two thousand and the other has zero thousand.

2	4	5	6
8	9	6	

COUNTING AND PATTERNS

Practice in skip counting helps decrease bumps in the counting road. Variations of the hundreds charts used in Chapter 7 can be a valuable tool in smoothing out the bumps.

Calculators are also useful in counting and pattern recognition. Seeing each value displayed on the calculator helps students develop important insight into what digits are changing and when. In the Classroom 7–4 provided a calculator counting activity. When children are counting by ones, they observe that the digit on the right (ones place) changes every time they "count," while the next digit (tens place) changes less frequently, and it takes much counting to change the third digit (hundreds place). On the other hand, Figure 8-9 develops an additional understanding

Counting by ones	Counting by tens	Counting by hundreds	Counting by thousands
284	284	284	284
285	294	384	1284
286	304	484	2284
287	314	584	3284

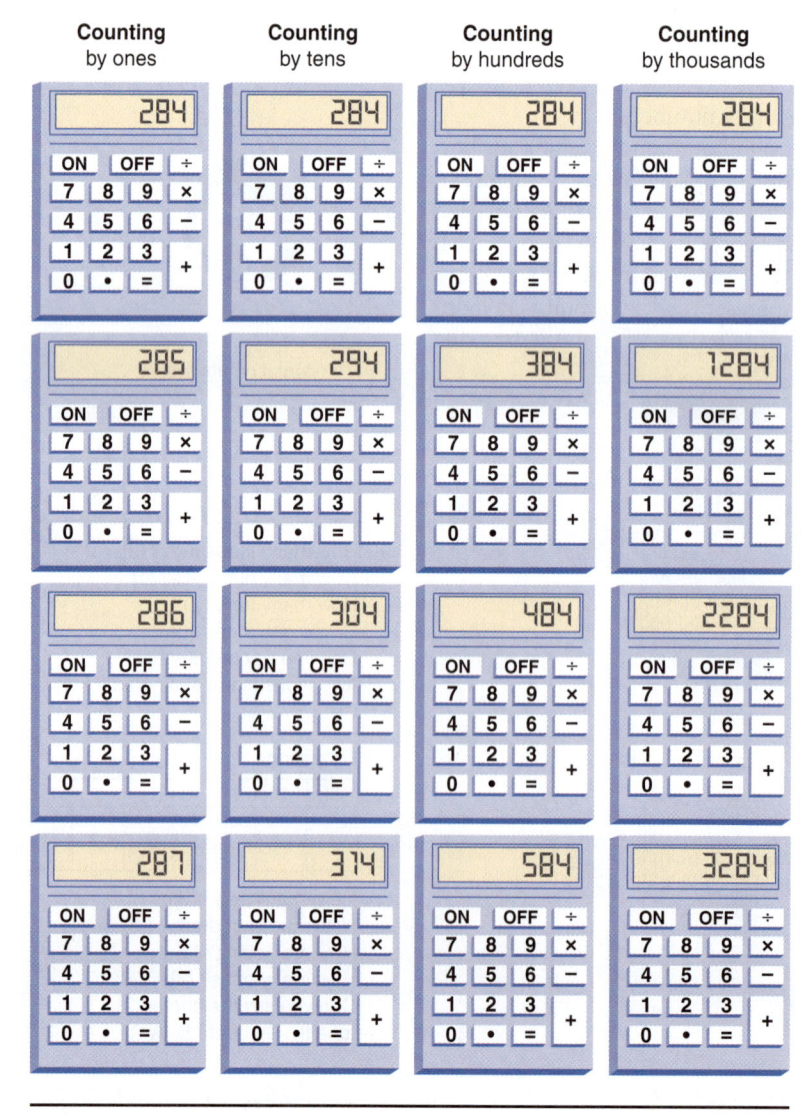

Figure 8-9 *Calculator counting to illustrate place-value patterns.*

Extending Place Value   167

of place value by displaying results of counting by ones, tens, hundreds, and thousands.

Calculator counting provides many opportunities to discuss patterns related to place value. Such counting can also contribute to a better grasp of large numbers, thereby helping to develop students' number sense. For example, having students count by ones to 100 and 1000 with the calculator helps them better understand the magnitude of these numbers. By expanding this activity, students will come to realize that it takes about the same amount of time to count to 1000 by ones on a calculator as to count to 1,000,000 by 1000 or to 100 by 0.1 on a calculator.

For example, one fourth grader, after doing this calculator counting, said, "That means there are as many thousands in one million as ones in one thousand." This is a profound observation—of the type that leads to a better understanding of both place value and large numbers, and it also reflects a growing sense of numbers.

As Figure 8-9 shows, counting on by tens, hundreds, or thousands never changes the ones place; however, when counting by tens, the tens place changes on each count and the next digit (hundreds) changes every 10 counts. Observing these patterns in counting larger and larger numbers helps students recognize place-value properties.

Counting back provides similar place-value patterns. The calculator facilitates counting back by fives, tens, and hundreds or whatever, and the resulting values when starting at 10 and counting back by fives, tens, and hundreds are shown in Figure 8-10. Place-value patterns resulting from counting back are apparent. Observing these patterns in counting larger and larger (or smaller and smaller) numbers helps students recognize that place-value properties are reflected in both positive and negative integers.

As you have seen, the hundred chart provides a useful model for counting and pattern recognition related to place value. Activities involving counting with multiples of 10 were illustrated in In the Classroom 8–1. In the Classroom 8–3 shows how similar ideas can be extended to a thousand chart. For example, consider this diagram and the thousand chart in In the Classroom 8–3:

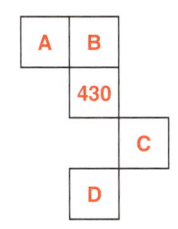

	A	B	
		430	
			C
		D	

A solution for this diagram requires the use of patterns:

This activity also leads to algebraic thinking when 430 is replaced by any number N.

A = N − 100 − 10
B = N − 100
C = N + 100 + 10
D = N + 200

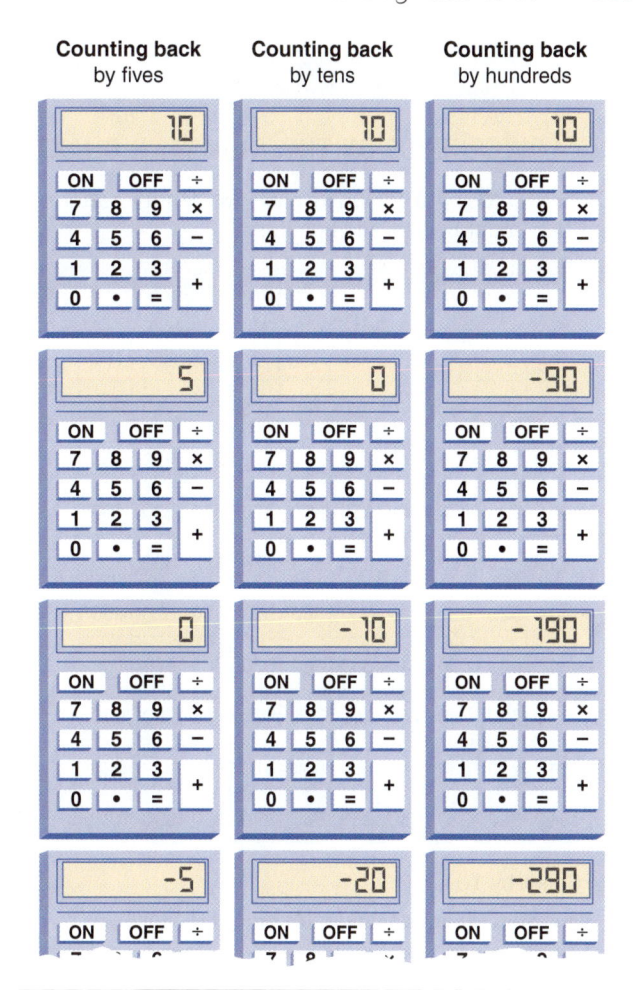

Figure 8-10 *Counting back to illustrate place value with negative integers.*

Such experiences develop students' number sense and provides practice in mental computation.

COMPOSING AND DECOMPOSING

Composing or decomposing happens when 6 tens 7 ones are considered as 5 tens 17 ones, or 245 is thought of as 24 tens 5 ones, or 40 pennies are traded for 4 dimes. The importance of clearly understanding the regrouping (i.e., composing and decomposing) process cannot be overemphasized (Goodrow and Kidd, 2008). Understanding is most likely to develop when children experience this bridging with physical models and practice trading and regrouping. Regrouping and place value are intertwined in later development of computation. Regrouping or composing happens whenever bridging occurs, as from one ten to another (such as 29 to 30) or from one hundred to another (such as 799 to 800).

Whenever regrouping or trading occurs, there are accompanying changes in how the number is recorded. Understanding this changed notation requires many experiences with problems involving trading and the related recording process. Figure 8-11 shows how regrouping affects digits and place value. Thus the ten pennies are regrouped and traded for one dime, resulting in $0.29 + $0.01 = $0.30.

In the Classroom 8-3

THE POWER OF 10 ON THE THOUSAND CHART

Objective: Using a thousand chart to identify patterns and skip count by tens and hundreds.

Grade Level: 2–3.

10	20	30	40	50	60	70	80	90	100
110	120	130	140	150	160	170	180	190	200
210	220	230	240	250	260	270	280	290	300
310	320	330	340	350	360	370	380	390	400
410	420	430	440	450	460	470	480	490	500
510	520	530	540	550	560	570	580	590	600
610	620	630	640	650	660	670	680	690	700
710	720	730	740	750	760	770	780	790	800
810	820	830	840	850	860	870	880	890	900
910	920	930	940	950	960	970	980	990	1000

- ▼ Count by 10:
  - Start on any square in the first three rows.
  - Count forward 10 squares, and tell where you stopped.
  - Start at a different square, and count forward 10 squares.
  - After you have done this several times, tell about a pattern that you found.
  - Describe a quick way to count "a hundred more" on this thousand chart.
- ▼ Count by 100:
  - Tell how you could use the thousand chart to add 300 to 240.
  - Tell how you could use the thousand chart to add 290 to 240.
- ▼ Connect the charts:
  - Tell how using the hundred chart helps you use the thousand chart.

Figure 8-11 *Nonproportional model illustrating relation between regrouping and place value.*

that continues to occur as new places are used. Children soon recognize that it becomes cumbersome to model large numbers with proportional models.

The calculator can be used with very large numbers. Minor variations of In the Classroom 7–7 (such as finding how long a calculator takes to count to one million or one billion) can help students develop a better grasp of large numbers. As the magnitude of numbers increases, physical models of them become more difficult to represent. Yet the need to continue to develop a sense of large numbers is a major goal of school mathematics (Kastberg and Walker, 2008).

READING AND WRITING NUMBERS

Reading and writing numbers are symbolic activities and should follow much modeling and talking about numbers. This recommendation is based on research with young children and alerts us to the danger of a premature focus on symbols (Verschaffel et al., 2007). A sustained development of number sense should accompany reading and writing numbers. This approach ensures that the symbols the students are writing and reading are meaningful to them.

Consider some ways in which understanding place value helps develop reading and writing numbers. Take the example of the number 123. The places (hundreds, tens, ones) as well as the value of each (1, 2, 3) are easy to identify. The 1 means 1 hundred. The 23 is both 2 tens 3 ones and 23 ones, and 123 is 1 hundred 2 tens 3 ones; 12 tens 3 ones; and 123 ones. These representations are equivalent and may be shown on the place-value mat:

Figure 8-12 further illustrates the regrouping process with two different models, one proportional and one nonproportional. Similar models with larger numbers should be used as soon as children have grasped the trading principles involving ones, tens, and hundreds. In fact, this extension process demonstrates the power of mathematical abstraction. Extending to thousands should be done with proportional models to illustrate the dramatic size increase

Reading and Writing Numbers   169

Abacus

Base-ten blocks

The largest one-digit number:

Add one.
9 ⟶ 10

The largest two-digit number:

Add one.
99 ⟶ 100

The largest three-digit number:

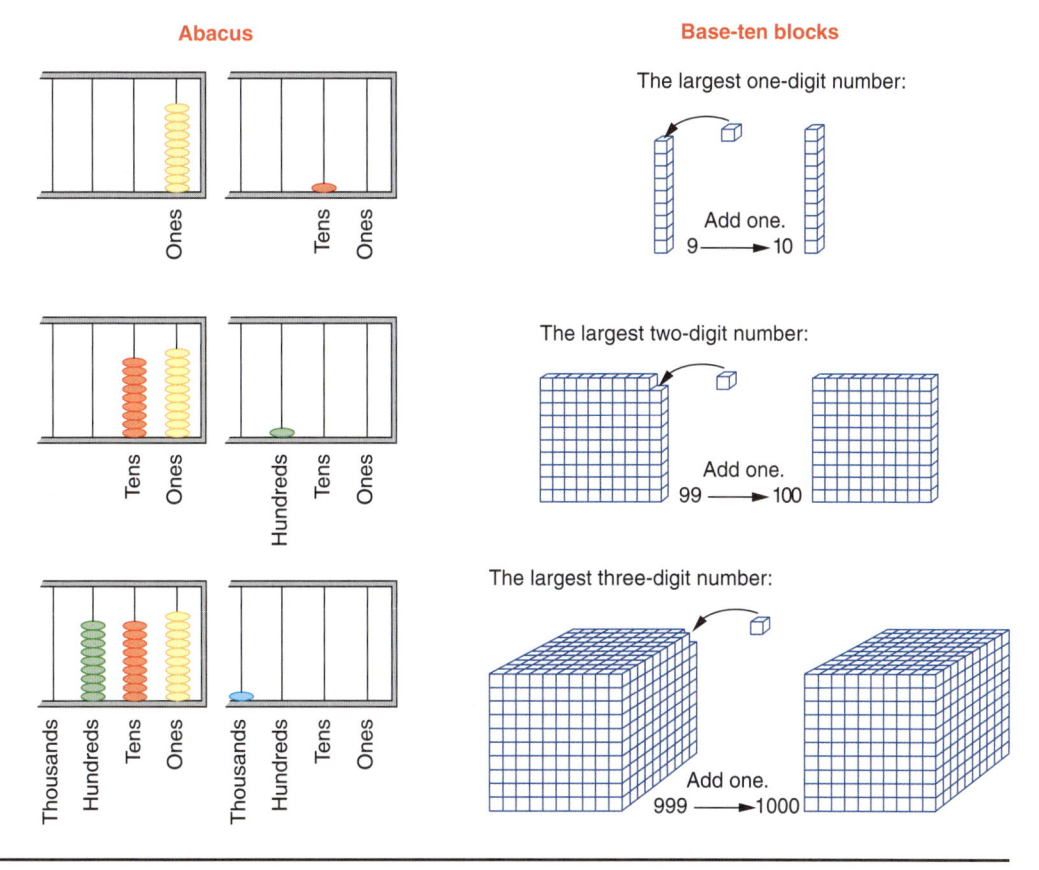

Add one.
999 ⟶ 1000

Figure 8-12 *Nonproportional and proportional models illustrating some relationships between regrouping and place value.*

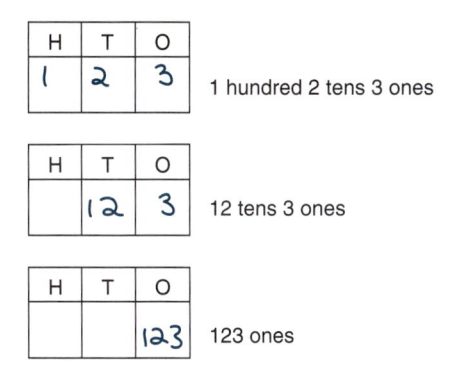

H	T	O
1	2	3

1 hundred 2 tens 3 ones

H	T	O
	12	3

12 tens 3 ones

H	T	O
		123

123 ones

In the Classroom 8–4 provides practice to further develop place value.

The skill of reading numbers in different ways (and understanding these representations) can be useful in many operations with whole numbers. For example, rereading provides a nice stepping stone to mental computation; more specifically, it leads to multiplying a number by 10:

H	T	O
3	8	0

Reads: 3 hundreds 8 tens 0 ones.
Realizes there are no ones.
Rereads: 38 tens, which is 380.

This idea extends naturally to larger numbers. For example, twenty three thousand is 23 (1000) or 23,000. Likewise, fifty million is 50 (1,000,000) or 50,000,000.

In the Classroom 8–4

PLACE VALUE AND THE POWER OF MOVING DIGITS

Objective: Using number cards to form different multi-digit numbers.

Grade levels: 1–3.

Use cards with 0, 1, 2, …9 and give student 2–5 cards.
Ask students to use any three of their cards to create a number that is

A. A three-digit number ___ ___ ___

B. The largest three-digit number possible with your cards ___ ___ ___

C. The smallest three-digit number possible with your cards

D. The largest three-digit even number possible with your cards ___ ___ ___

E. An even number in the tens place ___ ___ ___

F. More than 200 ___ ___ ___

G. Less than 600 ___ ___ ___

H. Between 100 and 400 ___ ___ ___

Although it seems logical to write number words as they sound, this procedure can lead to difficulty. If this were done, sixty-one would be incorrectly written as 601 and one hundred twenty-three as 100203. If a child made this mistake, the teacher could use the place 1-1 value mat as a model to demonstrate:

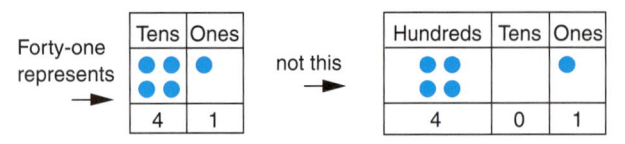

Forty-one represents →

Tens	Ones
●●●	●
●	
4	1

not this →

Hundreds	Tens	Ones
●●		●
●●		
4	0	1

Modeling several numbers on the mat can help clarify this notion.

Similar problems exist in naming and representing larger numbers. Open-ended questions such as those shown in Figure 8-13 encourage children to estimate and think about numbers. As they share their answers and talk about different ways of understanding millions and billions, their number sense grows. The Book Nook for Children identifies some great books, such as *Earth Day—Hooray!* by Stuart Murphy, *The Rajah's Rice: A Mathematical Folktale from India* by D. Berry, and *Can You Count to a Googol?* by Robert Wells. Each of these books provides real-world examples of larger numbers and connections to representing their place value symbolically. *Exploring Mathematics through Literature* (Thiessen ed., 2004) and *Math and Literature: Grades K–1 and 2–3* (Burns and Sheffield, 2004a) identify more books and ways to help children develop additional insight and appreciation of larger numbers.

It is helpful for students to link different models to larger numbers. For example, students might begin with a one-cubic-centimeter block. If they then make a cubic-meter box, that box will hold one million cubic centimeter cubes.

Base-ten blocks also can be used to help students make the connection between the concrete model and the symbolic representation, as in Figure 8-14. This model helps children mentally "see" that ten thousand is a long piece made up of 10 cubes, where each cube is one thousand. Although this model can be constructed physically, children quickly appreciate the power of constructing mental images to represent larger numbers. Once children begin to develop an intuitive grasp of larger numbers and begin to use millions and billions intelligently, they are ready to write and read these larger numbers. Place-value mats can be naturally expanded to represent larger numbers:

This provides students an opportunity to generalize that naming of numbers—as the hundreds, tens, ones—applies not only to three-digit numbers but to larger numbers as well. The only difference is that each block of three digits introduces different number period names as follows:

Ones
Thousands
Millions
Billions
Trillions

To develop facility in reading large numbers, children need to develop the correct vocabulary and practice in actually naming them aloud. For example, would you read 12,345,678 as "one ten million two million three hundred thousand four ten thousand" and so on? Certainly not! The period names would be read as "twelve million, three hundred forty-five thousand, six hundred seventy-eight." This example is a clear application of an organizing strategy: The digits within each period are read as hundreds, tens, and ones, as with "three hundred forty-five thousand." For this reason, children need to think of larger numbers (those of more than three digits) in blocks of three digits.

Recognition and understanding of the hundreds, tens, and ones pattern provide a powerful organizational strategy that can be called on in naming numbers. Only the key terms—*ones*, *tens*, and *hundreds*—along with recognition of the periods for thousands, millions, and billions are needed to name very large numbers.

In many countries, commas are not used to separate blocks of three digits. Instead, for example, the number 2346457 is written as 2 346 457. The blocks of digits remain visible but are separated by spaces rather than commas. Some newspapers, journals, and textbooks in the United States now print numbers this way. This change has instructional implications for both reading and writing numbers. In particular, children must become even more sensitive to the importance of writing numbers clearly and distinctly.

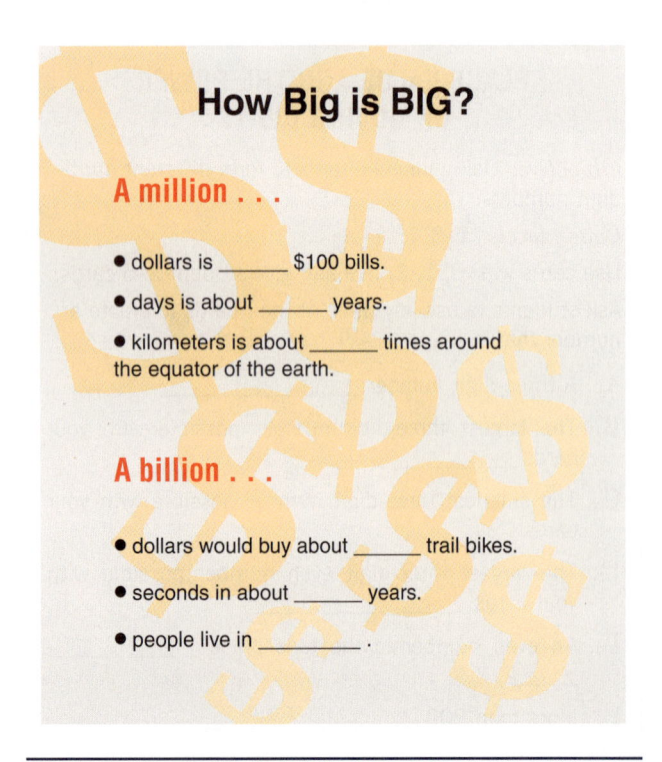

How Big is BIG?

A million . . .

- dollars is _____ $100 bills.
- days is about _____ years.
- kilometers is about _____ times around the equator of the earth.

A billion . . .

- dollars would buy about _____ trail bikes.
- seconds in about _____ years.
- people live in _____ .

Figure 8-13 *Questions to promote thinking about the size of numbers.*

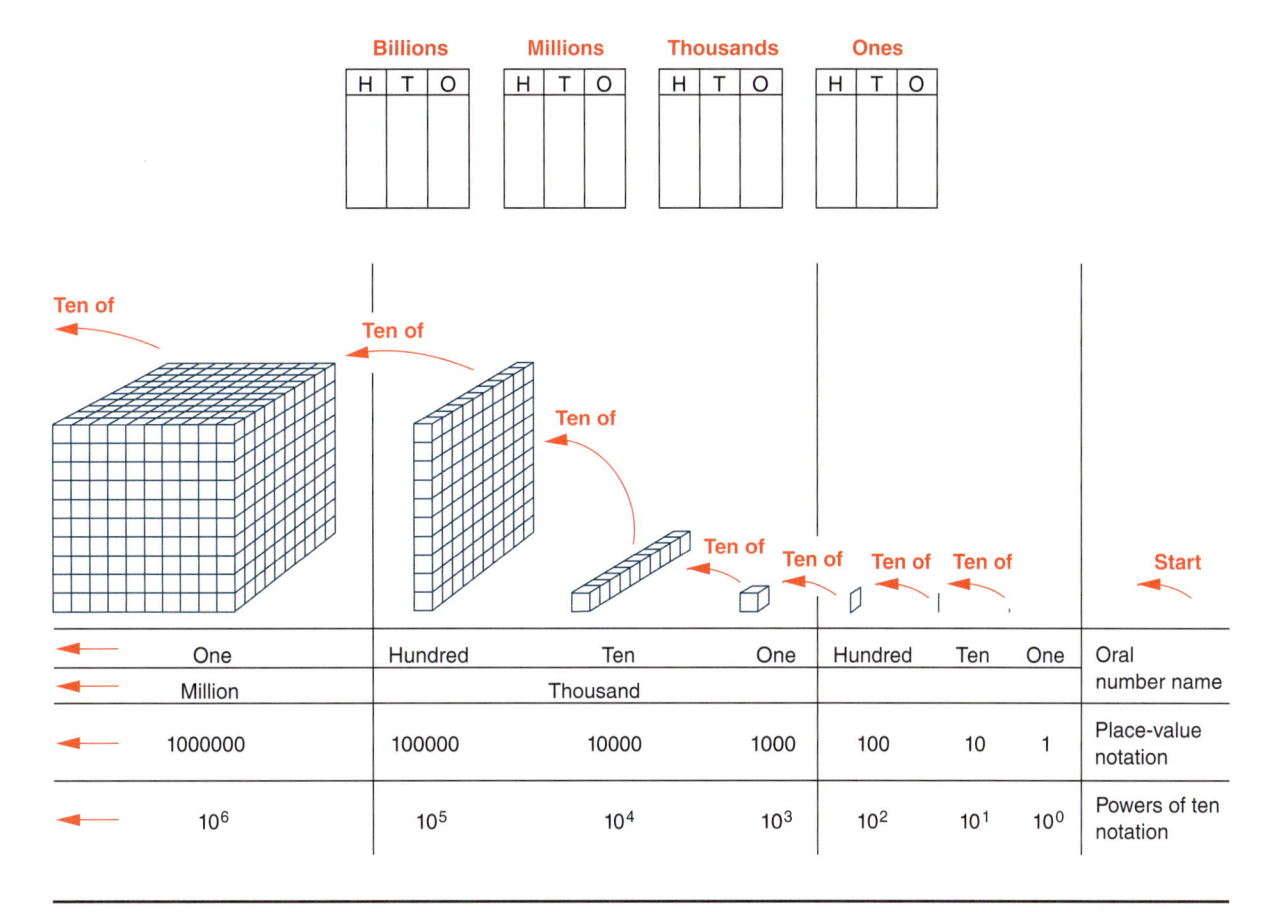

Figure 8-14 *Connecting the symbolic representation of one million with a concrete model.*

Newspapers provide a rich context to explore numbers of all sizes. Examine a newspaper and highlight all of the numbers reported in headlines and related stories. You may be surprised at the high frequency with which numbers occur. Naming numbers is clearly an important skill. Yet, with the widespread use of calculators, a more efficient way to read multidigit numbers is becoming common. For example, 32764 is read as "three two seven six four" and 4.3425 as "four point three four two five." Each of these readings is correct and much easier to say than the respective periods. There is the danger that children will say the digits without any realization of the value of the numbers involved, but such interpretations are not necessary at every stage of the problem-solving process. If it is desired to only copy a number displayed on a calculator, then a direct translation of digits is without a doubt the best way to read the number. Rather than requiring children to read numbers in a specific way, it is far better to recognize the value of each technique and encourage children to choose wisely—namely, to select the technique that is most appropriate for a given situation.

ROUNDING

An important aspect of developing number sense is recognizing that some numbers are approximate (such as our national debt) and some are exact (such as the number of people killed in a fatal airplane crash). Approximate values are associated with estimation, often involving rounding, and are encountered regularly in our daily lives.

Rounding integrates understanding of approximate values with place value and naming numbers. Numbers are usually rounded to make them easier to use or because exact values are unknown. How numbers are rounded depends on how they are used. For example, attendance at a major league baseball game may be 54321. Although the attendance could be rounded to the nearest ten (54320) or the nearest hundred (54300), it is more likely to be reported as "about 54000" or "over 50000" or "less than 60000" because these values are convenient and a little easier to communicate.

As children develop rounding skills, they should come to realize that rounding rules may vary and are not universal. For example, here are two different rules from current textbooks for rounding a number ending in 5:

1. Change the 5 to a 0 and increase the previous digit by 1.
2. Change the 5 to a 0. If the digit preceding the 5 is even, leave it alone. If the digit preceding the 5 is odd, increase it to the next even digit.

In either case, 75 would round to 80; however, 85 would round to 90 using rule 1 and 80 using rule 2. Neither rule is right or best, but this variability across textbooks can cause confusion. Checking to make sure students understand the specific rules of rounding that are to be used may avoid some confusion. Better yet, encourage children to round to numbers that are easier to work with and make sense to them.

The precision of the rounded numbers should make sense for the problem context. For example, a meter stick could serve as a number line. Consider this train of rods, with 7 decimeter rods and 4 centimeter rods:

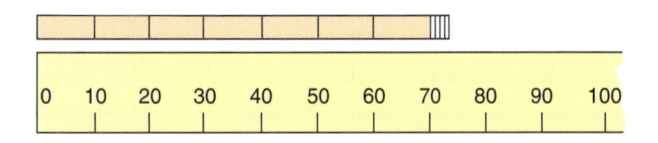

Is the train closer to 7 or 8 decimeters? [7 decimeters] Is it closer to 0 or 1 meter? [1 meter]

If you round to the nearest decimeter, then the length is 7 decimeters. If you round to the nearest meter, the length is 1 meter. How numbers are rounded depends on how the values will be used. For example, if you were to cut a strip of cloth to cover this train, it would be foolish to round to the nearest decimeter and cut a length of 7 decimeters; however, a meter of cloth would provide plenty of material to cover the train. Children must think about the problem context before rounding numbers and not just indiscriminately apply rounding rules.

Even in a given context, interpretation of rounded numbers is challenging, as illustrated by this national assessment question:

> The length of a dinosaur was reported to have been 80 feet (rounded to the nearest 10 feet). What length other than 80 feet could have been the actual length of the dinosaur?

Any answer from 75 to 85 was scored as correct in this open-ended question. (22% of the fourth graders provided a correct answer, 26% answered 90, and 48% gave another number.) This performance shows that interpreting a rounded result needs to be carefully addressed.

Base-ten blocks provide a natural method for developing rounding skills with larger numbers. Questions such as these focus attention on the quantity and the idea of *closer to*, which is essential in rounding:

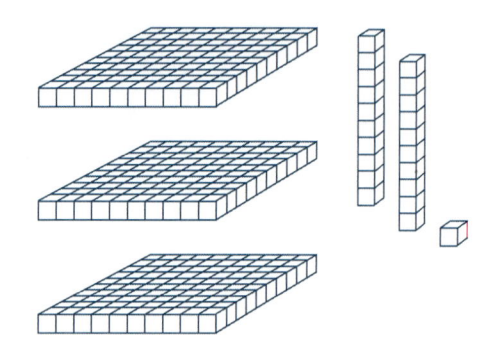

Is this more than three hundred? [yes]

Is this more than four hundred? [no]

Is this closer to three or four hundred? [closer to three hundred]

This model can also be extended to help children become more aware that 350 is halfway between 300 and 400.

A roller coaster model could be used to develop rounding skills. If children understand the number line, the roller coaster provides an effective tool for rounding.

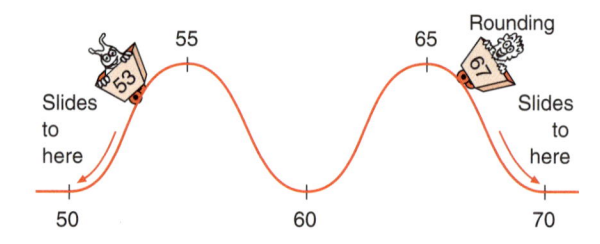

Children know what happens when the coaster stops at certain points. The model also suggests that something special happens at the top: The coaster could roll either way. This observation provides an opportunity to discuss a rule of rounding, such as "if the number ends in 5, you go over the hump to the next valley." In rounding, attention is given to the back-end digit or digits. Children should view rounding as something that not only makes numbers easier to handle but, more important, makes sense.

Encourage students to explore different ways to round. For example, consider these prices:

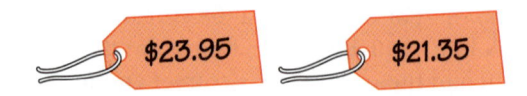

How would you round these prices if you wanted to round them to the same number? [to the nearest ten]

What would you round these prices to in deciding how much money is needed to make each purchase? [round up to $25 or $30]

Such real-world situations encourage students to think about both the advantages and consequences of rounding numbers. Meaningful rounding (knowing how much precision is necessary and what to round to) will improve through practice in many different problem contexts.

Cultural Connections

Counting to Ten

English	One	Two	Three	Four	Five	Six	Seven	Eight	Nine	Ten
Spanish	Uno	Dos	Tres	Cuatro	Cinco	Seis	Siete	Ocho	Nueve	Diez
Chinese	Yi	Er	San	Si	Wu	Liu	Qi	Ba	Jiu	Shi
Japanese	Ichi	Ni	San	Yo, Shi	Go	Roku	Nana, Shichi,	Hachi	Kyuu	Juu, Ju,

Eleven to Twenty

English	Eleven	Twelve	Thirteen	Fourteen	Fifteen	Sixteen	Seventeen	Eighteen	Nineteen	Twenty
Spanish	Once	Doce	Trece	Catorce	Quince	Diez y seis	Diez y siete	Diez y ocho	Diez y nueve	Veinte
Chinese	Shi Yi	Shi er	Shi san	Shi si	Shi wu	Shi liu	Shi qi	Shi ba	Shi jiu	Er shi
Japanese	Ju ichi	Ju ni	Ju san	Ju yon	Ju go	Ju roku	Ju nana	Ju hachi	Ju kyuu	Ni-ju

Some Larger Numbers

English	Twenty-one	Twenty-two	Twenty-three	Twenty-four	Twenty-five	Fifty	Hundred	Thousand	Ten thousand	Hundred thousand
Spanish	Veinte y uno	Veinte y dos	Veinte y tres	Veinte y cuatro	Veinte y cinco	Cincuenta	Cien	Mil	Dos mil	Cien mil
Chinese	Er shi yi	Er shi er	Er shi san	Er shi si	Er shi wu	Wu shi	Pai	Chien	Wen	Shi wan
Japanese	Ni-ju-ichi	Ni-ju-ni	Ni-ju-san	Ni-ju-yan	Ni-ju-go	Go-ju	Hyaku	Sen	Man	Ju-man

Figure 8-15 *Counting to 20 in different languages.*

CULTURAL CONNECTIONS

Young children have varying cultural, linguistic, and home experiences that shape their development of numbers. For example, patterns in naming numbers appear throughout Asian languages such as Japanese; yet in English and Spanish the number names don't always form consistent patterns. Figure 8-15 shows that in English and Spanish the number names from 11 to 20 do not explicitly name the number of tens as do the naming of numbers from 20 and larger. On the other hand, Figure 8-15 shows that in Japan and China the numbers from 10 to 20 clearly name the ten and a related number. In China and Japan, place-value patterns abound. For example, in Japan multiples of ten from 20 to 90 are (digit)*ju*; hundreds from 200 to 900 are (digit)*hyaku*; thousands from 2000 to 9000 are (digit) *sen*. These patterns consistently reinforce place value and facilitate naming numbers.

Figure 8-16 shows characters used to represent numbers in China. Notice how the larger numbers shown reflect place value and are additive. For example, 751 = 7 (100) + 5 (10) + 1. But 751 written with Chinese characters is 七五一, which literally means seven hundreds, five tens, and one. So if the Chinese characters are known, it is easy to express any Hindu-Arabic number and vice versa (Uy, 2003).

It is also interesting to note that in both Chinese and Japanese four groups of digits are used instead of three. In our number system, the unit changes every multiple of 1000, and we think ones, tens, hundreds and then move

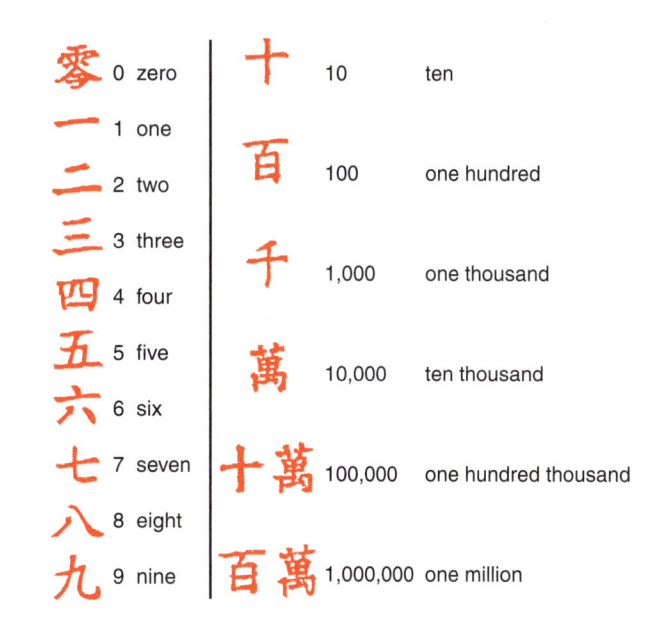

零 0 zero
一 1 one
二 2 two
三 3 three
四 4 four
五 5 five
六 6 six
七 7 seven
八 8 eight
九 9 nine

十 10 ten
百 100 one hundred
千 1,000 one thousand
萬 10,000 ten thousand
十萬 100,000 one hundred thousand
百萬 1,000,000 one million

Figure 8-16 *Chinese numeration system.*

to ones, tens, hundreds in the thousands place, then ones, tens, hundreds, in the millions place, and so on. In China and Japan, their number systems are based on ten thousand rather than one thousand. Thus, in Japan 10,000 is called a *man*. So in the United States, 3000 is read as "three thousand" whereas in Japan 30,000 is read *san-man*.

While mathematical concepts are universal across cultures, the language associated with these cultures is not. For example, in English we name the tens and then the ones for two-digit numbers larger than 20; thus, 35 is spoken as "thirty-five." In Arabic, numbers are named a bit differently. For example, for a two-digit number, say 45, the number is orally named "five forty." The number 345 is named three hundred "five forty"; 8345 is named "eight thousand three hundred five forty." As larger numbers are named, the digit in the ones place is named before the digit in the tens.

Naming large numbers in the United States and the United Kingdom can also become confusing. For example, in both the United States and the United Kingdom a million is a thousand thousands. However, a billion in the United States is a thousand million, whereas it is a million million in the United Kingdom. Thus 1,000,000,000,000 is a billion in England, but a billion in the United States is 1,000,000,000. These cultural differences in naming numbers underscores the need to know your students well and strive for similar understandings when numbers are named.

Math Links 8.3

Learning how different ancient cultures wrote numbers helps better understand and appreciate the qualities of number system. Here are some links that provide information on different systems of numeration, such as Babylonian, Mayan, and Roman:

http://www.math.wichita.edu/history/topics/num-sys.html#egypt

http://library.thinkquest org/J0112511/mayan_number.htm

http://mathforum.org/alejandre/numerals.html

These sites can also be accessed from this book's Web site.

www.wiley.com/college/reys

A GLANCE AT WHERE WE'VE BEEN

Children must have a clear understanding of our number system if they are going to be mathematically literate. They must be able to distinguish the four characteristics of our number system: the role of zero, the additive property of numbers, a base of ten, and place value. Many counting and trading experiences (particularly grouping by tens) are necessary.

As children develop their skills with the aid of various models (such as bean sticks, base-ten blocks, and an abacus), they need to learn how to organize the results in some systematic fashion and record them. Place-value mats serve as a visual reminder of the quantities involved and provide a bridge toward the symbolic representation of larger numbers. Establishing these bridges from the concrete to the abstract is particularly critical in developing place value, whose importance is second to none in all later development of number concepts.

Place-value concepts are developed over many years. Trading helps children compose and decompose numbers and begin to recognize equivalent representations. The power and importance of place value is developed, refined, extended, and expanded throughout the study of mathematics. We have seen how place value is reflected in the language and representation in different cultures. While language variations exist, the utilization of place value in these cultures is universal.

Understanding place value is essential to counting and facilitates operating with larger numbers. Place value is not taught and mastered over a few days, weeks, or grades. This means that place value is not completely developed before operations are introduced, because experiences with adding, subtracting, multiplying, and dividing whole numbers develops additional competence and understanding of place value. Work with decimals in later grades will further extend place-value concepts. For example, systematic work with smaller numbers, such as tenths, hundredths, and thousandths, is based on the same notion of trading, and similar patterns in naming numbers result. Thus place-value concepts initiated in the primary grades will be integrated and extended throughout elementary school.

Things to Do: From What You've Read

1. Identify four characteristics of our number system. Select one characteristic and make a visual representation of it.
2. Show how 201 and 120 would be represented with three different place-value models.
3. Distinguish between proportional and nonproportional models. Name an example of each.
4. Describe how trading provides opportunities for composing and decomposing numbers.
5. It has been suggested that centuries, decades, and millenniums can be used to demonstrate some notions of place value. Describe how this could be done.
6. Use your calculator. Enter the largest number possible in the display and name this number. How many digits does it have? Add one to this number and describe what your calculator does.

Things to Do: Going Beyond This Book

In the Field

1. *Place Value Concepts.* Tracking the number of school days provides a model for counting and place value that can be used throughout the year. Read the article by Goodrow and Kidd (2008), "Counting school days, decomposing number, and determining place value." Describe ways they involved children in counting and recording school days and some of the challenges their children encountered.

2. *Patterns on a Thousand Chart.* In the Classroom 8–4 provides an extension of the hundred chart. Show the chart to several children (grades 3–5) and say, "Please tell me about some patterns that you see." After each description, say, "Tell me about any other patterns you see." Be sure to allow them time to think and reflect; pattern recognition takes time. Make a list of the patterns for each child and compare their observations.

3. *Models for Large Numbers.* Different models exist for helping children develop a concept of large numbers. Several articles (Ellett, 2005; Joslyn, 2002; Kastberg & Walker, 2008; Losq, 2005; Nugent, 2006; Thompson, 2000) show and describe useful physical models. Examine one of these articles and describe the approach taken to help children better understand large numbers.

4. *Understanding Large Numbers.* Select either *Number Sense and Operations* (Burton et al., 1993), or *Number Sense: Simple Effective Number Sense Experiences Grades 1–2; 3–4; 4–6* (McIntosh et al., 1997). Choose an activity you think would be useful to help students develop a better understanding of large numbers and demonstrate how you would implement the activity.

With Technology

5. View the entire Annenberg video *Place Value Centers* excerpted in the Snapshot of a Lesson. Tell how measurement was used to explore place value. Describe the physical models being used to demonstrate place value. This first-grade lesson focused on two-digit numbers. Tell how these activities might be extended to larger numbers.

6. *Web surfing.* Use Google or some other web search engine to search for "100 charts" or "1000 charts." Identify several useful sites you found. Briefly describe three different resources available at these sites.

End Notes: Additional activities, suggestions, and questions are available in the field experience manual on the Student Companion site at www.wiley.com/college/reys

Note to Instructors: You can find additional resources, learning activities, and blackline masters in this text's accompanying Instructor's Manual, at www.wiley.com/college/reys.

Book Nook for Children

Berry, D. *The Rajah's Rice: A Mathematical Folk tale from India*. New York: W.H. Freeman, 1994.

When Chandra, an Indian village girl who bathes the raja's elephants, cures the beasts after they fall ill, the raja offers her jewels as a reward. She refuses, accepting only a measure of rice for the hungry villagers: two grains on the first square of a chessboard, four on the second, and so on, doubling the amount for each subsequent square. Although the amount seems insignificant at first, it grows at an alarming rate, since doubling has little effect on small numbers but an increasingly enormous effect as the numbers grow larger. The raja's storehouse is soon empty, and he must admit that he cannot fill her seemingly modest request.

Clements, A. *A Million Dots,* New York: Simon & Schuster, 2006.

The author shows the reader what a million looks like by using dots. Given examples of different things like shoeboxes, the author lets the reader think about how much a million looks like through the illustrations presented in the book. The reader will also learn about many different things presented in the book.

Murphy, S. J. *Earth Day—Hooray!* New York: HarperCollins, 2004.

The Maple Street School Save the Planet Club is having a can drive to recycle aluminum cans. As the club members collect the cans and sort them into different amounts, the reader will learn about place value. There are activities and games in the back of this book for parents and teachers to use.

Schwartz, D. M. *How Much Is a Million?* New York: Lothrop, Lee & Shepard Books, 1985.

The reader will see how much a million, a billion, a trillion can be. The pictures in the text help the reader better understand the size of these large amounts.

Schwartz, D. M. If *You Made a Million*. New York: Lothrop, Lee & Shepherd Books, 1989.

Marvelosissimo the Mathematical Magician and his team of cheerful kids (and their multitude of animal friends) take on some jobs. For each job, they'll be paid an appropriate amount of money. But soon the questions arise—what does that much money look like, and how can it be spent, saved, or used to pay off a loan? The fantasy cleverly introduces money from 1 penny to 1 million dollars.

Wells, R. E. *Can You Count to a Googol?* Morton Grove, IL: Albert Whitman, 2000.

This resource book will let children see how our decimal numeration system is developed. Starting with the number 1, the reader will see how our number system can develop many numbers using zeros that will contain the following numbers: billion, trillion, quadrillion, octillion, and finally a googol. Quantities and distances are related to the numbers.

CHAPTER 9

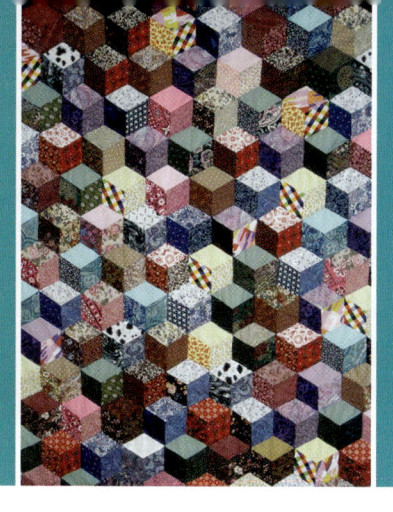

Operations: Meanings and Basic Facts

> "MATHEMATICS IS BUILT WITH FACTS AS A HOUSE IS BUILT WITH BRICKS, BUT A COLLECTION OF FACTS CANNOT BE CALLED MATHEMATICS ANY MORE THAN A PILE OF BRICKS CAN BE CALLED A HOUSE."
>
> — Poincaré (French mathematician, 1854–1912)

> SNAPSHOT OF A LESSON

KEY IDEAS

1. Have students develop familiarity and fluency with addition and subtraction basic facts (with whole number addends equal to or smaller than 7).
2. Have students recognize addition/subtraction relationships.

BACKGROUND

This lesson is based on the video *Window Puzzles*, from Teaching Math: A Video Library, K–4, from Annenberg Media. Produced by WGBH Educational Foundation. © 1995 WGBH. All rights reserved. To view the video, go to http://www.learner.org/resources/series32.html (video #13).

Terry Goens, a first-grade teacher at the Martha Fox Elementary School in Belgrade, Montana, engages her students in working in pairs on a number puzzle that involves sums and differences of one-digit numbers.

Ms. Goens begins the day's mathematics lesson by reviewing Windows, a puzzle the children worked on the previous week. This is an activity that Ms. Goens uses to engage her students in developing familiarity with basic facts for addition and subtraction. She provides the students with a sample window on the board.

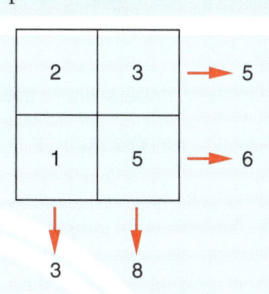

2	3	→ 5
1	5	→ 6
↓	↓	
3	8	

In this activity, the class starts by writing a number in each cell in the window (here, they've put in 2, 3, 1, and 5). Next, they find the row totals (here, 5 and 6) and the column totals (here, 3 and 8) and then they erase the numbers in the window cells and work to find other combinations of numbers that give the same row and column totals.

Ms. Goens asks Sam to describe how he might find other numbers that work for the puzzle.

Sam: Me and Rachel, we know that if 4 and 4 would equal 8, then we put those there [*in the cells in the right column*]. And if 1 and 4 would equal 5, we put that there [*put 1 in the top left cell*]. But 2 and 1 would equal 3 [*so put 2 in the bottom left cell*]. And if it was 2, and 4 would equal 6 [*checking that the bottom row sums to 6, as it should*].

Ms. Goens: OK, were we adding? Is that what we call it? Yes, we call that adding, don't we?

[Ms. Goens *clears the window by erasing all of the numbers.*]

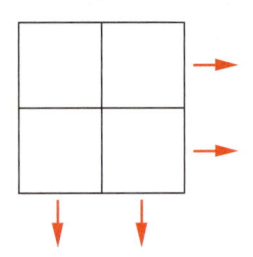

Ms. Goens: Well, boys and girls, let's create a window together in class. But I have one rule about putting tiles in these squares. You can put up to 7 tiles in these squares. That means you can put 7 or less, but you can't go over 7. OK? Does anyone have a number of tiles that we can put in the first square right here on the top?

The students offer numbers to place inside the four cells of the window. As the children suggest values, Ms. Goens places the corresponding number of tiles in each cell. The following window shows the result.

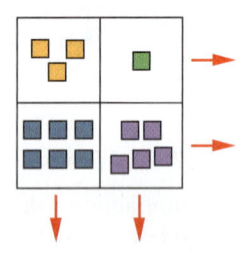

Ms. Goens: What do I do now?

Student: You write in the numbers where the arrows are.

Ms. Goens: And how do I get this number here? [*points to the top arrow*]

Student: With the tiles, you don't have to count them if you know the numbers. You can just put the numbers there, the first number up there and which—that first number up there would be 4.

Students continue to offer row and column totals, resulting in the complete window shown below.

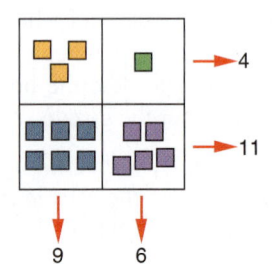

Ms. Goens: Do you think you could create a window at your desk? When you are done with creating your window, I have this recording sheet for you ... I want you to write all the different solutions you can come up with if you took all the tiles away and try to get these [*total*] numbers with different tiles.

The teacher removes all the tiles from the window displayed in the front of the class, leaving only the row and column totals. She asks the students to work in pairs to come up with other numbers that will make the puzzle true.

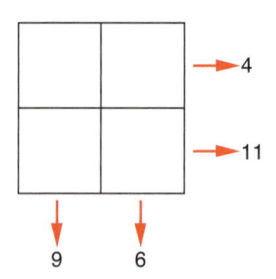

The teacher notes that this is a good lesson for the students because they can work at either the concrete or the symbolic levels of understanding—with tiles or with symbols, whichever makes them most comfortable. In order for students to complete the window, given both the column and row totals, they must consider the relationship between addition and subtraction.

As partners work on their own number windows, the teacher spends time with individual students and talks with them about their thinking and problem-solving strategies. The teacher reflects on her practices and indicates that her own mathematics skills are growing as she listens to the children.

FOCUS QUESTIONS

1. What prerequisites are important prior to engaging students in formal work on the four basic operations?
2. What general sequence of activities helps children develop meaning for the operations?
3. What three distinct types of situations lead to subtraction? What four types of structures lead to multiplication?
4. How should thinking strategies for the basic facts be taught?
5. Describe the key thinking strategies for learning basic facts for addition, for subtraction, for multiplication, and for division.

INTRODUCTION

This chapter's Snapshot of a Lesson incorporates several essential components of a well-planned classroom activity involving computation. First and foremost, it actively involves students in a problem situation that promotes reasoning and discussion. The problem is accessible to various levels of students. Those who need to count to find sums can use the blocks to find answers. Those who are already working on solidifying their knowledge of basic facts can work at a more abstract level. Note that the Window puzzle requires students to think about connections between addition and subtraction because the starting point is the row and column sums, so both operations are required to figure out what numbers could go in the cells. The problem is open-ended because multiple solutions are possible when row and column sums are the starting point.

These components are important in elementary school mathematics lessons, particularly as children develop understanding of the relevance and meaning of computational ideas. An understanding of addition, subtraction, multiplication, and division—and knowledge of the basic number facts for each of these operations—provides a foundation for all later work with computation. To be effective in this later work, children must develop broad concepts for these operations. This development is more likely to happen if each operation is presented through multiple representations using various physical models. Such experiences help children recognize that an operation can be used in several different types of situations. Children also must understand the properties that apply to each operation and the relationships between operations.

Learning the basic number facts is one of the first steps children take as they refine their ideas about each operation. By using these facts, plus an understanding of place value and mathematical properties, a child can perform any addition, subtraction, multiplication, or division with whole numbers. As the opening quote by the famous mathematician Poincaré makes clear, basic facts are an essential building block for arithmetic, although the broader field of mathematics consists of much more than just facts. Both understanding the operations and having immediate recall of basic facts are needed for doing pencil-and-paper computations proficiently with all sorts of numbers (whole numbers, fractions, decimals, and percents). But operation sense and basic fact knowledge are just as essential when a calculator is readily available because it is important to monitor the reasonableness of answers obtained by pushing buttons. Operation sense and basic facts also form the building blocks for estimating answers or doing exact computations mentally in many everyday situations where it would be inefficient to use either paper and pencil or a calculator. So, no matter what type of computation a child is doing—paper and pencil, calculator, mental exact, or mental estimation—both operation sense and quick recall of the basic number facts are essential.

Figure 9-1 lists the expectations related to operation meanings (addition, subtraction, multiplication, and division for whole numbers) and basic facts that are identified in the *Common Core State Standards for Mathematics* (CCSSI, 2010). We focus on these expectations throughout this chapter.

GRADE	CLUSTER HEADINGS AND SELECTED STANDARDS
Kindergarten	Understand addition as putting together and adding to, and understand subtraction as taking apart and taking from.
Grade 1	Represent and solve problems involving addition and subtraction. Understand and apply properties of operations and the relationship between addition and subtraction. Add and subtract within 20 (use strategies such as counting on, decomposing a number leading to ten, using the relationship between addition and subtraction, and creating equivalent but easier or known sums). Work with addition and subtraction equations. Use place value understanding and properties of operations to add and subtract.
Grade 2	Represent and solve problems involving addition and subtraction. Add and subtract within 20 **(by the end of grade 2, know from memory all sums of two one-digit numbers)**. Work with equal groups of objects to gain foundations for multiplication. Use place value understanding and properties of operations to add and subtract (fluently add and subtract within 100 using strategies based on place value, properties of operations, and/or the relationship between addition and subtraction).
Grade 3	Represent and solve problems involving multiplication and division. Understand properties of multiplication and the relationship between multiplication and division. Multiply and divide within 100 **(by the end of grade 3, know from memory all products of two one-digit numbers)**. Solve problems involving the four operations, and identify and explain patterns in arithmetic. Use place value understanding and properties of operations to perform multi-digit arithmetic.

Figure 9-1 Highlights from CCSSM relating to operations (with focus on operations meanings and fluency with number facts). (Source: From the *Common Core State Standards for Mathematics* (CCSSI, 2010).) This summary of the CCSSM standards was made by the authors of *Helping Children Learn Mathematics*. The full document (CCSSI, 2010) may be obtained at www.corestandards/the-standards/mathematics.

Math Links 9.1

A variety of Web resources for basic facts may be found at NCTM's Illuminations Web site under the Number and Operation Standard. They may be accessed from http://illuminations.nctm.org/swr/index.asp or from this book's Web site. Triangle Flashcards and Multiplication: An Adventure in Number Sense are particularly recommended.

www.wiley.com/college/reys

HELPING CHILDREN DEVELOP NUMBER SENSE AND COMPUTATIONAL FLUENCY

How many bottles?

We can count —
1, 2, 3, 4, 5, 6

How many bottles?

We could count, but it's more efficient to use a combination of counting (6 bottles in a carton, 5 cartons) and multiplication — $5 \times 6 = 30$.

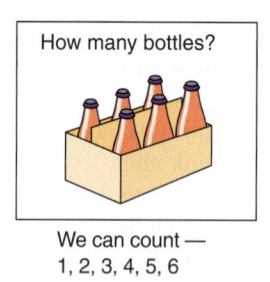

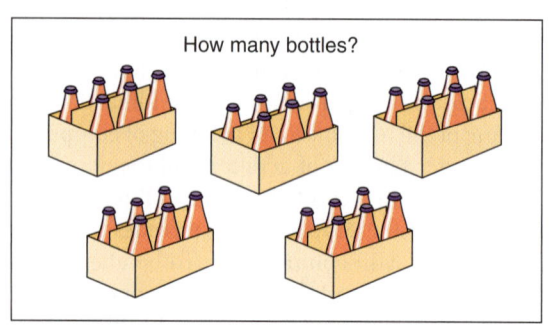

Figure 9-2 *An example showing the efficiency of using operations.*

Ultimately, the instructional goal is that children not only know how to add, subtract, multiply, and divide but, more important, know *when* to apply each operation in a problem-solving situation. Children also should be able to recall the basic facts quickly when needed.

How can teachers help children attain these skills and understandings? Begin by finding out what each child knows. Then capitalize on their knowledge while continuing to build on the number concepts they have already constructed (Kouba and Franklin, 1993, 1995; Russell, 2010). Most children entering school are ready in some ways and not ready in others for formal work on the operations. Four prerequisites for such work seem particularly important: (1) facility with counting, (2) experience with a variety of concrete situations, (3) familiarity with many problem-solving contexts, and (4) experience using language to communicate mathematical ideas.

FACILITY WITH COUNTING

Children use counting to solve problems involving addition, subtraction, multiplication, and division long before they come to school, as research has indicated (Baroody, Lai, & Mix, 2006; Clements & Sarama, 2007). Any problem with whole numbers can be solved by counting, provided there is sufficient time. Because it is not always efficient to solve problems by counting, children need to be able to use other procedures to cope with more difficult computations. Figure 9-2 illustrates this idea by comparing the counting method with multiplication, which is more efficient.

Counting nevertheless remains an integral aspect of children's beginning work with the operations. They need to know how to count forward and backward as well as how to skip count by 2s, 3s, and other groups (see Chapter 7).

They need to count as they compare and analyze sets and arrays. But they need more than counting to become proficient in computing.

Math Links 9.2

You can use electronic manipulatives, such as Number Line Arithmetic and Rectangle Multiplication, to help students visualize their basic facts. Visit http://nlvm.usu.edu/en/nav/vlibrary.html, or use the link from this book's Web site; then select Number and Operations and scroll through to find these and other useful manipulatives.

www.wiley.com/college/reys

EXPERIENCE WITH A VARIETY OF CONCRETE SITUATIONS

Children need to have many experiences in problem situations and in working with physical objects to develop understanding about mathematical operations. Understanding improves if they can relate mathematical facts and symbols to an experience they can visualize. Manipulative materials serve as a referent for later work with the operations as well as for constructing the basic facts. Materials also provide a link to connect each operation to real-world problem-solving situations. Whenever a child wants to be sure that an answer is correct, materials can be used for confirmation.

FAMILIARITY WITH MANY PROBLEM CONTEXTS

Problem situations are used in mathematics instruction for developing conceptual understanding, teaching higher-level thinking and problem-solving skills, and applying a variety of mathematical ideas. Research indicates the benefits of beginning with problem situations (Fuson, 2003a, 2003b). Word problems require reading, comprehension, representation, and calculation. Children generally have little difficulty with single-step word problems but have more difficulty with multistep, complex problems.

As with other mathematical content, a variety of problem contexts or situations should be used to familiarize students with the four basic operations, continuing all along the way until computational mastery is achieved. Children need to think of mathematics as problem solving—as a means by which they can resolve problems through applying what they know, constructing possible routes to reach solutions, and then verifying that the solutions make sense. Students must realize that mathematics is a tool that has real-life applications. Most children already know that computation is used in everyday life. Mathematics lessons need to be connected to those experiences, but students also need to realize that $6 \times 8 = n$ or $9 \times 2 = n$ also may be problems—ones that they can solve. They need to have the attitude, "If I don't know an answer, I can work it out."

EXPERIENCE IN TALKING AND WRITING ABOUT MATHEMATICAL IDEAS

Children need to talk and write about mathematics; putting experiences into words helps with making meaning. Both manipulative materials and problems can be vehicles for communicating about mathematics. All early phases of instruction on the operations and basic facts should reflect the important role that language plays in their acquisition. The *Principles and Standards for School Mathematics* (NCTM, 2000) discusses the roles of language in great depth in presenting the recommendations on communication and thus provides a valuable source of additional information.

Sometimes a move to symbols is made too quickly and the use of materials dropped too soon. Instead, the use of materials should precede and then parallel the use of symbols. Early on, children should be manipulating materials as they record symbols. As illustrated in the Snapshot of a Lesson that opened this chapter, when children talk and write about what is happening in a given situation, they are helped to see the relationship of the ideas and symbols to their manipulation of materials and to the problem setting.

The oral and written language that children learn as they communicate about what they are doing with materials helps them understand how symbolism is related to mathematical operations. By modeling, talking, and writing, the referent for each symbol is strengthened. Children should begin their work with operations after talking among themselves and with their teacher about a variety of experiences. They need to be encouraged to continue talking about the mathematical ideas they meet as they work with the operations. As soon as feasible, they need to put their ideas on paper—at first by drawings alone; but as soon as they are able to write, children should also be encouraged to write number sentences and narrative explanations of their thinking.

DEVELOPING MEANINGS FOR THE OPERATIONS

Children encounter the four basic operations in natural ways when they work with many diverse problem situations. By representing these problem situations (e.g., acting them out, using physical models, or drawing pictures), they develop meanings for addition, subtraction, multiplication, and division. Mastery of basic facts and later computational work with multidigit examples must be based on a clear understanding of the operations.

Thus, both computational proficiency and understanding of operations are desired outcomes of mathematics instruction. The following general sequence of activities is appropriate for helping children develop meaning for the four basic operations:

1. *Concrete—modeling with materials:* Use a variety of verbal problem settings and manipulative materials to act out and model the operation.

2. *Semiconcrete—representing with pictures:* Provide representations of objects in pictures, diagrams, and drawings to move a step away from the concrete toward symbolic representation.

3. *Abstract—representing with symbols:* Use symbols (especially numeric expressions and number sentences) to illustrate the operation.

In this way, children move through experiences from the concrete to the semiconcrete to the abstract, linking each to the others.

The four operations are clearly different, but there are important relationships among them that children will come to understand through modeling, pictorial, and symbolic experiences:

- Addition and subtraction are inverse operations; that is, one undoes the other:
$$5 + 8 = 13 \rightarrow 13 - 5 = 8$$
- Multiplication and division are also inverse operations:
$$4 \times 6 = 24 \rightarrow 24 \div 4 = 6$$
- Multiplication can be viewed as repeated addition:
$$4 \times \rightarrow 6 + 6 + 6 + 6$$

- Division can be viewed as repeated subtraction:

  $24 \div 6 = 4 \rightarrow 24 - 6 - 6 - 6 - 6 = 0$

These relationships can be developed through careful instruction with a variety of different experiences.

ADDITION AND SUBTRACTION

Figure 9-3 illustrates a variety of models (including counters, linking cubes, balance scale, and number line) that can be used to represent addition. Each model depicts the idea that addition means "finding how many in all." The models for addition can also be used for subtraction. Each model can be applied in the three different situations that lead to subtraction.

SEPARATION PROBLEMS *Separation*, or take away, involves having one quantity, removing a specified quantity from it, and noting what is left. Research indicates that this subtraction situation is the easiest for children to learn; however, persistent use of the words *take away* results in many children assuming that this is the *only* subtraction situation and leads to misunderstanding of the other two situations. This is why it is important to read a subtraction sentence such as $8 - 3 = 5$ as "8 minus 3 equals 5" rather than "8 take away 3 equals 5." Take away is just one of the three types of subtraction situations.

Peggy had 7 balloons. She gave 4 to other children. How many did she have left?

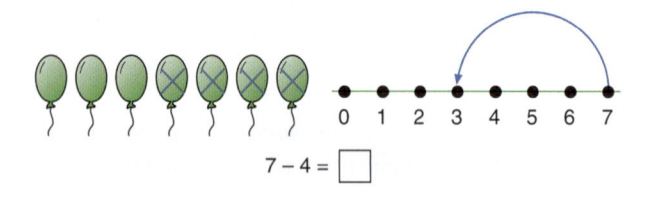

COMPARISON PROBLEMS *Comparison*, or finding the difference, involves having two quantities, matching them one to one, and noting the quantity that is the difference between them. Problems of this type can also be solved by subtraction, even though nothing is being taken away.

Peggy had 7 balloons. Richard had 4 balloons. How many more balloons did Peggy have than Richard?

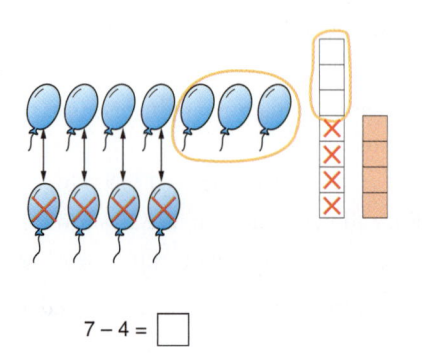

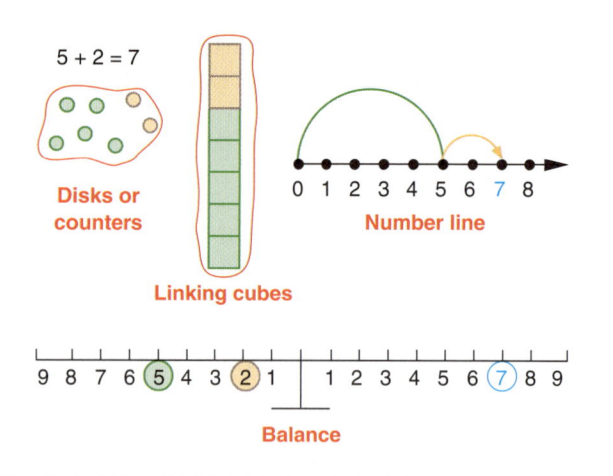

Figure 9-3 *Some models for addition.*

PART–WHOLE PROBLEMS The final type of subtraction situation is known as *part–whole*. In this type of problem, a set of objects can logically be separated into two parts. You know how many are in the entire set and you know how many are in one of the parts. You need to find out how many must be in the remaining part. Nothing is being added or taken away—you simply have a static situation involving parts and a whole.

Peggy had 7 balloons. Four of them were red and the rest were blue. How many were blue?

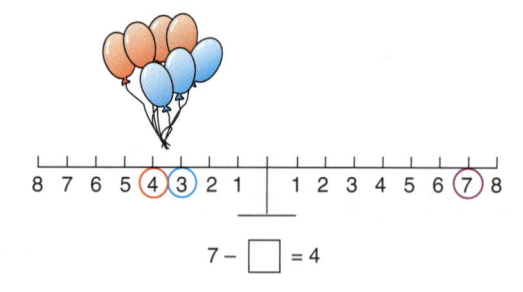

It is useful to help students recognize that many subtraction problems can also be thought of as "missing-addend" problems because the strategy "think addition" can be used to find the answer. For example, in the preceding problem, you could ask yourself: 4 plus what equals 7?

The importance of providing many varied experiences in which children use physical objects to model or act out examples of each operation cannot be overemphasized. The Snapshot of a Lesson at the beginning of this chapter provided one such experience. It illustrates a task that involves children in thinking about putting two piles of counters together as a lead-in to addition and subtraction. Some children might focus on the concrete, by actually placing chips in the window panes and counting. Other children might be just as comfortable drawing pictures or writing numerals as they work on the puzzle. All students are working toward the goal of being able to relate number sentences (such as $6 + 5 = 11$) with a row or column in the window pane. For students in the earliest stages of understanding, the situation need not be represented by written or spoken symbols. Moving tiles, counting, and thinking about additive and subtractive relationships are the important initial components of the activity. The teacher introduces symbols as a complement to the physical manipulation of objects. Symbols provide a way of recording what's happening with the materials. As work with numbers and operations progresses, the amount of symbolization that the teacher encourages and expects will increase.

> **Math Links 9.3**
>
> If you would like to see a variety of lesson plans that provide hands-on experiences and meaningful contexts for developing addition and subtraction concepts, visit NCTM's Illuminations Web site at http://illuminations.nctm.org/, or link from this book's Web site, then click on Lessons. Do It with Dominoes and Links Away are particularly recommended.
>
> www.wiley.com/college/reys

MULTIPLICATION AND DIVISION

The same sequence of experiences used for developing understanding of addition and subtraction—moving from concrete to pictorial to symbolic—should also be followed for multiplication and division. However, one important way that multiplication and division problems differ from addition and subtraction problems is that the numbers in the problems represent different sorts of things. For example, consider a problem such as "Andrew has two boxes of trading cards. Each box holds 24 cards. How many cards does he have altogether?" This problem may be written 2 × 24 = 48, where the *first factor* (2) tells us how many groups or sets of equal size (here, how many boxes) are being considered, while the *second factor* (24) tells us the size of each set (here, how many cards per set). The third number (48), known as the *product*, indicates the total of all the parts (here, the total number of cards). The old saying "you can't add apples and oranges" points out that in addition and subtraction problems, a common label must be attached to *all* the numbers involved. Apples and oranges can be added only if we relabel all the numbers in the problem with a common label such as "fruit." By contrast, in the trading card problem, the labels for the numbers are all different: in this problem we had boxes (2), cards per box (24), and cards (48).

Figure 9-4 illustrates some of the most commonly used models for illustrating multiplication situations: sets of objects, arrays, and the number line. Research indicates that children do best when they can use various representations for multiplication and division situations and can explain the relationships among those representations (Kouba and Franklin, 1993, 1995; Russell, 2010; Verschaffel, Greer, & De Corte, 2007).

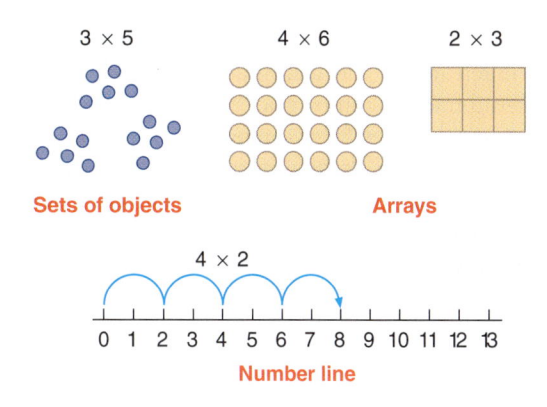

Figure 9-4 *Commonly used models for multiplication.*

Researchers have identified four distinct sorts of multiplicative structures: equal groups, multiplicative comparisons, combinations, and areas/arrays (Greer, 1992). Problems involving the first two of these structures are most common in elementary school, although students should eventually become familiar with all four. The four multiplicative structures are described here to help you, as the teacher, understand and recognize their variety. You are not expected to teach these labels to children, but you should try to ensure that they encounter a broad range of problem situations involving multiplication and division.

EQUAL-GROUPS PROBLEMS *Equal-groups problems* involve the most common type of multiplicative structure, where you are dealing with a certain number of groups, all the same size. When both the number and size of the groups are known (but the total is unknown), the problem can be solved by multiplication. The problem given earlier, about Andrew's trading card collection, is an example of an equal-groups problem. When the total in an equal-groups problem is known, but either the number of groups or the size of the group is unknown, the problem can be solved by division. Two distinct types of division situations can arise, depending on which part is unknown. These two types of division situations are known as measurement and partition division (described later).

COMPARISON PROBLEMS *Comparison problems* involve another common multiplicative structure. With comparison problems for subtraction, there are two different sets that need to be matched one-to-one to decide how much larger one is than the other. In similar fashion, comparison problems with multiplicative structures involve two different sets, but the relationship is not one-to-one. Rather, in multiplicative comparison situations, one set involves multiple copies of the other. An example of a multiplicative comparison situation might be: "Hilary spent $35 on Christmas gifts for her family. Geoff spent 3 times as much. How much did Geoff spend?" In this case, Hilary's expenditures are being compared with Geoff's, and the problem is solved by multiplication. The question does not involve "How much more?" (as it would if the problem involved additive/subtractive comparison). Instead, the structure of the problem involves "How many times as much?" If the problem is changed slightly to include information about how much Geoff spent but to make either Hilary's expenditures unknown or the comparison multiplier unknown, then the problem could be solved by division. Examples of these problem structures are: (a) Hilary spent a certain amount on Christmas gifts for her family, and Geoff spent 3 times as much. If Geoff spent $105, how much did Hilary spend? (b) Hilary spent $35 on Christmas gifts for her family and Geoff spent $105. How many times as much money as Hilary spent did Geoff spend?

COMBINATIONS PROBLEMS *Combinations problems* involve still another sort of multiplicative structure. Here the two factors represent the sizes of two different sets and the product indicates how many different pairs of things can be formed, with one member of each pair taken from each of the two sets. For example, consider the number of different sundaes possible with four different ice cream flavors and two toppings if each sundae can have exactly one ice cream flavor and one topping:

	Ice cream flavors			
	Vanilla	Cherry	Mint	Chocolate
Toppings Pineapple				
Butterscotch				

Other examples of combination problems are the number of choices for outfits if you have 5 T-shirts and 4 shorts, or the number of different sandwiches possible with 3 choices of meat, 2 choices of cheese, and 2 kinds of bread.

AREA AND ARRAY PROBLEMS Finally, *area and array problems* also are typical examples of multiplicative structure. The area of any rectangle (in square units) can be found either by covering the rectangle with unit squares and counting them all individually or by multiplying the width of the rectangle (number of rows of unit squares) by the length (number of unit squares in each row). Similarly, in a rectangular array—an arrangement of discrete, countable objects (such as chairs in an auditorium)—the total number of objects can be found by multiplying the number of rows by the number of objects in each row. The array model for multiplication can be especially effective in helping children visualize multiplication. It may serve as a natural extension of children's prior work in making and naming rectangles using tiles, geoboards, or graph paper.

These illustrations show a 2-by-3 or 3-by-2 rectangle.

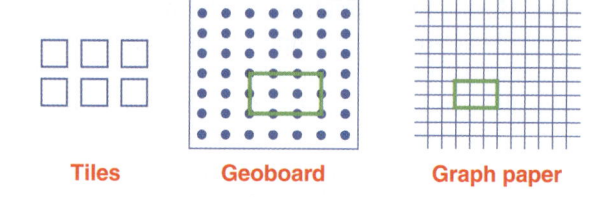

Tiles   **Geoboard**   **Graph paper**

Thus, each rectangle contains six small squares. Asking children to build and name numerous rectangles with various dimensions is a good readiness experience for the concept of multiplication. In the Classroom 9-1 illustrates several experiences designed for this purpose.

Just as sets of objects, the number line, and arrays are useful in presenting multiplication, they can also be useful in presenting division, with the relationship to repeated subtraction frequently shown. For division, however, two different types of situations must be considered: *measurement* and *partition*.

In the Classroom 9–1

RECTANGLES AND MORE RECTANGLES

Objective: Building rectangles with tiles to develop visual representations for multiplication facts.

Grade Level: 3.

- How many ways can you make a rectangle with this many tiles? Draw pictures. The first one is done for you.
- List the ways:

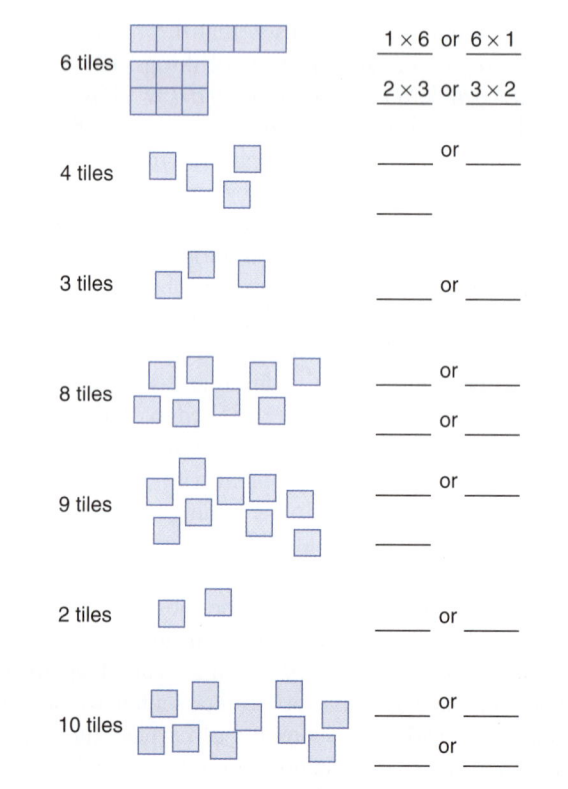

6 tiles 1×6 or 6×1
            2×3 or 3×2

4 tiles   ____ or ____
            ____

3 tiles   ____ or ____

8 tiles   ____ or ____
            ____ or ____

9 tiles   ____ or ____
            ____

2 tiles   ____ or ____

10 tiles   ____ or ____
            ____ or ____

MEASUREMENT (REPEATED-SUBTRACTION) PROBLEMS
In division situations of the *measurement* (or repeated-subtraction) type, you have equal sized groups, you know how many objects are in each group, and you must determine the number of groups.

> Jenny had 12 candies. She gave 3 to each person. How many people got candies?

Here, you can imagine Jenny beginning with 12 candies and making piles of 3 repeatedly until all the candies are gone. She is measuring how many groups of 3 she can make from the original pile of 12. (12 ÷ 3 = □, or □ × 3 = 12)

Person 1   Person 2   Person 3   Person 4

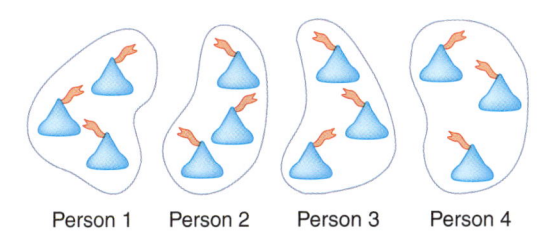

Another division example of the measurement (or repeated-subtraction) type is measuring how many 2-foot hair ribbons can be made from a 10-foot roll of ribbon. Imagine repeatedly stretching out 2 feet and cutting it off, thus measuring how many hair ribbons you can make. The hair ribbons are all the same size (2-foot) and you are figuring how many ribbons of that length can be made. (10 ÷ 2 = □, or □ × 2 = 10)

PARTITION (SHARING) PROBLEMS In division situations of the *partition* (or sharing) type, a collection of objects is separated into a given number of equivalent groups and you seek the number in each group. By contrast with measurement situations, here you already know how many groups you want to make, but you don't know how many objects must be put in each group.

> Gil had 15 shells. If he wanted to share them equally among 5 friends, how many should he give to each?

Here, imagine Gil passing out the shells to his five friends (one for you, one for you, one for you, etc., and then a second shell to each person, and so on) until they are all distributed, and then checking to see how many each person got. (15 ÷ 5 = □, or 5 × □ = 15)

Friend 1   Friend 2   Friend 3   Friend 4   Friend 5

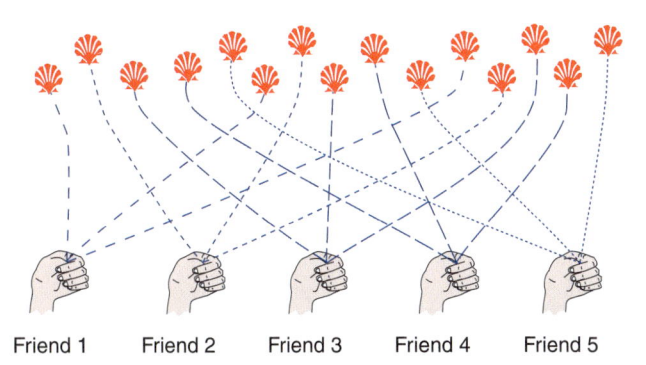

Partitioning (or sharing) is difficult to show in a diagram, but it is relatively easy for children to act out. Dealing cards for a game is another instance of a partition situation.

It is certainly not necessary for children to learn these terms or to name problems as measurement or partition situations. But it is important for you as a teacher to know about the two types of division situations so you can ensure that your students have opportunities to work with examples of each. It is vital that students be able to identify when a problem situation requires division, and that means being able to recognize both types of situations as involving division.

MATHEMATICAL PROPERTIES

An understanding of the mathematical properties that pertain to each operation (Table 9-1) is vital to children's understanding of the operation and how to use it. This understanding is not a prerequisite to work with operations, but it must be developed as part of understanding operations.

In elementary school, children are usually not expected to state these properties precisely or identify them by name. Rather, the instructional goal is to help children understand the commutative, associative, distributive, and identity properties and to use them when it is efficient. Table 9-1 gives the meaning of each property, states what children should understand, and provides examples to illustrate how the property can make learning and using the basic facts easier.

Understanding these properties implies knowing when they apply. For example, both addition and multiplication are commutative, but neither subtraction nor division is.

$$7 - 3 \text{ is } not \text{ equal to } 3 - 7$$

$$28 \div 7 \text{ is } not \text{ equal } 7 \div 28$$

Many children have difficulty with the idea of commutativity. They may simply "subtract the smaller number from the larger" or "divide the larger number by the smaller" regardless of their order. Care needs to be taken to ensure that children understand why order is important in subtraction and division.

OVERVIEW OF BASIC FACT INSTRUCTION

As children develop concepts of meanings of operations, instruction begins to focus on certain number combinations. These are generally referred to as the *basic facts*:

- *Basic addition facts* each involve two one-digit addends and their sum. There are 100 basic addition facts (from 0 + 0 up to 9 + 9; see Figure 9-5). To read off a fact (say, 4 + 9 = 13), find the first addend (4) along the left side and the second addend (9) along the top. By reading

186 Chapter 9 • Operations: Meanings and Basic Facts

TABLE 9-1 • Mathematical Properties for Elementary-School Children

Property	Mathematical Language	Child's Language	How It Helps
Commutative	For all numbers a and b: $a + b = b + a$ and $a \times b = b \times a$	If $4 + 7 = 11$, then $7 + 4$ must equal 11, too. If I know 4×7, I also know 7×4.	The number of addition or multiplication facts to be memorized is reduced from 100 to 55
Associative	For all numbers a, b, and c: $(a + b) + c = a + (b + c)$ and $(ab)c = a(bc)$	When I'm adding (or multiplying) three or more numbers, it doesn't matter where I start.	When more than two numbers are being added (or multiplied), combinations that make the task easier can be chosen; for example, $37 \times 5 \times 2$ can be done as $37 \times (5 \times 2)$ or 37×10 rather than $(37 \times 5) \times 2$
Distributive	For all numbers a, b, and c: $a(b + c) = ab + ac$	8×12 is the same as $8 \times 10 + 8 \times 2$. You can split numbers apart to multiply easy pieces, then add: $8 \times (10 + 2) = 8 \times 10 + 8 \times 2$	Some of the more difficult basic facts can be split into smaller, easier-to-remember parts; for example, 8×7 is the same as $(8 \times 5) + (8 \times 2)$ or $40 + 16$
Identity	For any whole number a: $a + 0 = a$ and $a \times 1 = a$	0 added to any number is easy; it's just that number. 1 times any number is just that number.	The 19 addition facts involving 0 and the 19 multiplication facts involving 1 can be easily remembered once this property is understood and established.

Figure 9-5 *The 100 basic facts of addition.*

horizontally across the 4-row from the left and vertically down the 9-column from the top, you find the sum (13).

- *Basic subtraction facts* rely on the inverse relationship of addition and subtraction for their definition. The 100 basic subtraction facts result from the difference between one addend and the sum for all one-digit addends. Thus the 100 subtraction facts are also pictured in Figure 9-5, the same table that pictures the 100 basic addition facts. You read off a basic subtraction fact (say, $13 - 4 = 9$) by finding the box in the 4-row that contains the sum 13, then reading up that column to find the difference (9) at the top of the table. Note that a sentence such as $13 - 2 = 11$ is neither represented in Figure 9-5 nor considered a basic subtraction fact because $2 + 11 = 13$ is not a basic addition fact (since 11 is not a one-digit addend).

- *Basic multiplication facts* each involve two one-digit factors and their product. There are 100 basic multiplication facts (from 0×0 up to 9×9).

- *Basic division facts* rely on the inverse relationship of multiplication and division, but there are only 90 basic division facts. Because division by zero is not possible, there are no facts with zero as the divisor.

Development and mastery of the addition and subtraction facts typically has begun in kindergarten or first grade and continued as multiplication and division facts were developed and practiced in third and fourth grades. However, researchers have documented that—at least until very recently—there has been considerable variation in when students were expected to have mastered their basic facts. Most state documents specified beginning attention to basic addition facts one year prior to when fluency was expected, which was most commonly at grade 2. The same was true for basic multiplication facts: Students usually began working with the facts about one year before they were expected to master them. Fluency with multiplication facts was most commonly expected at grade 4 (though a number of states specified grade 3). Table 9-2 provides an overview of state expectations for the timing of children's mastery of basic

TABLE 9-2 • Grade-Level Placement of Learning Expectations Related to Fluency with Basic Number Combinations for Each Operation[a]

Operation	Grade	Number of States	Operation	Grade	Number of States
Addition	1	8	Subtraction	1	7
	2	28		2	27
	3	2		3	3
	Not specified	1		Not specified	2
Multiplication	3	13	Division	3	6
	4	22		4	20
	5	1		5	3
	6	1		6	1
	Not specified	2		Not specified	9

[a]Analysis based on the 39 states that had standards covering at least K–6. (Adapted from Reys, 2006, p. 22).

facts. To avoid the confusion produced by 50 different state sets of standards, the *Common Core State Standards for Mathematics* offer recommendations for basic fact mastery to be followed by all states that choose to adopt the standards. The CCSSM recommend that basic addition/subtraction facts should be mastered by the end of grade 2, and basic multiplication/division facts should be mastered by the end of grade 3 (see Figure 9-1).

Why does it take some children years to master their basic facts? Sometimes the problem may be a learning disability that makes it impossible for a child to memorize the facts. Use of a calculator may allow such a child to proceed with learning mathematics. Or, as research by Clark and Kamii (1996) indicates for multiplication, a child may have trouble with multiplication facts because he or she has not developed the ability to think multiplicatively.

More commonly, children's difficulties in mastering basic facts may stem from one (or both) of the following two causes, and in these cases teachers can definitely provide help. First, the underlying numerical understandings may not have been developed. Thus, the process of remembering the facts quickly and accurately becomes no more than rote memorization or meaningless manipulation of symbols. As a result, the child has trouble remembering the facts. Second, the skill of fact retrieval itself may not be taught by teachers or understood by children, resulting in inefficient strategies. Teachers can do something about both of these problems by using a three-phase process for helping children learn basic facts:

1. Start where the children are.
2. Build understanding of the basic facts.
3. Focus on how to remember facts.

START WHERE THE CHILDREN ARE

Many children come to school knowing some basic facts. For instance, the chances are great that they can say "one and one are two," "two and two are four," and maybe even "five and five are ten." They may know that 2 and 1 more is 3, and that 6 and "nothing more" is still 6. But they probably don't know that 6 + 7 = 13, nor do they have a clear concept of the meanings of symbols such as + and =.

Similarly, they may know that if you have 3 and take away 2, you have only 1 left. But they might not know the meaning of the symbolism 3 − 2 = 1 or of the sentence "three minus two equals one." They may be able to figure out that buying three pieces of gum at 5¢ each will cost 15¢ but not know that 3 × 5 = 15. They may be able to figure out that 8 cookies shared fairly among 4 children means that each child gets 2 cookies, but they might not know that 8 ÷ 4 = 2.

In other words, young children can probably solve many simple problems involving facts, but they may not be able to either recognize or write the facts. Nor do many children understand that a number sentence such as 4 + 2 = n asks the same question as:

$$\begin{array}{r} 4 \\ +2 \\ \hline \end{array}$$

It is the teacher's task to help children organize what they know, construct more learning to fill in the gaps, and, in the process, develop meaning for basic facts and for the symbolism we can use to represent them.

You need to begin by determining what each child knows, using responses from group discussions, observations of how each child works with materials and with paper-and-pencil activities, and individual interviews.

Many teachers use an inventory at the beginning of the year, administered individually to younger children and in a questionnaire format in later grades. The purpose of such an inventory is to discover:

- Whether the children have the concept of an operation (Given a story problem involving simple addition, subtraction, multiplication, or division, can they model the situation? Solve the problem? Explain why their solution makes sense? Write a corresponding number sentence? Identify the operation they used?)
- What basic facts they understand (Given a number fact, can they draw a picture to illustrate it? Or given a picture, can they identify a related number fact?)
- What strategies they use to find the solution to combinations (Can they answer the question "How did you know 7 + 9 = 16?" beyond just saying "I know it"?)
- What basic facts they know fluently (Can they answer within about three seconds, without stopping to figure them out?)

Teachers use such information to plan instruction. Do some children need more work with manipulative materials to understand what multiplication means? Do some children need help in seeing the relationship of 17 − 8 and 8 + 9? Do the children recognize that counting on from a number is quicker than counting all? Which children need practice in order to move beyond counting to mastering the basic facts? You can group children to meet individual needs (as suggested in Chapter 3) and provide activities and direct instruction to fill in the missing links and strengthen understanding and competency.

BUILD UNDERSTANDING OF THE BASIC FACTS

Your emphasis in helping children learn their basic facts should be on aiding them to organize their thinking and to see relationships among the facts. Children should develop strategies for remembering the facts before they engage in drill to develop fluency. Researchers point out the following:

> The view on … how these number facts should be learnt and taught has drastically changed in the past decades. Whereas learning single-digit arithmetic was for a long time based mainly on memorizing those facts through drill-and-practice to the point of automatization … current instructional approaches put great emphasis on the gradual development of these number facts from children's invented and informal strategies.
>
> *(Verschaffel et al., 2007, p. 560)*

Generally, the facts with both addends or both factors greater than 5 are more difficult for most children, but what is difficult for an individual child is really the important point. Although many textbooks, workbooks, and computer programs emphasize practice on the generally difficult facts, many also encourage the child to keep a record of those facts that are difficult for him or her and suggest extra practice on those. The teacher should suggest or reinforce this idea.

How can the basic facts for an operation be organized meaningfully? Many textbooks present facts in small groups (e.g., facts with sums to 6: 0 + 6 = 6, 1 + 5 = 6, 2 + 4 = 6, etc.). Other textbooks organize the facts in "families" (e.g., facts in the "2–3–5 family" are 3 + 2 = 5, 2 + 3 = 5, 5 − 3 = 2, and 5 − 2 = 3). Still other textbooks organize the facts by "thinking strategies" (e.g., all facts where 1 is added, or facts involving "doubles" such as 7 + 7). No particular order for teaching the basic facts has been shown to be superior to any other order. Thus, the teacher can use professional judgment about what each group of children needs and can choose whether to use or modify the sequence given in the textbook.

A variety of thinking strategies can be used to recall the answer to any given fact. Thinking strategies are efficient methods for determining answers on the basic facts. The more efficient the strategy, the more quickly the student will be able to construct the correct answer for the sum, difference, product, or quotient of two numbers and, eventually, to develop fluency with the facts so he or she can quickly recall them.

Research has shown that certain thinking strategies help children learn the basic facts (Fuson, 2003a, 2003b). Understanding of the facts develops in a series of stages characterized by the thinking children use. Some of these thinking strategies involve using concrete materials or counting. Others are more mature because a known fact is used to figure out an unknown fact. Teachers want to help children develop these mature, efficient strategies to help them recall facts. The next section on thinking strategies for basic facts provides more detail on how these skills can be developed.

Many children rely heavily on counting—in particular, finger counting—and fail to develop more efficient ways of recalling basic facts. For example, a child might count 4 fingers and then 5 more to solve 4 + 5. This strategy is acceptable at first; however, this counting process should not have to be repeated every time 4 + 5 is encountered. The child needs to move beyond counting on from 4 (which is relatively slow and inefficient), to thinking "4 + 4 = 8, so 4 + 5 is 9" or using some other more efficient strategy. Eventually, the child must be able to recall "4 + 5 = 9" immediately and effortlessly. Some children discover efficient fact strategies on their own, but others may need explicit instruction. When the teacher is satisfied that children are familiar with a particular strategy (able to model it with materials and beginning to use it mentally), it is time to practice the strategy. In other words, before students start practicing fact retrieval, they should be able to

- State or write related facts, given one basic fact.
- Explain how they got an answer, or prove that it is correct.
- Solve a fact in two or more ways.

Overview of Basic Fact Instruction 189

> **Math Links 9.4**
>
> A variety of drill activities may be found on the Internet at http://funbrain.com/kidscenter.html, or link from this book's Web site. For example, you can try games such as Mathcar Racing, Soccer Shootout, Tic Tac Toe Squares, Line Jumper, and Power Football. A variety of operations and levels of difficulty may be selected.
>
> www.wiley.com/college/reys

FOCUS ON HOW TO REMEMBER FACTS

Consider this scene: Pairs of children are keying numbers on a calculator and passing it back and forth. Other pairs are seated at a table, some playing a card game and others playing board games. Several are busily typing numbers on computer keyboards. Still others are working individually with flashcards or number triangles. What are they all doing? They could be practicing basic facts.

If children are to become skillful with the algorithms for addition, subtraction, multiplication, and division and proficient at estimation and mental computation, they must know the basic facts with immediate, automatic recall. The goal is computational fluency and efficiency.

Here are some key principles for basic fact drill:

- Children should attempt to memorize facts only after understanding is attained.
- Children should participate in drill with the intent to develop fluency. Remembering should be emphasized. This is not the time for explanations.
- Drill lessons should be short (5 to 10 minutes) and should be given almost every day. Children should work on only a few facts in a given lesson and should constantly review previously learned facts.
- Children should develop confidence in their ability to remember facts fluently and should be praised for good efforts. Records of their progress should be kept.
- Drill activities should be varied, interesting, challenging, and presented with enthusiasm.

Computer software provides a natural complement to more traditional materials and activities, such as flashcards, games, and audiotaped practice, for establishing the quick recall of basic facts. For example, children may try games found in the Math Links. Most software for fact practice keeps track of the number of exercises attempted and the number answered correctly. Some programs display the time taken to give correct answers, thus encouraging students to compete against their own records for speed as well as mastery. Requiring short response time (within 3 or 4 seconds) is important because it promotes efficient strategies and encourages children to develop fluent recall. Many children enjoy computer software that displays a cumulative record of their individual progress. This feature allows children to diagnose for themselves the basic facts they know and don't know. It also provides a source of motivation because each student can compete against himself or herself, with the goal of complete mastery always in mind.

When using flashcards for practice, the child should start by going through the entire set and separating the cards into a pack of those known and a pack of those unknown. The known pack should be set aside so that time can be used efficiently to focus repeatedly on the facts in the unknown pack. The next time the child works with the cards, he or she should again review the cards from both packs, moving any newly learned facts to the known pack and moving any facts that have been forgotten to the unknown pack. Then practice begins again with the unknown pack. This approach makes progress evident and focuses attention on the facts that really need to be practiced.

Teachers need to take note of whether certain basic addition and multiplication facts occur more (or less) frequently in their elementary school textbook than others. If more difficult facts or facts involving larger numbers occur less frequently, then students are likely to get less practice with them, even though they might actually need more practice. In some textbooks, facts involving numbers larger than 5 may occur less frequently than facts involving numbers from 2 to 5; and problems involving 0 and 1 may be quite rare. If this is the case, it's important to ensure that more practice is provided with the facts that are not as commonly encountered in the textbook, especially the more difficult facts involving 6 through 9.

Several types of drill-and-practice procedures in the form of games are noted in In the Classroom 9-2. Children of all age groups find such games an enjoyable way to practice what they know. These activities supplement the many other drill-and-practice procedures that you will find in textbooks, journals, computer software, and other sources.

> **Math Links 9.5**
>
> Some basic fact games and puzzles also involve problem solving or require students to use critical thinking to develop a winning strategy. For example, visit http://nlvm.usu.edu/en/nav/vlibrary.html, or link from this book's Web site, then select Number and Operations and scroll through to find Number Line Bounce, Number Puzzles, and other challenges. Or go to http://illuminations.nctm.org, or link from this book's Web site, then select tools and scroll through to examine the Product Game and other interesting activities.
>
> www.wiley.com/college/reys

In the Classroom 9–2

GAMES FOR PRACTICING BASIC FACTS

Addition Bingo

Objective: Using a Bingo game to practice basic addition facts.

Grade Level: 1–2.

- Each player needs a different Bingo card and some buttons or macaroni for markers.

9	6	17	11	5
5	8	10	3	7
15	11	Free	13	14
7	9	2	1	12
2	13	11	18	17

The leader needs a pack of cards like these with all possible combinations (basic facts).

$\frac{9}{+5}$ $\frac{7}{+6}$ $\frac{8}{+4}$

- It's easy to play:
  - The leader draws a card and reads the addends on it.
  - Each player covers the sum on his or her Bingo card.

Not all sums are given on each card.

Some sums are given more than once on a Bingo card, but a player may cover only one answer for each pair of addends.

The winner is the first person with 5 markers in a row!

Multig

Objective: Using a game to practice basic multiplication facts.

Grade Level: 3–4.

- Use the playing board here or make a larger one on heavy construction paper. Each player needs some buttons, macaroni, or chips for markers.
  - Take turns. Spin twice. Multiply the 2 numbers. Find the answer on the board. Put a marker on it.

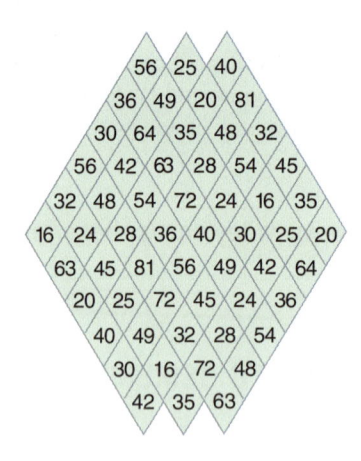

Don't forget the spinner. You can't play this game without it.

Spinner: 9, 4, 8, 6, 7, 5

Playing board (diamond of numbers):
56 25 40
36 49 20 81
30 64 35 48 32
56 42 63 28 54 45
32 48 54 72 24 16 35
16 24 28 36 40 30 25 20
63 45 81 56 49 42 64
20 25 72 45 24 36
40 49 32 28 54
30 16 72 48
42 35 63

- Score 1 point for each covered ♦ that touches a side or corner of the ♦ you cover.
- If you can't find an uncovered ♦ to cover, you lose your turn.
- Opponents may challenge any time before the next player spins.
- The winner is the player with the most points at the end of 10 rounds.

Zero Wins

Objective: Using a game to develop number sense and practice addition and subtraction acts.

Grade Level: 2–3.

- Make two identical sets of 19 cards with a number from 0 to 18 on each.

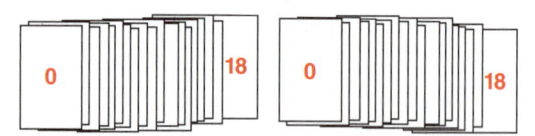

[Two stacks of cards showing 0 and 18]

- Follow these rules:
  - After shuffling, the leader deals 4 cards to each player and puts the remaining cards face down in the center of the table.
  - Players must add or subtract the numbers on their 4 cards so they equal 0. For example, suppose you had these cards:

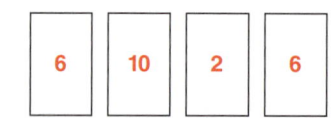

6	10	2	6

(continued on the next page)

- With these cards you could write

  $10 - 6 + 2 - 6 = 0$

  or

  $6 + 6 - 10 - 2 = 0$

  or various other number sentences.

- On each round of play, the players may exchange one card if they wish, and each player takes a turn being first to exchange a card on a round. To make an exchange, the first player draws a card and discards a card, face up. Other players can draw from either the face-down pile or the face-up discard pile.

The first player to get 0 on a round wins the round!

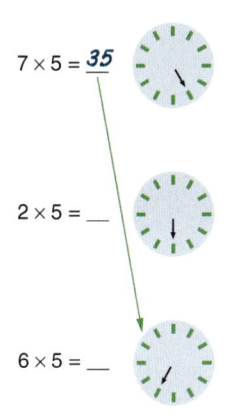

$7 \times 5 = \mathbf{35}$

$2 \times 5 = \_\_$

$6 \times 5 = \_\_$

$4 \times 5 = \_\_$

$5 \times 5 = \_\_$

Test Your Facts

Objective: Using a variety of activities and puzzles to practice multiplication and division facts.

Grade Level: 3–4.

- Fill in the empty box:

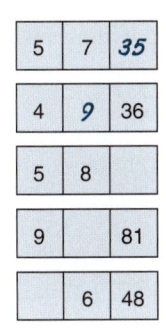

5	7	**35**
4	**9**	36
5	8	
9		81
	6	48

- ▼ Complete these five facts and match them with the clock minutes.

- Fill in each empty box to make the next number correct:

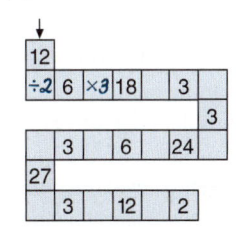

12				
÷2	6	×3	18	3
				3
	3		6	24
27				
	3		12	2

- Multiply each number in the middle ring by the number in the center:

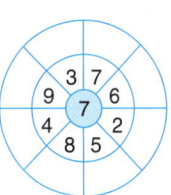

  (ring with 3, 7, 6, 2, 5, 8, 4, 9 around center 7)

THINKING STRATEGIES FOR BASIC FACTS

In the following four sections, we discuss thinking strategies for basic facts for addition, subtraction, multiplication, and division and illustrate ways to teach them.

THINKING STRATEGIES FOR ADDITION FACTS

The 100 basic facts for addition were shown previously in Figure 9-5. These facts should not simply be presented to children to memorize from this completed table; rather, the children should gradually and systematically learn the facts and may fill in or check them off on the table. However, having children examine a completed addition table to look

192 Chapter 9 • Operations: Meanings and Basic Facts

for patterns can be a challenging and educational activity (and can help them learn how to read the table, too). For example, you might ask the children questions such as the following:

- What patterns do you see in the table? (They will probably notice that each row or column shows numbers counting up by 1s, but they start at different starting points.)
- What is the largest sum? What are its addends? Can you write the addition sentence for this sum?
- Circle all the sums of 5. Why does 5 appear more than once in the table as a sum? Can you write all the addition sentences for this sum?
- What sum appears most often in the table? Can you write all the addition sentences for this sum?

Asking questions like this can help children recognize the orderliness of the basic addition facts. This overview can also help children see their goal as they begin to work on fluency with facts. Thinking strategies for teaching basic addition facts include commutativity; adding 1, 0, doubles, and near doubles; counting on; combinations to 10; and adding to 10 and beyond. For many facts, more than one strategy is appropriate.

COMMUTATIVITY The task of learning the basic addition facts is simplified because of the commutative property. Changing the order of the addends does not affect the sum. Children encounter this idea when they note that putting 2 blue objects and 3 yellow objects together gives the same quantity as putting 3 blue objects and 2 yellow objects together:

In work with the basic addition facts, children will see or write, for example:

$$\begin{array}{c} 2 \\ +5 \end{array} \quad \text{and} \quad \begin{array}{c} 5 \\ +2 \end{array}$$

or

$$2 + 5 = \Box \quad \text{and} \quad 5 + 2 = \Box$$

Students must realize that the same two numbers have the same sum, no matter which comes first. They need to be able to put this idea into their own words; they do not need to know the term *commutative property*. They need to use the idea as they work with basic facts, not merely parrot a term. A calculator can also help children verify that the order of the addends is irrelevant. Have them key into the calculators combinations such as "5 + 8 = " and "8 + 5 = ." Use a variety of combinations so the idea that the order does not affect the outcome becomes evident. Note that Figure 9-6 has exactly the same 45 numbers above and to the right of the diagonal as below and to the left. This mirror image helps you see that

+	0	1	2	3	4	5	6	7	8	9
0	0	1	2	3	4	5	6	7	8	9
1	1	2	3	4	5	6	7	8	9	10
2	2	3	4	5	6	7	8	9	10	11
3	3	4	5	6	7	8	9	10	11	12
4	4	5	6	7	8	9	10	11	12	13
5	5	6	7	8	9	10	11	12	13	14
6	6	7	8	9	10	11	12	13	14	15
7	7	8	9	10	11	12	13	14	15	16
8	8	9	10	11	12	13	14	15	16	17
9	9	10	11	12	13	14	15	16	17	18

Figure 9-6 *Addition facts derived by the commutative strategy.*

the commutative property reduces the number of facts to be learned by 45. Each box below the diagonal can be matched with a box above the diagonal with the same addends and sum. (If there are 100 facts altogether, why are there 55 distinct facts to learn, rather than just 50?)

ADDING ONE AND ZERO Adding one to a number is easy for most children. In fact, most children learn this idea before they come to school, and they only have to practice the recognition and writing of it rather than develop initial understanding. To reinforce their initial concept, experiences with objects come first, followed by such paper-and-pencil activities as these:

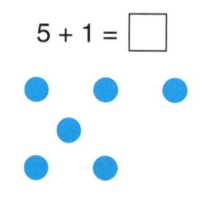

$$5 + 1 = \Box$$

Recognition of the pattern is then encouraged:

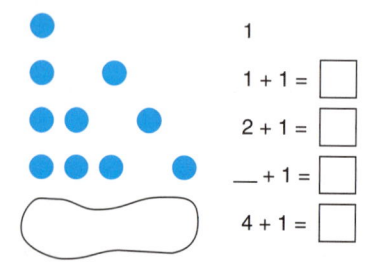

1
$$1 + 1 = \Box$$
$$2 + 1 = \Box$$
$$\_ + 1 = \Box$$
$$4 + 1 = \Box$$

The strategy for adding zero applies to facts that have zero as one addend. These facts should be learned, through experience, as a generalization: *Zero added to any number does not change the number.* This idea follows from many concrete examples in which children see that any time they add "no more" (zero), they have the same amount. Activities then focus on this pattern:

○   1 + 0 = ☐   0 + 1 = ☐

○○   2 + 0 = ☐   0 + 2 = ☐

○○○   3 + 0 = ☐   0 + 3 = ☐

○○○○   4 + 0 = ☐   0 + 4 = ☐

Although adding zero may seem easy to adults, this is actually one of the hardest strategies for children to learn. Concrete modeling of the situation is tricky because it is difficult to picture adding nothing. Therefore, explicit work on facts involving zero should be postponed until children have mastered some of the other fact strategies.

ADDING DOUBLES AND NEAR DOUBLES *Doubles* are basic facts in which both addends are the same number, such as 4 + 4 or 9 + 9. Most children learn these facts quickly, often parroting them before they come to school. Connecting doubles facts to familiar situations often helps students remember them (e.g., two hands shows 5 + 5 = 10, an egg carton shows 6 + 6 = 12, two weeks on a calendar shows 7 + 7 = 14). Students can profit from work with objects followed by drawings:

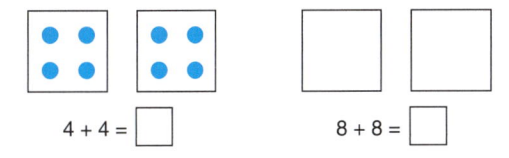

4 + 4 = ☐   8 + 8 = ☐

Another strategy, *near doubles*, can be used for the facts that are 1 more or 1 less than the doubles:

     Think

7 + 8 = ☐   7 + 7 = 14
               So 7 + 8 is one more
               7 + 8 = 15

     Think

7 + 6 = ☐   7 + 7 = 14
               So 7 + 6 is one less
               7 + 6 = 13

The four thinking strategies in the preceding two sections can be used with the addition facts shown in Figure 9-7.

COUNTING ON The strategy of *counting on* can be used for any addition facts but is most easily used when one of the addends is 1 or 2. To be efficient, it is important to count on from the larger addend. For example,

     Think

2 + 6 = ☐   6 ... 7 ... 8
               2 + 6 = 8

Initially, children will probably count all objects in a group, as noted in Chapter 7.

(a) Addition facts derived by the adding one and adding zero strategies

+	0	1	2	3	4	5	6	7	8	9
0	0	1	2	3	4	5	6	7	8	9
1	1	2	3	4	5	6	7	8	9	10
2	2	3								
3	3	4								
4	4	5								
5	5	6								
6	6	7								
7	7	8								
8	8	9								
9	9	10								

(b) Addition facts derived by the adding doubles and near-doubles strategies

+	0	1	2	3	4	5	6	7	8	9
0	0	1								
1	1	2	3							
2		3	4	5						
3			5	6	7					
4				7	8	9				
5					9	10	11			
6						11	12	13		
7							13	14	15	
8								15	16	17
9									17	18

Figure 9-7 *Addition facts derived by four strategies.*

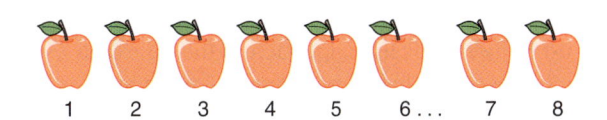

1   2   3   4   5   6 ...   7   8

They need to learn to start from the larger addend, 6, and count on, 7, 8. (Notice that understanding of the commutative property is assumed.) Often young children will count on, but not necessarily from the larger addend. Thus, the strategy must be taught to many children using activities such as this one:

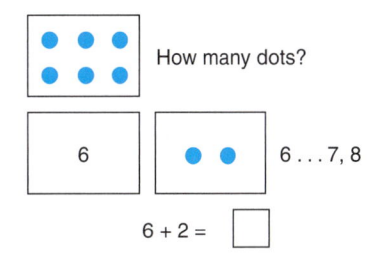

How many dots?

6   ••   6 ... 7, 8

6 + 2 = ☐

194   Chapter 9 • Operations: Meanings and Basic Facts

+	0	1	2	3	4	5	6	7	8	9
0										
1		2	3	4	5	6	7	8	9	10
2		3	4	5	6	7	8	9	10	11
3		4	5							
4		5	6							
5		6	7							
6		7	8							
7		8	9							
8		9	10							
9		10	11							

Figure 9-8 Addition facts derived by the counting-on (1 or 2) strategy.

The counting-on strategy can be rather efficiently used with the addition facts noted in Figure 9-8 (where at least one addend is 1 or 2); however, it is *not efficient* to use counting on when both addends are larger than 2 (e.g., for 6 + 8 = 14 or 5 + 8 = 13 or 9 + 8 = 17). Counting on more than one or two numbers is both slow and prone to error. Unfortunately, some children develop the habit of using counting on all the time and thereby fail to develop more efficient strategies, sometimes still relying on counting on even after they have graduated to middle school or high school. It is important to help students move beyond counting on because fluency (efficiency and accuracy) is extremely important for work with basic facts. When students are slowed down by inefficient strategies for basic fact retrieval, they are often hampered in doing more advanced work in mathematics.

COMBINATIONS TO 10 The combinations-to-10 facts are those nine pairs of numbers that together make 10: 1 + 9, 2 + 8, 3 + 7, 4 + 6, 5 + 5, 6 + 4, 7 + 3, 8 + 2, and 9 + 1. Note that recognition of the commutative property can reduce this list to just five facts (since each of the nine facts except 5 + 5 has a commutative partner-pair—for example, 8 + 2 and 2 + 8 are pairs, etc.). Of the nine combinations to 10, five have already been mentioned in previous fact strategies (either *counting on* or *doubles*). The "new" facts here are 3 + 7 (with 7 + 3) and 4 + 6 (with 6 + 4). As you will read in Cultural Connections at the end of this chapter, American students have traditionally been less likely than children in certain other countries to really use the combinations-to-10 strategy effectively. This is likely due, in part, to the structure of the English language, though it is probably also a reflection of how little American textbooks have emphasized these facts.

ADDING TO 10 AND BEYOND With the strategy of *adding to 10 and beyond*, one addend is broken apart so that one part of it can be used with the other addend to make 10. Then the remaining part of the first addend is added to the 10 to go beyond to the final teen sum. This strategy is used most easily when one of the addends is 8 or 9. However, if children are fluent with combinations to 10, then *adding to 10 and beyond* can be useful in many other situations too. Here is an example:

8 + 5 = ☐
Think
8 + 2 = 10 so I will use 2 from the 5
I'll think of 5 as 2 + 3
8 + 5 = (8 + 2) + 3 = 10 + 3 = 13

Recognizing that 8 + 2 = 10, the child mentally breaks 5 apart into 2 + 3 (to obtain the desired 2). The 2 is used with the 8 to "add to 10," and the 3 is used to go "beyond." A ten-frame can be helpful in teaching this strategy because it provides a visual image of adding to 10 and going beyond:

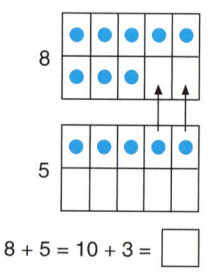

8 + 5 = 10 + 3 = ☐

The research of Funkhouser (1995) indicates that working with five-frames as a base and then moving to ten-frames may be particularly helpful for children with learning disabilities.

Children must know the combinations to 10 well in order to use the adding-to-10-and-beyond strategy efficiently. They also need to realize how easy it is to add any single digit to 10 to get a number in the teens, without having to count on. They should already be able to respond immediately to questions such as 3 + 10 = ?, 10 + 7 = ?, and 9 + 10 = ?.

The *combinations-to-10* and the *adding-to-10-and-beyond* strategies can be used with the addition facts shown in Figure 9-9.

+	0	1	2	3	4	5	6	7	8	9
0										
1										10
2									10	11
3								10	11	12
4							10		12	13
5						10			13	14
6					10				14	15
7				10					15	16
8			10	11	12	13	14	15	16	17
9		10	11	12	13	14	15	16	17	18

Figure 9-9 Addition facts derived by the combinations-to-10 and adding-to-10-and-beyond strategy.

+	0	1	2	3	4	5	6	7	8	9
0	0	1	2	3	4	5	6	7	8	9
1	1	2	3	4	5	6	7	8	9	10
2	2	3	4	5	6	7	8	9	10	11
3	3	4	5	6	7	8	9	10	11	12
4	4	5	6	7	8	9	10	11	12	13
5	5	6	7	8	9	10	11	12	13	14
6	6	7	8	9	10	11	12	13	14	15
7	7	8	9	10	11	12	13	14	15	16
8	8	9	10	11	12	13	14	15	16	17
9	9	10	11	12	13	14	15	16	17	18

Figure 9-10 *Addition facts derived by all thinking strategies for addition.*

In many cases, more than one strategy can be used to aid in recalling a fact. This point should be made with children. It encourages them to try different ways of recalling a fact, and it may strengthen their understanding of the relationships involved. Notice from Figure 9-10 that all 100 of the basic facts have been covered (with one or more strategies), once the strategies for adding one, adding zero, adding doubles and near doubles, counting on, combinations to 10, and adding to 10 and beyond have been learned. Children should be encouraged to look for patterns and relationships because almost all of the 100 basic addition facts can be developed from a variety of relationships with other facts.

It also should be noted that children might invent additional strategies of their own, such as

$6 + 7 = \square$

Think
I know $5 + 5 = 10$
6 is $5 + 1$
7 is $5 + 2$
So $10 + 3$
13

$6 + 8 = \square$

Think
I know my doubles
Move 1 from the 8 to the 6 to make the problem $7 + 7$
14

Encourage students' original ideas!

THINKING STRATEGIES FOR SUBTRACTION FACTS

For each basic addition fact, there is a related subtraction fact. In some mathematics programs, the two operations are taught simultaneously. The relationship between them is then readily emphasized, and learning the basic facts for both operations proceeds as if they were in the same family. Even when they are not taught simultaneously, however, the idea of a fact family is frequently used. Figure 9-11 shows several versions of materials for practicing facts via fact families. Students can practice with fact triangles by covering the number in one corner with their thumb and figuring it out by viewing the other two numbers. In a similar way, they practice with folding fact-family cards by folding one door shut to conceal one number, and figuring it out by viewing the other two numbers. The Snapshot of a Lesson at the opening of this chapter provides a clear example of a problem situation in which children work simultaneously on related addition and subtraction facts.

Think addition is the major thinking strategy for learning and recalling the subtraction facts. Encourage children to recognize, think about, and use the relationships between addition and subtraction facts. They can find the answers to subtraction facts by thinking about missing addends. For example,

$15 - 7 = \square$

Think
$7 + \square = 15$
$7 + 8 = 15$
So $15 - 7 = 8$

Other strategies for finding subtraction facts also can be taught: *subtracting one and zero, doubles, counting back,* and *counting on.*

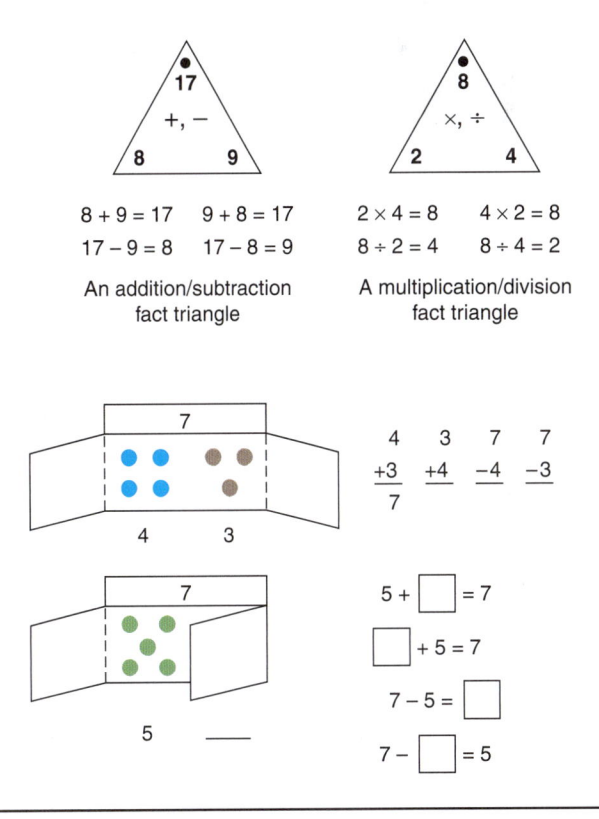

$8 + 9 = 17$ $9 + 8 = 17$
$17 - 9 = 8$ $17 - 8 = 9$

An addition/subtraction fact triangle

$2 \times 4 = 8$ $4 \times 2 = 8$
$8 \div 2 = 4$ $8 \div 4 = 2$

A multiplication/division fact triangle

$\begin{array}{cccc} 4 & 3 & 7 & 7 \\ +3 & +4 & -4 & -3 \\ \hline 7 & & & \end{array}$

$5 + \square = 7$
$\square + 5 = 7$
$7 - 5 = \square$
$7 - \square = 5$

Figure 9-11 *Examples of fact families.*

196 Chapter 9 • Operations: Meanings and Basic Facts

SUBTRACTING ONE AND ZERO Once they have learned strategies for adding 1 and adding 0, most children find it rather easy to learn the related subtraction facts involving 0 and 1. They can profit from work with materials and from observing patterns similar to those used for addition facts.

DOUBLES The strategy for doubles may need to be taught more explicitly for subtraction facts than for addition facts. It rests on the assumption that children know the doubles for addition. Here is an example:

$$16 - 8 = \Box \quad \begin{array}{l} \text{Think} \\ 8 + \Box = 16 \\ 8 + 8 = 16 \\ \text{So } 16 - 8 = 8 \end{array}$$

COUNTING BACK The strategy of counting back is related to counting on in addition. It is most efficient when the number to be subtracted is 1 or 2:

$$9 - 2 = \Box \quad \begin{array}{l} \text{Think} \\ 9 \ldots 8, 7 \\ 9 - 2 = 7 \end{array}$$

As for other strategies, use problems and a variety of manipulative materials to help children gain facility in counting back from given numbers.

COUNTING ON The strategy of counting on is used most easily and efficiently when the difference is 1 or 2 (i.e., when it is easy to see that the two numbers involved in the subtraction are quite close together). "Think addition" by counting on:

$$8 - 6 = \Box \quad \begin{array}{l} \text{Think} \\ 6 \ldots 7, 8 \text{ (that's two numbers} \\ \text{that I counted on)} \\ \text{So } 8 - 6 = 2 \end{array}$$

Thinking about the counting-on strategy can also be related to thinking about adding on: "How much more would I need?" Children should be encouraged to use known addition facts to reach a solution to the subtraction problem. This is particularly valuable with missing-addend situations, such as $6 + \Box = 8$. Research has shown that counting on is a very powerful subtraction strategy for many students (Fuson, 2003a, 2003b).

THINKING STRATEGIES FOR MULTIPLICATION FACTS

Multiplication is frequently viewed as a special case of addition in which all the addends are of equal size. The solution to multiplication problems can be attained by adding or counting, but multiplication is used because it is so much quicker.

×	0	1	2	3	4	5	6	7	8	9
0	0	0	0	0	0	0	0	0	0	0
1	0	1	2	3	4	5	6	7	8	9
2	0	2	4	6	8	10	12	14	16	18
3	0	3	6	9	12	15	18	21	24	27
4	0	4	8	12	16	20	24	28	32	36
5	0	5	10	15	20	25	30	35	40	45
6	0	6	12	18	24	30	36	42	48	54
7	0	7	14	21	28	35	42	49	56	63
8	0	8	16	24	32	40	48	56	64	72
9	0	9	18	27	36	45	54	63	72	81

Figure 9-12 *The 100 basic facts for multiplication.*

Instruction on multiplication ideas begins in kindergarten as children develop ideas about groups, numbers, and addition. In grades 1 and 2, children count by 2s, 5s, 10s, and later by other numbers such as 3s and 4s. These experiences provide a basis for understanding the patterns that occur with the basic multiplication facts. Use of the calculator, as described in Chapter 7, can aid teachers in developing ideas about these patterns of multiplication. Using the constant function on calculators, children realize that two 6s equal 12, three 6s equal 18, and so on.

The basic multiplication facts pair two one-digit factors with a product, as shown in Figure 9-12. The basic multiplication facts should not be given to children in the form of a table or chart of facts until they have been meaningfully introduced. Rather, the facts should be developed through problem situations, experiences with manipulatives and other materials, and various thinking strategies. The table becomes the end result of this process of developing understanding of operations and facts.

There has not been as much research done, to date, on the development of children's strategies for multiplying and dividing single-digit numbers as for adding and subtracting, but researchers report a growing body of literature in this area (Verschaffel et al., 2007). Thinking strategies for multiplication facts provide efficient ways for a child to attain each fact. These strategies include commutativity, skip counting, repeated addition, splitting the product into known parts, multiplying by 1 and 0, and patterns.

COMMUTATIVITY Commutativity applies to multiplication just as it does to addition. It is, therefore, a primary strategy for helping students learn the multiplication facts. In the Classroom 9-1 emphasizes this property. The calculator is also useful in reinforcing the idea. Children can multiply 4×6, then 6×4, for example, and note that the answer to both is 24. After they have tried many combinations, students should be able to verbalize that the order of the factors is irrelevant. Figure 9-13 (where each box in the upper right can be matched with a corresponding box in the lower left)

Thinking Strategies for Basic Facts   197

Figure 9-13 *Multiplication facts derived by the commutative thinking strategy.*

shows that, as for addition, there really are only 55 multiplication facts to be learned if you recognize the power of commutativity.

SKIP COUNTING The strategy of skip counting works best for the multiples children already know best, 2s and 5s, but it also may be applied to 3s and 4s (or other numbers) if children have learned to skip count by them.

Here is an example for 5:

	Think
$4 \times 5 = \square$	5, 10, 15, 20
	$4 \times 5 = 20$

Skip counting around the clock face (as you do when counting minutes after the hour) is a good way to reinforce multiples of 5 by skip counting. The facts that can be established with the skip-counting strategy for 2s and 5s are noted in Figure 9-14.

Figure 9-14 *Multiplication facts derived by the skip-counting strategy.*

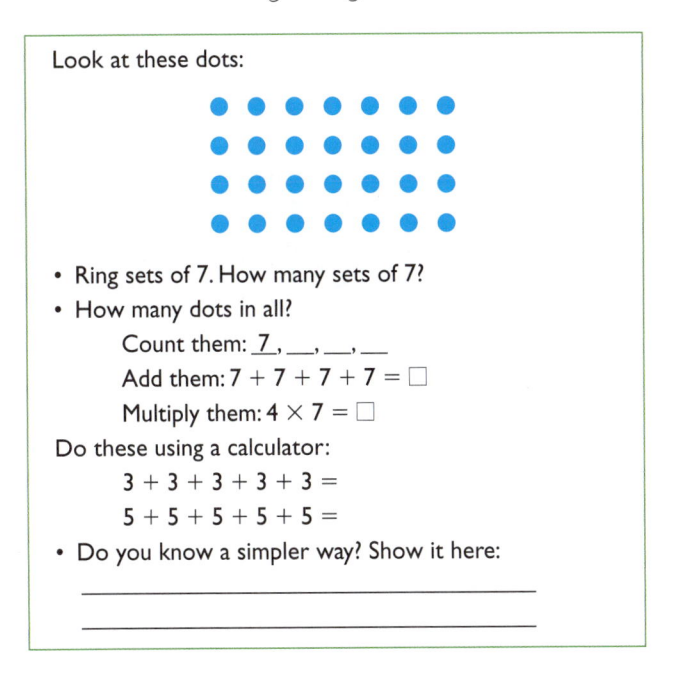

Look at these dots:

- Ring sets of 7. How many sets of 7?
- How many dots in all?
    Count them: _7_, __, __, __
    Add them: $7 + 7 + 7 + 7 = \square$
    Multiply them: $4 \times 7 = \square$
- Do these using a calculator:
    $3 + 3 + 3 + 3 + 3 =$
    $5 + 5 + 5 + 5 + 5 =$
- Do you know a simpler way? Show it here:
    _____
    _____

Figure 9-15 *Visualizing multiplication.*

REPEATED ADDITION The strategy of repeated addition can be used most efficiently when one of the factors is less than 5. The child thinks of the multiplication as repeated addition:

	Think
$3 \times 6 = \square$	$6 + 6 + 6 = 18$
	$3 \times 6 = 18$

Because this strategy is based on one interpretation of multiplication, children should have had many experiences with objects and materials. Drawings and explorations with a calculator can be used to provide additional experiences to help develop this strategy as well as the concept for the operation. Figure 9-15 illustrates these ideas. In Figure 9-16, note the facts that can be learned with this strategy.

SPLITTING THE PRODUCT INTO KNOWN PARTS As children gain assurance with some basic facts, they can use their known facts to derive others. The strategy, known as *splitting the product*, is based on the distributive property of multiplication. It can be approached in terms of "one more set," "twice as much as a known fact," or "known facts of 5."

- The idea of *one more set* can be used for almost all multiplication facts. If a certain multiple of a number, N, is known, the next multiple can be determined by adding the single-digit number, N. For example, to find 3×5 if doubles are already known, you can think of 2×5 (10) and add one more 5 (to get 15). The computation is slightly more difficult if the addition requires renaming (e.g., below, where 8×7 is found by adding one more set of 7 to the known fact $7 \times 7 = 49$):

198 Chapter 9 • Operations: Meanings and Basic Facts

×	0	1	2	3	4	5	6	7	8	9
0										
1										
2				4	6	8	10	12	14	16	18
3				6	9	12	15	18	21	24	27
4				8	12	16	20	24	28	32	36
5				10	15	20					
6				12	18	24					
7				14	21	28					
8				16	24	32					
9				18	27	36					

Figure 9-16 *Multiplication facts most appropriate for the repeated addition strategy.*

$8 \times 7 = \square$
   Think
   $7 \times 7 = 49$
   $8 \times 7 = 49 + 7 = 56$

Each fact can be used to help learn the next multiple of either factor. Illustrating this strategy using an array model can be helpful:

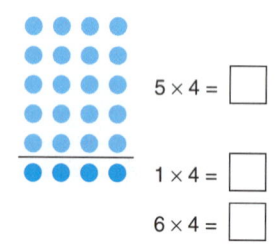

$5 \times 4 = \square$

$1 \times 4 = \square$

$6 \times 4 = \square$

Ask children to name each part of the array and write the multiplication fact for the whole array.

- *Twice as much as a known fact* is a variation of the foregoing strategy. It can be applied to multiples of 4, 6, and 8 because an array with one of these numbers can be split in half. The product is twice as much as each half:

$6 \times 8 = \square$
   Think
   $3 \times 8 = 24$
   6×8 is twice as much, or $24 + 24$
   So, $6 \times 8 = 48$

Note that some children may have trouble doing the computation mentally when renaming is required to do the doubling by addition. Again, using models helps provide a visual image of this strategy. In this case, children work with already divided arrays.

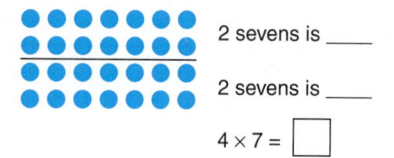

2 sevens is _____

2 sevens is _____

$4 \times 7 = \square$

As they progress, children can divide an array, such as the following; write about each part; and write the multiplication fact for the whole array.

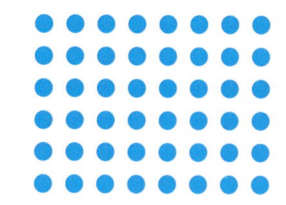

Some options for the array above (6×8) might be

$(5 \times 8) = 40$ and $1 \times 8 = 8$, so
$6 \times 8 = 40 + 8 = 48$

or $6 \times 4 = 24$ and 6×8 is twice as much,
so $6 \times 8 = 24 + 24 = 48$

- *Working from known facts of 5* also will aid children. It can be helpful for any problem with large factors but is most useful for multiples of 6 and 8 because both five 6s and five 8s are multiples of 10, so it is rather easy to add on the remaining part without renaming.

For example, here's how to figure 7×6, if you know $5 \times 6 = 30$.

$7 \times 6 = \square$
   Think
   $5 \times 6 = 30$
   $2 \times 6 = 12$
   So 7×6 is $30 + 12$, or 42

The facts that can be solved by splitting the product into known parts are shown in Figure 9-17.

×	0	1	2	3	4	5	6	7	8	9
0										
1										
2										
3										
4								24	28	32	36
5								30	35	40	45
6							24	30	36	42	48	54
7							28	35	42	49	56	63
8							32	40	48	56	64	72
9							36	45	54	63	72	81

Figure 9-17 *Multiplication facts derived by the strategy of splitting the product into known parts.*

Thinking Strategies for Basic Facts   199

MULTIPLYING BY ONE AND ZERO   The facts with one and zero are generally learned from experience working with multiplication. Children need to be able to generalize that "multiplying by 1 does not change the other number" and that "multiplying by 0 results in a product of 0." Figure 9-18 indicates the facts that can be learned with these strategies.

The preceding strategies account for all the multiplication facts. However, one more strategy can provide help with some difficult facts.

PATTERNS   Finding patterns is helpful with several multiplication facts. One of the most useful and interesting patterns concerns 9s. Look at the products of the facts involving 9. The digits of the products always sum to 9. Furthermore, the tens digit of each product is always one less than whatever factor was being multiplied times 9. (Why? Think about this. There's a reason.)

$1 \times 9 = 9$ $\qquad$ $0 + 9 = 9$
$2 \times 9 = 18$ $\qquad$ $1 + 8 = 9$
$3 \times 9 = 27$ $\qquad$ $2 + 7 = 9$
$4 \times 9 = 36$ $\qquad$ $3 + 6 = 9$

The tens digit is 1 less than 4.   The sum of the digits of 36 is 9.

So for 5×9,

Think

The tens digit is one less than 5   4
The sum of the digits is 9, so $4 + \square = 9$ 5
So, $5 \times 9 = 45$

Now try $7 \times 9 = \square$.

Challenge children to find this and other interesting patterns in a table or chart such as the one in Figure 9-12. They might note, for instance, that the columns (and rows) for 2, 4, 6, and 8 contain all even numbers, although the columns for 1, 3, 5, 7, and 9 alternate even and odd numbers. Why?

You will probably find that children also enjoy various forms of "finger multiplication" (see Figure 9-19), seemingly magical tricks for using finger manipulations to determine basic facts. Actually, though, finger multiplication works because of the patterns that are inherent in the basic fact table.

THINKING STRATEGIES FOR DIVISION FACTS

Teaching division has traditionally taken a large portion of time in elementary school. Now, with the increased use of calculators, many educators advocate reducing the attention accorded to it. Nevertheless, children still need an understanding of the division process and division facts. The facts help them to respond quickly to simple division situations and to understand better the nature of division and its relationship to multiplication.

Just as "think addition" is an important strategy for subtraction, "think multiplication" is the primary thinking strategy to aid children in understanding and recalling the division facts. Division is the inverse of multiplication; that is, in a division problem you are seeking an unknown factor when the product and some other factor are known. The multiplication table (Figure 9-12) illustrates all the division facts; you simply read it differently. For the division fact $54 \div 9 = \square$, look in the 9-row of the multiplication table for the number 54, then read up that column to find the other factor, 6. Students generally do not learn division facts separately from multiplication facts. Instead, they learn division facts such as $48 \div 6 = 8$ by remembering (and connecting with) multiplication facts such as $6 \times 8 = 48$.

It is important to realize that most division problems, in computations and in real-world problem situations, do not directly involve a multiplication fact that you have learned. For example, consider the computation $49 \div 6$. There is no basic fact involving 49 and 6. So what do you do? In this situation most people quickly and automatically mentally review the 6-facts that they know, looking for the facts that come closest to 49 ($6 \times 7 = 42$—too small, $6 \times 8 = 48$—just a little too small, $6 \times 9 = 54$—too big). From this mental review, you can conclude that $49 \div 6 = 8$, with 1 left over. Children need practice in thinking this way (mentally finding the answers to problems involving one-digit divisors and one-digit answers with remainders).

Just as fact families can be developed for addition and subtraction, they can also be useful for multiplication and division:

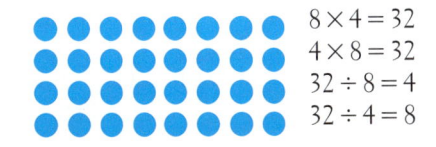

×	0	1	2	3	4	5	6	7	8	9
0	0	0	0	0	0	0	0	0	0	0
1	0	1	2	3	4	5	6	7	8	9
2	0	2								
3	0	3								
4	0	4								
5	0	5								
6	0	6								
7	0	7								
8	0	8								
9	0	9								

Figure 9-18   *Multiplication facts derived by the strategies for 0 and 1.*

$8 \times 4 = 32$
$4 \times 8 = 32$
$32 \div 8 = 4$
$32 \div 4 = 8$

Finger Multiplication for 9-Times Facts Only

1. Hold both hands in front of you with palms facing away.
2. To do 3 × 9, bend down the third finger from the left.
3. The fingers to the left of the bent finger represent tens. The fingers to the right of the bent finger represent ones. Read off the answer: 2 tens and 7 ones, or 27.
4. To do 5 × 9, bend down the fifth finger from the left, etc. Try this method for any 9-facts: 1 × 9 through 10 × 9.

Finger Multiplication for Facts with Factors 6–10 Only

1. Hold both hands in front of you with palms facing you.
2. Think of a multiplication in which both factors are 6 or more (up to 10). For example, you can do anything from 6 × 6 up to 10 × 10.
3. For each factor, find the difference from 10.
   - Let's try the example 8 × 9.
   - For the factor 8, think 10 − 8 = 2, and bend down 2 fingers on the left hand.
   - For the factor 9, think 10 − 9 = 1, and bend down 1 finger on the right hand.
4. Look at all fingers that are still up, and count by ten (10, 20, 30, 40, 50, 60, 70).
5. Look at the fingers that are bent down. Multiply the number of fingers bent down on the left hand by the number bent down on the right hand (2 × 1 = 2).
6. So 8 × 9 = 70 + 2 or 72.
7. Try this method for any multiplication facts from 6 × 6 up through 10 × 10.

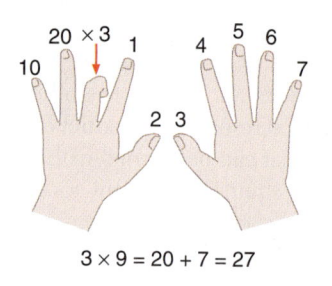

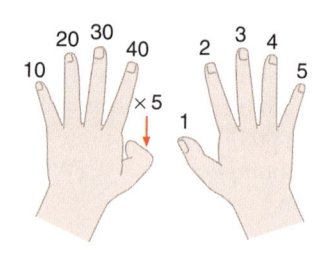

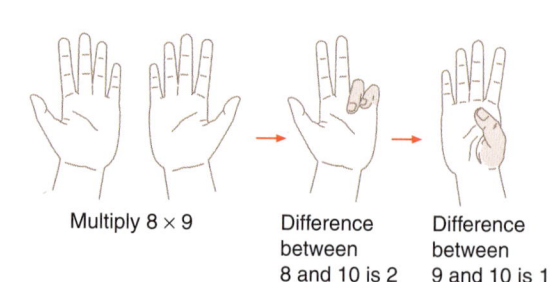

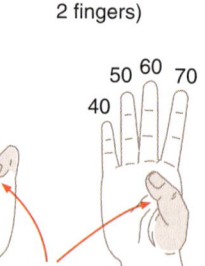

Figure 9-19 *Two forms of finger multiplication.*

Because of its relationship to multiplication, division can be stated in terms of multiplication:

$$42 \div 6 = \square \qquad 6 \times \square = 42$$

Thus, children must search for the missing factor in the multiplication problem. Because multiplication facts are usually encountered and learned first, children can use what they know to learn the more difficult division facts. Moreover, division is related to subtraction, and division problems can be solved by repeated subtraction (seeing how many times you must subtract to get down to zero):

What's 12 ÷ 3?
Since 12 − 3 − 3 − 3 − 3 = 0 (subtracting four 3s), then 12 ÷ 3 = 4

Four threes

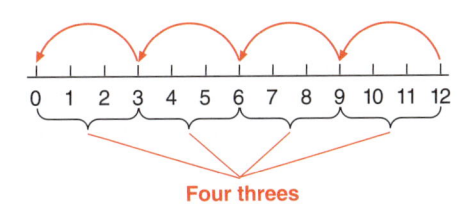

It is important for children to recognize that division can be thought of as repeated subtraction, just as multiplication can be thought of as repeated addition. Indeed, when children later encounter division involving multidigit numbers, they will need to understand division as equivalent to repeated subtraction, in order to appreciate why the division algorithm involves repeatedly subtracting.

However, as was mentioned earlier, counting backward is difficult for many children. Skip counting backward (to figure out a division problem mentally) is even more difficult than counting backward by one (to figure out a subtraction problem mentally). Here are two examples of skip counting backward:

$15 \div 3 = \Box$

Think
15 ... 12, 9, 6, 3, 0
I subtracted 3 five times.
So $15 \div 3 = 5$

$28 \div 7 = \Box$

Think
$28 - 7$
$21 - 7$
$14 - 7$ } 4 subtractions to get to 0
$7 - 7$
0
$28 \div 7 = 4$

A more productive approach is to "think multiplication" (just as it was productive to "think addition" in order to perform subtractions). If you know your multiplication facts, then $35 \div 7 = \Box$ is easily figured by thinking "what times 7 is 35?" For facts that are not immediately recognized, splitting the product into known parts is helpful. Obviously, this strategy relies heavily on knowledge of multiplication facts as well as on the ability to keep in mind the component parts.

$35 \div 7 = \Box$

Think
I know $2 \times 7 = 14$
I know $3 \times 7 = 21$
$14 + 21 = 35$
$2 + 3 = 5$
So $35 \div 7 = 5$

As with multiplication, work with arrays helps children relate the symbols to the action.

In general, children have little difficulty in dividing by one. But we all need to exercise caution when zero is involved in division. Dividing zero by some number and dividing some number by zero are very different situations. If you divide 0 by 6 ($0 \div 6$), the result is 0. Check this by multiplying your answer (0) by your divisor (6): $0 \times 6 = 0$.

But *division by zero is impossible*. For example, to solve $6 \div 0 = \Box$ would require a solution so that $0 \times \Box = 6$. However, there is no value for $\Box$ that would make this sentence true. Therefore, $6 \div 0$ has no solution. A similar difficulty arises if you try to divide zero by itself. That's why *division by zero is undefined* in mathematics. Just as you may have difficulty remembering which is possible, dividing zero by a number or dividing a number by zero, it is likely that children may find this difficult. They will need numerous experiences before they master the idea.

Thinking strategies for division are more difficult for children to learn than are the strategies for the other operations. There is more to remember and regrouping is often necessary. When using skip counting as a division strategy, for instance, the child needs to keep track of the number of times a number is named even as he or she simultaneously struggles to count backward. However, "think multiplication" is an extremely efficient strategy for division, which avoids the difficulties that other division strategies involve. It turns out that being able to recall multiplication facts quickly and accurately is an essential prerequisite to being able to divide efficiently.

CULTURAL CONNECTIONS

Researchers studying how people naturally learn basic addition facts in countries all around the world have noted some fascinating worldwide commonalities, as well as some significant differences. It appears there is a natural learning progression—across all cultures—for single-digit addition and subtraction. All learners very naturally build their later (more sophisticated) addition methods from their earlier methods "by chunking, recognizing and eliminating redundancies, using parts instead of entire methods, and using their knowledge of specific numbers" (Fuson, 2003a, 2003b, p. 73).

The first stages of this natural progression involve different forms of counting—which have already been discussed in Chapter 7. People initially add two piles of objects by counting each separately (1, 2, 3 and 1, 2, 3, 4), and then *counting all* (1, 2, 3, 4, 5, 6, 7). Later, they learn to *count on*—by simply stating the first addend (3) and counting up from there (4, 5, 6, 7). Soon, they learn to *count on* even without objects, keeping track of the numbers by using their fingers or auditory patterns.

In time, people begin to shortcut the counting process of addition by breaking up larger numbers into smaller, easy-to-remember number combinations. For example, they quite naturally begin to use doubles, sums involving 1 (learned by counting on), and other combinations of small numbers this way. For example, they think of 3 + 4 as 3 + 3 + 1 = 6 + 1 = 7.

In many parts of the world—especially Asian countries and some European countries—people progress next—often quite naturally—to the general "make-a-10" thinking strategy. This strategy is supported by the languages spoken in many of these countries. As was noted in Cultural Connections in Chapter 7, many languages other than English have teen numbers that transparently contain 10 and some extras; their counting into the teens sounds like "nine, ten, ten-one, ten-two, ten-three," and so forth. By contrast, our two-digit counting prior to 13 (ten, eleven, twelve) reveals absolutely no connection to ten! Two of the next three numbers (thirteen, fifteen) do suggest a connection to ten, but only if you recognize *teen* as meaning "ten" and if you also recognize the variants of the units words (*thir/three*, and *fif/five*). Even for fourteen, and numbers from sixteen up, the connection to ten is not immediately clear, since the word for ten (*teen*) comes after the word for the number of ones (fourteen, sixteen, seventeen, eighteen, nineteen).

The facts considered "basic" in some countries are not the same as those we call "basic facts" in the United States. For example, basic addition facts in Korea focus on combinations that total 10 or less. This is a reflection of their work with a ten-frame and decomposing larger numbers into combinations of 10. Thus, 7 + 8 is not considered a basic fact in Korea, but children there decompose one of the addends to create a basic fact. So, to add 7 and 8, Korean children would think "7 + 3 is 10" or "8 + 2 is 10," and then—in both cases—they would think "and 5 more is 15." This approach may seem different to you, but the high levels of performance that Korean children demonstrate is testimony to its effectiveness. With help from textbooks that stress this natural make-a-10 approach, most children in Asian countries are fluent in adding two single-digit numbers by the end of first grade.

Since U.S. and Canadian children are not naturally prompted by the English language to move to the more efficient make-a-10 addition strategy, one might think that American textbooks would attempt to help them along. However, "compared with other countries, the United States has had a very delayed placement of topics in the elementary curriculum" (Fuson, 2003a, 2003b, p. 74). American children often spend almost all of first grade learning addition and subtraction facts with sums below 10. Even in second grade, textbooks often review these simple facts, and many children struggle with learning the facts that sum to numbers higher than 10. It is not at all unusual to find children still counting on in the upper elementary grades.

Interestingly, research conducted in urban and suburban American classrooms, in English and in Spanish, has shown that it is feasible to teach American first graders to think about basic addition facts in a conceptual way so they can move beyond the limited approach of counting on. The studies showed that the entire range of basic facts were accessible to U.S. first graders, including those who were learning disabled and those with limited English proficiency (Fuson, 2003a, 2003b).

A GLANCE AT WHERE WE'VE BEEN

Problem-solving experiences help students develop meaningful skills in computation with whole numbers. In this chapter, we first considered how to help children develop and model meanings for the four basic operations (addition, subtraction, multiplication, and division). Experiences with counting, concrete situations, problem contexts, and talking and writing about mathematical ideas are important. These experiences should move from concrete modeling with materials to semiconcrete representations with pictures to abstract representations with symbols. Addition means finding out "how many in all?," but a variety of related situations lead to subtraction (separation, comparison, and part–whole). Multiplication situations include equal-groups, comparison, combinations, and areas/arrays. Measurement (repeated-subtraction) and partition (sharing) situations lead to division. Children should learn about mathematical properties at the same time that they are developing their understanding of the basic operations.

In the next part of the chapter we focused on the basic facts for each operation. A variety of thinking strategies for the basic facts for each operation help children move from counting to more mature, efficient ways of developing the facts. Specific suggestions for providing practice and promoting mastery help children master the basic facts for quick recall.

Things to Do: From What You've Read

1. Consider the Snapshot of a Lesson that opens this chapter. What prerequisites must children have before engaging in this lesson? Why did some students use tiles and others use symbols? How could this lesson be extended to explore other mathematical concepts? Discuss some reasons for using puzzles in learning mathematics.

2. Describe the thinking strategies a child might use with each of the following computations:

    $8 + 3 = \square$ $\quad$ $8 \times 5 = \square$ $\quad$ $7 + 8 = \square$
    $18 \div 3 = \square$ $\quad$ $16 - 7 = \square$

3. For each addition fact strategy (i.e., commutativity, adding one, adding zero, doubles, near doubles, counting on, combinations to 10, adding to 10 and beyond), list three different facts that could use that strategy and explain what a child would think when using that strategy.

4. When is counting back an effective strategy for subtraction? Give an example. Give an example of a fact where counting back is not such an effective strategy, and explain another strategy that would be better to use.

5. There are two types of division that arise in real-world problems: measurement (or repeated subtraction) division and fair-sharing (or partitive) division.
    a. Write a partitive (or fair-sharing) division word problem that would correspond to $24 \div 6$. Draw a picture to illustrate the solution to your problem and in simple language explain why your numerical answer makes sense.
    b. Write a measurement (or repeated-subtraction) division word problem that would correspond to $24 \div 6$. Draw a picture to illustrate the solution to your problem and, in simple language, explain why your numerical answer makes sense.

Things to Do: Going Beyond This Book

In the Field

1. Examine a recently published textbook for grades 2, 3, or 4. Analyze how basic facts for the four basic operations are presented in that text. How are the facts grouped? Are thinking strategies presented? If so, how do they compare with the strategies described in this chapter? How would you use this text in teaching basic facts? If you think you would probably want to supplement the text, explain what sorts of supplementary materials and experiences you would use and why.

2. [1]*Literature Context for Facts.* Find a popular trade book (perhaps from the Book Nook for Children) that would be useful as a motivator for a problem-based lesson revolving around one of the four basic operations (addition, subtraction, multiplication, or division). Identify the operation that the story evokes and write a problem that you might challenge students to work on after hearing the story. If possible, read the story to some children, have them try solving your problem, and analyze their written work for evidence of understanding of the basic operation the story involves.

3. [1]*Timed Tests.* Observe children as they take a timed test on basic facts. What behaviors and emotions can you document? How are the results of the test used by the teacher? Talk to at least five children or adults about their memories of timed tests on basic facts. What do they say about their experiences?

4. [1]*Basic Fact Difficulty.* Talk to a fifth- or sixth-grade teacher about what he or she does with students who are not fluent in basic facts. Do they receive special instruction in fact strategies? Do they use calculators routinely for classwork and homework? How many in the class have problems with basic facts? Which facts cause the most difficulties?

With Additional Resources

5. Read the brief research report on "The Empty Number Line in Dutch Second Grade" (Klein, Beishuizen, and Treffers, 2002) or the journal article entitled "The Empty Number Line: A Useful Tool or Just Another Procedure?" (Bobis, 2007). These papers describe the use of a number line with no zero point shown as a computational support for children doing multidigit addition and subtraction. What instructional and psychological reasons do the authors give for using the empty number line as a central model for addition and subtraction?

With Technology

6. Find a computer program on the Web that claims to help children learn basic facts and try it out. There are many such programs on the market. Most focus on encouraging development of quick responses, but

---

[1]Additional activities, suggestions, and questions are available in the field experience manual on the Student Companion site at www.wiley.com/college/reys.

a few encourage use of thinking strategies. Write a half-page review of the computer program (similar to those published in teacher journals). In your review you should identify the name and publisher of the program, the cost, and the type of hardware required to run it. Explain how the program works. Include answers to the following questions:

- Does the program help children learn strategies for basic facts?
- Does the program keep track of student progress?
- Does the program modify the facts presented to the student according to which ones he or she has gotten right or wrong in the past?
- Do you think the program would be effective in promoting quick recall of facts? Why?
- Do you think children would enjoy using this program? Why?
- How would you use such a program if your classroom had only one or two computers?
- Would you recommend this program to other teachers? Why?

Note to Instructors: You can find additional resources, learning activities, and blackline masters in this text's accompanying Instructor's Manual, at www.wiley.com/college/reys.

Book Nook for Children

Dahl, Michael. *Starry Arms: A Book about Counting by Fives.* Minneapolis, MN. Picture Window Books, 2005.

This book helps readers learn how to count by 5's using different ocean animals.

Dahl, Michael. *Toasty Toes: Counting by Ten.* Minneapolis, MN. Picture Window Books, 2006.

Readers see how groups of ten can be grouped together in different ways.

Hulme, J. N. *Counting by Kangaroos: A Multiplication Concept Book.* New York: Scientific American Books for Young Readers, 1995.

Multiplication is illustrated using animals from Australia. Groups of three squirrel gliders, four koalas, five bandicoots … , ten wallabies all crowd a house.

Merriam, E. *12 Ways to Get to 11.* New York: Simon & Schuster, 1993.

The book shows how to get to 11 by using different combinations of numbers. Many different pictures are used to show the different combinations of the number 11.

Murphy, S. J. *Spunky Monkeys on Parade.* New York: HarperCollins, 1999.

At the Monkey Day Parade, the reader sees monkeys marching as majorettes, cyclists, tumblers, and band members. They create a spectacle as they move in twos, threes, and fours. This counting book will let the reader see many different ways of counting. There are activities and games for teachers and parents to use at the back of the book.

Sierra, J. *Counting Crocodiles.* Orlando, FL: Gulliver Books, Harcourt Brace, 1997.

A clever monkey, using her counting ability, outwits the hungry crocodiles that stand between her and a banana tree on another island on a sea.

Wakefield, A. *Those Calculating Crows!* New York: Simon & Schuster Books for Young Readers, 1996.

Using different plans to prevent crows from invading his corn crop, Farmer Roy tries to scare off the crows. But the crows are crafty. Can they count? This is based on a true story by some hunters.

CHAPTER 10

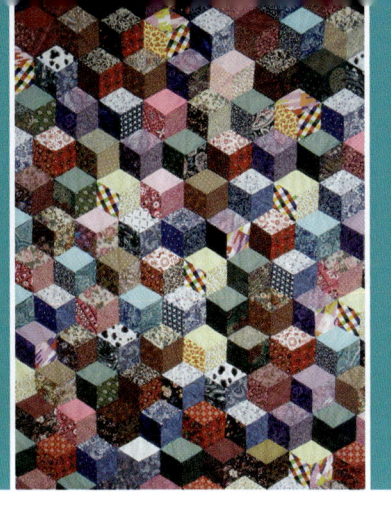

Computation Methods: Calculators, Mental Computation, and Estimation

> "IF YOU CAN DO THE ARITHMETIC IN YOUR HEAD, DO IT."
> —Joseph Ray (popular mathematics textbook author of 1800s)

> SNAPSHOT OF A LESSON

KEY IDEAS

1. Show students how to use different computational methods.
2. Help students understand estimating and checking whether answers are reasonable.

BACKGROUND

This lesson is based on the video *Choose a Method*, from Teaching Math: A Video Library, K–4, from Annenberg Media. Produced by WGBH Educational Foundation. © 1995 WGBH. All rights reserved. To view the video, go to http://www.learner.org/resources/series32.html (video #17).

Mary Holden teaches a combined class of fourth and fifth graders in Jeffersontown, Kentucky. She divides her class into three groups, assigning a different task to each group and rotating the students through the groups over a period of several days. One group, called "puzzles and games," works independently on developing tessellating designs, while a second group works on solving problems on computers. The third group meets with the teacher in an exploration activity involving base-ten blocks. This activity involves modeling operations with currency on base-ten manipulatives, which leads into a comparison of methods of computation.

Mrs. Holden: Once again today we're going to move back to another idea about these base-ten blocks. This is going to be one whole. [*holds up a flat*] So, this is a dollar. [*the flat*] What is this? [*holds up a rod*] Ten cents. And what is this? [*holds up a unit*]

Student: A penny.

Mrs. Holden: A penny, OK. Like, for instance, this would be a dollar. [*holds up one flat*] How much is this altogether? [*holds up one flat, one rod, and one unit*]

Student: A dollar and eleven cents.

Mrs. Holden: A dollar and eleven, OK.

The camera pans to show student-built base-10 structures. Mrs. Holden asks students to consider which student-built structure is "worth" the most. Students estimate the value of their classmates' structures.

Chris: Nataya's because it has two dollar bills and a lot of tens, and dimes.

Chrissy: I think it's Aubrey's because Nataya has a lot of ones in hers and, um, Aubrey's is built mainly of tens and wholes.

Student: I think Andy's would be more than the other ones. There's a lot on the outside. There's mostly tens and wholes.

Mrs. Holden: And how much do you think it's worth?

205

Student: I'd say $3.48.

Mrs. Holden: OK. [writes "Andy—$3.48" on chalkboard and also writes "Nataya—$2.73" and "Aubrey—$3.47"] OK, let's do the count. Go ahead, and as you trade them in, would you say out loud as you trade them in?

[*Students are shown in groups of about three, counting their base-ten blocks.*]

Mrs. Holden: Nataya, give me yours. How much is yours worth?

Nataya: $3.58.

Mrs. Holden: Look at the difference. This was our estimate. [*She refers to $2.73, already on board, and writes "$3.58" next to it.*] Look at the difference—this is our estimate, and this is what it really is. The truth. We were way off with hers, weren't we? OK. Aubrey's.

Aubrey: $4.20

Mrs. Holden: [*writes "$4.20" next to the estimate of $3.47*] Oh, look at the difference, though. If somebody told you that yours was worth $500.50, would that sound reasonable?

Students: No.

Mrs. Holden: No. Were we close enough to be reasonable?

Students: Yes.

Mrs. Holden discusses estimation with the students, suggesting that the students round values that are difficult to work with to values that are easier to work with.

Mrs. Holden: Is 23 cents, what kind of money is that close to?

Student: A quarter.

Mrs. Holden: A quarter. How much money would five quarters be?

Students: A dollar twenty-five.

Mrs. Holden: Is that close to $1.25?

Students: Yes.

After students become comfortable with the base-10 blocks, Mrs. Holden assigns a particular method of computation to each group member.

Mrs. Holden: OK, you're on paper–pencil, you're on calculator, you're on mental math, and you're on base ten. And we'll rotate. We'll go around. I'll give you some problems. Slushies are on sale today for 29 cents and you are going to buy everybody that sits at your table one. How many people are at your table?

Student: Four.

Mrs. Holden: Four.

Students work independently, using the assigned method of computation. After a few minutes, Mrs. Holden discusses their progress.

Mrs. Holden: Now, slushies cost 29 cents and you're buying four of them and it's going to cost 33 cents? What? [points to another student]

Student 1: A dollar sixteen.

Student 2: I'm totally off. I know that.

Student 3: What did you get?

Student 2: It's totally off. $2.48.

Student 3: I got 47 cents.

Mrs. Holden: Does that sound reasonable?

Student 3: Not from what everybody else has been saying.

Mrs. Holden: Twenty-nine cents and you're buying four of them?

Student 3: I kind of did like what the problem was on the board. I did it in my mind, and it didn't work.

Mrs. Holden: Let's think that through and I'll give you a little trick with the 29 ones. Twenty-nine is hard to work with, isn't it? If you think in your mind, four times 29, it's hard to do, isn't it? What's 29 close to?

Students: Thirty.

Mrs. Holden: Thirty. Then you do what?

Student: You take away a penny from each one.

Mrs. Holden: What do you come up with?

Student: A dollar sixteen.

Student: Say you want to equal up 29 times 4, you can use the tens first and then when you're done, you don't count up the tens, you count up the ones.

Mrs. Holden: OK, so like this is a ten, can I use my hands with this? [*Mrs. Holden counts by 10 on her fingers.*] 10, 20, is that one of them? 30, 40, is that another one of them? 50, 60, is that another one of them? 70, 80, OK is that all of the tens?

Mrs. Holden and the student discuss a strategy for combining the tens the teacher counted on her fingers and the ones from the original problem. The student adds that 4 times 9 is 36, arriving at a solution of $1.16. Mrs. Holden takes advantage of the opportunity to discuss the unique strategy and thanks the student for sharing his way of thinking with the group.

Mrs. Holden takes time at the end of the class to survey her students' opinions on the efficiency of the methods used in the group.

Mrs. Holden: Which method do you prefer when you're doing these kinds of problems? What would you rather do? I want to go around the circle and I want everybody to tell me what they think is the best method; paper–pencil, mental math, base tens, or calculator.

Student: I think it would be pencil and paper.

Student: Calculator.

Student: I think calculator.

Student: I think it would depend. If it's a harder question, it would have to be calculator. But if it were an easier one, I think it would be a lot easier with mental math.

Student: I think it would be mental math because you can round everything off to something.

Student: I think it would be mental math because if you go to a store when you buy a few things of the same, you're going to expect a reasonable answer. Maybe it will be a little bit more when they ring it up.

Mrs. Holden: OK, did we say that there's any method that's the best method?

Students: No, not really.

Mrs. Holden: No, it depends. Could you all agree on that? That it depends on what you're doing?

Mrs. Holden: [to camera] I think the best kind of math communication, the best kind of math writing, comes from feeling safe to express that in front of other people and to express wrong turns. And how when somebody makes a wrong turn in their thinking we can correct them again, but that it's safe to say that out loud in front of a group of peers and the teacher so we can get that corrected.

FOCUS QUESTIONS

1. Why is asking "Is the answer reasonable?" a good question to ask?
2. What are some myths and facts about using calculators?
3. Why are multiple strategies for mental computation important?
4. What are some different strategies for computational estimation?

INTRODUCTION

Computation evolved from counting to doing calculations in your head. As computation becomes more tedious, written algorithms were developed. More recently, the calculator has come into wide use. Wise use of calculators and written algorithms are important instructional goals. For example, we deplore the use of a written algorithm or calculator to compute 45×1000 because there are easy and efficient ways to produce this result. The admonition by Joseph Ray over a century ago, "If you can do the arithmetic in your head, do it," continues to be good advice.

Competence with different methods of doing computations is not merely useful; it is essential. This is reflected in the *Common Core State Standards for Mathematics* (CCSSM) that encourages mental computation strategies, written computation algorithms, and computational estimation techniques. For example, in Grade 2 it is recommended that students "Mentally add 10 or 100 to a given number 100-900, and mentally subtract 10 or 100 from a given number 100-900." Furthermore, the CCSSM state that students should "develop fluency in computations and make reasonable estimates of their results." This includes operations with whole numbers, fractions, and decimals. These are clear goals that require focused instruction over multiple years.

Historically, elementary school mathematics has emphasized written computation far more than other methods. This, together with the fact that learners are more likely to use methods with which they are familiar, means that students often tend to use written computation even when they could use a more efficient method. As a teacher, you will need to balance your instruction so that children learn about a variety of computational methods and learn when to use each.

All computation begins with a problem and with the recognition that computation is needed to solve the problem. Figure 10-1 shows there are two essential decisions in every computation. First, deciding on the type of result needed and, second, deciding on the best method for getting that result. The first decision involves the question of whether an estimate is appropriate or an exact answer is needed. The second decision involves such questions as whether a calculator would be helpful, or whether the numbers in the problem make mental computation possible, or whether paper-and-pencil calculation is most appropriate. Thus, your goals when teaching computation will be to help students do the following:

- Develop competence with each of the computational methods
- Choose a method that is appropriate for the computation at hand
- Apply the chosen method correctly
- Use estimation to determine the reasonableness of the result

In addition, by helping students choose and use computational methods appropriately and effectively, you will also help them develop their number sense.

In this chapter, we begin by discussing some of the issues in deciding how to find the right balance of instructional time devoted to different computational methods. We then consider the use of calculators in computation: in the process we dispel certain myths about the disadvantages of calculator use. Next we consider how to teach children mental computation, and we discuss the benefits that students derive from developing and doing computation in their head.

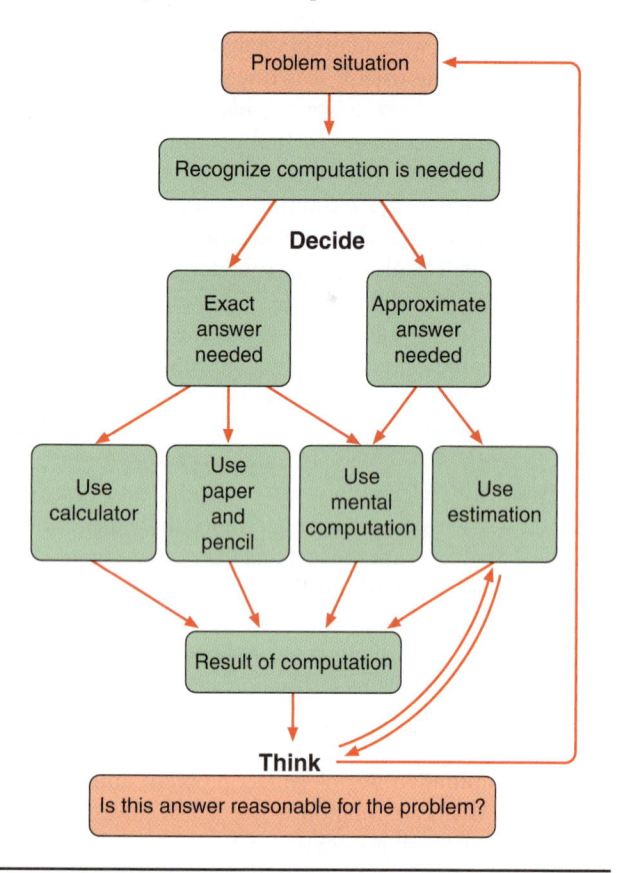

Figure 10-1 *Computational decisions and methods.*

of computational methods. For example, a report by the National Research Council suggests the following:

> Children should use calculators throughout their school work, just as adults use calculators throughout their lives. More important, children must learn when to use them and when not to do so. They must learn from experience with calculators when to estimate and when to seek an exact answer; how to estimate answers to verify the plausibility of calculator results; and how to solve modest problems mentally when neither pencil nor calculator is convenient.
>
> (*National Research Council*, (1989), *p.* 47)

Most people recognize the importance of each of the computational methods and suggest that a better balance of instructional time for the computational alternatives is needed than has historically occurred in U.S. elementary

Then we discuss ways of teaching children good strategies for making estimations and how to choose among these strategies. We finish by considering various cultural differences in the choice and use of different computational methods.

The next chapter is devoted exclusively to developing written computation.

BALANCING YOUR INSTRUCTION

As a teacher, you need to think about which computational methods are important enough to learn in school and how much time should be devoted to each. More than 80% of all mathematical computations in daily life involve mental computation and estimation, rather than written computation. Ironically, research also shows that, in the United States, 70–90% of the instructional time in elementary school mathematics directed toward computation has focused on written computation (Reys and Reys, 1998). The CCSSM continues this pattern. An analysis of the CCSSM documents that the lion's share of the "Number and Operation" standards (about 80%) focus on written computation, rather than mental computation or estimation.

Many proposals have been made regarding computation, ranging from prohibiting calculator use to eliminating the teaching of written algorithms. Educators have also called for helping children learn to make appropriate choices

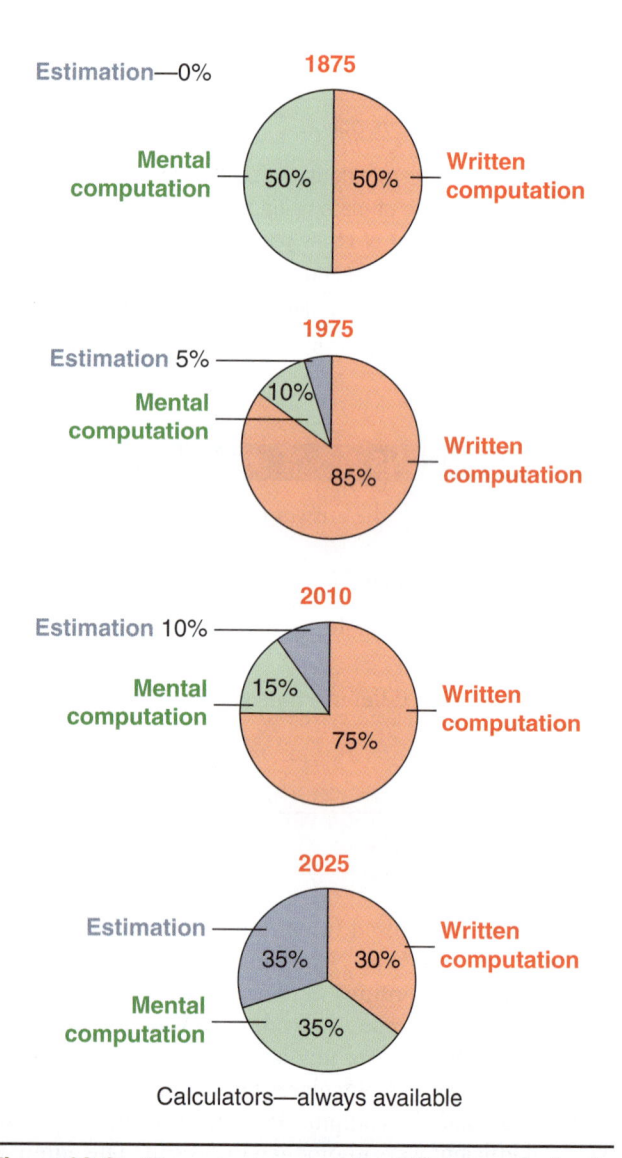

Figure 10-2 *Time spent on teaching different methods of computation in elementary school in the past, present, and, perhaps, the future.*

schools. What percentage of time should you allot for teaching each method? This remains an open question, and your answer clearly depends in part on the developmental levels of the students. As children develop more power and skill in mental computation and number sense, they are ready to apply that knowledge in estimating. Figure 10-2 shows the approximate percentages of instructional time spent on different computational methods in the past (1875 and 1975) and in the present (as of 2010), along with a proposal for a better balance in the future (2025).

Given the availability of technology (calculators and spreadsheets), no one spends much time doing tedious computations. Thus, a decline in instructional attention to written computation over the years is indisputable. A major benefit is that time previously devoted to developing proficiency with tedious written algorithms has become available for other uses. Much of this time can be dedicated to helping students develop their number sense and their ability to select and use other computational methods, including calculators, mental computation, and estimation. Reflecting on your experience and thinking about the future needs of your students will challenge you to develop ways of finding the right balance when you teach computation.

CALCULATORS

The NCTM position statement on calculators recommends that "appropriate calculators should be available to all students at all times." In a similar message the CCSSM says that "mathematically proficient students should consider the available tools when solving a mathematical problem" and calculators are one of those tools.

The CCSSM goes on to say that the tools used and the use made of them grows as students progress through elementary school. These statements suggest that as students' mathematical knowledge grows and changes, so does their need for calculators. Children outgrow calculators just as they outgrow shoes. The calculator needs of students in primary grades are different from those of students in intermediate grades and middle school. Electronics companies have responded to these changing needs with fraction calculators, scientific calculators, and graphing calculators.

As a teacher, you need to help students understand how to use calculators appropriately, which includes showing students that not all problems can be solved with a calculator. Also, as with any new tool, you need to give students time to explore with the calculator on their own, and you need to show them how to handle and care for calculators correctly.

In addition, you should communicate with parents so they know you are using calculators in the classroom and so they understand how and why you are using them. Often, a simple note like the one shown in Figure 10-3 is effective.

> **Math Links 10.1**
>
> The National Council of Teachers of Mathematics position paper on Computation, Calculators, and Common Sense is available at www.nctm.org or by going to this book's Web site.
>
> www.wiley.com/college/reys

Calculators have been used in schools for more than 30 years, but several myths, or incorrect beliefs, about using them persist (which is one reason why communicating with parents is important). In this section, we address these myths as follows:

- *Myth:* Using calculators does not require thinking.
- *Fact:* Calculators do not think for themselves. Students must still do the thinking.
- *Myth:* Using calculators lowers mathematical achievement.
- *Fact:* Calculators can raise students' achievement.
- *Myth:* Using calculators always makes computations faster.
- *Fact:* It is sometimes faster to compute mentally.
- *Myth:* Calculators are useful only for computation.
- *Fact:* Calculators are also useful as instructional tools.

Dear Parents:

We are using calculators in our second-grade class this year in many different ways. Sometimes we use them to develop skills in counting and recognizing patterns; other times to explore new topics, such as decimals; and often when solving problems that require tedious computations. We believe that a variety of early experiences with calculators will help young children make wise use of calculators both in and out of school.

If you are curious about how we are using calculators, please ask your child to tell you about some things he or she has been doing with this tool. If you have further questions, please let me know.

Figure 10-3 *A sample note to parents about using calculators in primary grades.*

USING CALCULATORS REQUIRES THINKING

Teachers must make it clear to students (and to parents) that calculators don't think—they only follow instructions. For example, suppose students were using a calculator to solve the following problem:

One bus holds 32 children. How many buses are needed to hold 1000 children?

Would the calculator decide which keys to press and in what order to press them? Would the calculator interpret the result? Clearly, the answer to both questions is no!

This problem is difficult for elementary students to solve with or without a calculator. To use the calculator effectively, students still must realize that they have to divide 32 into 1000, just as they would have to without a calculator. Then they must decide which calculator buttons to press in which order. And they must still interpret the result (31.25) properly. That is, does the result mean that 31 buses are needed or 32?

As you can see from this problem, calculators don't discourage thinking! In fact, properly used, calculators encourage thinking because they free students from tedious computations (picture using long division here) and give them more time to go through the important problem-solving processes that generally precede, and often follow, the computation.

USING CALCULATORS CAN RAISE STUDENT ACHIEVEMENT

Research has consistently indicated that the use of "calculators in concert with traditional instruction … can improve the average student's basic skills with paper and pencil, both in basic operations and in problem solving" (Hembree and Dessart, 1992). Moreover, "students using calculators possess a better attitude toward mathematics and an especially better self-concept in mathematics." Many parents and some teachers worry that students will become so dependent on calculators that they won't learn to compute. On the contrary, research shows the use of calculators improves students' attitudes toward mathematics and encourages students to persist longer when faced with problem solving situations (Ellington, 2006; Chval & Hicks, 2009). As a teacher, however, you have to respond to these concerns—not only by showing students how to make effective use of calculators but also by helping parents overcome their fears.

CALCULATORS ARE NOT ALWAYS THE FASTEST WAY OF DOING COMPUTATIONS

As we saw in relation to the bus problem above, calculators can make tedious computations fast and easy. But when it comes to simple calculations, the story is quite different. True, students can use a calculator to compute 2345×67 or $34{,}567 \div 89$ very quickly; however, calculations such as 4×40, $99 + 99 + 99 + 99$, $\frac{1}{2} + \frac{1}{4}$, and $23000 \div 100$ are much quicker to perform mentally than with a calculator, simply because of the time it takes to press the required keys. You can use an activity like the one shown in In the Classroom 10–1 to help children recognize which calculations are more quickly done mentally and which are faster with a calculator.

CALCULATORS ARE USEFUL FOR MORE THAN JUST DOING COMPUTATIONS

The most obvious use of calculators is to calculate, but the process of calculating can reveal many powerful mathematical ideas. For example, calculator counting forward and backward (see In the Classroom 7–4, 7–5, and 8–3) reveals patterns, gives students practice with important skills, and reinforces important knowledge such as basic multiplication and division facts. Older children can use calculators to see patterns such as powers of 10 (10, 100, 1000) and decimals (0.1. 0.01, 0.001). Resources (see Judd, 2007; Masalski and Elliott, 2005; Moss and Grover, 2007) are rich in ways to help students use calculators effectively.

As a teacher, you should consider using calculators for computation whenever computational skills are not the main focus of instruction. And you should keep in mind that calculators can even be helpful in learning those skills. Clearly, you should not let the use of calculators interfere with the establishment of basic facts, which remain an important goal of mathematics instruction in the primary grades. Basic facts are natural stepping stones to mental computation and estimation as well as to the development of number sense over the long term.

When is calculator use appropriate? Having students turn the calculator upside down to spell words or use the calculator to check written calculations may help motivate students to use calculators, but the educational value of such activities is questionable. They may communicate the message that a calculator is only for playing games or, even worse, that using a calculator is cheating, that students must do calculations by hand and just use a calculator to check their answers. Research shows that the more teachers use calculators in their classroom, the more they develop creative and productive ways to use them (Sutton and Krueger, 2002). There are two main uses for the calculator in the classroom: as a computational tool and as an instructional tool. A calculator should be used as a *computational tool* when it

- Facilitates problem solving
- Eases the burden of doing tedious computations
- Focuses attention on meaning
- Removes anxiety about doing computations incorrectly
- Provides motivation and confidence

A calculator should be used as an *instructional tool* when it

- Facilitates a search for patterns
- Supports concept development
- Promotes number sense
- Encourages creativity and exploration

Calculators also provide a safety net for students having particular difficulties in learning computational procedures. For example, if learning-disabled students find it difficult or impossible to use traditional computational procedures, the calculator can provide equity and allow them to engage in problem solving along with other students (Thompson & Sproule, 2005).

MENTAL COMPUTATION

Mental computation is computation done "all in the head"—that is, without tools such as a calculator or paper and pencil. This is certainly a natural way to do many computations, as shown by the Snapshot of a Lesson at the beginning of this chapter as well as by the fact that many young children develop ways of computing before they are able to write. In fact, research has documented a wide variety of mental computation techniques that children have created on their own and that make sense to them (Fuson, 2003a; Reys & Barger, 1994).

STRATEGIES USING COMPATIBLE NUMBERS AND DECOMPOSITION

Mental computation builds on the thinking strategies used to develop basic facts (see Chapter 9) and naturally extends children's mastery of basic facts. Extending basic facts is an early step in developing more powerful strategies and skills in mental computation. Extending basic facts takes different forms. Students might:

- Extend $4 + 5 = 9$ to $40 + 50 = 90$ and $400 + 500 = 900$.
- Use place value in doing a subtraction such as $675 - 200 = 475$ or $675 - 50 = 625$.
- Combine basic facts with decomposition and place value and recognize that $47 + 16 = 47 + 10 + 6$ and $47 + 10 = 57$ and $57 + 6 = 63$.
- Decompose (break up) numbers to make them easier to handle. For example, to calculate $18 + 17$, the student could decompose 17 into $2 + 15$; thus, $18 + 17 = 18 + 2 + 15$, and $18 + 2 = 20$, and $20 + 15 = 35$. Alternatively, the student could decompose 18 into $10 + 8$ and

---

In the Classroom 10–1

CALCULATORS VERSUS MENTAL COMPUTATION—WHEN TO USE WHICH

Objective: Help students recognize which types of computations are best done with calculators and which types are best done mentally.

Grade Level: 2–4

- Divide the class into pairs of students. In each pair, one student has a calculator and must use it to do computations. The other student doesn't have a calculator and must do computations mentally.
- Write computation problems on the board, one at a time.
- In each pair, the student who first has an answer raises his or her hand. The other student finishes the computation but doesn't raise his or her hand.
- After everyone has finished the computation, ask the students with their hand raised for their answer and whether they used a calculator. If their answer is correct, mark it in the appropriate column of the chart below.

Computation	First and Correct Using a Calculator	First and Correct Using Mental Computation
1. 4×100	☐	☐
2. $400 + 50 + 8$	☐	☐
3. $99 + 99 + 99 + 99$	☐	☐
4. $1000 - 200$	☐	☐
5. $50 + 50 + 50 + 50 + 50 + 50$	☐	☐
6. $½ + ¾$	☐	☐
7. $6300 \div 7$	☐	☐
8. $10 \times 10 \times 10 \times 10$	☐	☐
9. $0.75 + 0.25 + 0.50$	☐	☐
10. $\$1.00 - \$0.25$	☐	☐

- Write two lists on the board—one list showing the computations that more students did correctly and faster using a calculator, the other list showing the computations that more students did correctly and faster using mental computation.
- Discuss the differences in the two lists.

decompose 17 into 10 + 7; thus, 18 + 17 = 10 + (8 + 7) + 10, and 8 + 7 = 15, and 10 + 15 = 25, and 25 + 10 = 35. Both solutions use basic facts and place value, while using different ways of decomposing numbers to make them easier to work with mentally.

As you can see from the last two examples above, mental computation is often done by using compatible, or "friendly," numbers. These are numbers that can be combined to make numbers that are easy to compute with, such as 10 (e.g., 8 and 2 are compatible numbers because they add up to 10). Having children work with the ten-frame will help them see how to link different combinations to make 10:

8 + __ = 10   __ + 5 = 10   7 + __ = 10

Using compatible numbers facilitates a mental computation such as 8 + 7 + 22 + 5 + 13. For example, a student could recognize that 8 and 22 are compatible numbers (8 + 22 = 30) and that 7 and 13 are compatible, too (7 + 13 = 20).

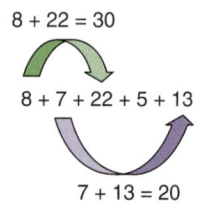

8 + 22 = 30
8 + 7 + 22 + 5 + 13
7 + 13 = 20

Then the student could finish the computation by thinking: 30 + 20 = 50, and 50 + 5 = 55.

Another student could do this same computation differently, by thinking: 8 + 7 = 15, and 15 + 5 = 20, and 22 + 20 = 42, and 13 = 10 + 3, and 42 + 10 = 52, and 52 + 3 = 55. There are many other ways of doing this same mental computation, but they all would reflect good number sense and show the importance of fluency with basic facts and with adding 10 or multiples of 10. Understanding the different strategies children use and helping the children

Figure 10-4 *Find two or more adjacent numbers in a row or column with a sum of 10. (Try to find at least eight different combinations.)*

Figure 10-5 *Find two or more adjacent numbers in a row or column with a sum of 50. (Try to find at least eight different combinations.)*

understand each other's strategies make the development of mental computation exciting.

The first step in developing proficiency with using compatible numbers is learning to recognize them. Figures 10-4 and 10-5 show one way you can give children practice in this. Such practice also demonstrates that compatible numbers vary and that they can include combinations that add up to 10, 50, or whatever other numbers are easy to work with mentally.

There are many useful strategies for mental computation. Just as teachers can help children develop a repertoire of problem-solving strategies, they can also help children develop a repertoire of mental computation strategies. Figure 10-6 shows several common strategies for whole-number addition. Chances are you have used some of these strategies yourself.

Problem	Strategy	How I Did It
43 + 48	Adding from the left	40 plus 40 is 80, 3 plus 8 is 11, 80 plus 11 is 91.
43 + 48	Counting on	I'll count by tens. 48 . . . 58 . . . 68 . . . 78 . . . 88. Then I'll count by ones. 89 . . . 90 . . . 91.
43 + 48	Making tens	48 plus 2 is 50, 50 plus 40 is 90, 90 plus 1 more is 91.
43 + 48	Doubling	48 plus 48 is 96. Since 43 is 5 less than 48, 96 minus 5 is 91.
43 + 48	Making compatibles	43 and 7 are compatible because they make 50, 50 plus 40 is 90, 90 plus 1 more is 91.
43 + 48	Bridging	I'll break up a number and add the parts. 43 plus 8 is 51, add 40 more is 91.

Figure 10-6 *Some common strategies for whole number addition using several common strategies.*

Mental computation encourages flexible thinking, promotes number sense, and encourages creative and efficient work with numbers. Consider some of the strategies that children could use to do different types of computations mentally:

- Children could solve a problem in whole-number addition such as 165 + 99 in different ways. For example,

  "I subtracted 1 from 165 and added it to 99. Then I added 164 plus 100 to get 264."

  "I added 165 plus 100 and got 265, then I subtracted 1 and got 264."

- For multiplication problems such as 4 × 600 and 4 × 6000, children could extend the basic multiplication fact 4 × 6 = 24 and combine that with their understanding of place value. For example,

  "4 × 600 is 4 × 6 hundreds, which is 24 hundreds, or 2400."

  "4 × 6000 is 4 × 6 thousands, which is 24 thousands, or 24000."

- And for a division problem such as 1200/4, children could combine place value with extending the basic division fact 12/4 = 3. For example,

  "1200 is 12 hundreds, so 1200/4 is 12 hundreds/4, or 3 hundreds, or 300."

The strategy of combining basic facts and place-value concepts will suggest different patterns and relationships to students. And by combining mental computation with calculators, students can explore a wide enough range of patterns and relationships to stimulate them to make conjectures and discoveries. For example, students could explore patterns and relationships among multiplications such as 4 × 6, 4 × 60, 4 × 600, and 4 × 6000, and can tell how the results are alike and how they are different. A slightly different exploration could be used to challenge students to tell why the results are the same: 4 × 600, 40 × 60, 400 × 6, and 4000 × 0.6. Doing mental computation rewards thoughtful analysis, and it encourages flexible thinking because students see that they can often get the correct results by using very different techniques, as illustrated in Figure 10-7. Each strategy is meaningful to the student using it, even if it seems strange to others. By encouraging children to talk with each other about the different ways they do problems in their heads, you help them learn to think more freely and flexibly. This sharing and explaining is an important part of instructional activities that promote the development of mental computation skills, and it often helps children learn new strategies.

ENCOURAGING MENTAL COMPUTATION

You should make sure that you give students a range of activities that encourage and reward the wide variety of skills they develop when doing mental computation. Rather than have everyone use the same strategy to solve a problem, encourage students to use the strategies that make sense to them individually. Here are some things you should encourage children to do:

- Always try mental computation before using paper and pencil or a calculator.

- Use numbers that are easy to work with:

The problem	*One way to think about it*
397 × 4 = ?	400 × 4 = 1600
	3 × 4 = 12
	1600 − 12 = 1588
$6.98 + $7.98 + $9.98 = ?	7 + 8 + 10 = 25
	3 × 2 cents = 6 cents
	$25 − 6 cents = $24.94

- Look for an easy way:

The problem	*One way to think about it*
2 × 3 × 7 × 5 = ?	2 × 3 = 6
	6 × 5 = 30
	30 × 7 = 210
	OR
	2 × 5 = 10
	3 × 7 = 21
	10 × 21 = 210
6 × 8 × 19 × 0 = ?	0 is a factor, which means the product is zero

- Use logical reasoning:

The problem	*One way to think about it*
15 × 120 = ?	That's halfway between 10 × 120 and 20 × 120
	Halfway between 1200 and 2400 is 1800
	OR

Thought bubbles in Figure 10-7:

"8 × 25 is double 4 × 25... or double 100..., that's 200. 20 × 200 is 4 with three zeros.. that's 4000."

"8 × 20 = 160, 160 is 100 plus 60, 100 × 25 is easy, that's 2500, so 50 × 25 is 1250, 10 × 25 = 250, so it's 2500 + 1250 + 250, that's 4000."

"20 × 25 = 2 × 25 × 10..., that's 500: 8 × 500 = 4000"

Figure 10-7 *The volume of the box is 25 × 8 × 20. Each student finds the answer in a different way.*

That's 10 × 120 plus half of
10 × 120
1200 + 600 = 1800

- Use knowledge about the number system:

The problem	One way to think about it
56 − 24 = ?	50 − 20 = 30
	6 − 4 = 2
	30 + 2 = 32
	OR
	54 − 24 = 30
	so 56 − 24 = 32

Why emphasize mental computation? There are several reasons for helping students develop mental computation skills:

- *Mental computation is very useful.* Adults do more than three-fourths of all their calculations mentally. You should encourage and reward the early development of this practical skill that students will use for the rest of their lives.

- *Mental computation is the most direct and efficient way of doing many calculations.* For example, research has shown that the computation 200 − 5 is better done mentally than with paper and pencil. Students are more likely to make sensible judgments about the result and less likely to come up with 205 (a common error when writing). In the middle grades, a computation such as $\frac{3}{4} - \frac{1}{2}$ is easier for children to do mentally than by writing (McIntosh, Reys, and Reys, 1995).

- *Mental computation is an excellent way to help children develop critical-thinking skills and number sense and to reward creative problem solving.* Students must figure out a strategy and use it. In this process, they become aware that there is more than one way to perform most calculations mentally, and they are encouraged to seek simple and economical methods that make sense to them.

- *Proficiency in mental computation contributes to increased skill in estimation.* Mental computation is the foundation for estimation because it helps children see alternative algorithms and nonstandard techniques for finding answers.

MENTAL COMPUTATION VERSUS WRITTEN ALGORITHMS As we have seen, there are many efficient techniques for mental computation. Moreover, in the process of becoming proficient at mental computation, children gain mathematical insights and develop skills that will serve them well all their life. Unfortunately, however, when children above grade 2 tried mental computation, their dominant strategy was to apply written algorithms mentally (McIntosh et al., 1997). For example, when they were asked to compute 165 + 99 mentally, a typical response was: "I added 5 plus 9 and got 14. I carried the 1, and 5 plus 9 plus 1 is 15. I carried that 1 and got 2. It's two six four, or 264."

Many students think of mental computation this way, as applying written algorithms in their heads. In fact, extensive practice with written computation increases the likelihood that students will try applying written algorithms mentally. This is an international dilemma. In Canada and the United States, for instance, research showed that the range of strategies used for mental computation was far greater than in Japan. However, in all three countries, when students were asked to do a computation mentally, the dominant strategy was to apply a written algorithm mentally. The mental application of written algorithms most likely reflects the emphasis given to written algorithms in school; unfortunately, this strategy appears to inhibit the development of flexible and more efficient strategies for mental computation (Reys and Yang, 1998; Shigematsu et al., 1994). As a teacher, you should increase your emphasis on flexible, student-generated strategies that enable students to compute mentally without relying on written algorithms.

GUIDELINES FOR DEVELOPING MENTAL COMPUTATION SKILLS Here are some guidelines for helping students develop mental computation skills:

- *Encourage students to do computations mentally.* Make it clear that when mental computation is possible, it is not only acceptable but preferable to written computation. Students report that mental computation is not encouraged and is often even discouraged in school. When teachers say things like "show your work," students may interpret this to mean that mental computation is unacceptable. Thus, instead of quickly and easily computing 1000 × 945 mentally, students might use paper and pencil, thinking they are required to apply the unnecessary and inefficient written algorithm:

$$\begin{array}{r} 945 \\ \times 1000 \\ \hline 000 \\ 000 \\ 000 \\ 945 \\ \hline 945000 \end{array}$$

Metaphorically, this is using a sledge-hammer to kill a fly, and is a classic example of using a written algorithm inappropriately.

- *Learn which computations students prefer to do mentally.* Research suggests that, when given the choice, students from grade 4 and up prefer to use written computations rather than calculators or mental computation. For example, most fifth graders chose to do 1000 × 945 with either a calculator or paper and pencil, even though this computation is most easily done mentally (McIntosh & Sparrow, 2004). This choice typically reflects the student's lack of experience and confidence in doing computations mentally. In the Classroom 10–2 illustrates one way to explore the computational preferences of students with whole numbers, and In the Classroom 10–3 does the same

Mental Computation   215

In the Classroom 10–2

HOW WOULD YOU DO IT?

Objective: Encourage wise use of computational alternatives.

Grade Level: 3-4.

	In Your Head	With a Calculator	With Paper/Pencil
60 × 60	☐	☐	☐
945 × 1000	☐	☐	☐
450 × 45	☐	☐	☐
24 × 5 × 2	☐	☐	☐
	☐	☐	☐
	☐	☐	☐
4 × 15	☐	☐	☐
50 × 17 × 2	☐	☐	☐

Follow Up

- Write a computation YOU would solve with a calculator.
- Write a computation YOU would solve mentally.
- Write a computation YOU would solve with paper and pencil.

In the Classroom 10–3

HOW WOULD YOU DO IT?

Objective: Encourage wise use of computational alternatives.

Grade Level: 4-5.

	In Your Head	With a Calculator	With Paper/Pencil
$\frac{1}{2} + \frac{1}{4}$	☐	☐	☐
$1 - \frac{1}{3}$	☐	☐	☐
$\frac{3}{4} + \frac{3}{4}$	☐	☐	☐
$\frac{1}{5} + \frac{1}{6}$	☐	☐	☐
$\frac{1}{2} + \frac{5}{6}$	☐	☐	☐
$1\frac{1}{2} + 2\frac{3}{4}$	☐	☐	☐
$2 - \frac{3}{4}$	☐	☐	☐
$\frac{1}{2} - \frac{1}{3}$	☐	☐	☐

Follow Up

- Write a computation YOU would solve with a calculator.
- Write a computation YOU would solve mentally.
- Write a computation YOU would solve with paper and pencil.

---

with fractions. If you see that students prefer using a calculator or paper and pencil to do computations that are better done mentally, you can focus on helping them develop mental computation skills that would lead them to change their preferences.

- *Find out if students are applying written algorithms mentally.* Ask students to tell how they did a computation in their heads so that their strategies and thinking become clear to you.

- *Plan to include mental computation systematically and regularly as an integral part of your instruction.* Systematic attention and practice will improve your students' mental computation skills. You should focus on helping students develop strategies and thinking patterns that make sense to them. Students may develop these strategies and patterns on their own or may adopt ones they have learned from others.

- *Keep practice sessions short, perhaps 10 minutes at a time.* Many teachers use an activity such as Follow Me while children are waiting in line ("3 × 4 + 10 − 4 + 20 … " where the teacher calls out "three times four" and the first student in line says "twelve" and the teacher says "plus ten" and the next student says "twenty-two," etc.). Another useful activity is Today's Target, where the date serves as a target that students have to "hit" using specific types of computations (see Figure 10-8). You can vary this activity by changing the target number to something besides a date—for example, you might say, "Today's target is 100," and then ask students to think of different ways to make 100. Or you could vary the questions, for example, by asking them to "use addition and subtraction" or to "do it an unusual way." This type of activity gives students practice with mental computation, and the solutions give you insight into students' thinking—both into the strategies they use and into the level of computational complexity at which they are operating.

- *Develop children's confidence.* Pick numbers that are easy to work with at first (e.g., 3 × 99 or 1/2 of 84), and then increase the difficulty (e.g., 5 × 75 or 2/3 of 96). Working with compatible numbers helps children develop confidence at mental computation and improves their number sense.

- *Encourage inventiveness.* There is never just one right way to do any mental computation, but certain ways may be more efficient and interesting than others. Asking students "How did you do that?" can reveal highly ingenious mental computation strategies. For example, students who computed 60 × 15 mentally reported the following strategies:

216 Chapter 10 • Computation Methods: Calculators, Mental Computation, and Estimation

Today's target is May 24	
Try to hit this target by	
Adding three numbers	**Subtracting one number from another number**
Some possible answers	*Some possible answers*
8 + 8 + 8	30 − 6
20 + 2 + 2	25 − 1
10 + 10 + 4	100 − 76

Multiplying two numbers	**Using a fraction**
Some possible answers	*Some possible answers*
4 × 6	$23\frac{9}{10} + \frac{1}{10}$
8 × 3	$48 \times \frac{1}{2}$
24 × 1	$24\frac{1}{2} - \frac{1}{2}$

Figure 10-8 *Today's target and some ways to hit it.*

"10 times 60 is 600. 5 times 60 is 300. 600 plus 300 is 900."
"60 times 10 is 600, and half of 600 is 300, so it is 600 plus 300, or 900."
"60 is 4 times 15, so that is 4 times 15 times 15. 15 squared is 225, times 4 is 900." (This one was reported by an eighth grader.)

As we have noted, children often develop their own strategies for mental computation; the power of such self-developed, out-of-school techniques has been documented (McIntosh and Sparrow, 2004).

- *Mental computation or estimation?* Make sure children are aware of the difference between estimation (in which answers are approximations) and mental computation (in which answers are exact).

ESTIMATION

Computational estimation is a process of producing answers that are close enough to allow for good decisions without performing elaborate or exact computations. The video *Choosing a Method*, excerpted in this chapter's Snapshot of a Lesson, nicely illustrates that there are many different real-world situations for using estimation. Computational estimation is typically done mentally. Figure 10-9 illustrates how students can use estimation to monitor their computations at three different points in the process:

- *Before starting exact computation*, students can use estimation to get a general sense of what to expect.
- *While doing computation*, students can use estimation as a check to determine if the computation is moving in the right direction.

(a) Before computing

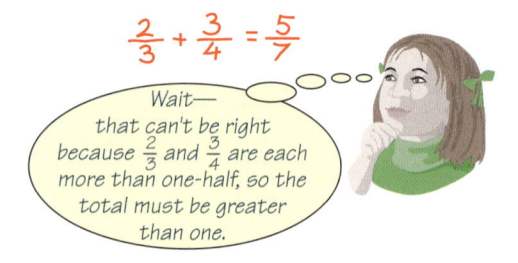

(b) During computing

(c) After computing

Figure 10-9 *Different times to estimate.*

- *After completing a computation*, students can use estimation to reflect on their answer and decide if it makes sense.

As children become more aware of these different uses of estimation, they develop a greater respect for its power and view it as an essential part of the total computational process.

BACKGROUND FOR ESTIMATING

Students who are proficient at written computations are not necessarily good estimators. Research suggests that memorizing rules and procedures discourages many students from using estimation (Alajmi and Reys, 2007; Reys and Yang, 1998). Good mental computation skills and number sense provide the foundation for the successful development of estimation skills. You can significantly improve students' estimation skills by paying systematic attention to estimation in your instruction. Although some dramatic improvements can occur quickly, the development of good estimation skills is a process that takes years. Thus, students must be given repeated opportunities to estimate.

> **Virtual Classroom Observation**
>
> Go to www.wiley.com/college/reys,
> Access the Wiley Resource Kit
> Click on the Virtual Classroom Observations Section
> Data Analysis and Probability: Measures to Center
> Teaching Examples:
> Video: Guessing versus estimating
>
> What are some strategies used as students moved from guessing to estimating?

Learning about estimation gives students their first encounter with an area of mathematics that does not focus on exact answers. As young children talk about estimation in mathematics, their mathematical vocabulary will expand to include words such as *about, almost, just over,* and *nearly*. As children grow older, they will add words such as *approximate, reasonable,* and *unreasonable* to their vocabulary for describing mathematical answers, as well as phrases such as *in the ballpark* and *close enough*. Becoming comfortable and confident in using language to describe the inexactness found in the real world contributes not only to developing number sense but also to developing better estimation skills.

You should begin by making students aware of what estimation is about so that they acquire a tolerance for error. Estimation involves a different mindset from the mindset that says only an exact answer will do. You have to help children change their exact-answer mindset before beginning to teach specific estimation strategies. This change begins when students recognize that estimation is an essential and practical skill. Estimation should be developed and emphasized along with procedures for exact computation.

Also important is giving students immediate feedback on their estimates and not being overly critical initially. Ask students to explain how they obtained their estimates, because such discussion helps clarify procedures and may even suggest new approaches to estimating with a given problem. Make sure students understand that estimation, which produces an approximate answer, is not the same as mental computation, which produces an exact answer.

One of the keys to helping children develop good estimation strategies is encouraging them to be flexible when thinking about numbers. For example, suppose you want students to estimate 418 + 349. One child might think, "400 + 300 is 700, and 49 + 18 is less than 100, so the sum is between 700 and 800." This approach uses the leading, or front-end, digits. Another child might think, "418 is about 400 and 349 is almost 350, so the sum is about 750." This child rounded to numbers that are easy to work with and used number sense to make adjustments.

As with mental computation, it is important that children develop different strategies for estimating different computations. That is, they must learn to think about the problem, the operations, and the numbers involved and not rely on a fixed set of rules to produce an estimate. Many adults think only of rounding when considering estimation; however, as with mental computation, you must help children develop a repertoire of strategies from which they can choose. In the following sections, we discuss some of these strategies.

FRONT-END ESTIMATION

The front-end strategy for estimation is a basic yet powerful approach that can be used in a variety of situations. Front-end estimation involves checking (1) the leading, or front-end, digit in a number and (2) the place value of that digit. Figure 10-10 shows how front-end digits are used to obtain an initial estimate. The unique advantage of the front-end strategy is that the original problem shows all the numbers that students have to work with. This enables students to reach an estimate quickly and easily. The front-end strategy also encourages students to use number sense as they think about the computations.

Make sure students understand that a front-end estimate is a lower bound and that students can use the remaining digits to adjust this estimate, as we discuss in the next section.

ADJUSTING

Number sense has many dimensions, one of which is recognizing when an estimate is a little more or a little less than the exact answer would be. As students develop their estimation skills, it becomes natural for them to refine an initial estimate

Figure 10-10 *Front-end estimation in action.*

218 Chapter 10 • Computation Methods: Calculators, Mental Computation, and Estimation

In the Classroom 10–4

FRONT-END ESTIMATION

Objective: Using front-end digits and adjusting to estimate sums.

Grade Level: 3-5.

- Front-end estimation is quick, reasonable, and done in your head.

Estimate

$4.47
$2.19
$1.89

My estimate is $4 + $2 + $1 = $7

I adjusted my estimate. It is over $7 or $7+

- Try These

1. $1.03
   $2.51
   $1.59

   Front-End ___
   Adjust to Get Closer
   Over $8 ___
   Under $8 ___

2. $8.53
   $6.57
   $3.95

   Front-End ___
   Adjust to Get Closer
   Over $20 ___
   Under $20 ___

3. $5.19
   $0.41
   $0.59
   $1.44

   Front-End ___
   Adjust to Get Closer
   Over $7 ___
   Under $7 ___

4. $0.27
   $6.57
   $5.95
   $0.19
   $8.53

   Front-End ___
   Adjust to Get Closer
   Over $20 ___
   Under $20 ___

---

by adjusting it, or compensating. Students can do this no matter which strategy they used to make their initial estimate.

For example, consider how the student in Figure 10-10 could adjust a front-end estimate of 13000 for the sum 4219 + 7912 + 2446, by thinking: "The answer would be exactly 13000 if all the digits except the front-end ones were zeros. But they're not zeros, so the answer must be more than 13000 … "The student could then go on to consider the digits in the hundreds place and raise the estimate accordingly. More specifically, adding the hundreds (2 + 9 + 4) would raise the estimate to 14600.

Figure 10-11 illustrates how this type of adjusting would work in solving a real-world problem such as deciding if you have enough money to buy lunch. In the Classroom 10-4 presents an activity that will give students practice in using the front-end estimation strategy and then adjusting.

When you teach estimation, be sure students understand that they should adjust estimates regardless of their initial estimation strategy and regardless of the operations involved in the problem.

COMPATIBLE NUMBERS

As in mental computation, using compatible numbers—numbers that go together naturally and are easy to work with mentally—is often helpful in estimation. For example, 14/31 + 2/7 are messy fractions to add, but 14/31 is almost ½, and 2/7 is about ¼. By changing these fractions to compatible numbers that are about the same makes an estimate of ¾ quick and easy.

Figure 10-12 illustrates how compatible numbers can make estimation easier with a variety of different operations.

Using compatible numbers for estimation means rounding the numbers in the problem to numbers that are easier to work with, *given the operations needed to solve the problem.* Notice in Figure 10-12 that 64 and 8 are not compatible for multiplication (64 × 8 is hard to compute mentally) but are compatible for division (64/8 is easy to compute mentally). Similarly, 60 and 8 are not compatible for division but are compatible for multiplication. Thus, compatibility depends on the operation as well as on the numbers themselves.

The compatible-numbers strategy is particularly powerful for division. Children given the problems shown in Figure 10-13, for example, need to be encouraged to think about why 7 and 2800 or 6 and 3000 are compatible

I've got $5 to pay for this. Will that be enough?

Burger $2.19
Soda $1.29
Fries $1.17

This makes $4.

Burger $2.19
Soda $1.29
Fries $1.17

You need to adjust! This means taking a second to look at the estimate to see if something should be added or subtracted to make a better estimate.

Figure 10-11 *Front-end estimation with adjustment.*

Estimation   219

8000 is a lot more than 7029, so the estimate of 400 is probably pretty rough.

- 75 is evenly divided by 25, so 75 and 25 are compatible for division. 7500 divided by 25 is 300, and 7500 is fairly close to 7029, and 25 is close to 23, so an estimate of 300 is probably close.

Compatible numbers are also often used to estimate addition problems. In the following example, the student first used the front-end strategy and then used compatible numbers to adjust the initial estimate:

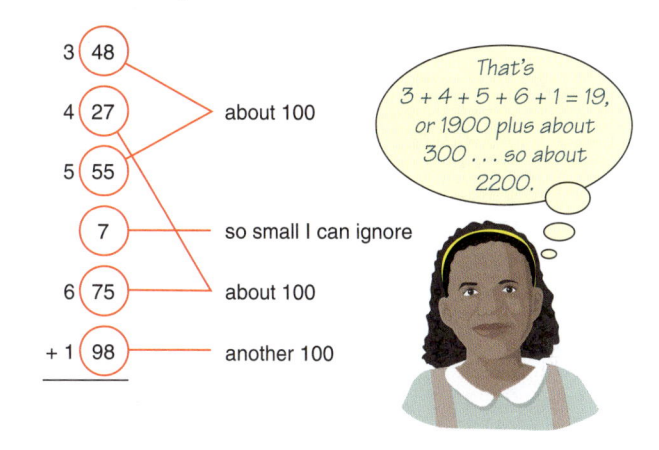

Compatible numbers are numbers that are easy to work with for the given operation.

Look! 16/31 and 4/7 aren't speaking to each other!

- These are not compatible numbers:
- These are compatible numbers:
- Using compatible numbers gives these estimates

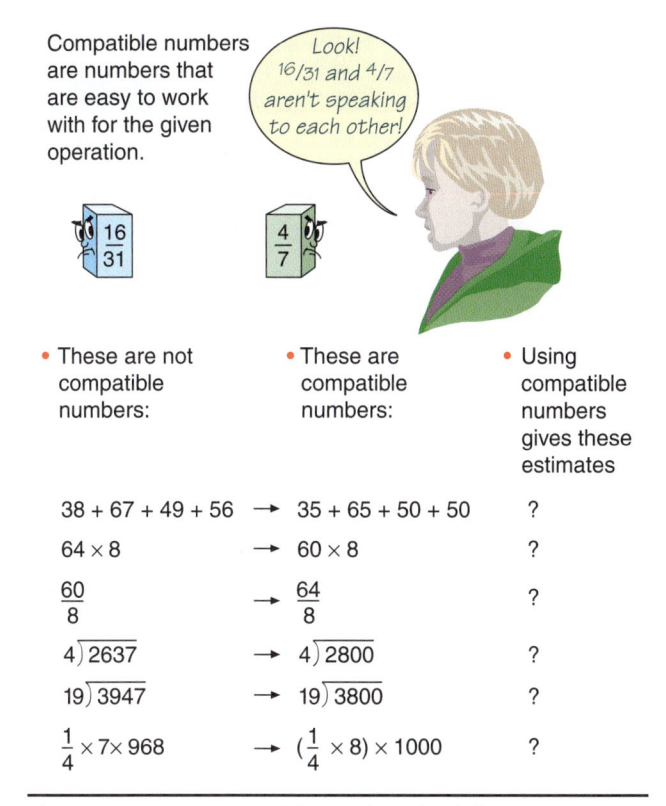

38 + 67 + 49 + 56	→ 35 + 65 + 50 + 50	?
64 × 8	→ 60 × 8	?
$\frac{60}{8}$	→ $\frac{64}{8}$	?
4)2637	→ 4)2800	?
19)3947	→ 19)3800	?
$\frac{1}{4}$ × 7 × 968	→ ($\frac{1}{4}$ × 8) × 1000	?

Figure 10-12   *Compatible numbers for different operations.*

numbers for the problem. Asking children to think of "compatible-numbers pairs" for different problems in division and to tell why the numbers are compatible will help them come up with thoughtful ideas and will give you insight into their number sense. For example, if you ask students to think of some compatible-numbers pairs to use in estimating 7029 divided by 23, they might come up with ideas like these:

- 7029 is fairly close to 6900, and 69 is divided evenly by 23, so 23 and 69 are compatible for division. 6900 divided by 23 is 300, which is probably a pretty good estimate.
- 8000 is divided evenly by 20, so 20 and 8000 are compatible for division. 8000 divided by 20 is 400, but

3) 48
4) 27   ⎫ about 100
5) 55   ⎭
7   so small I can ignore
6) 75   about 100
+ 1) 98 another 100

That's 3 + 4 + 5 + 6 + 1 = 19, or 1900 plus about 300 . . . so about 2200.

FLEXIBLE ROUNDING

Rounding to get numbers that are easier to work with is a very useful estimation strategy. Rounding is a more sophisticated strategy than front-end estimation because rounding changes, reformulates, or recomposes the numbers in the original problem. When using rounding for estimation, students should use *flexible rounding*, as shown in Figure 10-12. Flexible rounding results in numbers that are close but are also compatible. Flexible rounding is appropriate for all operations with all types of numbers.

Figure 10-14 shows how students started with the same multiplication problem but rounded differently. Notice that each approach produced a different result, but each is a

6)2800 . . ?
6)2700 . . ?

Estimate:
6)2775

Make it easy . . choose compatible numbers to work with. These are numbers that are easy to mentally compute.

7)2800 is easy, and so is 6)3000.

Remember . . . look for something close, but easier to mentally compute.

Estimate:
18)371

You can change one of the numbers or both of them, . . so you could try 18)360 or 19)380.

Figure 10-13   *The compatible numbers strategy for division.*

220 Chapter 10 • Computation Methods: Calculators, Mental Computation, and Estimation

Estimate the result
29
× 24

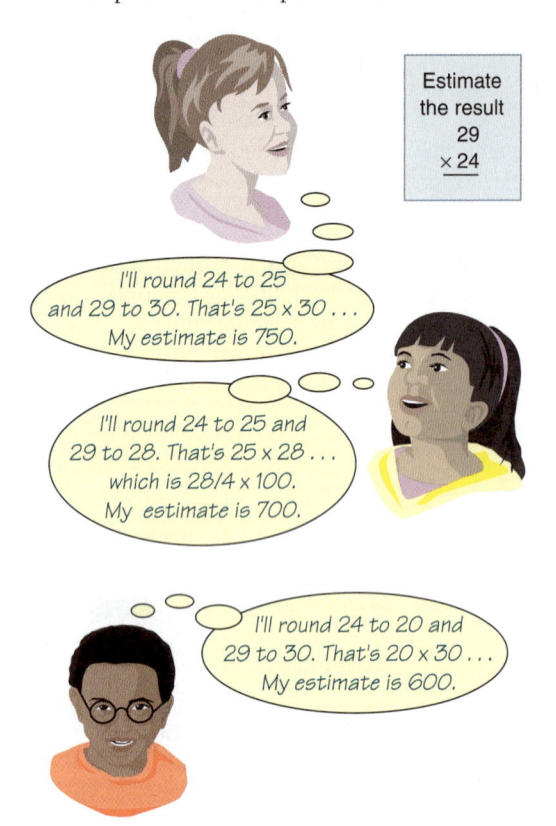

I'll round 24 to 25 and 29 to 30. That's 25 × 30 ... My estimate is 750.

I'll round 24 to 25 and 29 to 28. That's 25 × 28 ... which is 28/4 × 100. My estimate is 700.

I'll round 24 to 20 and 29 to 30. That's 20 × 30 ... My estimate is 600.

Figure 10-14 *Flexible rounding and the different estimates that result.*

reasonable estimate. It is important for students to know they have the freedom to choose different estimation strategies, but they also need to realize that different strategies produce different estimates. Accepting a range of reasonable estimates, rather than a single best estimate, fosters number sense and encourages students to choose and use their own strategies.

Problems like the one shown in Figure 10-14 need to be accompanied by questions that go beyond asking what the estimate is to encouraging thought about the process—for example:

How did you change the numbers?

Why did you change the numbers?

Children should be encouraged to think of possible substitutions and to reason about which substitutions would be best. The choices illustrated in the following example lead naturally to such questions as, "Which pair would you choose to make the estimate? Why?"

40   50
× 90   × 80
40   45
× 80   × 90

Estimate
43
× 88

Problem	Estimate with Rounding	Adjusted Estimate
42 × 61	40 × 60 = 2400	A little more than 2400
39 × 78	40 × 80 = 3200	A little less than 3200
27 × 32	30 × 30 = 900	About 900

Figure 10-15 *Different ways of estimating products using rounding and then adjusting.*

Also, encourage children to adjust estimates that they get by rounding, in order to compensate for how the numbers were rounded. In the following example, the student makes an initial estimate by rounding:

38 × 47 = 2000

The nearest tens are 40 and 50. 40 × 50 is 20 with two 0's or 2000.

Then the student adjusts the initial estimate:

38 × 47 = { a little under 2000

It's gotta be less than 2000 because I rounded up on both numbers. 2000 is an overestimate.

Figure 10-15 shows some other examples. For the first two examples, the rounding procedure clearly indicates whether the result is an underestimate or an overestimate, but the last example is not clear in this regard. Students must think more deeply about the numbers used to produce their estimate.

As students become more skillful at mental computation, they become more flexible in their use of rounding to make estimates. They learn to rely more on common sense than on traditional rounding rules, so they round to numbers that are easy for them to work with mentally. This means that different people may round the same numbers in a problem differently, depending on what is easiest for them, as we saw in Figure 10-14.

CLUSTERING

The *clustering*, or averaging, strategy uses an average for estimating a sum. Whenever a group of numbers clusters around some value, students can use this two-step process to estimate the sum of all the numbers in the group:

Estimation 221

Speech bubbles (Figure 10-16):
- "Estimate the average, and then use it to estimate the total."
- "The average is about $3. So 6 × 3 is..."
- Price tags: $3.42, $2.12, $2.98, $3.78, $2.50, $3.79
- "Use clustering! When a lot of numbers need to be totaled, you can sometimes make it simpler by estimating an average for them."
- Estimate the total: 82, 90, 87, 97, 94
- "These numbers cluster around 90. So 90 × 5 = 450."

Figure 10-16 *The clustering strategy for addition.*

1. Estimate the average value of the numbers—that is, the value that the numbers cluster around.
2. Multiply by the number of numbers in the group.

Figure 10-16 shows two examples of this strategy in action. Questions such as the following can help students understand this strategy:

What value do all the numbers cluster around?
Why is this value the estimated average?
Why is the estimated average multiplied by the number of numbers in the group?

Asking what the estimated total is for a range of problems similar to those in Figure 10-16 will help students become adept at using this strategy.

Clustering is a limited strategy, since it is appropriate only for quickly estimating the sum of a group of numbers that aren't too different from one another (i.e., that cluster around some value). The strength of this strategy is that it eliminates the mental computation involved in adding up a long list of front-end digits or rounded numbers.

CHOOSING ESTIMATION STRATEGIES

In many cases, different estimation strategies can work for the same problem. The choice of strategies depends mainly on the student, as well as on the specific numbers and operations involved. A challenge for you as the teacher is to help your students become aware of the various estimation strategies and to help them develop confidence in their ability to use any strategy successfully. Here are some useful guidelines for meeting that challenge:

- *Give your students problems that encourage and reward estimation.* For example, you can expect your students to be able to compute 78 + 83 mentally and get an exact answer, but giving them 78342 + 83289 will encourage them to use estimation. Make sure the numbers are messy enough that students will want to estimate the answer, rather than compute an exact answer.

- *Make sure students are not computing exact answers and then rounding to produce estimates.* Research has documented that students frequently use this technique (McIntosh and Sparrow, 2004). Unfortunately, this often goes undetected. You should talk with and observe students to see if they are truly estimating.

- *Ask students to tell how they made their estimates.* Individual students often develop unique approaches to estimation. By sharing their different approaches, students develop an appreciation of each other's strategies. Research has shown that asking students to share and compare strategies used in producing their estimates significantly enhances their learning (Star, Kenyon, Joiner, and Rittle-Johnson 2010).

Over or Under

1. An SUV usually uses about 14L of gas to travel about 100 km. If your gas tank is one quarter full and it is 132 km to the next service station, is it a good time to over- or underestimate your gas mileage?

2. You are talking with a car dealer about buying a car, and you ask about the gas mileage. Do you think the car dealer will over- or underestimate the gas mileage of the car?

3. You have $60 to spend on groceries for a group picnic. As you place each item in the grocery cart, is this a good time to over- or underestimate the cost of each item?

4. You are forecasting the lava speed of an active volcano. As the lava is moving down the mountainside toward a town, a decision needs to be made when to evacuate the town. Should you over- or underestimate the speed of the lava?

5. Your plane is scheduled to leave at 4:00 and it usually takes about one hour to get to the airport. Should you over- or underestimate the time needed to get to the airport when deciding what time to leave?

Figure 10-17 *Real-world situations requiring underestimation or overestimation.*

- *Fight the one-right-answer syndrome from the start.* Don't let students fall prey to the mentality of "one right" or "best" estimate. This requires helping students realize that several different estimates for the same problem could be equally acceptable. One way to do this is to let students suggest how close to the exact answer an estimate has to be, so all estimates within that range are good estimates. Not only will this counter the idea of one right answer, but it will also generally reveal the students' different estimation strategies and help everyone learn about other strategies. Such experiences help students become more comfortable with the notion that there are several different but reasonable estimates.

- *Encourage students to think of real-world situations that involve making estimates.* Doing this will help students sharpen their critical thinking skills as they decide when to overestimate or underestimate, depending on the situation. Figure 10-17 describes several situations that could encourage such thinking.

The National Research Council underscores the importance of systematically attending to mental computation and estimation in elementary school. The NRC recommends that students should have opportunities "to develop and use techniques for mental arithmetic and estimation as a means of promoting a deeper number sense" (Kilpatrick et al., 2001).

CULTURAL CONNECTIONS

For decades, students from Asian countries have excelled at written computation on international mathematics assessments (Lemke et al., 2004), and their performance in mental computation has also been strong. These results reflect the instructional attention given to developing mental and written computation skills throughout elementary school.

In Japan and Taiwan, mental and written computation are an integral part of mathematics programs, but computational estimation has received little attention in their national curricula (Yoshikawa, 1994). In fact, estimation is often held in low esteem by Japanese and Taiwanese students because of the emphasis on exact computation. An uncomfortableness in dealing with situations where a range of answers can be correct prompted one Japanese girl to say, "I don't like estimation. I think it is an evil method" (Yoshikawa, 1994). This serves as a reminder of the challenges associated with teaching estimation, not just in Japan or Taiwan, but in many different countries, including the United States.

Beware of assuming that students who are proficient at written computation are also good at estimating; much research shows otherwise. For example, Taiwanese students who showed considerable skill at tedious written computations performed much more poorly at estimating the identical computations (Reys and Yang, 1998), as shown in Table 10-1. Notice that both of the problems in the table involve numbers that are tedious to compute in writing but that lend themselves well to estimation.

Nearly two-thirds of the Taiwanese sixth graders were able to correctly compute the sum of 12/13 + 7/8. Each of these fractions is slightly less than 1, so the sum should be less than 2. Yet only one-fourth of these same students recognized two as a best estimate, while over one-half of them choose 19 or 21 as a good estimate for the sum. The other computation (534.6 × 0.545) shown in Table 10-1 is another complex computation using a written algorithm, but recognizing that 0.545 is about equal to one-half clearly indicates that the reasonable choice for an estimate is 291.357 (one-half times a number between 500 and 600 is between 250 and 300, so 291.357 is the only reasonable answer of the three choices). Nevertheless, nearly 90% of the Taiwanese students selected 29.1357. Similar results were found among eighth graders in Kuwait (Alajmi & Reys, 2007). Clearly, 29.1357 is an unreasonable answer, but it is one that might result by applying the rule "count the digits to the right of the decimal point." The results in Table 10-1 are vivid reminders that teachers must look beneath correct answers to check on students' number sense and their ability to produce reasonable estimates.

Low performance on computational estimation is common across many different countries, including Australia, Kuwait, Japan, Sweden, Taiwan, and the United States (Reys et al., 1999). Much needs to be done in helping children everywhere become competent at estimation. This represents a real opportunity for you to bring about dramatic change in the way your students view estimation and improve their performance.

A GLANCE AT WHERE WE'VE BEEN

Teachers must help students develop confidence and skill in choosing and using different methods of computation. Thus, teachers must find the right balance in the amount of time they spend on each method.

The NCTM recommends that appropriate calculators should always be available to all students. Teachers must work against myths regarding calculator use and must help students and parents understand that calculators can be very useful tools in elementary mathematics instruction. The important points are these: Using a calculator requires thinking; using a calculator can raise student achievement; a calculator is not always the fastest method of computation; the calculator is useful both for computation and as an instructional tool.

Mental computation is a natural way of doing many computations. Mental computation techniques include extending basic facts, using place value, and decomposing numbers. Children often develop their own efficient techniques for mental computation, and it is important for teachers to encourage and reward this. Mental calculation is the most direct

TABLE 10-1 • Performance of Taiwanese Sixth Graders on Written Computation and Parallel Estimation Items

Written Computation Items Requiring an Exact Answer		Percentage Correct by Sixth Graders
12/13 + 7/8		61
Estimation Items Requiring a Correct Choice		
Without calculating an exact answer, circle the best estimate for 12/13 + 7/8	A. 1	10
	B. 2[a]	25
	C. 19	36
	D. 21	16
	E. I don't know	13
This multiplication has been carried out correctly except for placing the decimal point: 534.6 × 0.545 = 291357 Place the decimal point using estimation	A. 29.1357	87
	B. 291.357[a]	11
	C. Other answer	2

[a] Denotes correct answer choice.

and efficient way of doing many calculations, it helps children develop their critical thinking skills and number sense, and it contributes to the development of estimation skills.

Estimation produces an approximate rather than an exact answer. Estimation is useful before, during, and after exact computations, to get a sense of what kind of answer to expect, to check that the computation is moving in the right direction, and to check on the reasonableness of results. Teachers must help children change their exact-answer mindset so they can see the power and usefulness of estimation. Estimation strategies include front-end estimation, using compatible numbers, flexible rounding, and clustering. Regardless of the strategy used, children should learn to adjust their result to get a better estimate. Guidelines for helping children choose and use estimation strategies include giving problems that encourage and reward estimation, making sure students are not computing exact answers and then rounding to produce estimates, getting students to talk about how they made their estimates, helping students overcome the one-right-answer syndrome, and encouraging students to think of real-world situations where estimation is needed. Studies across cultures show that students who excel at written computation and mental computation are not necessarily good at estimation.

Things to Do: From What You've Read

1. Here is an actual quote from a fourth grader telling what she thinks her teacher wants:

    "I think she doesn't like us to use mental (computation) because she can't see the writing. And she doesn't like us to use the calculator a whole lot because then we'll get too used to it and we won't want to learn and stuff …. And she likes us to use written [computation] because she can see what we're doing, and if we're having problems she can see what we're doing wrong."

    Suppose you had an opportunity to talk with the student's teacher. What would you tell her teacher?

2. How would you respond to the following statement by a parent? "Calculators should not be used in school, because if students use them, they will never have to think."

3. Look again at Figure 10-2. Decide how you think instructional time devoted to computation should be allocated. Be prepared to defend your proposal.

4. Offer some reasons why student performance on computational estimation is lower than on written computation in many countries.

5. A fifth grader described how she computed 7 × 499 in her head: "I put the 499 on top and the 7 on bottom, then I get 3 and carry the 6 and 9 carry the 6 and 7 times 4 is 28 plus 6 is 34." Would you say this student was doing mental computation the best way? Explain why. Describe some other strategies she could have used to do this computation mentally.

6. Discuss why it is important for students to develop a tolerance for an acceptable range of answers when doing computational estimation.

7. Discuss why the following is not a good assignment: "I want you to estimate the answers to these problems, then compute the correct answers, and see how far off your estimate was from the correct answer."

Things to Do: Going Beyond This Book

In the Field

1. [1]*Do It Mentally.* Ask at least two students to compute 99 + 165 mentally. Ask them how they did it. Describe the strategies that they used.

2. *About How Much?* Ask several adults to estimate 29 × 24 and tell how they made their estimate. Describe their strategies and results. Did any of them use the rounding strategies shown in Figure 10-14?

In Your Journal

3. *Assessing Mental Computation and Estimation.* Select a standardized achievement test. Review the test and determine how much attention is given to assessing mental computation and computational estimation. In your journal, discuss the strengths and weakness of test items that address mental computation or estimation.

With Additional Resources

4. Review a mental math book (Hope, Leutzinger, Reys, and Reys, 1988; Hope, Reys, and Reys, 1987). Select a lesson and highlight the key ideas for your classmates.

5. Read the article "Calculators as Learning Tools for Young Children's Explorations of Number" (Huinker, 2002). Summarize this article and describe an activity that uses the calculator to expand young children's opportunities to explore number concepts.

6. Read the article "Counting is for the birds" (Jones, 2010). Describe how the activities engage students in estimation, and how mental computation facilitates the estimation process.

With Technology

7. *Calculators and Order of Operations.* Some calculators use arithmetic logic in performing operations. For example, if you entered 3 + 5 × 4, the calculator would display 32 = (3 + 5) × 4. Other calculators use algebraic logic and would display 23 = 3 + (5 × 4). Explain how you would solve these two problems using each type of calculator: (a) (253 − 85) / 4; (b) 253 − 85 / 4).

8. View the entire video *Choose a Method*, and then answer these questions:

    - Did you see students doing mental computation and computational estimation? Describe an example of each.
    - How did the students get compatible numbers to make their computation easier?
    - Do you think the students themselves chose their computational methods? Tell why.
    - If you were doing this lesson, would you do anything differently? Explain.

---

[1]Additional activities, suggestions, and questions are available in the field experience manual on the Student Companion site at www.wiley.com/college/reys.

Note to Instructors: You can find additional resources, learning activities, and blackline masters in this text's accompanying Instructor's Manual, at www.wiley.com/college/reys.

Book Nook for Children

Goldstone, B. *Great Estimations.* New York: Henry Holt and Company, 2006.

The teacher will see how to present and teach ways to es-timate different things in everyday life. By using base ten, the reader can see or the teacher can show how to estimate using base ten with different objects. The book shows everything from money to swimmers and how to go about estimating just how many there could be. A book all classrooms should have to teach estimation.

Murphy, S. J. *Coyotes All Around.* New York: HarperCollins, 2003.

Many coyotes try to determine how many road-runners and other creatures of the desert are in their area. Different coyotes have different methods of counting, but Clever Coyote rounds off and estimates. There are activities and games at the back of the book for parents and teachers to use.

Neuschwander, C. *Amanda Bean's Amazing Dream.* New York: Scholastic, 1998.

This mental computation book shows the reader that Amanda Bean loves to count. She has a dream that helps her realize that being able to multiply will help her count things faster.

Pinczes, E. J., and MacKain, B. *One Hundred Hungry Ants.* Boston, MA: Houghton Mifflin, 1993.

This is a great book for teaching multiplication, division, and grouping strategies. It begins with ants looking for a picnic, but they discover that spending too much time organizing can result in not getting the product! The ants march in one group of 100, two groups of 50, four groups of 25, and so on. Encourage the students to model the ants, draw new patterns, and write a happier ending for the story.

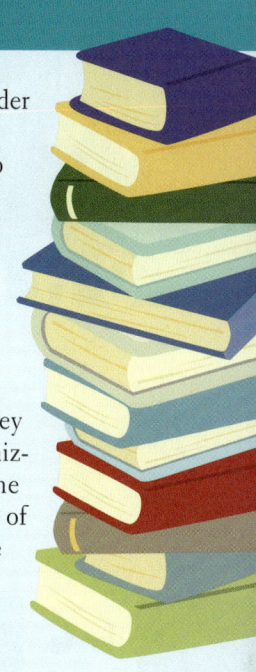

CHAPTER 11

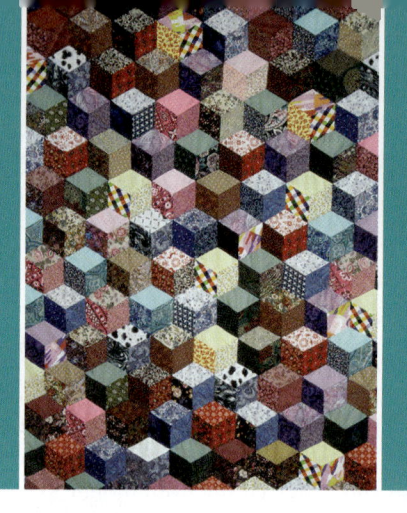

Standard and Alternative Computational Algorithms

> "SHE WAS WISE, SUBTLE, AND KNEW MORE THAN ONE WAY TO SKIN A CAT."
>
> — Mark Twain (A Connecticut Yankee in King Arthur's Court, 1889)

>> SNAPSHOT OF A LESSON

KEY IDEAS

1. Understand that children can invent strategies for multidigit computation if they understand how to take numbers apart and put them back together.
2. Realize that different children will think of different strategies.
3. Understand that children's invented strategies help them develop understanding of multidigit computation.

BACKGROUND

This lesson is based on the video *Bean Sprouts*, from Teaching Math: A Video Library, K–4, from Annenberg Media. Produced by WGBH Educational Foundation © 1995 WGBH. All rights reserved. To view the video, go to http://www.learner.org/resources/series32.html (video #15).

Malia Scott's second-grade class is beginning a new lesson, exploring strategies for multi-digit subtraction. Ms. Scott's class already is familiar with the concept of subtraction. They have been working on addition and subtraction since the beginning of the school year, and it's now the spring semester. They know their basic facts, and they can do simple mental subtraction and estimation involving larger numbers. But they have not yet learned any formal methods for subtracting multi-digit numbers. On this day, they are working on two subtraction problems in context: 56 − 27 = ? and 56 − 38 = ?.

Ms. Scott: [*to the class*] I'm going to ask for a volunteer who would like to read this first problem. Jack?

Jack: [*reading aloud*] Last year the second graders in Miss Scott's class planted 56 bean seeds. On Tuesday the class counted 27 beans sprouting. How many plants had not sprouted yet?

Ms. Scott: What's a bean sprout? What does it mean when they're saying beans are sprouting? Es?

Es: When the bean just pops out of the dirt, and it begins to grow its leaves and all.

Ms. Scott: What is the difference between the root and the sprout?

Boy #1: The root grows down and the sprout grows up.

Ms. Scott: [*narrating*] They had a familiar context with the beans that we've been studying in our life-cycle unit. As they worked through this problem, they were able to constantly refer back to the problem itself to help them get understanding of what they were trying to find out.

Ms. Scott: [*to the class*] Look at the numbers that are involved in this problem. Think about what's happening in this problem, and tell me what you think would be a reasonable estimate. Maria?

Maria: 56 ... hmm ... 56 minus 20. I think it's going to be around 30.

Boy #2: I think it's going to be around 39.

Boy #3: I think it's also going to be around 39.

Ms. Scott: OK. Tell me why.

Boy #3: Because if you took away 20 from 50, it's, it's ... Whoa! Actually I think it's going to be around 30!

Ms. Scott: Tell me why.

Boy #2: Because 50 take away 20 is 30.

Ms. Scott: I want you to keep those estimates in your head as we are going back and problem solving. That will help you to figure out if you come up with a reasonable answer.

I'm going to give you one sheet for you and your partner. So you're going to need to work together to come up with strategies that will work for that problem. I'll ask you to come up with two strategies for each problem. One person will record one strategy, one person will record the other. If you have questions, first talk with your partner and see if you can figure out that problem together.

Ms. Scott: (*narrating*) I really build in, from day one, the expectation that they will share their thinking and communicate their understanding. And along the way, I really try to build in as many supports as possible, to help them to feel comfortable and confident.

Ms. Scott circulates around the room as the children work in pairs. The video shows different pairs of students. They are using a wide variety of strategies to determine the exact answer to this multi 1 = 1 digit subtraction problem. (Only some of the examples are included in this video snapshot. If you watch the entire video, you can see more of the classroom work.)

Girl #1: [*talking aloud, explaining her work, as shown below*] 56 take away 27. And then I put a line for equals. 6 take away 7 is negative 1. And 50 take away 20 equals 30. And then I took the 30 take away 1 is 29.

$$\begin{array}{r} 56 \\ -\ 27 \\ \hline 30 - 1 = 29 \end{array}$$

This girl is using a very interesting strategy that is accurate but nontraditional. She writes her work in vertical format, so it starts out looking like traditional two-digit subtraction. Also, she subtracts the ones digits first, then the tens digits, as we do traditionally. But she doesn't need to regroup when subtracting a larger digit from a smaller digit, because she uses negative numbers when the digit on the bottom is larger than the digit on the top.

Ms. Scott: [*offering narration about another student's work*] One student was working with the hundreds board. He was able to start at 56, so he was able to conserve that number in his head as a whole. But then he counted backwards by ones [*pointing at the numbers on the hundreds board*]. And he was really looking at those numbers as individual units. He wasn't in any way chunking those numbers together, which is ideally where I would like to bring him, to help him to reach his answer more efficiently.

One pair of boys are working on the second problem from the day's worksheet: "The next day, Wednesday, the class counted again, and a total of 38 beans had sprouted. How many plants had not sprouted yet?" The boy who is writing explains his thinking aloud as he writes.

Boy #4: I'm doing 38 plus how many equals 56? So I borrow from here [56] and make this 40 and 16. And then 30 plus 10 equals 40. And 10 plus 8 equals 18.

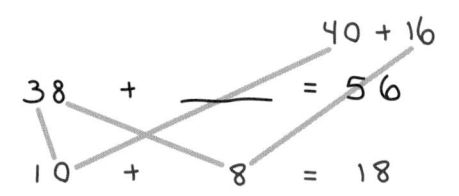

$$\begin{array}{c} 40 + 16 \\ 38\ +\ \underline{}\ =\ 56 \\ 10\ +\ 8\ =\ 18 \end{array}$$

A girl is working on the same problem. She explains her work:

Girl #2: I made 56 circles and crossed over 38, and I counted those that aren't crossed over, and it equals 18.

Ms. Scott: [*narrating*] I think it is so important for students to be able to communicate their thinking in a clear and effective manner. Not only does this help to sort of clarify their own thinking, but it also helps them to hear how other students are thinking about the problems as well.

Ms. Scott brings the class back together. She asks a few of the children to write their solutions on the board and to explain their thinking for the entire class.

Ms. Scott: I heard so many great strategies as you were working today. And what I'd like to do is, I'd like to take a few minutes to share those strategies with the rest of the class.

Ms. Scott: [*narrating*] It's really been a year-long process, to get them to the point where they are today with their sharing. We really began at the beginning of the year to build up their self-confidence so they were comfortable getting up in front of the class and sharing their thinking.

During sharing time, we see students present two more approaches to multi-digit subtraction. One boy works with tens first, and then with ones to compute 56–27. He uses "save boxes" to remind himself of the numbers he hasn't yet used. He comfortably writes horizontal number sentences to show his thinking.

$$56 - 27$$

$$6 \diagdown \diagup 7$$

$$50 - 20 = 30$$
$$30 - 7 = 23$$
$$23 + 6 = 29$$

Ms. Scott asks Es, one of the girls in the class, to explain, in her own words, why this solution makes sense.

Es: Well, he separated the ones from the tens. And he started with the tens because they are easier, and he subtracted the tens (*well, the twenty*). And then he subtracted the 7 from the 30 (*which was the answer from the 50 – 20*). And that was 23. And then he added the 6 to the 23, and that was 29.

Next, Melina and Kate show how they computed 56 – 38. They chose to keep the 56 intact, subtracting the 30 first, and then 8. Kate reports that it was easier for her to think of it that way.

$$56 - 38$$
$$56 - 30 = 26$$
$$26 - 8 = 18$$

Ms. Scott: This approach for teaching subtraction has really provided me with wonderful insight into my students' learning and given my students an opportunity to really work their way through their understanding of this very, very complex function. They are given an opportunity to share their strategies, which helps me as a teacher to get a stronger understanding of what they are thinking and where their understanding is, and start to help me to plan where I want to take them next.

FOCUS QUESTIONS

1. What is a computational algorithm? How and why are manipulative materials useful in helping children develop understanding of algorithms?
2. How can teachers help children develop the addition algorithm? Do all children need to use the same addition algorithm?
3. What are two standard subtraction algorithms and how did they develop?
4. How does the distributive property support the development of the multiplication algorithm?
5. What is the partial-products algorithm for multiplication, and how is it related to the traditional multiplication algorithm?
6. Why is the traditional division algorithm the most difficult for children to master?

INTRODUCTION

For hundreds of years, computational skill with paper-and-pencil procedures (or *computational algorithms*) has been viewed as an essential component of children's mathematical education. The teaching of computation has become more exciting in the past few years. Instead of teachers merely presenting algorithms, showing children just what to do, step by step by step, the focus has shifted to more attention on what children construct or develop for themselves. Computation has become a problem-solving process, one in which children are encouraged to reason their way to answers, rather than merely memorizing procedures that the teacher says are correct.

This change also means that children can explore alternative algorithms. Children who have thought about more than one way to perform multi-digit calculations, and who have made sense of various approaches, are likely to be more flexible and efficient. As suggested by the Mark Twain quote that opens this chapter, it can be powerful to know "more than one way to skin a cat." The algorithms that teachers have traditionally presented in mathematics class have been refined over the centuries. They are highly efficient, but they do not necessarily reflect the way children think as they compute. Accuracy and efficiency are still important today, but efficiency is of lesser importance because, with the advent of calculators and computers, no one needs to carry out pages and pages of calculations by hand anymore. Instead, it is important that today's classrooms give serious attention to students' abilities to reason and think about computation.

As mentioned in previous chapters, NCTM's *Principles and Standards for School Mathematics* (2000) includes a Number and Operations standard that lists expectations for operation sense, basic facts, and computation by grade band: PreK–2, 3–5, and 6–8. In the years after NCTM's standards were published, states throughout the nation developed their own curricular standards or frameworks. Those documents often made specific recommendations, by grade level, about when certain concepts, facts, or skills should be mastered. Confusingly, there was not much consensus across the states about when children should master the four basic operations (addition, subtraction, multiplication, and division) with whole numbers (Reys, 2006). In some states, children were expected to begin adding and subtracting multi-digit numbers as early as kindergarten, while in other states this might be recommended as late as third grade. In fact, the target for mastering addition/subtraction of multi-digit numbers ranged from grade 1 to grade 6 across the states. Multi-digit multiplication and division were typically a focus of study at grades 3 or 4, with mastery expected about a year later (in grades 4 or 5), although some states didn't expect mastery until sixth grade. (For example, one state recommended that addition should be mastered by the end of first grade, while fifteen states said this should be accomplished by the end of fourth grade and three states recommended this by the end of sixth grade.) In response to this disconcerting patchwork of state-level expectations, NCTM published its *Focal Points* (2006), which specified "grade-level expectations" for the most important mathematics concepts and skills, including computation. Then in 2010, the development of the *Common Core State Standards for Mathematics* offered states a single set of grade-level expectations, bringing the potential for greater consistency of expectations across the entire USA (CCSSI, 2010). Tables 11-1 and 11-2 show the CCSSM recommendations for development of computational methods and mastery of computational algorithms for addition/subtraction and multiplication/division, respectively.

> **Math Links 11.1**
>
> You will find detailed information about the *Common Core State Standards for Mathematics* at http://www.corestandards.org/ or by linking from this book's Web site. The CCSSM document provides a learning trajectory for developing computational skills.
>
> www.wiley.com/college/reys

TABLE 11-1 • Highlights from CCSSM Relating to Use of Relating to Use of Computational Algorithms for Addition and Subtraction[a]

Grade	Cluster Headings (and Selected Standards) Relating to Addition and Subtraction
Grade 1	Use place value understanding and properties of operations to add and subtract.
	• Add within 100 . . . using concrete models or drawings and strategies based on place value, properties of operations, and/or the relationship between addition and subtraction; relate the strategy to a written method and explain the reasoning used.
	Subtract multiples of 10 (10-90) from multiples of 10 (10-90) using concrete models or drawings and strategies based on place value, properties of operations, and/or the relationship between addition and subtraction; relate the strategy to a written method and explain the reasoning used.
Grade 2	Use place value understanding and properties of operations to add and subtract.
	• Fluently add and subtract within 100 using strategies based on place value, properties of operations and/or the relationship between addition and subtraction.
	• Add up to four two-digit numbers using strategies based on place value and properties of operations.
	• Add and subtract within 1000, using concrete models or drawings and strategies based on place value, properties of operations, and/or the relationship between addition and subtraction; relate the strategy to a written method. Understand that in adding or subtracting three-digit numbers, one adds or subtracts hundreds and hundreds, tens and tens, ones and ones; and sometimes it is necessary to compose or decompose tens or hundreds.
	Explain why addition and subtraction strategies work, using place value understanding and properties of operations.
Grade 3	Use place value understanding and properties of operations to perform multi-digit arithmetic.
	• Fluently add and subtract within 1000 using strategies and algorithms based on place value, properties of operations, and/or the relationship between addition and subtraction.
Grade 4	Use place value understanding and properties of operations to perform multi-digit arithmetic.
	• Fluently add and subtract multi-digit whole numbers using the standard algorithm.
Grade 5	Perform operations with multi-digit whole numbers and with decimals to hundredths.
	• Add and subtract decimals to hundredths, using concrete models or drawings and strategies based on place value, the properties of operations, and/or the relationship between addition and subtraction; relate the strategy to a written method and explain the reasoning used.
Grade 6	Compute fluently with multi-digit numbers.
	• Fluently add and subtract multi-digit decimals using the standard algorithm.

[a] From the *Common Core State Standards for Mathematics* (CCSSI, 2010). This summary of the CCSSM standards was made by the authors of *Helping Children Learn Mathematics*. The full document (CCSSI, 2010) may be obtained at www.corestandards/the-standards/mathematics.

TABLE 11-2 • Highlights from CCSSM Relating to Use of Computational Algorithms for Multiplication and Division

Grade	Cluster Headings (and Selected Standards) Relating to Multiplication and Division
Grade 3	Use place value understanding and properties of operations to perform multi-digit arithmetic. • Multiply one-digit whole numbers by multiples of 10 in the range 10–90 using strategies based on place value, and properties of operations.
Grade 4	Use place value understanding and properties of operations to perform multi-digit arithmetic. • Multiply a whole number of up to four digits by a one-digit whole number, and multiply two two-digit numbers, using strategies based on place value and the properties of operations. Illustrate and explain the calculation by using equations, rectangular arrays, and/or area models. • Find whole number quotients and remainders with up to four-digit dividends and one-digit divisors, using strategies based on place value, the properties of operations, and/or the relationship between multiplication and division. Illustrate and explain the calculation by using equations, rectangular arrays, and/or area models.
Grade 5	Perform operations with multi-digit whole numbers and with decimals to hundredths. • Fluently multiply multi-digit whole numbers using the standard algorithm. • Find whole number quotients of whole numbers with up to four-digit dividends and two-digit divisors, using strategies based on place value, the properties of operations, and/or the relationship between multiplication and division. Illustrate and explain the calculation by using equations, rectangular arrays, and/or area models. • Multiply and divide decimals to hundredths, using concrete models or drawings and strategies based on place value, the properties of operations, and/or the relationship between addition and subtraction; relate the strategy to a written method and explain the reasoning used.
Grade 6	Compute fluently with multi-digit numbers and find common factors and multiples. • Fluently divide multi-digit numbers using the standard algorithm. • Fluently multiply, and divide multi-digit decimals using the standard algorithm for each operation.

[a] From the *Common Core State Standards for mathematics* (CCSSI, 2010). This summary of the CCSSM standards was made by the authors of *Helping Children Learn Mathematics*. The full document (CCSSI, 2010) may be obtained at www.corestandards/the-standards/mathematics.

Math Links 11.2

Within the full-text version of NCTM's Principles and Standards for School Mathematics (PSSM) (available for 90-day free access at www.nctm.org), you can read an all-grades overview of NCTM's Number and Operations Standard in Chapter 3. You can also click on PreK–2, 3–5, or 6–8 for more detailed information by grade level. Access electronic interactive examples tied to the standards by going to http://standards.nctm.org/document/eexamples/index.htm or by linking from this book's Web site.

www.wiley.com/college/reys

No matter what grade level is recommended for mastery of computational skills, and in spite of the fact that calculators are readily available these days to relieve the burden of computation, the ability to use paper-and-pencil algorithms is still considered essential for all children. However, the NCTM *Standards* stress the need for children to be able to choose computational techniques that are appropriate for problems at hand and also to determine whether the answers they obtain are reasonable.

A balance between conceptual understanding and computational proficiency is essential for developing computational fluency. To develop such an enhanced ability, recommendations for the teaching of computation include the following:

- Fostering a solid understanding of and proficiency with simple calculations
- Abandoning the teaching of tedious calculations using paper-and-pencil algorithms in favor of exploring more mathematics
- Fostering the use of a wide variety of computation and estimation techniques—ranging from quick mental calculation, to paper-and-pencil work, to using calculators or computers—suited to different mathematical settings
- Developing the skills necessary to use appropriate technology and then translating computed results to the problem setting

- Providing students with ways to check the reasonableness of computations (number and algorithmic sense, estimation skills)

> ### Math Links 11.3
>
> A variety of Web resources for whole-number computation, which can be used by teachers and children, may be found at NCTM's Illuminations Web site under the Number and Operations Standard. It may be accessed from http://illuminations.nctm.org/swr/index.asp or from this book's Web site.
>
> www.wiley.com/college/reys

Over the years, educators have developed several models for teaching the algorithms for the four basic operations. Underlying the development of these models is the idea that children must be actively involved in constructing their own mathematical learning. Children may be more comfortable using a standard operational algorithm or an alternative variation; the choice is theirs. In this chapter, we emphasize the importance of teaching children to choose an appropriate calculation procedure, depending primarily on whether an exact or approximate answer is needed.

> ### Virtual Classroom Observation
>
> Go to www.wiley.com/college/reys,
> Access the Wiley Resource Kit
> Click on the Virtual Classroom Observations Section
>
> Module 4: Pre-Algebra: Patterns and Functions.
>
> 1. Watch a video: Teaching Examples #3: Finding Patterns in Numbers
> 2. Think and discuss: How does being challenged to explain why numeric patterns make sense help children develop their abilities generalize, to make sense of algorithmic procedures, and to extend that understanding to learning computational algorithms?

TEACHING ALGORITHMS WITH UNDERSTANDING

Computational fluency with addition, subtraction, multiplication, and division includes the ability to use various methods for computing and to recognize the relationships among the various methods. This ability requires that children learn the procedures for particular algorithms; however, they must also gain conceptual understandings that support their procedural understanding. As teachers focus their instruction on algorithms, they need to support children in gaining "a balance and connection between computational proficiency and conceptual understanding" (NCTM, 2000, p. 35). Two instructional considerations are important in ensuring that children do not just learn algorithmic procedures by rote, but that they learn them with understanding.

> ### Math Links 11.4
>
> Working with electronic manipulatives can help children develop deeper understandings of multiplication and division. Visit http://nlvm.usu.edu/en/nav/vlibrary.html, or link from this book's Web site, then select Number and Operations and scroll through to see all the choices available. Rectangle Multiplication and Rectangle Division are particularly recommended.
>
> www.wiley.com/college/reys

USING MATERIALS

The use of manipulative materials in developing understanding of the algorithms is essential. Materials form a bridge between the real-life problem situation and the abstract algorithm, helping to forge the recognition that what is written down represents real objects and actions. Children must be given sufficient time to handle the materials and make the transition to pictures and symbols. Unifix cubes, base-ten blocks, rods, Popsicle sticks, bean sticks and loose beans, buttons, and myriad other materials, either derived from the problem setting or representative of that setting, help children construct an understanding of when and how an algorithm works.

USING PLACE VALUE

Each of the algorithms for whole-number computation is based on place-value ideas, many of which were discussed in Chapter 8. Children need to have a firm understanding of these ideas before they can work effectively with the algorithms. Linking place-value ideas directly with renaming ideas is a necessary step as children explore and develop algorithms for each operation. Providing numerous trading activities is accompanied by renaming activities.

Regroup 138 ones into tens and ones using base-ten blocks:

138 is ___ tens ___ ones

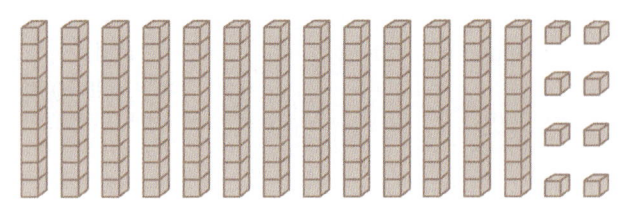

Show 46 with bundles of tens and ones, using Popsicle sticks and rubber bands to bundle them:

46 is ___ tens ___ ones

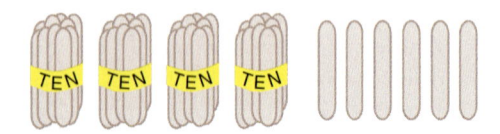

- Rename 7 tens 16 ones as 8 tens ___ ones

Tens	Ones
7	16

T	O
8	6
7̶	1̶6̶

- Rename 35 as 2 tens ___ ones

Tens	Ones
3	5

T	O
2	?
3̶	5̶

- Rename 5 tens 3 ones to show more ones:

Tens	Ones
5	3

T	O
4	13
5̶	3̶

- Write 6 tens 5 ones in five different ways
- Write 37 in all the different ways you can

Don't be surprised at the many different ideas that arise in response to this last activity. Students frequently can handle challenges that you thought were beyond them.

As you saw in this chapter's Snapshot of a Lesson, developing algorithms that work with multi-digit numbers evolves from students' understanding of place value, which

> ... involves building connections between key ideas of place value—such as quantifying sets of objects by grouping by ten and treating the groups as units—and using the structure of the written notation to capture this information about groupings
>
> *(Wearne and Hiebert, 1994, p. 273).*

ADDITION

Fluency with basic addition is a goal for the preK–2 grades, but gaining fluency depends on many diverse experiences in grouping and counting. As children work with multiple objects, they develop a concrete understanding of addition. They move objects together and use a counting strategy to identify the total number of objects. Initially, they may begin by recounting all the objects; later, this strategy often yields to a more efficient strategy of counting on. A closer look at how children make sense of addition reveals a wide variety of estimation and computation strategies.

Suppose you are teaching first grade. After a brief activity to review renaming numbers (e.g., renaming 52 as 5 tens 2 ones and also as 4 tens and 12 ones, and so on), you pose a problem to the children:

> Jill and Jeff both collected baseball cards. Jill had 27 cards and Jeff had 35. How many did they have together?

To focus attention on the reasonableness of the answers they will find, you might begin with such questions as:

> Would they have more than 50 cards? Why do you think so?
>
> More than 100 cards? Why do you think so?

Then you might give actual cards (or slips of paper to represent cards) to each group of four or five children, with the direction to figure out an answer, show it on paper, and be ready to tell why it is right. Instead of providing them with cards, you might simply encourage students to use whatever aids they are accustomed to using for math problem solving, such as tiles, base-ten blocks, an abacus, or drawing pictures. At other times, instead of working in groups, the children might be asked to solve the problem individually and to write an explanation of what they did. They need time and repeated opportunities to explore and invent strategies and to make connections between different ways of thinking about or doing the same process. After each group or individual has reached an answer, the whole class may meet to share the ways in which they solved the problem. You might record on the board the essence of each explanation.

- Group 1 counted out 27 cards and counted out 35 cards, and then counted all the cards together, one by one: 27 + 35 = 62

- Group 2 made 2 piles of 10 cards, with 7 extras, and 3 piles of 10 cards, with 5 extras:

  2 tens + 7
  3 tens + 5
  5 tens + 12, which is renamed as 6 tens + 2, or 62

- Group 3 recognized that 20 and 30 make 50, and 7 and 5 make 12:

$$\begin{array}{r} 20 + 7 \\ +30 + 5 \\ \hline 50 + 12, \text{ or } 62 \end{array}$$

- Group 4 relied on what Kim's older brother had shown her, first adding 5 + 7, to get 12, writing the 2 ones and "carrying 1" ten, then adding it to 2 tens and 3 tens to get 6 tens, or 62 altogether. The teacher showed another way to record this way of thinking about the problem.

$$\begin{array}{r} \overset{1}{2}7 \\ +35 \\ \hline 62 \end{array} \qquad \begin{array}{r} 27 \\ +35 \\ \hline 12 \\ 50 \\ \hline 62 \end{array}$$

- Group 5 did almost the same thing, but because they thought about adding 20 + 30 before adding 7 and 5, the teacher proposed recording their work a little differently:

$$\begin{array}{r} 27 \\ +35 \\ \hline 50 \\ 12 \\ \hline 62 \end{array}$$

- Group 6 decided to work with a multiple of ten that is easier to think about, and then compensated. They checked their answer by doing the problem a different way, too!

$$\begin{array}{r} 27 + 3 = 30 \\ +35 - 3 = 32 \\ \hline 62 \end{array} \qquad \begin{array}{r} 27 - 5 = 22 \\ +35 + 5 = 40 \\ \hline 62 \end{array}$$

- Group 7 added 27 + 5 = 32, and then 32 + 30 = 62. Some of the students in the class pointed out that you could also think 27 + 30 = 57, and then 57 + 5 = 62.

Everyone was sure they were right and could tell why—and they were sure that the other groups were right, too! Is 62 correct just because it is the answer every group reached? "No," retorted a number of children! It is a reasonable answer because 27 is almost 30, and 35 more would make it 65, but 27 is 3 less than 30, so 62 is right. Note that some of the students worked from left to right, as research has indicated children frequently do when constructing their own algorithms (Kamii, Lewis, and Livingston, 1993). Not all children used the manipulative materials, but those who did were able to describe or write an algorithm that modeled what they had done with the materials. Or they used procedures that they had invented outside of school. If children have had frequent opportunities to think about how numbers fit together, and how our base-ten system can help with mental computation, they will be more likely to be flexible in their thinking about computation. For example, In the Classroom 11-1 shows a game that helps students develop the ability to add and subtract 10 and multiples of 10 from any number.

STANDARD ADDITION ALGORITHM

Perhaps it occurred to you when reading the Snapshot of a Lesson that opened this chapter that some of the procedures the children used were as plausible for mental computation as for paper-and-pencil computation. Children need to recognize this, too, as well as that much addition involving one- and two-digit numbers can and should be done mentally. Lots of activities with renaming have given these children a good base from which to tackle addition algorithms. Note that only the group that relied on what Kim's brother had told them came up with the standard algorithm:

	Think
$\overset{1}{2}7$	7 + 5 = 12
+35	Write 2 in the ones column
	and 1 in the tens column.
62	2 + 3 + 1 = 6 tens

That most students did not use the standard algorithm should not surprise you. That algorithm is the result of centuries of refinement. It is commonly used because it is efficient (requires less writing than some other algorithms), but it is also less obvious. Someone probably needed to tell you about placing the one ten from 12 in the tens column. When students attempt to use the standard algorithm without understanding why it works, they may be more prone to errors than when they do addition in ways that intuitively make sense to them. For this reason it is important for the students to develop this idea from activities such as those in Figure 11-1. Here children experience the "creation and placement of the one" (or regrouping) by trading 10 units for 1 ten using the base-ten blocks.

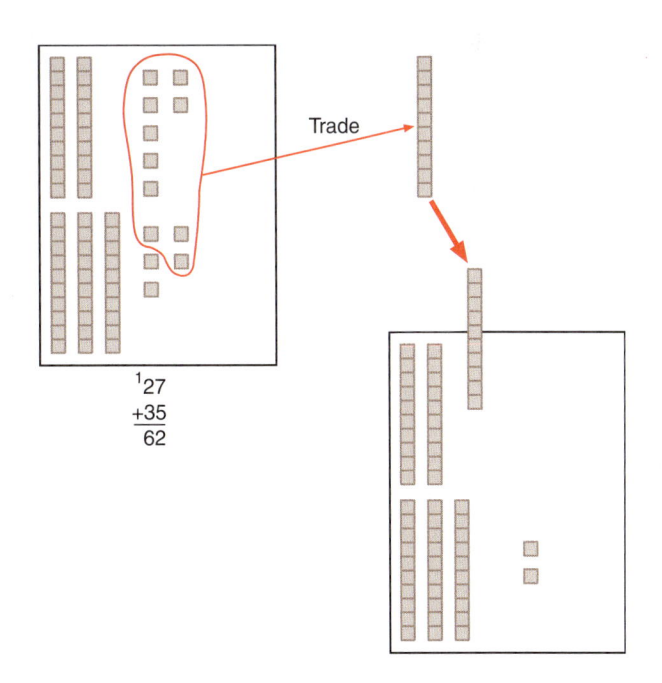

Figure 11-1 *Base-ten blocks build the ideas of regrouping in the standard addition algorithm.*

In the Classroom 11–1

CAPTURE

Objective: Using a chart game to teach addition and subtraction of multiple of tens and ones.

Grade Level: K-3.

Materials:

1. A chart divided into 100 squares numbered from 1 to 100
2. A deck of 52 change cards numbered + 1, − 1, +2, −2, +3, −3, + 5, − 5, + 10, −10, + 20, − 20, + 30, − 30 (Each of the these numbers appears 4 times.)
3. Student sheets for recording moves
4. 12 markers of the same color
5. 1 game piece per person or group

How to play

1. Place each marker on any number on the chart.
2. Place your game piece on any number without a marker. That is where you will begin playing.
3. Place the change cards face down.
4. Take 5 cards from the deck of change cards.
5. Spread them on your table in the order in which you want to use them.
6. A number with a minus (−) sign tells you to take steps backwards. A number with a plus (+) sign tells you to go forward.
7. You can use one or more than one card at a time.
8. If your last count is on a marker, capture that marker.
9. You can use your card or cards only once. After you have used them, place them face down on a discard pile and replace them from the change card pile.
10. If your card or cards are not good, replace them with new ones from the deck and reshuffle the deck.
11. If you replace a card or cards, you lose your chance to play.
12. Begin the game by placing markers on the squares shown in the chart at the right.

Activity

1. Jane's game piece was on 9 and she captured a marker on 25. She wrote her equation as: 9 + 3 + 10 + 2 + 1 = 25. How many spaces did Jane move? Explain how you found this out.

   Rewrite Jane's equation to show how far she moved: 9 + _____ = 25.

2. Lee's game piece was on 52 and he captured a marker on 85. He wrote his equation as: 52 + 3 + 20 + 10 = 85. How many spaces did Lee move? Explain how you found this out.

   Rewrite Lee's equation to show how far he moved: 52 + _____ = 85.

3. Kanye's game piece was on 10 and he captured a marker on 33. He wrote his equation as: 10 + 20 + 2 + 1 = 33. How many spaces did Kanye move? Explain how you found this out.

   Rewrite Kanye's equation to show how far he moved: 10 + _____ = 33.

4. Alona's game piece was on 72 and she captured a marker on 59. She wrote her equation as: 72 − 10 − 1 − 2 = 59. How many spaces did Alona move? Explain how you found this out. Rewrite Alona's equation to show how she moved.

   Rewrite Alona's equation to show how far she moved: 72 − _____ = 59.

5. Ramona's game piece was on 97 and she captured a marker on 56. She wrote her equation as: 97 − 30 − 10 − 3 + 2 = 56. How many spaces did Ramona move? Explain how you found this out.

   Rewrite Ramona's equation to show how far she moved: 97 − _____ = 56.

6. Partner with one of your classmates. Now place the 12 markers wherever you want and start a new game of your own.

1	2	3	4	5	6	7	8	9	10
11	12	13	14	15	16	17	18	19	20
21	22	23	24	25	26	27	28	29	30
31	32	33	34	35	36	37	38	39	40
41	42	43	44	45	46	47	48	49	50
51	52	53	54	55	56	57	58	59	60
61	62	63	64	65	66	67	68	69	70
71	72	73	74	75	76	77	78	79	80
81	82	83	84	85	86	87	88	89	90
91	92	93	94	95	96	97	98	99	100

(Source: Adapted from Economopoulos, K., and Russell, S. J. Putting Together and Taking Apart: Addition and Subtraction. A Unit in the *Investigations in Number, Data, and Space Curriculum*. White Plains, NY: Dale Seymour, 1998, pp. 112–117 and pp. 180–184.)

Addition **235**

> **Math Links 11.5**
>
> Practice using electronic base-ten blocks to add whole numbers (or even decimals). See if you can use them to illustrate a variety of problems. Visit http://nlvm.usu.edu/en/nav/vlibrary.html or link from this book's Web site, then select Number and Operations and scroll through to see all the variations of base-ten blocks activities available.
>
> www.wiley.com/college/reys

As groups share their different ways of working, children learn from each other. At some point, you will probably want to share the standard algorithm with the children, noting that it is shorter than some of their methods and that this is probably the way that their parents and other adults learned to add. Using chips and a place-value mat, as in Figure 11-2, provides a natural progression in building the notion of regrouping before introducing the standard algorithm.

PARTIAL-SUM ADDITION ALGORITHM

The algorithm used by group 5 in the Snapshot of a Lesson is similar to the standard algorithm but less obscure because all the partial sums are shown separately, so there is less chance for errors with regrouping. As group 5 used it, they added from left to right (tens first, then ones), which is more natural than working right to left. But this same partial-sum algorithm works equally well starting with ones, then tens, and so on (as shown by group 4). The partial-sum algorithm can be used as an alternative algorithm for addition (an end goal for students) or it can be useful as a transitional algorithm (intermediate step on the way to learning the standard algorithm). Ideally, however, children should be encouraged to work with whichever procedure they find easiest to understand.

Suppose the next problem posed to the children involved the addition of 13 and 54. Until recently, teachers might have introduced this problem, which does not require regrouping, far earlier than the one involving 27 and 35, which does involve regrouping. Teachers have come to realize, however, that this type of artificial control of problem situations can lead students to misconceptions about the operation and the algorithm. It is more realistic and better pedagogically for children to work, from the outset, on addition computations stemming from problems that involve both regrouping and no regrouping.

The addition of

$$\begin{array}{c}437\\+25\end{array} \quad \text{or} \quad \begin{array}{c}437\\+521\end{array} \quad \text{or} \quad \begin{array}{c}254\\+283\end{array} \quad \text{or} \quad \begin{array}{c}672\\+188\end{array}$$

is simply an extension of the regrouping procedure to the hundreds place. Children need enough experience with numbers with more than two digits so they realize that there are no hidden difficulties; if they had no calculators, they could do the computation. Most practice, however, focuses on two-digit numbers, with which they do need to become proficient.

Adding multiples of 10, 100, and so on is another simpler case. This is an important skill for students to develop, for use both in mental computation and estimation and as a building block in performing written computations:

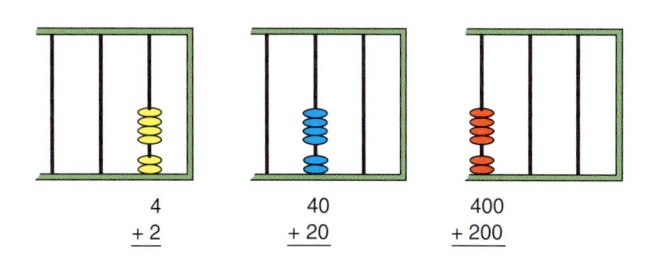

$$\begin{array}{c}4\\+2\end{array} \qquad \begin{array}{c}40\\+20\end{array} \qquad \begin{array}{c}400\\+200\end{array}$$

As children work with addition (and other operations), they must be encouraged to estimate in order to ascertain whether the answer they reach is approximately correct. Use of *compatible numbers*—numbers that are easy to compute mentally—is a powerful estimation strategy. Some problem-solving activities to practice mental computation and sharpen estimation skills, as well as deepen understanding of addition algorithms, are found in In the Classroom 11-2.

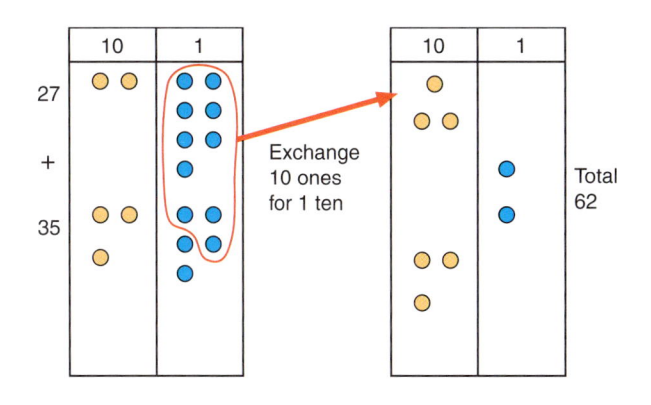

Figure 11-2 *Use of chips and a place-value mat for modeling the standard addition algorithm.*

In the Classroom 11–2

STARTERS

Objective: Using compatible numbers to develop mental computation and estimation skills involving addition.
Grade Level: 1–3.

▼ Find each missing digit:

```
  52        29       ■8       452
 +16       +36      +21      +■7
 ─────     ─────    ─────    ─────
  6■       ■5        79      489
```

▼ Use only the digits given in the cloud to make a problem with the sum shown:

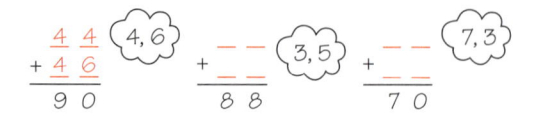

```
  4 4           — —           — —
 + 4 6         + — —         + — —
 ─────         ─────         ─────
  9 0           8 8           7 0
```
(clouds: 4, 6) (3, 5) (7, 3)

▼ Use 2, 4, 6, and 8 for these problems:
- Use each digit once to make the smallest sum possible.

  ```
    ■ ■
  + ■ ■
  ─────
  ```
- Use each digit once to make the largest sum possible.

  ```
    ■ ■
  + ■ ■
  ─────
  ```
- Use each digit once to make a sum as near 100 as possible.

  ```
    ■ ■
  + ■ ■
  ─────
  ```

▼ Use only these numbers:

| 24 | 40 | 22 | 15 | 31 | 14 |

- Name two numbers whose sum is 64. _____
- Name two numbers whose sum is more than 70. _____
- Name two numbers whose sum ends in 8. _____
- Name three numbers whose sum is 93. _____

---

attention on the relationship of 9 + 5, 19 + 5, 29 + 5, and so on. As a result of this activity, children realize the following:

- In each example, the sum will have a 4 in the ones place because 9 + 5 = 14, and the tens place will always have 1 more ten.
- The sum of 9 + 5 is 4 more than 10, so the sum of 19 + 5 is 4 more than 20 and the sum of 29 + 5 is 4 more than 30.
- For 59 + 5, there will be a 4 in the ones place and 6 (5 + 1) in the tens place.

In the Classroom 11–3

LOOK FOR PATTERNS

Objective: Using observed patterns to develop mental computational skills with addition.
Grade Level: 1–2.

▼ Complete the sums in these addition problems:

```
 3 + 6 =  9
13 + 6 = 19
23 + 6 = 29
53 + 6 = ☐
83 + 6 = ☐
```
Find a pattern in the answers.

▼ How does 4 + 8 help you to add 14 + 8?

```
  4       14
 +8      +8
 ──      ──
 12      22
```

▼ Your turn:

```
  9    19    29    39    79    89
 +5    +5    +5    +5    +5    +5

  7    17    27    37    47    87
 +3    +3    +3    +3    +3    +3
```

A more detailed plan for this activity is available on the Student Companion site at www.wiley.com/college/reys.

Children should learn to perform higher-decade additions automatically, without needing to think about adding ones and then tens and without counting on. Children who have already developed good number sense and facility with basic-fact addition strategies (such as those described in Chapter 9) should have little difficulty thinking through higher-decade addition problems. Looking for patterns such as those in In the Classroom 11-3 and explicitly discussing them with classmates and with the teacher not only encourage mental computation but also greatly increase the skill of higher-decade addition.

HIGHER-DECADE ADDITION

Combinations such as 17 + 4 or 47 + 8 or 3 + 28, called *higher-decade combinations*, are used in a strategy sometimes referred to as "adding by endings." Note that the two-digit number may come either before or after the one-digit number.

The need for higher-decade addition arises in many real-life problems; for instance, adding 6¢ tax to a purchase of 89¢. The skill is also necessary in multiplication. The strategy of counting on will probably occur to some children, and it is clearly one way of solving this type of problem, but counting on is not efficient. In the Classroom 11-3 focuses

SUBTRACTION

Like addition, subtraction of multidigit numbers requires prerequisite knowledge of basic facts and of place value. Experiences with trading or regrouping are important to develop the idea that 10 in any place-value column can be traded for 1 in the column to the left, and vice versa. Children need to become comfortable with the idea, for example, that 37 can be thought of as 3 tens and 7 ones, or as 2 tens and 17 ones, or as 1 ten and 27 ones, or just as no tens and 37. If they can think this way, they have developed the flexibility of thought needed to develop strategies for subtraction with meaning. Similarly, prior to learning subtraction algorithms, children benefit from plenty of experiences with problem-solving situations involving subtraction (solved in any way they can). In this chapter's Snapshot of a Lesson, you read about a variety of strategies that children used to solve two-digit subtractions. The Web-based video that the snapshot comes from contains even more examples, so you are encouraged to watch the entire video.

STANDARD SUBTRACTION ALGORITHM

The standard subtraction algorithm taught in the United States for the past 50 or 60 years is the *decomposition algorithm*. It involves a logical process of decomposing or renaming the sum (the number you are subtracting from). In the following example, 9 tens and 1 one is renamed as 8 tens and 11 ones:

$$\begin{array}{r} \overset{8\,1}{9\,1} \\ -2\,4 \\ \hline 6\,7 \end{array} \quad \begin{array}{l} \textit{Think} \\ 11 - 4 = 7 \text{ ones} \\ 8 \text{ tens} - 2 \text{ tens} = 6 \text{ tens} \end{array}$$

Children often naturally develop different algorithms for subtraction, just as they do for addition. They may try to work from left to right, and at this point some confusion may begin. But if they think carefully about the numbers, they may construct alternatives such as the following:

$$\begin{array}{r} 74 \\ -58 \\ \hline 20 \\ -4 \\ \hline 16 \end{array}$$

or

$$\begin{array}{r} 70 + 4 \\ -50 - 8 \\ \hline 20 - 4 = 16 \end{array}$$

or

$$\begin{array}{r} 74 \\ -58 \end{array} \rightarrow \begin{array}{r} 58 \\ + \\ \hline 74 \end{array} \rightarrow \begin{array}{r} \overset{1}{58} \\ +\,6 \\ \hline 74 \end{array} \rightarrow \begin{array}{r} \overset{1}{58} \\ +16 \\ \hline 74 \end{array} \rightarrow \begin{array}{r} 74 \\ -58 \\ \hline 16 \end{array}$$

To learn to use algorithms with understanding, children can think about connecting the steps used to solve a problem with manipulatives to the steps used in the written algorithm. Questions that encourage children to focus on the connections between manipulatives and symbols are important. Consider the following example using a problem posed by Rathmell and Trafton (1990, p. 168):

> There were 61 children who did not sign up for hot lunch. Of these, 22 went home for lunch. The rest brought a cold lunch. How many brought a cold lunch?

What are we trying to find? (How many brought cold lunch?)

Do you think it will be more or less than 60? (Less, because there are 61 children who didn't have a hot lunch, but some of those went home for lunch.)

How many did not sign up for hot lunch? (61)

How many went home? (22)

How can we show this? (Use cubes to model 61, then separate out 22 cubes.)

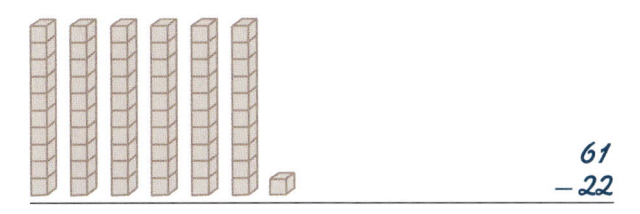

$$\begin{array}{r} 61 \\ -22 \end{array}$$

Let's assume we want to subtract off the ones first. Are there enough ones to subtract 2 ones? (No.)

How can we get more ones? (We can trade a stick of tens.)

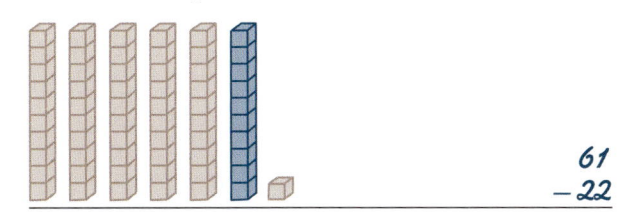

$$\begin{array}{r} 61 \\ -22 \end{array}$$

If we trade in a 10 for ones, how many ones do we get? (10)

How many ones do we have now? (11)

And how many tens? (5)

How much is 5 tens and 11 ones? (61)

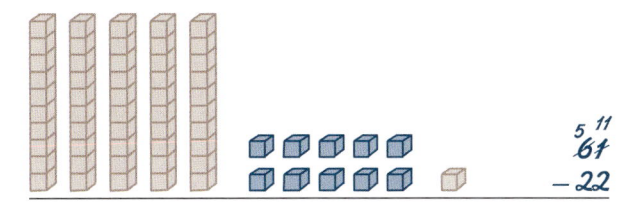

$$\begin{array}{r} \overset{5\,\,11}{6\,1} \\ -2\,2 \end{array}$$

Now are there enough ones to subtract 2? (Yes, we have 11 ones. 11 is more than 2.)

What is 11 − 2? (9)

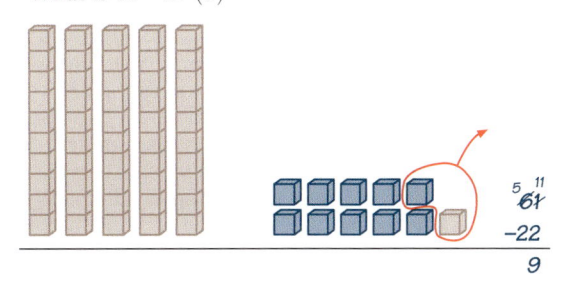

$$\begin{array}{r} \overset{5\,\,11}{6\,1} \\ -2\,2 \\ \hline 9 \end{array}$$

238 Chapter 11 • Standard and Alternative Computational Algorithm

Now what do we subtract? (the tens)

What is 5 tens minus 2 tens? (3 tens)

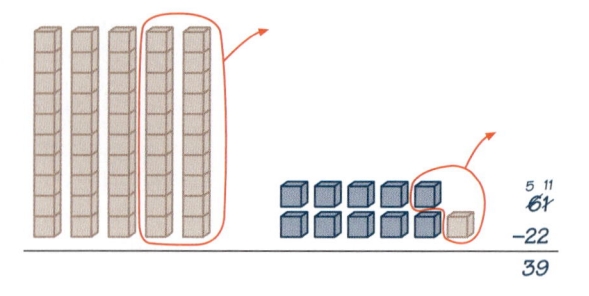

$$\begin{array}{r} \overset{5\ 11}{\cancel{6}\cancel{1}} \\ -22 \\ \hline 39 \end{array}$$

So, if we have 9 ones and 3 tens, how many children are eating a cold lunch? (39)

Does this answer make sense? Why? (Yes, 39 makes sense because it is close to 40. There were about 60 kids who didn't have a hot lunch. There were about 20 kids who went home. 60 − 20 = 40. So there should be about 40 kids who brought a cold lunch. Because 39 is close to 40, it makes sense.)

Experiences with regrouping from tens to hundreds or in both places are also helpful. In the Classroom 11–4 (Racing with Base-Ten Blocks) describes two games played with base-ten blocks that can help develop students' understanding of regrouping for addition or subtraction.

The presence of zeros in the sum demands special attention when the standard algorithm is used. In this example, with zero in the ones place only, 50 is renamed as 4 tens and 10 ones:

$$\begin{array}{r} \overset{4\ 10}{850} \\ -237 \end{array}$$

When zero appears in the tens place, the problem is slightly more difficult, especially when regrouping in the ones place is also necessary. Two ways to approach the regrouping (using either one step or two steps) are shown in Figure 11-3.

In the Classroom 11–4

RACING WITH BASE-TEN BLOCKS

Objective: Developing students' understanding of regrouping for addition or subtraction using base-ten blocks.

Grade Level: 1–5.

Number of Players: 2–5.

Materials: Base-ten blocks, place-value mat (full-sized paper divided into three columns) for each player, and one pair of dice.

Place-Value Mat

Flats	Longs	Units

Race-to-a-Flat

Before the Game Begins: Choose one player to be the base-ten "banker."

How to Play: Players take turns rolling the dice, adding the numbers on the tops of the two dice, and then asking the banker for that number of units. For example, if you roll a 4 and a 5 as shown on the dice, ask for 9 units.

Place the pieces on your place-value mat. You may never have 10 or more pieces in any portion of your place-value mat. Whenever you can, you must trade units for a long (or longs for a flat). When all the trading is complete, it is the next player's turn to roll the dice, collect units, and trade pieces. The first player to trade up to a flat is the winner.

Give-away-a-Flat

Before the Game Begins: Choose one player to be the base-ten "banker." The banker distributes one flat to each player, who places it in the flats column on her or his place-value mat.

How to Play: Players take turns rolling the dice, finding the sum of the numbers on the tops of the two dice, and *giving away* that number of units to the banker. Clearly, trades will have to be made during each player's first turn and possibly again during subsequent turns. The first player to return all of her or his base-ten pieces to the banker is the winner.

Questions for Give-away-a-Flat

1. Is it possible to give away one long with one roll of the dice?

2. Is it possible to give away two longs with one roll of the dice?

3. Can you win the game with one roll of the dice?

4. What is the fewest number of rolls required to give away all of the pieces?

5. What is the largest number of rolls required to give away all of the pieces?

Variations: To challenge fifth and sixth graders, race to a cube (1000) or give away a cube. For games involving larger goals, students can multiply the two dice instead of adding.

A more detailed plan for this activity is available on the Student Companion site at www.wiley.com/college/reys.

Subtraction **239**

A subtraction involving numbers with more than one zero is generally even more difficult. One alternative is multiple renaming—renaming from hundreds to tens, then renaming again from tens to ones:

$$\begin{array}{cccc} & & \overset{9\,10}{} & \overset{9\,10}{} \\ & \overset{4\,10}{} & \overset{4\,\cancel{10}}{} & \overset{4\,\cancel{10}}{} \\ 500 & \cancel{500} & \cancel{500} & \cancel{500} \\ -257 & -257 & -257 & -257 \\ \hline & & & 243 \end{array}$$

If students learn a procedure like this rotely, without understanding, it is likely that they will make errors but be unable to notice or correct them. Place-value experiences are important in preparing children to cope with these problems. They must clearly understand that 500 can be renamed as 4 hundreds and 10 tens, or as 4 hundreds, 9 tens, and 10 ones. They will then find it easier to recognize the need for multiple regrouping when they see multiple zeros and will be able to do all the renaming at once:

$$\begin{array}{ccl} & \overset{4\,9\,1}{} & \\ 500 & \cancel{500} & 500 = 4 \text{ hundreds, 9 tens, 10 ones} \\ -283 & -283 & \end{array}$$

Alternately, if they understand that 500 can be thought of as 50 tens, they can think of renaming it directly as 49 tens and 10 ones. In Figure 11-3, for the problem 207 − 39, 207 can be renamed as 19 tens and 17 ones in one step. The need to do this is not readily recognizable by all children, however; those who need to do the double renaming should be allowed to do so.

PARTIAL-DIFFERENCE SUBTRACTION ALGORITHM

A common error made by children who have been taught to subtract without understanding is to simply do the subtraction in each column separately, always subtracting the smaller number from the larger, without recognizing when regrouping is required. For example, a student might incorrectly write

$$\begin{array}{r} 523 \\ -385 \\ \hline 262 \end{array}$$

If your students make this sort of mistake, it's probably important to have them return to modeling some simple problems (such as 23 − 19 or 34 − 15) so they can see why the answers here are not 16 and 21.

However, an interesting correct variation of this approach is the following strategy, which we have named the "partial-difference subtraction algorithm" because it bears some resemblance to the previously mentioned partial-sum addition algorithm. Here, the subtractions in each column can be done in any order—left to right, right to left, or even out of order. You first write down the result of each subtraction, column by column, and then combine those partial differences to get your final answer. The key is that when the bottom number is larger than the top number, you do not worry about regrouping—instead, you simply make note of how much more needs to be subtracted. (Note that Girl #1

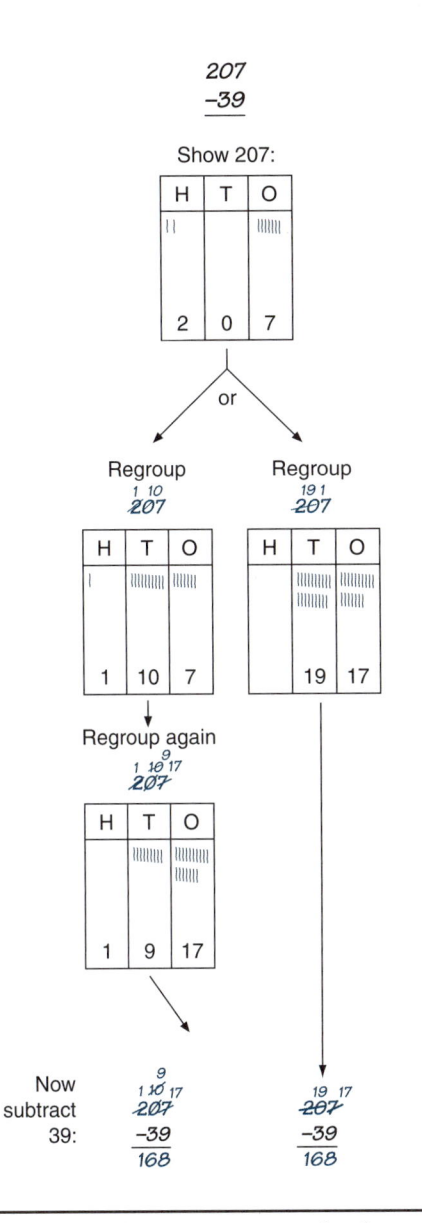

Figure 11-3 *Place-value representations for the subtraction model.*

Math Links 11.6

Practice using electronic base-ten blocks to subtract whole numbers (or even decimals). See if you can use them to illustrate the hot lunch problem. Experiences with problems that require regrouping from hundreds to tens (or even multiple regroupings) are also helpful. Visit http://nlvm.usu.edu/en/nav/vlibrary.html, or link from this book's Web site, then select Number and Operations and scroll through to see all the variations of base-ten blocks activities available.

www.wiley.com/college/reys

used this subtraction strategy in the opening Snapshot of a Lesson.) For example, when doing 20 – 80, you write –60. (Adults may think of this as "negative 60," but children are more likely to think something like "Starting with 20, I can get to 0 by subtracting 20, but I will still need to subtract 60 more." The second step in the procedure is to combine the partial differences. If a child has developed good number sense for basic subtraction facts and for working with multiples of 10 and 100, it will not be difficult for him or her to do each part of the second step mentally (e.g., doing 200 – 60 = 140 mentally and, finally, 140 – 2 = 138 mentally). The Capture game from In the Classroom 11–1 provides valuable practice in exactly this skill.

```
   523
  –385
   200 ⎫
 –  60 ⎬
 –   2 ⎭
   140
 –   2
   138
```

MULTIPLICATION

Before children tackle multiplication algorithms, they must have a firm grasp of place value, expanded notation, and the distributive property, as well as the basic facts of multiplication. As with the other operations, it is wise to review each of these prerequisites before beginning work with multiplication algorithms. It is also wise to develop situations or problems for which children compute mentally, without concern for the paper-and-pencil forms.

MULTIPLICATION WITH ONE-DIGIT MULTIPLIERS

Children should be encouraged to talk about their thinking, at the same time that the meaning of multiplication is reinforced with materials:

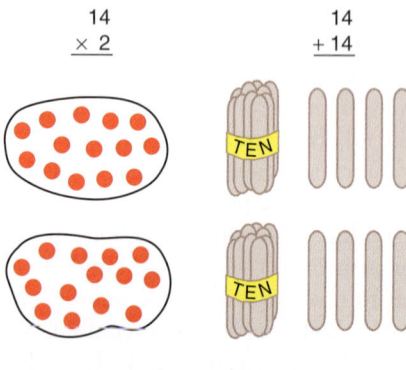

$2 \times 14 = 14 + 14$

Looking at these models, we can see not only that 14 can be represented by one bundle of tens and 4 ones, but also that 2×14 can be represented by *two* bundles of tens and *two* groups of 4 ones. This points out the use of the distributive property:

$$2 \times 14 = 2 \times (10 + 14) = (2 \times 10) + (2 \times 4)$$
$$= 20 + 8 = 28$$

Arrays can also be used to develop meaning and visual images of multiplication:

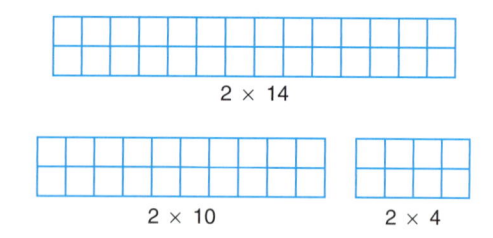

2×14

$2 \times 10 \qquad 2 \times 4$

Place-value ideas are noted along with materials:

T	O		T	O
1	4		1	4
1	4	×	2	
2	8		2	8

Expanded algorithms can be developed or constructed by the children

```
  14        1 ten  4 ones        14
×  2       ×         2         ×  2
           2 tens  8 ones         8    2 × 4  =  8
                              + 20     2 × 10 = 20
                                28     20 + 8 = 28
```

PARTIAL-PRODUCTS MULTIPLICATION ALGORITHM The partial-products algorithm for multiplication (Figure 11-4) is one form of the expanded algorithm, where all the individual products produced by the multiplication are recorded on separate lines, then summed. Just as the partial-sums algorithm for addition is a useful alternative or transitional algorithm, the partial-products algorithm can be useful for multiplication.

```
   57                          452
 ×  3                        ×   7
   21  (3 × 7)                  14   (7 × 2)
 +150  (3 × 50)                350   (7 × 50)
  171                        +2800   (7 × 400)
                              3164
3 × 57 = (3 × 50) + (3 × 7)
                         7 × 452 = (7 × 400) + (7 × 50) + (7 × 2)
```

Traditional   **Partial products**

```
   372                     372
 ×  28                   ×  28
  2976                      16   (8 × 2)
   744                     560   (8 × 70)
 10416                    2400   (8 × 300)
                            40   (20 × 2)
                          1400   (20 × 70)
                          6000   (20 × 300)
                         10416
```

Figure 11-4 *Multiplication comparing the traditional multiplication algorithm with the partial-products algorithm.*

The traditional American multiplication algorithm is a streamlined version of the partial-products algorithm. In the traditional algorithm, some of the partial products are added mentally before their sum is written down. In some cases this traditional method is more efficient (because it requires less writing). But it is not as easy to understand step-by-step and it may be more susceptible to error (because steps are combined and done mentally).

The distributive property helps children understand the relationship between the partial-products algorithm and the traditional algorithm (see Figure 11-4). Whenever students see a multiplication such as 7 × 124, they should readily think that it could be computed by finding (7 × 100) + (7 × 20) + (7 × 4). Similarly, they can think of 4 × 59 as (4 × 50) + (4 × 9). Or, if the multiplication is to be done mentally, they might think of 4 × 59 = (4 × 60) − (4 × 1). In the Classroom 11-5 includes several types of activities that focus on using the distributive property.

LATTICE MULTIPLICATION ALGORITHM Lattice multiplication is an interesting algorithm, though not many people today are familiar with it. For each multiplication problem, you draw a rectangular lattice of boxes with diagonals drawn from bottom left to upper right in each interior square. You write the numbers to be multiplied on the top and right of the rectangle, one digit per square, and the boxes in the lattice help you organize where to write digits of the partial products. Some people really like this method because using the lattice eliminates the need for thinking about place value or for writing lots of zeros. Lattice multiplication is a very old approach. In fact, historians believe it has Hindu origins in India more than 900 years ago, though this method first appeared in print in the 1400s in Italy. In Figure 11-5 you can see the same multiplication problem (26 × 368) done with the partial products algorithm (Method A) and with the lattice multiplication algorithm (Method B). Where do you see the answer in Method B? Look at the digits in each column in Method A and compare them with the digits found along each diagonal in Method B. For example, can you see which diagonal in Method B corresponds with each column (ones, tens, hundreds, thousands) in Method A? Can you explain to someone else how Method B works?

Method A
(Partial Products Algorithm)

```
   368
 × 26
    48
   360
  1800
   160
  1200
  6000
  9568
```

Method B
(Lattice Multiplication Algorithm)

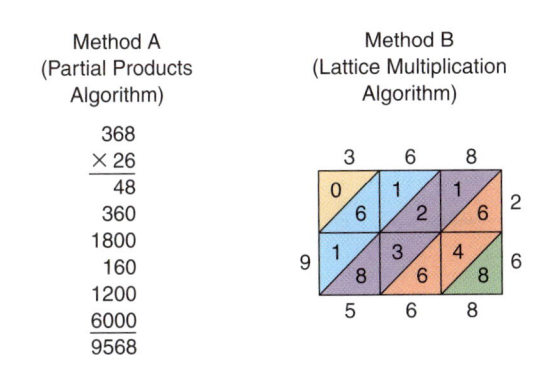

Figure 11-5 *Comparing partial-products multiplication with lattice multiplication.*

In the Classroom 11-5

WHAT'S MISSING?

Objective: Using number puzzles to develop estimation skills with multiplication.

Grade Level: 3–5.

▼ Guess the numbers that will go into the circles and boxes.

Write your number and then check it on a calculator. Score 2 points if correct on the first try and 1 point if correct on the second try.

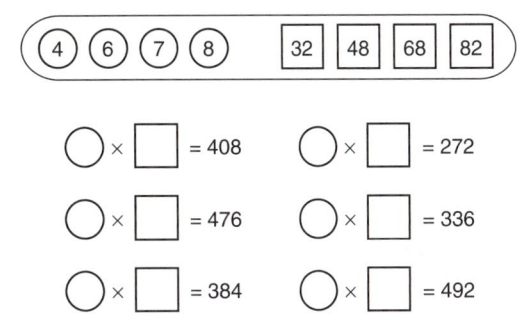

(4　6　7　8)　　[32　48　68　82]

○ × □ = 408　　○ × □ = 272

○ × □ = 476　　○ × □ = 336

○ × □ = 384　　○ × □ = 492

▼ Try these using only 2, 3, and 4. (You may use the same numeral more than once.)

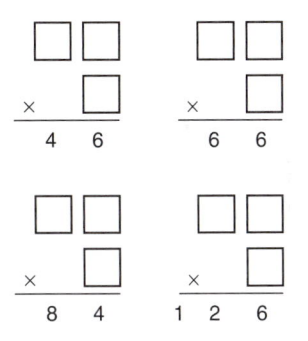

　　□□　　　　□□
× □　　　　× □
4　6　　　　6　6

　　□□　　　　□□
× □　　　　× □
8　4　　　　1　2　6

- Now use only 4, 6, 8, and 9.
- Make the largest product. (Use each numeral only once.)

□□□
× □

- Make the smallest product: (Use each numeral only once.)

□□□
× □

MULTIPLICATION BY 10 AND MULTIPLES OF 10

The ability to work flexibly with powers of 10 is an important prerequisite to handling multiplication algorithms with understanding. Multiplying by 10 comes easily to most children, and is readily extended to multiplying by 100 and 1000 as children gain an understanding of larger numbers. Anytime a number is multiplied by 10, all its digits move one place to the left—for example, in $10 \times 763 = 7630$, the 700 becomes 7000, the 60 becomes 600, and the 3 becomes 30. Similarly, all the digits move two places to the left if a number is multiplied by 100. Notice that it makes much more sense to think about the digits moving to the left, than to think about the decimal point moving to the right (Flores, 2007). Children can be shown a series of examples, after which they are asked to discuss and generalize, noting where the digits have moved to, and discussing why.

Multiplying by 20, 30, 200, 300, and so on is an extension of multiplying by 10 and 100. Emphasize what happens across examples and generalize from the pattern. For example, have children consider 3×50:

$3 \times 5 = 15$

3×5 tens $= 15$ tens $= 150$

$3 \times 50 = 150$

Then have them consider 4×50:

$4 \times 5 = 20$

4×5 tens $=$ ___ tens $=$ ___

$4 \times 50 =$ ___

A similar sequence can be followed for examples with hundreds, such as 2×300:

$2 \times 3 = 6$

2×3 hundreds $=$ ___ hundreds $=$ ___

$2 \times 300 =$ ___

MULTIPLICATION WITH ZEROS

When zeros appear in the factor being multiplied, particular attention needs to be given to the effect on the product or partial product. Many children are prone to ignore the zero. Thus, for 9×306, their answer may (incorrectly) be

$$\begin{array}{r} \overset{5}{306} \\ \times\ \ 9 \\ \hline 324 \end{array}$$

When an estimate is made first, children have a way of determining whether their answer is in the ballpark. Writing each partial product separately, rather than attempting to add them mentally, can also help avoid place-value errors due to zeros. In the previous problem, a student might actually write the partial products 54 and 0 and 2700. Use of a place-value chart may also help students understand what the correct procedure must be, as will expanded notation:

Estimate

$9 \times 3 = 27$
$9 \times 300 = 2700$

$9 \times 306 = 9 \times (300 + 6)$
$= (9 \times 300) + (9 \times 6)$
$= 2700 + 54$
$= 2754$

MULTIPLICATION WITH TWO-DIGIT MULTIPLIERS

When children begin to work with two-digit multipliers, the use of manipulative materials becomes cumbersome. Arrays or grids offer one way to bridge the gap from concrete materials to symbols and also help illustrate, once again, why the partial-products algorithm makes sense. As shown in Figure 11-6, a grid can also provide entirely new ways of viewing and writing about a multiplication example. Materials such as base-ten blocks can be used to tie this work to previously learned procedures, but increasingly the emphasis will shift to working with symbols, as shown in Figure 11-7. This approach is possible because students already have a conceptual foundation built on the use of materials and on understanding of the distributive property.

It is always important for students to connect new ideas to earlier understandings, continually building on a foundation of previous knowledge. Children should see that the process of multiplying by a two- or three-digit number is just an extension of the procedure for multiplying by a one-digit number. Similarly, illustrating this process by using an array is essentially the same no matter what the multiplier. Moreover, in the middle grades, when students begin to learn about symbol manipulation in algebra, they should be able to connect the new idea of multiplying binomials to their prior understanding of multidigit multiplication. In Figure 11-8, the illustration of the "foil" method for multiplying

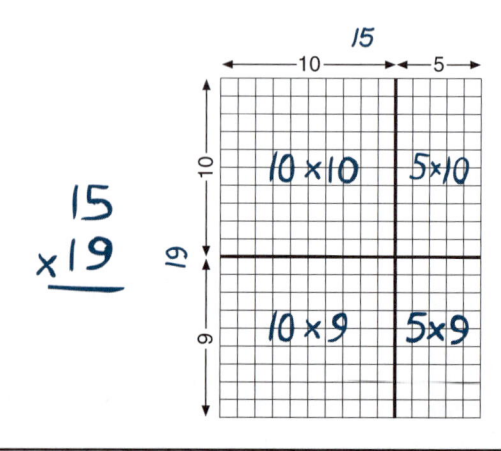

Figure 11-6 *Use of a grid to solve multiplication with two-digit multiplier.*

Division

MULTIPLICATION WITH LARGE NUMBERS

How can we use a calculator to solve computations involving numbers that appear too big for the calculator? As children experiment with using a calculator for multiplication, there will come a time when they overload the calculator. Sometimes the number to be entered contains more digits than the display will show. At other times the factors can be entered, but the product will be too big for the display. For instance, if this example is entered in a four-function calculator, an error message results:

$$\begin{array}{r} 2345678 \\ \times\ 4003 \\ \hline \end{array}$$

When this happens, children should be encouraged to estimate an answer and then use the distributive property plus mental computation along with the calculator.

<u>Estimate</u>

$4000 \times 2000000 = 8000000000$

(4 followed by 3 zeros) times (2 followed by 6 zeros) yields 4×2 or 8 followed by 9 zeros, since $3 + 6 = 9$)

<u>Calculate</u>

$4003 = 4000 + 3 = (4 \times 1000) + 3$
$4 \times 2345678 = 9382712$ (with the calculator)
$9382712 \times 1000 = 9382712000$ (mentally)
$3 \times 2345678 = 7037034$ (with the calculator)

Calculate $9382712000 \times 7037034$ (by hand, or with the calculator by temporarily leaving off the 938 at the front, and replacing it afterwards)

Such examples show how an understanding of multiplication, plus problem-solving skills, can be used with calculators to reach a solution. Use of the calculator, mental computation, and paper-and-pencil recording may be integrated to solve a single problem. This sort of approach reminds us that calculator algorithms differ from the currently used paper-and-pencil algorithms.

DIVISION

The traditional division algorithm is without doubt the most difficult of the algorithms for children to master, for several reasons:

- Computation begins at the left, rather than at the right as for the other operations.
- The algorithm involves not only the basic division facts but also subtraction and multiplication.
- There are several interactions in the algorithm, but their pattern moves from one spot to another.
- Trial quotients, involving estimation, must be used and may not always be successful at the first attempt—or even the second.

Figure 11-7 *Multiplication with two-digit multiplier.*

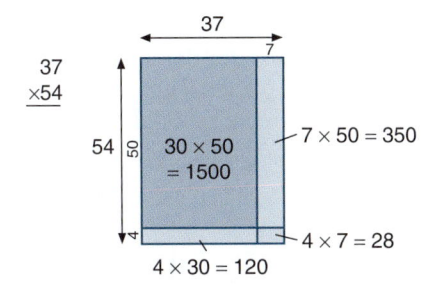

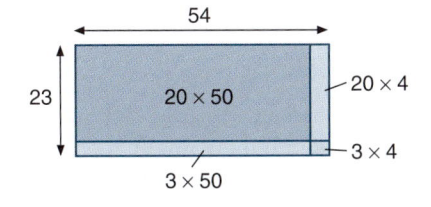

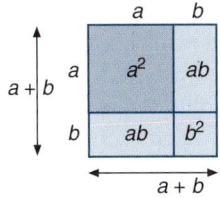

Figure 11-8 *Extending multiplication ideas to algebra.*

two binomials should look familiar because it is the same as the illustration for multiplying two two-digit numbers together; the simple explanation is that the distributive property governs both of these parallel procedures.

Many teachers struggle to teach the long-division algorithm, and many children struggle to learn it. Since handheld calculators became readily available more than 30 years ago, making it possible for everyone to calculate answers to division problems quickly and easily, mathematics educators have suggested that too much emphasis on the division algorithm can be a waste of scarce instructional time. As *An Agenda for Action* (NCTM, 1980, p. 6) indicated, more than 30 years ago:

> For most students, much of a full year of instruction in mathematics is spent on the division of whole numbers—a massive investment with increasingly limited productive return.... For most complex problems, using the calculator for rapid and accurate computation makes a far greater contribution to functional competence in daily life.

On the National Assessment of Educational Progress (NAEP), some exercises are given both with and without the use of calculators. For example, on the fourth NAEP, third graders were given the problem $3\overline{)42}$. When calculators were not used, about 20% of the answers were correct; when calculators were used, about 50% were correct (Lindquist et al., 1988). On one item on the sixth NAEP, students who reported using a calculator scored 82% correct, and those who reported they did not use a calculator scored only 35% (Kenney and Silver, 1997). Even more dramatic were results from the second NAEP (Carpenter et al., 1981). One exercise involving division by a two-digit divisor was given to 9-, 13-, and 17-year-olds. All the 9-year-olds were tested with calculators available (perhaps because they presumably had not yet been taught division with a two-digit divisor). For comparison purposes, some students in each of the two older groups were tested with calculators, while others were tested without calculators. Without calculators, 13-year-olds scored 46% and 17-year-olds scored 50% without calculators. With calculators, their scores rose dramatically, to 82% and 91%, respectively. Fifty percent of the 9-year-olds obtained the correct answer (with calculators available). Considering that half of the 17-year-olds could not perform the division even after years of practice, is it not reasonable to let them use the tool with which they are successful—and which they will use anyway for the rest of their lives?

The NCTM's *Standards* stress that in today's technological society, teachers should rethink how computation is taught and how it is performed. Helping students understand *when* division is appropriate and *how to estimate* answers to division problems is extremely important. In terms of actually performing computations, division with one-digit divisors should be the focus of instruction, followed by some work with two-digit divisors so that students understand how such division is done. Performing more complex division—perhaps any that takes more than 30 seconds to do—with paper and pencil is a thing of the past for adults. Schools should not demand that children spend countless hours mastering an antiquated skill. Teachers can't afford the instructional time for this task. Instead, estimation skills should be used to define the bounds of the quotient so that the reasonableness of calculator answers can be determined—just as such estimates should be used with division with one-digit divisors and, in fact, with all addition, subtraction, and multiplication.

DIVISION WITH ONE-DIGIT DIVISORS

It is essential to begin division instruction with real-world problems. The most important thing is for students to develop an understanding of what division is all about. There is no reason for children not to encounter division problems with remainders from the outset of their work with division. Children should be encouraged to use blocks or counters to model division situations and to explain their thinking. The teacher should help students see how their thinking can be clearly and efficiently recorded on paper. Thus, learning a paper-and-pencil algorithm essentially turns into learning a way to record what is done concretely.

For example, Burns (1991) posed problems and had third-grade students explain their answers and the methods they used to reach them. One problem called for four children to divide 54 marbles equally. Group 1 counted out 54 cubes (to represent the marbles), divided them into four groups, and wrote:

> We think we get 13 each. We think this because we took the cubes and dealed out them one by one. We had two left. Bryce was sick so we gave the extra to him.
> (Burns, 1991, p. 16)

Other groups wrote:

Group 2: We had 54. We gave each person 10 because we thout if there was 40 there would be 4 tens and 14 would be left. Each person gets 3, witch leves 2. Each person gets 13.

Group 3: We drew 54 marbles and then I numbered them 1, 2, 3, 4, and then we counted the ones and then we knew that each child gets 13 marbles and we lost the other two.

Group 4: I wrote down 54. I took away 12. I got 42. I took another 12 away. I got 30. I took another 12 away. I got 18. I took another 12 away. I got 6. Then I took away 4. I got 2. I chiped each of them into halfs that made 4 halfs. Each person got 13 and a half.
(Burns, 1991, pp. 16–17)

Burns noted that, from listening to each others' methods, the children not only heard different approaches to solving division problems, but they also learned that division can be done in a variety of ways. Their teacher might have helped the class see how each of their different explanations could be recorded symbolically in a slightly different way, but always with the same final result. Refer to Figure 11-9 to see how each group's work might be recorded.

Division 245

When using the distributive algorithm for this problem, first share as many of the 9 hundreds as possible (8 hundreds can be shared, 2 to each of the 4 people). Record how many were given to each person above (2 hundreds). Record the total number shared below (8 hundreds). Be careful to preserve place value by writing these numbers in the proper columns. "Subtract and bring down" and note that you still have 1 hundred and 5 tens to be shared (or 15 tens). Begin the process again, sharing as many of the tens as possible, recording the number for each person above, recording the total shared below, and subtracting to see how much more still remains to be shared. Share 12 of the 15 tens, 3 to each of the 4 people, so write 3 above (in the tens place) and 12 below. Now, "subtract and bring down" again, noting now that you still have 3 tens and 4 ones to be shared (or 34 ones). Once again, share as many as possible, this time writing 8 above and 32 below, and then subtracting. You now have just 2 ones left. Starting with $954, you were able to give $238 to each of 4 people, with $2 left over.

SUBTRACTIVE ALGORITHM

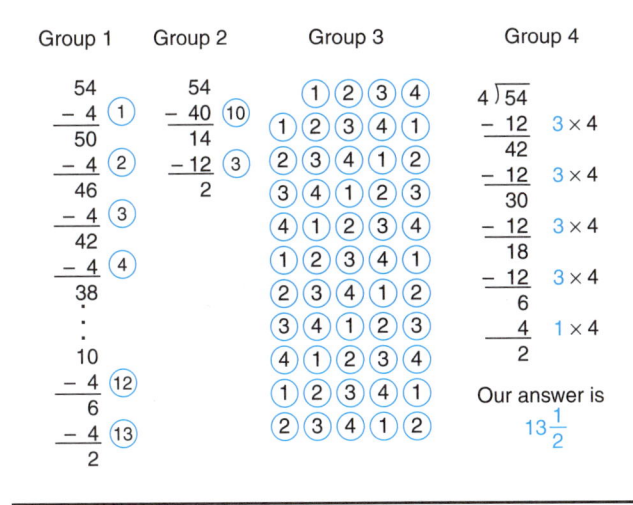

The *subtractive algorithm*, although not as familiar as the distributive algorithm, provides an intuitive and straightforward method for helping children learn to divide. This algorithm is easier than the standard algorithm for most children to learn because it is flexible enough to be made to correspond to a wide variety of approaches to thinking about division. Refer to Figure 11-10b for an example of how the subtractive algorithm might be used to solve this same division problem. You have $954 to share among four people. Perhaps begin by giving each person $50 (that's 200 shared out so far). Subtract and see that you still have $754 to share. Obviously, you didn't need to start with such a small share. Now choose to give each person $100 (that's 400 to share). You still have $354 to share. You might give each person $50 again, leaving $154 still to share. You don't have enough to share $50 again, so maybe give each person $30. That leaves $34. Now give each person $8, leaving $2 as remainder. The total each person received was $50 + $100 + $50 + $30 + $8, or $238, the same answer you obtained above with the standard algorithm. The choices made here in using the subtractive algorithm were arbitrary. You might instead have chosen to repeatedly distribute $30, over and over, and in that case the problem would take a long time to complete (see Figure 11-10c). On the other hand, if you were quite efficient, the solution would look very much like the standard algorithm solution (see Figure 11-10d).

The thinking process used in the subtractive algorithm is also clearly illustrated by the work of groups 2 and 4 in the previous problem involving marbles. Both groups were computing 54 ÷ 4. Group 4 began by recognizing that 12 marbles could be shared fairly among four people (three to each). So they first shared just 12 marbles (leaving 42 still to be shared). Then they repeatedly shared 12 more each time (leaving 30, then 18, then 6 still to be shared). Finally, they

Figure 11-9 *Solutions of groups 1–4.* (Source: Burns (1991).)

Figure 11-10 *Examples of the distributive division algorithm (a) and the subtractive division algorithm (b–d).*

DISTRIBUTIVE ALGORITHM Clearly, there are many ways to figure and record the answer to a division computation, and certain methods are more efficient than others. Two algorithms have been used for division most frequently; both are effective. The *distributive algorithm* is most common and familiar and is considered the standard division algorithm by most Americans. Consider the problem of sharing $954 among 4 people (refer to Figure 11-10a).

shared four of those final six, leaving two marbles (which, amusingly, they cut in half to complete the sharing). Because each person was awarded three marbles four distinct times (for a total of 12 marbles), and also was awarded 1 more marble plus 1/2 marble, the solution can be written 54 ÷ 4 = 13½. By contrast, group 2 initially thought about sharing 40 of the 54 marbles among the four people (10 to each). That left 14 still to be shared. Those 14 could be distributed 3 per person. Thus each person got 10 + 3 marbles (or 13), and 2 marbles were left over.

In using the subtractive algorithm, the child can choose to subtract off any multiple of the divisor at each step. Although the process may take more or less time, depending on how efficient the thinking is, the answer will be correct no matter how many steps are required. If children are extremely efficient, the subtractive algorithm can look almost exactly like the traditional division algorithm. Thus, the subtractive algorithm can serve as a transitional algorithm (a stage on the way to learning to divide using the traditional method), or children may learn it as an alternative algorithm.

Although teachers can help children understand algorithms such as those described previously, they also need to let them explore. As with the other operations, children are likely to develop their own algorithms. Many students will use number sense, reasoning in terms of the numbers involved, as the following examples indicate:

52 ÷ 7 52 divided by 7 is close to 49 divided by 7, or 7. But there are 3 left over. So the answer is 7 remainder 3.

85 ÷ 5 20 divided by 5 is 4. There are four 20s in 80, so 80 ÷ 5 is 4 times as much as 20 ÷ 5.4 × 4 = 16. And then there is one more 5 in 85. So the answer is 17.

69 ÷ 4 Think 60 + 9. 60 divided by 4 is 15, and 9 divided by 4 is 2, with a remainder of 1. So the answer is 17 remainder 1.

Have you begun to believe that the possibilities for constructing ways to find an answer are myriad? Just when you think you've seen all the possible algorithms, another student will probably come up with a new one. Enjoy this process. Be an explorer right along with the children. (Some short, but informative articles about division algorithms and making sense of division include those by Boerst, 2004; Fuson, 2003a; Gregg and Gregg, 2007; Hedges, Huinker, and Steinmeyer, 2005; Martin, 2009; and Sellers, 2010.) But, as a teacher, you also have another role. Just because all these algorithms that we've been exploring are correct doesn't mean that children never come up with incorrect algorithms. They do. You need to watch for this possibility—and you need to make children aware of the need to prove that a procedure really works across examples. You also need to be aware of consistent error patterns that crop up in children's work—error patterns that usually indicate some difficulty such as lack of knowledge about place value,

or a lack of mastery of some basic facts, or some other misunderstanding. Fortunately, there are sources of information, such as Ashlock (2010), to which you can turn. This source not only helps you identify the patterns but also provides specific suggestions for how to help the child get onto the right track.

As with the other operations, it is important that children work with manipulative materials as well as place-value charts. They need to keep in mind the problem situation and not forget that it is the process of sharing that is of concern, not merely working their way through the algorithm. A variety of experiences, such as those in In the Classroom 11–6 and 11–7, may help children gain facility with the division algorithm.

In the Classroom 11–6

THE MISSING DIGITS GAME

Objective: Using a game to develop logical reasoning and number sense involving division.

Grade Level: 3–5.

▼ Try this game with several friends who have been divided into two teams.

- Teams take turns selecting the leader for a round of the game.
- The leader starts a round by working out a division problem and putting the pattern on the board.
- Players on each team take turns asking about the digits that go in the boxes. For example, a player on Team A might ask if there are any boxes containing 2s; then a player on Team B might ask if there are any containing 7s.
- If a digit appears in the problem, the leader writes it in the appropriate box(es).
- A team scores a point for each box that gets filled in during its turn. If there are no boxes for a digit, the team scores 0 for that turn.
- The winner for the round is the team with the highest score. (If a team guesses the problem before all the boxes are filled in, it wins the round and gets bonus points for the number of boxes that are still empty.)

▼ Play several rounds to determine which team wins the game!

In the Classroom 11–7

MAKING EXAMPLES

Objective: Using number puzzles to develop number sense involving division.

Grade Level: 3–5.

▼ Make (and work out) a division example with
  - A dividend of 47 and a divisor of 3
  - A dividend of 81 and a divisor of 5

▼ Now make (and work out) a division example with
  - A quotient of 6 r2
  - A quotient of 10 r4
  - A quotient of 23 r5

▼ Try these:
  - A divisor of 6 and a quotient of 15 r3
  - A divisor of 3 and a quotient of 25 r2
  - A dividend of 83 and a quotient of 11 r6

I know I can do this..

Determining a reasonable answer can be particularly difficult with the division algorithms, but questions can help students make sense of them. A necessary first step in determining a quotient is to estimate the size of the quotient. Consider 839 ÷ 6:

Are there as many as 10 sixes in 839? [Yes, 10 sixes are only 60.]

Are there as many as 100 sixes? [Yes, 100 sixes are 600.]

Are there as many as 200 sixes? [No, 200 sixes are 1200.]

So the quotient is between 100 and 200—and probably closer to 100.

Consider another example, 187 ÷ 3:

Are there enough hundreds that we could divide them into three piles? [No. We have only 1 hundred.]

Are there enough tens? [Yes, 18 tens—from 1 hundred and 8 tens, and 18 tens ÷ 3 = 6 tens.]

So the quotient has two digits [and there are 6 tens].

So you know the answer is between 60 and 70.

Developing the habit of mentally asking these sorts of questions helps students recognize the range for a quotient, and mental questions such as these also help students make sense of the division algorithm, while developing valuable estimation skills.

DIVISION WITH TWO-DIGIT DIVISORS

Work with two-digit divisors should aim toward helping children understand what the procedure involves but not toward mastery of an algorithm. The calculator does the job of multi-digit division for most adults, so there is little reason to have children spend months or years mastering it. Other mathematics is of more importance for children to learn.

The development of division with two-digit divisors proceeds through stages from concrete to abstract, paralleling the work with one-digit divisors. Much practice may be needed with the symbolic form if proficiency is the goal. Use of the calculator is interwoven into the activities, as indicated in In the Classroom 11–8, where the calculator is used to strengthen understanding of the relationships between numbers.

In the Classroom 11–8

EASY DOES IT!

Objective: Using calculators to develop number sense involving division.

Grade Level: 3–5.

- Use your calculator to find the missing numbers:

  45) 159 r13 ___

  23) 69783 ___ r ___

  ___) 7683 98 r39

  37) 16972 ___ r ___

  ___) 4745 46 r7

  73) ___ 89 r 3

MAKING SENSE OF DIVISION WITH REMAINDERS

Children should encounter division problems involving remainders from the time they begin to work with division ideas (just as they should encounter addition involving regrouping from the outset). It is artificial and counterproductive to begin teaching division with problems that always come out even, saving problems involving remainders for later. As long as problems remain on a concrete level, the concept of remainder is rather easy. Here are some examples of varied situations:

- Pass out 17 candies to 3 children. (Each child receives 5 candies with 2 candies left over. Or, if the candies can be cut into pieces, each child could have 5 candies plus 2/3 of a candy.)
- To make each valentine's card, you need 3 pieces of lace. You have 17 pieces of lace. How many cards can you make? (You can make 5. You'll probably have to discard the two extra pieces of lace, or save them for another project.)
- If 17 children are going on the class trip and 3 children can ride in each car, how many cars are needed? (You will need 6 cars. With 5 cars you could seat only 15 children, with 2 children still waiting for a ride. With 6 cars, you can seat all 17 children, with 1 seat left over.)
- Separate a class of 17 children into 3 teams. (There can be 5 children on each team, and the 2 children left over can be the scorekeepers, or perhaps you could make two teams of 6 and 1 team of 5.)

Note that the remainder is handled differently in each of these real-world problems. In the candy problem, both the whole number (5) and the remainder are reported as parts of the answer. In the valentine problem, the answer involves only the whole-number part (the remainder must be discarded or ignored). In the class trip problem, the quotient is rounded up because the remainder cannot be ignored (no matter how small). In the team problem, children need to find some other use for the remainder (or they may abandon the attempt to form equal-sized groups if they want everyone to be able to play).

Thus, it is important for children to think about and be able to deal with remainders as they appear in real-world problems. With calculators, the results of inexact division are expressed in decimal form, and children need to learn how to interpret the remainder when it is not an integer (see Chapter 13, and also see the Snapshot of a Lesson in Chapter 5).

Initially, children are taught to write the remainder in one of the following ways:

$$6)\overline{13} \quad \text{with 1 left over}$$
$$\underline{12}$$
$$1$$

$$5)\overline{27} \quad \text{remainder 2}$$
$$\underline{25} \quad \text{(later shortened to R2 or r2)}$$
$$2$$

It is important to emphasize the real-life situations from which examples arise and to decide whether a "remainder of 2" makes sense in that situation. Activities such as the game in In the Classroom 11–9 provide practice in identifying the remainder.

Calculators can be used to solve problems with remainders, but the correct answer depends on the thinking a child does. Consider this problem:

A bus holds 36 children. If 460 children are being bussed to a concert, how many buses do we need?

In the Classroom 11–9

THE REMAINDER GAME (FOR 2 TO 4 PLAYERS)

Objective: Using a game to practice mental computation involving division.

Grade Level: 3–5.

Materials:
- A copy of this gameboard on heavy paper
- Four cards of each numeral 1 to 9
- A counter for each player

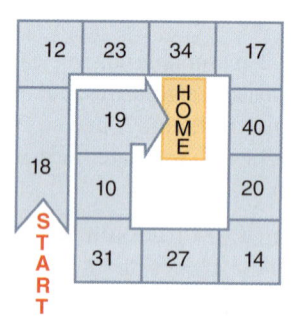

12	23	34	17
18	19	HOME	40
	10		20
START	31	27	14

Game Rules:

1. Place the cards face down in a pile.
2. The first player draws a card.
   - For a player's first turn, divide the first number on the board by the number drawn.
   - For a player's additional turns, the dividend is the number on the space where his or her counter landed on the previous turn.
3. Move the counter forward by the number of spaces indicated by the remainder. If the remainder is 0, no move is made.
4. Each of the other players go in turn.
5. To get "home," a player must be able to move the exact number of spaces left. The first person home wins!

A more detailed plan for this activity is available on the Student Companion site at www.wiley.com/college/reys.

Research shows this problem to be difficult for children to solve with or without a calculator. An answer of 12 remainder 28 is often reported without a calculator, and 12.777 buses is often reported with a calculator. The children seem more intent on producing an answer than on deciding if their answer makes sense. A sensible answer would be 13 buses (12 buses will be full and the last bus will have some empty seats).

CHECKING

Just as it is important to estimate before computing, it is important to check after computing. Ordinarily, addition and subtraction are used to check each other, as are multiplication and division. Unfortunately, checking does not always achieve its purpose of ascertaining correctness. In fact, if you talk with children as they perform the computational check, you may find that they frequently force the check (i.e., make the results agree without actually performing the computation). Children need to understand the purpose of checking as well as what they must do if the solution in the check does not agree with the original solution.

The calculator can serve many other functions, but its use in checking has not been overlooked by teachers. Nevertheless, the calculator should not be used primarily to check paper-and-pencil computation. It insults students to ask them to spend large amounts of time on a paper-and-pencil computation and then use a machine that does the computation instantly. Encourage estimation extensively, both as a means of identifying the ballpark for the answer *and* as a means of ascertaining the correctness of the calculator answer. The words "use your calculator to check" should generally be used only when the calculator is to perform the computation following an estimate.

CHOOSING APPROPRIATE METHODS

As the NCTM *Standards* and the discussion in Chapter 10 make very clear, children must learn to choose an appropriate means of calculating. Sometimes paper and pencil is better; sometimes mental computation is more efficient. Other times use of a calculator is better than either, and sometimes only an estimate is needed. Encouraging students to defend their answers often yields valuable insight into their thinking. Children need to discuss when each method or tool is appropriate, and they need practice in making the choice, followed by more discussion, so that a rationale for their choice is clear. They need to realize that this is a personal decision. A problem that one child chooses to do with a calculator may be done with mental computation by another.

BUILDING COMPUTATIONAL PROFICIENCY

Computational fluency with addition, subtraction, multiplication, and division is an important part of mathematics education in the elementary grades; however, "developing fluency requires a balance and connection between conceptual understanding and computational proficiency" (NCTM, 2000, p. 35). Research has demonstrated that extensive practice without understanding is often either forgotten or remembered incorrectly (Hiebert, 1999; Kamii and Dominick, 1998). On the other hand, practice is important to help students gain the computational fluency desired. Teachers must carefully provide practice opportunities intertwined with the development and maintenance of conceptual understanding. If students use reasoning, their practice is more likely to help them develop computational fluency.

Many kinds of computer software are available on the market. The challenge is to identify programs that require more than mindless practice with the algorithms. You will find that most programs are typically visually appealing, and they use a variety of characters in a story format; however, few programs involve problem solving or estimation. Some involve the use of games to motivate the child's participation. The question you should ask is whether the software offers advantages over the materials you already have available. Your search for useful computer software must be guided by the mathematics, not by the "bells and whistles" features.

Worksheet activities are available for motivating students to practice their skills with computation. Some worksheets feature a mystery for the students to solve using the answers obtained by completing the computations. In the Classroom 11–10 illustrates one example of this type of worksheet. In this example, the answers to some of the multiplication problems provide clues to identifying the mystery number. The student must organize these clues to help identify the mystery number. While these types of worksheets provide needed practice, overuse can lead to mindless practice rather than actually helping students gain computational fluency.

CULTURAL CONNECTIONS

The computational algorithms typically taught in the United States are not universal. In different parts of the world, people have different ways of doing paper-and-pencil computation. If you have children in your classroom who have attended school in other countries or whose parents were educated outside the United States, particularly in Europe or in South America, you may find confusion arising when those parents attempt to help their children with computation. It helps if you, as the teacher, are well aware of the existence and validity of alternative algorithms. Asking students to explain their methods can help you uncover the reasons for errors that may arise if students confuse the steps from different algorithms. Sometimes you may need to do some careful thinking to figure out how an unfamiliar algorithm works and why it is valid.

For example, the long division problem 360 ÷ 8 is typically taught this way in American schools:

$$\begin{array}{r} 45 \\ 8\overline{)360} \\ \underline{32} \\ 40 \\ \underline{40} \\ 0 \end{array}$$

This same computation might look like this in some other countries:

$$\begin{array}{r} 360\ \overline{)8} \\ 40\ \ 45 \\ 0 \end{array}$$

Sometimes this method is known as *Cuban division* because it was particularly noticed in the United States when many Cubans immigrated to Florida in the 1960s, although this same notation is also used in other cultures, including Arabic, Lao, and Vietnamese. And people in the Netherlands and in Poland also use a similar algorithm (Hirigoyen, 1997).

Children schooled in other countries may also use different methods for subtraction. In particular, it is helpful to be familiar with the *equal-additions* or *add-tens-to-both* method used in many European and Latin American countries. In fact, before the now-standard decomposition algorithm for subtraction gained prominence in this country, the *equal-additions algorithm* was taught here, too. The equal-additions algorithm is still taught and used in many other countries around the world (Ellis and Yeh, 2008; Ron, 1998; Whiteford, 2009/10). In the equal-additions subtraction algorithm, both the sum (the number you are subtracting from) and the known addend (the number you are subtracting) are renamed. In this example, ten is added to each number involved in the subtraction: here 10 ones have been added to the sum (91) and 1 ten has been added to the known addend (24):

$$\begin{array}{r} \overset{1}{9}1 \\ \underset{3}{-2}4 \\ \hline 67 \end{array}\quad \begin{array}{l} \textit{Think} \\ 11 - 4 = 7 \text{ ones} \\ 9 \text{ tens} - 3 \text{ tens} = 6 \text{ tens} \end{array}$$

The computation has been changed from 91 – 24 to 90 – 30 and 11 – 4 (which is easier to do). This algorithm works because the difference between the numbers (91 and 24) remains the same after both have been increased by the same amount [i.e., 91 – 24 = (91 + 10) – (24 + 10)]. It may help to picture a number line. The distance between 91 and 24 on the number line is what you are finding when you subtract. If you move both numbers up the number line the same amount (10), the distance between them remains the same. The reason for adding 10 to each number (but adding it in different forms) is to make the subtraction possible without any borrowing or regrouping. Although this algorithm may seem strange to many Americans, people who have learned to subtract this way generally find it easy and natural to use. What is important is that individuals understand *what* they are doing and *why* when they learn an algorithm. A classic research study established that both the typical American decomposition algorithm and the equal-additions algorithm are effective in terms of speed and accuracy when taught meaningfully (Brownell, 1947). A third subtraction algorithm is used by the Japanese, and it also involves doing some addition as part of the procedure for subtracting. To do 531 – 398

In the Classroom 11–10

WHAT NUMBER AM I?

Objective: Practicing with multiplication to solve a mystery.

Grade Level: 3–4.

- **Directions**: Solve each of the multiplication problems. Some of the solutions can help you identify the clues to help you find the mystery number. Circle the answers that are solutions to some of the problems; cross out the other clues. Use the clues to find the mystery number.

1. 15 ×9	2. 51 ×21	3. 63 ×24	4. 68 ×21
5. 77 ×21	6. 18 ×15	7. 22 ×14	8. 69 ×66
9. 18 ×9	10. 72 ×9	11. 88 ×18	12. 34 ×31

Answer	Clue
162	I am less than 100.
189	I am more than 100.
1502	I am an even number.
1071	I am an odd number.
1085	I am divisible by 3.
1308	I am divisible by 4.
280	I am divisible by 5.
647	I am a multiple of 3.
4554	I am a multiple of 5.
1607	I am a multiple of 10.
208	The sum of my digits is an even number.
1428	The sum of my digits is an odd number.

using the Japanese method, you first determine the "nines complement" of 398, then add that number to 531, then make an adjustment to get the final answer. The "nines complement" of any number is found by subtracting each of its digits from 9, so the "nines complement" of 398 is 601 (because 999 − 398 = 601). Add 601 to 531, and you get 1132. Remove the 1 from the front (thousands place), and add 1 to the ones place to get 133, and this is the answer to the desired computation: 531 − 398 = 133. (Why does this work? This question is left for you to ponder. *Hint*: At first you are actually figuring 999 − 398 = 601. What happens next in the algorithm?)

Many other interesting algorithms are used throughout the world. Recognizing their validity and acknowledging children's rights to use them is important in supporting the cultural backgrounds of all students. (For further reading about this idea, see articles by Civil, Planas, and Quintos (2005), Ellis and Yeh (2008), Philipp (1996), and Whiteford, 2009/10).

A GLANCE AT WHERE WE'VE BEEN

Although computational skill is viewed as an essential component of children's mathematical achievement, its role in the curriculum and the methods of teaching it are changing. Helping children learn computational algorithms involves encouraging them to reason their own way through problems requiring basic computations. Important prerequisites to developing computational fluency include knowing basic facts, understanding place value, and using manipulative materials and drawings as models. We have not only provided suggestions for developing the standard (familiar) algorithms for each of the four operations but also described many transitional and alternative algorithms. The use of calculators has been interwoven with activities that use the algorithms as well as activities using other materials.

Things to Do: From What You've Read

1. What prerequisites must a child have in order to succeed in the chapter's opening Snapshot of a Lesson?

2. Analyze each of the following student-invented algorithms. Do they work all the time? If so, do two more examples using that method. If not, explain why the method does not work.

   a. Subtraction—uses missing addend addition:

   $$\begin{array}{r}3452\\-1784\\\hline\end{array} \rightarrow \begin{array}{r}1\\1784\\+8\\\hline 3452\end{array} \rightarrow \begin{array}{r}11\\1784\\+68\\\hline 3452\end{array} \rightarrow \begin{array}{r}111\\1784\\+668\\\hline 3452\end{array} \rightarrow \begin{array}{r}111\\1784\\+1668\\\hline 3452\end{array}$$

   b. Subtraction—subtract from 10 when regrouping, then add on:

   $$\begin{array}{r}410\\3452\\-1784\\\hline 6\\+2\end{array} \rightarrow \begin{array}{r}10\\3410\\3452\\-1784\\\hline 26\\+42\end{array} \rightarrow \begin{array}{r}10\,10\\2\,3\,4\,10\\3452\\-1784\\\hline 326\\+342\end{array} \rightarrow \begin{array}{r}10\,10\\2\,3\,4\,10\\3452\\-1784\\\hline 1326\\+342\\\hline 1668\end{array}$$

   c. Subtraction—work left to right, crossing out when regrouping is needed:

   $$\begin{array}{r}3452\\-1784\\\hline 2\end{array} \rightarrow \begin{array}{r}1\\3452\\-1784\\\hline \cancel{2}7\\1\end{array} \rightarrow \begin{array}{r}11\\3452\\-1784\\\hline \cancel{2}\cancel{7}7\\16\end{array} \rightarrow \begin{array}{r}111\\3452\\-1784\\\hline \cancel{2}\cancel{7}\cancel{7}8\\166\end{array}$$

   d. Division—uses distributive property to subtract twice for each multiplication by two-digit divisor:

   $$\begin{array}{r}2058r20\\23\overline{)47354}\\40\\7\\6\\135\\100\\35\\15\\204\\160\\44\\24\\20\end{array}$$

3. Sketch a diagram to show how 12 × 17 can be modeled with an array on grid paper. Relate the model to the partial-products algorithm for 12 × 17, showing how each of the partial products in the array relates to the model.

4. Consider problems involving subtraction with regrouping, such as 52 − 25, 91 − 79, and 47 − 18. How would you approach these problems if you were teaching second grade? What do you think children need to understand or be able to do before they can start learning subtraction with regrouping? (You might refer to this chapter's Snapshot of a Lesson for ideas.)

5. Assume you are a sixth-grade teacher. You notice that several of your students are making the same mistake when multiplying multidigit numbers. When computing 15 × 23 or 173 × 234, they seem to forget to "move over the numbers" on each line. They are writing:

$$\begin{array}{r}23\\ \times 15\\ \hline 115\\ 23\\ \hline 138\end{array} \quad \text{instead of} \quad \begin{array}{r}23\\ \times 15\\ \hline 115\\ 23\\ \hline 345\end{array} \quad \text{and} \quad \begin{array}{r}234\\ \times 173\\ \hline 707\\ 1638\\ 234\\ \hline 2574\end{array} \quad \text{instead of} \quad \begin{array}{r}234\\ \times 173\\ \hline 707\\ 1638\\ 234\\ \hline 40482\end{array}$$

What will you do to help these students? How will you help them correct their mistake?

6. For the problem 4578 ÷ 7, find the quotient by using the subtractive algorithm. Then do the problem again, using at least three more steps or three fewer steps than you used the first time to illustrate the flexible nature of this algorithm.

7. Write two story problems involving 35 ÷ 6 where the answers appropriate for the real world would actually not be the exact answer to the computation ($5\frac{5}{6}$). For your first story problem involving 35 ÷ 6, the appropriate answer should be 5. For your second story problem involving 35 ÷ 6, the appropriate answer should be 6.

Things to Do: Going Beyond This Book

In the Field

1. [1]*Computation in the Textbook and Classroom.* Choose a textbook for the grade level of a classroom you can visit and analyze a lesson plan on computation in the teacher's guide. What stages from concrete to abstract are involved in the lesson? Talk to the teacher about how he or she teaches computation. Is the textbook approach supplemented? How? What manipulatives, if any, are used? How?

2. [1]*Addition with Regrouping.* Develop three evaluation items that would assess students' understanding of adding two two-digit numbers with regrouping. Explain why you chose the examples that you did. Try your items with several children and analyze their responses.

3. [1]*Transitional Algorithms.* Ask a classroom teacher if you can spend some time working with a student who is having trouble with paper-and-pencil computation (addition, subtraction, multiplication, or division, depending on age and grade level). See if you can figure out what the child's problem is. You might try helping him or her by suggesting the use of one of the transitional algorithms described in this chapter.

In Your Journal

4. Some people wonder why it makes sense to encourage students to invent their own computational algorithms, since students are unlikely to discover the standard algorithms on their own. The standard algorithms, which are generally quick and efficient, were refined through centuries of invention and modification. Why not just teach the standard (traditional) algorithms directly to students? Answer this question in your journal—either agreeing or disagreeing that standard algorithms should be taught directly, and providing rationale for your position.

5. Consider the order in which the four operations have been introduced in this chapter. Do you think the order in which they are taught to students matters? Why or why not? Analyze and write about the relationship among the four operations.

6. "Now that we have calculators, long-division algorithms should be abolished." Write a convincing argument, pro or con.

With Additional Resources

7. Choose a textbook for grade 3 or 4. Trace the development of multiplication algorithms. How are they introduced? What steps do children go through? Or choose a textbook series and look through the books for grades 3–6 to trace the development of division algorithms. At what grade level do whole number division algorithms first appear? What is the highest grade level where instruction in whole-number division is included? Does this textbook series include attention to student-generated algorithms or to algorithms other than the traditional long-division algorithm? If so, describe.

8. Read "Creative Arithmetic: Exploring Alternative Methods" (Ellis and Yeh, 2008)—which shows examples of some different ways to subtract and multiply. Create a two-digit subtraction problem and a three-digit subtraction problem, and show how you would do them using the method described in this article. Create a two-digit multiplication problem, and show how you would do it using the method described in the article. Ask a parent, grandparent, neighbor, or friend if he or she knows another way to do subtraction or multiplication. Have him or her show you an example. Create another problem, and do it using that method. Why does that method work?

9. Read "Computational Fluency, Algorithms, and Mathematical Proficiency: One Mathematician's Perspective," an article by the noted mathematician,

Hyman Bass (2003). Using the five qualities that Dr. Bass considers important in evaluating the usefulness of computational algorithms, write an analysis of the relative usefulness of three multiplication algorithms: the partial-products method, the lattice multiplication method, and the traditional method.

With Technology

10. *Drill and Practice Software*. Find a drill-and-practice software program that is available on the Web for computations of your choice (+ , − , ×, ÷). Learn to use this program and describe how you might use this with a class when teaching algorithms such as those developed in this chapter.

11. *Calculators*. In division, the calculator provides a decimal remainder. For example, for 476÷23= 20.69565217. Describe two different ways you could find the fractional remainder using a simple calculator.

12. *Internet Software* for Worksheets. Many Web sites are available to help teachers create math worksheets. Find one of these sites where you can create a worksheet for students to practice computational skills. Explain when and why you would recommend use of these worksheets, and how you would decide which numbers to use.

Note to Instructors: You can find additional resources, learning activities, and blackline masters in this text's accompanying Instructor's Manual, at www.wiley.com/college/reys.

---

[1]Additional activities, suggestions, and questions are available in the field experiences manual on the Student Companion site at www.wiley.com/college/reys.

Book Nook for Children

Anno, M. *Anno's Magic Seeds.* New York: Philomel Books, 1999.

Jack meets a wizard who gives him two magic seeds and instructs him to eat one, which will sustain him for a full year, and to plant the other. The following spring, the plant bears two seeds. Jack eats one and plants the other, as he does for several years until he decides to plant both seeds. The next year he has two plants each bearing two seeds; he eats one seed and plants the other three. Six seeds! As the years go by, he marries, raises a family, plants many crops, endures a flood, and saves enough seeds to feed his family and start planting again.

Calvert, P. *Multiplying Menace: The Revenge of Rumpelstiltskin.* (Illustrated by Wayne Geehan). Watertown, MA: Charlesbridge Publishing Co., 2006.

After ten years of being tricked, Rumpelstiltskin returns to the royal family to wreak vengeance using multiplication (by whole numbers). Fractions are used in the story, too.

Dodds, D. A. *The Great Divide.* Cambridge, MA: Candlewick Press, 1999.

Eight people begin to race the cross-country race using bicycles. As they race, the great divide begins. New challenges occur and each time they do, the number of racers is divided in half.

Murphy, S. J. *Too Many Kangaroo Things To Do*. New York: HarperCollins, 1996.

A kangaroo is getting ready to have a birthday. His friends are planning a surprise party for the kangaroo. The reader will see how to add and multiply many different things that they will have at the party. There are activities and games in the back of the book for teachers and parents to use.

Murphy, S. J. *Divide and Ride.* New York: HarperCollins, 1997.

A group of 11 friends go to a carnival. At the carnival, some of the rides take two people to a seat and, therefore, they must divide themselves for the ride. The book uses different algorithms to show how they divide themselves for the rides as well as for tickets and other carnival activities. There are activities and games to use in the back of the book for teachers and parents to use.

CHAPTER 12

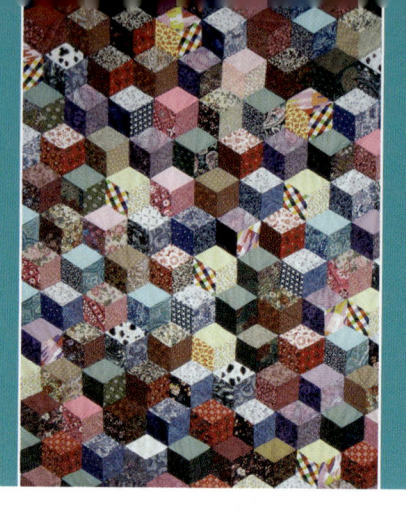

Fractions and Decimals: Concepts and Operations

> "FIVE OUT OF EVERY FOUR AMERICANS HAVE DIFFICULTY WITH FRACTIONS."
>
> — Unknown

>> SNAPSHOT OF A LESSON

KEY IDEA

1. Extend students' concept of fractions to include regions with equal areas.

BACKGROUND

This lesson is based on the video *Fractions with Geoboards*, from Teaching Math: A Video Library, K–4, from Annenberg Media. Produced by WGBH Educational Foundation © 1995 WGBH. All Rights reserved. To view the video, go to http://www.learner.org/resources/series32.html (video #37). Constance Richardson is one teacher of a class of 40–50 students in Tucson, Arizona. In this video, the class is examining various ways of representing the halves of a large square on a 5×5 geoboard.

Ms. Richardson: What do we mean by halves?

After eliciting students' responses such as "two of the same size," "two equal parts," and "two equal lengths," Ms. Richardson asks the students how they would show halves on a geoboard. The students give the first example shown. They discuss how to tell that the two rectangles are equal parts of the large square. One student focuses on the number of pegs, others say they look alike, while others consider the size of the rectangles. Ms. Richardson then asks what they mean by the same size, and this leads to counting the small squares. They establish that the area of each part (the halves) is 8 small squares.

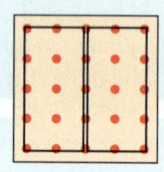

The students' next example involves cutting the square on the diagonal to make two triangles. This leads to an interesting discussion of whether or not the area of each triangle is 8 small squares. Watch the entire video to see how the students establish this.

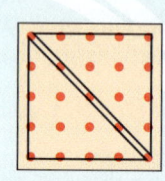

Ms. Richardson then challenges the students to find as many ways as they can to show halves. One group comes up with this example, saying, "The two parts looked the same." Ms. Richardson has them think about area. They see that each part consists of 6 small squares and 2 triangles. They put the 2 triangles together and see that the 2 triangles are equivalent to 2 small squares.

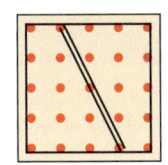

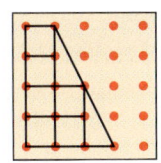

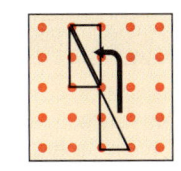

Students come up with several other ways of showing halves of the large square, as illustrated below. Ms. Richardson has the students record their solutions on geopaper (see Appendix A) and then try to convince the class that their solutions do, in fact, show halves.

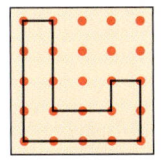

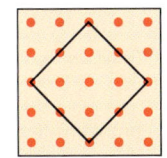

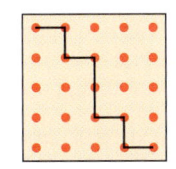

FOCUS QUESTIONS

1. What are three meanings of fractions, and what are models of the part–whole meaning?
2. How can you help children make sense of fractions, and how can you use concrete and pictorial models to develop children's understanding of ordering fractions and equivalent fractions?
3. Describe how children can use estimation strategies for adding and subtracting by rounding to whole numbers or using benchmark numbers to determine the reasonableness of answers to fraction and decimal problems.
4. How can models assist the development of children's conceptual understanding of adding, subtracting, multiplying, and dividing fractions or decimals?

INTRODUCTION

The opening quote, "Five out of four Americans have difficulty with fractions" is indicative of the reaction to fractions by many people. Fractions and decimals have long been a stumbling block for many students. One reason may be that curricula tend to rush to symbolization and operations without developing the strong conceptual understandings that children must have for fractions and decimals. Therefore, much of this chapter is devoted to helping children develop the concepts they need for working with fractions and decimals.

There is some controversy over the importance of studying fractions, since technology uses decimals almost exclusively; however, common fractions are important in daily life and necessary for further study in mathematics, so current recommendations are that the development of computational fluency should include that with common fractions (Kilpatrick et al., 2001). The recent *Core Content State Standards for Mathematics* (CCSSM) place more emphasis on fractions at Grades 3-6, but less in the lower grades than do many present standards in the states. An overview of the CCSSM standards will be given in the various sections of this chapter. You can reference the entire statements as indicated in Math Links 12.1.

Math Links 12.1

The full text of the *Common Core State Standards for Mathematics* can be accessed at http://www.corestandards.org/ or from this book's Web site.

www.wiley.com/college/reys

In this chapter, the initial concentration on fractions does not imply that the entire study of fractions precedes the study of decimals. Teachers can introduce decimal notation after a beginning foundation has been built with fractions. In fact, adding and subtracting decimals are easier than the corresponding operations with fractions and can be taught meaningfully before children learn to add and subtract fractions.

CONCEPTUAL DEVELOPMENT OF FRACTIONS

The *Principles and Standards for School Mathematics* (NCTM, 2000) emphasizes that students should be given the opportunity to develop concepts as well as number sense with fractions and decimals. A careful examination of the items on the mathematics assessment of the National Assessment of Educational Progress (NAEP) reveals that fractions concepts are not well developed by U.S. students (Wearne and Kouba, 2000).

The concepts associated with fractions are complex; however, two rather simple but powerful ideas—*partitioning* and *equivalence*—can help tie many of the concepts together. Partitioning refers to sharing equally (sharing a cake equally among eight people means partitioning the cake into eight equal portions; sharing six candy bars equally among four people means partitioning six candy bars into four equal portions). Two fractions are equivalent if they represent the same amount.

Figure 12-1 is an overview of the recommended expectations in the CCSSM; we will relate to these recommendations in the remainder of this section.

We begin this section by examining three meanings of fractions and models associated with one of these meanings. This is background information for you; we follow with details of how to help children make sense of fractions. We then examine comparing and ordering fractions which leads naturally to equivalent fractions. Although we close

	Number and Operations-Fractions Domain
Grade 2	Partition circles and rectangles into two, three, or four equal shares, describe the shares using the words *halves, thirds, half of, a third of*, etc., and describe the whole as two halves, three thirds, four fourths. Recognize that equal shares of identical wholes need not have the same shape. (from Geometry Domain)
Grade 3	Understand a fraction *1/b* as the quantity formed by 1 part when a whole is partitioned into *b* equal parts; understand a fraction *a/b* as the quantity formed by a parts of size *1/b*. Represent 1/b and a/b as numbers on a number line. Understand two fractions as equivalent if they represent the same amount. Compare fractions. Emphasis is on using visual models and like denominators. Recognize that comparisons are valid only when the two fractions refer to the same whole. Record the results of comparisons with the symbols >, =, or <, and justify the conclusions.
Grade 4	Extend understanding of comparing fractions with different denominators. Explain why a fraction *a/b* is equivalent to a fraction $(n \times a)/(n \times b)$ by using visual fraction models. Use this principle to recognize and generate equivalent fractions. Compare two fractions with different numerators and different denominators, e. g., by creating common denominators or numerators, or by comparing to a benchmark fraction such as 1/2.
Grade 5	Interpret a fraction as division of the numerator by the denominator. Solve word problems involving division of whole numbers leading to answers in the form of fractions or mixed numbers.

Figure 12-1 *Summary of the CCSSM standards relating to fraction concepts.* This summary was made by the authors of *Helping Children Learn Mathematics*. The full document (CCSSI, 2010) may be obtained at www.corestandards.org/the-standards/mathematics.

this section by examining improper fractions and mixed numbers, we recommend that the study of these fractions be used as the other concepts are developed.

THREE MEANINGS OF FRACTIONS

Three distinct meanings of fractions—part–whole, quotient, and ratio—are found in most elementary mathematics programs, but the focus is usually on the part–whole meaning, with little development of the other two meanings. Ignoring these other meanings may be one source of students' difficulty with fractions.

PART–WHOLE The part–whole meaning of a fraction indicates that a whole has been partitioned into equal parts—for example, the fraction $\frac{3}{5}$ indicates that a whole has been partitioned into five equal parts and that three of those parts are being considered. This fraction may be shown with a region model:

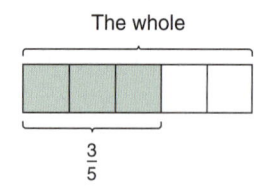

The CCSSM approach makes much use of unit fractions (fractions with numerators of 1) and being able to also look at this as three one-fifths. You will see how we use this subtle, but powerful idea in helping children make sense of fractions and operate with them.

QUOTIENT The fraction $\frac{3}{5}$ may also be considered as a quotient, 3 divided by 5. This meaning also indicates partitioning. For example, suppose you had 20 big cookies to share equally among 5 people. If you gave each person 1 cookie, then another, and so on until you had distributed all the cookies, you could represent this process mathematically by the quotient $20 \div 5$—each person would get four cookies.

But what if you had just 3 big cookies and you still wanted to share them equally among 5 people? Now the quotient would be $3 \div 5$. How much would each person get? Would anyone get a whole cookie? One way to solve this problem would be to use pictures of the cookies, as shown in Figure 12-2.

Begin with 3 cookies. Cut each into 5 parts

Each person gets $\frac{1}{5}$ of each cookie.
Thus, each person gets

$\frac{1}{5} + \frac{1}{5} + \frac{1}{5}$ or $\frac{3}{5}$ or $3 \div 5 = \frac{3}{5}$

Figure 12-2 *The quotient interpretation of a fraction.*

This meaning of fractions is used when a remainder in a division problem is expressed as a fraction. It is also the interpretation that is needed to change any fraction to a decimal (i.e., $\frac{5}{8}$ is equivalent to 0.625). The CCSSM recommends that this meaning be developed in grade 5. The basis of the lesson in the Snapshot of a Lesson in Chapter 2 is the problem posed in *The Doorbell Rang* (Hutchins, 1986) as grade 4 students are introduced to the division meaning of fractions. It does take time for students to realize these two different meanings of 3/5 (3 parts of 5 parts of a whole and 3 divided by 5).

RATIO The fraction notation $\frac{3}{5}$ may also represent a ratio—for example, in a class of 6 boys and 10 girls, the ratio of boys to girls is 6 to 10, which is equivalent to $\frac{3}{5}$. That is, for every 3 boys, there are 5 girls. What else could be the make up of a class with this ratio?

The ratio interpretation of fractions is conceptually different from the part–whole and quotient interpretations, since it doesn't involve the idea of partitioning. In this chapter, we focus on the part–whole interpretation and bring in the quotient interpretation as relevant (see Chapter 13 for discussion of the ratio interpretation). You may read about the measurement meaning (Flores, Samson, and Yanik, 2006), but we have not made a special section on this meaning since it is closely aligned with the measurable attributes discussed here.

MODELS OF THE PART–WHOLE MEANING

Attributes such as region, length, set, and area are useful concrete and pictorial models for the part–whole meaning of a fraction. You can also model the quotient meaning using any of these attributes; we have chosen to introduce the models with the part-whole meaning. Other attributes, such as capacity, volume, and time, also can model both meaning of fractions.

These models are not only used to develop the meanings of fractions but also to find equivalent fractions, to order fractions, and to solve problems. Often students do not realize that these models can be helpful in solving problems (Wearne and Kouba, 2000).

REGION The region is the most commonly used model for this interpretation of fractions because it is the most concrete and the most easily handled by children. In the region model, which is a special version of the area model, the region is the whole (the unit), and the parts are *congruent* (same size and shape). The region may be any shape, such as a circle, rectangle, square, or triangle. You should use a variety of shapes when presenting the region model so that the children do not think that a fraction is always a particular shape—for instance, "part of a pie" (if the region is a circle).

Some Types of Regions

Circle: Easy to see it is a whole; difficult to partition.

Rectangle: Easy to partition; difficult to know if it's a whole.

Triangle: Difficult to partition; difficult to know if it's a whole.

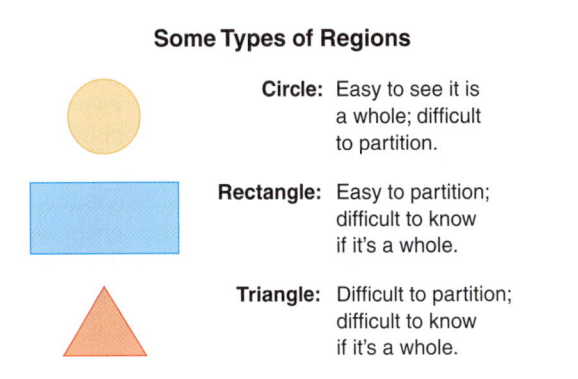

Figure 12-3 *Types of regions.*

A rectangle is the easiest region for children to draw, and as Figure 12-3 indicates, it is also the easiest region to partition. (Try partitioning each of the shapes shown into three equal parts to see which is easiest.) The circle has the advantage of being easy to see as a whole, but this does not outweigh the advantages of the rectangle, which we use extensively throughout this chapter.

LENGTH Any unit of length can be partitioned into fractional parts of equal length. Children can begin by folding (partitioning) a long, thin strip of paper into halves, fourths, and so on. Later, you can use length partitioning to represent fractions as points on a number line. For example, as shown in Figure 12-4, you could partition unit lengths into thirds to help children realize that the number $4\frac{1}{3}$ is $4 + \frac{1}{3}$.

Keep in mind, however, that young children need plenty of prior experience with the number line before they can understand it as a model for fractions. Fewer than one-third of fourth graders could identify the point on a number line that represented a given fraction, even when the number line was partitioned into the same number of parts as the denominator (Kloosterman et al., 2004). Because of the many subtleties involved in using a number line and the importance that the CCSSM recommendations place on the number line, we will discuss it more thoroughly in the section on making sense of fractions.

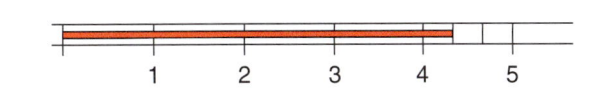

Figure 12-4 *Length model of $4\frac{1}{3}$.*

SET The set model uses a set of objects as a whole (the unit). This model sometimes causes children difficulty, partly because they have not often considered a set of, say, 12 objects as a unit. In the 1996 NAEP, about three-fourths of fourth-grade students chose the correct region model for a given fraction, whereas only about half chose the correct set model (Wearne and Kouba, 2000). It is encouraging to note that there was a significant improvement on both these tasks from 1996 to 2000 (Kloosterman et al., 2004). Another, more obvious reason why children have difficulty with the set model is that they have not had much experience with physically partitioning the objects in a set into thirds, fourths, fifths, and so on.

Without mentioning fractions, you should give children experiences with physically partitioning sets. This will give them a background for both division and fractions. For example, you could ask a student to share 12 toys equally among 4 children. Later, you should focus on whether a given number of objects can be partitioned equally among a given number of people. For example, you can ask whether 15 toys can be partitioned (shared) equally among:

5 people? [yes]

4 people? [no]

3 people? [yes]

2 people? [no]

Once children develop the understanding to deal with these questions, you can use the set model for fractions (e.g., finding fifths by partitioning the set into five equal parts). Figure 12-5 shows a set of 15 marbles that has been partitioned into 5 equal parts. Each part is one-fifth of the whole set. From this modeling, children develop the set meaning of fractions and can answer questions such as these:

What is one-fifth of 15? [3]

Two-fifths of 15? [6]

Three-fifths of 15? [9]

Experience of this type allows children to solve many practical problems and prepares them for multiplication of fractions. It also provides background experience for multiplicative reasoning as students determine the number of marbles, for example, in 4 groups of 3 marbles. This is an appropriate activity to include in grade 3 as children are building their concepts both of fractions and of multiplication and division.

AREA The area model is a more general version of the region model, in which the parts must be equal in area but not necessarily congruent. Before using this model, you must be sure that your students have some idea of when two different shapes have equal areas. This makes the model more appropriate for older children (about third and fourth grades) than for younger ones. The Snapshot of a Lesson at the beginning of this chapter

The whole:

15 Marbles

Partitioned into 5 equal parts:

Each part is one-fifth $\left(\frac{1}{5}\right)$.

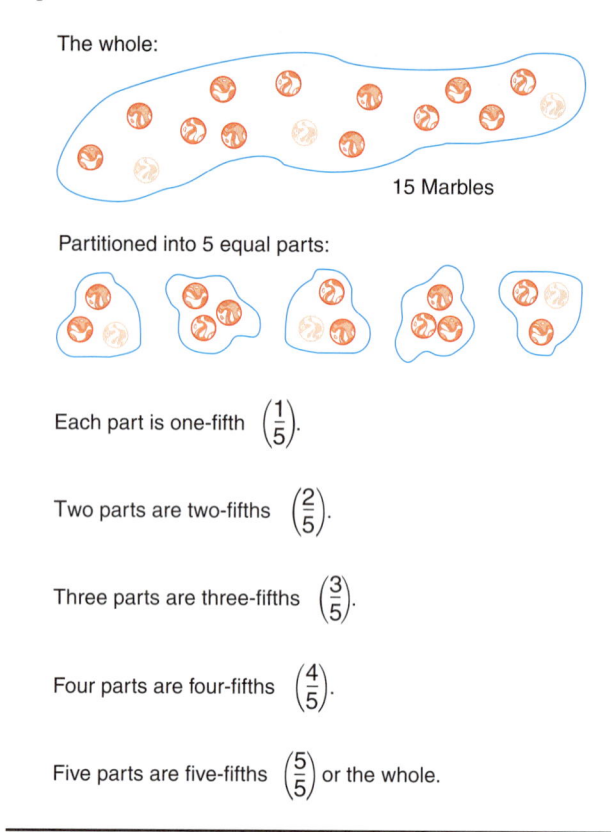

Two parts are two-fifths $\left(\frac{2}{5}\right)$.

Three parts are three-fifths $\left(\frac{3}{5}\right)$.

Four parts are four-fifths $\left(\frac{4}{5}\right)$.

Five parts are five-fifths $\left(\frac{5}{5}\right)$ or the whole.

Figure 12-5 *Working with a set model.*

shows students in grades 4 and 5 using the area model for halves.

In the Classroom 12-1 shows examples and nonexamples of three-fourths. It provides a way to assess children's understanding of the different models—region, length, set, and area. Use it first as a self-assessment, and then try it with elementary students. Which do you expect to be the easiest examples for students?

A more detailed plan for this activity is available on the Student Companion site at www.wiley.com/college/reys.

MAKING SENSE OF FRACTIONS

Begin to build children's conceptual understanding of fractions by starting with the simplest meaning and the simplest model: the part–whole meaning and the region model. After you introduce this model and the language and symbols associated with fractions, you can show the children other models of the part–whole (length, set, and area). Many materials, such as fraction bars and pattern blocks, are available to assist you in helping children develop concepts about fractions (see Appendix for paper versions of these commercial materials). Some of the most effective materials, such as construction paper and counters, are commonly found in classrooms.

Conceptual Development of Fractions 259

In the Classroom 12–1

THREE-FOURTHS

Objective: To assess children's understanding of three-fourths.

Grade Level: 3–5.

▼ In which of the following models does the shading show three-fourths of the whole? Why or why not?

A. Whole: large rectangle

B. Whole: box of marbles

C. Whole: large rectangle

D. Whole: large rectangle

E. Whole: box of shapes

F. Whole: large rectangle

G. Whole: box of candy

H. Whole: large rectangle

I. Whole: large triangle

J. Whole: large rectangle

A more detailed plan for this activity is available on the Student Companion site at www.wiley.com/college/reys.

PARTITIONING Partitioning, making equal shares by separating a whole into equal parts, is an important idea if fractions are to be understood. The whole is whatever is specified as the unit. At first, you should use a whole that will be obvious to children (e.g., a rectangle). If the whole is a region, then the parts must be equal in size and the same shape; if it is a length, then the parts must be equal lengths; if it is a set, then the parts must have the same number of objects in each; and if it is an area, then the parts must be equal in area. Whatever the model, children must learn to partition the whole into equal parts and to describe those parts with fractional names.

Let the children do the partitioning. For example, each child could be given a "candy bar" (a piece of paper the size of a large candy bar) to share with a friend. Have them fold the "candy bar" to show how they would share it. Talk about whether a fold like the one sketched here would be a "fair share" for two people.

Move on to sharing equally among 3, 4, 6, and 8 people. For some of the more difficult sharing problems, such as with 5 people, you may want to use a strip of paper that has already been marked. When children become familiar with the process of sharing by partitioning regions into equal parts, describing the parts with words, and counting the parts, it is time to move to the other models. Children especially like to share collections of objects (set model)—for example, sharing 12 marbles between 2 friends is appealing to children because they can share fairly, with each friend getting half the marbles.

WORDS FOR THE EQUAL PARTS As children develop the idea of equal parts, introduce the words *halves*, *thirds*, *fourths*, and so on. The words will be familiar to students since they are usually the plural of the ordinal words: third, fourth, and so on. This connection can help, but you need to make sure that students hear the *s* and that they connect the word with the number of parts of the whole. Help them make that connection by asking such questions as "How many fifths make a whole? How many eighths? How many twenty-fourths?" On the NAEP mathematics assessment, only half of the fourth-grade students correctly responded that 4 fourths make a whole (Wearne and Kouba, 2000).

COUNTING Once children are familiar with the words for fractional parts, begin counting the parts. This process should not be any more difficult than counting apples, but the children need to know what fractional parts they are counting. Counting is a powerful step in helping children make sense of fractions.

In the Classroom 12-2 has a few ideas for getting children to practice counting fractional parts.

Counting can help develop the concept that 3/5 is three parts of size 1/5 as suggested by the expectations in the CC-SSM expectations (see Figure 12-1). Counting the number of parts also can lead to questions about ordering and equivalent fractions—for example, Which is more—3 fourths or 5 eighths? How many halves are equal to 4 eighths? Counting fractional parts can also help children see how much more is needed to make a whole. For example, if children have counted to 3 fifths with a model, they can see that 2 more fifths would make the whole. Counting fractional parts can lead to developing improper fractions and mixed numbers. For example, if three children are counting all the fourths they have, the class can hear the third child counting "9 fourths, 10 fourths, … " and can see that the three children have 2 wholes and 1 fourth, 2 wholes and 2 fourths, and so on. When they hear "12 fourths," they will see the

260 Chapter 12 • Fractions and Decimals: Concepts and Operations

---

In the Classroom 12–2

FRACTION BARS

Objective: To develop an understanding of equal sharing and counting fractional parts using fraction bars.

Grade Level: 2–5.

Materials: 4 construction paper strips (3″ × 9″) of 4 different colors for each child.

Preparation of Fraction Bars: Each child should fold a green strip into "halves," mark the fold with a dark line, and write halves on the back.

Make fraction bars for thirds, fourths, and sixths in other colors in the same way.

Activities:

- Ask each child to count the parts on a fourths bar: [one-fourth, two-fourths, …]

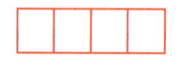

  1, 2, 3, 4 fourths

- Ask each to count the parts of another bar.
- Ask a pair of children to count the fourths in two bars:

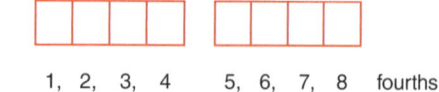

  1, 2, 3, 4   5, 6, 7, 8 fourths

- Make sure the class sees that 8 fourths is 2 wholes, 6 fourths is 1 whole and 2 fourths, and so on.
- Ask a group of 4 students to count all the sixths in their group's sixths-bar.
- Challenge all the children to tell how many strips it would take to show 11 sixths or 23 sixths. Let them experiment in groups of 4.

In the Classroom 12–3

SHARING RODS

Objective: To investigate sharing rods equally and naming the fractional parts and the amount in each equal part.

Grade Level: 1–2.

Materials: Unifix cubes or other connecting cubes.

Launch: Have each child make a Unifix rod of 12 cubes and name it the 12-rod. Tell the children that we are going to investigate making halves, thirds, fourths, and sixths from the 12-rod.

Questions:

1. Have the children "break" their 12-rod into two rods of equal length. (Make 2 shares.) Ask the children: "What is each share (fractional part) called?" [halves] "How many halves make the 12-rod?" [2] "How many cubes in each half?" [6] "What is one-half of a 12?" [6]

2. Have the children make 3 equal shares from their 12-rod. Ask: "What fractional part is one share?" [thirds] "How many thirds make the 12-rod?" [3] "How many cubes in each third?" [4] "What is one-third of 12?" [4]

3. Have the children make 6 equal shares from their 12-rod. Ask: "What fractional part is 1 share?" [sixths] How many sixths make the 12-rod? [6] How many cubes in each sixth? [2] What is one-sixth of 12? [2]

4. Have the children make 4 equal shares from their 12-rod. Ask: "What fractional part is one share?" [fourths] How many fourths make the 12-rod? [4] How many cubes in each fourth? [3] What is one-fourth of 12? [3]

Extend:

1. Ask the children how many cubes in two-thirds and three-thirds of their 12-rod. Continue with other fractional parts, such as two-fourths and five-sixths.

2. Use other rods, such as a 10-rod (halves and fifths) or a 16-rod (halves, fourths, eighths).

---

3 wholes. This also is a natural introduction to adding and subtracting fractions with like denominators.

After developing the initial ideas of partitioning, the words for fractional parts, and counting parts, you can introduce other fractional models. For example, In the Classroom 12-3 uses a length model and connects it to the set model while still using only the fractional words without symbols.

SYMBOLS Symbols and written words should be used together or alternately until students understand the meaning of the symbols.

The following quote summarizes much of research and meaningful classroom practice concerning the development of symbols:

> Using the written fraction symbol is appropriate only after children can name, count, and compare using oral language with facility…. Written symbols are developed in the same way as oral language. The same kind of questioning can be used, but now the model and the oral language are connected with the symbol.
>
> (Payne, Towsley, and Huinker, 1990, p. 185).

Conceptual Development of Fractions 261

Connections can be depicted as follows:

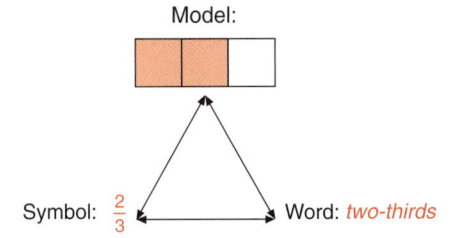

Model:

Symbol: $\frac{2}{3}$ Word: *two-thirds*

When children can match the words with the model, it is time to tell them that is the symbol for *two-thirds*. Then they need many opportunities to make all three kinds of connections between the different representations:

Connection 1. Model and words
Connection 2. Model and symbol
Connection 3. Words and symbol

Textbooks usually concentrate on part of connection 2 (going from a model to a symbol), so you will find plenty of examples of these. However, they often do not go from a symbol to having the students produce the model. Connection 3 requires oral work, so you need to give children opportunities to solve problems and describe the solutions. When you read fractions, be sure to use the fraction words (e.g., *three-fifths*). Children who hear only "3 over 5" often think of two numbers rather than a single number, a fraction.

DRAWING A MODEL Children who have modeled fractions by folding paper or by choosing a picture should also be able to draw a picture of fractional parts. The rectangle is probably the easiest shape to use to show a good approximation of a fractional part. Encourage the children to be as accurate as possible, and do not worry too much about other aspects of their drawings. For example, which of these two drawings would you prefer as a picture of two-thirds?

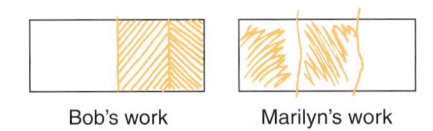

Bob's work Marilyn's work

Bob's work is neater than Marilyn's, but he seems to have missed the point that the three parts must be equal. You might help Marilyn be a little neater, but she does seem to have the idea of two-thirds. Being able to produce your own model is essential in using models to solve problems.

EXTENDING THE MODEL Table 12-1 shows a question on the second national mathematics assessment and how children responded to it (Carpenter et al., 1981). Little has changed in the over 30 years since this item was given; the performance level on a similar question on the 2003 assessment was almost identical for fourth-grade students (age 9). There is no doubt that this model is more complicated than those described in the previous section, since the number of parts (12) is greater than the denominator (3) of the simplest fraction corresponding to it $\left(\frac{1}{3}\right)$. However, this model is useful for introducing equivalent fractions and for teaching children how to order fractions.

TABLE 12-1 • Results of National Assessment Question Using the Region Model
What fractional part of the figure is shaded?

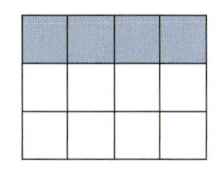

Responses	Percent Responding Age 9	Age 13
Acceptable responses $\frac{1}{3}$, $\frac{4}{12}$, .33	20	82
Unacceptable responses $\frac{1}{4}$, .25	5	4
Top 4 or $\frac{4}{3}$	36	6
Other	15	6
I don't know	17	1
No response	7	1

You can also use paper folding to introduce the model in Table 12-1 as well as to introduce equivalent fractions. Give each child a sheet of plain paper to fold into thirds and to shade two-thirds. Then fold the paper in half the other way. Ask how many parts and what kind of parts [6, sixths]. Then ask what part is shaded, and make it clear that both $\frac{2}{3}$ and $\frac{4}{6}$ are correct answers. Tell the children that $\frac{2}{3}$ and $\frac{4}{6}$ are called *equivalent fractions* because they represent the same amount.

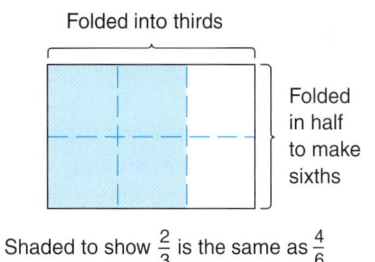

Folded into thirds

Folded in half to make sixths

Shaded to show $\frac{2}{3}$ is the same as $\frac{4}{6}$

Move to rectangle models such as the one below. Make certain the students can identify the way the paper was folded in both directions. Have them show thirds and fourths. Then have them identify $\frac{1}{4}$, $\frac{2}{4}$, $\frac{3}{4}$, $\frac{1}{3}$, and $\frac{2}{3}$, as well as $\frac{4}{12}$, and $\frac{8}{12}$. Some children will need to color in the parts on different models to focus their attention on the specific part.

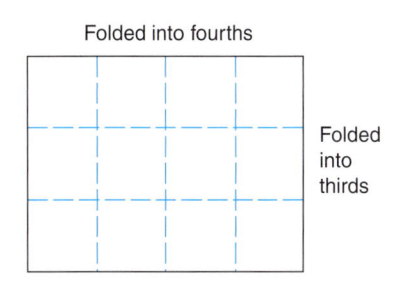

Folded into fourths

Folded into thirds

262 Chapter 12 • Fractions and Decimals: Concepts and Operations

After doing some more examples and getting more practice with folding paper, children should be ready for drawing. Reinforce the connection between folding and drawing by having the children make a step-by-step record of the folding process on a picture of a rectangle. Look at the CCSSM overview in Figure 12-1 and see which grade is recommended for this activity.

A CLOSER LOOK AT THE SET MODEL Look at the two examples below:

The child who responds that $\frac{3}{4}$ of the doughnuts are small is correct but who says $\frac{3}{4}$ of the rectangle is yellow is incorrect. In the doughnut example, the child has learned that "3 out of the 4" doughnuts are small; thus, the fraction is $\frac{3}{4}$. Isn't it logical that the child could also reason that "3 out of the 4" pieces are yellow?

We unwittingly encourage this thinking by over using *out of*. The *out of* encourages the above actions as well as thinking of the fraction only as two whole numbers. Siebert and Gaskin (2006) have a thoughtful discussion of this difficulty. It certainly does not help children understand the set model and how partitioning is related to it. It also does not help students focus on what attribute is being considered. In fact, the child who says $\frac{3}{4}$ of the rectangle is yellow is correct if the attribute is not a region but is a set of 4 pieces. If we could make one change in curricula and assessment, it would be to throw out all the early references to *out of*.

A CLOSER LOOK AT THE NUMBER LINE Many children use whole-number ideas when working with fractions or they think of a fraction as two numbers, the numerator and the denominator. When they do not know what to do, they revert to thinking of only whole numbers. A common error when children were asked to place unit fractions on a number line is shown in the example (adapted from Petit et al., 2010).

Child A has relied on his or her knowledge of whole numbers and put the denominators in order without regard for the fractions. Child B has the correct order but does not have them placed reasonably on the number line. Child C realizes the placement of one-half as being halfway between 0 and 1. It is not clear whether he or she knew that one-fourth is halfway between 0 and one-half. It is evident that he or she knew that is $\frac{1}{3}$ is more than $\frac{1}{4}$ and less than $\frac{1}{2}$ but did not realize that $\frac{1}{3}$ is nearer to $\frac{1}{4}$. Child D relied on a length model to place the decimals. It is very difficult to place decimals on a number line that is not subdivided. Children can get more accurate with the drawings, but one has to question how much accuracy is needed.

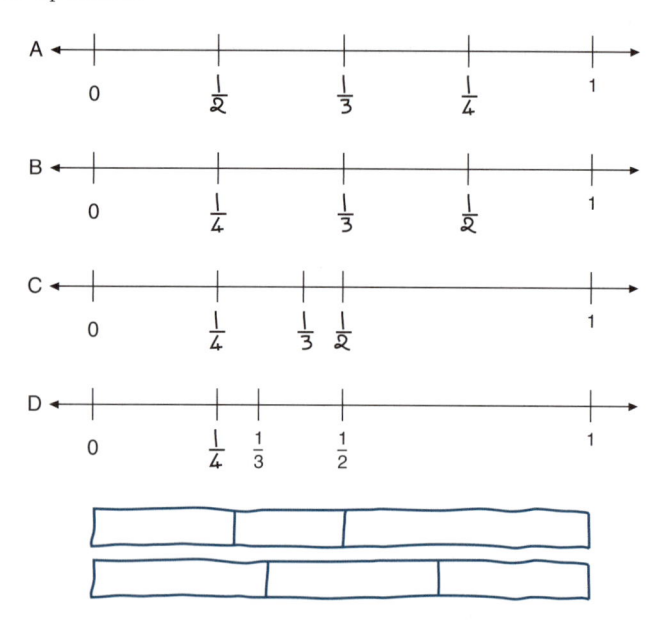

It is also not clear what Child D is thinking. Look at the drawing D. The bottom left rectangle represents one-third. The child has successfully transferred this model to the length (distance) model, realizing the end mark is one-third of the distance from 0 to 1.

We have shown a number line between only 0 and 1, but students also need to see the number line including negative integers and numbers greater than 1. Often when children see a number line from 0 to 2 and are asked to mark $\frac{1}{2}$, they put the mark at 1. They are answering another question—what is halfway between 0 and 2, not where is $\frac{1}{2}$ located. They need to realize that the unit on a number line is 1, or the distance from 0 to 1. When negative numbers are used, students often locate fractions to the left of 0 and reason (incorrectly), "A fraction is little so it must be less than 0."

For more information about children's understanding of the number line and ways to support students' learning, read the chapter by Saxe and colleagues (2007) or the chapter by Petit and colleagues (2010). Both clearly show that it takes skillful teaching and time for students to learn to use this model of fractions.

BENCHMARKS Initially, children need to develop facility with friendly fractions such as halves, thirds, and fourths so that they can relate these to fractions that look more intimidating, such as sevenths, elevenths, and twenty-fifths. If you help children build benchmarks, they will be able to reason with these more difficult fractions. For example, when you consider the following fractions, you can easily see that they are all near one-half:

$$\frac{6}{11} \quad \frac{11}{23} \quad \frac{19}{35} \quad \frac{44}{90} \quad \frac{354}{700} \quad \frac{800}{1605} \quad \frac{100001}{200000}$$

Moreover, you can quickly say which are more than $\frac{1}{2}$ and which are less than $\frac{1}{2}$. Children can build this same intuitive feel for fractions near the benchmark $\frac{1}{2}$ by beginning with smaller numerators and denominators.

Students also need to be able to tell if a fraction is near 0 or near 1. This will let them use the benchmarks 0, $\frac{1}{2}$, and 1 to put in order a set of fractions such as $\frac{13}{25}, \frac{2}{31}, \frac{5}{6}, \frac{4}{11}$, and $\frac{21}{20}$, which would be a time-consuming task if they tried to find a common denominator. Using benchmarks as shown on the following number line makes the task rather easy: $\frac{13}{25}$ is a little more than $\frac{1}{2}$, $\frac{4}{11}$ is a little less than $\frac{1}{2}$, $\frac{2}{31}$ is near zero, $\frac{5}{6}$ is a little less than 1, and $\frac{21}{20}$ is a little more than 1. Thus, the correct order is $\frac{2}{31}, \frac{4}{11}, \frac{13}{25}, \frac{5}{6}, \frac{21}{20}$.

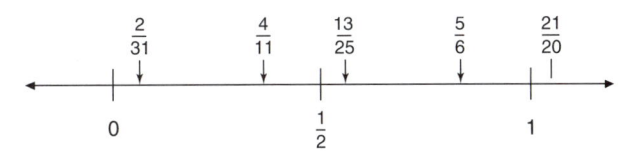

Children can also use benchmarks to check the reasonableness of computations with fractions and to estimate answers. For example, if they know that $\frac{3}{5}$ and $\frac{6}{7}$ are each more than $\frac{1}{2}$, then they know that $\frac{3}{5} + \frac{7}{6}$ is not $\frac{9}{12}$—the sum must be more than 1.

> ### Virtual Classroom Observation
>
> Go to www.wiley.com/college/reys,
> Access the Wiley Resource Kit
> Click on the Virtual Classroom Observations Section
>
> 1. Module 3: Numbers and Operations: The Magnitude of Fractions.
> 2. Watch a video: Teaching Examples #1: If I Only Had Half a Cube.
> 3. Think and discuss: What strategies did the students use to determine whether $\frac{5}{18}$ is closer to the benchmark fraction, $\frac{1}{4}$ or $\frac{1}{3}$? How would you determine this?

> ### Virtual Classroom Observation
>
> Go to www.wiley.com/college/reys,
> Access the Wiley Resource Kit
> Click on the Virtual Classroom Observations Section
>
> 1. Module 3: Numbers and Operations: The Magnitude of Fractions.
> 2. Watch the accuracy commentary on Teaching Examples #1: If I Only Had Half a Cube.
> 3. Think and discuss: The commenter explains the reason the teacher did not push the students to give an exact answer to whether $\frac{5}{18}$ was closer to $\frac{1}{4}$ or $\frac{1}{3}$. Why?

GOING FROM A PART TO A WHOLE Most students can recognize a simple representation of a fractional part of a whole. However, asking students to figure out a whole when they are shown a fractional part will often reveal that children's understanding of fractions is limited. The following exchange from a fifth-grade classroom illustrates the kind of question you should ask (Mrs. May is the teacher):

MRS. MAY: If I told you that this picture represents three-fifths of a cake, could you draw a picture of the whole cake?

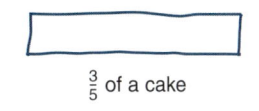

$\frac{3}{5}$ of a cake

ROSA LEE: It sure would be a skinny cake.

CAMILLE: That depends where the two-fifths was cut off.

MRS. MAY: You are both right; you won't be able to tell exactly what the shape of the cake was before it was cut. Let's pretend you can see that it was cut here. [pointing to the right side] What do you know?

GILBERT: You have three of the five equal pieces. If we divide this into three equal parts, we know how large one-fifth is.

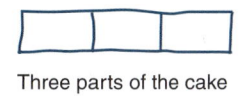

Three parts of the cake

OLAV: Now all you have to do is add the missing two-fifths. Rosa Lee, it is a skinny cake!

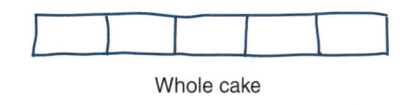

Whole cake

Mrs. May continued the classroom discussion by giving the children other tasks—for example: Given $\frac{2}{7}$ of a stick of licorice (to represent a length) or $\frac{2}{5}$ (12 pieces) of a box of candy (set model), find the whole. Children should work independently on several problems and then share their solutions. These types of tasks should also include an improper fraction (e.g., $\frac{8}{5}$) and a mixed number (e.g., $1\frac{2}{3}$), so you can see which students have made sense of fractions greater than 1.

ORDERING FRACTIONS AND EQUIVALENT FRACTIONS

Part of understanding fractions is realizing that they are numbers and therefore can be ordered, added, subtracted, multiplied, and divided. The goal is to have children compare two fractions or order more than two fractions and find equivalent fractions symbolically, but you can help children connect the concrete to the symbolic. Many problems involving ordering capture children's interest because they want to know which is more, which is shorter, which is larger, and so on. Problems about equivalent fractions are central to understanding fractions and being able to operate (add, subtract, multiply, and divide) with them.

Math Links 12.2

A wide variety of electronic fraction manipulatives (try the fraction bars) may be found at the National Library of Virtual Manipulatives for Interactive Mathematics Web site, which you can access from http://www.nlvm.usu.edu/en/nav/vlibrary.html or from this book's Web site.

www.wiley.com/college/reys

CONCRETE MODELS Children can make concrete models and then use them to order fractions or to find equivalent fractions. For example, fraction bars (see In the Classroom 12-2) are concrete models that children can make and then use to find out which is larger, $\frac{2}{3}$ or $\frac{3}{4}$. They can fold the thirds bar so that it is $\frac{2}{3}$ long and the fourths bar so that it is $\frac{3}{4}$ long and then compare the models to see which is longer.

Fraction bars can also be used as concrete models to find equivalent fractions. For example, by comparing a thirds bar and a sixths bar, children can readily see that $\frac{2}{3}$ is the same as $\frac{4}{6}$. For simplicity, our examples have focused on the rectangle and on paper models. Keep in mind, however, that children need to work with multiple representations. Pattern blocks (see Appendix) are another model accessible to young children. In the Classroom 12-4 shows how to plan a lesson using pattern blocks for ordering fractions and finding equivalent fractions.

PICTORIAL MODELS Children can also use pictorial models to order fractions. For example, the accurate scale of the fraction bars in the Appendix lets children compare lengths to decide which is larger, $\frac{7}{9}$ or $\frac{8}{10}$. These bars can also be used to identify equivalent fractions—for instance, children can see that $\frac{1}{2}, \frac{2}{4}, \frac{3}{6}, \frac{4}{8}$, and $\frac{6}{12}$ are all the same length and therefore represents the same amount.

You should also make sure that children are able to draw their own pictorial models to represent fractions so that they can solve problems for which no ready-made models are available. For example, suppose you give students this problem: "Jacqui ate $\frac{1}{3}$ of the cake. Tom ate $\frac{2}{7}$ of the cake. Who ate more?" You could have students approach this problem as follows:

1. Have the children draw two equal-sized pictures of the cake, partitioning one into thirds horizontally and the other into sevenths vertically and then shading $\frac{1}{3}$ of one and $\frac{2}{7}$ of the other, as shown on the right:

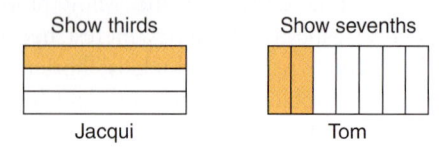

Show thirds — Jacqui
Show sevenths — Tom

In the Classroom 12–4

COMPARING MODELS

Objective: To use concrete models to represent fractional parts of a whole.

Grade Level: 3–5.

Materials: Pattern blocks.

The four basic shapes of the pattern blocks that you will use are the hexagon, triangle, rhombus, and trapezoid. Each child or pair of children will need multiple copies of each shape.

Let children have a chance to explore by making designs and seeing how they could make the hexagon with the other pieces.

Idea One:

Use the hexagon as the whole. What is the size of each of the other pieces? Use the pieces to show that

$$\frac{2}{6} = \frac{1}{3} \quad \frac{3}{3} = 1 \quad \frac{3}{6} = \frac{1}{2} \quad \frac{2}{3} > \frac{1}{2} \quad 2\frac{2}{3} = \frac{8}{3}$$

Idea Two:

Use the figure made from two hexagons as the whole.

Use ▱ to cover the yellow shape. How many does it take? What fractional part of the whole is the rhombus? What fractional part of the whole is the trapezoid?

Use the shape to show that $\frac{3}{4}$ is greater than $\frac{2}{3}$.

Idea Three:

Give the shape of a fractional part and have children construct the whole as in the lesson snapshot

If ▱ is $\frac{2}{3}$, what is the whole?

If ▽ is $\frac{3}{4}$, make the whole.

If △ is $\frac{1}{5}$, make the whole.

If △ is $\frac{2}{5}$, draw the whole.

2. Have the children recognize that it is difficult to tell which shaded part is larger.

3. Have the children draw two more equal-sized pictures of the cake, but this time portioning both into thirds horizontally and into sevenths of the other, as show below:

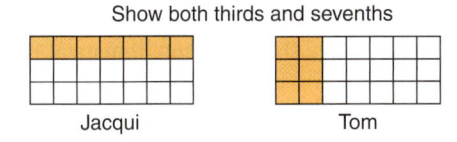

Show both thirds and sevenths

Jacqui   Tom

4. Now the children can see that Jacqui ate 7 pieces and Tom ate 6 pieces, so Jacqui ate more (i.e., $\frac{1}{3}$ is more than $\frac{2}{7}$).

Your students may find other ways to solve the problem, but this is one approach you could have in your repertoire. Also, you can use the picture of the cake partitioned into thirds horizontally and into sevenths vertically to help children understand a symbolic way of comparing fractions as well as to help them understand a way of adding and subtracting fractions.

SYMBOLIC REPRESENTATION It is easier to compare two measurements given in the same unit (e.g., 78 meters and 20 meters) than two measurements given in different units (e.g., 83 meters and 231 feet). Similarly, it is easier to compare two fractions that are symbolically represented by the same partitioning (e.g., $\frac{3}{5}$ and $\frac{2}{5}$; the denominators are the same) than two fractions that are represented by different partitioning (e.g., $\frac{3}{5}$ and $\frac{5}{8}$; the denominators are different). In mathematics, we often try to deal with difficult cases by changing them into simpler ones. With fractions, making the difficult case with different denominators simpler means expressing each fraction as an equivalent fraction so that both fractions represent the same partitioning—that is, so they have a common denominator.

Now we can see how to use the cake partitioning shown above to compare $\frac{1}{3}$ and $\frac{2}{7}$ by expressing them as equivalent fractions with a common denominator. The rectangle that has been partitioned into both thirds and sevenths is partitioned into 21 parts—that is, into twenty-firsts. Children can easily recognize that $\frac{1}{3}$ is 7 of the 21 parts, or $\frac{7}{21}$, and that $\frac{2}{7}$ is 6 of the 21 parts, or $\frac{6}{21}$. Since $\frac{7}{21}$ is greater than $\frac{6}{21}$, $\frac{1}{3}$ is greater than $\frac{2}{7}$.

Students who have considered various models showing equivalent fractional parts such as $\frac{2}{3}$ and $\frac{4}{6}$ will be familiar with the concept of equivalent fractions, but they may not have developed many of the skills associated with finding equivalent fractions symbolically. Finding an equivalent fraction rests on the generalization that both the numerator and denominator of a fraction may be multiplied (or divided) by the same number without changing the value of the fraction. This is stated in the CCSSM as a/b = (n × a)/(n × b). One way to develop this generalization is to start with a paper-folding model and symbolically describe what is happening.

Fold into fourths

Fold in half to make eighths

1. Make a model of $\frac{3}{4}$ by folding a piece of paper in fourths and shading three of the fourths. Write down how many fourths there are (4) and how many are shaded (3). Then fold the paper in half the other way to make eighths.

2. Have the children look at the fourths. Ask what happened to the fourths when the paper was folded in half. You created twice as many equal parts (2 × 4 = 8).

A   B

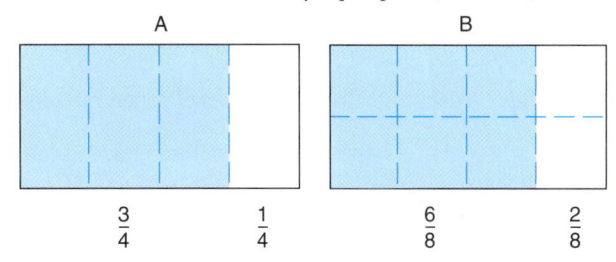

$\frac{3}{4}$ $\frac{1}{4}$ $\frac{6}{8}$ $\frac{2}{8}$

3. Notice that there are also twice as many shaded parts in B (2 × 3 = 6).

4. Express the model symbolically:

$$\frac{2 \times 3}{2 \times 4} = \frac{6}{8}$$

After investigating more examples like this one, children should be able to make the generalization that multiplying the numerator and the denominator by the same number results in an equivalent fraction. Later when they have learned about multiplying fractions, they will realize they were multiplying the original fraction by 1. Thus, they have used a version of the multiplicative identity property ($a \times 1 = a$).

Note that you could also begin with the model of eight parts and describe how to get to four parts by dividing the eight parts by two (8 ÷ 2) and dividing the six shaded parts by two (6 ÷ 2). In this case, you would express the model symbolically as:

$$\frac{6}{8} = \frac{6 \div 2}{8 \div 2} = \frac{3}{4}$$

Examples of this type should lead children to make the generalization that dividing the numerator and denominator by the same number also results in an equivalent fraction.

Another understanding that needs to be built is whether or not a fraction can be written as an equivalent fraction with a specified denominator. For example, can fourths be expressed as eighths, as tenths, or as twelfths? Drawing models such as rectangles showing, say, fourths drawn vertically, students can explore whether they can draw horizontal lines than $\frac{1}{3}$ that may be expressed as equivalent fractions with a denominator of 48 [e.g., $\frac{1}{4} = \frac{12}{48}$ or $\frac{3}{16} = \frac{9}{48}$].

COMPARING FRACTIONS USING VARIOUS REASONING STRATEGIES Although we have looked at various models (concrete, pictorial, and symbolic) for comparing fractions, children with a firm foundation can often use adaptive reasoning, one of the five strands of mathematical proficiency (see Chapter 1, Figure 1-3). How would you tell which is the larger fraction in each pair?

1. $\frac{3}{7}$ and $\frac{5}{7}$
2. $\frac{1}{5}$ and $\frac{1}{7}$
3. $\frac{4}{5}$ and $\frac{6}{7}$
4. $\frac{3}{8}$ and $\frac{3}{5}$
5. $\frac{9}{11}$ and $\frac{2}{13}$
6. $\frac{18}{25}$ and $\frac{9}{13}$

Listen to the reasoning strategies used by students in elementary school (these are reasoning strategies from students answering other pairs in this way).

1. Ali: That is easy—you know that 5 of something is more than 3 of that same thing—the thing is sevenths.
2. Bob: I thought of a pie. Fifths means that 5 people share and sevenths mean that 7 people share—so those pieces are smaller. So the fifths are bigger and so one-fifth is larger than one-seventh.
3. Carl: It is like the last one except someone ate a piece out of each pie. The person that ate the one-fifth ate more—so more is left in the sevenths pie. Six-sevenths is more. (Some students still needed to see a sketch to see what Carl was saying.)
4. Dana: The numerators are the same. So 3 of the smaller pieces (eighths) is less than 3 of the larger pieces (fifths).
5. Emma: Nine-elevenths is way over one-half and two-thirteenths is way under. So it's nine-elevenths.
6. Fred: I used the "get the same numerator"—so I had 18 of each—twenty-fifths and twenty-sixths. Twenty-fifths is larger, so 18 of them would be more.

FINDING FRACTIONS BETWEEN TWO FRACTIONS Between any two whole numbers, there may or may not be another whole number. For example, between 3 and 8 there are four whole numbers (4, 5, 6, and 7), but between 5 and 6 there is no whole number. Between any two fractions, there are an infinite number of fractions. This is called the **density property** of fractions.

First students need to see that between any two fractions, there is another fraction. Eventually, they will see that there are many, many fractions between any two given fractions. You may want to connect this to the number line

---

In the Classroom 12–5

WHOLE HOG

Objective: To compare fractions.

Grade Level: 3–5.

Materials: A copy of the H grid; squares of paper; markers

- Cut 10 squares of paper. Write one of these fractions on each square:

  $\frac{1}{2}$ $\frac{1}{3}$ $\frac{1}{6}$ $\frac{2}{6}$ $\frac{1}{9}$ $\frac{2}{9}$ $\frac{3}{9}$ $\frac{1}{18}$ $\frac{2}{18}$ $\frac{3}{18}$

Game Rules:

1. Shuffle the fraction cards and put them in a pile face down.
2. Each player picks a card from the top of the pile and turns it face up.
3. The player with the smaller fraction colors in that fractional part of her or his H.
4. Put both cards at the bottom of the pile.
5. If the player with the smaller fraction cannot color the fractional part shown on the card (because not enough of the H is left uncolored), both players put their cards at the bottom of the pile and pick again.
6. Continue playing until one player colors the whole H. That person is the first to go Whole Hog and loses the game.

- Make a new H and play again.

A more detailed plan for this activity is available on the Student Companion site at www.wiley.com/college/reys.

---

to show eighths, tenths, or twelfths. They will soon learn that there must be a multiplicative relationship. In this example, if the denominator is a multiple of four, then an equivalent fraction, with a whole number numerator, can be found.

Children who can order fractions symbolically may enjoy the quick and easy game from In the Classroom 12-5. If they have difficulty playing, let them use a concrete or pictorial representation of the fractions. If the game is too easy, use a gameboard with 48 squares and fractions less

and two given points, say A and B. Students can see that there are many points between these two given points.

```
  •———•——→
  A   B
```

Suppose A is $\frac{3}{8}$ and B is $\frac{6}{8}$. Ask students what fractions are between these two fractions. It is rather clear that $\frac{4}{8}$ and $\frac{5}{8}$ are between $\frac{3}{8}$ and $\frac{4}{8}$. See if they can find another fraction between $\frac{4}{8}$ and $\frac{5}{8}$. Continuing this line of questions and showing the associated points on the number line will help students build the idea that there are many fractions between any two fractions.

Students who have a firm conceptual understanding of fractions will find many different strategies for finding fractions between two given fractions depending on the pair. Have students try to find a fraction between the following pairs. See if the students use different strategies such as concrete or pictorial models, decimal equivalents, or equivalent fractions.

Pair 1: $\frac{3}{5}$ and $\frac{4}{5}$

Pair 2: $\frac{1}{4}$ and $\frac{1}{2}$

Pair 3: $\frac{3}{7}$ and $\frac{4}{7}$

Pair 4: $\frac{3}{8}$ and $\frac{3}{7}$

MIXED NUMBERS AND IMPROPER FRACTIONS

Through models, you can lead children naturally into mixed numbers and improper fractions, even as they are learning the initial concepts of fractions. We began the ideas when we were counting fractional parts—one-fourth, two-fourths, three-fourths, four-fourths, five-fourths (oh! we have a whole and another fourth). This leads naturally to examining the equivalence between improper fractions (a fraction greater than 1) and a mixed number (a whole number and fraction).

A mixed number is a natural symbolic representation of the following model:

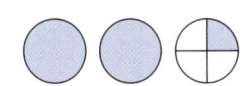

You can add partitions in the model to show all the fourths, so children can see that the initial counting is 9 fourths, or the improper fraction $\frac{9}{4}$.

This helps them understand that $2\frac{1}{4}$ and $\frac{9}{4}$ are equivalent representations. To help children gain experience with mixed numbers and improper fractions, use the models as much as possible and ask them to write *both* types of numbers to represent the models. After children understand this process, they need practice in changing from one form to the other without the use of models. Do not rush to teach them a routine,

however, but encourage them to think problems through. In the following example, see if you can follow Cerise's thinking as she considers how to change $8\frac{2}{3}$ to an improper fraction.

Note that her thinking depends on understanding that a whole can be equivalently expressed in terms of fractional parts (e.g., 5 fifths and 6 sixths each make a whole).

I know that 1 is 3 thirds...

Cerise: I know that 1 is 3 thirds. So I need to figure how many thirds in 8 wholes. That is 8 groups of 3 thirds, which is 24 thirds. Now I have to add on the 2 thirds, so it is 26 thirds, or $\frac{26}{3}$.

Children also must be able to think through the opposite process—changing an improper fraction to a mixed number—for example, changing $\frac{17}{6}$ to $2\frac{5}{6}$. In this case, their thinking could go like this: 6 sixths make a whole, so 17 would make 2 wholes, and there would be 5 sixths more. So $\frac{17}{6}$ is $2\frac{5}{6}$. If children understand the quotient meaning of fractions, then students can also think of $\frac{17}{6}$ as $17 \div 6$, which they can calculate as $2\frac{5}{6}$.

OPERATIONS WITH FRACTIONS

The key to helping children understand operations with fractions is to make sure they understand fractions, especially the idea of equivalent fractions. When children are solving problems that involve operations with fractions, be sure the children can use what they know to make sense of the operations. Whenever possible, they should be able to extend to fractions what they know about operations with whole numbers. Keep in mind, however, that some operations with whole numbers do not extend directly to fractions. For example, multiplication of two fractions is not repeated addition as it is with whole numbers. There are other differences, too—for instance, when you multiply

268 Chapter 12 • Fractions and Decimals: Concepts and Operations

two positive whole numbers, the product is always larger than either factor; but when you multiply two positive fractions, less than 1 the product is always less than either factor. Children often have the idea that if you divide, you make something smaller. However, if you divide a whole number by a proper fraction, the quotient will be larger than the whole number.

Children gain a better understanding of operations with fractions if they learn to estimate answers by using whole numbers and benchmarks such as $\frac{1}{2}$. For example, before computing the answer to $3\frac{2}{3} + 4\frac{5}{6}$, they should estimate the answer by considering, first, that the sum must be more than 7 (3 + 4) and, second, that $\frac{2}{3}$ and $\frac{5}{6}$ are each more than $\frac{1}{2}$, so the answer must be more than 8. Developing this type of number and operation sense makes it easier to figure out whether answers to problems are reasonable. Having the conceptual understanding to check answers for reasonableness is also important because many calculators can perform operations with fractions.

Your goal is to help children make sense of adding and subtracting fractions. One of the most famous examples from NAEP (Carpenter et al., 1981, pp. 36) was a multiple-choice item given to students at ages 13 and 17.

Estimate the answer to $\frac{12}{13} + \frac{7}{8}$.

You will not have time to solve the problem using paper and pencil.

The choices for answers were 1, 2, 19, 21, and "I don't know." At age 13, a greater percentage of students chose 19 (28%) and 21 (27%) than the correct answer of 2 (24%). Students actually did better when asked to add fractions with paper and pencil, which shows they had little understanding of reasonable answers. A more recent study (Petit and others, 2010) showed that on a similar problem, students who could add the fractions still could not estimate or give a justification for their answers.

ADDITION AND SUBTRACTION

Figure 12-6 is an overview of the recommendations in the CCSSM document.

To help with building understanding, we are suggesting that instead of beginning addition and subtraction of fractionss with a symbolic sentence such as $\frac{2}{3} + \frac{1}{4}$, begin with problems that involve joining and separating. Such problems, together with pictorial models, can

- Help children see that adding and subtracting fractions can solve problems similar to problems involving whole numbers
- Help children develop an idea of a reasonable answer
- Help children see why a common denominator is necessary when adding or subtracting fractions with unlike denominators

Use what children know about solving whole-number addition and subtraction word problems to begin examining

	Number and Operations-Fractions Domain
Grade 4	Add and subtract fractions (both proper and mixed numbers) with like denominators. Understand the a/b is the sum of fractions 1/b. Decompose a/b into the sum of fractions with denominator of b in more than one way. Understand addition and subtraction of fractions as joining and separating parts referring to the same whole. Solve word problems involving adding and subtracting of fractions with like denominators.
Grade 5	Add and subtract fractions with unlike denominators using equivalent fractions. Solve word problems involving addition and subtraction of fractions referring to the same whole. Use benchmark fractions and number sense of fractions to estimate mentally and assess the reasonableness of answers.

Figure 12-6 *Summary of the CCSSM standards relating to addition and subtraction of fractions.* This summary was made by the authors of *Helping Children Learn Mathematics*. The full document (CCSSI, 2010) may be obtained at www.corestandards.org/the-standards/mathematics.

how to add and subtract fractions. If children understand how to count fractional parts (e.g., 3 sevenths and 2 more sevenths is 5 sevenths) and how to compare two fractions using a pictorial model, they may find their own ways to solve joining and separating problems with fractions. The experience suggested in the CCSSM standards of decomposing a fraction in many ways such as seeing that $\frac{6}{9}$ is $\frac{2}{9} + \frac{4}{9}$, $\frac{5}{9} + \frac{1}{9}$, $\frac{3}{9} + \frac{2}{9} + \frac{1}{9}$, $\frac{3}{9} + \frac{3}{9}$ (or $\frac{1}{3} + \frac{1}{3}$) is helpful background for adding and subtracting fractions.

Look at the word problems in Figure 12-7. Even very young children have no difficulty with A, and those with a strong conceptual background in fractions should be able to solve B without being taught a rule. Have them show how they solved each one, encouraging them to tell how the two problems are alike (e.g., both problems

---

A
Cyrilla ate 2 apples, and Carey ate 1 apple. How many apples did they eat altogether?

B
Cyrilla ate $\frac{2}{5}$ of a pie, and Carey ate $\frac{1}{5}$ of a pie. How much pie did they eat altogether?

C
Cyrilla ate $\frac{2}{3}$ of a cake, and Carey ate $\frac{1}{4}$ of the cake. How much of the cake did they eat altogether?

Figure 12-7 *Three joining problems.*

involve adding 2 and 1; in problem A, you have to add Cyrilla's 2 apples to Carey's 1 apple, and in problem B, you have to add Cyrilla's 2 fifths of the pie to Carey's 1 fifth). Also, have children draw a picture to illustrate the first two problems, even if they find it quite simple to do so. Then let them try problem C. Discuss how problem C is like the other two problems and how it is different. If the children need help, encourage them to draw a rectangle to represent the cake.

Children may come up with various ways to approach the problem. If they have drawn pictures to compare fractions, they may draw pictures like these:

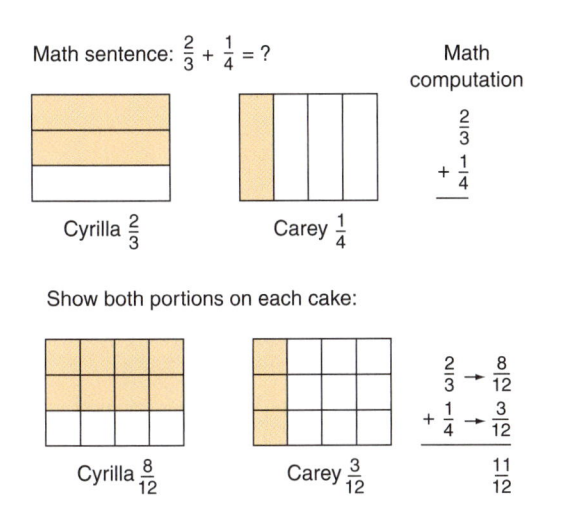

Math sentence: $\frac{2}{3} + \frac{1}{4} = ?$ Math computation

$\begin{array}{r} \frac{2}{3} \\ + \frac{1}{4} \end{array}$

Cyrilla $\frac{2}{3}$ Carey $\frac{1}{4}$

Show both portions on each cake:

Cyrilla $\frac{8}{12}$ Carey $\frac{3}{12}$

$\begin{array}{r} \frac{2}{3} \rightarrow \frac{8}{12} \\ + \frac{1}{4} \rightarrow \frac{3}{12} \\ \hline \frac{11}{12} \end{array}$

Note how the picture above could also be used to see the result if Carey had eaten $\frac{3}{4}$ of the cake and Cyrilla had still eaten $\frac{2}{3}$. The picture, shaded differently, could show that Carey had eaten $\frac{9}{12}$ of a cake. Together, therefore, they would have eaten $\frac{17}{12}$ or one whole cake and $\frac{5}{12}$ of another.

This highlights the point that, when adding fractions, the whole (the unit) is assumed to be the same. That is, in this example, where Cyrilla ate $\frac{2}{3}$ of a cake and Carey ate $\frac{3}{4}$ of a cake, not only would there have to be two cakes, but the cakes are assumed to be the same size. If Cyrilla ate $\frac{2}{3}$ of a very small cake and Carey ate $\frac{3}{4}$ of a large cake, you would need to know the relative sizes of the cakes before you could determine how much they ate altogether or who ate more.

Adding mixed numbers is no more difficult than adding proper fractions. Children who have made sense of adding proper fractions can put this together with what they know about adding whole numbers to handle a computation such as $5\frac{2}{3} + 3\frac{3}{4}$. Adding the whole numbers gives 8, and adding the fractions gives $\frac{17}{12}$, as we saw above. Putting these together gives $8\frac{17}{12}$, and children can then use their knowledge that $\frac{17}{12}$ is express the total as $9\frac{5}{12}$.

You can approach subtracting two proper fractions in the same way that you approached adding. Begin with subtraction problems that involve comparison. For example, how much more cake did Cyrilla eat than Carey in problem C (Figure 12-7)? You could use the same model to find the answer. Note that knowing how to order fractions is important in real-life situations. In problem C, it would not make sense to ask how much more cake Carey ate than Cyrilla. Subtracting improper fractions or mixed numbers is often more difficult, partly because children have not worked on changing a mixed number to another mixed number and partly because they lack understanding of regrouping. Using a model such as the fraction bars in the Appendix can help children think about how fraction equivalents are useful for subtracting when regrouping is necessary. For example, the symbolic subtraction problem $6\frac{3}{7} - 2\frac{4}{7}$ requires regrouping $6\frac{3}{7}$ to express it as $5\frac{10}{7}$. You can help children understand this regrouping by getting them to see that 1 is $\frac{7}{7}$; so 6 is $5 + \frac{7}{7}$.

Thus, $6\frac{3}{7}$ is $5 + \frac{7}{7} + \frac{3}{7}$, or $5\frac{10}{7}$.

If children are having difficulty with regrouping fractions, a model like the following can help them see, for example, that $6\frac{3}{7} = 5\frac{10}{7}$.

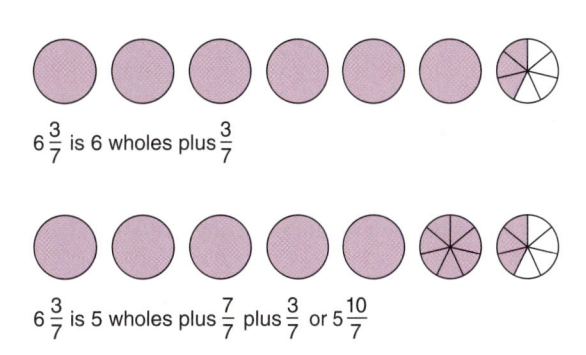

$6\frac{3}{7}$ is 6 wholes plus $\frac{3}{7}$

$6\frac{3}{7}$ is 5 wholes plus $\frac{7}{7}$ plus $\frac{3}{7}$ or $5\frac{10}{7}$

Students often find subtraction more difficult than addition because subtraction often involves more steps. As in addition, students must express the fractions as equivalent fractions with a common denominator. However, in subtraction, they may also have to regroup if the fractional part being subtracted is larger than the other fractional part. The previous example ($6\frac{3}{7} - 2\frac{4}{7}$) requires children to regroup $6\frac{3}{7}$ to $5\frac{10}{7}$ since they cannot subtract $\frac{4}{7}$ from $\frac{3}{7}$.

Students who approach subtraction problems by trying to follow a rule have a great deal of difficulty with problems like $2 - 1\frac{3}{5}$. In contrast, students who have a mental picture of this subtraction and who see that $\frac{2}{5} + \frac{3}{5}$ makes a whole will be able to think through this problem rather than compute mindlessly. They lack the flexibility of procedural fluency, mentioned in Chapter 2 as an essential part of being mathematically proficient. Children who have a firm understanding of fractions will find ways to add and subtract fractions without having to rely solely on the traditional algorithms.

MULTIPLICATION AND DIVISION

Figure 12-8 is an overview of the expectations presented in the CCSSM document.

MULTIPLICATION   The algorithm for multiplication of fractions is one of the simplest: Multiplying the numerators gives the numerator of the product, and multiplying the

	Number and Operations-Fractions Domain
Grade 4	Multiply a fraction by a whole number. Understand that a/b is a multiple of 1/b (3/4 is 3 × ¼) and a multiple of a/b is also a multiple of 1/b (5 × ¾ is 15 × ¼). Solve word problems by using visual fraction models and equations to represent the problem.
Grade 5	Multiply a fraction by a fraction. Understand the size of result of multiplying a fraction by a fraction less than one and greater than one. Divide whole numbers by a unit fraction and unit fractions by whole numbers. Solve problems involving multiplication and division with these types of numbers.
Grade 6	Interpret and compute quotients of fractions, and solve word problems involving division of fractions by fractions, e.g., by using visual fraction models and equations to represent the problem.

Figure 12-8 *Summary of the CCSSM standards relating to multiplication and division of fractions.* This summary was made by the authors of *Helping Children Learn Mathematics*. The full document (CCSSI, 2010) may be obtained at www.corestandards.org/the-standards/mathematics.

denominators gives the denominator of the product. Students can learn this algorithm in minutes (and forget it in seconds unless they practice it a great deal); however, learning this algorithm does not give students insight into why it works or when to use it. That is why we suggest teaching multiplication of fractions in a way that helps students understand the meaning of multiplication, helps them get an idea of the size of an answer, and helps them see why the algorithm works before they are presented with the algorithm itself.

In the following sections, we examine three different cases of multiplication: a whole number times a fraction, a fraction times a whole number, and a fraction times a fraction. In each case, we tie multiplication with fractions to a meaning of multiplication with whole numbers—for example, using the knowledge that 3×4 means three groups of four and that 3×4 is the area of a 3-by-4 rectangle.

Whole Number Times a Fraction Begin with a problem like this one:

> You have 3 pans, each with $\frac{4}{5}$ of a pizza. How much pizza do you have?

Ask your students how this problem is like having three bags, each with four marbles. See if this helps them understand the similarity between multiplying whole numbers and multiplying a whole number times a fraction. How do we find three groups of four? We could put out three groups of four marbles and count them. Similarly, to find three groups of $\frac{4}{5}$, we could put out three pans of pizza, each with $\frac{4}{5}$ of a pizza, and count the number of fifths. This approach is shown by a model like the following:

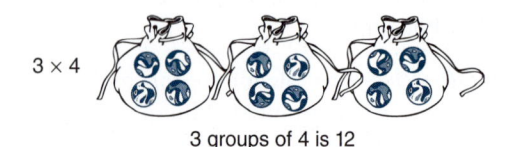

3×4

3 groups of 4 is 12

$3 \times \frac{4}{5}$

3 groups of 4 fifths is 12 fifths

If we symbolize three groups of four as 3×4, it makes sense to symbolize three groups of $\frac{4}{5}$ as $3 \times \frac{4}{5}$.

Then we can consider 3×4 as repeated addition: $4 + 4 + 4 = 12$. We can approach $3 \times \frac{4}{5}$ in the same way:

$$\frac{4}{5} + \frac{4}{5} + \frac{4}{5} = \frac{12}{5}$$

After students have solved problems like this one with pictures, see if they can solve them without pictures, maybe with repeated addition. Be sure to place a strong emphasis on the words. For example, in discussing the problem $5 \times \frac{2}{3}$, have them listen carefully as you read:

$5 \times \frac{2}{3}$ is 5 groups of two-thirds, which is

5 groups of 2 *thirds* or 10 *thirds*—which is $\frac{10}{3}$.

Fraction Times a Whole Number Again, begin with a problem—for example:

> You have $\frac{3}{4}$ of a case of 24 bottles. How many bottles do you have?

If children have worked with the set model for fractions, they have the background to solve this problem using physical objects. Now they must move to solving it symbolically and tying it to multiplication.

First, help the children look at why it makes sense to consider this problem a multiplication problem. Ask them: "If you had 5 cases with 24 bottles in each, what would you do to find out how many bottles you have?" Get them to see that they would multiply 5×24. Similarly, for 20 cases they would multiply 20×24, and for 53 cases they would multiply 53×24. Thus, for $\frac{3}{4}$ of a case it is logical that they should multiply $\frac{3}{4} \times 24$.

Then have children tell how to find $\frac{3}{4}$ of 24. Lead them to see that they would first partition (or divide) the set into four equal parts (each part would have 6); in other words, first find $\frac{1}{4}$ of 24, which is 6, and then look at 3 of the parts—that is, $\frac{3}{4}$ is 3 times as many as $\frac{1}{4}$, or $3 \times 6 = 18$. Thus, $\frac{3}{4} \times 24 = 18$. This thinking is summarized below:

$\frac{3}{4} \times 24 =$ 	**Think**
one-fourth of 24 is 6
three-fourths is 3 times as many
or 3×6
$3 \times 6 = 18$

At first glance, this procedure looks different from the usual algorithm for multiplying numerators and multiplying denominators. First, you divide the number in the set (24) by 4 and then multiply that by 3. When you perform the standard algorithm, you multiply first (3×24) and then divide by 4.

Another way to approach this type of problem (fraction times a whole number) is to use the idea of commutativity. That is, point out to the children that $\frac{3}{4} \times 24$ is the same as $24 \times \frac{3}{4}$, so they can multiply using the procedure discussed above (whole number times a fraction). However, children do not readily understand that $\frac{3}{4}$ of a case of 24 is the same as having 24 groups of $\frac{3}{4}$. Although this approach may be easier, the opportunity to present the "of" meaning of multiplication is lost.

Fraction Times a Fraction Consider this problem:

You own $\frac{3}{4}$ of an acre of land, and $\frac{5}{6}$ of this is planted in trees. What part of the acre is planted in trees?

Children need to understand why this is a multiplication problem. Showing them pictures like the ones below may help them see how this problem is related to the area of a rectangle and thus to multiplication. In the first picture, the acre is partitioned into fourths and the amount you own ($\frac{3}{4}$) is shaded.

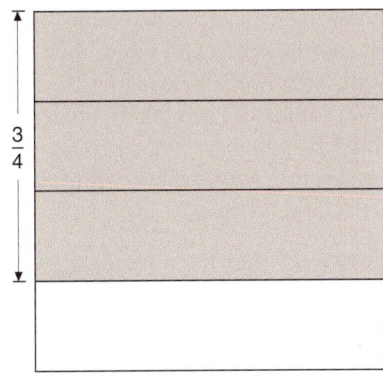

$\frac{3}{4}$ of an acre (the amount you own)

In the next picture, the acre is partitioned into sixths, with little drawings of trees marking the $\frac{5}{6}$ of your land that is planted in trees. Help the children see that they need to identify the size of each of the small rectangles in order to find what part of the whole acre is planted in trees. Since there are 24 small rectangles, each is $\frac{1}{24}$ of an acre. And since 15 of the small rectangles are planted in trees, $\frac{15}{24}$ of the acre is planted in trees.

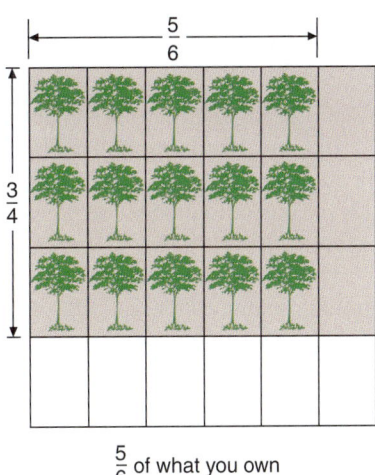

$\frac{5}{6}$ of what you own

Pictures like these can help the children see that the tree-planted rectangle is $\frac{3}{4}$ by $\frac{5}{6}$, so its area is found by multiplying $\frac{3}{4} \times \frac{5}{6}$.

You need to develop this model slowly and only after children have had experience with finding areas of rectangles. After children use a model for problems like this, make a list of fraction times fraction products (do not simplify the products). See if the children notice a pattern for multiplying two fractions.

This model can be used with children who have had experience with finding area to help them understand the algorithm for multiplying fractions (multiply the numerators, multiply the denominators). You can refer to the diagrams to show why the procedure works. The acre is partitioned into fourths one way and sixths the other way, so there are $4 \times 6 = 24$ equal parts (24 is the denominator). Trees were planted in three rows of five of these parts, so the numerator is $3 \times 5 = 15$. Symbolically, write

$$\frac{3}{4} \times \frac{5}{6} = \frac{3 \times 5}{4 \times 6} = \frac{15}{24}$$

At this point, return to earlier examples and have the children verify that this algorithm works for those examples also. Make sure the children understand that any whole number can be represented by a fraction (e.g., $6 = \frac{6}{1}$); this will help them see that they can use the algorithm of multiplying numerators and denominators when one of the factors is a whole number.

DIVISION Consider this problem:

How many 2-foot hair ribbons can be made from a 10-foot roll of ribbon?

Students should understand that they are being asked to find how many 2-foot pieces (the size of the group) are in the whole, or how many twos are in ten. A picture like this would show them how they might do this in real life:

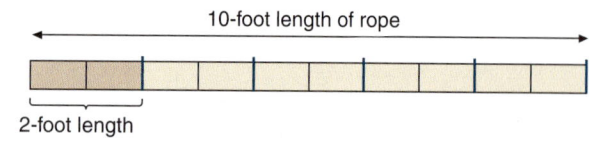

10-foot length of rope

2-foot length

Students can readily see that the 10-foot roll contains five 2-foot hair ribbons. This is symbolized by $10 \div 2 = 5$.

Turn the attention now to dividing by a unit fraction (see Figure 12-8). you could modify the previous picture by marking off $\frac{1}{2}$-foot sections instead of 1-foot sections to show students how to find the number of $\frac{1}{2}$-foot hair ribbons they could make from this same 10-foot roll (i.e., what is $10 \div \frac{1}{2}$?). The picture would look similar to the one Susie drew when she was finding the number of $\frac{1}{4}$-foot hair ribbons she could make from a 3-foot roll:

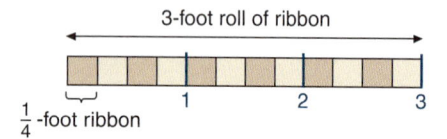

3-foot roll of ribbon

$\frac{1}{4}$-foot ribbon

In this problem, we want to know how many one-fourths are in 3, or $3 \div \frac{1}{4}$. Students can see that there are twelve $\frac{1}{4}$-foot ribbons in a 3-foot roll. You should also have students notice that the picture shows that there are 3 groups of 4 (3×4) hair ribbons.

Now ask students how they would draw a picture to find out how many $\frac{3}{4}$-foot pieces there are in a 6-foot roll of ribbon? They should see that, in this case, they would mark off $\frac{3}{4}$-foot lengths, as shown here:

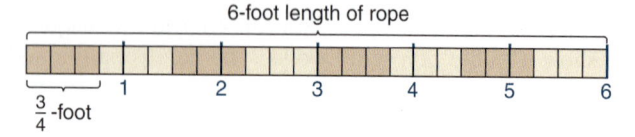

6-foot length of rope

$\frac{3}{4}$-foot

Point out that fourths are marked off first, and then the fourths are grouped by threes. That is, we multiplied by 4 and divided by 3 (or multiplied by $\frac{1}{3}$). When you were in school, you probably learned this as inverting the divisor and then multiplying, as indicated by this symbolic representation:

$$6 \div \frac{3}{4} = \frac{6}{1} \times \frac{4}{3} = \frac{6 \times 4}{1 \times 3} = \frac{24}{3}, \text{ or } 8$$

In these types of problems, students (even adult students!) often have difficulty with ribbon being left over after they cut as many pieces of the specified length as possible. You can help students understand such cases by posing a problem like this:

> Suppose that you have a 2-foot roll of ribbon and you want to cut as many $\frac{3}{4}$-foot ribbons as you can. You know that you can cut at least two because you can get one from each foot of ribbon. Can you cut three?

Students can approach this problem by looking at this picture.

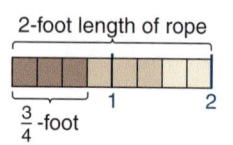

2-foot length of rope

$\frac{3}{4}$-foot

Students can see that they can cut two $\frac{3}{4}$-foot ribbons (indicated by the two shaded parts) but that the leftover (unshaded) part is not enough for another $\frac{3}{4}$- foot ribbon. Ask them what part of another they can make. They should see that the unshaded part is $\frac{2}{3}$ of a $\frac{3}{4}$-foot ribbon. From this, children should be able to conclude that they can make $2\frac{2}{3}$ hair ribbons (each $\frac{3}{4}$-foot long).

Doing examples like the last one ($2 \div \frac{3}{4}$) will help children reason through problems in which the divisor is larger than the dividend—for example, how many $\frac{1}{2}$-foot ribbons could be cut from a strip of ribbon that is $\frac{1}{3}$-foot long ($\frac{1}{3} \div \frac{1}{2}$). Children should recognize that $\frac{1}{2}$ is larger than $\frac{1}{3}$, so they could not cut even one $\frac{1}{2}$-foot ribbon. The question is, what part of a $\frac{1}{2}$-foot ribbon could they cut? Have the children draw a picture to convince themselves that they could cut $\frac{2}{3}$ of a $\frac{1}{2}$-foot ribbon.

In the Classroom 12-6 consists of a few rhymes that incorporate questions about fractions. Even older children like their silliness. Be sure, however, that you balance fanciful activities like this with activities involving situations from students' everyday life. Use the diversity in your class as a resource by making up problems that relate to students' different customs.

CONCEPTUAL DEVELOPMENT OF DECIMALS

As students develop decimal concepts, they should relate decimals to what they already know about common fractions and place value. For ease of discussion, we consider these relationships separately, but you should keep in mind that you will need to weave them together in your teaching. You should also keep in mind that it is important to relate decimals to everyday experiences familiar to the students, since contextual situations often help students understand the mathematics. The video described in Math Link 12.3 shows young children becoming familiar with the use of decimals in their lives.

Math Links 12.3

To see young children discussing the use of decimals in their lives, view *Everyday Decimals*, video 35 from the Annenberg series, Teaching Math: A Video Library, K–4, which you can access from http://www.learner.org/resources/series32.html or from this book's Web site.

www.wiley.com/college/reys

Conceptual Development of Decimals

	Number and Operations-Fractions Domain
Grade 4	Understand decimal notation for fractions with denominators of 10 and 100. Compare decimal fractions involving tenths and hundredths. Understand the place value meaning of tenths and hundredths. (0.34 is 3/10 + 4/100 or 34/100).
Grade 5	Extend the understanding of place value, recognizing that a digit in one place represents 10 times as much as it represents in the place to its right and 1/10 of what it represents in the place to its left. Read and write (standard and expanded form) decimals to thousandths using base-ten numerals. Compare decimals to thousandths. Use place value understanding to round decimals to any place.

Figure 12-9 *Summary of the CCSSM standards relating to multiplication and division of fractions.* This summary was made by the authors of *Helping Children Learn Mathematics*. The full document (CCSSI, 2010) may be obtained at www.corestandards.org/the-standards/mathematics.

The conceptual development of decimals recommended in the CCSSM standards are summarized in Figure 12-9. Note that in Grade 4 the meanings are closely tied to fractions and decimals.

RELATIONSHIP TO COMMON FRACTIONS

Decimals are just another notation for fractions—specifically, for tenths, hundredths, and other powers often parts of a unit. Thus, basic to understanding decimals is understanding these fractional parts—an understanding that you should help students build when you are developing the meaning of fractions.

TENTHS Before introducing the decimal notation for tenths, review what students should know about tenths from their prior study of fractions. They should know that partitioning a unit into tenths results in 10 equal parts. They also should be able to make the connection between a model of a fraction, the word for the fraction, and the fraction written symbolically. And they should know that 10 tenths make a whole, that 7 tenths is less than a whole, and that 27 tenths is more than a whole.

With this background, children should be ready to learn that, for example, 0.3 is a symbol for $\frac{3}{10}$. You should help children link their knowledge about place value to this new notation (see the discussion in Chapter 8 on place value). You also need to stress that 0.3 and $\frac{3}{10}$ are both said as "three-tenths." Then have them look at a number such as $2\frac{7}{10}$. Show them that this is written 2.7 in the new notation and said as "two and seven-tenths." Point out that the word *and* is said for the decimal point. Saying decimals aloud in

---

In the Classroom 12–6

A NEW TWIST ON OLD RHYMES

Objective: To verify whether or not computational statements about fractions are true.

Grade Level: 4–6.

- Have the students decide if the following rhymes are true and show how they know by using models or drawing pictures.

Humpty Dumpty sat on a wall.
"Wow, I have 45 balloons in all."
The wind blew $\frac{2}{5}$ of them to the sky.
"Now, I have 30 balloons, oh my!"

Little Miss Muffet
Sat on a tuffet
And ate $\frac{1}{2}$ of a candy bar.
Along came a spider
And ate $\frac{1}{5}$ right beside her.
The remaining $\frac{1}{10}$ won't go far.

Jack and Jill went up the hill
To fetch a pail of water.
Jack filled $\frac{2}{3}$ without fail.
Jill added $\frac{1}{4}$ to the pail.
And $\frac{11}{12}$ spilled as Jack caught her.

Little Jack Horner
Sat in the corner
Eating a pizza pie
He ate $\frac{1}{2}$ and $\frac{1}{2}$ of the rest
"None left," said he, "I am the best."

- Have the students make up their own rhymes using fractions or change these rhymes with other fractions.

this way helps students relate what they know about fractions to decimals. In the Classroom 12-7 shows a quick game that children can play to practice writing in decimal and fraction notation.

Before introducing hundredths, you should make certain that students have made all three connections shown by this triangle (model and symbol, symbol and word, word and model).

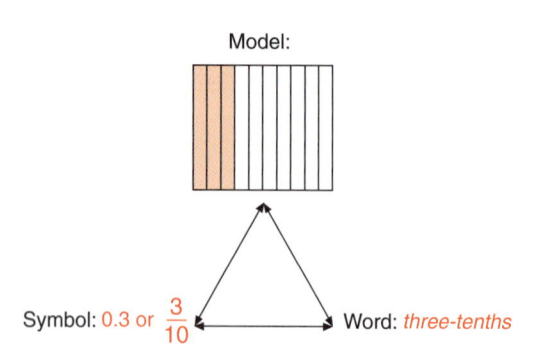

Model:

Symbol: 0.3 or $\frac{3}{10}$ ⟷ Word: *three-tenths*

> **In the Classroom 12–7**
>
> ### CAN YOU BEAT THE TOSS?
>
> *Objective:* Write numbers as fractions or decimals.
>
> *Grade Level:* 3–5.
>
> *Materials:* Paper and pencil for children; a penny for the teacher.
>
> Directions:
> - ▼ The teacher reads a number.
> - ▼ Each child writes the number in either decimal or fraction notation.
> - ▼ Each child receives 1 point if he or she writes the number correctly.
> - ▼ The teacher tosses a coin.
>   - If it's *heads*, each child who wrote a correct decimal receives one point.
>   - If it's *tails*, each child who wrote a correct fraction receives one point.
>
> Suggested Numbers:
>
> | | |
> |---|---|
> | two-tenths | one and three-tenths |
> | seven-tenths | twenty-two and five-tenths |
> | five and seven-tenths | thirty-four and no-tenths |
> | eleven and four-tenths | |
>
> Challenge Numbers:
>
> | | |
> |---|---|
> | sixteen-tenths | five and eleven-tenths |
>
> A more detailed plan for this activity is available on the Student Companion site at www.wiley.com/college/reys.

HUNDREDTHS Give each child multiple copies of the model of hundredths (in Appendix, the decimal or percent paper contains six copies of the 10-by-10 grid that you can use for this model). Make certain the children know that the unit (the entire square grid) is one and that each part (each small square) is one hundredth. Ask them to shade $\frac{3}{100}$, $\frac{7}{100}$, $\frac{10}{100}$, and $\frac{21}{100}$ on the different copies of the model.

Then have the children shade the first and second columns on the hundredth model and three more small squares. Ask what part of the whole is covered. Elicit both responses: $\frac{2}{10} + \frac{3}{100}$ as well as $\frac{23}{100}$. Then write the following:

$$\frac{23}{100} = \frac{2}{10} + \frac{3}{100}$$

Now connect the place-value interpretation and the decimal notation 0.23. You should also use the model to show that 0.2 = 0.20. Continue having children make the connections shown in the triangle (i.e., by connecting the model, the symbol, and the word). These examples are only a brief introduction to hundredths. You will need to return to models as you continue helping students make sense of hundredths and how they relate to tenths. See Chapter 8 for other place-value models. You can change these models and use them with decimals. The dollar is also a natural model for hundredths. However, it should not be the only model, since children do not necessarily relate the cents (whole numbers) to a fractional part of a dollar. The site referenced in Math Link 12-2 also has decimal models.

THOUSANDTHS AND OTHER DECIMALS Older students often do not understand small decimals such as 0.234 or 0.0003, partly because there is less emphasis on these decimals but also because teachers often expect children to generalize from hundredths to all the other places. Most of your work with smaller decimals should be done through the place-value interpretation because the fractions and models become unwieldy. However, you should make sure children understand that a thousandth is one-tenth of a hundredth, and you should develop this understanding by using a model. Students often have difficulty hearing the difference between "thousands" and "thousandths"—careful pronunciation of all the fractional names (tenths, hundredths, thousandths, etc.) is essential.

WRITING FRACTIONS AS DECIMALS If decimals have been introduced carefully, students should be able to write the fraction notation for any decimal and the decimal notation for any fraction expressed in tenths, hundredths, and so on. For example, they should be able to write $\frac{2}{10}$ and $\frac{23}{100}$ as 0.2 and 0.23, no matter whether they read the written fractions or hear them read as "two tenths" and "twenty-three hundredths."

However, the fact that students can write fractions as decimals when the fractions are expressed in tenths, hundredths, and so on does not mean that they will be able to write other fractions as decimals. Students try to make sense of the relationship between fractions and decimals, but they often do not have a firm understanding of the quotient meaning of fractions. For example, they do not understand that $\frac{4}{5}$ means 4 divided by 5. They should realize that the quotient is less than 1. In context, this is evident. For example, if you have 4 cookies to divide among 5 children, each will receive less than 1 cookie. To divide 4 by 5, children must know that 4 is equivalent to 4.0 and must be able to divide decimals. If you can help students gain these understandings, they can use a calculator or the traditional algorithm to see that $\frac{8}{10}$ is 0.8. Children who also understand equivalent fractions should know that $\frac{4}{5}$ is $\frac{8}{10}$ or 0.8. Having children see that both methods lead to the same decimal is important. They can use the method of division to change any fraction to a decimal (or a decimal approximation, if the division doesn't terminate in zeros).

Children can explore many interesting patterns. Some can be investigated with a calculator—for example, the decimal approximations for ninths:

$$\frac{1}{9} = 0.11111$$

$$\frac{2}{9} = 0.22222$$

$$\frac{3}{9} = 0.33333$$

You should do a few of these calculations by hand to show that the pattern continues forever. If you only use a calculator, students may think that $\frac{1}{9} = 0.11111111$ exactly (i.e., the number of places on their calculator display).

Similarly, you should be looking at the meaning of $\frac{1}{3}$ to help students see that $\frac{1}{3}$ does not exactly equal 0.33, 0.333, or 0.3333. For instance, you can point out that if $\frac{1}{3} = 0.33$, then each of three equal parts of a whole would be 0.33, and thus the whole would be 0.33 + 0.33 + 0.33, or 0.99, which is not 1. Similarly, you can show students that $\frac{1}{3}$ does not equal 0.3333 (although this is a closer approximation to $\frac{1}{3}$).

RELATIONSHIP TO PLACE VALUE

The place-value interpretation of decimals is most useful for helping children understand computation with decimals. Let's look now at one way to develop this interpretation and consider how to use it in teaching children how to order and round decimals.

What do children know about place value? Consider the number 2463. By third grade, children can identify the places (ones, tens, hundreds, and thousands) and the number in each place (3, 6, 4, and 2). They know, for example, that the 4 means 4 hundreds. They also have learned how the places are filled: beginning with 1 as a unit, grouping 10 of these to form a new unit (tens), grouping 10 of these to form a new unit (hundreds), and so forth.

In introducing place-value ideas with decimals, again begin with 1 as the unit. But instead of grouping by tens, group by one-tenth of the one to form the new unit of tenths. To indicate this new unit in the place-value system, use a decimal point after the ones place. Help students understand that 10 of the tenths make a one (just as 10 of any unit make the next larger unit). They should also be able to identify the tenths place in a number. Be sure to integrate the place-value interpretation of decimals with the interpretation of decimals as fractions.

When introducing hundredths, you should again focus on the place-value interpretation. Given a number such as 51.63, a child should be able to tell what number is in the tenths place and what number is in the hundredths place, as well as the relationships between the places (a hundredth is one tenth of a tenth, or 10 hundredths is 1 tenth). After you introduce thousandths in a similar way, the children should be able to generalize to any decimal place.

Be sure to read decimals aloud in a way that emphasizes the tie between decimals and fractions. For example, read 24.09 as "twenty-four and 9 hundredths," not as "two four point zero nine" or "twenty four point oh nine." The words *tenths*, *hundredths*, and so on help students maintain that tie to the meaning.

Often, you can use a place-value grid to help students having difficulty with decimals. Consider, for example, this grid for the number 32.45:

T	O	tth	hth
3	2	4	5

Point out that the decimal can be seen both as *32 and 45 hundredths* and as *32 and 4 tenths and 5 hundredths*. It can also be read as 3242 hundredths. What other ways could you express this number in words?

These children will also benefit from being asked questions like the ones in Figure 12-10 and organizing their thoughts on a decimal grid.

Children can use a grid in as rote a way as a rule unless you continue to help them explain their thinking and relate the grid to the model. It is certainly not enough to fill in zeros on the grid anywhere there is not a number. It is more important to discuss why, for example, 4.3 is equivalent to 4.30.

ORDERING AND ROUNDING DECIMALS

Children should be able to understand the ordering and rounding of decimals based directly on their understanding of decimals and their ability to order and round whole numbers. For decimals, this understanding must include being able to interpret the decimals in terms of place value and being able to think of, for example, 0.2 as 0.20 or 0.200.

Here is an example of two children discussing which is larger, 23.61 or 23.9.

CARTER: I think that 23.61 is larger because it has more places, just like 2361 is larger than 239.

JEANE: But both of these (23.61 and 23.9) are 23 and some more, so all we have to do is compare the decimal parts.

Write 8 hundredths:

T	O	tth	hth
	•	0	8

Children should realize there are 0 tenths.

What is 29 hundredths?

T	O	tth	hth
	•	2	9

Children should realize there are 2 tenths and 9 hundredths.

What is 29 tenths?

T	O	tth	hth
	2 •	9	

Children should realize there are 2 ones and 9 tenths.

Now look at 4.3 on the grid. How many tens?

T	O	tth	hth
	4 •	3	

Children should realize there are no tens.

Tell the children that we could write a 0 in the tens place, but this is not customary.

How many hundredths?

T	O	tth	hth
	4 •	3	0

Children should realize that 4.3 is equivalent to 4.30.

Figure 12-10 *Using a decimal grid with place value.*

CARTER: OK, it's not like whole numbers.

JEANE: Well, sort of. We begin with the largest place—the tens are the same, the ones are the same, so we need to look at the tenths. The 9 is larger than the 6, so 23.9 is larger than 23.61.

CARTER: I follow you until that last part. Now you have a 9 and a 61, and 61 is larger.

JEANE: But the 9 is 9 tenths and the 61 is 61 hundredths. Want to look at a model?

CARTER: I get it—I remember the model. Nine tenths is 9 strips, or 90 little squares. It is 90 hundredths, so it is larger.

As you see from this conversation, if children really understand the notation and meaning of decimals, they can Figure out ways to compare decimals.

In rounding a decimal such as 24.78 to the nearest tenth, children need to ask themselves the same types of questions as with whole numbers. Children must also understand that 24.7 = 24.70.

Questions	Expected Responses
What "tenths" is 24.78 between?	It's between 24.7 and 24.8 (or 24.70 and 24.80)
Is it nearer 24.7 or 24.8?	It's nearer to 24.80
How will you round it?	To 24.8

OPERATIONS WITH DECIMALS

One advantage of decimals over fractions is that computation is much easier, since it basically follows the same rules as for whole numbers. It thus makes sense that little emphasis (two expectations) is placed on decimal computation in the CCSSM (see Figure 12-11).

In teaching the algorithms for decimals, you should build on the place-value interpretation of decimals and the corresponding whole-number algorithms. (Adapt the ideas in Chapters 10 and 11 for use with decimals operations.) Given the wide availability of calculators, it is important that you spend as much time seeing whether answers are reasonable as you spend on the algorithms. Thus, estimation skills are crucial (see Chapter 10).

In the Classroom 12-8 shows an activity that requires estimation skills. Encourage students to use whole numbers that are compatible when trying to answer the questions. You can present questions of this type to students before they learn about decimal computation. For example, England's population is about 50 million and Scotland's is about 5 million, so England has about 10 times as many people as Scotland.

	Number and Operations-Fractions Domain
Grade 5	Add, subtract, multiply, and divide decimals to hundredths, using concrete models or drawings and strategies based on place value, properties of operations, and/or the relationship between addition and subtraction.
	Explain patterns in the number of zeros of the product when multiplying a number by powers of 10, and explain patterns in the placement of the decimal point when a decimal is multiplied or divided by a power of 10.
Grade 6	Fluently add, subtract, multiply, and divide multi-digit decimals using standard algorithms.

Figure 12-11 *Summary of the CCSSM standards relating to multiplication and division of fractions.* This summary was made by the authors of *Helping Children Learn Mathematics.* The full document (CCSSI, 2010) may be obtained at www.corestandards.org/the-standards/mathematics.

ADDITION AND SUBTRACTION

Children who have a good concept of decimals have little trouble extending the whole-number algorithms for addition and subtraction to decimals. They realize that they need to

- Add or subtract like units (tenths with tenths, hundredths with hundredths, etc.).
- Regroup in the decimal places as they did with whole numbers.

Difficulty with adding or subtracting decimals arises mainly when the values are given in horizontal format or in terms of a story problem and the decimals are expressed to a different number of places (e.g., 51.23 + 434.7). To deal with this difficulty, have the children first estimate by looking at the wholes (about 480). Some children may need help in lining up the like units and so may benefit from using a grid:

	H	T	O	t^{th}	h^{th}
		5	1 .	2	3
+	4	3	4 .	7	
	4	8	5 .	9	3

MULTIPLICATION AND DIVISION

Multiplying and dividing a decimal by a whole number are conceptually easier than multiplying two decimals or dividing a decimal by a decimal. Children can build on this to understand the more sophisticated computation. Begin by having children consider a problem like this:

> Six tables are lined up end to end. Each table is 2.3 meters long. How long is the line of tables?

From their previous work with multiplication, students should realize that this is a multiplication problem. Of course, they should also see that they can solve the problem by adding decimals. That is: $6 \times 2.3 = 2.3 + 2.3 + 2.3 + 2.3 + 2.3$, which equals 13.8.

However, just as children moved away from repeated addition to find the product of two whole numbers, they need to do so when multiplying a decimal by a whole number.

Children with a firm understanding of the place-value interpretation of decimals can think about multiplication with decimals in another way. For example: 2.3 is 23 tenths, so 2.3×6 is

$$\begin{array}{r} 23 \text{ tenths} \\ \times 6 \\ \hline 138 \text{ tenths, which is } 13.8 \end{array}$$

Similarly, for the problem 2.37×6, children could realize that 2.37 is 237 hundredths, so:

$$\begin{array}{r} 237 \text{ hundredths} \\ \times 6 \\ \hline 1422 \text{ hundredths, which is } 14.22 \end{array}$$

Working with a decimal grid can help students remember that 138 tenths is 13.8. In using this method, you should first have the students decide on a reasonable answer. Ask whether 6×2.37 is more than 12 and whether it as much as 138?

To help students make sense of multiplying a decimal by a decimal, rather than only remembering a rule about counting decimal places, see if they can model this problem: "What is the area of a porch that is 3.2 meters by 1.9 meters?"

In the Classroom 12–8

WHAT'S YOUR ANSWER?

Objective: To use estimations about decimal quantities to answer questions.

Grade Level: 4–5.

- Use the data in this table. These are data about the parts of the United Kingdom.

Parts	Area (Thousands of km²)	Population (Millions)
England	130.422	50.714
N. Ireland	14.144	1.733
Scotland	78.772	5.103
Wales	20.761	2.977

▼ Answer these questions:

- About how much larger in area is England than Scotland? ___
- About how large in area is the United Kingdom? ___
- About what is the population of the United Kingdom? ___
- Do twice as many people live in Wales than live in N. Ireland? ___
- Is England more than six times larger in area than Wales? ___

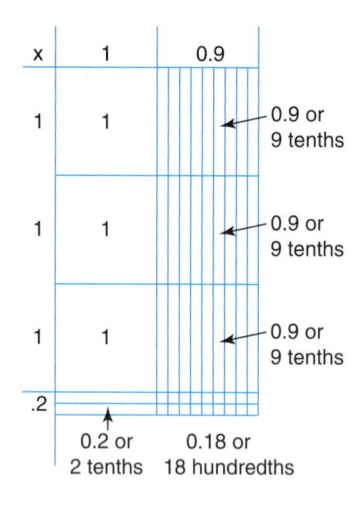

278 Chapter 12 • Fractions and Decimals: Concepts and Operations

```
 x   |  1  | 0.9
-----+-----+-----
  1  |  1  | 0.9 or 9 tenths
  1  |  1  | 0.9 or 9 tenths
  1  |  1  | 0.9 or 9 tenths
 .2  | 0.2 | 0.18 or
     | or  | 18 hundredths
     | 2 tenths
```

Add the parts:
1 + 1 + 1 = 3
9 tenths + 9 tenths + 9 tenths = 27 tenths or 2.7
Now find the total:
   3.
   2.7
   0.2
   0.18
   ————
   6.08

Students need to discuss whether 6.08 square meters is a reasonable answer. Ask them: "What would be the area of a porch that is 3 meters by 2 meters?"

Children can use the distributive algorithm for dividing whole numbers to divide a decimal by a whole number. Consider this problem:

A vinegar company distributed 123.2 million liters of vinegar equally among eight customers. How much vinegar did each customer receive?

First, ask for reasonable answers.

Did each customer get more than 10 million liters? [Yes, that would be only 80 million liters.]
Did each get more than 20 million liters? [No, that would be 160 million liters.]
What is an estimate for the answer? [Between 10 and 20 million liters.]

Then, talk through the division, as in Figure 12-12.

Later, students will need to develop division of a decimal by a decimal (why do you move the decimal points in the divisor and dividend) as well as multiplying any decimals (where and why do you place the decimal point in the product)? The development of fluency with multiplication and division of decimals takes more practice than with addition and subtraction, partly because children's skills with multiplication and division of whole numbers is not as firm as their skills with addition and subtraction.

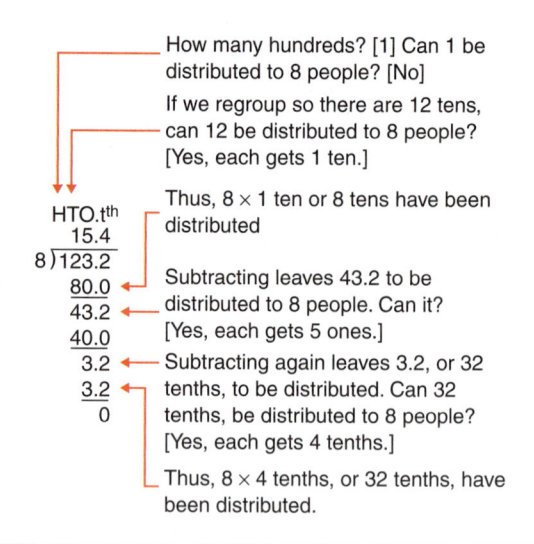

```
       HTO.tth
       15.4
    8)123.2
       80.0     ← How many hundreds? [1] Can 1 be distributed to 8 people? [No]
                  If we regroup so there are 12 tens, can 12 be distributed to 8 people? [Yes, each gets 1 ten.]
                  Thus, 8 × 1 ten or 8 tens have been distributed
       43.2     ← Subtracting leaves 43.2 to be distributed to 8 people. Can it? [Yes, each gets 5 ones.]
       40.0
        3.2     ← Subtracting again leaves 3.2, or 32 tenths, to be distributed. Can 32 tenths, be distributed to 8 people? [Yes, each gets 4 tenths.]
        3.2
        0         Thus, 8 × 4 tenths, or 32 tenths, have been distributed.
```

Figure 12-12 *Talking through division of a decimal by a whole number.*

CULTURAL CONNECTIONS

The Egyptians as early as 3500 B.C. were one of the first cultures in recorded history to make use of fractions. However, in their hieroglyphs, they wrote all fractions as the sum of unit fractions. For example, $\frac{3}{4}$ would have been written as $\frac{1}{2} + \frac{1}{4}$, and $\frac{2}{3}$ as $\frac{1}{2} + \frac{1}{6}$. They could do all four operations with fractions using this notation. (And you thought *our* system is cumbersome!)

Decimal fractions are found in the *Book of Chapters on Indian Arithmetic*, written by Abul Hassa Al-Uqlidisi in 900s. The author said the answer to successively halving 19 five times was 0.59375. The vertical mark over 0 indicates that the decimal fraction begins to the right. In America and Britain, this would be written 0.59375, however, other European countries use a comma instead of a period, and it would be written 0,59375. We often think of numbers as being an international language, but there are differences in notation that we use of which we should be cognizant.

At grade 4 there are considerable differences among countries with regard to the emphasis placed on fractions (Mullis, Martin, Gonzalez, and Chrostowski, 2004). Japan expects only half of its fourth graders to have a firm background in concepts involving fractions; students do little with fractions until about grade 5, when they are expected to add and subtract fractions with like denominators. The United States expects most of fourth-grade students to have some conceptual background in fractions as well as experience in adding and subtracting like fractions. Other countries, such as Singapore, expect all students to have experience in both of these areas. If you have students from other countries, you can expect a wide variety of skills and experience in dealing with fractions.

Japanese (Wantabe, 2006) students begin the study of fractions in the fourth grade and focus on developing concepts of proper fractions, improper fractions, and mixed numbers. Students begin developing fractions in a measurement context, which emphasizes that a fraction is an amount. This contrasts with the view of many students in the United States that a fraction is two numbers, which often leads to whole-number thinking rather than fractional thinking. In grade 5, they focus on equivalent fractions, comparison of fractions, fractions as quotients, and the relation of fractions to decimals. Operations with fractions are left until grade 6.

Some countries teach operations with decimals before operations with fractions, while others do the opposite. Many countries use a comma instead of a period for the decimal point, and this, along with the way numbers are displayed on a calculator, is one reason you often see whole numbers written without commas. Some countries say the denominator before they say the numerator. Be alert to these differences; students will adjust to the U.S. system if you are aware of the situation.

A GLANCE AT WHERE WE'VE BEEN

Three meanings of fractions are part–whole, quotient, and ratio (this chapter focuses on the part–whole meaning). Useful models of the part–whole meaning include such attributes as region, length, set, and area. The region model is the easiest for children to understand, and the rectangle is the easiest type of region. To help children make sense of fractions, begin with the important idea of partitioning (making equal shares) and then introduce words for fractional parts and counting the parts. Help children make connections among words, symbols, and models. Make sure children can figure out a whole from a fractional part and that they understand the idea of equivalent fractions. Use concrete models, pictorial models, and symbolic representations to build children's ability to order fractions and find equivalent fractions. Make sure children understand how to use benchmarks for estimating and checking the reasonableness of answers. Use models to develop children's understanding of mixed numbers and improper fractions.

To teach addition and subtraction of fractions, begin with problems and pictorial models that help children (1) see how adding and subtracting fractions can solve problems similar to problems with whole numbers, (2) develop an idea of a reasonable answer, and (3) see why a common denominator is necessary. Teach multiplication of fractions in a way that helps children (1) understand the meaning of multiplication and (2) estimate the size of an answer. This will help them see why the algorithm for multiplication works (multiply numerators and denominators). Begin with a whole number times a fraction, then a fraction times a whole number, and then a fraction times a fraction. For division, relate the way partitioning works with whole numbers to the way it works with fractions, using pictorial models.

Help children develop the concept of decimals by relating decimals to common fractions. Emphasize the place-value interpretation of decimals to help children in ordering and rounding decimals and to prepare them for operations with decimals.

Students in different countries learn about fractions and decimals at different grade levels; they also have different ways of writing and saying decimals. Teachers should be aware of these differences so they can deal appropriately with diverse students.

Things to Do: From What You've Read

1. The part–whole meaning of a fraction can be developed with different attributes, such as length and sets. Draw part–whole models for four different attributes to show $\frac{3}{4}$.
2. Illustrate three different meanings of $\frac{3}{4}$.
3. What is partitioning? How is it related to division and to fractions?
4. Verify that the last three examples in the Snapshot of a Lesson are halves. Then, show six different ways to make fourths. (Use the geoboard paper in Appendix; you may want to copy it so you will have a master for geoboard paper when you are teaching.)
5. Illustrate how two fractions can be compared using the pictorial model described on page 264. Show how this model could also be used to see if two fractions are equivalent.
6. With pictures, show why $2\frac{3}{5}$ is equivalent to $\frac{13}{5}$; show why $2\frac{3}{5}$ is equivalent to $1\frac{8}{5}$.
7. What is a benchmark for fractions? How are benchmarks used? Propose a problem ordering fractions that could be solved using benchmarks for fractions.
8. How is multiplication of fractions different from multiplication of whole numbers?
9. Describe how decimals and common fractions are alike and different.
10. How would you help children who have difficulty with this: 25.03 − 8.459?

Things to Do: Going Beyond This Book

In the Field

1. [1]*Three-fourths.* Interview at least three children using the tasks from In the Classroom Feature 12-2. Which of the models for fractions were the most difficult for children to understand?
2. [1]*Comparing Models.* Develop two activities for fractions using pattern blocks (see In the Classroom 12-4) that children could do at centers.
3. [1]*Can You Beat the Toss?* Play the game from In the Classroom 12-7 with children. Include hundredths if the children are familiar with these decimals.

In Your Journal

4. Give three different sets of two different decimals for students to compare. Tell why you chose each pair, that is, what you would expect to learn about children who did these exercises. Tell what you would do if the children encountered difficulties.
5. Describe how you were taught fractions in elementary school. Did those experiences help you make sense of fractions?

With Additional Resources

6. Read the article by Cramer and Whitney (see references). What points do they make that are also made in this chapter?
7. "Fraction Action" (Moone and de Groot) in *Teaching Children Mathematics* (see references) describes students learning about fractions involving the set model. Describe the basic approach to these lessons.
8. Create five story problems that ask children to compare fractions using a story from one of the books (or another) in the Book Nook for Children.

With Technology

9. Experiment with a calculator that has a fraction key. What fraction skills and understandings would be important for children to have if they were to use such calculators?
10. Watch the video *Fractions with Geoboards*, summarized in the Snapshot of a Lesson. What are three things from the video you would do if you were the teacher, and what are three things that you would change?

Note to Instructors: You can find additional resources, learning activities, and blackline masters in this text's accompanying Instructor's Manual, at www.wiley.com/college/reys.

---

[1]Additional activities, suggestions, and questions are provided in *Teaching Elementary Mathematics: A Resource for Field Experiences* (Wiley, 2010).

Book Nook for Children

Dodds, D. A. *Full House: An Invitation to Fractions.* Somerville, MA: Candle Wick Press, 2007.

Miss Bloom, the hostess of Strawberry Inn, welcomes a cast of hilarious characters until all the rooms are taken. In the middle of the night, Miss Bloom realizes that something is wrong at her inn, The guests are eating her cake! One delicious cake divided by five hungry guests and one hospitable hostess equals a perfect midnight snack at the Strawberry Inn and a good way to emphasis equal sharing and fraction names.

Hutchins, P. *The Doorbell Rang.* New York: Greenwillow Books, 1987.

This book is an excellent source to use when introducing the concept of partitioning. Children are challenged to share cookies fairly in what can be an interactive story.

Murphy, S. J. *Jump, Kangaroo, Jump!* New York: HarperCollins, 1999.

This book is about Australian animals planning a field day of events at a camp. Kangaroo and his friends must divide themselves up into different groups for the events that are taking place. There are activities and games in the back of the book for teachers and parents to use.

Napoli, D. J. *The Wishing Club: A Story About Fractions.* New York: Henry Holt, 2007.

Four siblings wish on a star but get only a fraction of their wishes. Then they decide to combine their wishes to get a new pet for their family.

Pallotta, J., and Bolster, J. *The Hershey's Milk Chocolate Fraction Book.* Needham, MA: Title Wave Press, 1999.

This book uses the whole candy bar, partitioned into various smaller squares, to present problems involving fractional parts. Vocabulary is stressed throughout the book.

CHAPTER 13

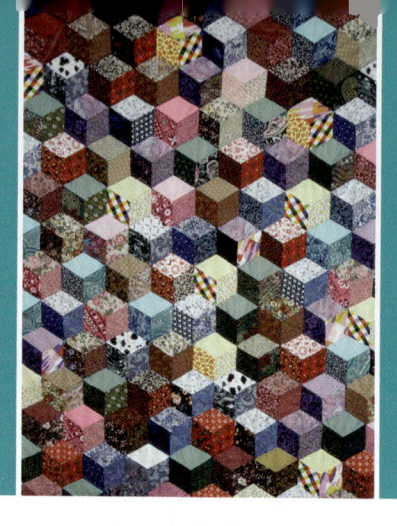

Ratio, Proportion, and Percent: Meanings and Applications

> "WE DON'T KNOW A MILLIONTH OF ONE PERCENT ABOUT ANYTHING."
>
> —Thomas Edison

>> SNAPSHOT OF A LESSON

KEY IDEAS

1. Help students develop ratios in a real-world context.
2. Have students engage in measurement and use equivalent ratios.
3. Guide students to represent ratios in different ways.
4. Help students connect ratios to fractions and decimals.

BACKGROUND

Carey Tuckey is a seventh-grade teacher. She is using the story of *Gulliver's Travels* to introduce ratios. The lesson has children measuring, recording ratios, and then connecting their ratios to fractions and decimals. This excerpt is from the beginning of the lesson that leads to the children's constructing a model of a shirt for Gulliver. This lesson is based on the video *Body Ratios*, from the Modeling Middle School Mathematics Project. To view the video, go to http://www.mmmproject.org/br/mainframe.htm.

Ms. Tuckey: Who knows the story of *Gulliver's Travels*? Can anybody tell us about *Gulliver's Travels*? Allison?

Allison: Well, there's this guy and they think he's crazy, but then he goes sailing away and he is on a strange land and he's like really big and these little people, called Lilliputians, they tie him down and they want to make a new shirt for him.

Later in the lesson, Ms. Tuckey has a student read from the book.

Ms. Tuckey: And Allison, would you read for us what they decided would work in order to measure Gulliver's shirt?

Allison: "The seamstresses took my measure. As I lay on the ground one standing at my neck and another at my mid leg with a strong chord extended, that each, held by the hand, while the third measured the length of the cord with a rule of an inch long. Then they measured my right thumb and desired no more, for by a mathematical computation twice around the thumb is once around the wrist and so on, to the neck and the waist. And by the help of my old shirt, which was displayed on the ground before them for a pattern, they fitted me exactly."

Ms. Tuckey: If we knew that the Lilliputians were 6 inches tall, and Jonathan Swift who wrote this said Gulliver was about 6 feet tall, what is the ratio of one Gulliver to a Lilliputian? How many Lilliputians is it going to take to make a Gulliver? Pablo?

Pablo: Um … 12.

Ms. Tuckey: And what ratio is that?

Pablo: 1 to 12.

Ms. Tuckey: That means the Lilliputians were pretty tiny compared to Gulliver. I have a little friend here that I found who is exactly $\frac{1}{12}$ of my height. And if we put him down on the ground next to me, can you imagine him having to measure to make a shirt for me? I'm pretty huge compared to him, aren't I?

Ms. Tuckey: The Lilliputians were smart. They had a formula. And their formula used ratios. They said that twice round the thumb is once round the wrist and so on to the neck and the waist. What did they mean by "and so on to the neck and the waist"? David what does that mean?

David: Twice around the wrist equals around the neck and twice around the neck equals twice around the waist.

Ms. Tuckey: So we have a ratio, don't we? And if we look at the ratio of thumb to wrist, everybody, how many thumbs did they say it took to make a wrist?

Class: Two.

Ms. Tuckey: Do you believe that?

Class: No.

Ms. Tuckey: You don't? What we're going to do is see if it is real. In the box on your desk, you have some string and you have some scissors.

The students begin the measuring activity.

Ms. Tuckey: [*to the camera*] The kids are very skeptical as to whether or not once around your thumb was indeed half of your wrist measurement. When we moved into the first exploration, we took string and they used rulers and all kinds of things to verify that ratio. They were very surprised to find out that Mr. Swift did know what he was talking about.

Student: She measured my thumb and then she put it on this [*indicates ruler*], then she measured my wrist and she put it on this [*ruler*] and then we said, since we have to do two times around that's 8, and then 8 times 2 is 16 and then this is 16 … almost 16.

Later, the teacher has students share their findings.

Ms. Tuckey: So how many of you found out that the Lilliputians were correct? You were pretty close. Almost all of you. They were pretty smart, weren't they? Let's suppose then that we have a classmate whose thumb measures 5 centimeters. What would we estimate his wrist would measure. Laurie?

Laurie: 10 centimeters.

Ms. Tuckey: And how did you figure 10 centimeters?

Laurie: Because if you go twice around your thumb, the Lilliputians say it would equal your wrist, so twice 2 times 5 is 10.

Ms. Tuckey: So if we know that the thumb is 5 centimeters and the wrist measures 10 centimeters, how do we say that as a ratio?

Class: 5 to 10.

Ms. Tuckey: Would you please take your dry erase boards out and write that ratio two different ways?

The students comply with the teacher's request.

Ms. Tuckey: I'm glad to see that everybody remembered that not only can we write a ratio as a fraction, we can write a ratio using a colon and we can write a ratio using the word *to* and all of these are read the same way. How is that?

Class: 5 to 10.

Ms. Tuckey: Now sometimes we find that it's much easier to compare ratios if we make them into decimals. We know that you can take a fraction and you can easily make into a decimal by doing what? Sara?

Sara: By dividing them.

Ms. Tuckey: Good, and are we going to divide 5 by 10 or 10 by 5?

Sara: 5 by 10.

Ms. Tuckey: So if we take 5 divided by 10, what decimal will we end up with?

Class: Five-tenths.

Ms. Tuckey: And how do we write that as a decimal?

Class: 0.5.

Ms. Tuckey: 0.5—good. Let's see if the Lilliputians knew what they were doing when they were talking about ratios.

We're going to use our string again, and we're going to see if one time around our wrist is half of our neck. So is our neck indeed twice our wrist? We're going to do the same thing going from our neck to our waist. Now, as we measure this time, I'm going to ask you to use centimeters. When we talked about the Lilliputians and Gulliver, we were using inches and feet. But one thing that's nice about ratios is no matter what unit of measure you use, the ratio is going to stay the same.

FOCUS QUESTIONS

1. How is a ratio different from a proportion?
2. How can teachers use student intuition to develop students' thinking about proportions?
3. Why do students mistakenly use an additive method to find proportions rather than the correct multiplicative method?
4. What models help develop the concept of percents?

INTRODUCTION

"Twice around your thumb is equivalent to once around your wrist."
"Family income this year increased by 10%."
"Ian did only half the work Angela did."

"Her salary is three times my salary."
"The cost of living tripled during the last 8 years."
"Your chances of winning the lottery are less than one in a million."
"I can purchase a 12-ounce bottle of soda for $1 or a 16-ounce bottle for $1.50. Which bottle is the better buy?"

These statements show ratio, proportion, and percent in action. They demonstrate that much of quantitative thinking is relational. In such thinking, what is important is the multiplicative relationship between numbers, rather than the actual numbers themselves. By emphasizing multiplicative relationships, you will help your students take advantage of the many opportunities to use these relationships in real-world situations.

The *Common Core State Standards for Mathematics* clearly state that students in elementary school and middle school must be able to express appropriate relationships using fractions, ratios, proportions, and percents. In Figure 13-1, note that although the main focus on ratio, proportion, and percent takes place at grade 6, there is a developmental progression from grade 3 to grade 6 of the necessary underlying concepts.

Here is an example to get you started thinking about ratios, proportions, and percents. Consider the prices of three carpets:

CARPET SALE

A $9.00 per sq. yd.
B $18.00 per sq. yd.
C $27.00 per sq. yd.

The difference in price between carpet A and carpet B is $9 per square yard, and so is the difference in price between carpet B and carpet C. Yet B is twice as expensive as A while C is three times as expensive and 50% more expensive than B. Whether one considers the difference in the prices or the ratio of the prices is essential to relational thinking that we will explore.

RATIOS

Ratios involve comparing things. In *Gulliver's Travels*, comparing the distance around the thumb to the distance around the wrist to the distance around the neck illustrates a series of ratios. Some of children's earliest experiences with comparisons involve ratios or rates. For example, if a child pays one dime for three stickers, the rate is three stickers for one dime.

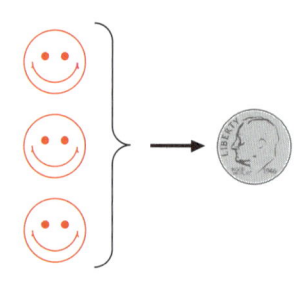

This ratio may be read as "three to one" and recorded as 3 to 1, 3:1, or $\frac{3}{1}$. Any of these forms is acceptable.

Ratios and Proportional Relationships
Understand ratio concepts and use ratio reasoning to solve problems.

- Understand the concept of a ratio and use ratio language to describe a ratio relationship between two quantities. For example, "The ratio of wings to beaks in the bird house at the zoo was 2:1, because for every 2 wings there was 1 beak." "For every vote candidate A received, candidate C received nearly three votes."
- Understand the concept of a unit rate a/b associated with a ratio a:b with b≠0, and use rate language in the context of a ratio relationship. For example, "This recipe has a ratio of 3 cups of flour to 4 cups of sugar, so there is ¾ cup of flour for each cup of sugar." We paid $75 for 15 hamburgers, which is a rate of $5 per hamburger."
- Use ratio and rate reasoning to solve real-world and mathematical problems, e.g., by reasoning about tables of equivalent ratios, tape diagrams, double number line diagrams, or equations.
  - Make tables of equivalent ratios relating quantities with whole-number measurements, find missing values in the tables, and plot the pairs of values on the coordinate plane. Use tables to compare ratios..
  - Solve unit rate problems including those involving unit pricing and constant speed. For example, if it took 7 hours to mow 4 lawns, then at that rate, how many lawns could be mowed in 35 hours? At what rate were lawns being mowed?
  - Find a percent of a quantity as a rate per 100 (e.g., 30% of a quantity means 30/100 times the quantity); solve problems involving finding the whole, given a part and the percent.
  - Use ratio reasoning to convert measurement units; manipulate and transform units appropriately when multiplying or dividing quantities

Figure 13-1 *Ratio and proportion standards for grade 6 from CCSSM.* This excerpt of the CCSSM standards was made by the authors of *Helping Children Learn Mathematics*. The full document (CCSSI, 2010) may be obtained at www.corestandards.org/the-standards/mathematics.

We can also form ratios to report the number of stickers to nickels as 3 to 2 or stickers to pennies as 3 to 10:

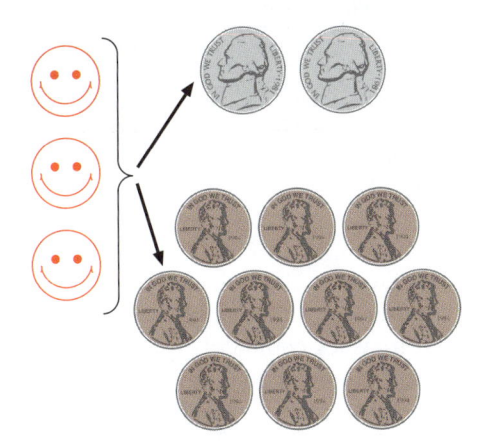

This ratio could also be expressed in reverse, as 10 pennies for every 3 stickers, which we could write as 10 to 3 or as a quotient, $\frac{10}{3}$. You could use the quotient form to help children see that the cost per sticker is more than three pennies, which illustrates a powerful application of ratios—knowing the cost of multiple items lets us calculate the cost of a single item.

Early experiences with ratios should enable children to think of two related numbers simultaneously. As children use manipulatives such as coins, or draw pictures, they should be encouraged to think about ordered pairs of numbers, such as (3 stickers, 10 pennies). The models help link the operation of multiplication directly to ratios.

Working with facts about money (the number of pennies in a nickel, nickels in a quarter, and so on) gives children natural and meaningful experiences with ratios. In the Classroom 13-1 shows how organizing such facts in a ratio table not only visually displays many ratios but also helps students realize that a ratio is a multiplicative relationship between two or more numbers in a given order. The ratio table also contains patterns that encourage students to explore relationships, generate formulas, and engage in algebraic thinking.

Prices—for instance, two cans for 99¢, 3 pounds for $1.99, and 88¢ per dozen—provide meaningful contexts for ratios. The money model also lets you naturally extend ratios to more than two numbers—for example, the ordered triple (2, 10, 50) relates 2 quarters, 10 nickels, and 50 pennies. Other real-life examples of ratios are shown in *Each Orange Had 8 Slices: A Counting Book* (Giganti, 1992).

Patterns often lead to ratios. For example, in Figure 13-2, the ratio of red links to green links in chain A is 2 to 3. Extending the chain would reveal a pattern where two red links are followed by three green links, followed by two red links, and so on. Extending chain B would show a different pattern—four red links, followed by six green links, followed by four red links, and so on. In chain B, the ratio of red to green links is 4 to 6. Chain C shows how doubling chain A forms a chain with the same ratio of red to green links as chain B (4 to 6). Chains B and C are the same

---

In the Classroom 13–1

KNOW YOUR COINS

Objective: Using value of coins in a ratio table to develop patterns and proportions and to examine algebraic relationships.

Grade Level: 4–6.

- Use a ratio table to help complete these patterns:

Number of Quarters	1	2		4	5	6
Number of Nickels	5	10	15	20		30
Number of Pennies	25	50	75		125	150

- Answer these questions and fill in the blank:
  - Describe a pattern you found in each row.
  - Write a ratio for the number of quarters to nickels. ___
  - Write a ratio for the number of nickels to pennies. ___
  - Write a ratio for the number of quarters to pennies. ___
- Try these:
  - How many nickels will be needed for 8 quarters? ___
    Tell two different ways to decide.
  - How many pennies will be needed for 10 quarters? ___
    Tell two different ways to decide.
  - Give three numbers (not shown in the table) that could go
    in the nickels row ___
    in the pennies row ___
  - Give three numbers that could *not* go
    in the nickels row ___
    in the pennies row ___
  - How many quarters would you have when the number of quarters plus the number of nickels plus the number of pennies (all in one column of the ratio table) first exceeds $30? ___

Chain A—with 2 red links and 3 green links—RRGGG

Chain B—with 4 red links and 6 green links—RRRRGGGGGG

Chain C—with 2 red links and 3 green links—RRGGGRRGGG

Figure 13-2 *Chains showing equivalent ratios.*

286   Chapter 13 • Ratio, Proportion, and Percent: Meanings and Applications

length (10 links), and each chain has the same number of red links (4) and green links (6). This shows that 2:3 and 4:6 are equivalent ratios (though they make a different pattern). Children may better connect the representation by writing "3 green links to 2 red links" when expressing the ratio 3 to 2, 3/2 or 3:2.

So far, we've been discussing the chains in Figure 13-2 from a part-to-part perspective—that is, we've been discussing the ratio of one part of the chain (the red links) to another part of the chain (the green links). Now let's examine these chains from a part-to-whole perspective. For example, consider the ratio between the number of red links (the part) and the total number of links (the whole) in each chain. In chain A, this ratio is 2 to 5, whereas it is 4 to 10 in both chains B and C, which shows that these two ratios are also equivalent. Note that we have used the same model to show different ratios, depending on the perspective (part-to-part was 2 to 3 and part-to-whole was 2 to 5). You must be very clear about what is being compared when discussing ratios.

Children encounter ratios in many different forms: "three video games for a dollar," "twice around," "half as much," and so on. Showing children real-world examples of ratios helps them develop a greater awareness and understanding of ratios and how they work. For instance, adjusting recipes to serve more or fewer people provides an opportunity to work with ratios involving multiple values. Other examples might include a can of mixed nuts with a ratio of cashews to almonds to peanuts of 3:5:10 or a sack of lawn fertilizer with "8–12–20" printed on it, indicating the ratio of the percentages of phosphorus, nitrogen, and potash in the fertilizer. You can also use examples like these to remind students of the order of the numbers in a ratio and the importance of understanding what each number represents—for instance, the ordered pair (5, 25) from ratio table in In the Classroom 13-1 could represent 5 quarters and 25 nickels or 5 nickels and 25 pennies. Ratios have little meaning unless you know what the numbers represent. Using physical models such as those shown in In the Classroom 13-1 and Figure 13-2 helps children learn to make sense of ratios.

Figure 13-3 shows a ratio table with the ratios between the numbers of wheels on different types of vehicles, as well as between the number of vehicles and the number of wheels on those vehicles. The ratio table contains many different patterns and gives children a visual

	Number of vehicles					
	1	2	3	4	5	6
Number of wheels on bicycles	2	4	6	8	10	12
Number of wheels on tricycles	3	6	9	12	15	18
Number of wheels on wagons	4	8	12	16	20	24

Figure 13-3 *Ratio table comparing the number of vehicles with number of wheels for bicycles, tricycles, and wagons.*

way of understanding different ratios. For example, the table lets children see that the ratio of wheels on bicycles to wheels on tricycles could be expressed as 2:3, 6:9, or 10:15; it also lets them see that the ratio of the number of wagons to the number of wheels on those wagons could be expressed as 1:4, 4:16, or 6:24.

Students are sometimes confused by the different symbols used to write ratios. Linking the symbols to models and promoting class discussion can help minimize this confusion. For example, round chips or cut-out pictures of wheels could be used to model the wheels in Figure 13-3 before children try writing the ratios. And ratio tables help organize the information. After children record the ratios in writing, you can help them talk about the mathematics by guiding them to discuss the information in the table and the situation represented in the ratios. For the ratios in Figure 13-3, for example, you could give children early practice in verbalizing and describing some of the mathematics surrounding the concept of ratio by asking questions such as the following:

> If I have three bicycles and six wheels, is the ratio three to six or six to three?
>
> If the ratio of bicycles to wheels is four to eight, how many bicycles do I have?
>
> If I have six bicycles, how many wheels are there? What is the ratio of bicycles to wheels?
>
> If, instead of six bicycles, I have half as many bicycles, how many wheels? What is the ratio of bicycles to wheels?

In addition to understanding that ratios can be constructed in different ways (e.g., part-to-part and part-to-whole), children need to understand that combining ratios is not the same as adding fractions. For example, in Figure 13-4, chain A has 2 blue links and 3 orange links (a ratio of $\frac{2}{3}$),

Chain A	ratio of blue to orange is 2 to 3 or 2/3
Chain B	ratio of blue to orange is 4 to 5 or 4/5
Chain C	ratio of blue to orange is 2 to 3 or 6/8

Figure 13-4 *Combining ratios is different from adding fractions.*

and chain B has 4 blue links and 5 orange links (a ratio of $\frac{4}{5}$). We could join together chains A and B to form chain C, with 6 blue and 8 orange (a ratio of $\frac{6}{8}$). However, the sum of the fractions $\frac{2}{3} + \frac{4}{5}$ is not $\frac{6}{8}$ (it's $1\frac{7}{15}$). Remember that operations, such as addition, can be performed on fractions but not on ratios. Of course, it is important that you help students understand the equivalence of different representations of ratios (e.g., 2 to 3, 2:3, and $\frac{2}{3}$) and that you help them link these representations to real-world models. But far more important is helping students understand the relationships signified by ratios. Young children with a good understanding of numbers often use expressions such as "twice as much" or "half as much." When children use these expressions, you should take advantage of the opportunity to help them make connections and develop reversible thinking.

Math Links 13.1

Session 8, Rational Numbers and Proportional Reasoning, from the Annenberg Learning Math Series provides some excellent examples that you can use to explore part–part interpretation, part–whole interpretation, and absolute and relative comparisons. It may be accessed from http://www.learner.org/channel/courses/learningmath/number/session8/index.html or from this book's Web site.

www.wiley.com/college/reys

PROPORTIONS

A proportion is a statement that two or more ratios are equal or "the same"—for example, that the ratio 2:3 is the same as the ratio 4:6. Alternatively, you could say that two ratios are proportional (e.g., 2:3 and 4:6) or not proportional (e.g., 2:3 and 4:8). Understanding the relationships in proportions and working with these relationships is termed *proportional reasoning* and has been called the "capstone" of elementary school mathematics (Cramer, Post, and Currier, 1993). It is called the capstone because proportional reasoning requires a high level of thinking, and it provides a direct link to many algebraic topics, including direct and inverse relationships. To get a sense of what proportional reasoning means, consider the following problem, which is illustrated in Figure 13-5.

A month ago, plant A was 12 cm tall and plant B was 15 cm tall. Now plant A is 18 cm tall and plant B is 21 cm tall. Which plant grew the most?

You could compare the absolute amounts that the plants grew, which might lead you to conclude that the two plants

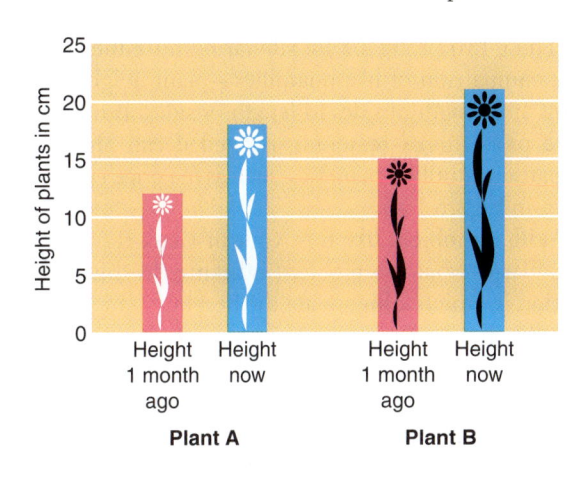

Figure 13-5 *Which plant grew the most?*

grew equally, since both plants grew 6 cm. Or you could make a multiplicative, or relative, comparison by looking at how much each plant grew in relation to its original height. From this perspective, the plant A grew $\frac{6}{12}$ of its original height, whereas plant B grew just $\frac{6}{15}$ of its original height, which might lead you to conclude that plant A grew more. Of course, both conclusions are correct—they just reflect a different basis of comparison, absolute versus relative. But it is the ability to make comparisons in relative rather than absolute terms that characterizes proportional reasoning, and it is this ability that is the hallmark of the formal operational level of cognitive development in Piaget's system (see Chapter 2).

Now let's look at another example. The two rectangles shown below are similar (i.e., have the same shape) but of different sizes.

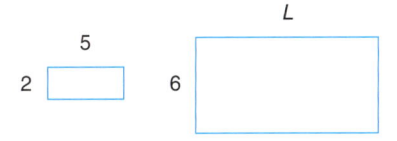

When asked to find the length of the side L, students frequently come up with 9, instead of 15, because they add when they should multiply. That is, they see that the height of the large rectangle is 4 units more than the height of the small rectangle, so they add the same 4 units to the length of the small rectangle to get $L = 9$. Instead, they should find the ratio of the height of the large rectangle to the height of the small rectangle $\frac{6}{2} = 3$ and then multiply the length of the small rectangle by that same ratio to find L ($5 \times 3 = 15$). The fact that the two rectangles are the same shape means that their corresponding sides are in the same ratio—that is, they are in proportion. Research shows that this type of error reflects students' misconceptions about ratio and proportion and is not simply the result of carelessness (Lamon, 2007; Lesh, Post, and Behr, 1998).

288 Chapter 13 • Ratio, Proportion, and Percent: Meanings and Applications

These two problems are similar because they both involve comparison of like quantities—plant height to plant height and length of sides to length of sides. But problems where proportional reasoning is needed can also involve comparison of unlike quantities, such as the price comparison shown in Figure 13-6.

With a problem like this, your first task is to help students clarify what the problem is asking. In this case, the question is, which lollipops are the best buy, which cost less per lollipops? Next, you have to help students determine a method for approaching the problem. For instance, you could encourage them to construct ratio tables like these:

Number of lollipops	1	3
Cost		51¢

Number of lollipops	1	4
Cost		64¢

Alternatively, you could suggest that double number lines (also called two-sided number lines) would be a powerful model:

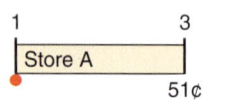

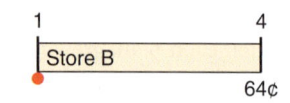

Store A: 1 — 3, 51¢
Store B: 1 — 4, 64¢

With either of these models, your goal would be to help students find equivalent ways of representing the cost of a lollipop at the two stores. Both models let students come up with these equivalent representations:

At store A: 3:51¢ = 1:C or $\frac{3}{51¢} = \frac{1}{C} \rightarrow \frac{1}{17¢} = \frac{1}{C}$, so C = 17¢

At store B: 4:64¢ = 1:C or $\frac{4}{64¢} = \frac{1}{C} \rightarrow \frac{1}{16¢} = \frac{1}{C}$, so C = 16¢

Thus, at store A a lollipop costs 17¢, while at store B a lollipop costs 16¢. This means that the best buy is at store B.

SALE Lollipops 3 for 51¢

I wonder if I should buy them down the block at 4 for 64¢

Figure 13-6 *Comparing prices of lollipops in two places.*

(Of course, this answer ignores issues such as the quality of the lollipops as well as whether to walk a block to save a penny!)

Students will frequently encounter situations involving comparisons of like and unlike quantities, both in problems they are given at school and in their everyday lives. Generally speaking, children have an easier time working with proportions involving unlike quantities, as they are less likely to make the error of adding instead of multiplying.

You can also solve the lollipops problem by comparing fractions with common denominators. That is, you can represent the ratios as the quotients $\frac{1}{17}$ and $\frac{1}{16}$, which can be converted into quotients with the common denominator 272 (16 × 17 = 272):

$$\frac{1}{17} = \frac{16}{272} \text{ compared to } \frac{1}{16} = \frac{17}{272}$$

This again shows that store B offers the better buy because you can get 17 lollipops for $2.72, compared to 16 for the same cost at store A.

> **Math Links 13.2**
>
> Go to http://www.fi.uu.nl/wisweb/en or this book's website and click on applets and you will find a wide range of applets. Several applets (Enlargement and Radio Table) provide dynamic opportunities to explore proportions.
>
> www.wiley.com/college/reys

Now let's examine another problem that requires proportional reasoning:

Suppose 10 horses need 4 acres of pasture to live comfortably. How many acres would 15 horses need to live comfortably?

Rather than teach students an algorithm for solving a problem like this, you want them to approach it using their intuition or their existing skills. Figure 13-7 shows five student responses to this problem. Many students, like Karen, draw pictures (or use manipulatives); others, like Tom, make a list showing a pattern; some make a ratio table and use their knowledge of equivalent fractions, as Carrie did; and some find a unit rate, as Jose did. Still others may use an incorrect strategy, like Bob.

Note that Bob's approach reflects a misconception that students commonly bring to solving proportion problems: "When in doubt, add" (Miller and Fey, 2000). The concept of proportion is closely linked to multiplication, yet it is often difficult for students to make the necessary connections. Having students share their solutions—both correct and incorrect—on the board may help them grasp this important idea.

Proportions **289**

How much will a dozen balloons cost if 3 balloons cost 48 cents?

Here are visual models with the double number line for two popular thinking strategies upper elementary children use to solve this problem:

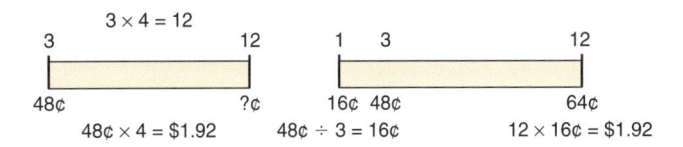

$3 \times 4 = 12$

3	12
48¢	?¢

$48¢ \times 4 = \$1.92$

1	3
16¢	48¢

$48¢ \div 3 = 16¢$

	12
	64¢

$12 \times 16¢ = \$1.92$

a. By package-per-price strategy
  3 balloons cost $0.48
  so 12 balloons is 4×3
  so $4 \times \$0.48$ is $\$1.92$.

b. By unit-price strategy
  3 balloons cost $0.48
  so each balloon is $0.16
  12 balloons would be $12 \times \$0.16$ or cost $\$1.92$.

Middle grade students will be able to write an algebraic equation in a variety of ways:

$\dfrac{3}{48} = \dfrac{12}{\square}$ Thinking $\dfrac{3 \text{ balloons}}{\$0.48}$ as $\dfrac{12 \text{ balloons}}{\square}$

(where the symbol $\square$ is the cost of the balloons)

$\dfrac{3}{12} = \dfrac{48}{\square}$ Thinking $\dfrac{3 \text{ balloons}}{12 \text{ balloons}}$ as $\$0.48$ is to $\square$

$\dfrac{48}{3} = \dfrac{\square}{12}$ What would students be thinking in this case?

$\dfrac{12}{4} = \dfrac{\square}{48}$ What would students be thinking in this case?

Although students should recognize the equivalence of these statements, they should feel free to use the model or form of their choice.

The pennies–nickels comparisons shown in In the Classroom 13-1 contain many equivalent ratios (1:5 = 2:10, 2:10 = 3:15, and so on). The concept of equivalent ratios is important and can be anchored in different ways. For example, Figure 13-8 shows a table displaying the result of 5 pennies for 1 nickel as $y = 5x$. These tabular data provide opportunities to discover patterns, explore an algebraic relationship, and observe that the ratio of all of these proportions is constant or invariant. This understanding of invariance is an essential component of good proportional reasoning (Lobato and Ellis, 2010). Students need to see that even though the values of x and y are different in each pair, the ratio of y to x remains invariant.

Another approach to help students make sense of proportions is to use estimation skills. Think about the following problem.

KAREN: *Using a picture or a diagram* I drew the picture and then counted the acres to be 6.

TOM: *Using patterns* I noticed:
- 10 horses need 4 acres of pasture . . . so
- 20 horses need 8 acres of pasture

Then I split the difference and got 15 horses need 6 acres of pasture.

CARRIE: *Using a ratio table* I did it this way:

Number of horses	10	15
Number of acres	4	?

I thought 10 is to 4 as 5 is to 2 and 15 is to 6. So they need 6 acres of land

JOSE: *Using a unit rate or density* I thought, 4 divides into 10 two times with a remainder of $\frac{2}{4}$ or $\frac{1}{2}$. So there are $2\frac{1}{2}$ horses per acre. Then I saw how many acres it would take for 15 horses. I divided 15 by $2\frac{1}{2}$. Or $15 \div \frac{5}{2}$ which is the same as $15 \times \frac{2}{5}$. That means that it would be 6 acres.

BOB: *Using an incorrect addition/subtraction strategy* I figured there are six more horses than acres (10 − 4).
So I subtracted 6 from 15 and the number of acres is 9.

"I drew the picture and then counted the acres to be 6."

Figure 13-7 *Methods children use to solve the horse and pasture problem.*

We have considered three different types of problems where proportional reasoning is needed (Kilpatrick et al., 2001; Miller and Fey, 2000). The lollipops problem (see Figure 13-6) illustrates *numerical-comparison problems*; the horses per acre problem (see Figure 13-7) illustrates *missing-number problems*; and the problem with similar rectangles illustrates a *scaling* problem.

Once students have made sense of proportions by solving many problems in a variety of ways, you can introduce them to a more algebraic method of solving missing-number problems. For example, consider the following problem:

290 Chapter 13 • Ratio, Proportion, and Percent: Meanings and Applications

X	y
1	5
2	10
3	15
4	20
5	25
6	30
7	35

Figure 13-8 *Table comparing number of nickels (X) to number of pennies (Y).*

A trout swam 5.8 meters in 3 seconds. At this rate, how far could it travel in 19 seconds?

Students could think:

That's about 6 meters in 3 seconds or 2 meters per second. And 2 meters per second times about 20 seconds would give 40 meters.

Another student could say:

19 seconds is about 6 times the 3 seconds, so approximately 6 meters times 6 would be about 36 meters.

Although each estimate is different, each of them is reasonable. This type of estimation thinking is very productive. It should be both encouraged and rewarded. It uses estimation along with ratios to produce ballpark answers. Frequent experiences similar to this one will improve students' judgment, making them less likely to fall victim to unreasonable answers resulting from indiscriminate number crunching. Suppose the student mistakenly set up the proportion:

$$\frac{5.8}{3} = \frac{19}{\square}$$

And then correctly computed that n = 9.8

What mistake did the student make in this proportion? How could he determine that he was making an error in his thinking? What would a correct proportion be? Using estimation to produce a ballpark result might help the student realize that 9.8 is an unreasonable answer and encourage the student to reflect on the procedures used.

Of course, using a double number line model provides a model for a correct ratio, $\frac{3}{5.8} = \frac{19}{\square}$, but also allows the rejection of 9.8 as being an unreasonable answer.

Figure 13-9 shows an item given to fourth graders on a national assessment (Kenney, Lindquist, and Heffernan, 2002). Examples of two correct and incorrect responses are also shown. The incorrect answers demonstrate how some children considered this an additive rather than multiplicative problem. The overall performance was low, as only 6% of the students gave a correct answer with a correct explanation, and another 7% gave a correct answer without an explanation or made an error using a correct method. The majority (86%) of the fourth graders gave an incorrect response.

Consider using other real-life applications of proportions with which children are familiar (Reeder, 2007). This is illustrated by comparisons of thumb to wrist or other body parts of Gulliver as described in this chapter's Snapshot of a Lesson. Other real-world applications include recipes, photographs of different sizes, maps, and scale drawings or models. Using a zoom in or out on a computer screen or when photocopying both illustrate proportions between the original and its new image. The concepts of ratio and proportionality can be

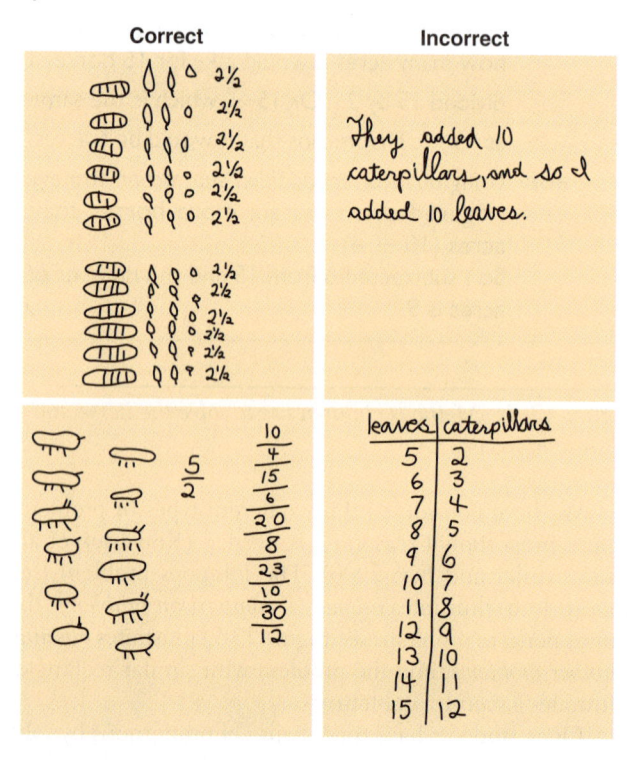

Figure 13-9 *Samples of students' solutions to a problem involving proportion.*

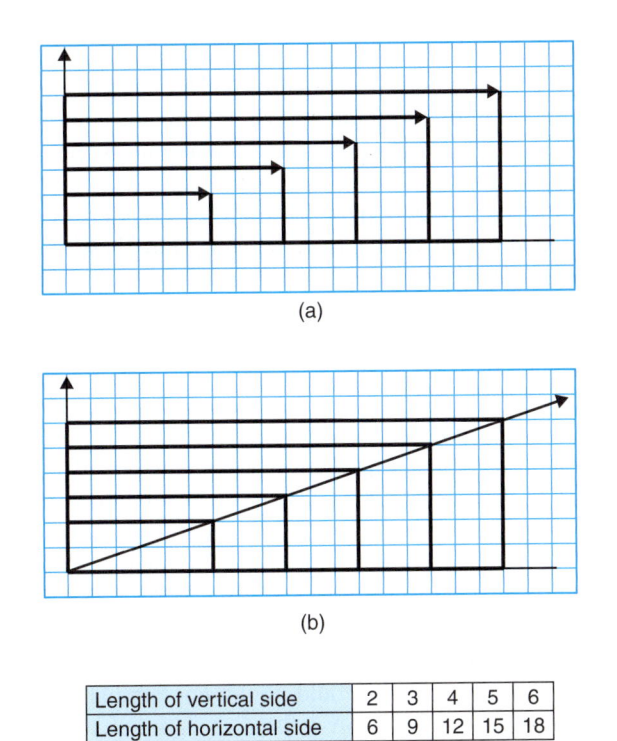

Figure 13-10 *Drawing similar rectangles and ratio table.*

PERCENTS

You only need to read a newspaper or watch television to be reminded that percent is one of the most widely used mathematical concepts:

Save 23 to 55%
MONEY MARKET CERTIFICATES
14.70% YIELDS
16.07%

City Asks For 40% Budget Increase

Serious crimes drop 5 percent

Wholesale Price Index Up 0.6 Pct.

38 Percent Plunge In Profits

Understanding of percent is taken for granted, although there is plenty of evidence to the contrary. Incorrect usage of percent is common among secondary students and adults. Flagrant errors abound, suggesting that often the most basic ideas are unclear (Gay and Aichele, 1997). For example, about one-third of the 17-year-olds and adults missed the following question on the first NAEP:

If 5% of the students are absent today, then 5 out of how many are absent?

An error on such a fundamental idea suggests they did not know that 100 is the comparison base for percent.

Misconceptions, distortions, and confusion surrounding percent are surprisingly easy to find. Here are some examples:

- "Prices reduced 100%." If this advertisement were correct, the items would be free. Probably, the prices were reduced 50%. If an item that originally cost $400 was on sale for $200, then the ad based the 100% on the sale price, when it should have been based on the original price.
- "Of all doctors interviewed, 75% recommended our product." This type of claim could be an effective advertisement for a company. If, however, the ad said "3 out of the 4 doctors we interviewed recommended our product," the consumer reaction might be different. Percents can often be used to disguise the number involved. Thus they can be misused. Percents allow for easy comparisons because of the common base of 100, but they may appear to represent a larger sample than actually exists.

Discussing the following questions and providing real-life examples can help students develop a number sense for percents:

- Can you eat 50% of a cake?
- Can you eat 100% of a cake?
- Can you eat 150% of a cake?
- Can a price increase 50%?

naturally connected to geometry problems, as similarity is based on proportions. Two figures are similar if their respective sides are in the same ratio (i.e., proportional). Thus, all squares are similar, but all rectangles are not. Figure 13-10 shows drawings of similar rectangles. Notice that in each of the five rectangles, the ratio between the vertical side and the horizontal size is constant. For example, in the smallest rectangle, the ratio is 2:6 and in the largest rectangle the ratio is 6:18. The rectangles in Figure 13-10b show that a single line passes through the vertex of each of these rectangles, and this demonstrates that the slope of the diagonals (i.e., the ratio of the vertical height to the horizontal length) of each of these rectangles is the same. This is visible from Figures 13-10a and 13-10b. The ratio table shown in Figure 13-10c shows these lengths and the equivalence of each of the ordered pairs shown. The pattern can be extended in the graphs as well as in the ratio table. An examination of these ratios, such as 2:6, 3:9, or 5:15, confirms that each of them is equivalent, and the resulting slope is $\frac{1}{3}$.

Children's books, such as *Anno's Math Games* (Anno, 1987), provide different yet interesting contexts to explore ratios and proportions and connect them to percent. Other excellent ideas on using literature to explore proportional reasoning are available (Thompson, Austin, and Beckmann, 2002; Martinie and Bay-Williams, 2003).

- Can a price increase 100%?
- Can a price increase 150%?
- Can a price decrease 50%?
- Can a price decrease 100%?
- Can a price decrease 150%?

Ironically, the understanding of percent requires no new skills or concepts beyond those used in mastering fractions, decimals, ratios, and proportions. In fact, percent is not really a mathematical topic, but rather the application of a particular type of notational system. The justification for teaching percent in school mathematics programs rests solely on its social utility.

As is true with decimals and fractions, percents express a relationship between two numbers. Percents are special ratios based on 100 and without a doubt are the most widely used of all ratios. Percent is derived from the Latin words *per centum*, which mean "out of a hundred" or "for every hundred." The origin of percent and its major uses are closely associated with ratio, fractions, and decimals. Thus, 25% is the ratio 25:100 or $\frac{25}{100}$, which then connects to decimals (0.25) and the fraction $\frac{1}{4}$.

When is percent understood? Students understand percent when they can use it in many different ways. For instance, if a child understands 25%, he or she can do the following tasks.

- Find 25% in various contexts:
    Cover 25% of a floor with tiles.
    Determine 25% off the price of a given item.
    Survey 25% of the students in class.

In many such situations, estimates of 25% are not only appropriate but essential.

- Identify characteristics of 25%:
    25% of the milk in a glass is less than half.
    If 25% of the milk in a glass is spilled, then 75% remains.

- Compare and contrast 25% with a range of other percents and numbers such as 5%, 50%, 100%, one-fourth, one-half, and 0.25.
    25% is half as much as 50%, one-fourth as much as 100%, five times as much as 5%, less than one-half, and the same as one-fourth and 25 hundredths.

Early development of percent can be facilitated with literature connections. For example, *The Grizzle Gazette* illustrates percent in several different ways (Murphy, 2003).

UNDERSTANDING PERCENTS

Percents should be introduced only after students thoroughly understand fractions and decimals. Percent is not studied extensively in elementary school, although it is typically introduced in fifth or sixth grade. The *Common Core State Standards for Mathematics* state that students in grade 6 should be able to "find a percent of a quantity as a rate per 100 (e.g., 30% of a quantity means 30/100 times the quantity)." Students should also be able to "solve problems involving finding the whole, given a part and the percent" (Figure 13-1).

Initially, students need a variety of experiences with the fundamental concepts of percent, and these experiences should be connected to various concrete models and real-world contexts. Computation applying percent in problem-solving situations is generally reserved for later. Students who understand that percent means parts out of one hundred and have a good pictorial representation of percent are more successful in solving percent problems than those who do not. Research has found that students naturally use benchmarks to make initial judgments about percent situations (Lembke and Reys, 1994). Helping students develop the concepts for common percent benchmarks and their fraction and decimal equivalents—such as 10%, 25%, 33%, 50%, and 75%—helps them apply their percent number sense to problem-solving situations.

Initial instruction should build on familiar models (Gay and Aichele, 1997). A dollar is made up of 100 cents; therefore, it provides a natural connection among percents, fractions, and decimals. For example, 25 cents is $0.25, one-fourth of a dollar, and also 25% of a dollar.

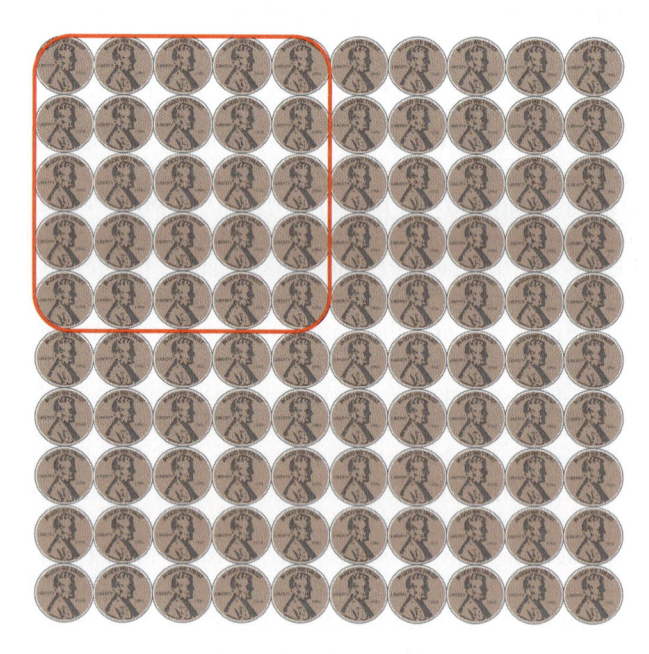

This model should be expanded to illustrate a wide range of percents, such as 50%, 90%, 5%, 100%, 99%, 1%, and 200%.

These percents should be illustrated with a variety of different situations and models. For example, a rope with 100 disks arranged in multiples of 10 in alternating colors shows 50% of each color.

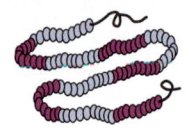

These two rectangular grids also show 50%:

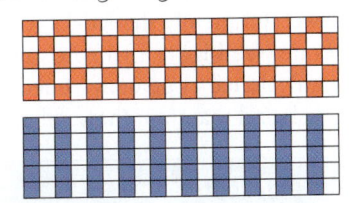

Percent bars can be used to show percent. This bar shows 75%:

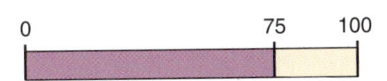

0   75   100

A meterstick provides another easily accessible and effective model. Cover part of the meterstick with paper and ask children to estimate the percent of the meterstick that is covered. Also, to remind them of the notion of 100%, be sure to ask: About what percent of the meterstick is not covered with paper? This model allows many different situations (25%, 50%, 1%, and so forth) to be presented and discussed quickly. Patterns may also emerge as students realize that the sum of the covered and uncovered portions always totals 100%. The percent bars can also be used to show connections between percents and decimals:

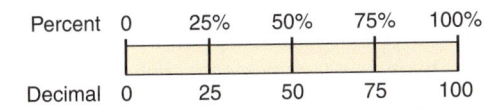

Percent   0   25%   50%   75%   100%
Decimal   0   25   50   75   100

Shaded fraction circles (Appendix) can also be used to provide multiple representations and help students see connections between fractions and percents.

These early experiences with percent should be followed by activities that center on direct translation experiences involving 100. The NAEP data revealed that only 31% of seventh graders could correctly write the decimals 0.42 and 0.9, respectively, as percents. These data are reminders of the fragile nature of students' understanding and the importance of connecting the representation of percents to decimals and fractions. Having students use percent bars, base-ten blocks, or share decimal paper (see blackline master in Appendix) provides another concrete model for percents and helps students make connections between decimals and percents.

Figure 13-11 illustrates how the same diagram can be represented symbolically by a fraction, a decimal, and a percent. It should be emphasized that each small square represents 1% and the large square represents 100%. Practice activities using this model to convert percents to fractions and decimals, and vice versa, should be plentiful. For example, Figure 13-11 provides a visual reminder that 17% can also be thought of as $17 \times \frac{1}{100}$. Also, 17% can be thought of as $\frac{17}{100}$, so the concept of ratio is reinforced. This approach helps students feel comfortable with different interpretations of percent. Calculators that convert fractions, decimals, and percents are useful when students are discovering patterns and relationships.

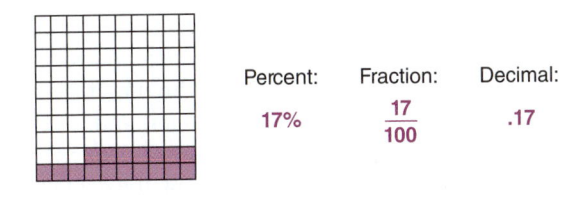

Percent: Fraction: Decimal:
17% $\frac{17}{100}$ .17

Figure 13-11 *Models and symbolizations of 17%.*

The importance of establishing 100 as the base for percent cannot be overemphasized. Also, 50% should be recognized as the fraction $\frac{50}{100}$ or the product $50 \times \frac{1}{100}$. It is also important that students know that an infinite number of equivalent fractions $[\frac{1}{2}, \frac{2}{4}...]$ also represent 50%.

In the Classroom 13-2 provides a natural means of developing some important ideas in an informal and yet meaningfully structured way. Each of the four activities

---

In the Classroom 13–2

USING PERCENTS

Objective: Using different models to relate quantities to percents.

Grade Level: 4–5.

A. Color this circle:
- 25% red
- 50% blue

What percent is uncolored? _____

B. Use 3 colors:
- Blue
- Green
- Yellow

Color each log with only one color.
Color all logs.

	Number Colored	Percent Colored
Blue	_____	_____
Green	_____	_____
Yellow	_____	_____

C. Put 100 pennies in a box.
Shake the box. Count the number of heads, and fill in the blanks.

	Number	Percent
Heads	_____	_____
Tails	_____	_____

D. Here are 20 poker chips. Count the number of each color.

	Number Colored	Percent Colored
Blue	_____	_____
Green	_____	_____
Yellow	_____	_____

should further develop children's concept of percent. Activities B, C, and D require some collection and recording of data before reporting the percents. The use of three different base numbers (10 logs, 100 pennies, and 20 chips) helps strengthen the link between ratio and percent. Even though answers for each activity depend on the data recorded, some patterns will emerge. A few questions from the teacher should trigger some stimulating discussion. For example, do the percents in each item total 100%? Why does this happen? Can you think of a time when it would not?

One particularly troublesome aspect of percents involves small percents between 0% and 1%. For example, percent, as in "percent milk fat," is not well understood. A visual representation, as in Figure 13-12, can help show that percent may indeed be less than 1%. Understanding rests on the earlier agreement that each small square represents 1%. As is true with all percents, this percent also can be shown symbolically as a fraction or a decimal. In the opening chapter quote, Thomas Edison said "We don't know a millionth of one percent about anything." Clearly this is a small value, and connecting this percent to a fraction or decimal produces

$$\frac{1}{1,000,000} \text{ of } 1\% \text{ is } \frac{1}{1,000,000} \text{ of } \frac{1}{100}, \text{ which is } \frac{1}{100,000,000}, \text{ or } (0.00000001)$$

While this computation provides a nice exercise, it is more important in elementary school to establish the intuitive notion of relative size of small percents than to devote extensive time to the algebraic gymnastics of showing the fraction and decimal equivalents.

Development of percents greater than 100% also is challenging and should be illustrated with models. Once the idea is established that a given region represents 100%, more than one such region can be used to represent percents greater than 100%. For example, 234% could be represented by two completely shaded large squares and a partially shaded one (see Figure 13-13). Using every opportunity to show equivalence of percents, fractions, and decimals helps establish and maintain these relationships.

> **Math Links 13.3**
>
> Go to the National Library of Virtual Manipulatives at http://nlvm.usu.edu/en/nav/vlibrary.html, or access it from this book's Web site. There you will find Percentages and Pie Chart under the Number and Operations for grades 3–5. Each of these interactive activities provides visual connections between part–whole relationships and some connections to ratios, fractions, and decimals.
>
> www.wiley.com/college/reys

APPLYING PERCENTS

Early experiences with percent should help children establish important benchmarks, such as 0%, 50%, and 100%. Then other benchmarks, such as 10% and 90%, should also be examined. Familiarity with benchmarks provides conceptual anchors to use when percents are encountered. Children come to realize that 10% is small, 50% is one-half, and 90% is nearly the entire unit. In the Classroom 13-3 provides opportunities to apply these percents by making visual estimates. In In the Classroom 3A, Al's stack of coins serves as 100% and provides a gauge to estimate the others. The notion of 100% is implicit when the glass is full, as shown in In the Classroom 13-3B, whereas a relative comparison is required in In the Classroom 13-3C. In all of these activities, estimation rather than exact counts should be used to make decisions.

In elementary school, students should solve percent problems meaningfully and avoid rushing toward symbolic methods. Whenever possible, students should be asked to discuss how to solve percent problems mentally, using what they know about common percent benchmarks. Even when a formal method is required, using informal methods first to obtain an estimate promotes sense making and will help students focus on the reasonableness of formal results.

Although percents are regularly encountered in many real-life problem-solving situations, only three basic types of problems involve percents. Several different formal methods can be used to solve percent problems; two methods typically found in elementary and middle school textbooks are equation and ratio, both of which utilize algebraic thinking. The effective use of these methods requires a firm understanding of the concepts of percent and ratio as well as the ability to solve simple equations and proportions. Such skills are developed over a period of several years and need not be rushed.

In the equation method, the following equation is used:

$$\text{percent} \times \text{total} = \text{part}$$

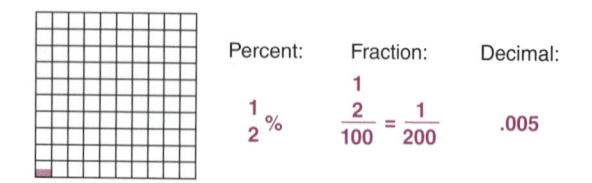

Figure 13-12 *Model and symbolizations of 1/2 %.*

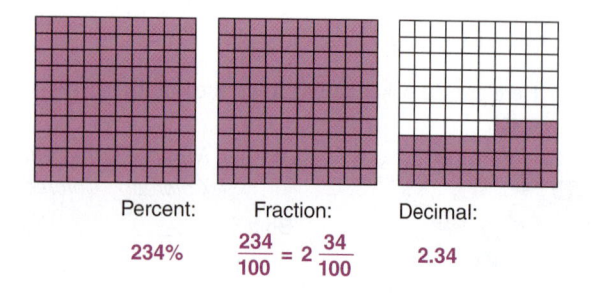

Percent: $\frac{1}{2}\%$ Fraction: $\frac{1}{2}\cdot\frac{1}{100} = \frac{1}{200}$ Decimal: .005

Figure 13-13 *Model and symbolizations of 234%.*

Percent: 234%   Fraction: $\frac{234}{100} = 2\frac{34}{100}$ Decimal: 2.34

In the Classroom 13-3

ESTIMATING PERCENTS

Objective: Connecting benchmarks of 10%, 50%, and 90% to physical quantities.

Grade Level: 4–5.

A. Al has 100 pennies. Estimate to decide who has about 90 pennies, 50 pennies, and 10 pennies.

        Al   Barb   Rustin   Whitney

B. Estimate to decide which of these glasses are closest to 10%, 50%, and 90% full.

D. Here are 20 poker chips. Count the number of each color.

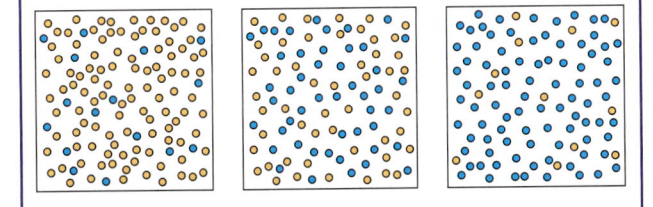

The two known values are placed in the equation and then students solve the equation for the third, unknown value.

In the ratio method, a (part) is b (percent) of c (total). The variables are set up as equal ratios or a proportion:

$$\frac{b(\text{percent})}{100} = \frac{a(\text{part})}{c(\text{total})}$$

Students place the three known values into the proportion and solve for the fourth, unknown value.

Now let's take a brief look at each of the three types of percent problems and how they may be solved using both informal and formal methods:

- Finding the percent of a given number.
  Lucas receives $60 per month for a paper route. Next month he will get a 10% raise. How much will his raise be?

The context of this problem suggests that the raise will be something considerably less than $60. The situation might be solved mentally or modeled as shown in Figure 13-14.

The rectangular array shown in Figure 13-14 is very similar to the percent bar model and can be used to construct several different proportions:

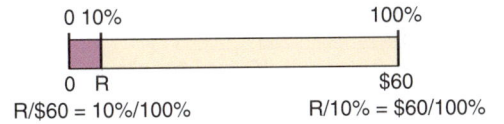

R/$60 = 10%/100%   R/10% = $60/100%

Either of these proportions leads to R = $6. The computation is simple and may disguise the level of difficulty this type of problem presents. Only about one-half of eighth graders were successful at solving this type of problem on the seventh NAEP (Silver and Kenney, 2000).

- Find what percent one number is of another number.
  The Cardinals won 15 of their 20 games.

What percent did they win? Intuitively, it is clear that the Cardinals did not win all of their games, so the answer must be less than 100%. Similarly, they won more than half of their games, so it must be more than 50%. The situation could be solved mentally or modeled as shown in Figure 13-15.

▼ **Mental strategy:**

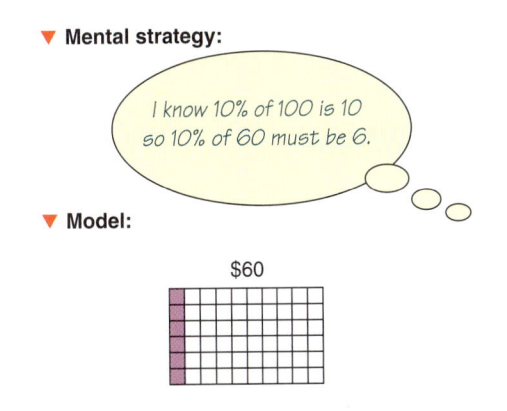

I know 10% of 100 is 10 so 10% of 60 must be 6.

▼ **Model:**

$60

Figure 13-14 "How big a raise is 10% of $60?" Use a mental strategy or a model to find a percent of a number.

296 Chapter 13 • Ratio, Proportion, and Percent: Meanings and Applications

▼ **Mental strategy:**

I know the fraction 15/20 is the same as the fraction 3/4. I also know that one-fourth is 25% so three-fourths is 75%. They won 75% of their games.

▼ **Model:**

20 games

15

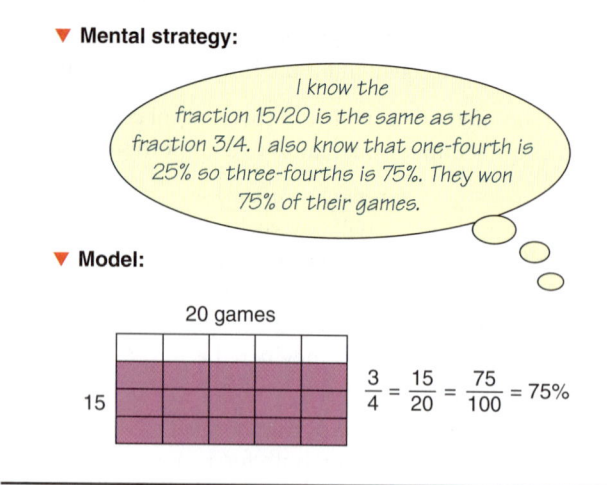

$$\frac{3}{4} = \frac{15}{20} = \frac{75}{100} = 75\%$$

Figure 13-15 *"The Cardinals won 15 of their 20 games—what percentage did they win?"* Use of a mental strategy or model for finding what percent one number is of another number.

Here are two other ways to find the solution:

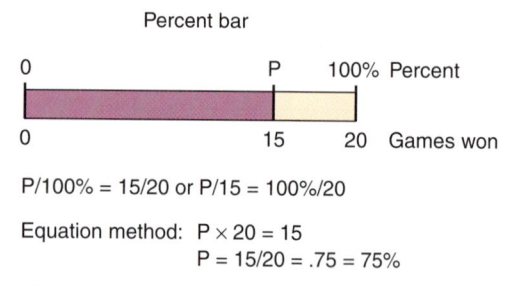

Percent bar

0 — P — 100% Percent
0 — 15 — 20 Games won

P/100% = 15/20 or P/15 = 100%/20

Equation method: P × 20 = 15
P = 15/20 = .75 = 75%

The percent bar provides a model to help children construct different proportions that lead to the same result. Once again, the computation is easy, but the NAEP provides a reminder of the difficulty students have with percent. For example, 43% of seventh graders correctly answered "30 is what percent of 60?" but only 20% correctly answered "9 is what percent of 225?" (Silver and Kenney, 2000).

- Find the total (100%) when only a percent is known. The sale price on a coat was $40 and it was marked down 50%. What was its original price?

Common sense suggests that the original price should be more than $40. Guess-and-test is often a very effective strategy in solving this type of problem. For example, if children guess an original price of $60, then the sale price of $30 is too low. Still, they are on the right track, and if they continue this approach, it will eventually lead to the correct price of $80. The problem could be solved mentally or modeled as shown in Figure 13-16.

The percent bar model provides a base for either of these solutions, where *OP* is the original price:

Ratio method: $\frac{50\%}{100\%} = \frac{\$40}{OP} = \$80$

Equation method: 50% of *OP* = $40

$$OP = \frac{\$40}{.50} = \$80$$

▼ **Mental strategy:**

I know 50% is the same as one-half. If $40 is half of the original price and 40 is half of 80, the original price must be $80.

▼ **Model:**

Percent bar

0 — 50% — 100%
0 — $40 — OP
   Sale   Original
   price   price

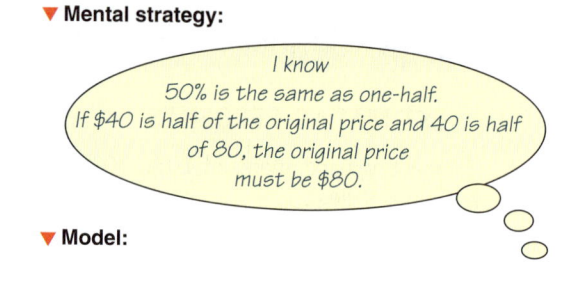

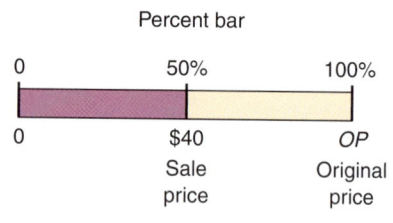

Figure 13-16 *"A coat is on sale for $40 after being marked down 50%—what was the original price?"* Use of a mental strategy or model for finding the total when only a percent is known.

Research has shown that this type of problem is typically more difficult to solve (Silver and Kenney, 2000). The consistently poor performance on percent problems means that instruction must become more meaningful. If emphasis is placed on a particular method before the problem is thought through and well understood, the result will probably be confusion and poor performance. Instructional emphasis in elementary school must be on thinking aloud and talking about what should be done and what a reasonable answer would be. These teacher-led discussions should occur before any serious efforts are made to solve the problem with pencil and paper. Early emphasis on writing a solution to a percent problem forces many students to operate mechanically (without any conscious thinking) on the numbers to produce an answer. Students should be encouraged to think quantitatively in solving problems involving percent. Research does not support the teaching of a single method. You should present a variety of problems involving percent and then follow student leads flexibly toward solutions. Encourage students to verbalize their solutions as they engage in the actual problem-solving process. This verbalization helps clarify what was done (right or wrong). It also encourages them to reflect on the reasonableness of an answer.

This less formal, intuitive approach lacks the structure and security of emphasizing a particular method, but it has several important advantages. In particular, it encourages students to understand the problem in their own minds, along with possible solutions, and decreases the likelihood of their applying a method blindly.

CULTURAL CONNECTIONS

Ratio, proportion, and percent are core elements in mathematics programs in countries throughout the world. International assessments have documented comparatively low performance on items involving proportional thinking, yet the cognitive processes required in developing these concepts are consistent across countries. For example, much of the work done by Piaget involved French children, and the levels of thinking he proposed to describe their cognitive development have been documented to apply across different cultures, including the United States. Studies in different countries have also demonstrated that children reflecting high levels of success with proportional thinking are consistently operating at a formal operations level (Kilpatrick et al., 2001; Lamon, 2007).

Percent is a common topic as well, and research has demonstrated that children in other countries who are not proficient in computational procedures often reflect a lack of some underlying principles. For example, only about half of the sixth graders tested in Japan correctly reported that 100% of 48 is 48 (Reys, 1993). It is fair to conclude that the topics of ratio, proportion, and percent are challenging to teach and learn regardless of where teachers and children are in the world.

A proportion provides a relationship between two or more ratios. Together with proportion, ratios provide an opportunity to practice many computational skills as well as strengthen problem-solving skills. Proportions provide a way to find answers to problems where the numbers are relational. These relationships may be considered additive (absolute) or multiplicative (relative). Children often apply additive procedures to proportions that require multiplicative thinking, particularly in dealing with like quantities. The ability of children to apply multiplicative thinking or proportional reasoning is an important instructional goal and is one of the reasons proportional reasoning has been called the capstone of elementary school mathematics. The use of the double number line is a powerful model for helping children engage in proportional reasoning.

Proportional reasoning is complex, both in terms of the mathematics and of the developmental experiences it requires. Yet it is an important skill for students to gain because it facilitates algebraic thinking. Ratios and proportions also provide a natural means of studying percent, which has a comparison base of 100. Because few mathematical topics have a more practical usage than percents, it is essential that meaningful and systematic development of percent be provided. Instruction should use concrete models to support foundational percent concepts. Key among those concepts are benchmarks of 100% and 50% as well as 90%, 10%, and 1%. These benchmarks provide anchors for students to solve a wide range of problems involving percent and to gauge the reasonableness of their answer.

A GLANCE AT WHERE WE'VE BEEN

Ratios compare two or more numbers. They take different forms and have many applications; money (pennies for nickels), measurements (12 inches in a foot), consumer purchases (3 for 29¢), scale drawings, and blueprints are but a few. Ratio tables provide an excellent model for exploring patterns involving ratios.

Things to Do: From What You've Read

1. Describe how estimating the number of heartbeats in an hour or in a lifetime uses ratios. How could you use a calculator or a spreadsheet to solve the problem?
2. Describe how proportions could be used to compare two products: If a 24-ounce can of pears sells for $1.90 and a 32-ounce can sells for $2.65, which one is the better buy?
3. If a dinner bill was $48, tell how you would estimate a 15% tip. Explain how knowing a way to find 10% of a number mentally would be useful.
4. Rose was making $45,000 per year. She received a 10% raise. Later in the year, the company started losing money and reduced all salaries by 10%. Rose said, "I'm making less money than last year." Is her thinking correct? Tell why.
5. Describe how you would think through a solution to this problem: "The population of a city increased from 200,000 to 220,000. What is the percent increase?" Demonstrate two different ways to solve the problem.

Things to Do: Going Beyond This Book

In the Field

1. [1]*Finding Out About Proportions.* Ask several fifth or sixth graders this question: "A design is 2 cm wide and 2.4 cm long. If the picture is to be enlarged so that it is 5 cm wide, how long will it be?" Ask them to explain their reasoning and answer. Check to see if students tended to be "adders" (i.e., added 3 cm to the dimensions, thus getting 5 cm wide and 5.4 cm long). If they modeled the solution correctly and used multiplication, report the thinking they used.
2. [1]*Finding Out About Percents.* Ask fifth or sixth graders some of these questions: "What is 50% of $80? What is 100% of 50? What does the % on a milk carton mean? Is it possible to get a raise of 200%?" In each question, ask them to explain their thinking. Do you think they have a good understanding of percent? Tell why.

In your Journal

3. Ratios may be used to determine the best buy. Use local newspaper advertisements to find information and set up ratios to compare prices for different sizes

of the same product. Respond in your journal to this question: Can you refute the claim "the larger the quantity, the lower the unit price"?

With Additional Resources

4. Read the articles "Three Balloons for Two Dollars: Developing Proportional Reasoning" (Langrall and Swafford, 2000), "Proportional Reasoning" (Miller and Fey, 2000), "Improving Middle School Teachers' Reasoning About Proportional Reasoning" (Thompson and Bush, 2003), "Assessing Proportional Reasoning" (Bright, Joyner, and Wallis, 2003), or any article from the *2002 NCTM Yearbook*. Describe instructional suggestions they offer from research to help students better understand proportions.

5. Read the articles "Our Diets May Be Killing Us" (Shannon, 1995), "Diet, Ratios, Proportions: A Healthy Mix" (Telese and Abete, 2002), or "Making Sense of Percent" (Metz, 2003). Describe how you could use ideas like those to make a ratio or percent lesson.

6. Read the articles "The Legend of Paul Bunyan" (Bohl, Oursland, and Fince, 2003) or "Using Literature as a Vehicle to Explore Proportional Reasoning" (Thompson, Austin, and Beckmann, 2002). Discuss some ways these stories might be used to help students understand ratio and proportion.

With Technology

7. If you were able to view the tape *Body Ratios*, excerpted in this chapter's Snapshot of a Lesson, here are some questions:

   Did the children believe that some of the ratios of body parts discussed in *Gulliver's Travels* were true? Do you?

   How is ratio used to make connections with fractions and decimals?

   Explain how the teacher used estimation to help children deal with some messy ratios.

   How could the ratios experienced in this lesson be used to develop proportions?

8. Read "Map scale, proportion, and Google earth" (Roberge & Cooper, 2010). Describe three different activities and tell how each engaged students in proportional reasoning.

---

[1]Additional activities, suggestions, and questions are provided in *Teaching Elementary Mathematics: A Resource for Field Experiences*, found on this text's accompanying website, at www.wiley.com/college/reys.

Note to Instructors: You can find additional resources, learning activities, and blackline masters in this text's accompanying Instructor's Manual, at www.wiley.com/college/reys.

Book Nook for Children

Anno, M. *Anno's Math Games.* New York: Philomel Books, 1987.

This book uses proportion to compare the relative numbers of sugar cubes in different-sized containers. A wealth of ideas appear in this book that can be extended to a number of situations; for this reason, it may be desirable to explore one chapter at a time.

Bair, S. Rock, Brock, and Savings Shock. Morton Grove, IL: Albert Whitman & Co. 2006.

Gramps teaches his twin grandsons the value of saving money when he pays each a dollar a week to help with summer chores, then matches every dollar each boy saves.

Giganti, Jr., P. *Each Orange Had 8 Slices: A Counting Book. New York: Greenwillow Books,* 1992.

If each orange has eight slices and each slice has two seeds, how many seeds are there in all? You can multiply, add, or count your way through the math puzzles hiding in the world all around you.

McCallum, A. *Beanstalk: The Measure of a Giant.* Watertown, MA: Charlesbridge Publishing Co., 2006.

This is a story about Jack, who climbs a giant beanstalk and meets a lonely giant boy. Using ratios and proportions, he makes toys that both can use.

Murphy, S. J. *The Grizzly Gazette.* New York: HarperCollins, 2003.

Camp Grizzly is having a parade and must elect a mascot to wear the Grizzly bear outfit. Candidates have started to campaign to wear the outfit. The reader will see how the candidates are keeping track of the daily polls by using percentages in circle graphs. As the last day approaches, each candidate gives a speech; the candidate who won wrote a new camp cheer, and she won with 50% of the vote. There are activities and games in the back of the book for teachers and parents to use.

Swift, J. *Gulliver's Travels* was originally published in 1726.

A number of different adaptations of this masterpiece have been published. Many contemporary versions exist, and the one by James Riordan, published by Oxford University Press in 1998, is excellent. Following Gulliver on the miniature island of Lilliput provides a rich context for discussing and exploring ratio and proportion.

Walton Pilegard, V. *The Warlord's Puppeteers.* Gretna, LA: Pelican Publishing Company, 2003.

Chuan is traveling back to the warlord's palace in China. Chuan joins a troupe of puppeteers and learns about puppets and their proportions.

CHAPTER 14

Algebraic Thinking

> "I WOULD ADVISE YOU, SIR, TO STUDY ALGEBRA, ... YOUR HEAD WOULD BE LESS MUDDY."
> — Samuel Johnson (1709–1784)

>> SNAPSHOT OF A LESSON

KEY IDEAS

1. Help students explore and describe growing patterns.
2. Encourage students to think algebraically.

BACKGROUND

This lesson is based on the video *V Patterns*, from the Modeling Middle School Mathematics Project. To view the video, go to http://www.mmmproject.org and click on the video matrix.

Mr. Dauray, a sixth-grade teacher, is using the mathematics program, *Mathematics in Context* (Romberg and de Lange, 1998), in a school in Rhode Island.

Mr. Dauray: The other day, I come out of my house and ... I heard this [*sound of geese honking*]. It was a sure sign of fall. Does anybody know what that might be?

Jesus: It would probably be the birds that fly in a pattern, with the V-line. I forgot the name.

Student: Geese.

Jesus: Yeah, geese.

Mr. Dauray: I've got these geese right here. Can someone come up to the board and show me how you might see them flying in the air? Say I had just three geese. Can someone come up and put them on the board?

The students add geese to continue the pattern as shown in the dots below. Then Mr. Dauray explains the notation they will use.

V1   V2   V3

Mr. Dauray: This happens to be the first V. This is number 1, OK, the first V. Another V-pattern, this is the second V; we'll call it V2. And then we get this next one; this is the third V-pattern. And it looks like this. How many birds are in the first V-pattern? Jenny?

Jenny: There are 3 birds in the first V-pattern.

After discussing the first three V-patterns, the notation, and the number of birds in each, Mr. Dauray has the students work in groups.

Mr. Dauray: I want you to copy this chart, and I want you to extend the chart for a V-pattern of 4, 5, and 6. And I want you to tell me what it looks like. Discuss in your groups, what do you think you see happening?

The students talk among themselves about the patterns. One student thinks there is something about two and two times a number. Watch the video to see how his fellow students help him with his misconception.

Another group focuses on odd numbers. Watch the video to see if you think they justified their conjecture that there would always be an odd number of geese in the V pattern.

We rejoin the video when the class is back as a whole.

Mr. Dauray: Can someone tell me the number of geese that would be in the V-number 10, the 10th pattern along?

Oscar: I got 21.

Mr. Dauray: And how did you find that information?

Oscar: I multiplied 10 by 2.

Mr. Dauray: Why did you do that?

Oscar: Because I was thinking that every time the two extra birds come, they come and then they have a group … . Then I multiplied it by 2 and it comes out to 20, and then I added a 1, the one in the middle, then it came out to 21.

Jenny: What they are saying is that there are two geese in a pair and there are 10 pairs, so he would try and multiply it by 2 × 10 and then add the leader would be 21.

Mr. Dauray: Does everybody agree? Anybody want to try it out? Let's try it out for a V-number of 15. How many geese would that be?

Ashley: In V-pattern number 3, the V means that there are three groups, or three pairs, plus that extra one. So let's say there's 15 pairs in the 15th pattern, plus this one here.

Mr. Dauray continues with larger and larger numbers to see if students were making the generalization of multiplying the V number by 2 and adding 1.

FOCUS QUESTIONS

1. What does algebraic thinking mean for elementary students?
2. How do patterns help children develop algebraic thinking and ideas?
3. How can you encourage algebraic thinking as children solve problems?
4. How can you use children's understanding of relations to help develop the concept of variables?
5. How do representing, generalizing, and justifying intertwine?

INTRODUCTION

Samuel Johnson, the British author and lexicographer, made the remark in the opening quote to a warehouse packer in the late 1700s. He saw the importance of everyone studying algebra to prevent muddled thinking. I hope that, in studying this chapter, you will see how algebraic thinking can help children clarify much of their thinking about arithmetic.

Traditionally, children do not begin to study algebra until they have a solid foundation in arithmetic. Thus, many students who have not done well with arithmetic never have the opportunity to learn algebra. Research in the latter part of the twentieth century indicated that students in elementary school should receive instruction that prepares them to study algebra. More importantly, it showed that these students built a deeper understanding of arithmetic concepts and skills and became better problem solvers (Carpenter, Franke, and Levi, 2003). Algebra also can reach diverse learners, providing challenges as well as helping students make sense of arithmetic (Schifter, Russell, and Bastable, 2009).

Substitute "algebra" for "mathematics" in the five views of mathematics in Chapter 1, and you will see reasons and ways to include algebra in the elementary school:

1. *Algebra is a study of patterns and relationships.* Patterns and relationships are important in their own right, and studying them leads naturally into algebraic ideas.

2. *Algebra is a way of thinking.* Algebra provides strategies for analyzing representations, representing situations, generalizing ideas, and justifying statements.

3. *Algebra is an art, characterized by order and internal consistency.* Children who are involved in algebraic thinking gain a better understanding of the underlying structure and properties of mathematics (Carpenter and Levi, 2000).

4. *Algebra is a language that uses carefully defined terms and symbols.* These terms and symbols enhance children's ability to communicate about real-life situations and mathematics itself.

5. *Algebra is a tool.* In the past, many algebra courses presented algebra as a tool. But too often, students only learned the intricacies of the tool, without learning how to use it in any practical or meaningful way. The NCTM's *Principles and Standards* (2000) do not recommend this limited approach to algebra, nor is it the approach we will present here. Algebra is a tool, but one that should be meaningful and useful.

Bringing algebra to elementary school does not mean just adding one more topic to a crowded curriculum. It does mean using the mathematics in the curriculum to help children develop algebraic thinking or reasoning. As

Carraher and Schliemann (2010, p.27) write "The early mathematics curriculum abounds with opportunities for promoting algebraic reasoning."

This chapter begins with a discussion of topics—problems, patterns, and relations—that will be used in the remainder of the chapter to build algebraic thinking. After discussing the algebraic language and symbols appropriate for elementary students, we will look at ways to help children develop algebraic thinking through representing, generalizing, and justifying.

> ### Virtual Classroom Observation
>
> Go to www.wiley.com/college/reys,
> Access the Wiley Resource Kit
> Click on the Virtual Classroom Observations Section
> Module 4: Pre-Algebra: Patterns and Functions.
>
> 1. Watch a video: Teaching of Algebra
> 2. Think and discuss: What turned this teacher's ideas of mathematics around? What lessons do you take from listening to her talk about her experience in learning and teaching mathematics?

PROBLEMS, PATTERNS, AND RELATIONS

The teaching of algebra in elementary school should build on ideas that are an essential part of the curriculum. Problems, patterns, and relations are each an essential part of elementary school mathematics. Throughout this chapter, we will develop these three topics to illustrate ways they tie to algebraic thinking. As you read this section, you should work the problems, extend the patterns, answer the questions, and ask yourself how these topics can be used to develop algebraic thinking. In subsequent sections, we will discuss the connections of these three topics to algebraic thinking.

PROBLEMS

Both routine and nonroutine types of problems provide good opportunities for developing algebraic concepts and thinking. We list a few such problems, which we will view from an algebraic perspective later in this chapter.

ROUTINE PROBLEMS Routine problems are often considered exercises for practicing computation, but they can also be used to build algebraic understanding. You will find routine problems like these in every textbook:

Routine Problem A: Before her birthday, Jane had 8 toy trucks. At her birthday party, she was given some more trucks. That night she counted all her trucks and found she had 15. How many trucks was she given at her party?

Routine Problem B: Bill had 24 pencils to put in 4 boxes. He put the same number of pencils in each box. How many pencils did he put in each box?

Routine Problem C: Pat took 4 pails to the beach. She had 3 shovels and 2 rakes in each pail. How many tools did she take to the beach?

NONROUTINE PROBLEMS Many nonroutine problems, especially look-for-a-pattern problems and number puzzles, lend themselves well to an algebraic approach.

Here are the *look-for-a-pattern problems* that we will use in this chapter:

Non-Routine Problem 1: *Geese*. The geese problem in the Snapshot of a Lesson at the beginning of this chapter can be solved by examining the numerical pattern of the number of geese in any V-formation. It is the type of problem that can lead to generalization in a very natural way. Young children can solve this problem for a given V-formation by making a table, but algebra is needed to represent the general solution for all V-formations.

Non-Routine Problem 2: *Sidewalk garden*. A family is building a rectangular garden that is always 1 foot wide surrounded by a sidewalk made of 1-foot by 1-foot tiles. For a 1-foot by 1-foot garden, they need 8 tiles. For a 1-foot by 2-foot garden, they need 10 tiles. Draw a picture to verify the number of tiles needed for these two gardens. How many tiles do they need for a 1-foot by 8-foot garden?

Non-Routine Problem 3: *Cube painting*. A large cube is made of small cubes. The large cube is 4 small cubes high, 4 small cubes wide, and 4 small cubes deep. The surface of the large cube is to be painted. How many of the small cubes are painted on 6 faces, 5 faces, 4 faces, 3 faces, 2 faces, 1 face, and no face?

Here are the *number puzzles* we will use in this chapter. Try them. Can you explain why each works?

Puzzle 1: *Same number*. Pick a number. Add 4. Double the sum. Subtract 6. Divide by 2. Subtract 1. What did you get? Why does it work?

Puzzle 2: *Calendar*. Choose any 3 consecutive days (e.g., Tuesday, Wednesday, and Thursday) for any 3 consecutive weeks in a month (e.g., July 8–10, 15–17, and 22–24). How can you easily find the sum of all nine dates?

PATTERNS

Patterns are an important part of mathematics because they help children organize their world and understand mathematics. Figure 14-1 overviews the expectations set forth in the CCSSM. In Chapters 7-9, we discussed how you can use patterns to help children think about numbers. Here we will use two types of patterns, repeating and growing,

	Operations and Algebraic Thinking Domain
Grade 3	Identify arithmetic patterns (including patterns in the addition table or multiplication table), and explain them using properties of operations.
Grade 4	Generate a number or shape pattern that follows a given rule. Identify apparent features of the pattern that were not explicit in the rule itself.
Grade 5	Generate two numerical patterns using two given rules. Identify apparent relationships between corresponding terms. Form ordered pairs consisting of corresponding terms from the two patterns, and graph the ordered pairs on a coordinate plane.

Figure 14-1 *Excerpt of the CCSSM standards relating to patterns.* This excerpt was made by the authors of *Helping Children Learn Mathematics*. The full document (CCSSI, 2010) may be obtained at www.corestandards.org/the-standards/mathematics.

for building algebraic ideas. Although the CCSSM does not emphasize repeating patterns, we have included them because we want to emphasize that often there are many ways, unless the pattern is well-defined, to look at a pattern. It is the justifying by students of their thinking that make them part of algebraic reasoning.

REPEATING PATTERNS A repeating pattern has a core element that is repeated over and over. One of the simplest repeating patterns is an alternating pattern such as the following:

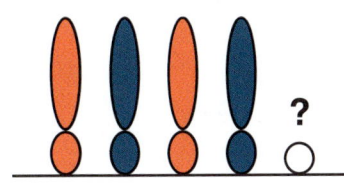

This simple pattern raises many questions about how children work with patterns. What color should come next? The answer to this question is not as simple as it appears, since there are many ways to think about the pattern. Some children may say that the next color should be red because the pattern is just alternating red and blue (i.e., the core element is a red followed by a blue). Others might say that the next color should be blue because the pattern is a red with one blue, followed by a red with two blues, followed by a red with three blues, and so on (i.e., the core grows by one blue each time it repeats). It is often difficult to get into the minds of children, so having them explain their thinking is crucial. Also, sharing ideas helps children consider alternative ways of looking at patterns. As children become familiar with patterns like this one, they can extend other patterns or make their own patterns.

Research has shown that children should begin recognizing and thinking about patterns by describing and extending patterns with pictorial models, as illustrated by In the Classroom 14-1. As children become able to identify the missing pieces in the patterns, encourage them to look at the patterns in different ways that would lead to different ways of extending the patterns. Ask questions such as "How can you describe this pattern?" or "How can you extend this pattern?" Questions like these help children begin to think algebraically.

In the Classroom 14–1

THINKING ABOUT PATTERNS

Objective: Recognizing, describing, and extending patterns.

Grade Level: 1–2.

▼ Provide children with a variety of patterns like these, asking them questions like:
  - How do you describe this pattern?
  - What shape (or object) goes in the empty space? Why?
  - Are some of these patterns alike? How?

1. What pattern block shape should come next?

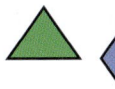

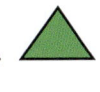

2. One of the shapes is missing in this string of shapes. What shape and what color would you insert? Why?

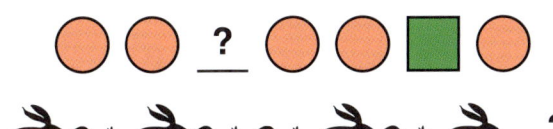

3. What animal should be in the next place? Why?

304 Chapter 14 • Algebraic Thinking

Can you see the cores in the repeating patterns below?

Pattern A: ■■◆■■◆■■◆■

Pattern B: ●■●■■●■●■■●■●■■

Pattern C: ♣♣♣♣♣♣♣♣♣♣♣♣♣♣♣

In pattern A, the core consists of three elements: two red squares followed by one blue rhombus. This is repeated over and over. In pattern B, the core consists of 5 elements: one pink circle, one green square, one pink circle, two green squares. Do you see that the core in pattern C is the same type as in pattern B?

Research from several mathematics assessments of the National Assessment of Educational Progress (NAEP) indicates that fourth-grade students often have more difficulty identifying a missing element in a pattern (as in Problem 2 from In the Classroom 14-1) than identifying the element that continues the pattern. This difficulty may simply be the result of having been asked to continue patterns more often than to fill in missing parts. Be sure to ask children both types of questions.

Math Links 14.1

Watch kindergarten children work with patterns on the video *People Patterns* (video 38) from the Annenberg series, Teaching Math: A Video Library, K–4, which you can access from http://www.learner.org/resources/series32.html or from this book's Web site.

www.wiley.com/college/reys

GROWING PATTERNS The geese problem in the Snapshot of a Lesson involves a growing pattern. You can present such patterns verbally, through pictures only, or with symbols only. Both the geese problem and the sidewalk-garden problem, for example, could be presented just with pictures, and you could ask young children to continue the pattern:

Problem 1: *Geese problem*. What would be the next three pictures (terms 4, 5, and 6)?

Term 1   Term 2   Term 3

Problem 2: *Sidewalk-garden problem*. The green squares represent the garden, and the blue squares represent the tiles. What would be the next picture (term 4)? Draw pictures to illustrate terms 5 and 6.

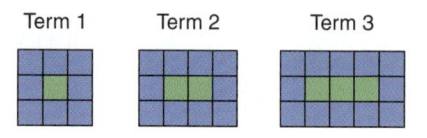

Term 1   Term 2   Term 3

You can use pictorial representations like these as powerful links to analyzing number patterns that are growing patterns. Mathematics curricula often include problems where children are asked to continue growing number patterns or sequences of numbers. For example, the beginning terms of the geese problem could be symbolized as 1, 3, 5, 7, ... , and the sidewalk-garden problem could be symbolized as 8, 10, 12, ... Do you see how these growing patterns differ from repeating patterns? In growing patterns like these, each successive term changes by the same amount from the preceding term (in both the geese problem and the sidewalk-garden problem, this change is 2 more). Traditionally, children in elementary school have been given only numerical patterns like these, where the difference between successive terms is some constant amount. However, numerical patterns need not be so simple. Growing patterns, like repeated patterns, can be thought of in different ways, and we need to help children look at different ways in which patterns can grow. For example, consider the numerical patterns below:

Pattern A: 1, 2, 4, ...

Pattern B: 1, 4, 9, 16, 25, ...

Pattern C: 1, 1, 2, 3, 5, ...

In a fourth-grade classroom working with pattern A, Andy suggested that the next number in the sequence is 5. The teacher asked what the next number after 5 would be, and Andy answered 7 (1, 2, 4, 5, 7, ...). Do you see what Andy may have had in mind? Some children said: "Andy writes two numbers, skips one, and writes the next two," "Andy skips all the threes—3, 6, 9, and all those 3-times numbers," and "Andy adds one, then two, then one, then two." Sue suggested a different pattern from Andy: "I think that the pattern goes 8, 16, and on. It just doubles." Argon said, "I think it goes 1, 2, 4, 6, 9, 12, and so on." Can you figure out what rule Argon was using?

This episode illustrates several key ideas to consider as you work with children. First, do not expect children to see patterns the way you do. It is important to have them explain their thinking. It is also important to have them realize that there is often more than one way to look at a pattern and to explain it.

Pattern B is the sequence of perfect squares. Wells and Coffey (2005) gave this sequence to second graders, who most likely were not familiar with perfect squares. One child did see that the differences between consecutive numbers were 3, 5, 7, 9 and used this to continue the pattern. The next difference would be 11, so the next number in the sequence would be 36. Another child just wrote the next differences: 11, 13, and 15. Another child used the 5 elements as the core of a repeating pattern and wrote 1, 4, 9 for the next three entries. This is a correct answer to another question. Think about how you could pose the problem so the child would not think a repeating pattern was a possibility. The fourth child wrote 34, 43, and 52 for the next three entries. What was this child thinking?

Pattern C is the Fibonacci sequence. The next numbers are 8 and 13 (1, 1, 2, 3, 5, 8, 13, …). Can you figure out the number after 13? Once you know the rule, it is easy to generate the sequence, but many people (adults as well as children) do not immediately see the rule. Children need practice in recognizing and analyzing patterns before we can start teaching them to represent patterns algebraically.

RELATIONS

We will explore two types of relations that we can use to help children build algebraic concepts and thinking: *properties of numbers* and *functions*. Another relation — the *equality relation* — underpins both these types of relations as well as all of algebra.

Young children have a good sense of equality in terms of the same size or amount. Even young children know when they do not get a cookie that is the same size as yours, when one block tower is not the same height as another, or when they have three marbles just as you do. However, very few programs build on the idea of presenting equality as being "the same as" in size or amount and describing the relationship of two quantities as being the same (equal) or different (unequal). A balance provides a model of how to relate two quantities in this way (Mann, 2004). In the Classroom 14-2 shows some questions that children in primary school can investigate.

PROPERTIES OF NUMBERS Table 9-1 in Chapter 9 reviews some of the mathematical properties of numbers (commutative, associative, distributive, and identity). In this chapter, we will discuss how to use number properties to promote algebraic thinking. Here are examples of relations based on the mathematical properties and meanings of operations that you could have students investigate. Which statement is not true? Why?

1. If you add a number to a given number and then subtract that same number from the sum, then you get the given number.
2. If you subtract 0 from a given number and add the given number to the difference, then the sum is twice the given number.
3. If you multiply any two numbers, the product is larger than each of the two numbers.
4. If you divide a positive whole number by a proper fraction, the quotient is larger than the positive whole number.

In the Classroom 14–2

BALANCE ME

Objective: To explore what balances a balance scale.

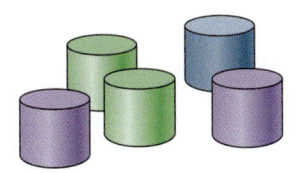

Grade Level: 1–3.

Materials: Weights (1–10 ounces or 1–10 washers), or labeled containers weighing 1–10 units (2 of each)

(Note: You can fill containers with Unifix cubes to make weights.)

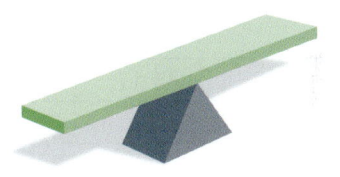

Balance scale

Questions To Ask Students:

A. I am going to put some weights on each side of the balance. Which will balance?

Left side	Right side
4	4
3	8
2 and 3	5
3 and 4	6
1, 2, and 3	6

B. If I put a 5 weight on the left side, what could I put on the right side to balance it?

C. If I put a 3 weight on one side and a 7 weight on the other side, what would I need to add to the side with the 3 weight to make the scale balance?

D. If I put an 8 weight on the left side and a 2 weight on the right side, can I balance the scale by adding a 6 weight to the left side?

E. I have the scale balanced with a 6 weight on each side. What could I add to each side to keep it balanced?

F. I only have four weights: 3, 4, 5, and 7. Could I put all or some of the weights on the scale to have it balance?

G. Could I balance the scale by using one of each of the 10 different weights?

A more detailed plan for this activity is available on the Student Companion site at www.wiley.com/college/reys.

FUNCTIONS Suppose two sets of numbers are related in such a way that each number in the first set is related to one and only one number in the second set. A function is a way of expressing that relation. Young children can explore ways that functions work in a very informal way. For example, they can explore how the number of boys in a group is related to the number of hands. Begin by having students tell the number of hands of 3, 4, and 10 boys; then write the numbers in a table, as shown below.

Number of boys	3	4	10	6	11	5	11
Number of hands	6	8	20				

Have students complete the table. You want the students to focus on the relationship between the number of boys and the number of hands—that is, to focus on the function. For this reason, it is best not to write the number of boys in order from 1 to 12 in the table, so students cannot easily see the pattern of the number of hands increasing by 2 with each additional boy in the group (see the table below).

Number of boys	1	2	3	4	5	6	7	8	9	10
Number of hands	2	4	6	8	10	12	14			

Another informal way for children to explore functions would be to consider the relationship of the number of girls to the number of triangles as the girls make triangles with string as shown below.

Unlike the 1-to-2 relationship of boys to hands, this is a 3-to-1 relationship (3 girls for each triangle). Some children may have difficulty describing this relationship in general terms because it calls for division rather than multiplication. Again, begin by writing the numbers in a table (in this case, it may be useful to write the number of girls in order, to help students see the relationship):

Number of girls	3	6	9	12	15	18	21	24
Number of triangles	1	2	3	4	5	6	7	8

The article (Johanning and others, 2010) gives an example of using literature, *The Polar Express*, with young students to examine the relations between number of cars and number of wheels and windows.

One technique to look at relations is to use a "machine" that outputs a number for each number input. The challenge for the students is to decide what the machine does and how to describe it. These machines can be simple, such as one that always gives an output of two more than the number input, twice as much as the number input, or two less than the number input. A more complicated machine might relate input and output as shown in the Function Table below. The relationship involves two rules. Can you determine the function?

Function Table

Input	3	6	7		20	53
Output	7	13	15	29	41	

You need to help children focus on the change from the input to the output. Some children will focus only on the change from one input value to the next, but this does not help them understand the input–output relationship. Two items with this same relationship (double the input and add one) as in the function table above were given to fourth-grade students on the 2005 and 2007 NAEP mathematics assessments (see sample items at http://www.nces.ed.gov/nationsreportcard). In one item, the children had to chose a rule that related the input and output, and in the other they had to produce an output for a given input. In both items, students tended to answer considering only the relation between the first input and output. For example, they would have chosen the option of 57 as the output for 53 since 7 is 4 more than 3. The other choices, including the correct answer, were selected at a chance level. Children at this grade level often think additively (the first output is 4 more than the input). They need experience with multiplicative functions. They also tend to make judgments based on one instance. You need to ask them if their rule would work for the next input–output pair (and all the given pairs).

In the Classroom 14-3 shows an activity with a "function machine" that you can modify to suit the level of your students. Children love making a machine (a big box with a child in it or behind it). One student puts the number in the box; the child in the box uses the rule and writes the

In the Classroom 14–3

IN–OUT MACHINE

Objective: To analyze the actions of a function machine.

Grade Level: 3–4.

Dr. de Zine invented an in–out machine. The machine adds, subtracts, multiplies, or divides markings on animals as they go through it. He has kept a record of what happened each day as animals went in and came out of the machine. Your job is to fill in the blanks in his record and to figure out how the machine worked each day.

	IN	OUT
M	skunk with 3 stripes	skunk now has 12 stripes
O	leopard with 12 spots	leopard now has 48 spots
N	zebra with 5 stripes	zebra now has _____ stripes
T	dog with 5 spots	dog now has 15 spots
U	cat with 2 stars	cat now has 12 stars
E	rabbit with 3 marks	rabbit now has _____ marks
W	horse with 12 lines	horse now has 6 lines
E	pig with 8 spots	pig now has 4 spots
D	cow with 0 spots	cow now has _____ spots
T	elephant with 32 stars	elephant now has 16 stars
H	rhino with 24 diamonds	rhino now has 8 diamonds
U	hippo with 41 spots	hippo now has _____ spots
F	frog with 4 spots	frog now has 11 spots
R	toad with 3 spots	toad now has 9 spots
I	salamander with _____ stripes	salamander now has 15 stripes

A more detailed plan for this activity is available on the Student Companion site at www.wiley.com/college/reys.

output. The other children guess the rule. Do not hesitate to give a rule involving multiplication or two simple operations.

LANGUAGE AND SYMBOLS OF ALGEBRA

Children can learn the language and symbols associated with algebra as they are learning about numbers. In this brief section, we look only at equality and inequality, variables, and expressions and equations, because these are the ideas that are central to the algebra that is appropriate for elementary students.

The CCSSM places emphasis on representing problems and solving equations as shown in Figure 14-2. In order to make this possible, one has to develop the language and symbols of algebra. Research shows that children adapt to using letters for numbers and amounts when given the opportunity.

Of course, there are many other algebraic terms and symbols, but we will use them only when what is meant is clear from the context. For example, you probably understood what was meant by the three dots that we used in earlier sections of this chapter to represent the continuation of number sequences (e.g., 1, 4, 9, 16, …). When you are helping children learn to think algebraically, you should use algebraic terms and symbols as you would if you were teaching children a foreign language. That is, you should weave the terms and symbols naturally into your discussions, as the teacher did in this chapter's Snapshot of a Lesson. For instance, the teacher used the terms V_1, V_2, and V_3 in a natural manner that made sense to the students.

	Operations and Algebraic Thinking Domain
Kindergarten	Represent simple addition and subtraction problems with objects, pictures, and equations.
Grade 1	Represent addition and subtraction problems (sums ≤ 20) with equations using an unknown symbol. Solve such equations. Understand the meaning of the equal sign.
Grade 2	Represent addition and subtraction problems (sums ≤ 100) with equations using an unknown symbol. Solve such equations.
Grade 3	Represent multiplication and division problems (products ≤ 100) with equations using and unknown symbol. Solve such equations. Represent two-step word problems with equations using a letter to stand for the unknown quantity.
Grade 4	Represent multiplication and division problems with equations using and unknown symbol. Represent multi-step word problems with equations using a letter to stand for the unknown quantity.
Grade 5	Use parentheses, brackets, or braces in numerical expressions, and evaluate expressions with these symbols. Write simple expressions that record calculations with numbers, and interpret numerical expressions without evaluating them.
Grade 6	Write, read, and evaluate expressions in which letters stand for numbers. Apply the properties of operations to generate equivalent expressions and identify equivalent expressions. Understand solving an equation or inequality as a process of answering a question: which values from a specified set, if any, make the equation or inequality true? Represent and solve inequalities. Use variables to represent two quantities in a real-world problem that change in relationship to one another; write an equation to express one quantity, thought of as the dependent variable, in terms of the other quantity, thought of as the independent variable. Analyze the relationship between the dependent and independent variables using graphs and tables, and relate these to the equation.

Figure 14-2 *Excerpt from the CCSSM standards relating to representing and solving problems. This excerpt was made by the authors of Helping Children Learn Mathematics.* This summary was made by the authors of *Helping Children Learn Mathematics.* The full document (CCSSI, 2010) may be obtained at www.corestandards.org/the-standards/mathematics.

When children begin writing expressions for mathematical situations, they often use a mixture of symbols and words. This is natural and many of these expressions communicate well. For example, a child may describe the rule in the previously discussed function table as "double the input + 1." Eventually, if the input was labeled x and the output was labeled y, as shown in the table below, you could write that the change was $2x + 1$ or that $y = 2x + 1$. This symbolic or abstract version is the goal at this level, but not a place for children to begin.

x	y
3	7
5	11
6	13
	29
53	

Math Links 14.2

If you have questions about algebra, turn to "Ask Dr. Math." This service is available from the Math Forum. It can be accessed from http://www.mathforum.org/math_help_landing.html or from this book's Web site.

www.wiley.com/college/reys

EQUALITY AND INEQUALITY

As we have noted, young children understand sharing equally and balancing a scale. Nevertheless, they are baffled by the equal sign. It is well known that many children think that the equal sign means "get an answer." For example, many children in grades 1–6 respond that 12 or 17 is the number that should be placed in the box to make the following symbolic sentence true: $8 + 4 = \Box + 5$. Research showed that only 5% of students in grades 1–2 and 9% of students in grades 3–4 responded with the correct answer, 7 (Carpenter and Levi, 2000). Typically, children in these grades have not learned what "equals" means and have not learned how to interpret the structure of symbolic sentences. However, after instruction similar to that recommended in this chapter, children's results were greatly improved (66% correct in grades 1–2 and 72% correct in grades 3–4). For further information and research about developing meaning for the equal sign, read the article by Molina and Ambrose (2006).

Math Links 14.3

A variety of Web resources for algebra that may be used by teachers and children may be found at NCTM'S Illuminations Web site under the Algebra Standard. In particular, the balance activity gives students an opportunity to explore equality. You can access this from http://illuminations.nctm.org/ActivityDetail.aspx?id = 26 or from this book's Web site.

www.wiley.com/college/reys

You can help children grasp the meaning of the equal sign (and the unequal sign) by introducing it in an activity like the one in In the Classroom 14-2. You can tell children that two quantities are equal if the scale balances. Thus, the relationships in part A of In the Classroom 14-2 would be represented as shown in the following table.

You can help children use the inequality symbols ($<$ and $>$) to describe larger or smaller quantities. Children can describe the relationship using words and then use the symbols. For example, they can say "3 is less than 8" and then write $3 < 8$; similarly, they can say "12 is greater than 4" and then write $12 > 4$.

Left side	Right side	Representation
4	4	$4 = 4$
3	8	$3 \neq 8$
2 and 3	5	$2 + 3 = 5$
3 and 4	6	$3 + 4 \neq 6$
1, 2, and 3	6	$1 + 2 + 3 = 6$

In everyday life, the equal sign is used in different ways. For example, on a spreadsheet an equal sign signals the computer to do an operation. In the classroom, it is often carelessly used in carrying out a string of operations. Look at this statement:

$$5 + 3 = 8 - 6 = 2 \times 5 = 10 + 14 = 24$$

What does this mean? The equal sign was used to signal steps performing operations. It would be better to use another symbol such as an arrow. Certainly, $5 + 3$ does not equal 24.

VARIABLES

You should be aware of the three different uses of *variable*. The most common use in elementary school is as a *placeholder*—for example, in the sentence $3 + \Box = 7$, the box is a placeholder for the number that makes the sentence true. This is no different from using a letter, as in $3 + a = 7$; children easily adjust to using both boxes and letters to represent variables. Variables are also used in *generalizations*—for example, the generalization "any number subtracted from itself is equal to zero" may be represented as $a - a = 0$. In this generalization, the variable a represents any number, in contrast to the single number represented by a variable used as a placeholder.

Variables are also used in *formulas* and in *functions*. For example, the function relating the number of boys (B) to the number of hands (H) could be represented by $H = 2 \times B$, where H and B are the variables. The function relating the number of girls (g) to the number of triangles is $t = g \div 3$. If the value of one variable changes, then the value of the other variable also changes. Note also that the value of g must be a multiple of 3 (or the triangle would not be complete).

The following table summarizes these three uses of variables:

Use of Variable	Representation	Characteristics
Placeholder	$3 + a = 7$	Specific value for a
Generalization	$a - a = 0$	All values of a make the sentence true
Function	$H = 2 \times B$	Each value of B produces one and only one value of H

EXPRESSIONS AND EQUATIONS

Think of an expression as representing a phrase and an equation as representing a complete sentence. For example, the expression $n + 3$ represents the phrase "three more than n" (or "n plus three" or any other equivalent phrase). The equation $n + 3 = 7$ represents the complete sentence "three more than n is 7" (or "n plus 3 is 7" or "n plus 3 equals 7" or any other equivalent sentence). Knowing how to represent phrases will help students represent complete sentences. In many elementary programs, an equation such as $n + 3 = 7$ or $\Box + 3 = 7$ is called an *open sentence* because there is an unknown. In contrast, the equation $4 + 3 = 7$, which does not have an unknown, is referred to as a *closed sentence*.

Children often think there is only one number that will satisfy the open sentence. This NAEP item was given to fourth-grade students on the 2005 mathematics assessment. The students had numbered tiles (0–9) that would fit the open box, which was larger than shown here. Otherwise, they might have thought that the boxes must be replaced by the same number.

Jan entered four numbers less than 10 on his calculator. He forgot what his second and fourth numbers were. This is what he remembered doing.

$$8 + \Box - 7 + \Box = 10$$

List a pair of numbers that could have been the second and fourth numbers. (You may use the number tiles to help you.)

List a different pair that could have been the second and fourth numbers.

About one-third of the students gave two solutions, and another 22% gave one solution to the open sentence. A challenge for students would be to list all ten solutions. Try it. What is the relationship between the two numbers of every pair? This task is a good example of how, with simple modifications, you can reach the needs of different children in your class.

Using only the digits 0–9 (you may want students to use the number cards in the Appendix), what are the solutions to the following open sentences?

A. $8 - \Box = 2$
B. $3 < \Box$
C. $0 > \Box$
D. $5 + \Box = \Box + 5$

Notice that there is only one number that is the solution to A, there are several whole numbers (4–9) that are solutions to B, there are no whole numbers from 1 to 9 that satisfies the inequality in C, and all the numbers satisfy D. Presenting children with a variety of types of open sentences such as A–D prevents them from thinking that every open sentence has one and only one number that satisfies it.

REPRESENTING, GENERALIZING, AND JUSTIFYING

The NCTM's Algebra Standard calls for students to represent and understand quantitative relationships, to represent and analyze mathematical situations and structures using algebraic symbols, and to analyze change in various contexts. In this section, we will examine how the processes of representing, generalizing, and justifying relate to each of these topics in elementary school mathematics.

ROUTINE PROBLEMS

The following discussions show how you can use representing and justifying to help children think algebraically about routine problems. Such problems, however, do not lend themselves to generalizing, because the equations corresponding to those problems contain variables used only as placeholders, not to make general statements about numbers.

REPRESENTING Many elementary mathematics curricula include the use of equations to solve routine problems, but all too often, these curricula fail to emphasize helping children understand where these equations come from and how the equations make mathematical sense. For example, consider this routine problem:

Routine Problem A. Before her birthday, Jane had 8 toy trucks. At her birthday party, she was given some more trucks. That night she counted all her trucks and found she had 15. How many trucks was she given at her party?

Young children can represent this situation with physical materials (concrete representations). For instance, they can put out 8 trucks and keep adding trucks until they reach 15 trucks. (A potential difficulty is that many young children will lose track of how many trucks they have added on and will have to repeat the process many times before they realize that it would be helpful to keep the original 8 trucks separated from the others. Therefore, you should begin representing routine problems with simpler problems that involve just joining or separating existing groups of materials.) After representing with physical materials, children can move to drawing pictures (semiabstract representations). At first, the pictures may be realistic on picture. The child draws 8 red trucks and then draws green ones, one at a time, to determine that 7 more are needed to make 15.

With encouragement, children will move to more abstract drawing of the situations (showing only the original 8 trucks) such as those below:

The child who drew A needed to keep some semblance to a truck, while child B was happy with just using t for truck. Note that the child in C has simplified the drawing to a mark. All of these are more abstract than the truck drawings and certainly take much less time. If you model such behavior, children will move to these stages.

Often, the purpose of representing, whether with physical materials or with pictures, is to find an answer. However, we want to turn children's attention to the idea of representing a situation with an open sentence or with an equation (abstract representation). Second-grade students who were already familiar with using open sentences to represent problems involving simple joining or separating explained their thinking about this more complicated problem as follows:

> Agnes, who wrote $8 + \square = 15$, explained: "Jane started with 8 trucks. She got some more; some more is a box because we don't know how many she got. She ended up with 15."
>
> Baylor, who wrote $15 - 8 = \square$, explained: "Jane had 15 trucks after the party. If you took away the ones she had before, you could tell how many she got at the party, and that is what we want to know."
>
> Candiza, who wrote $15 = 8 + \square$, explained: "I thought about the balance. I put 15 here [*pointing to the left side of a make-believe balance*] and 8 on the other side. That is sort of like the before and after. I have to change the before side, the 8, to make it equal to the 15. Need to add on to the 8 to make it equal the 15."
>
> Julian asked Candiza: "Why didn't you take away from the 15 to make it balance?" Couldn't you write $15 - \square = 8$?"

What would you ask Julian? The teacher, who had been working with her students on writing open sentences to represent problems, asked Julian what the open box stood for in his sentence. She was satisfied that his sentence made sense to him when he explained it was the number of trucks Jane was given. He said, "Jane had 15, if she gave back those she got (the box), she would have 8. That is what my sentence says."

This type of discussion does not happen the first time students write sentences. Look at the understanding of the problem and the open sentences that this group developed: They understood the structure of both addition and subtraction sentences, the meaning of the equal sign, and that many different sentences could represent the same problem.

Now consider the second routine problem:

> Routine Problem B. Bill had 24 pencils to put in 4 boxes. He put the same number of pencils in each box. How many pencils did he put in each box?

How would you represent this with an open sentence or an equation? How would you explain each of the following possibilities, where p stands for the number of pencils in each box?

Sentence A: $4 \times p = 24$
Sentence B: $24 \div 4 = p$
Sentence C: $p + p + p + p = 24$
Sentence D: $24 - p - p - p - p = 0$

Sentence A focuses on the multiplicative nature of the problem. To understand this sentence, a child would have to understand that it is a sentence of the form $G \times N = T$, where G stands for the number of groups, N stands for the number of items in each group, and T stands for the total number of items. Sentence B also focuses on the multiplicative nature of the problem but puts it in terms of division. Sentences C and D focus on the additive/subtractive nature of the problem. Students writing these sentences were probably thinking: "I have to put p pencils in one group and then in another group and so on until "I have 4 groups, so $p + p + p + p$ must equal the 24 pencils." At grades 3 and 4, you need to help children move from additive to multiplicative thinking. That is, they see the pencil problem as involving multiplication, not just addition.

These two problems and children's responses to them illustrate that different children often look at problems, even routine ones, in different ways. Therefore, you need to be careful about giving tests that ask questions like "What operation would you use to solve this problem?" As the pencil problem shows, there will often be more than one correct answer to such questions.

You should ask children not only to write open sentences that represent problems but also to write problems that correspond to given open sentences. For example, here are four different problems that correspond to the equation $28 - \square = 12$:

> Jennifer had 28 cookies. She gave some to friends for lunch. After lunch, she had 12. How many cookies did she give away at lunch? (separating)
>
> Jill has 28 books and Amanda has 12. How many more books does Jill have than Amanda? (comparing)
>
> How many 1-ounce weights need to be removed from the left side to balance the scale? (equalizing)
>
> James has 28 stickers. Twelve of them are blue and the rest are green. How many are green? (part–whole)

Note that these problems differ not only in the objects involved (cookies, books, weights, and stickers) but also in the type of activity involved (separating, comparing, equalizing, and part–whole). You should make sure that your students have experience with writing all four types of problems. Many mathematics programs do not require students to write problems of the equalizing type, but we think they are important because of their close tie to the idea of balancing and the meaning of equality. They give children another way of understanding abstract representations and help them make sense of equations.

312 Chapter 14 • Algebraic Thinking

JUSTIFYING Having children justify their answers or their approaches to problems can help them understand the mathematics and gain confidence in their knowledge and skills. You should apply this principle when children are working with open sentences and ask them to justify their solutions. For example, if a student says the solution to the open sentence $35 + a = 51$ is 16, ask, "How do you know?" Note that you should do this even when the solution is correct, as it is in this case. Often, because of time pressure, teachers ask students to justify only when their solutions are incorrect. Think how this could affect a child's confidence!

Figure 14-3 shows some open sentences and how students explain the thinking behind their solutions. Having students explain their thinking is a way to encourage reasoning and sense making as they justify their answers or method. The students' solutions depend on understanding equality and the abstract representation of the equation. See if you can follow each student's explanation.

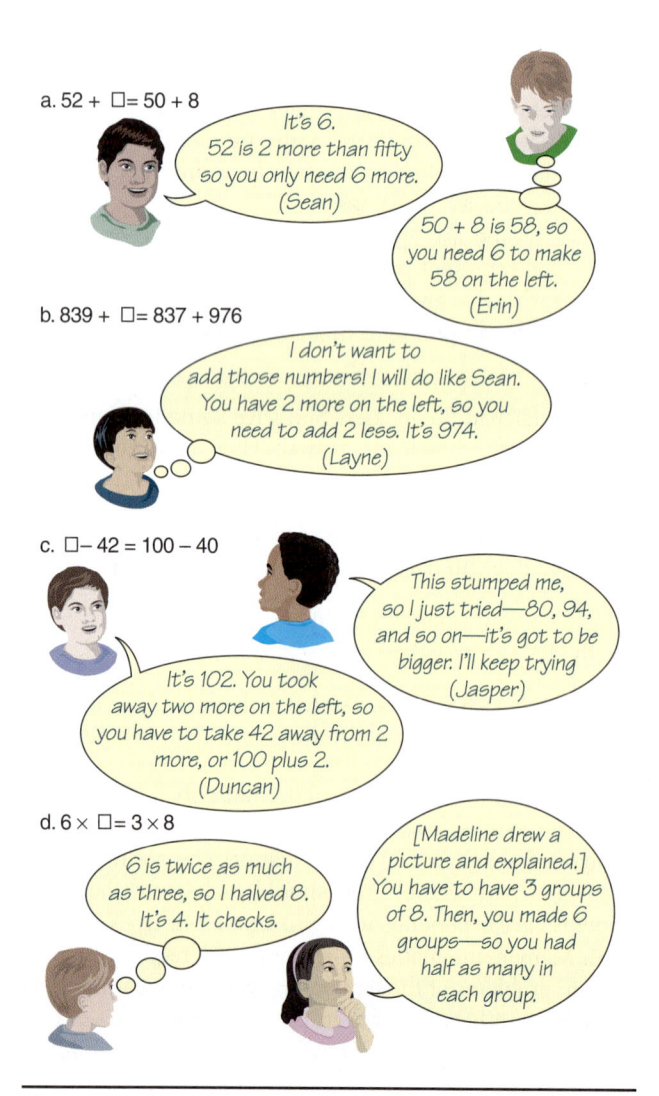

a. $52 + \square = 50 + 8$

It's 6. 52 is 2 more than fifty so you only need 6 more. (Sean)

50 + 8 is 58, so you need 6 to make 58 on the left. (Erin)

b. $839 + \square = 837 + 976$

I don't want to add those numbers! I will do like Sean. You have 2 more on the left, so you need to add 2 less. It's 974. (Layne)

c. $\square - 42 = 100 - 40$

This stumped me, so I just tried—80, 94, and so on—it's got to be bigger. I'll keep trying. (Jasper)

It's 102. You took away two more on the left, so you have to take 42 away from 2 more, or 100 plus 2. (Duncan)

d. $6 \times \square = 3 \times 8$

6 is twice as much as three, so I halved 8. It's 4. It checks.

[Madeline drew a picture and explained.] You have to have 3 groups of 8. Then, you made 6 groups—so you had half as many in each group.

Figure 14-3 *Explaining solutions to open sentences.*

PATTERNS

REPRESENTING Sometimes, you may ask students directly to describe a growing pattern, and at other times, you may give students a problem whose solution requires them to use a growing pattern. In either case, representing the pattern with a table showing every term will help students organize their thinking and describe the pattern.

The table below summarizes the information that the students in this chapter's Snapshot of a Lesson found from representing the geese problem (Non-Routine Problem 1) with paper geese and then with pictures.

Geese V-term	1	2	3	4	5	6	...	V
Number of geese (n)	3	5	7	9	11		...	?

The teacher's task would be to help students generalize the pattern so they can represent it symbolically. There are two ways to consider the generalization: as a *recursive expression* or as an *explicit equation*.

A recursive expression is an expression that tells how to find the value of a term given the value of the previous term. In the geese problem, it is clear that if you know the value of a term, the value of the next term will be two more (i.e., two geese join the pattern each time). Thus, if you know that there are 101 geese in the pattern, you know that the next step in the pattern would have 103 geese. Symbolically, we could express this as "if n is the value of one term, then $n + 2$ is the value of the next term." The recursive expression is easier for young children.

An explicit equation lets you calculate the value of one term given the number of the term. In the geese problem, an explicit equation would let you calculate the number of geese given the number of the V-term. The students in the Snapshot of a Lesson expressed this as "double the term number plus one," which can be represented by the equation $n = (2 \times V) + 1$. An equation is a more powerful representation than a recursive expression because you do not need any information about values of previous terms but can calculate the value of any arbitrary term.

Now let's consider the sidewalk-garden problem (Nonroutine Problem 2) again. Below is the picture showing the first three gardens along with a table of values that give the number of tiles needed for the first four terms.

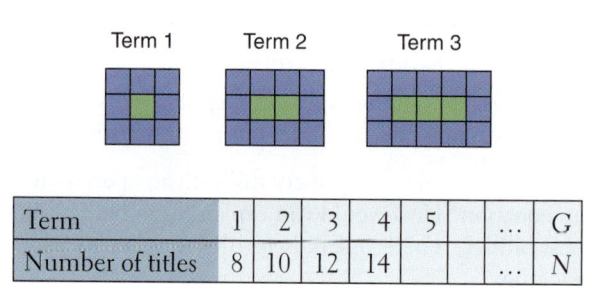

Term 1   Term 2   Term 3

Term	1	2	3	4	5	...	G
Number of titles	8	10	12	14		...	N

Using the table, young children would find it relatively easy to tell how many tiles would be needed for garden numbers 5 and 6. As in the geese problem, the recursive expression is $n + 2$ (2 more tiles each time); thus, if garden 3 has 12 tiles, garden 4 would have 14 tiles, garden 5 would have 16 tiles, and garden 6 would have 18 tiles. Young children with patience could extend this to tell how many tiles for garden 10 or even, say, garden 33! But what if they wanted to know how many tiles for garden 200? An explicit equation would be more useful.

Look again at the picture of the growing garden above. There are several ways to look at the number of tiles without just counting them. Note that, no matter how long the garden is, there are three tiles at each end. Also note that the numbers of tiles above and below the garden are both the same as the term number (G)—that is, in term 2, there are 2 tiles above and 2 tiles below; in term 3, there are 3 tiles above and 3 tiles below; and so on. Thus, we can describe the sum (N) of all the tiles as $N = 3 + 3 + G + G$ (i.e., 3 tiles at one end plus 3 tiles at the other end plus G tiles above plus G tiles below).

The article by Earnest and Balti (2008) looks at a growing pattern with students in Grade 3 and presents three instructional strategies that help young children to move to representing such patterns.

> **Virtual Classroom Observation**
>
> Go to www.wiley.com/college/reys,
> Access the Wiley Resource Kit
> Click on the Virtual Classroom Observations Section
> Module 4: Pre-Algebra: Patterns and Functions.
>
> 1. Watch the videos: Teaching Examples #1- #2.
> 2. Think and discuss: How does the teacher develop growing patterns from the first explorations (video #1) to the more general case? How does she build on different ideas that her students have?

JUSTIFYING If you have students working on these problems, trying to describe or represent the pattern, it is also important to have them explain their thinking. At this level, you are having them justify their representation. For instance, in the sidewalk-garden pattern, one might say $N = 6 + (2 \times G)$, another might say $N = (2 \times 3) + (2 \times G)$, and still another might say $N = 2 + 2(G + 2)$. Look at the picture and see if you can tell how these students might explain their thinking to justify their solutions. Of course, the power of algebra would let us show that all these expressions are equivalent, but children will find that out later in their study of algebra. At the elementary levels, the important thing is that they should think about the representations and patterns and explain their thinking.

When you show elementary students number patterns that are not in a context, such as a problem or a geometric figure, encourage them to explain the rule, either in terms of a recursive expression that relates one term to the next or, possibly, in terms of a closed sentence that expresses the value of one specific term. For example, students presented with the pattern 2, 4, 6, 8, ... could be encouraged to come up with a recursive expression such as "each number is two more than the previous one" or to come up with a closed sentence such as "the tenth number will be 20."

GENERALIZING Finding a rule that generates the pattern is a type of generalization. One of the most powerful ways to tie repeating patterns to algebra is to number the terms as you would do for growing patterns. This simple strategy can really help children begin to recognize and generalize the pattern. For example, consider the repeating pattern of geometric shapes in Figure 14-4. Questioning guides student thinking about patterns in a more general way. Ask questions like:

- What shape is the ninth term? The twelfth term? (They are both squares.)
- What is the sixteenth term? (a hexagon) How do you know? (The 15th term is a square, the hexagon follows the square.)
- What can you say about the terms of all the squares? (they are all multiples of 3.)
- How many of each shape are needed to extend the pattern to 20 terms? (5 hexagons, 5 trapezoids, and 4 squares)

Children in the upper elementary grades can begin to describe the terms of all the squares as multiples of 3. They may also see that the hexagons are always two places before the squares and the trapezoids are one place before the squares. They can now begin to use symbols to describe their thoughts. The square's place can be described by $3 \times s$, where s represents the number of the squares thus far in the pattern; thus, the first square ($s = 1$) is in the third place. This means that the place of the hexagons can be described as $(3 \times s) - 2$ (i.e., hexagons are two places before squares). The first hexagon is in the first place and the second hexagon is the fourth place. The place of the trapezoids (one place before the squares) can be described by $(3 \times s) - 1$. These generalized rules written with symbols will come later, but telling what shape is in any place is well within the reach of students familiar with multiplication and division.

Figure 14-4 *Repeating pattern with terms numbered.*

NONROUTINE PROBLEMS

We separated the discussion of representing, generalizing, and justifying in the previous section, but these processes are so intertwined that we will discuss each when appropriate for a particular problem, function, or number property.

In the Classroom 14-4 shows an expanded version of the cube-painting problem (Non-routine Problem 3), which we presented earlier in this chapter. This problem involves several different patterns to analyze and describe. The activity can be done with children in grades 5–8, but it is a good one for you and your peers to try as well. Go through the steps of modeling with actual cubes, find a systematic way of counting the painted faces, record your counts in the table, find a general rule for the counts in each column, and justify your generalizations. To see the patterns, you may need to model and count up to a $5 \times 5 \times 5$ cube and visualize the counts for a $6 \times 6 \times 6$ cube. Why does it help to figure out the total number of small cubes in each large cube?

In this example, students who carefully count each of the cubes can often find the number in the next larger cube by extending the number pattern, but they often do not have the algebraic language to express the number of cubes that are painted on one or two faces since the general rule involves quadratic and cubic expressions. However, they all should be able to justify why each cube has exactly 8 smaller cubes that are painted on three sides and the number of cubes painted on only one face for any cube.

Figure 14-5 shows a number puzzle (Puzzle 1) that also was presented earlier in the chapter. Number puzzles like this can help children think about generalizing. First, have children do the puzzle with many different numbers chosen by the children. Then let them use materials to represent the steps. As they use the materials, you should be able to see whether they have strongly developed concepts about the operations. If you discover, for example, that some children do not know how to represent "multiply by two" or "divide by two," then you can reinforce the meanings.

In the Classroom 14-4

PAINTING CUBES

Objective: Modeling and generalizing.

Grade Level: 5–8.

Materials: Small cubes.

Problem:

If you painted the surface of a large cube made from smaller unit cubes, how many of the smaller cubes would be painted on no (0) face, 1 face, 2 faces, 3 faces, 4 faces, 5 faces, and 6 faces?

Represent:

Fill in the table below for the **1 × 1 × 1** cube. Make a **2 × 2 × 2** cube and refer to it to fill in the table below. Do the same with a **3 × 3 × 3** cube and a **4 × 4 × 4** cube. Some of the numbers have been filled in for you. Why is there no column for 4 faces or 5 faces?

Generalize:

What pattern do you see? Can you predict the numbers for a **5 × 5 × 5** cube? (You may have to build a model for the **5 × 5 × 5** cube to verify your numbers.) How about for an **N × N × N** cube? Justify your answers.

Cube Size	Painted Faces					Total Small Cubes
	None	One	Two	Three	Six	
1×1×1	0				1	1
2×2×2	0			8	0	8
3×3×3	1		12			
4×4×4						
5×5×5					0	125
N×N×N					0	

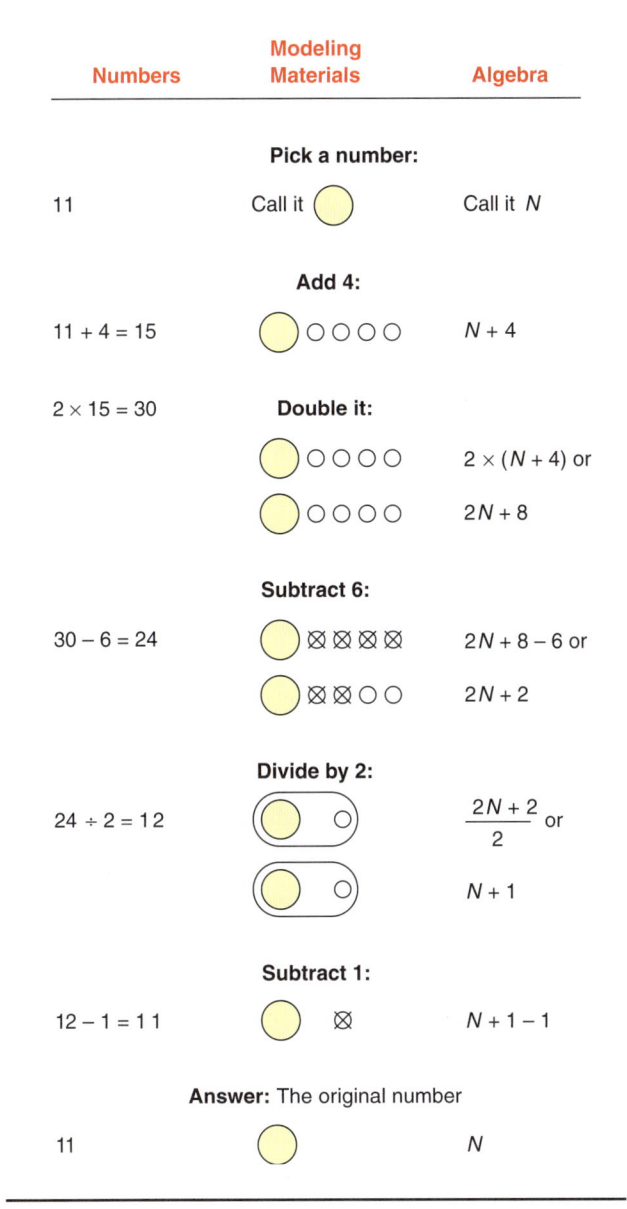

Representing, Generalizing, and Justifying 315

Numbers	Modeling Materials	Algebra
	Pick a number:	
11	Call it ⬤	Call it N
	Add 4:	
11 + 4 = 15	⬤ ○○○○	N + 4
	Double it:	
2 × 15 = 30	⬤ ○○○○	2 × (N + 4) or
	⬤ ○○○○	2N + 8
	Subtract 6:	
30 − 6 = 24	⬤ ⊗⊗⊗⊗	2N + 8 − 6 or
	⬤ ⊗⊗○○	2N + 2
	Divide by 2:	
24 ÷ 2 = 12	⬤ ○	$\frac{2N+2}{2}$ or
	⬤ ○	N + 1
	Subtract 1:	
12 − 1 = 11	⬤ ⊗	N + 1 − 1
	Answer: The original number	
11	⬤	N

Figure 14-5 *A number puzzle.*

You can move to a symbolic representation of the puzzle, as shown in the third column of Figure 14-5. After the students represent the actions with materials, you can ask them to describe the steps using symbols. Notice the concepts they can gain by doing this. For example, they can easily see that 2 × (N + 4) is equal to (2 × N) + 8 because both of these symbolic representations describe the same step.

Most children love number puzzles of all types, especially puzzles that let them predict a number, such as someone's favorite number. Find puzzles like these and first let children just enjoy them. Then challenge them to figure out how the puzzles work, and ask them to justify their solutions, by using either materials or symbolic representations.

You can also use calendars (Puzzle B) as a context for number patterns that will engage students in algebraic thinking. Figure 14-6 uses the calendar for a September as the context for the calendar number puzzle that we presented earlier. Have younger children look at how the numbers on the calendar change as you go across a row for a week. Also have them look at how the numbers change as you go down a column for a specific day of the week. The children may be able to recognize patterns such as that every other date is an even number or that the date of the next Friday after September 12 is 7 more, or September 19. Have students find the sum of all the numbers in a 3 × 3 square of dates (as shown in Figure 14-6). After finding this sum in several different squares of the same size, students may realize that the sum is always 9 times the number in the center. Older students should be able to see why this is the case. As shown in Figure 14-6, using a letter (*m*) for the number in the center and using expressions with that letter for the numbers in the other squares can help. Note that the sum of the expressions in the nine squares is 9 × *m*. Children in the upper elementary grades should not find it difficult to make this generalization.

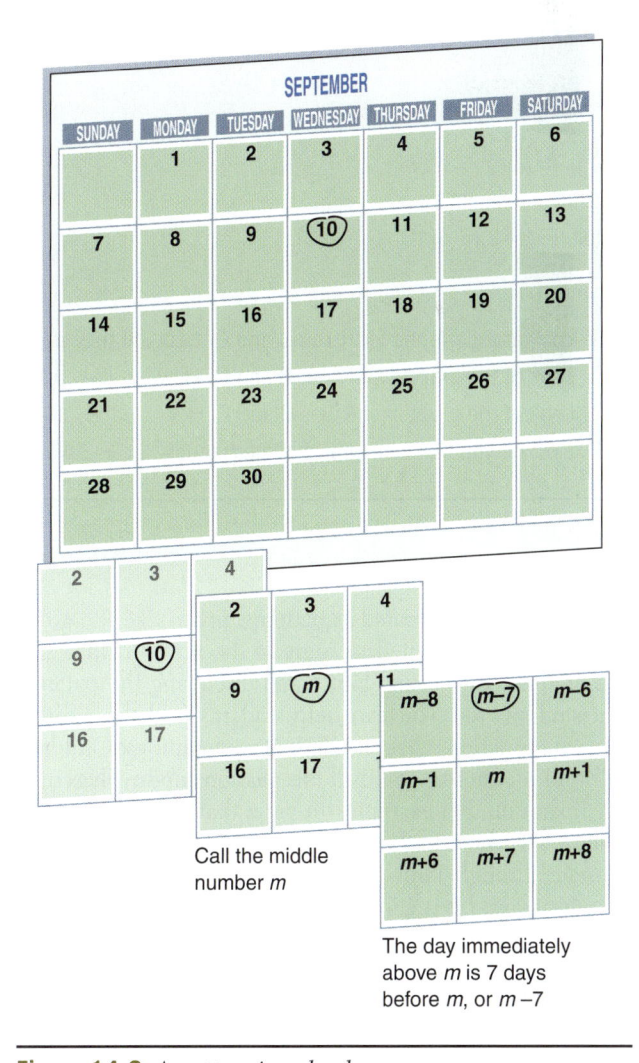

Figure 14-6 *A pattern in calendars.*

In the Classroom 14–5

FROM A RULE TO A GRAPH

Objective: To graph a function given a rule. To analyze the graphs.

Grade Level: 4–6 (graphing) 5–6 (analyzing).

Materials: Graphing paper.

▼ Make a table of values for each of these rules. Let x be 0, 1, 2, 3, 4, and 5.

Rule A
$y = 2x + 1$

Table

x	y
0	1
1	3
2	5
3	7
4	9
5	11

Graph

Rule B: $y = x$

Rule C: $y = 2x$

Rule D: $y = 4$

Rule E: $y = 2x + 2$

Rule F: $y = 2x + 4$

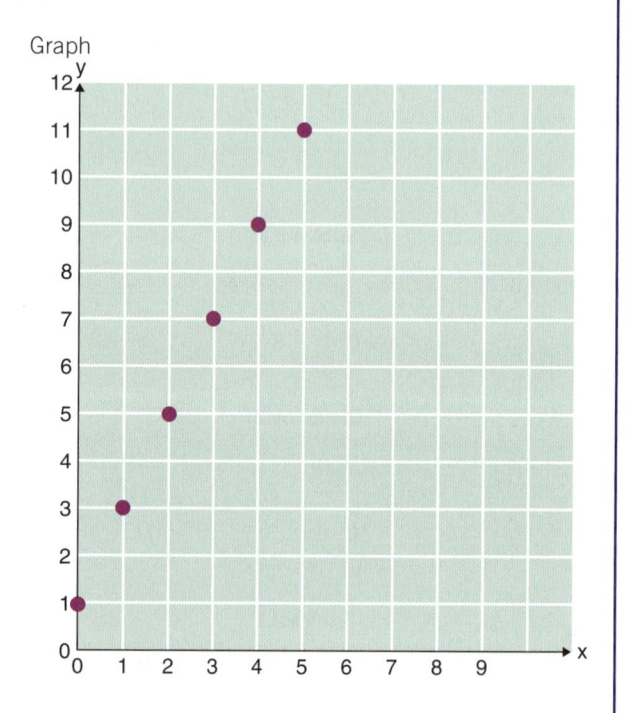

1. Look at the graphs for Rules B and C. How are they alike? How are they different?
2. Look at the graphs for Rules B and D. How are they alike? How are they different?
3. Look at the graphs for Rules C and E. How are they alike? How are they different?
4. How do you think the graph of the rule $y = 2x + 3$ would compare with Rule E and Rule F? Check your conjecture.

RELATIONS: FUNCTIONS

Functions may be treated as patterns, much like the growing patterns represented in some of the previous tables in this chapter. By making tables and discussing the patterns that students see, you can help students find the general rule (the function) that relates the two quantities. Children will find it easier to do this if the relationship involves only one operation. For example, the rules that describe how the machine works in In the Classroom 14-3 are easy to state for Monday through Thursday, as shown below (R is the result, or output, and I is the input I):

Monday: $R = I \times 4$
Tuesday: $R = I + 10$
Wednesday: $R = I \div 2$
Thursday: $R = I - 16$

What about Friday? Remember, the machine does the same thing all day long. Student reactions tend to be like this: "It looks like it added 7 spots to the frog. But that doesn't work for the toad. It's 3 times for the toad." Here are some questions you can ask students to start them on the right track:

- Do the animals have more or fewer spots when they come out than when they went in? (more)
- Does adding the same number work? (no)
- Does multiplying by the same number work? (no—it's 3 times for the toad, but it's not quite 3 times for the frog and the snake)
- What if the machine used two operations? What if it multiplied and then added? (let the students try to work with just this hint)

Do not rush the students in their thinking. Later in their study of algebra, they will learn specific techniques for solving such problems, but at this point it is important for them to make conjectures and try to verify them.

In recent TIMSS and NAEP assessments, fourth-grade students were asked questions about how pairs of quantities were related. Often, they found a rule that related one pair but not the others. They did not realize that every pair had the same relation and that they were supposed to determine it. Activities like In the Classroom 14-3 are needed to help make children aware of this general idea of expressing patterns in terms of functions.

You should also ask children to "undo" their thinking. For example, when doing In the Classroom 14-3, you would tell children the output (how the animal came out) and ask them to tell how many marks it had going in. Educators call this *reversibility of thinking* (Driscoll, 1999) and look at it as one of the crucial skills for children to develop as they learn to think algebraically.

Functions also are a good way to study change in subject areas other than mathematics. For example, in science lessons, students often collect data about how plants grow or how other natural phenomena change, and the patterns in which such changes occur can often be expressed as functions. Such changes may not always fit a simple rule, but students can represent the changes by graphing the data and then look at the graph to see if that helps them determine the rule. In the Classroom 14-5 helps children see the relationship between rules (functions), numerical values of pairs of variables, and graphs.

Connecting equations (the rules), the table of values, and the graphs gives students ways to analyze situations. The questions asking the students to compare the graphs are a first step in having them look at how changing one representation (the rule) changes the other (the graph).

Math Links 14.4

Electronic manipulatives (such as algebraic tiles and graphing) that may be used to build algebraic concepts are available from the National Library of Virtual Manipulatives for Interactive Mathematics Web site, which you can access from http://nlvm.usu.edu/en/nav/index.html or from this book's Web site.

www.wiley.com/college/reys

RELATIONS: PROPERTIES OF NUMBERS

Investigating properties such as distributivity can help students develop their ability to make and justify generalizations. Students may make conjectures entirely on their own, or you can initiate the process by asking thoughtful questions. Some teachers keep a written record available to the class, showing students conjectures along with the name of the child who stated each conjecture. Students can revisit these conjectures as they gather evidence as to whether or not they are true of all numbers. Remember that only one example is needed to prove that a conjecture is not true for all numbers. Consider, for example, the conjecture "When you multiply two numbers, the product is always larger than either number." Although this is true for positive whole numbers, it is not true for all whole numbers. One example, $0 \times 6 = 0$, shows that the product (0) is smaller than one of the numbers (6). Thus, the conjecture is not true for all numbers and therefore the conjecture is false. Disproving conjectures is as important as proving them, and often it is simpler.

Young children use many of the mathematical properties intuitively because "they work." For example, most first-grade students will find $2 + 9$ by counting on 2 from 9. Actually, they are using the commutative property for addition, which can be stated like this: $a + b = b + a$, for any two numbers a and b. Of course, this language and symbolization are too formal for very young children, but they can justify it by representing it with concrete objects. Figure 14-7 shows how third- and fourth-grade students used linking cubes to represent $8 \times 5 = 5 \times 8$, which is a specific example of $a \times t = t \times a$ (the commutative property for multiplication) (Carpenter and Levi, 2000).

This figure illustrates a particular case, but the students went on to argue that the representation shows that all you have to do is turn one of the array of cubes on its side, so it would work for any array. Does this convince you that the commutative property holds for all whole numbers?

The distributive property states that $a \times (b + c) = (a \times b) + (a \times c)$ for any three numbers a, b, and c. Again, students use this property naturally without realizing it. For example, consider a problem that we presented earlier in the chapter:

Routine Problem C. Pat took 4 pails to the beach. She had 3 shovels and 2 rakes in each pail. How many tools did she take to the beach?

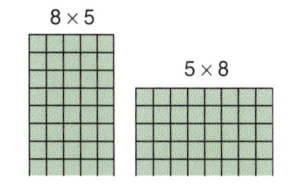

Figure 14-7 *A representation of* $8 \times 5 = 5 \times 8$.

Generally, children approach this problem in one of two ways:

1. They might add the number of shovels and number of rakes (3 + 2) first, to find out the number of tools in each pail, and then multiply the sum by 4. You could represent this by (3 + 2) × 4 or use the commutative property to represent it as 4 × (3 + 2).
2. They might find the total number of shovels (4 × 3) and the total number of rakes (4 × 2) and then add (4 × 3) + (4 × 2).

The fact that both approaches give the correct answer (20) demonstrates the commutative property:

$$4 \times (3 + 2) = (4 \times 3) + (4 \times 2)$$

How would you represent the distributive property with physical materials? or justify it more generally?

It is important to give children examples illustrating mathematical properties to see if they can generalize those properties. First show them closed sentences and have them verify that the sentences are true; then have them solve similar open sentences; finally, encourage them to make conjectures for generalizing the properties illustrated by the example sentences. Do not expect children to immediately come up with good generalizations; it may help to give them some additional example sentences or to have the students themselves give more examples. The wording of the children's conjectures will probably need to be revised; a class discussion about how to word the conjectures so they make sense is important in helping children learn to write mathematically. The following table shows the kinds of closed and open sentences and corresponding generalizations that you can use with students:

rule for inserting fractions between two fractions and how she helped him prove it for the easier cases of one and two fractions being inserted.

ANOTHER LOOK AT REPRESENTING, GENERALIZING, AND JUSTIFYING

In this section, we have shown examples of problems, patterns, and relations that can be used to encourage representing, generalizing, and justifying. These processes are often done together, but we often focus on one of these. For example, you can represent without justifying the representation or making a generalization, or conjecture a generalization without representing or justifying it. However, the power of mathematics comes when we do include justification in whatever we do. This is a way to see if students are making sense of mathematics.

"The arguments that students typically use to justify generalizations fall into three basic categories: appeal to authority, justification by example, and deductive arguments" (Carpenter and Romberg 2004, p. 28). These educators then go on to point out that appealing to authority is really avoiding justification. For example, a student who says "My teacher told us to divide fractions, you just flip the one after the division sign and multiply" is not providing a mathematical justification for doing so and is not showing an understanding of division of fractions. Children need to learn that, in mathematics as in most subject areas, they should not do something a certain way just because someone tells them to; rather, they need to understand why doing it that way makes sense (or doesn't make sense).

Closed Sentences	Open Sentences	Generalizations (Mathematical Properties)
79 + 0 = 79	37 + ☐ = 37	When you add 0 and any number, you get that number.
0 × 54 = 0	0 × ☐ = 0	When you multiply any number by 0, you get 0.
33 − 0 = 33	83 − ☐ = 83	When you subtract 0 from any number, you get that number.
67 − 67 = 0	3456 − ☐ = 0	When you subtract any number from itself, you get 0.
24 × 1 = 24	1002 × ☐ = 1002	When you multiply any number times 1, you get that number.

The wording of the generalizations in this table is informal but correct. Children will also use informal wording, and you should help them focus on whether or not their wording is clear and correct. For example, if a student says, "When you subtract any number and 0, you get that number," encourage the class to discuss why this is not clear ("subtract any number *and* zero" doesn't make sense).

Often you can help older children explain their discoveries with algebra. Thus, you need to be familiar with using algebra. In the Classroom 14-6 gives an example of how a teacher questioned Robin to help him make a more general

In this chapter, we have mainly focused on justification by examples and by informal explanations (i.e., explanations that aren't as tightly logical as formal deductive arguments). We have encouraged you to have children make representations, move on to symbolic representations when they can, and discuss how they arrived at their generalizations. These are important steps in understanding and fully justifying generalizations, and they may be as much as young children can handle.

We close this section with a summary of an interview with a second grader, Susie. Her teacher was in a study group investigating how to encourage algebraic thinking in

Cultural Connections

In the Classroom 14–6

FRACTIONS DISCOVERY

Objective: To discover a rule for finding fractions between two fractions.

To prove the rule algebraically.

Grade Level: 6 and above.

Robin found a rule for inserting one number of fractions between two special types of fractions. Answer these questions to see if you can discover the rule.

1. Robin's rule worked for these fraction pairs:

    A. $\frac{1}{3}$ and $\frac{2}{3}$ B. $\frac{3}{5}$ and $\frac{4}{5}$ C. $\frac{4}{11}$ and $\frac{5}{11}$

    It would not work for $\frac{3}{7}$ and $\frac{4}{9}$.

    How would you describe the pairs of fractions that worked?

2. Robin said, "If I want to put one number between the pair, I multiply the numerators and denominators by 2." For B, I multiply like this:

    $\frac{3 \times 2}{5 \times 2}$ $\frac{4 \times 2}{5 \times 2}$

    That gives me $\frac{6}{10}$ and $\frac{8}{10}$, $\frac{7}{10}$ is between.

    Try this rule with A and B.

3. Robin's teacher asked, "What do you think you would do if you want to put 2 fractions in between these pairs." Robin conjectured, "Since for 1 between, I multiplied by 2, I think I will try 3." Check his conjecture for all three pairs of fractions.

4. Write your own conjecture for putting 3 fractions between these pairs. Check it.

5. Try more examples, and write a conjecture for putting n fractions between these pairs.

    Challenge A. Prove that the rule for 1 and 2 fractions between works for this type of pairs. Hint: If one fraction is $\frac{a}{b}$, how would you describe the second fraction?

    Challenge B. Prove the rule for any number (n) of fractions for pairs like this.

    Challenge C. What if the denominators are not the same? Can you discover a method?

---

elementary school. The class had been making and justifying conjectures such as the one Susie was asked to justify: $a + b - b = a$. The video of the interview (Carpenter and Romberg, 2004) shows Susie thinking about the conjecture and the interviewer, Ms. L., encouraging her to delve deeper.

Susie initially proposes to justify the conjecture by trying a lot of different examples with different types of numbers. She recognizes that she cannot try all the numbers, and that she cannot be sure the generalization is true for all numbers by trying a few examples.

Ms. L realizes that Susie had moved from her initial strategy of operating on numbers to using properties appropriate for a more general proof. For example, Susie says "I just know if you have a number and you minus the same number, it gets you to zero." She then uses two conjectures that her class has discussed: $b - b = 0$ and $a + 0 = a$. She argues that if these conjectures are true, then her conjecture of $a + b - b = a$ must be true.

(Carpenter and Romberg, 2004, pp. 30–31).

This is a powerful example of a young child beginning to make a convincing deductive argument. When asked if her way is better than trying numbers (the way she began), she answers: This way, because it would take your whole life, and you still wouldn't have tried all the numbers. And this way you only have to do one thing, and you proved it." (Carpenter and Romberg, 2004, pp. 31).

CULTURAL CONNECTIONS

Algebra has it roots in many cultures. The word comes from the Latin variant of the Arab word "al jabr" meaning restoration or reunion. The Indians, Babylonians, Egyptians, and Greeks were cultures that contributed to the early development of algebra; the Moors are credited with bringing algebra to Europe through Spain.

Math Links 14.5

More information on the origin of the word algebra can be found at http://www.und.edu/instruct/lgeller/algebra.html or from this book's Web site.

www.wiley.com/college/reys

Development of classical algebra, which is often considered a generalization of arithmetic, began in the fourth century. It was not until the nineteenth century, however, that algebra began to be studied in secondary schools around the world. During the twentieth century, algebra became a central subject for those who expected to advance in their study of mathematics. But only recently has it become a subject that all students in the United States are expected to study; this is in contrast to most other developed countries, where algebra long has been a core subject for all students.

Today, topics in algebra that are commonly taught in fourth grade or before in many countries around the world appear in the latest TIMMS 2011 Mathematics Framework (Mullis and others, 2009). These include representing situations with expressions or number sentences, solving equations, extending or finding missing terms in a well-defined pattern, describing relationships between adjacent terms in a sequence and between the sequence number of the term and the term, writing or selecting a rule for a relationship given some pairs of whole numbers satisfying the relationship, and generating pairs of whole numbers following a given rule. All of these topics are reflected in recent CCSSM.

A GLANCE AT WHERE WE'VE BEEN

We have considered problems, patterns, and relations as topics to develop algebraic thinking. These topics occur in all strands of the mathematics curriculum as well as in other subjects. Helping children to think algebraically does not mean adding another topic; it does mean we probably need to teach those areas differently. We can help children develop algebraic concepts and habits of algebraic thinking through questioning; helping students represent problems, patterns, and relations; encouraging them to generalize; and expecting them to justify their thinking and statements. We have emphasized the need for teachers to ask questions and provide opportunities for children to discuss and defend their ideas with any of the mathematics they are learning.

Things to Do: From What You've Read

1. Why should algebra be studied in elementary school? How would you describe that algebra?
2. What are topics in the elementary curriculum that can be used to develop algebraic concepts and thinking?
3. What are two common types of patterns? How do they differ? How are they alike? In particular, how are they alike and different in developing the algebraic thinking discussed in this chapter?
4. What algebraic language and symbols are essential for elementary school students?
5. What is the difference between an open sentence and a closed sentence?
6. How does representing routine problems promote algebraic thinking?
7. Name and give examples of three processes were used to describe algebraic thinking?
8. What are the different meanings of *variable*?
9. Complete the cube-painting problem (In the Classroom 14-5).
10. Justify the statements listed in the section Properties of Numbers.

Things to Do: Going Beyond This Book

In the Field

1. [1]*Equality*. Ask students about the truth of mathematical sentences or to solve sentences such as the ones in the chapter that showed initial misunderstandings. What did you learn about students' understanding of equality, mathematical sentences, and justifying?
2. [1]*In–Out Machine*. Use the activity sheet in In the Classroom 14-3 or one that you design. Do one or two other examples with children. Do you notice that children want to make a judgment of the rule after one in–out? Can they find the number but not describe the rule?
3. [1]*Algebraic thinking*. Ask several students in grade 5 or 6 to examine the sidewalk garden problem (see page 313) with you. Encourage the students to draw the next few gardens and sidewalks (1 by 3 and 1 by 4) if they are having difficulty generalizing to the 1 by

---
[1]Additional activities, suggestions, and questions are available in the field experience manual on the Student Companion site at www.wiley.com/college/reys.

8 garden. Encourage them to describe what patterns they see between the garden and the number of tiles. See if that helps them to generalize to larger gardens or if they still need to draw a picture. For students who have a good grasp of the problem, ask how many tiles would be needed for a 1 by x garden.

In Your Journal

4. What was your experience in algebra classes? Do you think that if you had the background recommended in this chapter that it would have been different? Give ways and examples to explain your comments.

5. Describe why equality is an important idea in algebraic thinking.

With Additional Resources

6. Select an instructional idea in a current textbook series that is listed as an algebra lesson. Briefly summarize the lesson and tell how it supports the ideas or how it is contrary to the ideas in this chapter.

With Technology

7. Explore the pan balance on the NCTM's Illuminations Web site (Math Links 14-3). You can use this interactive pan balance to "weigh" numeric or algebraic expressions. Develop a lesson plan that would use this activity.

8. Watch one of the mathematics videos from *Powerful Practices* (Carpenter and Romberg, 2004). What ideas are in both the video and this chapter?

Note to Instructors: You can find additional resources, learning activities, and blackline masters in this text's accompanying Instructor's Manual at www.wiley.com/college/reys.

Book Nook for Children

Barry, D. *The Rajah's Rice*. New York: W. H. Freeman, 1994.

A smart peasant girl outsmarts the rajah and ensures the village plenty of food forever after the rajah agrees to give the girl the rice that would cover his chessboard when he puts one grain on the first square, and doubles it for the next square, and then doubles the amount on that square, and so on for all 64 squares.

Calmenson, S. *Dinner at the Panda Palace*. New York: HarperCollins, 1991.

This book for grades 1 to 3 counts dinner guests as they arrive at the Panda Palace. Each group is one guest larger than the group before. And children could predict how money guests would be in the tenth group or the fifteenth group. They might be challenged to make up a similar story with a different expanding pattern. The class can find out how many groups it would take to fill the restaurant.

Geringer, L. *A Three Hat Day*. New York: Harper & Row, 1987.

A man who loves hats has a collection of 12 different hats. One day he decides to wear three hats and it brings him good luck. The students figure out how many different combinations of three he could make with the hats he has. The book is most appropriate for third and fourth graders.

Murphy, S. J. *Beep, Beep, Vroom, Vroom!* New York: HarperCollins, 2000.

Kevin keeps his toy cars in a pattern on the shelf. Suddenly, the cars fall from the shelf and Kevin's sister, Molly, puts the cars back on the shelf in a different pattern. This happens again, but this time the pattern is the same as the original. There are activities and games for parents and teachers to use.

Swinburne, S. R. *Lots and Lots of Zebra Stripes*. Honesdale, PA: Boyds Mills Press, 1998.

This book explains with colorful illustrations how patterns encompass our world. The author explains that the patterns can have a purpose or be seasonal.

CHAPTER 15

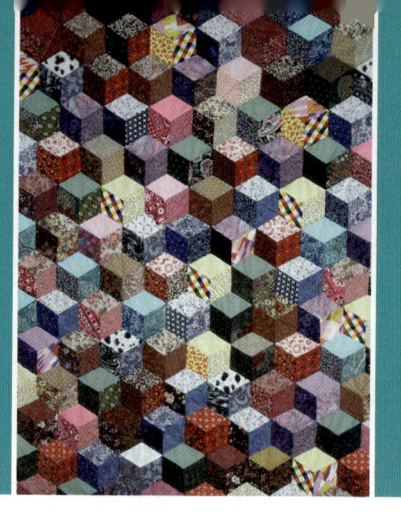

Geometry

> "AND SINCE GEOMETRY IS THE RIGHT FOUNDATION OF ALL PAINTING, I HAVE DECIDED TO TEACH ITS RUDIMENTS AND PRINCIPLES TO ALL YOUNGSTERS EAGER FOR ART."
>
> — Albrecht Durer (artist)

>> SNAPSHOT OF A LESSON

KEY IDEA

1. Help students make and identify basic geometric shapes.
2. Identify sides of a shape.

5 × 5   First fold   "Airplane" fold   Final: open

BACKGROUND

This lesson is based on the video *Shapes from Squares*, from Teaching Math: A Video Library, K–4, from Annenberg Media. Produced by WGBH Educational Foundation © 1995 WGBH All rights reserved. To view the video, go to www.learner.org/resources/series32.html (video #20).

Mr. Ramirez is a teacher in a bilingual second/third-grade classroom in Tucson, Arizona. In the video, the students are investigating what shapes they can make by folding a square. Mr. Ramirez presents step-by-step directions on folding a 5" × 5" construction paper square to his class and then challenges them to work in pairs to find as many shapes as they can by folding only on the creases. [Before you watch what the children did, follow the abbreviated directions below and make the folding square.]

1. Fold the square in half.
2. Make airplane folds from each corner to the fold line.

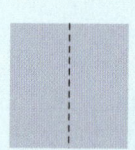

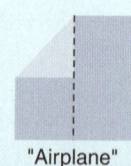

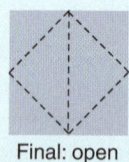

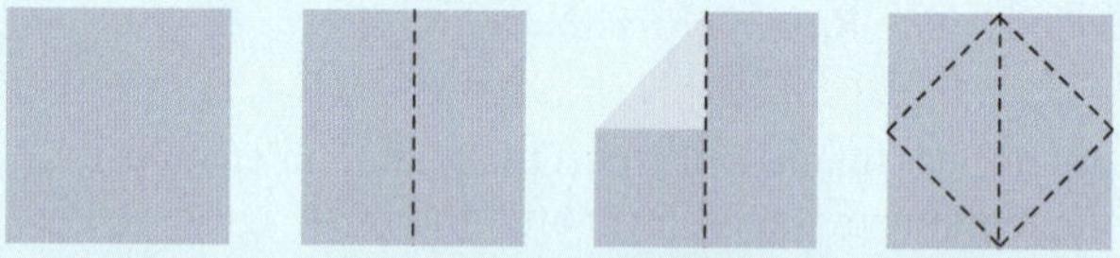

The children work in pairs at tables making different shapes and recording their findings. Mr. Ramirez talks with various groups, having them identify the shape and the number of sides. One pair made a shape with 5 sides. When Mr. Ramirez asked what the name was, they replied fiveagon. They called their 6-sided shape a sixagon, but one student said six is the same as hex, so it is a hexagon. [Make the hexagon: Do you see why the students may not have related it to the hexagon in the pattern blocks? Paper pattern blocks are in the Appendix.]

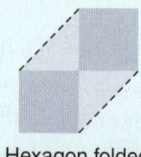

Hexagon folded from the square   Hexagon from pattern blocks

322

A boy, we'll call him Antonio, made the shape below and recorded it by drawing a sketch.

Mr. Ramirez: What is this shape?

Antonio: This is a triangle and this is a triangle. [pointing to the two triangles in his drawing]

Mr. Ramirez: So the two triangles make up this shape. [pointing to the whole shape] How many sides does it have?

Antonio: Five. [He counts the five, saying the inside line is a side of the triangle.]

After some discussion and counting sides of other shapes, Antonio persists in counting the inside lines.

Mr. Ramirez: Let's look at your hand. [Antonio places his hand flat on the table.] Where are the sides? [Antonio points to an edge of his hand. Mr. Ramirez points to a line on Antonio's palm. Antonio responds that is a palm, not a side.]

Mr. Ramirez returns to a folded shape but Antonio persists in counting the inside lines.

Mr. Ramirez: Open up your paper shape. What is this?

Antonio: It is a square.

Mr. Ramirez: How many sides does it have?

Antonio: Four.

Mr. Ramirez: Why didn't you count these? [pointing to the inside lines]

Antonio: Those are folds.

Mr. Ramirez: How about this one. [back to the original shape under discussion]

Antonio: Four sides. That other one is a fold.

Mr. Ramirez brings the class back together to share what they had found. He sees this as just one of many opportunities for his students to work with shapes.

FOCUS QUESTIONS

1. Why should the elementary mathematics program include geometry?
2. What should early childhood and elementary students learn about geometric shapes?
3. How do the van Hiele levels guide the development of geometric experiences for elementary children?
4. What types of explorations with geometry help build elementary children's spatial reasoning and visualization skills?
5. What roles do location and movement have to play in geometry?

INTRODUCTION

You may find yourself working with some teachers who respond to geometry in the following ways:

"Oh, I could never do proofs."

"The children don't understand it, so why do it?"

"We do it if we finish everything else first."

There are many reasons for such responses. Some are based on personal experiences such as an unsatisfactory geometry course in high school. Some are based on inappropriate geometry curriculum materials for elementary students that use, for example, an abstract, definitional approach. Some are based on a historical emphasis on computation—even though various professional groups have recommended geometry be included in the curriculum for over a hundred years.

You may find that you are working with teachers who respond like this:

"It amazes me who is good in geometry; it's not always my best arithmetic students."

"What a joy it is to see a child's eyes light up as she discovers ..."

"Some of my students could work on a geometry problem for hours."

"Geometry gives me an opportunity to work on communication skills and to help children follow instructions."

"The change in the spatial ability of children after they work with geometric shapes always surprises me."

"I love to learn with my students; I never liked geometry before."

Why should you include geometry in elementary school? The previous responses certainly give several reasons, including being a mathematics topic that engages children differently both in performance and persistence. It is a natural site for including other skills, such as following directions and reasoning about shapes and their attributes. Children can make and verify conjectures about geometric figures. For example, by folding models of isosceles triangles on the line of symmetry, children can see that two sides are congruent and two angles are congruent. It is also a topic that will help you teach many other mathematical topics. For example, many representations are geometric in nature—models for fractions, area models for multiplication, and patterns that lead to algebraic expressions. The German Durer, the author of the opening quote, understood that knowing geometry was essential to him as an artist as do the two architects shown in Math Links 15.1.

The *Principles and Standards for School Mathematics* (NCTM, 2000) calls for geometry to be an integral part of the mathematics programs of all elementary students in the Geometry Standard. In fact, it is seen as the second most important area in pre-K–2 and in grades

	Geometry Summary
Kindergarten	Identify, describe, and name common three- and two- dimensional shapes. Describe relative position of objects. Compare, create, and compose shapes.
Grade 1	Focus is on attributes: distinguish between defining attributes (e.g., triangles are closed and have three sides) versus non-defining attributes (e.g., orientation and color); build and draw shapes to possess defining attributes. Compare shapes and compose shapes to make other shapes (e.g., two triangles to make square).
Grade 2	Recognize and draw shapes having specified attributes, such as a given number of angles or a given number of faces.
Grade 3	Understand that shapes in different categories (e.g., rhombuses, rectangles, and others) may share attributes (e.g., having four sides), and that the shared attributes can define a larger category (e.g., quadrilaterals). Recognize rhombuses, rectangles, and squares as examples of quadrilaterals, and draw examples of quadrilaterals that do not belong to any of these subcategories.
Grade 4	Emphasis is on identifying points, lines, line segments, rays, angles (right, acute, obtuse), and perpendicular and parallel lines in two-dimensional figures. Classify two-dimensional figures based on the presence or absence of parallel or perpendicular lines, or the presence or absence of angles of a specified size. Recognize and draw a line of symmetry for a two-dimensional figure.
Grade 5	Understand that attributes belonging to a category of two-dimensional figures also belong to all subcategories of that category. Classify two-dimensional figures in a hierarchy based on properties.
Grade 6	Draw polygons in the coordinate plane given coordinates for the vertices. Represent three-dimensional figures using nets made up of rectangles and triangles.

Figure 15-1 *Overview of standards relating to geometric concepts.* Source: From the *Common Core State Standards for Mathematics* (CCSSI, 2010). This summary was made by the authors of *Helping Children Learn Mathematics*. The full document (CCSSI, 2010) may be obtained at www.corestandards.org/the-standards/mathematics.

3–5 (see Figure 1-1). In an effort to streamline the curriculum, the CCSSM does not place as much emphasis on non-measurement geometry as it does on those attributes that you can measure such as length, area, and volume which are examined in the next chapter. An overview of the CCSSM recommendations is given in Figure 15-1.

This chapter is organized into four sections—shapes, space, transformations, and visualization—as was the geometry in the NCTM *Standards*. In this time of transition, you will still encounter these topics in the elementary curricula and assessments, and should be familiar with them in order to make wise decisions about their inclusion in your teaching. We place the most emphasis on the geometry of shapes as does the CCSSM (K-6).

SHAPES

Understanding the attributes of objects and the relationships among different geometric objects is an important part of elementary mathematics. Before discussing how to help children learn and use geometric ideas, we introduce some of the research on children's learning of these ideas.

The research of Dina van Hiele Geldof and Pierre Marie van Hiele, a Dutch couple, has provided teachers with guidelines for the growth of geometric understanding and thought. They proposed and studied a model of levels of understanding. Clements and Battista (1992) have studied this model and offer a modification of the original van Hiele levels, as shown in Figure 15-2. Although this model will not tell you what to do, it will help you analyze activities. Too often, we have students only recognize and name shapes. It is not difficult to see why our students do not do well in advanced geometry classes (Kloosterman et al., 2004), since many geometry experiences at the elementary level require students to be only at the visual level (Level 1). We need to help them move to the next two levels in elementary and middle school.

Very young children often attend to only visual cues that are salient to them. Later, children focus on visual cues—it is a rectangle because it looks like a rectangle (Level 1). This is much like—it is a dog because it looks like one. Just as very young children can identify many dogs when they see one, children can identify rectangles. Where can they see an object shaped like a rectangle? A door, a book, and a tabletop—the examples are plentiful in our world. Even in kindergarten, children can begin to look at shapes more critically, such as describing a rectangle as having four sides and square corners.

Level	Description	Sample Responses of Children
Level 0 (Prerecognition)	Children only focus on some visual cues.	Shown a triangle, children may focus on straightness and say it is a square.
Level 1 (Visual)	Children view a geometric shape as a whole. They can describe attributes based on visualization but not based on analysis of the attributes.	Children may say it is a rectangle because It "looks like a door." It has four sides because they can count them.
Level 2 (Descriptive/Analytic)	Children focus on the relationships between parts of a shape and defining attributes.	Children may describe a rectangle is a four-sided figure with opposite sides equal and parallel with four right angles.
Level 3 (Abstract/Relational)	"Students interrelate geometric attributes, form abstract definitions, distinguish among necessary and sufficient sets of attributes for a class of shapes ..." (Battista, 2007)	Students know that a sufficient definition of a rectangle is "a quadrilateral with two pairs of parallel sides and a right angle."
Level 4 (Formal Axiomatic)	Students use deduction to prove statements. This is the level needed to be successful in a formal, high school geometry class.	Students, given axioms, can write a deductive proof.

Figure 15-2 *Levels of geometric thought (modified van Hiele levels by Clements and Battista).*

Math Links 15.1

Two female landscape architects show an award-winning project that uses many geometry terms mentioned in this chapter. You can access this from www.thefutureschannel.com/dockets/hands-on_math/landscape_architects/index.php or from this book's Web site.

www.wiley.com/college/reys

With many experiences that specifically require them to describe the attributes of the figures (Level 1, Visual), children begin to recognize and more carefully describe attributes of all rectangles and move to Level 2 (Descriptive/Analytic). At this level they are moving toward more a precise description of classes of shapes. For example, they might describe all rectangles as having four sides; two pair of parallel, congruent sides; and four right angles. However, they do not realize that it is it is sufficient to define a rectangle as a quadrilateral that has two pairs of parallel sides and one right angle. This more abstract thought is at Level 3 (Abstract/Relational). At this level students establish relationships among attributes and among figures.

As students are doing the same activity, you can often see them working at different levels and, through questioning, you can help them develop a deeper understanding. We will return to these levels throughout this section.

THREE-DIMENSIONAL SHAPES

We live in a three-dimensional world that can be represented and described geometrically. We begin with three-dimensional shapes because this is the world of young children— a world that they can explore in the geometric sense. It is important to begin their geometry explorations from their perspective with familiar three-dimensional objects, such as balls and blocks. They may roll the ball and perhaps try to roll the block. Different experiences help them describe these solid, three-dimensional shapes. A three-dimensional shape such as a ball is often referred to as a *solid* even though the object may be hollow; we will refer to the shapes as solids in this chapter.

So that you can see the development of ideas about three-dimensional shapes, we discuss them together in this section. This does not mean that we recommend doing everything discussed here before considering two-dimensional shapes. Students need concepts from two-dimensional shapes in order to more completely describe three-dimensional objects.

Models play an important role in all of geometry, but especially in three-dimensional geometry. If wooden or plastic solids are not available, you can make models, as suggested later in this section. You and your students could also collect real objects that have particular geometric shapes, including spheres (balls), cylinders (cans), prisms (boxes), and cones, as well as solids that may not have geometric names.

Studying geometric attributes of three-dimensional objects also provides an opportunity to emphasize the process goals of geometry. This section is built around some of these processes: describing and sorting, constructing, exploring, and discovering.

DESCRIBING AND SORTING Children need to be able to describe attributes of three-dimensional objects to tell how two or more objects are alike or different geometrically. Describing and sorting are processes that are begun in early childhood and should be continued throughout elementary school. New and more complex attributes can be added to children's repertoire. Describing and sorting activities help children develop thinking at Level 1 and lead toward the Level 2 thinking described by the van Hiele model. They are essential to develop the abstract/relational (Level 3) thinking that is a precursor to high school geometry.

In the activities that follow, vocabulary and attributes appropriate for beginning, intermediate, and more advanced students are used. Older children who have not been exposed to three-dimensional activities will benefit from activities such as those described in the beginning activities. Their responses, of course, will be more sophisticated.

BEGINNING ACTIVITIES Children are often taught the names of the geometric shapes, but they do not develop the discriminating power they need to use the names with meaning. In these beginning activities you should build on the children's own vocabulary, adding new words as appropriate. Although the names of the solids can be used, they need not be formally introduced until children have participated in activities like these. Notice that they are not just asking questions like "mark all the pictures of cones"; they require children to describe and to justify their answers. These activities are at Level 1 (see Figure 15-2), but pushing children toward Level 2 as they justify their answers.

1. *Who am I?*
   Put out three objects (such as a ball, a cone, and a box). Describe one of them (it is round all over, it is flat on the bottom, or its sides are all flat). Ask the children which object you are describing and have them tell why it is that object and not one of the others.

2. *Who stacks?*
   Provide a collection of solids for children in small groups to sort according to which will stack, which will roll, and which will slide. A more sophisticated sorting is one that requires three sets of objects: solids that can be stacked no matter what face is down, solids that can be stacked if placed in some ways but not in other ways, and solids that cannot be stacked in any way.

3. *How are we alike or different?*
   In a whole-class discussion, hold up two solids such as the following:

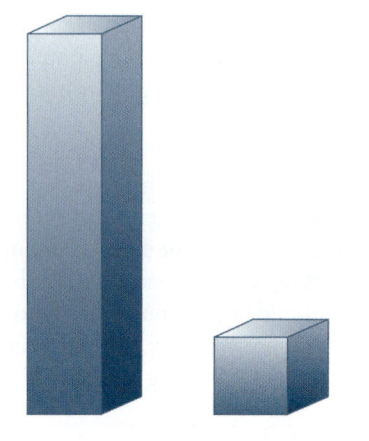

Ask children to tell how they are alike or different. For example, children may compare the two solids shown as follows:

"They are both flat all over."

"One is tall."

"One has bigger sides."

"They both have some square faces."

"They both have six faces."

4. *Who doesn't belong?*
   For another whole-class activity, put out three solids such as these:

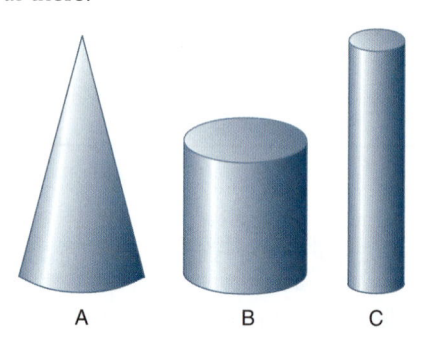

   A   B   C

   Ask the children which does not belong with the other two. Since there are many ways to answer this question, be ready to encourage lively discussion. For example, some children may say A doesn't belong because it has a point. Others may say B doesn't belong because it's short, or C doesn't belong because it's skinny or has a smaller bottom.

5. *How many faces do I have?*
   A *face* is a flat side of a solid object. Have children in small groups count the number of faces on solids of various shapes. Then ask them to collect objects with six faces (boxes, books), with two faces (cans), and with zero faces (balls). You will be surprised at what they find! Some children have difficulty counting the number of faces on many-faced solids. You can help by using tape and numbering them as they count.

INTERMEDIATE ACTIVITIES The following activities introduce the names of some solids and consider sizes (or measurable attributes) as well as edges, faces, and vertices. Most of the thinking is at Level 2; the vocabulary and ideas are more mathematical than those in the beginning activities.

1. *Edges, vertices, and faces: Who am I?*
   After introducing the concepts of *edge* (a straight segment formed by two faces) and *vertex* (a point at which three or more edges come together), have children in small groups solve these riddles:

   I am a solid with:

   Eight edges—who am I?

   Six edges and four faces—who am I?

   Five corners—who am I?

The same number of vertices as faces—who am I?

No faces (no corners)—who am I?

One face and no corners—who am I?

2. *Classifying solids.*
Introduce each type of solid—cube, cone, pyramid, cylinder, and sphere—by putting out solids that are examples and nonexamples of each type of solid. As illustrated below, show two or three cylinders and two or three solids that are not cylinders. Then have the children select which are cylinders and which are not from the third set. Since it may be difficult to collect examples of such cylinders, you could use pictures. The important part of this activity is the discussion, having children say why the object is or is not a cylinder. For example, many children think A is too "thin" to be a cylinder. They may be thinking at Level 0—it does not look like one. Whole-class discussions about what is common about all cylinders (a circular cylinder has two faces that are congruent circles joined by a curved surface) can help them move to a higher level of thought.

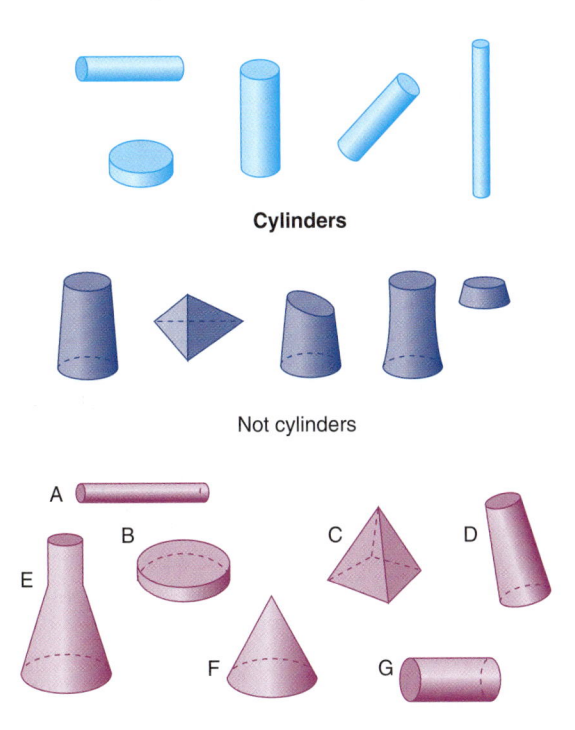

Cylinders

Not cylinders

Which are cylinders?

3. *Searching for solids.*
Make up a set of activity cards that children can use in small groups to search out solids according to the size and shape of the faces and the length of the edges. The clues you give will depend on the solids you use, but here are some samples to get you started:

Search for a solid with:

Exactly two faces that are the same size and shape (congruent)

Exactly three faces that are the same size and shape

All edges the same length

Edges of three different lengths

ADVANCED ACTIVITIES These activities focus on the attributes of parallel and perpendicular faces and edges as well as more careful definition and classification of the solids. These activities are at Level 2 and use more sophisticated geometric ideas than in the previous activities. They also consider more than one attribute, leading students to see relationships between attributes and helping them move toward Level 3.

1. *Parallel faces.*
This activity can be done after parallel lines and parallel faces have been introduced. It consists of questions about real objects and why faces are parallel. A few sample questions that you can use in a whole-class discussion are given here to start you thinking:

Why are the top and bottom of soup cans parallel?

Why are shelves of a bookcase parallel to the floor?

Why are roofs of houses in cold climates usually not parallel to the ground?

Why is the front side of a milk carton parallel to the backside?

2. *Perpendicular edges.*
In the Classroom 15-1 provides clues about particular solids, focusing on perpendicular edges, and asks students to construct the solids from sticks and connectors. See the Edge Models section that follows for a variety of materials that can be used to model the mystery objects. Students enjoy making up their own mysteries for other students to solve. This experience provides an excellent opportunity to work on developing written descriptions that are clear, precise, and not contradictory.

3. *Right prisms.*
This activity introduces the definition of right prisms and how to name prisms. Show examples and nonexamples of right prisms (as in the second of the intermediate activities), and ask students to describe the bases and faces, ultimately encouraging them to come up with the definition of *prism*—a solid that has congruent and parallel bases (top and bottom) joined by rectangular faces. Then have students in a whole-class activity discuss how prisms are named. See if they can determine a way to distinguish between prisms. For example, if the base is a triangle, it is a triangular prism.

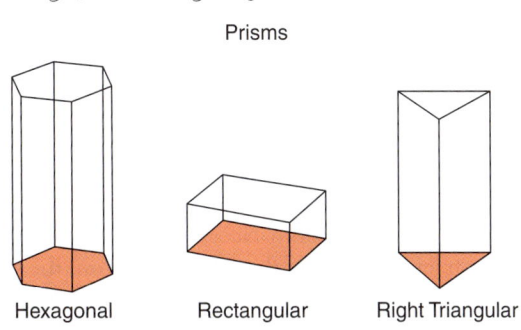

Prisms

Hexagonal   Rectangular   Right Triangular

328 Chapter 15 • Geometry

In the Classroom 15–1

SOLID MYSTERY

Objective: To build three-dimensional objects with specified conditions about the perpendicularity of the edges.

Grade Level: 4–8.

- Solve each of these mysteries by constructing a "suspect" from sticks and connectors. If you think there is more than one suspect, look at Clue 2.

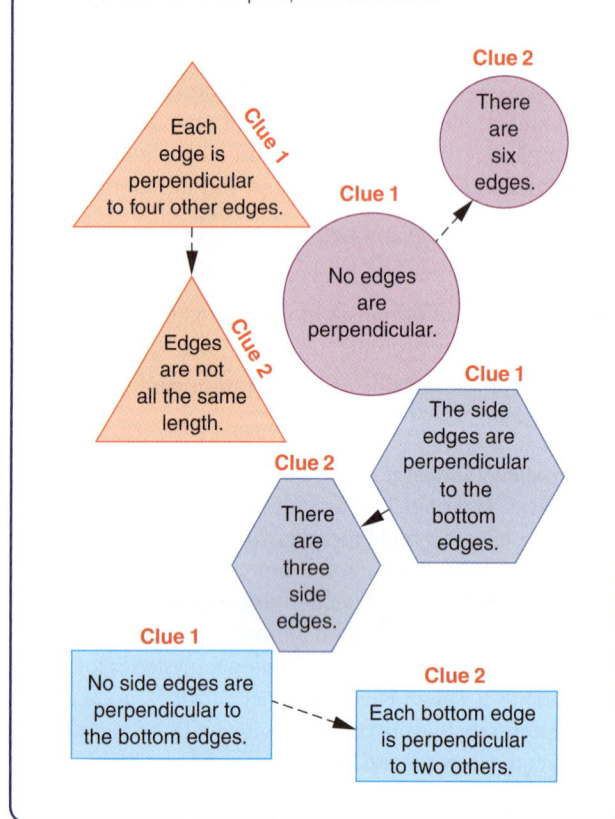

Clue 1: Each edge is perpendicular to four other edges.
Clue 2: Edges are not all the same length.

Clue 1: No edges are perpendicular.
Clue 2: There are six edges.

Clue 1: The side edges are perpendicular to the bottom edges.
Clue 2: There are three side edges.

Clue 1: No side edges are perpendicular to the bottom edges.
Clue 2: Each bottom edge is perpendicular to two others.

Use construction paper and masking tape to construct these tubes. Fold and tape each as shown.

- Prism with three congruent faces:

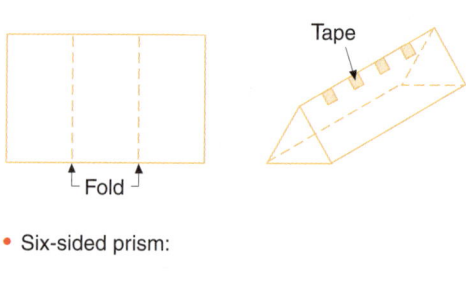

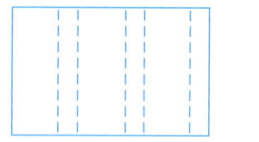

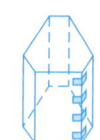

- Six-sided prism:

- Truncated prism:

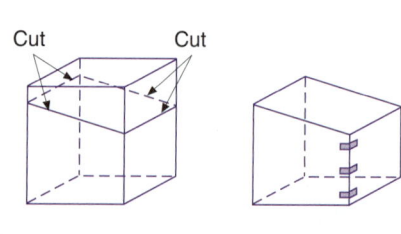

Figure 15-3 *Constructing tubes to think about attributes of prisms.*

and taping it to the tube. As this figure suggests, many variations of prisms can be made, and these can be cut (truncated) to create many strange shapes.

Any solid whose faces are all polygons is a polyhedron. Prisms and pyramids are polyhedra (plural of *polyhedron*), but spheres, cones, and cylinders are not. The exploration in In the Classroom 15-2 deals with a special case of Euler's formula. [Euler's (pronounced "oilers") formula relates the number of edges (E), faces (F), and vertices (V) of a polyhedron to each other: $V + F = E + 2$.] In this activity, students investigate the relationship of the number of edges, faces, and vertices in open-ended prisms (called tubes). The formula for these tubes is $V + F = E$. Have your students look for other relationships between two of these variables. For example, they may see that the number of edges is always three times the number of faces (not counting the top and bottom faces). This is another opportunity to use an algebraic description: $E = 3 \times F$.

Triangular or square grid paper (see the Appendix) can be used to make polyhedra. A challenging activity is to find all the networks consisting of six squares and determine which will fold into cubes. Since there are 35 different networks, this is a good small-group project. If this is too advanced for your students, try using only five squares (12 different networks), and fold them into open cubes. Students often do not have the opportunity to work with the triangular grid paper. A larger triangle consisting of four smaller triangles

---

CONSTRUCTING TO EXPLORE AND DISCOVER One of the difficulties that children have with three-dimensional geometry is visualizing the solids. It is essential to have models of the solids, but they are often expensive to purchase. There are many ways for children to make three-dimensional models. We have included suggestions of two ways to make "face models" and two ways to make "edge models." Face models (construction paper tubes and polyhedra shapes) emphasize the faces, while edge models (toothpicks and newspaper) focus on edges. As children are making models, they often discover many attributes of solids. However, for some attributes they may need structured investigations like those suggested here.

FACE MODELS Some of the easiest and most versatile models can be made from heavy construction paper. Figure 15-3 shows how to make open-ended paper tube models of prisms. The top and bottom faces may be added to these models by tracing the top of the tube, cutting out the shape,

In the Classroom 15-2

WHAT CAN YOU DISCOVER?

Objective: To explore the relationships among the number of vertices, edges, and faces of prisms.

Grade Level: 4–5.

It's easy to complete this table if you've made the tubes in Figure 15-3.

	Tube 1	Tube 2	Tube 3
Faces	3		
Edges			
Vertices			

A. Count the number of faces of Tube 1. (Remember there is no top or bottom, so don't count them.)

B. Count the number of edges of Tube 1. (Don't forget the top and bottom edges.)

C. Repeat A and B for Tube 2.
- Do you see an easy way to tell how many edges if you know the number of faces? _____
- Write your conjecture: _____

D. Count the vertices of Tube 1 and Tube 2.
- Do you see an easy way to tell how many corners if you know the number of faces and edges? ____
- Write your conjecture: _____
- Check your conjecture with Tube 3.

---

will fold into a triangular pyramid. Can you arrange the four smaller triangles in any other way to make a pyramid? Experiment with other numbers of triangles. K. Jeon in an article in *Teaching Children Mathematics* entitled "Mathematics Hiding in the Nets for a Cube" presents children's thinking and other ideas you can use (Jean, 2009).

EDGE MODELS Models can be made from straws, pipe cleaners, toothpicks, or other "sticks" that can be connected with clay or tape. There are also reasonably priced, commercial materials designed for this purpose. Young children enjoy building stick models from toothpicks and gumdrops. In the Classroom 15-3 uses this stick model to build specific objects given the number of gumdrops (vertices) and the number of edges (toothpicks). If your students need more of a challenge, have them make the shapes that are described in In the Classroom 15-1.

In the Classroom 15-4 illustrates how to make a three-dimensional figure from newspaper sticks and tape. Rolling the paper tightly takes some practice for some children; however, as with many geometry activities, often you will be surprised at who can do this easily. When building three-dimensional polygons, the children can investigate the rigidity of triangles, squares, or other polygons. They should find that the pyramid they built is a rigid structure, but the cube needs bracing to be sturdy. This is why we often see angle braces added to furniture.

TWO-DIMENSIONAL SHAPES

There are several general things that you should keep in mind as you work with children in extending their knowledge of two-dimensional shapes. Children first recognize shapes in a holistic manner (Level 1)—that is, a triangle is a triangle because it looks like a shape that someone has called a triangle. If an equilateral triangle with its base parallel to the bottom of the page is always used, then this shape will be the children's image of a triangle. A teacher could not understand why all her first graders were coloring triangles green (example of a non-defining attribute of a triangle). Then she realized that she had only been using the pattern blocks (see the Appendix) and this was their only image of a triangle.

Children should begin to recognize types of shapes through examples and nonexamples, not through formal definitions. In the Classroom 15-5 gives examples and nonexamples of triangles. Do you see why the different shapes were included? If a child says that C, B, F, or G is a triangle, what attribute of triangles do you think is being ignored in each case? If a child fails to realize that D is a triangle, what do you think may be the reason?

Children need to be able to recognize geometric shapes as models for real objects. For example, you might have young children write a "book" about circles. What is shaped like a circle? Let them find examples and draw pictures. Older children can be challenged to tell why certain objects are shaped in a certain way:

Why are walls rectangular?

Why are support braces triangular?

Why are most buttons shaped like a circle?

Why is paper rectangular?

The article (Whitin & Whitin, 2009) entitled "Why Are Things Shaped the Way They Are?" provides many more examples and shows children's thinking about this question.

Children also should know the names of common shapes such as *triangle, square, rectangle, circle,* and *parallelogram*. They also should be aware of other words that are used with shapes (e.g., children should be able to identify the *center, radius, diameter,* and *circumference* of a circle or the *sides, vertices,* and *angles* of a polygon). Build vocabulary gradually, tying the new vocabulary to other words. For example, how is a triangle like a tricycle and a tripod? The most important thing with all vocabulary is that, after it is introduced, it is used.

In the Classroom 15–3

GUMDROPS AND TOOTHPICKS

Objective: To construct three-dimensional figures with given number of edges and faces.

Grade Level: 4–6.

Materials: Gumdrops and toothpicks (about 20 of each for each child). Clay can be substituted for gumdrops.

Launch: Explore making three-dimensional shapes with the gumdrops and toothpicks. Here are a few examples. How many gumdrops and toothpicks did each take?

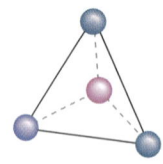

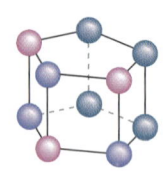

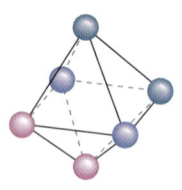

Activity: Now, try some with the given number of gumdrops and toothpicks. Three rules to follow:

1. Two gumdrops can be used on each toothpick, one at each end.
2. Toothpicks should be used to make only the outline of a face (no extra ones stuck across the face).
3. The figure should be closed. Imagine it is a container with no openings.

Figure out which ones (A–L) are possible and why. If you can make a shape, tell whether it is a prism, a pyramid, or another shape.

Letter	Gumdrops	Toothpicks	Possible? (Yes or no)	Name of Shape
A	6	9		
B	5	8		
C	8	5		
D	6	10		
E	7	12		
F	8	12		
G	10	15		
H	12	18		
I	8	14		
J	10	3		
K	6	12		
L	5	9		

1. What is your conjecture about the ones that could not be made?
2. Which ones make pyramids? What is the relationship of the number? Using the models, explain why.
3. Which ones make prisms? What is the relationship of the numbers? Using the models, explain why.

A more detailed plan for this activity is available on the Student Companion site at www.wiley.com/college/reys.

Children are often taught the geometric names without being given much opportunity to explore the attributes or to solve problems. A recent national assessment shows that students in the United States know the names of geometric shapes but have difficulty with complex geometric attributes (Kloosterman et al., 2004). This section discusses each attribute separately, but the aim is to have children respond as in Figure 15-4. Notice the differences in the geometric vocabulary and attributes on which students focus. All the descriptions, however, could be based on visual clues that are only at the Level 1 or beginning Level 2. Asking children to explain or justify their answers would have them delve deeper into the attributes.

Shapes **331**

In the Classroom 15–4

BUILD YOUR OWN PYRAMID

Objective: Explore pyramids through construction.

Grade Level: 5–6.

- Follow these easy steps to construct a "stick" pyramid from newspaper. Use masking tape for the connectors.

   Step 1: Take three sheets from a newspaper and roll tightly from corner to corner.

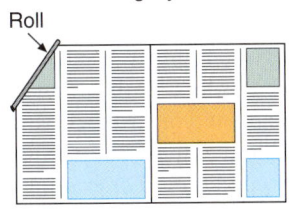

   Step 2: Tape to hold rod.

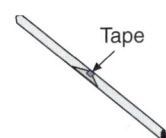

   Step 3: Make several rods and tape together as shown. Cut off tails.

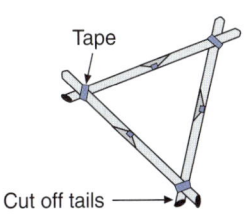

   Step 4: Put them together to make a pyramid.

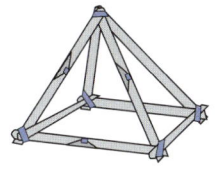

▼ Your turn:

Use the same method to construct other three-dimensional shapes, such as a cube.

In the Classroom 15–5

FIND ME

Objective: To identify examples and nonexamples of triangles.

Grade Level: 2–6.

▼ These are triangles:

▼ These are not triangles:

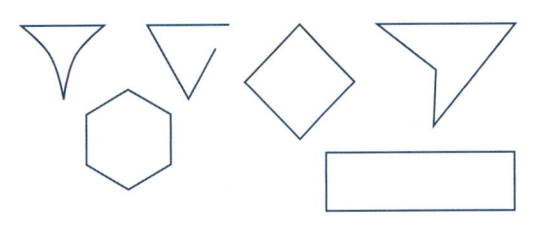

▼ Which are triangles? Tell why the others are not triangles.

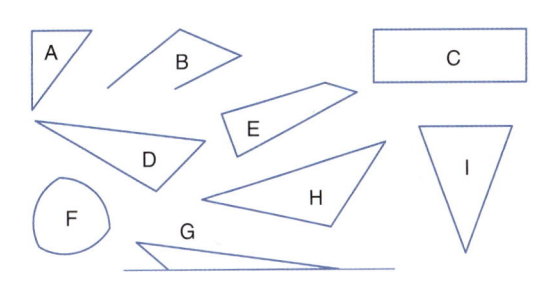

A more detailed plan for this activity is available on the Student Companion site at www.wiley.com/college/reys.

NUMBER OF SIDES AND CORNERS One of the first attributes children focus on is the number of sides. They readily count the number of sides (line segments) of a shape, unless a shape has many sides. Then they may need to mark the place where they begin counting. Or, as you no doubt noticed in the Snapshot of a Lesson, children may have difficulty knowing what counts as a side. As you have children participate in activities such as those that follow, they will begin to make many conjectures about shapes and learn vocabulary.

1. *How many sides?* This simple activity uses pattern blocks (see the Appendix). Each child needs only one shape for this activity.

   - Call a number and ask children who have a shape with that number of sides to stand. Call numbers such as two and seven, for which no one will stand.

   - Have a search for all the different shapes that have three sides, four sides, five sides, six sides, and zero sides (the circle).

   - Put a sample of each different shape somewhere within view of all children, and call on children to tell how the four-sided shapes differ (some are bigger than others, some are skinny, and some are

332 Chapter 15 • Geometry

Compare these shapes:

A   B

Primary responses

*They have four sides.
A is a rectangle. B isn't.
A can be folded to match.
They each have corners.
B looks lopsided.*

Intermediate responses

*They each have 4 sides and 4 corners.
A is a rectangle.
B is a parallelogram.
A has perpendicular sides.
B has 2 pairs of parallel sides; so does A.
The angles of A are equal: they are right angles.
The length of opposite sides are equal.
They aren't congruent.*

Middle school responses

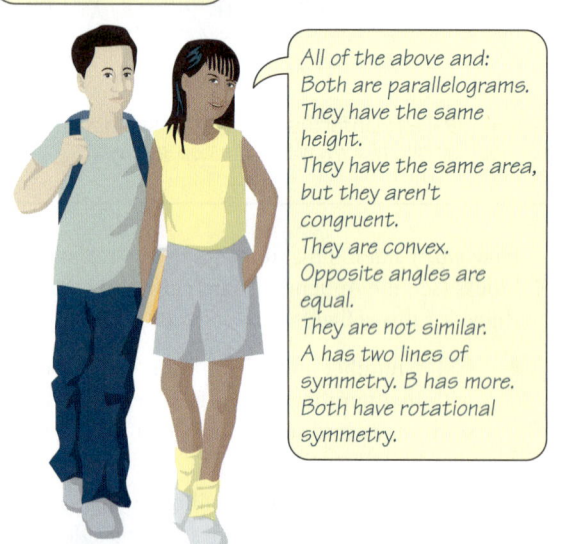

*All of the above and:
Both are parallelograms.
They have the same height.
They have the same area, but they aren't congruent.
They are convex.
Opposite angles are equal.
They are not similar.
A has two lines of symmetry. B has more.
Both have rotational symmetry.*

Figure 15-4 *Examples of children's responses that show their knowledge of geometric terms.*

slanty—accept their everyday words at this point). It is important for them to realize that the number of sides does not determine the shape.

2. *Less is best.* In the Classroom 15-6 provides a more advanced activity in which children put pattern blocks together to make new shapes with as few sides as possible. Two or more children choose three pattern blocks. Watch to see that they are counting the sides of the new shape, not the sides of the pieces. For example, if Marrietta said ten, it is likely that she counted the sides of the individual pieces.

3. *Can you make?* In this challenging activity, children make a figure of a given number of sides on a geoboard. (See the Appendix for a geoboard model.) Give each child a geoboard and one rubber band. Begin by asking the children to make simple shapes, and gradually add other conditions. For example:

- Can you make a four-sided figure?
- Can you make a four-sided figure that touches only four pegs?
- Can you make a four-sided figure that touches six pegs?
- Can you make a four-sided figure that has two pegs inside (not touching) it?

Closely related to counting the numbers of sides is counting the number of corners. Children will soon realize that any polygonal figure has the same number of sides as corners if they count both on each shape. The activities suggested for counting the number of sides can be modified for counting corners.

SYMMETRY Two types of symmetry—line or reflectional symmetry and rotational symmetry—may be used to describe geometric shapes as well as objects in the real world. To introduce line symmetry, ask children to compare two snowmen:

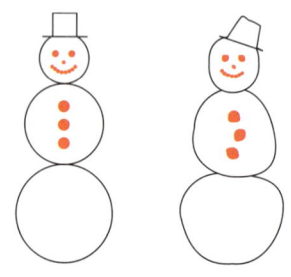

When they say one looks lopsided, show them how they can fold the drawings in the center to see if the sides match. A child's first perception of symmetry is visual (Level 1). Use this visual perception to help build the idea of folding to match the sides or edges. You may also want the children to explore with mirrors or Miras to bring in the idea of reflection. A figure has line or reflectional symmetry if, when

Shapes **333**

In the Classroom 15–6

LESS IS BEST
(A GAME FOR TWO OR MORE)

Objective: To make a shape with fewest sides.

Grade Level: 1–3.

▼ Choose a partner to play this game:
- Put an assortment of pattern blocks in a box
- Without looking, each player chooses 3 blocks and puts them together to make a new shape.
- Count the number of sides of each of the new Whoever has the shape with the fewest side best!

▼ Make a table to record your scores, and play sever determine the winner.

Cos: I'm good at this game — only 4 sides

Marrietta: I've got 4 sides, too!

| Cos | 4 | | | |
| Marrietta | 4 | | | |

A more detailed plan for this activity is available on the Student Companion site at www.wiley.com/college/reys.

In the Classroom 15–7

HOW MANY LINES?

Objective: To identify lines of symmetry.

Grade Level: 4–5.

- Draw the lines of symmetry on each shape.

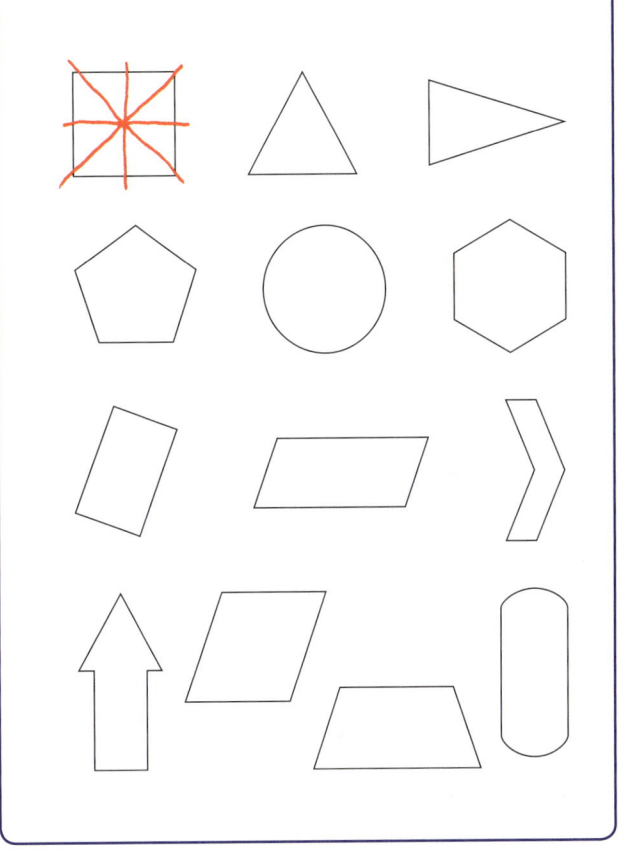

---

reflected over a line, the resulting image coincides with the original figure.

Have students find the line(s) of symmetry of geometric shapes. Be sure to let them try folding a square (four lines of symmetry), an equilateral triangle (three lines of symmetry), and a circle (an infinite number) to find lines of symmetry before moving to activities such as those in In the Classroom 15-7. Older children can often see the lines of symmetry without folding, but some shapes (such as parallelograms) are misleading. Many children say a parallelogram has two lines of symmetry. Try it yourself to see when this is true and when it is false. Children notice symmetry in things around them.

Together, you might make a bulletin board of pictures of things that are symmetric. Children also enjoy making symmetric shapes. One way to make a symmetric shape is to fold a piece of paper and cut the folded piece, leaving the fold intact. (Can you figure out how to make a shape with two lines of symmetry?) There are many activities on the computer in which children can investigate symmetry. Battista (2003) describes children's activities and learning about line symmetry using the *Geometer's Sketchpad*. And many additional ideas may be found on the Web such as the one suggested in Math Links 15.2.

Math Links 15.2

A video with activities introducing line symmetry in real life, symbols, and geometric shape can be found at http://www.linkslearning.org/Kids/ or accessed from this book's Web site.

www.wiley.com/college/reys

LENGTHS OF SIDES Many of the definitions of geometric shapes, as well as classification schemes, depend on the lengths of sides. Help children focus on the length by having them find the shape with the longest side, find the shortest side of a given shape, and measure lengths of sides. In the Classroom 15-8 presents an example of a more advanced activity in which children make shapes on a geoboard according to certain specifications about the lengths of the sides. Try the items yourself and classify each as to whether it is easy, medium, or challenging.

In the Classroom 15–8

SHOW MY SIDES

Objective: To make shapes with a specified number of congruent sides.

Grade Level: 5–8.

▼ Use a geoboard to show these figures:

1. Can you make a 4-sided figure with exactly two equal sides?
2. Can you make a 12-sided figure with all sides equal?
3. Can you make a 3-sided figure with three equal sides?
4. Can you make an 8-sided figure with four sides of one length and the other four of another length?
5. Can you make a 5-sided figure with exactly four equal sides?
6. Can you make a 4-sided figure with two pairs of equal sides that is not a parallelogram?
7. Can you make a 3-sided figure with two equal sides?
8. Can you make a 7-sided figure with no equal sides?

A more detailed plan for this activity is available on the Student Companion site at www.wiley.com/college/reys.

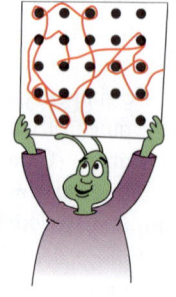

SIZES OF ANGLES There are many ways to examine the angles of geometric figures. A more complete introduction to angles may be found in Chapter 16. Here are some of the attributes related to angles that older children may discover:

- The sum of the angles of a triangle is 180 degrees.
- The sum of the angles of a quadrilateral is 360 degrees.
- The base angles of an isosceles triangle are equal.
- Opposite angles of a parallelogram are equal.
- A polygon with more than three sides can have equal sides without having equal angles.
- The angle opposite the longest side of a triangle is the largest.

A guided activity such as the one in In the Classroom 15-9 will help students discover for themselves that the angles of a quadrilateral sum to 360 degrees.

PARALLEL AND PERPENDICULAR SIDES In addition to examining parallel and perpendicular sides in geometric shapes, children need to be able to identify parallel lines and perpendicular lines in a plane and, later, in space. Two lines in a plane are parallel if they never intersect. (Remember, a line can be extended indefinitely in either direction.) Another useful definition states that two lines are parallel if they are always the same distance (perpendicular distance) apart. Two lines are perpendicular if they intersect at right angles.

It is important that children recognize perpendicular and parallel lines in the world around them. Have them search for perpendicular and parallel lines in the room. You might start a list together on the board, letting children add to it as they find other examples. Here is a start:

Parallel lines

- Opposite sides of a book
- The horizontal lines in E
- The top and bottom of the chalkboard

Perpendicular lines

- Adjacent edges of a book
- The vertical line and horizontal lines in E
- The edge of the wall and the edge of the floor

You may also have the children identify parallel and perpendicular sides on the pattern blocks. Ask them to find all the pieces that have one pair of perpendicular sides (none), the pieces with more than one pair of perpendicular sides (square), and the pieces with one pair of parallel sides (trapezoid). Use the puzzles in In the Classroom 15-10 to challenge children to arrange pattern blocks to make shapes with a specified number of parallel sides.

CONVEXITY AND CONCAVITY Often children are exposed only to convex shapes (any polygon with all angles less than 180 degrees). Many of the activities suggested thus far have included concave shapes. When children are making shapes, concave examples will often give interesting variety.

Show children two shapes such as these and have them describe how they are alike and different:

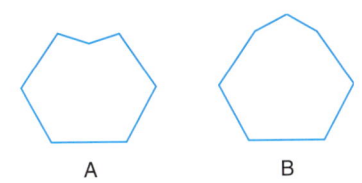

A   B

They will probably express the idea that shape A "comes back" on itself or "caves in" (concave). Introduce the terms *concave* and *convex*. After children classify shapes as convex or concave, you might have them investigate questions such as the following:

- Can you draw a four-sided (five-sided, six-sided, seven-sided) figure that is concave?
- Can you draw a five-sided (six-sided, seven-sided) figure that is concave in two places (or that has two angles greater than 180 degrees)?
- Can you draw a six-sided (seven-sided, eight-sided) figure that is concave in three places?

It is challenging to try these exercises with *Logo* on the computer because the turtle turns the external angles of shapes. Have the children keep a record of the steps it takes to make each figure.

ALTITUDE The altitude (or height) of a geometric shape depends on what is specified as the base. Identifying and measuring the altitude is essential in finding the area of geometric figures. In the Classroom 15-11 is designed to

---

In the Classroom 15–9

HOW MANY DEGREES IN A QUADRILATERAL?

Objective: To explore the sum of the degrees of the angles in quadrilaterals and triangles.

Grade Level: 5–8.

▼ Try this method for finding the sum of the angles in a quadrilateral:

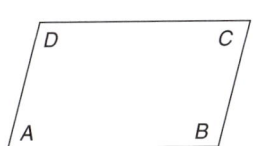

Trace and cut out. 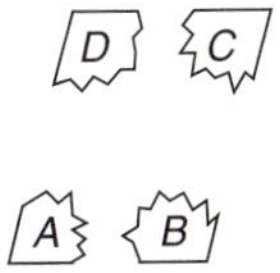   Tear off the corners.    Put the angles around a point.

Name of shape *parallelogram*

Number of degrees of A + B + C + D   *360°*

▼ Use the same method for these quadrilaterals. (Reminder: there are 360 degrees around a point.)

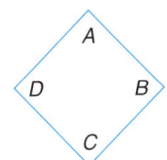

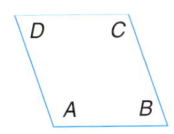

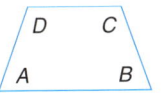

Name of shape _____   Name of shape _____
Number of degrees _____   Number of degrees _____

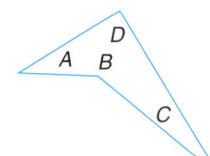

Name of shape _____   Name of shape _____
Number of degrees _____   Number of degrees _____

▼ Now
- Try some more four-sided figures.
- Try a five-sided figure.
- Try a three-sided figure.
- What do you conclude?

> ### In the Classroom 15–10
>
> ## PIEZLES
>
> *Objective:* To build shapes with parallel sides.
>
> *Grade Level:* 4–5.
>
> *Materials:* Pattern blocks.
>
> ▼ Solve these piezles (puzzles) using pattern blocks. Draw a sketch of the shape you made.
>
> 1. Use two different pieces: make a shape with
>   - Exactly 2 pairs of parallel sides
>   - Exactly 1 pair of parallel sides
>   - No parallel sides
>
> 2. Use three different pieces; make a shape with
>   - Exactly 3 pairs of parallel sides
>   - Exactly 2 pairs of parallel sides
>   - Exactly 1 pair of parallel sides
>   - No parallel sides
>
> 3. What is the largest number of pairs of parallel sides of a shape you can make from
>   - 2 pieces
>   - 3 pieces
>   - 4 pieces
>
> 4. Can you put all the pieces together to make a shape with no parallel sides?

> ### In the Classroom 15–11
>
> ## WHAT'S MY ALTITUDE?
>
> *Objective:* To identify and measure the altitude of a triangle.
>
> *Grade Level:* 4–5.
>
> ▼ Make a triangle from a stiff piece of paper. Cut a strip 2 cm by 20 cm. Mark off segments of 9 cm, 4 cm, and 7 cm and label them *A*, *B*, and *C*, respectively. Fold and tape as shown:
>
> 1. Set the triangle on side *A*. This is the base.
>   - How long is the base? _____
>   - What is the altitude? _____
>   - How long is the altitude? _____
>
> 2. Set the triangle on side *B*. This is the base now.
>   - How long is the base? _____
>   - What is the altitude? _____
>   - How long is the altitude? _____
>
> 3. Set the triangle on side *C*. This is the base now.
>   - How long is the base? _____
>   - What is the altitude? _____
>   - How long is the altitude? _____

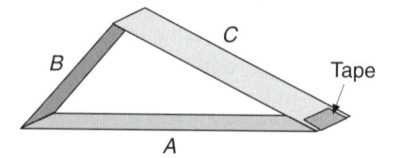

help children realize that a geometric object has different heights, or altitudes. After students do this with the triangular model, have them draw a triangle on paper and again measure each side and each altitude.

CLASSIFICATION SCHEMES What makes a parallelogram a parallelogram? When is a rhombus a square? What is a regular polygon? When students begin to understand the defining attributes of two-dimensional shapes, they are moving into Level 3. We take a brief look at classifying triangles, quadrilaterals, and polygons and some of the defining attributes.

Triangles Triangles are classified either by sides or by angles:

By Sides	By Angles
Equilateral three congruent sides	**Acute** all angles less than 90 degrees
Isosceles at least two congruent sides	**Right** one angle equal to 90 degrees
Scalene no sides congruent	**Obtuse** one angle greater than 90 degrees

After children have learned to identify triangles by sides and by angles, the two classification schemes may be put together. For example, can you make an isosceles, right triangle? A scalene, obtuse triangle? An equilateral, right triangle?

What defines a triangle? Here are some definitions that students wrote.

a. A triangle has three sides.
b. A triangle has three sides and three angles.
c. A triangle is a polygon with three sides.

Although all three statements are true, only the last defines the triangle. Can you think of a figure that has three sides and is not a triangle? (Hint: try a non-closed figure.) How about an example of b that is not a triangle?

Quadrilaterals There are many special names for quadrilaterals. Common names are parallelograms, rectangles, squares, rhombuses, trapezoids, and kites. These classes of quadrilaterals are not disjoint; one shape may fall into several categories depending on the definition.

For example, a rectangle is also a parallelogram. This type of classifying process is more difficult for children than partitioning the whole set into disjoint classes, as is the case of classifying triangles by lengths of sides. It requires more than just recognizing examples of figures; it requires understanding the defining attributes (Level 3). For example, a parallelogram is a quadrilateral with two pairs of parallel sides. Assuming you know that a quadrilateral is a four-sided, closed, simple figure, can you identify which of the following are parallelograms? What other names do they have?

You are correct: They, all are parallelograms. Thus, a square, a rhombus, and a rectangle are all special types of parallelograms.

Discuss these questions with a peer. A *rhombus* is a parallelogram with all sides congruent. Does that mean that a square is a rhombus? Why? A *rectangle* is a parallelogram with right angles. Does that mean a square is a rectangle? Why? Through the centuries, definitions of shapes have changed. At one period, we had oblongs and squares. Oblong was a rectangle that was not a square.

How do you begin to teach such relationships? Children must first begin to verbalize many attributes of the figure. For example, they must be able to describe a square:

As a closed, four-sided figure (attribute 1),
With opposite sides parallel (attribute 2),
All right angles (attribute 3), and
All sides congruent (attribute 4).

Attributes 1 and 2 make it a parallelogram; attributes 1, 2, and 3 make it a rectangle; attributes 1, 2, and 4 make it a rhombus; attributes 1, 2, 3, and 4 make it a square. In the Classroom 15-12 helps students with this idea.

Polygons   Polygons are named according to the number of sides:

3 sides: triangles
4 sides: quadrilaterals
5 sides: pentagons
6 sides: hexagons
7 sides: heptagons
8 sides: octagons
9 sides: nonagons
10 sides: decagons

This classification scheme is not difficult, but often children are shown only regular polygons. Thus, among the shapes shown here, a child sees only the first as a hexagon, instead of realizing they are all hexagons.

In the Classroom 15–12

CLASSIFY ME

Objective: To classify and name quadrilaterals.

Grade Level: 4–5.

▼ Mark each of the figures with a

1. if it is a quadrilateral
2. if it has two pairs of parallel sides
3. if it has all right angles
4. if it has all congruent sides

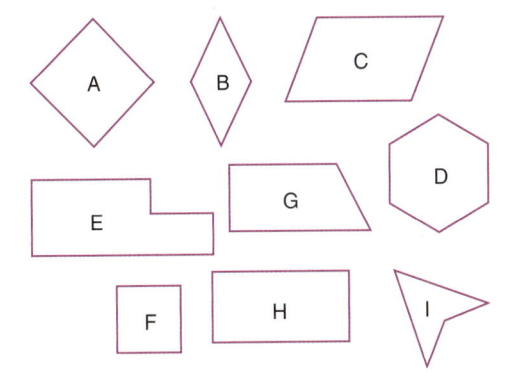

- Any figure marked 1 and 2 is a _____.
- Any figure marked 1, 2, and 3 is a rectangle as well as a _____.
- Any figure marked 1, 2, and 4 is a _____ as well as a _____.
- Any figure marked 1, 2, 3, and 4 is a _____ and a _____ as well as a square.

A more detailed plan for this activity is available on the Student Companion site at www.wiley.com/college/reys.

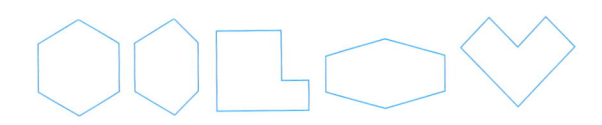

Children should be encouraged to think of real objects that are shaped like these. A fun hunt for children at any age is to find different-shaped polygons in their neighborhood. If the children live in a rural area, have them draw all the different-shaped pentagons they see (the sides of many roofed buildings). Buildings in cities often use wonderful geometric shapes.

The names *heptagon*, *nonagon*, and *decagon* are not widely used, so in doing activities you may have to remind children of these names. In the Classroom 15-13 uses the names as well as other attributes; students are asked to see shapes within other shapes, a task that is difficult for some. The geometric design paper in the Appendix may be used for similar searches.

338　Chapter 15 • Geometry

Children often have difficulty seeing figures within other figures. In the 2007 mathematics assessment of NAEP, the following two items were given. Students in the fourth grade assessment were provided three copies of the grid in Figure 15-5. They were asked to outline a square in each grid and were told that the squares must not be the same size. The other item was the same except the figures were to be triangles. The squares that are turned gave the students more difficulty than the two squares that are not turned. Less than half of the fourth-grade students outlined three of the four, and fewer students outlined the three triangles. Even more difficult was a third item that asked students to draw a four-sided shape that was not a rectangle (or square). This should be a reminder of three things:

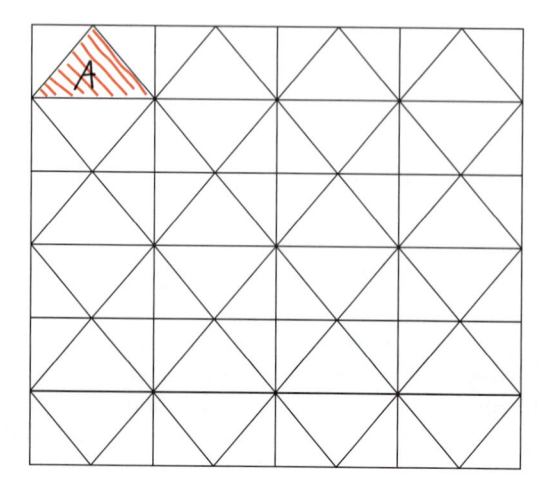

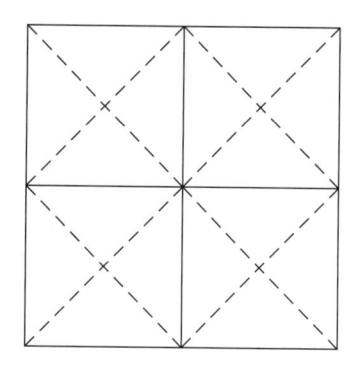

Figure 15-5　*Grid from NAEP 2007.*

1. Students need to see many different examples and nonexamples of shapes in many sizes and orientations.
2. Students need to make shapes that illustrate given attributes rather than always identifying attributes in shapes.
3. Students need experiences with seeing and making shapes within shapes.

In the Classroom 15–13

CAN YOU FIND IT?

Objective: To shade polygons with given attributes.

Grade Level: 5–8.

▼ See if you can find each of these in the design. Fill in the shape, and mark it with the matching letter.

- A. triangle—isosceles
- B. triangle—scalene
- C. quadrilateral—not symmetric
- D. quadrilateral—4 lines of symmetry
- E. pentagon—concave
- F. pentagon—convex
- G. hexagon—exactly 2 pairs of parallel sides
- H. hexagon—symmetric
- I. heptagon (7 sides)—symmetric
- J. heptagon—not symmetric
- K. octagon

SPACE

Typically geometry in elementary school focuses on shapes. Yet another part of geometry deals with space—locations and movements that describe direction, distance, and position. Where are you? Are you above or below the floor? Are you in front of or behind your desk? Are you between the cabinet and the computer? Where would you be if you moved five steps forward? Starting at your desk, how far from your desk will you be if you move five steps forward and three steps backward?

Examining locations and movements gives children a way to describe their world and give order to their surroundings, and provides an opportunity to build mathematical concepts such as that of positive and negative numbers (forward and backward) and skills connected to other subjects, such as map skills. Give students opportunities to move by describing directions and distances traveled and where they must identify positions of an object through several movements. You might begin by taping to the floor of the classroom a map of several city blocks, as in Figure 15-6. Ask students to describe the directions and the distances in blocks between specific locations, such as between school and home. Have the students describe a set of directions from home to school where they would use the crosswalks provided at stoplights and stop signs. Have the students describe a set of directions from school to the playground.

Students were asked on the 2005 NAEP mathematics assessment to give directions on a simple grid to go from a point marked school to a point marked park. They were asked to use the appropriate directions north, south, east,

Space **339**

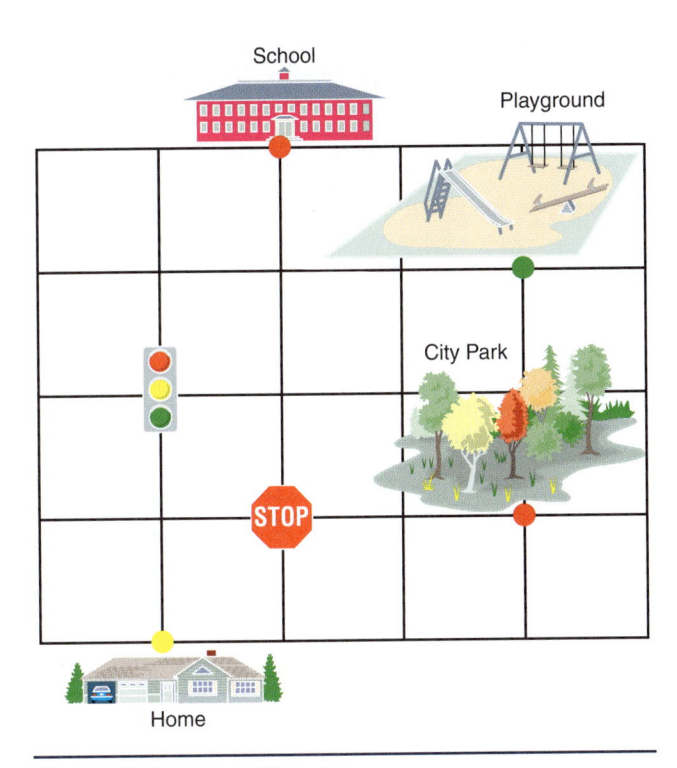

Figure 15-6 *Map of locations.*

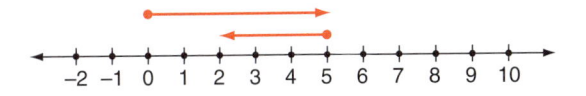

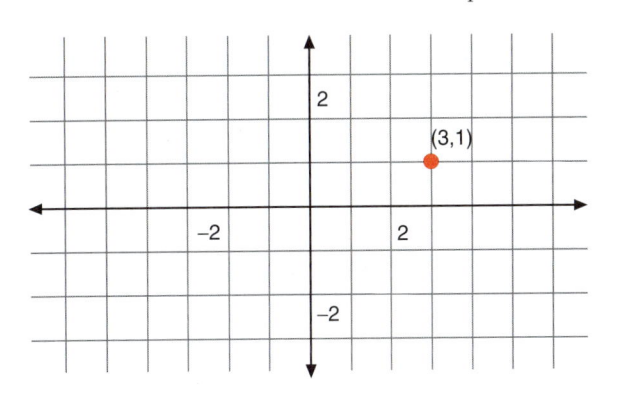

Figure 15-7 *Coordinate system for describing spatial relationships.*

and west which were illustrated in the picture. The first direction, go 2 blocks east was given. Over two-thirds of the fourth-grade students gave the remaining two directions (4 blocks north and 3 blocks east). The others did not give the distance needed. This is a practical skill well in reach of students at this and earlier grades.

Students can formalize their navigational thinking by examining a coordinate system as in Figure 15-7. The coordinate system in Figure 15-7 uses all four quadrants. Although many programs do not introduce negative numbers in early childhood, this is a natural way for children to use them. You may want to begin with only the positive quadrant, similar to map grids. Tie movement and location together. For example, describe a movement and ask for a location: If I am at point (3, 1) and move 2 spaces, where could I be? (Many answers.) Describe two locations [e.g., (3, 2) and (5, 6)] and ask what movements would take you from one point to the other. (Many answers.)

The number line is a representation that involves location and can be used to represent movements on a line rather than in two dimensions. Movements like move forward five steps and back three steps can be represented on the number line. Suppose you begin at 0, where would you end? The number line below shows this movement:

There are three parts to problems like this: the beginning point, the movement, and the ending point. Knowing any two, you can figure out the other. The movement, however, could occur in many ways. For example, suppose you begin at 2 and end at 8, what are ways you could have moved? The shortest movement is forward 6, but consider forward 4, back 3, forward 5, back 4, and so forth. Will you get there? How many separate movements will it take? How far would you have moved in all?

Logo programs and other software for computers offer a wide variety of activities for children to explore movement and location. In *Logo*, an angle is considered as a turn and a figure is a path created as the turtle travels with its pen. With only a few simple commands (FD for ForwarD, BK for BacK, RT for RighT, etc.), children can begin drawing pictures and hypothesizing about geometric figures. An important feature of *Logo* is the angle that is created when the turtle is instructed to turn. Figure 15-8 shows some turtle commands and the resulting figure that is drawn. What is the measure of the angle the turtle needs to turn in order to create an equilateral triangle where each of the angles is 60 degrees?

Children in the early grades begin exploring *Logo* through the *Logo* simulation such as the two, Hiding Ladybug and Turtle Pond suggested in Math Links 15.3.

Math Links 15.3

For Pre-K-2 students at hiding the turtle and for older students have them move the turtle around a pond. You can access these from the NCTM Web at http://www.nctm.org/standards/content.aspx?id=25009 and http://illuminations.nctm.org/ActivityDetail.aspx?ID=83 or from this book's Web site.

www.wiley.com/college/reys

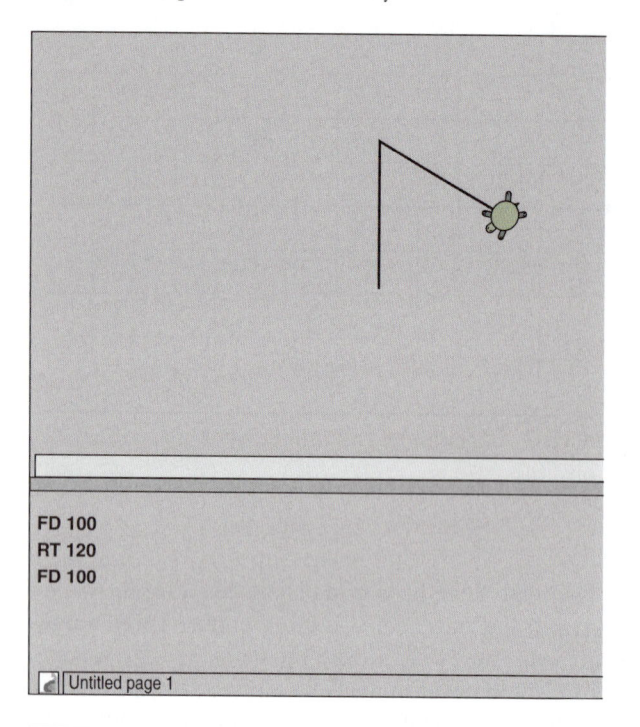

Figure 15-8 *Logo commands: instructing the turtle to begin to create an equilateral triangle.*

Children love to plot points on a coordinate system and connect the dots to make pictures. This certainly is one way to have them practice plotting points, but it is a rather low-level activity. It is more challenging if they draw a picture and create their own directions. There are many problem-solving activities that involve coordinates and geometric shapes. In the Classroom 15-14 has challenging problems involving shapes and changing shapes. If you are teaching younger children, then consider having them draw shapes with given descriptions related to the coordinate grid. For example, you may ask them to draw a rectangle in the first quadrant (the one with all positive coordinates) with one vertex at (3, 2) where the length of the longer sides is 4 and shorter sides is 3.

TRANSFORMATIONS

Geometry is often studied in terms of patterns. Think of wallpaper patterns that you have seen or look at the three border patterns in Figure 15-9. What basic design is repeated? How is the basic design moved? Describe the moves in your own words. Did you use words such as *turn*, *slide*, or *flip*?

Although the CCSSM does not call for study of rigid geometric transformations until after grade 6, you should be familiar with them: slide (translation), turn (rotation), and flip (reflection), illustrated in Figure 15-10.

In the Classroom 15–14

SHAPES ON A GRID

Objective: To explore various shapes on a coordinate grid.

Grade Level: 4–6.

Materials: Grid paper (see the Appendix).

Launch: Use these as a problem of the day (or week). Have students share the different solutions. Only rule: The vertices must be on the grid.

1. Two vertices of a square are (2, 0) and (4, 2). What are the other two vertices? (3 solutions)

2. Two vertices of a right triangle are (0, 0) and (3, 0). What could be the third vertex? How many possibilities are there? Give at least 5 examples.

3. Parallelogram ABCD has vertices at (0,0), (0,4), and (1,3). What are the coordinates of the fourth vertex shown in the picture? Find other parallelograms with these three vertices. Give the fourth vertex.

4. The octagon below is in a 3-by-3 grid. What other octagons can you draw in a gird this size? Remember: The vertices must be on the grid.

Which of your octagons are
a. symmetric?
b. convex?

Which has
a. the largest area?
b. the largest perimeter?

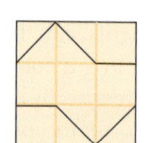

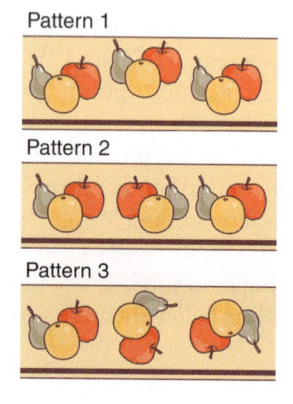

Figure 15-9 *Wallpaper borders.*

1. Slide (Translation)

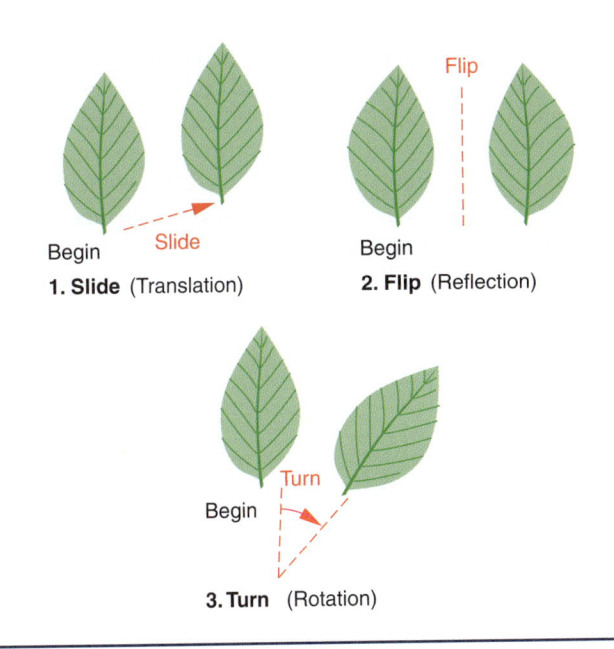

2. Flip (Reflection)

3. Turn (Rotation)

Figure 15-10 *Motions for determining congruence.*

If students are asked to make conjectures and justify their thinking, they develop a deeper understanding of the transformations. For example, older children may be asked to investigate what one motion would be the same as two flips over parallel lines, first over line *m* and then over line *l*:

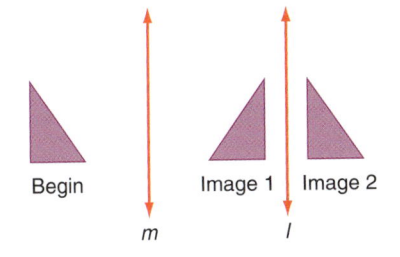

What if the lines are not parallel? What if the lines are perpendicular? What if you first flipped it over line *l* and then over line *m*?

Some questions about transformations can be answered just by looking, but for others children will need to develop techniques. For example, children can fold on a line of reflection as they have done on lines of symmetry to see if the image is a reflection over that line. They can trace and turn or slide shapes to see the rotation or the slide. Computer programs such as the *Geometer's Sketchpad* provide a way even for elementary school children to investigate many of the characteristics of transformations.

One of the most interesting applications of transformations for students involves tessellations. A simple transformation of a shape can produce an interesting design. You can find many ideas on the Web and a simple example in the article "Tessellating T-Shirts" (Shockey and Snyder, 2007).

CONGRUENCE Two shapes are said to be *congruent* if they have the same size and the same shape. Young children grasp this idea when they see that one shape can be made to fit exactly on the other. If the two shapes are line segments, they are congruent if they have the same length. If the two shapes are two-dimensional and they are congruent, then they have the same area. The converse is not true. Two shapes with the same area may not be congruent. Children have difficulty with this concept, and many middle school students would respond that the parallelogram shown here is congruent to the rectangle:

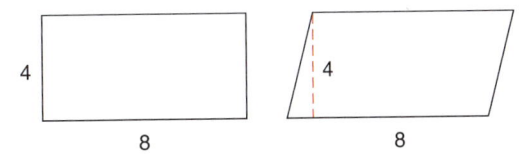

Young children have little difficulty identifying figures with the same shape and size, such as with these two right triangles. They can cut the shapes out and show that one fits exactly on top of the other to demonstrate the idea. Thus, the task becomes one of asking young children to match figures to see if they are the same size and shape and gradually introducing and using the word *congruence*.

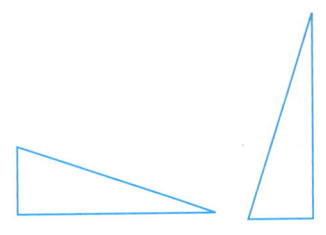

Congruence is often investigated through transformations. If two shapes are congruent, they can be made to fit by one or more of the three rigid motions. What motion or motions could you use to move one of the triangles above onto the other? In terms of transformations, why are the rectangle and parallelogram shown above not congruent?

SIMILARITY Another type of transformation is that of enlargement (or reduction). You have probably used these terms when using a copy machine. What happens to a triangle on a sheet of paper when you reduce or enlarge it by a certain percentage? We say the two triangles are similar.

Similar is a word used in everyday language to mean alike in some way such as same shape. In mathematics *similar* has a special meaning or definition: Two figures are *similar* if corresponding angles are equal and corresponding sides are in the same ratio. We have also used *same shape* to mean, for example, that all the figures are all rectangles. But not all rectangles are similar. So part of your task in teaching similarity will be to refine intuitive notions about similarity to fit the mathematical definition stated above.

This definition is too formal for a beginning. Instead, you can begin by using a geoboard and geopaper. Children can make a design on the geoboard and transfer it to smaller geopaper, or they can copy designs from one size of graph paper to another. Older students can explore doubling, tripling, and so on the lengths of the sides of geometric shapes to see what effect this has on the perimeter and area on the electronic geoboard (see Math Links 15.4). After students have been introduced to ratio, they can investigate similarity in a more rigorous way. (Similar triangles are discussed in Chapter 13 as an example of proportions.)

> **Math Links 15.4**
>
> A wealth of electronic geometry manipulatives may be found at the National Library of Virtual Manipulatives for Interactive Mathematics Web site, which you can access from http://nlvm.usu.edu/en/nav/index.html or from this book's Web site.
>
> www.wiley.com/college/reys

VISUALIZATION AND SPATIAL REASONING

If you have been doing the activities in this chapter thus far, you have probably been visualizing, using spatial reasoning or modeling to solve problems. Geometric modeling using number lines, coordinate grids, and rectangular arrays for multiplication has been used throughout the book. In this section we consider several aspects of visualization and spatial reasoning that we have not emphasized in the previous discussions.

Battista (2007, p. 483) defines spatial reasoning as the "ability to 'see,' inspect, and reflect on spatial objects, images, relationships, and transformations. Spatial reasoning includes generalizing images, inspecting images to answer questions about them, transforming and operating on images, and maintaining images in the service of other mental operations."

USING GEOMETRIC PHYSICAL AND PICTORIAL MATERIALS

One way to develop visualization and spatial reasoning is to use physical and pictorial materials that can be moved, subdivided, put together, and otherwise changed. Research shows that the use of physical materials, including paper models, plastic or wooden geometric shapes, geoboards, and solids, can be helpful in developing geometric representations, but they must be used wisely (Clements, 2004). The ability to see shapes rearranged to make other shapes or seeing shapes within shapes is important to develop. The more that students work with materials to do this, the more ways they have to visualize geometric shapes and attributes.

The opening Snapshot of a Lesson used simple paper folding to make shapes from a square. The tangram puzzle (see Appendix A) provides pieces for endless explorations of ways to put the shapes together to make animals, figures, and other shapes. Children love to make designs and fill in pictures with the pattern blocks. All of these types of activities are a beginning, but students need to be encouraged to take these activities further—to make predictions, to investigate various ways to make the same shape, to describe the pictures they make with geometric terms, and to make shapes with given attributes. In the Classroom 15-15 uses the pattern blocks (see the Appendix) to present some problems of combining shapes to make other shapes.

> **In the Classroom 15–15**
>
> ### SHAPE MAKER
>
> *Objective:* To make shapes with given characteristics.
> *Grade Level:* 3–5.
> *Materials:* These pattern blocks.
>
> ▼ Use as many of the blocks as you need. Record your answers with drawings showing the blocks used.
>
> 1. How many different ways can you make the hexagon?
> 2. Can you make a triangle with 2 pieces? With 3 pieces? With 4 pieces? With 5 pieces? With 6 pieces?
> 3. Can you find two different ways to make a triangle with 4 pieces?
> 4. Make a rhombus using one hexagon and other pieces.
> 5. Make a convex hexagon using one hexagon and other pieces. Find another way.
> 6. Make a concave hexagon using one hexagon and other pieces. Find another way.
> 7. Make a nonregular hexagon (all sides are not congruent) with 3 pieces, with 4 pieces, with 5 pieces, with 6 pieces.
> 8. Make a large triangle only using the green triangles. How many did it take?
> 9. Make a larger rhombus using only the blue rhombuses. How many did it take?
> 10. Make a larger trapezoid using only the red trapezoids? How many did it take?
> 11. You cannot make a larger hexagon using only yellow hexagons. Can you explain why?

Pictures or diagrams also are important tools. Can you imagine learning what a rectangle is without seeing examples? You should think about this if you have students who have visual difficulties. What can you do to help those students with pictures and diagrams? Think about using models with raised interior lines (a piece of string glued) that they can feel. It is not only these students who have difficulty; many students see diagrams and pictures differently than intended. It is important to have them create their own pictures and diagrams from work with the physical materials.

Computers have provided a tool that opens up many possibilities for many students. Students who have trouble moving the pattern blocks into place and recording their work can often use one of the shape-making computer programs. Computers also allow students to create many more examples of shapes they are trying to explore. For example, the exploration of altitude and bases (In the Classroom 15-11) is done with only one triangle. On a computer, students could extend this to many examples and realize that to find the area of a triangle, it does not matter which side is the base. Many of the textbooks you will be using will have special computer programs; see what is available in your school to use.

USING MENTAL IMAGES

Although you can help children learn geometric ideas by using physical materials, you also need to help them develop visual images of geometric shapes. As your students mature, you can help them manipulate these images in their minds. For example, visualize a rectangle that is longer than it is wide. Now turn it (in your mind) to the right 90 degrees. What does it look like now? Draw a line from the top right corner to the middle of the bottom side. What two figures did you create?

One of the classic tasks to encourage developing mental images is to cover up most of the shape and reveal it slowly, asking students to predict what shape is covered. See the example below. There are questions to get you started, but let children come up with what they think it is before you ask them these specific questions.

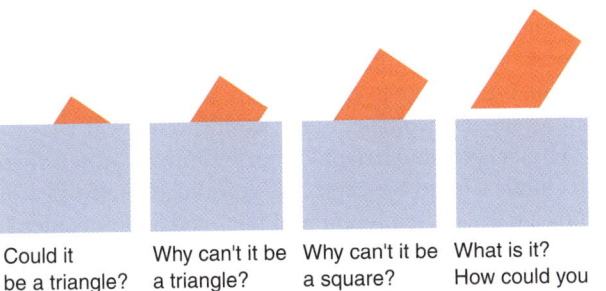

Could it be a triangle? square? rectangle? trapezoid? other?

Why can't it be a triangle? Could it still be a square?

Why can't it be a square? Could it still be a rectangle?

What is it? How could you change the bottom so it would not have been a trapezoid?

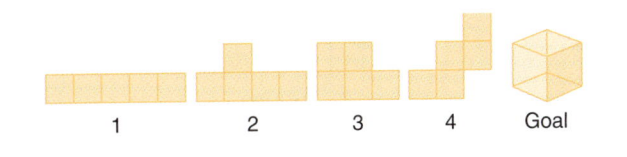

Figure 15-11 *Which figure can be folded into the shape of the open box?*

Another version of this type of activity can be found in the overlap problem (Tayeh, 2006). Student solutions are given to the problem of overlapping two transparent squares and asking what shape the overlap is. Before reading the article, try this yourself. Are you as flexible as the students?

Visualizing two-dimensional patterns that can be folded to make three-dimensional patterns is an important aspect of geometric thinking. In the process, you may visualize the folding, or you may use spatial reasoning to help you decide. For example, imagine trying to fold an open cube (no top face) from five connected squares. Which of the configurations in Figure 15-11 could be folded into an open cube? Which would be the bottom square? Being able to make these visual manipulations mentally is enhanced by experience in doing such manipulations physically.

Visualizing three-dimensional objects from different perspectives is another aspect of geometry. What happens if you view a solid from different perspectives? Pretend you are a fish, swimming below a pyramid that is sitting on top of the water. You can see the entire bottom of the pyramid. What do you see? Begin with putting the children in easy locations where they see only one face and then move to ones that show more than one face. Pretend you are a bird and you are looking down on the point of a (square) pyramid. Can you describe what it would look like? Then have children put out solids and ask them to draw what they see from different views (down, under, directly in front, front when the solid is turned to see two faces).

Students need to work with models to help them make these mental images. Young children can begin with the basic solids and describing the faces. Make imprints in play dough (homemade works great) of the faces of wooden geometric solids. Have children match each solid with a face. The faces can be traced instead of making imprints, but younger children can work easily with imprints.

One way to help students see shapes from different perspectives is to look at structures built with cubes. Have students build a structure with cubes and then draw the different faces on square grid paper. There are several variations of this task. Give the children the drawings of the faces and see if they can construct the building. Or have the children draw the structure using the isometric paper as shown below. (A master sheet of isometric paper can be found in the Appendix.)

Front

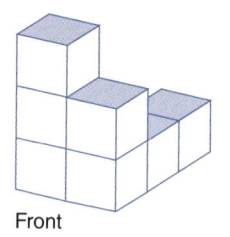

Another visualization task involves cross sections of solids. Cutting paper tubes (see Figure 15-3) helps children to see the cross sections. Using other objects that can be cut (e.g., oranges, carrots shaped in cones, or zucchini shaped in different solids), you can help the children see the cross sections. The next time you eat cheese, cut some small cubes of cheese and then try slicing each cube so the slice shows different geometric shapes. Can you slice the cube to see a triangle? A different triangle? How can you slice the cube to make a rectangle that is not a square? Try a pentagon and a hexagon.

CULTURAL CONNECTIONS

Geometry opens the doors to many cultures. Literally, a search on the Web for doors from other countries can be an exciting adventure. We found African doors from the thirteenth century, including a beautifully carved wooden door with an intricate geometric pattern around the edge and a pitcher of wheat in the center. Mexican doors often have many rectangular panels, some plain and some decorated with geometric designs; Chinese doors have elaborate screen designs. Of course, the doors of every country differ by the period when they were made, the availability of materials, and the socioeconomic status of the owners. Have you ever thought about what cultures did not have doors? What cultures have doors that are not rectangular? Why are most doors rectangular? What geometry do you see in the doors in your neighborhood?

Cultures have their own designs for their arts and crafts—quilts, rugs, pottery, jewelry—the list is endless. Many of these designs at some period in any culture's history are geometric. Homes and other structures also have distinctive geometric attributes. In the United States, homes are some variation of rectangular prisms with roofs that are a mixture of triangular prisms. Look around your neighborhood or in house magazines for homes that are made up of rectangular prisms. What other shapes do you see?

In the nineteenth century, Orson Squire Fowler claimed that octagonal houses were the most economical (less siding material for the same area of floor space) and the healthiest (more windows and ventilation). Although he promoted his idea through several books and his own designs, there are only a handful of octagonal buildings. What other shapes are used for the structure of buildings or roofs? A classic article (Zaslavsky, 1989) looks at round houses.

Geometry has played different roles in schools of other countries. The French curriculum concentrates more on geometry in middle grades than we do, but it does much less before those grades. "Although there are 11 geometry topics in the TIMSS fourth grade mathematics assessment, their inclusion in participants' curriculum varies more widely than any other strand, as does the percentage of students taught each of the topics" (Mullis et al., 2004, p. 213). Most of the topics are included in other grades, mainly in higher grades.

While in the United States we have spent the majority of the time in elementary school teaching geometric vocabulary, the curriculum in Japan expects students to learn the vocabulary naturally as they use it in solving geometric problems. Realize that the geometric vocabulary in many countries is more closely aligned with other words. For example, *tri* (from *triangle*) is a variation of *tres* (Spanish) or *trois* (French). Also the names of many geometric shapes are often descriptive. In English, this would be like using "six sides" for hexagons.

In the diverse classrooms of today, you will find students with many different backgrounds. The visual nature of geometry helps you accommodate these differences in many of the geometric activities.

A GLANCE AT WHERE WE'VE BEEN

Geometry is a topic that is often neglected in elementary school, yet it has many benefits for children if it is presented in an intuitive, informal manner. This chapter presented a variety of sample activities that provide this type of informal experience. The discussion of three-dimensional shapes included ways to have children describe and sort solids as well as ways to make three-dimensional models. The discussion of the two-dimensional shapes focused on attributes of shapes. It emphasized moving from simple recognition of geometric shapes (modified van Hiele Level 0) to the higher levels.

Geometry is used to help us represent the space in which we live and to describe the movements and the relationships between objects in space. Ideas were presented for young children and more formal ideas using a coordinate system. Geometric transformations also can be used to describe our world and to solve problems. Building children's visual skills and reasoning is an important aspect of geometry. The ultimate aim is to be able to use geometry to solve problems and to appreciate the geometry in the world around you.

This chapter touched on only a few things you can do to help children build concepts and skills in geometry, as well as only a few ways to present problems and apply geometry. There are many other fascinating topics and activities that you can use. Begin collecting and using these in your teaching. Your job is to use these ideas wisely, helping students deepen their understanding of geometry from simple recognition to analyzing and justifying geometric statements and to solving problems involving geometry.

Things to Do: From What You've Read

1. Name three attributes of solids that children should learn at each of these levels: beginning, intermediate, and advanced.
2. Explain how the children's answers in Figure 15-4 differ from primary to intermediate. What more is shown by the middle school students than the primary and intermediate students? What level of the modified van Hiele scheme do you think they are operating?
3. Give the reason for including each example and nonexample of a triangle in In the Classroom 15-6.
4. Design a sorting activity for intermediate students based on one of the ideas in the beginning activities with solids.
5. Explain how quadrilaterals can be classified. Why are squares a type of rectangle? How could rectangles be defined so that squares would not be rectangles?
6. Complete the activity in In the Classroom 15-8. Build a list of at least 10 objects in the world around you that have lines of symmetry.
7. Do one of the challenging activities from In the Classroom 15-11, 15-14, or 15-15.

Things to Do: Going Beyond This Book

In the Field

1. [1]*Solid Activities*. With children, try one of the three-dimensional activities at the beginning, intermediate, or advanced levels in the section on describing and sorting objects.
2. [1]*Find Me or Classify Me*. In the Classroom 15-6 provides examples and nonexamples of triangles. Ask young children to tell which of the mixed examples are not triangles and to justify their selections. Or use In the Classroom 15-13 with older children to see what they understand about the classification of quadrilaterals.
3. [1]*Compare Shapes*. Interview a range of K–8 students, asking them to compare the two shapes in Figure 15-4. Compare their responses with those in the text.
4. [1]*Gumdrops and Toothpicks* In the Classroom 15-3 is one of the activities with building three-dimensional shapes. As you try this with children, observe which shapes children make and whether they can tell the number of edges and vertices. Can they make shapes with a given number of edges and vertices?
5. [1]*Less Is Best or Show My Sides*. In the Classroom 15-6 is a game for younger children, and In the Classroom 15-10 has challenging puzzles. Try one of these with students and observe how they react to games or puzzles.

In Your Journal

6. Defend spending time in a classroom constructing models of solids or defend spending time teaching geometry in elementary school.

With Additional Resources

7. Read about the use of geometry in everyday life in another country or culture. Describe how it is different and how you could use this information in your teaching.
8. Look at the geometry in a textbook at a given grade level. Make a list of the activities from this chapter that would complement the text.
9. Read one of the children's books listed in the Book Nook for Children or another book of your choice. Describe how you would use the book in teaching geometry.

With Technology

10. Choose one of the geometric materials from the Virtual Manipulatives for Interactive Mathematics (see Math Link 15-4). Teach one of your peers to use the material and together design an activity for children.

---

[1]Additional activities, suggestions, or questions are available in the field experience manual on the Student Companion site at www.wiley.com/college/reys.

Note to Instructors: You can find additional resources, learning activities, and blackline masters in this text's accompanying Instructor's Manual at www.wiley.com/college/reys.

Book Nook for Children

Burns, M. *The Greedy Triangle*. New York: Scholastic, 1994.

This story is about a shape that wants to be another shape. A shapeshifter turns the triangle into another shape, but the triangle still is not happy. This continues until the triangle realizes that he really wants to be a triangle. There are activities for children, parents, and teachers.

Flournoy, V. *The Patchwork Quilt*. New York: Four Winds Press, 1991.

Tanya loves listening to her grandmother talk about the quilt she is making from pieces of colorful fabric from the family clothes. When Grandma becomes ill, Tanya decides to finish Grandma's masterpiece with the help of her family. This book can lead to discussions about different shapes used in the quilt pieces.

Friendman, A. *A Cloak for the Dreamer*. New York: Scholastic, 1994.

A tailor has been asked by the archduke to make new clothes for a very important journey that he will take. The tailor asks his sons—Ivan, Alex, and Misha—to help in making the clothes. The tailor is concerned when Misha, the dreamer, uses geometric shapes in his tailoring designs. There are extended activities to use in the classroom.

Grifalconi, A. *The Village of Round and Square Houses*. New York: Little, Brown, and Company 1986.

This is based on a true story of how people in Central Africa live in a village called Tos in Cameroon. The women live in round houses and the men live in square houses. A young girl listens to her grandmother as she tells the story of a volcano erupting and burning everything but one round house and one square house.

Hoban, T. *Spirals, Curves, Fanshapes and Lines*. New York: Greenwillow Books, 1992.

Intriguing shapes—spirals, curves, fanshapes, and lines—are all around. This book helps to heighten children's attention to the geometry in their own world.

Tompert, A. *Grandfather Tang's Story*. New York: Crown, 1990.

A tangram adventure in which the animals change into different animals. Just rearrange the seven tangram pieces to form a rabbit, a dog, a squirrel, a hawk, and a crocodile. Using the tangram pattern (see Appendix A in this text), children can explore different shapes.

Neuschwander, C. *Mummy Math*. New York: Henry Holt, 2005

The Zills family is summoned to Egypt to help find the hidden burial chamber of an ancient pharaoh. When Matt and Bibi get trapped in the pharaoh's pyramid, they use their knowledge of geometric solids to find their way out of the burial chamber.

CHAPTER 16

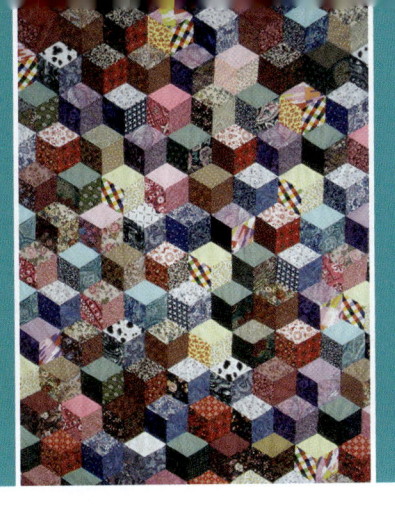

Measurement

> "WE MUST MEASURE WHAT IS MEASURABLE AND MAKE MEASURABLE WHAT CANNOT BE MEASURED."
>
> — Galileo

> SNAPSHOT OF A LESSON

KEY IDEA

1. Help students apply area concepts and skills to a practical situation.

BACKGROUND

This lesson is based on the video *Pencil Box Staining*, from *Teaching Math: A Video Library, K–4*, from Annenberg Media. Produced by WGBH Educational Foundation © 1995 WGBH. All rights reserved. To view the video, go to http://www.learner.org/resources/series32.html (video #27).

Mr. Levy is a teacher of fourth-grade students in Lexington, Massachusetts. In this video, each member of the class has made a wooden pencil box. The task today is to determine how much stain is needed to cover all the boxes.

Mr. Levy: This is an example of our boxes we are making. This one has been stained on the inside and outside. If I go to the hardware store and say, "We need some stain," what is Mr. Cuttingham going to ask?

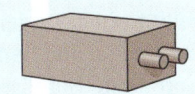

Danny: Did you measure what you are staining?

Mr. Levy: Right, and here is the chart to tell us the amount of stain:

1 pint	About 25 square feet	$6.30
1 quart	About 50 square feet	$8.40
1 gallon	About 200 square feet	$28.60

After introducing the problem, Mr. Levy has the students work in groups of 4. Each student in a group is assigned a specific role (measurer, recorder, calculator, director); together they are to figure out how much stain is needed.

(We have provided a glimpse into what is happening in four of the groups. If you watch the video you will get a more thorough understanding of the groups at work.)

Group 1: We should begin with the big pieces. Do we know if any are the same size? Mr. Levy, do we need to find how much for our box or all the boxes? [*After Mr. Levy asks the student who posed the last question what she thinks, she realizes that the stain was for all the boxes, so they need to find out the total area for 26 boxes. They then begin the task.*]

Group 2: This is $\frac{1}{2}$ inch—OK, that is 0.5—and 0.5 × 9 is 4.5. [*Later, we see this group adding the areas of all the rectangles.*] The total is 162 square inches. Oh no, we need square feet. Do we have to do it again? [*They figure out they can divide 162 by 144. Using calculators, they find this is about 1.1 square feet. They conclude that 26 boxes would be about 28.6 square feet.*]

Group 3: [*This group made a model of 1 square foot and covered the model with the pieces to get an estimate of the area.*]

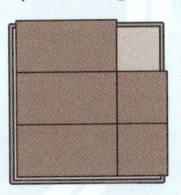

347

Group 4: [*This group made one large rectangle (as nearly as possible) from all the pieces, and found the area of that rectangle and doubled it, since they were staining both sides.*]

After the students had worked, Mr. Levy gathers the class together to report their findings. He records the findings of the groups who had completed the task on the board:

65 sq. ft.	1 pt. 1 qt.	$14.70
29 sq. ft. (28.6 rounded)	1 qt.	$8.40
39 sq. ft.	1 qt.	$8.40
86 sq. ft.	2 qt.	$16.80

Mr. Levy: What in the world am I going to tell Mr. Cuttingham?

The group that answered 39 square feet realized they had figured out stain for only one side, so they doubled their answer to 78 feet, the amount to 2 quarts, and the price to $16.80.

Mr. Levy asked the students how they could resolve this. One student answered that they could do it together the next day. Another suggested he buy 1 quart to stain one side and see if they needed more for staining the next day.

Although this is a practical solution, there were widely diverse findings on the amount of area to be painted. If you were the teacher, what would you do the next day to resolve these differences?

FOCUS QUESTIONS

1. Why should measurement be included in school mathematics?
2. What is the measurement process and why is it important?
3. Why are concepts related to the unit of measurement important to develop?
4. How is measuring one attribute like measuring another attribute?
5. How can you use estimation to strengthen measurement skills?

INTRODUCTION

The Snapshot of a Lesson provides a glance into a fourth-grade classroom learning about measurement. What experiences should the children have before this lesson? How do you help children have these experiences? What concepts and skills are important for the children to learn? We will examine these and other questions in this chapter, but first let's consider why measurement should be included in school mathematics.

Reason 1. Measurement is useful in everyday life. Stop and think about how you have used numbers in the past few days. Did you tell someone how long it took you to drive to school, how many calories are in a piece of chocolate cake, how far it is to the nearest store, or how many cups of coffee you drank? All of these are measurements. The opening quote made in the early 1600s indicates that Galileo saw the importance of measurement. What did he mean by "we must … make measurable what cannot be measured?" (Hint: Think about what Galileo was trying to measure at that time.)

Reason 2. Measurement uses and can be used to develop many mathematics concepts and skills. For example, children may count the number of grams it takes to balance an object on a scale, multiply to find a volume, divide to change minutes to hours, subtract to see how close an estimate was to the actual amount, or add to find the perimeter of a triangle. To report the number of units, children may use whole numbers, common fractions, decimals, or negative numbers.

Many of the numeration models are based on measurement. For example, the number line can be connected to length. An important model for multiplication is the area of a rectangle. Also, there are concepts and procedures that underlie both measurement and number ideas. As shown in Figure 16-1, measuring to the nearest unit is similar to rounding to a given unit showing how measurement ideas may be used to complement numerical ideas.

Reason 3. Measurement is useful in other areas of the curriculum. If you are trying to connect mathematics with other subjects, consider the ways that measurement is used in art, music, science, social studies, and language arts. Much of the practice with measurement skills could be accomplished as you teach these other subjects.

Reason 4. Measurement is an effective way to engage many types of students. Some students whom you may have difficulty reaching through other topics, often see the usefulness of the task as it relates to them personally. Think about the Snapshot of a Lesson at the beginning of this chapter. Were you fortunate enough to have such lively math classes when you were in fourth grade?

Research from international studies has often shown that measurement is a strand on which U.S. students perform less well than their counterparts in other countries. This was also true for fourth-grade students on the TIMSS 2007 assessment, being the only strand on which U.S. students were not above the international average. Through the years, the data from the mathematics assessments of the National Assessment of Educational Progress (NAEP) have given reason for concern about students' performance on measurement items. As you read about these results in this chapter, think about your responsibility to help students develop a better understanding of measurement and better measuring skills.

The Common Core State Standards for Mathematics sets expectations for students to learn measurement concepts and skills and makes a strong case for using measurement to develop other mathematical ideas and to solve both real-world and mathematical problems. Figure 16-2 has a summary of these expectations, but more detail can be found at the Web site referenced in Math Links 16.1.

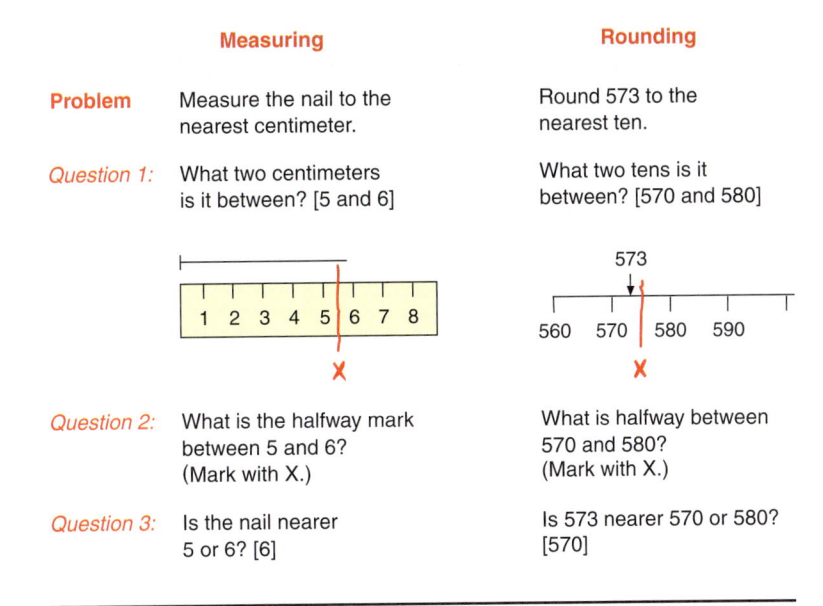

	Measuring	**Rounding**
Problem	Measure the nail to the nearest centimeter.	Round 573 to the nearest ten.
Question 1:	What two centimeters is it between? [5 and 6]	What two tens is it between? [570 and 580]
	(ruler with nail between 5 and 6, X below)	(number line 560 570 580 590 with 573 marked, X below)
Question 2:	What is the halfway mark between 5 and 6? (Mark with X.)	What is halfway between 570 and 580? (Mark with X.)
Question 3:	Is the nail nearer 5 or 6? [6]	Is 573 nearer 570 or 580? [570]

Figure 16-1 *The similarity between measuring and rounding.*

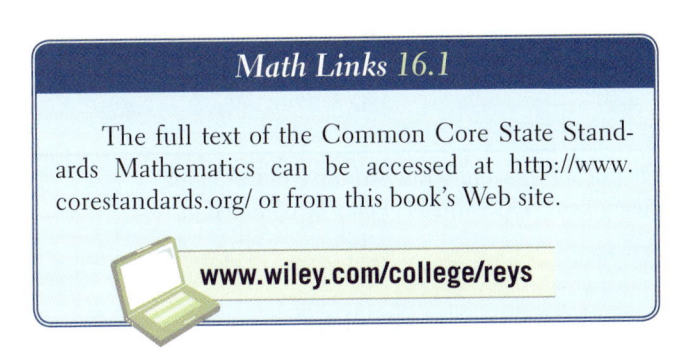

Math Links 16.1

The full text of the Common Core State Standards Mathematics can be accessed at http://www.corestandards.org/ or from this book's Web site.

www.wiley.com/college/reys

THE MEASUREMENT PROCESS

Much of the research about how children measure and think about measurement has focused on children's conceptual development. However, there have also been classroom studies (Lehrer, 2003) that shed light on what students can do when given the opportunity. Wilson and Osborne (1988, p. 109) gave the following recommendations that are as appropriate for today as they were then:

- Children must measure frequently and often, preferably on real problems rather than on textbook exercises.
- Children must develop estimation skills with measurement in order to develop common referents and as an early application of number sense.
- Children should encounter activity-oriented measurement situations by doing and experimenting rather than by passively observing. The activities should encourage discussion to stimulate the refinement and testing of ideas and concepts. See Math Links 16.2 for activities that encourage this.
- Instructional planning should emphasize the important ideas of measurement that transfer or work across measurement systems.

Measurement is a process by which a number is assigned to an attribute of an object or event. Length, capacity, weight/mass, area, volume, time, and temperature are the measurable attributes considered in most elementary mathematics programs. Although each of these attributes is different, the process of measuring is common to all of these attributes. The following outline, based on the measuring process, can be used to plan instruction:

I. Identify the attribute by comparing objects
   A. Perceptually
   B. Directly
   C. Indirectly through a reference
II. Choose a unit
   A. Nonstandard
   B. Standard
III. Compare the object to the unit by iterating the unit
IV. Find the number of units by
   A. Counting
   B. Using instruments
   C. Using formulas
V. Report the number of units

	Measurement and Data Domain (Measurement only)
Kindergarten	Describe and compare measurable attributes such as length and weight.
Grade 1	Order three objects on length, compare lengths indirectly and measure length by iterating units.
Grade 2	Measure length by selecting and using appropriate tools such as rulers, yardsticks, meter sticks, and measuring tapes. Estimate lengths using units of inches, feet, centimeters, and meters. Use different units to measure the same length to examine the role of the unit.
Grade 3	Measure lengths using rulers marked with halves and fourths of an inch. Develop the concept of area and a unit square. Measure areas with standard square units (both metric and non-metric). Solve problems involving perimeters and investigate rectangles with the same perimeter and different areas or with the same area and different perimeters. Use area to investigate fractional parts.
Grade 4	Know relative sizes of measurement units within one system of units including km, m, cm; kg, g; lb., oz.; l, ml; hr., min., sec. Express measurements of a larger unit in terms of a smaller unit. Solve problems involving distances, intervals of time, liquid volumes, mass, and money. Apply the area and perimeter formulas for rectangles.
Grade 5	Convert like measurement units within a system. Recognize volume as an attribute of solid figures and understand volume concepts. Measure volumes by counting unit cubes, using cubic cm, cubic in., cubic ft., and non-standard units. Solve real-world and mathematical problems involving volume. Focus on right rectangular prisms and related formulas.
Grade 6	Find the area of triangles, special quadrilaterals, and polygons by composing into rectangles or decomposing into triangles and other shapes. Find the volume of a right rectangular prism with fractional edges. Find the length of vertical and horizontal line segments on a coordinate grid. Find the surface area of three-dimensional figures by using their nets. Apply all techniques to solving problems.

Figure 16-2 *Excerpt of CCSSM standards relating to measurement concepts and skills. This excerpt was made by the authors of Helping Children Learn Mathematics. The full document (CCSSI, 2010) may be obtained at www.corestandards.org/thestandards/mathematics.*

If a new attribute is being introduced, students should cycle through the measurement process. Focus first on comparison, and then use nonstandard units and counting to report the number of units. The next time through, use standard units and counting, and finally introduce instruments or formulas. This cycling may take place over several years for the first attributes studied, but after several attributes have been introduced, the time spent on the cycle can and should be shortened.

Math Links 16.2

A variety of Web resources for measurement useful for teachers, parents, and children, may be found at the NCTM's Illuminations Web site. Check the activities, lesson plans, and connections to other sites at http://illuminations.nctm.org/Default.aspx or from this book's Web site.

www.wiley.com/college/reys

I. IDENTIFYING ATTRIBUTES BY COMPARING

To measure with understanding, children should know what attribute they are measuring. Consider the attribute of attitude which we often measure by attitude scales. These scales are difficult to interpret because we really do not understand the underlying concept of attitude. For young children, measuring the area of an object can also be difficult if they do not understand the concept of area. Older children, who have difficulty measuring an angle, often do not have a solid understanding of an angle. Thus, one of your first tasks is to build an understanding of measurable attributes.

Three types of comparisons can build understanding of attributes: comparing two objects perceptually (they look the same or they look different), comparing two objects directly (they are placed next to each other), and comparing two objects indirectly (a third object is used to compare objects). As children make these types of comparisons, not only are they gaining an understanding of the particular attribute and the associated vocabulary, but they are also learning procedures that will help them in assigning a number to a measurement. We consider each attribute commonly found in elementary mathematics programs because of the importance of having children participate in such comparison experiences for each attribute. As you read, see if you can tell the differences among the three types of comparison and how each can be used to develop an understanding of that attribute.

LENGTH   Length is one of the most easily perceived attributes of objects. Very young children have some concept of length and some vocabulary associated with it; however, they often have what adults may consider misconceptions about length. For example, they may say that a belt is shorter when it is curled up than when it is straight. These misconceptions disappear as children develop cognitively and are involved in constructive experiences. In the Classroom 16-1, 16-2, and 16-3 illustrate comparison activities that are appropriate for early childhood from preschool through grades 1 and 2.

In the Classroom 16-1 involves comparisons made perceptually and is designed to be done with very young children. Many children in preschool have no difficulty with this task, yet some will have difficulty when the objects are not so perceptually different. In the beginning of this activity, all irrelevant perceptual attributes have been masked (i.e., the objects are the same except for length). This allows children to build the concept of length as an attribute of long, thin things. Note that you do not ask which is longer, but allow the children to tell you what they observe and know. In the second part of this activity, objects that differ on many attributes (color, function, etc.) are used, but the focus remains on length. You can assess how your students are answering these questions. Do they focus on length? Do they know the vocabulary associated with length? You may need to gear down for some students who have not had the opportunity to learn the vocabulary (especially *shorter*) by asking more questions and letting students learn from each other. You can gear up by asking questions about a set of three to five objects: Which object is the longest, which object is the shortest, or what object is shorter than the longest object?

In the Classroom 16-2 is designed to be used at learning centers after introducing the procedure of placing two objects side by side on a common baseline (comparing directly). Present a problem of comparing two objects that are not perceptually different in length. Ask children which object is longer and how they can justify their answer. If no one suggests putting the two objects side by side on a baseline, do so. This activity may be extended to seriating objects by length (i.e., arranging them from shortest to tallest). Seriating could be another center for students who need. For the young child, this task is more difficulwt because multiple comparisons must be made. When children are using rulers, they are directly comparing two lengths—the object to the ruler.

In the Classroom 16-3 presents the problem of indirectly comparing two objects when they cannot be placed side by side. Children must represent the lengths to help them

---

In the Classroom 16–1

PERCEPTUAL COMPARISON OF LENGTHS

Objective: To compare lengths of physical objects perceptually.

Grade Level: K–1.

Materials: Collections of long, thin objects such as rods, spaghetti, pencils, and crayons.

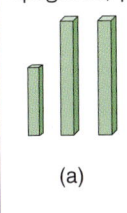

(a)
▼ Hold up two long objects and one short object.
- Ask children to tell which is different.
- Use vocabulary of *shorter, taller,* and *longer.*
- Repeat with other objects that differ only on the length and are obviously different in length.

(b)
▼ Hold up a long pencil and a short crayon.
- Ask how they are different. Expect answers such as color, type of object, paper wrapper, eraser.
- If no one says the crayon is shorter or the pencil is longer, ask "Which is longer?"

---

make the comparison. Notice that in launching this activity, the students are presented with a problem to solve. For example, which is longer—the height of the teacher's desk or the width of the door? Children can represent the height of the desk and the width of the desk with string or connecting link chains. They can then directly compare the strings or link chain to determine which object is longer. Later, a ruler will be used to represent length, and the objects will be compared indirectly using numbers.

Through activities such as these, children begin to develop an understanding of length as an attribute of long, thin objects; however, length is used in other ways. For example, length is the distance around your waist. Young children can use string to compare their waists or to compare their wrists with their ankles. In the Classroom 16-4 is an activity appropriate for older children in which they estimate the distance around a can and the height of a can, and then check their conjecture by comparing the lengths with a representation (a piece of string).

The *distance between* two points is also measured by length. Distance is often more difficult to perceive than the length of a long, thin object, because you have to imagine

In the Classroom 16–2

DIRECT COMPARISON OF LENGTH

Objective: To compare lengths of physical objects directly.

Grade Level: K–1.

Materials: A box of long, thin objects and three large sheets of paper (labeled shorter, same, longer).

Launch:

Present the class with a problem by holding up two pencils that look like they are the same length. Ask the students which is longer? You should have some disagreement. Have a child compare the lengths by putting the pencils upright on a table.

After children have had the opportunity to compare objects directly (their height with a partner, the length of two shoes, etc.) using a baseline, set up the following center.

Center:

Place the box of objects at a center and choose one object to be the reference. Place it on the sheet labeled "Same". Have the students sort the other objects as shown below.

Students can then change the reference object and re-sort.

Gear Down:

Work with individual students to help them with the words and with the idea of a baseline. Have them verify their sort by showing you why they placed an object in that category. First ask about an object they placed correctly.

Gear Up:

Challenge the students to pick a reference object so one category will have no objects. See if they can generalize that if they choose the shortest (longest) object, there will be no objects in the shorter (longer) category.

In the Classroom 16–3

INDIRECT COMPARISON OF LENGTH

Objective: To indirectly compare two objects on length.

Grade Level: 2–3.

Materials: String or connecting links, index cards, tape, and objects in the room selected for children to represent.

Description: Choose two lengths that are about the same on objects that cannot easily be moved to compare directly. For example, the height of your desk with the width of your desk. Ask the children which is longer, and let them pose a solution of how they can tell.

- Discuss the ways they propose and then show them how you would use the links (or string) to represent each of the lengths and then compare.
- After this discussion and demonstration, give each pair of children a string and have them represent one of the objects you have marked.
- Make a "graph" of all the lengths for easy comparison.

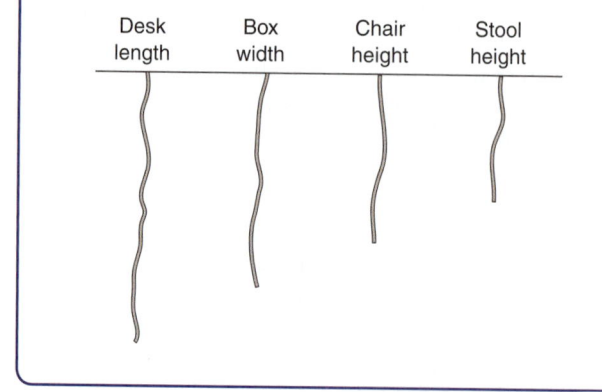

---

the straight path between the two endpoints. Words such as *nearer* and *farther* may be used when comparing two distances. There is also other vocabulary associated with length, such as *higher* and *lower* as well as the superlatives of all these words.

Perimeter—the distance around a region—is a special type of length. Children should be given the opportunity to measure the distance around a region with string or a measuring tape if they are familiar with a ruler. Later, they can add the lengths of the various sides of the region to find the perimeter. There are many everyday examples of problems that require finding the perimeter, such as finding the distance around a bulletin board to determine how much border is needed or how far it is around a city block or around a track on a field. *Circumference* is a special word for the distance around a circle.

In the Classroom 16–4

COMPARING HEIGHT AND CIRCUMFERENCE

Objective: To compare the distance around an object with its height.

Grade Level: 3–4.

Materials: Cans of various sizes (at least one for each pair of students) and string. Label the cans and make a recording sheet as suggested below. (If you cannot find cans, roll construction paper to represent cans of various sizes.)

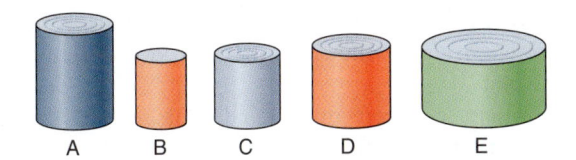

Launch: Ask students, showing them a can, which they think is longer: the distance around the can or the height of the can. Students are often surprised about the comparison. Have them verify their conjecture by using string to mark each of the distances.

Activity: Have pairs of children work together. They should first decide which is longer (the distance around or the height) by looking and then verify their guess by measuring each with a piece of string. Have them record their guess and verification on a chart like the following:

Can	My Eyes Said	The String Showed
A	height	distance around
B		
C		
D		

Discussion: Which cans were obvious? Which cans fooled your eyes?

A more detailed plan for this activity is available on the Student Companion site at www.wiley.com/college/reys.

CAPACITY Capacity is considered the attribute that tells "how much can a three-dimensional container hold." Although perceptual comparisons can be made between two containers, young children often make the comparisons based on length (height) rather than on capacity. When asked which holds more—a tall container or a short container—most children will choose the taller container even if the shorter one may actually hold more. Thus it is probably best to begin the study of capacity by using direct comparisons.

Some type of filler is needed to make direct comparisons. Water and sand are perhaps messy but easy for even preschool children to use. Given a variety of containers, they can fill one and pour it into the other to see which holds more. After children have experimented with direct comparison (filling one and pouring it into another container), activities involving perceptual comparisons are possible. For example, children greatly enjoy guessing contests in which they guess which container holds more, and then you can check the results together. Figure 16-3 displays a possible comparison.

Indirect comparisons are used when two containers cannot be compared perceptually or directly. For example, suppose you have two containers with small openings that make it difficult to pour from one into another. By pouring the filler in each into a pair of identical large-mouth containers, the capacities can be compared. Note that this activity is similar to what you do when you use graduated cylinders in a science lab to identify amounts of liquid.

WEIGHT/MASS Weight and mass are different: Mass is the amount of a substance, and weight is the pull of gravity on that substance. On Earth, the weight and mass of an object are about the same. In contrast, the weight of on object on the moon is about one-sixth of its mass. For example, if you weigh 120 pounds on Earth your weight on the moon would be about 20 pounds. Your mass would be the same on both the Earth and Moon.

In everyday usage, we often refer to the mass of an object as its weight. You need to know there is a difference between the two attributes, but this difference does not need to be stressed with young children. When we live on the moon, we will here to adjust our thinking! The same comparison words are used with children for both attributes.

The weight/mass of two objects are compared perceptually by feel by lifting the two objects. Children should be given a variety of pairs of objects (one of which is much heavier than the other) and asked to hold one in each hand.

Figure 16-3 *Which container holds more seeds? Tell why you think the one you chose holds more.*

Children often think that a larger object weighs more. These children need opportunities to compare small and heavy with large and light objects. Children should learn that to find which is heavier, they must do more than look at the object. An easy way to provide this experience is by having children compare two identical containers with lids (such as cottage cheese containers) but filled with different amounts of objects. For example, you could put 3 washers in one container and 14 washers in the other.

If you cannot feel the difference between the weights/masses of two objects, you need a balance scale to assist in the comparison. To introduce the balance, choose two objects that differ greatly in weight so children can see that the heavier object "goes down" on the balance. Figure 16-4 shows that even though one rock is larger, it is not the heavier of the two rocks.

Many activities may be set up for children to compare the weights of two objects. One challenging activity is to compare five identical containers, with lids secured and filled with different amounts of objects, and to put them in order from lightest to heaviest. If the masses are not perceptually different, then this task requires multiple comparisons on the balance.

Indirect comparisons are not necessary until units of weight are introduced, because whenever each of two objects could be compared to a third on the balance, it would be much simpler to compare the two objects directly.

AREA Area is an attribute of plane regions that can be compared by sight (perceptually) if the differences are large enough and the shapes similar enough. You can modify the length activity (In the Classroom 16-1) for area. Begin with three sheets of construction paper, two of which are the same size and one that is larger. Ask the children which is different and to describe the difference. They will say that one is "bigger." Begin to build the vocabulary: this region is "larger in area."

The first direct comparisons should be made with two regions, one of which fits within the other. This is simple for students, so begin to challenge them to compare regions like A, B, and C. If the regions can be cut out, it is fairly easy to compare regions B and C by placing one on top of the other. It is more difficult to compare either B or C with A. When children have some idea of *conservation of area*—that a region can be cut and rearranged without changing the area—this experience can be meaningful. However, you will be surprised at the solutions that young children will propose; they are often more perceptive than older students, who often rely on formulas.

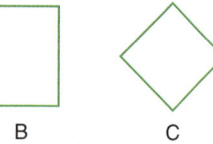

A   B   C

If objects cannot be moved to place one on top of the other, children can trace the objects and use these representations to make an indirect comparison. Choose two objects in the room that are about the same area and cannot be moved. For example, the electric switch cover and the door handle plate.

To help children understand that regions can be rearranged without affecting the area, many experiences with geometry activities are helpful. Notice that in In the Classroom 16-5 children are asked to discuss the area of the shapes; this type of discussion helps focus on the objective of comparing shapes indirectly, using the right triangles as a representation to help make the comparisons. Although this activity was designed for younger children, it is appropriate for children in upper elementary who have not had such experiences. The CCSSM recommend area as a focus in

Figure 16-4 *Use a balance to show which rock weighs more.*

In the Classroom 16–5

ARE WE THE SAME SIZE?

Objective: To indirectly compare areas.

Grade Level: 2–3.

Materials: Cut out the four right, isosceles triangles below.

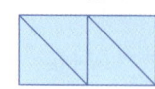

Are We the Same Size?

1. Draw a line to show how you placed two triangles to make each of the shapes A, B, and C.

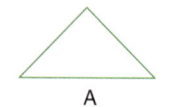

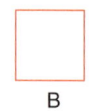

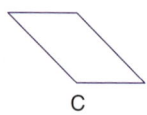

   A   B   C

2. Which of shapes A, B, or C looks the smallest to you? Explain to your classmates why each shape actually is the same size in terms of area. What is different?

3. What shapes can you make from three of the triangles? Draw a sketch of each.

4. What is the area (use the triangle as the unit) of each shape you made? What would be the area if the unit was the square shown in B above?

grade 3. However, it is important to help children understand the attribute of area through direct and indirect comparisons prior to the more formal work at this level.

VOLUME Volume is defined as "how much space a three-dimensional object takes up," so it is difficult to make anything but perceptual comparisons before units are introduced. Thus, volume should receive little formal attention until fifth grade. However, because of the close connection between volume and capacity, some background can be provided for young children if containers are filled with non-liquids, such as blocks, balls, or other objects. Figure 16-5 uses blocks to fill a box, allowing for the units counted to make volume comparisons with other boxes. Challenge your students to see if they can figure out how many cubes it would take without filling the entire box. Experiences like this are essential to help children make sense of volume formulas. The CCSSM document recommends that volume be a focus of grade 5, especially for rectangular solids.

Figure 16-5 *How many blocks will fit?.*

ANGLE Students are not often given the opportunity to compare angles before they are taught to measure them with a protractor. If an angle is considered a turning (such as the clock hands), then even young children can compare perceptually two angles (the amount of turning). Young children can also compare angles directly by comparing the amount of space the turn would make. Look at the direct comparison of these two angles:

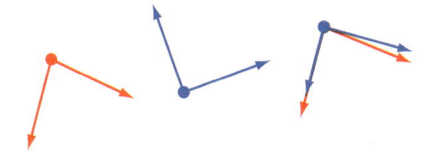

Students can move to comparing the angles indirectly, tracing one and then comparing the tracing with the other angle. Later, they will use an instrument such as a protractor to make these comparisons.

TEMPERATURE You can certainly sense (perceptually) great differences in temperatures. Before introducing reading the thermometer, you can have children compare to see which of two objects is colder (or warmer). You can also talk about things (or times) that are hot or cold; however, there are few other comparisons you can make without an instrument (thermometer).

TIME Time is a very abstract attribute; like temperature, it is not an attribute of objects that can be seen. There are two attributes of events (rather than of objects) that can be measured: *time of occurrence* and *length of duration*. You can begin describing the time of occurrence: We went to the gym yesterday, we have rug time in the morning, or we collected fall leaves in October. Young children need to develop the vocabulary of days, months, and seasons of the year. They also need to hear occurrences described in hours and minutes: school begins at 9 o'clock, you eat lunch at 12 o'clock, or you have 30 minutes to play. A lot of this learning takes place naturally if children are made aware of time. Look for opportunities in all your teaching to use this type of vocabulary.

Children can tell which of two events takes longer (duration) if their lengths are greatly different. Does it take longer to brush your teeth or read a story? If the events are similar in duration, children can tell which lasts longer if they both begin at the same time (note the similarity to deciding which object is longer when both are placed on a baseline). You can think of many contests that use this idea: who won the race, whose paper plane flew longer, whose eyes were shut longer, or who hopped for a longer time.

OTHER ATTRIBUTES There are many other measurable attributes appropriate for upper elementary students; for example, speed and density. One way to look at speed is found in the activity referenced in Math Link 16.3. You will find other interactive sites or science experiments in which students investigate speed and density.

Math Links 16.3

A three-part electronic lesson plan for students in grades 3–5, Competing Coasters, may be viewed at the NCTM Web site. In these lessons, children use the Internet to learn about roller coasters from around the country. They predict which one is faster, higher, and takes longer. Then they compare their estimates with actual data. You can access it at http://illuminations.nctm.org/LessonDetail.aspx?id=L241 or from this book's Web site.

www.wiley.com/college/reys

II. CHOOSING A UNIT

After children have begun to develop a firm concept of an attribute through comparison activities, it is important to help them move through the rest of the measurement process (II–V). To answer the question "How long is the pencil?" you can say "It's longer than my thumb" or "It's shorter than my arm." These are relative statements that give a range of possibilities for length but do not do an accurate job of describing it. To be accurate, you need to compare the pencil to a unit. You can use a nonstandard unit such as a paper clip and report that the pencil is 7 paper clips long, or you can use a standard unit and report that it is 16 centimeters long. First you must choose a unit. In so doing, you consider all the concepts listed below.

These concepts develop over time. A single activity as described for each concept will not suffice; you need to include similar activities. Look for opportunities within any measurement activity, whether it is with nonstandard, customary, or metric units, to help children develop these nine concepts. Children will learn these concepts as they proceed through the measurement process.

1. *A unit must remain constant.*
   Young children (Clements and Sarama, 2007) do not understand that units must be of equal size. They may measure a length using pencils of different sizes and report the measurement as 6 pencils or use different shapes to cover an area and report that the area is 8. They also will compare measurements made with different units and see no conflict, as discussed previously. They still look at measuring as counting only.

2. *A measurement must include both a number and the unit.*
   How many times do you remember a teacher telling you to be certain to write inches? Probably because you were only measuring in inches, you did not see why this was necessary. When children measure the same object with many different nonstandard units, they are more likely to see the need to report the unit. Having children measure the length of a book with paper strips, erasers, or cubes, or weigh an object with washers, pennies, or paper clips, is the type of task that will encourage younger children to write (or draw) the unit. In the Classroom 16-6 provides an activity that encourages children to focus on the importance of reporting the unit.

3. *Two measurements may be easily compared if the same unit is used.*
   This may be an obvious concept to you, but comparing the results of measuring with different units will shed light on children's understandings. Young children often rely only on numbers to make comparisons. For example, if one pencil is 6 blue strips long and another pencil is 2 red strips long, some children will say that the pencil that measures 6 is longer. They may not yet have reached the stage where they can coordinate the number with a unit. Students with special needs will need more experiences of measuring the same object with different units to realize that different numbers may occur if different units are used.

In the Classroom 16–6

MEASURING LENGTH WITH ARBITRARY UNITS

Objective: To measure lengths with different arbitrary units; to compare those measurements.

Grade Level: 2–3.

Materials: Erasers, paper clips, books, boxes, and other objects that can be used as arbitrary units of measure

Description:
- Give each pair of children an arbitrary unit and the name of what they are to measure.
- Have the children record their measurements at a place everyone can see. Observe to see whether they record the unit (if they do not, it will lead to the discussion below).
- Ask children to use the information about the lengths to compare the lengths of the different objects.
- Ask children why the unit is necessary. Ask the children which comparisons are easy to make (ones measured with the same unit) and what they can tell about those made with different units.

If students realize that it is easy to compare sizes if the same unit is used, they will understand the need to convert to the same unit when comparing measurements such as 3.4 meters with 1459 millimeters or when comparing $\frac{4}{5}$ with $\frac{2}{3}$.

4. *One unit may be more appropriate than another to measure an object.*
   The size of the unit chosen depends on the size of the object and on the degree of accuracy desired. If children are allowed to choose the unit rather than always being told what unit to use, the idea of the unit's size depending on the size of the object develops naturally. To encourage this concept, ask questions such as "Should you use the edge of a book or a paper clip to find the height of the door? Should you use a thimble or a coffee mug to find out how much water the big bucket will hold? Should you use a paper clip or a pencil to find out the length of the book?"

5. *There is an inverse relationship between the number of units and the size of the unit.*
   When measuring the same object with three different units, children realize that the larger the unit, the fewer are required. For example, you could ask each child to weigh an object with pennies, washers, and cubes. If each child makes a graph of the results, a pattern becomes apparent when all the graphs are compared. As depicted in Figure 16-6, it always takes more cubes than washers to weigh each object. So these cubes are lighter than these washers. When rulers are introduced, you can use rulers of different units (Clarkson et al., 2007) to reinforce this and the other ideas. The article cited gives more detail about activities and how to access such rulers.

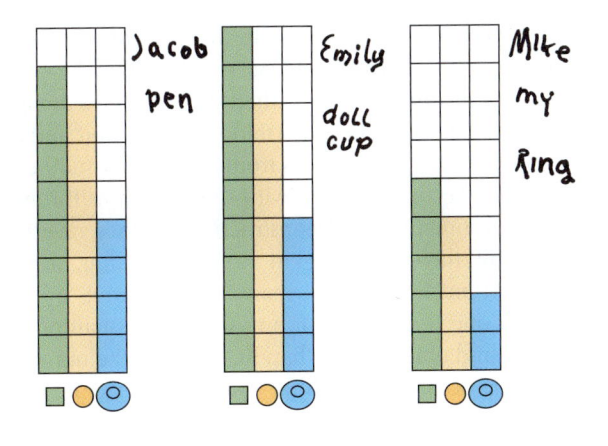

Figure 16-6 *Children's graphs of weights of objects.*

6. *Standard units are needed to communicate effectively.*
Many concepts about units can be developed with nonstandard units. At the same time, you will be teaching procedures of measuring with units (e.g., how to line up units, use a balance, cover a region, keep track of how many units). However, at some point children need to learn about standard units. Standard units are either customary (e.g., inch, pint, pound) or metric (e.g., meter, liter, gram). Unfortunately, in the United States both types of units are used. It is clear from recent national assessments that children have a better understanding of customary units (Blume, Galindo, and Walcott, 2007).

Stories and activities that demonstrate the difficulty in communicating sizes when there is no standard of measurement are one way to present the necessity of a standard unit. The classic story retold in *How Big Is a Foot?* (see Myller in the Book Nook for Children) is an interesting and amusing source to use in helping children see the necessity for a standard unit. Another enjoyable activity for children is making a recipe of powdered drink such as lemonade, using a very large cup for the water and a very small spoon for the powder. (Be sure to have enough of the powder so the children can make it tasty.)

7. *A smaller unit gives a more exact measurement.*
All measurements are approximate. If children do a lot of measuring of real objects, they are reporting approximate measurements, but perhaps without being fully aware of it. A practice of saying "about 6 inches," "more than 6 inches," or "between 6 and 7 inches" helps develop this idea.

To set up the need for a more precise measurement, give one child a strip of paper that is 28 centimeters long and another child, sitting across the room, a strip that is 29 centimeters long. Give each child one-decimeter (10 centimeters) strip with which to measure the paper. After they report the measurement, ask the class which strip is longer (no fair comparing the strips directly). Next, have them measure with centimeters. Discuss with the class why a smaller unit was needed.

8. *Units may be combined or subdivided to make other units.*
When students are given the experience of making their own instruments, they often combine units (say, inches) to make a larger unit (feet). Seldom are they given the opportunity to subdivide or partition units to make smaller units. This can be done by folding to make fractional parts of a length unit or by using fractional parts of square units when covering areas.

9. *Units must match the attribute that is being measured.*
Length is measured with a length unit, area with a square (or other shape that will cover the region) unit, and volume with a cubic unit that "fills" the same space. That is, we compare the object to be measured to a unit of the same type. Children often say the area is so many inches because they have only learned to use the length of the sides to calculate area. They have missed the concept of square units in the rush to formulas.

III. AND IV. COMPARING AN OBJECT TO A UNIT AND FINDING THE NUMBER OF UNITS

In the section on identifying attributes, we considered ways that children could compare two objects on a specified attribute. Measuring with units is very similar to this. Instead of comparing two objects, you compare an object with a unit and find how many units would be equal to that object. In all attributes, you must iterate the unit to make the comparison. For attributes such as length, weight, area, capacity, and volume, you can take copies of the physical unit and "fill the space." In measuring area, for example, you can compare a region to a square unit of area and find how many of these square units cover the region.

There are three techniques that are used to find the number of the units. The first is counting units; this may be done by merely counting the units or by using addition or multiplication to assist in that counting. The second is using an instrument such as a ruler that tells the number of units. Lastly, there are formulas that produce the number of units.

COUNTING UNITS Length is the first attribute that most children measure. The beginning activities should be with multiple copies of a nonstandard unit that are put end to end long enough to represent the object they are measuring. From previous experiences with direct comparisons of two objects, children should know when the lengths are the same. Some children do not understand that they need to line up the units in a straight line with no gaps and no overlaps. Do not expect too much precision with young children, but keep encouraging this skill. Give your students problems to solve: "How long is your desk in strips?" Observe to see how they are lining up the strips. Students with special needs and all students enjoy measuring with units that connect, such as connecting cubes or links. These are good units because they are easy to handle and line up without gaps or overlaps.

Later, children can take one unit and move it along the object (an *iteration*). This more advanced skill is needed for proper use of a ruler when measuring objects longer than the ruler. It is a skill that should not be pushed too early. If you find that children cannot move the unit and mark the end, then postpone this skill until later.

If children have been measuring with nonstandard units, then using standard units should be easy. They will have a good understanding of the process of measuring, so the purpose should be to give them a feel for the standard unit. We have illustrated this with the decimeter because it is the easiest unit of length for young children to handle.

Give each child a paper strip that is 10 centimeters (1 decimeter) by 2 centimeters. Strips can easily be cut from a 4-by-6 index card, or you could use 10-rods from a Cuisenaire set. Here are some simple ways to have children compare, estimate, or construct to develop a mental image of a decimeter:

- *Decimeter list-up.* (Comparing to a decimeter.) Pose the problem for children to find something that is the same length as a decimeter. Put a list on the board of objects children find. Also have each child find a decimeter length on his or her hand. Perhaps it is from the tip of the thumb to the wrist. This is a handy reference.

- *Decimeter hold-up.* (Estimating a decimeter length.) Pair the children and have one child try to hold two forefingers one decimeter apart—vertically, horizontally, and obliquely. The partner should check each time; then roles should be reversed. Or you could have them try to write their name or draw a picture that is one decimeter long.

- *Decimeter stack-up.* (Constructing a decimeter length.) Set up centers with pennies, chips, clips, beans, cubes, and the like. At one center, children stack the chips 1 decimeter high. At another center they make a line of chips 1 decimeter long.

After children are somewhat familiar with the length of a decimeter, they can begin to measure with decimeters. A good way to begin is by asking them to estimate the lengths of objects and then measure. As children are measuring with one-decimeter strips, they (and you) should notice that putting down strip after strip is not the easiest way to measure. Have them tape their strips together, end to end, alternating colors.

Now they have a "decimeter ruler," except they still have to count the units. After they have done some counting, see if anyone suggests numbering the strips. This activity helps children understand how rulers are made and that they are counting units. Most children will place the number in the middle of the unit. Where is it on a ruler?

| 1 | 2 | 3 | 4 | 5 | 6 | 7 | 8 | 9 | 10 | 11 | 12 |

Once children are familiar with the decimeter measurement of small objects, the next stage is measuring something very long. Have each child make a 10-decimeter ruler. Tell the children that this unit is called a *meter* and is used to measure longer distances. Activities similar to those suggested with decimeters will help them become familiar with meters. New units should be related either by combining or subdividing the units that children have already used. This will help them understand the new unit and assist them in making conversions from one unit to another.

Children become familiar with standard units through comparing, estimating, and constructing. As they measure with that unit, they will also gain an understanding of it. It is also important that not too many standard units are introduced at one time, that the unit is not too small or too large for a child of that age to handle, and that the numbers generated are not too large. Table 16-1 is a guide to the most common standard units used in elementary school and approximate grade levels at which it is appropriate to introduce them. Although we have only concentrated on length and only on a few units, these ideas can be generalized to many of the other attributes. Think about how you could introduce a square inch, a liter, a pound, or a square centimeter.

> ### Virtual Classroom Observation
>
> Go to www.wiley.com/college/reys,
> Access the Wiley Resource Kit
> Click on the Virtual Classroom Observations Section
> Module 5: Geometry: Calculating the Area of a Triangle.
>
> 1. Watch the video: Teaching Examples #2: How Do We Express Area.
> 2. Think and discuss: Find your own examples of the square units. What else csould you do to help students think about the size of these units?

USING INSTRUMENTS Instruments are used to measure some attributes. In elementary school, the more common instruments are rulers, scales, graduated containers, thermometers, protractors, and clocks. Other attributes (such as area and volume) are assigned a measurement by the use of a formula after an instrument has been used to measure some dimensions. Later, other attributes are derived from measurements of more than one attribute (e.g., speed is derived from distance and time).

Much of the emphasis in the elementary curriculum is on instruments and formulas, and some children encounter difficulty with both. One probable source of difficulty is that the children do not understand what they are measuring and what it means to measure. The activities and suggestions presented so far in this chapter have dealt with building this understanding. Here we will look at some common problems students with special needs have with particular instruments and some ways to assist in developing the appropriate skills.

Ruler A ruler automatically counts the number of units, but children must realize what unit they are using and line up the ruler properly. Children focus on the unit being used if the scale on the ruler has only that unit. For example, if the unit is centimeters, choose a ruler marked only in centimeters, not in centimeters and millimeters. Or, if the unit is inches, choose a ruler without the markings of fourths or eighths.

TABLE 16-1 • Standard Units for Elementary Students (with Approximate Grade Level)

Attribute	Metric Units	Customary Units
Length	Decimeter (1–2) Centimeter (2–3) Meter (2–3) Millimeter (3–4) Kilometer (4–5)	Inch (1–2) Foot (2–3) Yard (2–3) Mile (4–5)
Weight	Kilogram (2–3) Gram (4–5)	Ounce (2–3) Pound (2–3)
Capacity	Liter (1–2) Milliliter (4–5)	Quart (2–3) Cup (1–2) Gallon (2–3)
Area	Square centimeter (3–5) Square meter (4–6)	Square inch (3–5) Square foot (4–5) Square yard (4–5)
Volume	Cubic centimeter (5–6) Cubic meter (5–6)	Cubic inch (5–6) Cubic foot (5–6) Cubic yard (5–6)
Temperature	Celsius degree (2–3)	Fahrenheit degree (2–3)
Time		Hour (1–2) Minute (1–2) Second (3–4) Day (K–1) Week (1–2) Month (K–1)

It is important that children measure real objects with the ruler. Make certain that you include activities in which children measure objects longer than one ruler. Can they move the ruler (iterate), and do they have the addition skills to add the units? For example, suppose the children have a 25-centimeter ruler and they are going to measure something that is 43 centimeters long. Can they add 25 and 18? Of course, they could use counting or techniques that rely on their place value and counting background: 25, 35, 36, 37, …, 43. Let them try; they will surprise you if they are given a problem to solve. Your role is to ask questions to assist them.

Older children may have difficulty in measuring to the nearest fourth, eighth, or sixteenth of a unit. One cause of this difficulty may be the smallness of the unit (there is more room for error), but more probably it is their lack of confidence and understanding of fractions, their lack of understanding of the unit (fourths), and their lack of consciousness of how to measure to the nearest unit. In measuring to a unit, they tend to decide by simply looking at the nearest larger unit. The example previously given in Figure 16-1 shows how to make this process more explicit. The rulers in the Appendix marked in halves, fourths, and eighths provide a good intermediate step before moving to a standard ruler marked in sixteenths. It is also helpful to emphasize the units by asking, "What two units is the nail between—two-fourths and three-fourths or three-fourths and four-fourths?" Children must be able to answer this question before deciding which it is nearer. A firm foundation of fractions and fractional parts of lengths helps with measuring to the nearest unit.

Figure 16-7 illustrates an item for using a ruler similar to a question that has appeared in various versions of the NAEP. What do you think was the most common answer of fourth-grade students? Most said the segment was 5 inches long.

Children begin counting with 1, so it is natural that they often begin measuring at 1. Activities like making a ruler and seeing that the 1 is placed at the end of the first unit are helpful. If children are having difficulty, return to directly comparing the object with the ruler (use a baseline as described in the section on direct comparison). You also could use separate units or just mark the inches on the segment and count the units.

Although we have emphasized difficulties that some children have in learning to use a ruler, do not be discouraged. Children love to use rulers, and overall they do quite well with them. Make certain you have children not only measuring objects but also constructing line segments or objects of a given length. Research shows that children have more difficulty drawing a line segment of a given length than in measuring. Part of this difficulty may be the lack of experience in producing a line segment.

Scaled Instruments Instruments such as bathroom scales, graduated cylinders, and thermometers are easy to read if they are digital or if each unit is marked. They cause children some trouble when each unit is not marked. On national assessments, students have always had difficulty reading thermometers on which the markings represent 2 degrees. On the 2003 NAEP assessment, only 40% of fourth-grade students correctly reported that such a thermometer showed 84°. The same percentage said it showed 82°, reading each marking as one degree. This has been a long-term difficulty for students. As number lines become more prevalent in elementary schools, it will be helpful if some are marked in units other than units of one.

One way to help children become more aware of the markings on a scale is to have them make their own instruments. They can make graphs, for example, using different scales, or they can mark their own graduated cylinders. In the Classroom 16-7 gives instructions on how to make the "cylinder." Children can make such a measurement instrument and then use it to measure the amounts other containers hold. Note that this activity is also good practice for finding multiples of a number; in this case, children are working with multiples of 6, since it took 6 spoons to fill the small container.

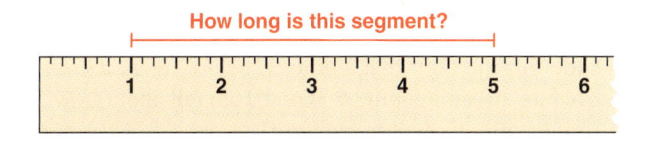

Figure 16-7 *Assessing understanding of a ruler.*

360 Chapter 16 • Measurement

> **In the Classroom 16–7**
>
> **MAKE YOUR OWN**
>
> *Objective:* Make a measuring container.
>
> *Grade Level:* 3–5.
>
> *Materials:* Large glass, Spoon, Small container, Masking tape, Water, Felt-tip pen.
>
> - Now follow these directions:
>   1. Put a piece of tape on the side of the glass.
>   2. Fill the small container with spoonfuls of water. Count how many it takes. Empty the small container into the glass. Mark the level of the water and the number of spoonfuls on the tape.
>   3. Fill the small container again. Empty into glass and mark.
>   4. Continue to the top of the glass.
>
> *Now that you have your own measure, use it to see how much other containers hold.*

Clocks The ordinary dial clock is one of the most complicated instruments to read—and yet it is often one of the first instruments to be taught. Not only are there two or more ways to read the scale on a clock (hour, minute, and second), but also the hands (indicators of the measures) move in a circular fashion. Children who easily read linear scales (rulers, for example) may be confused by this circular arrangement. There is no set age at which children appear ready to learn how to tell time; you will often notice a wide range of ability within a class. The following list of skills associated with telling time is not necessarily in the order in which children may develop them:

- Identify the hour hand and the minute hand and the direction they move.
- Orally tell time by the hour (noting that the minute hand is on the 12) and move the hands of a clock to show the hourly times.
- Identify the hour that a time is "after" (e.g., it's after 4 o'clock).
- Count by 5s to tell time to the nearest 5 minutes and report it orally (e.g., as 4 o'clock and 20 minutes after).
- Count on by 1s from multiples of 5 to tell time to the nearest minute (e.g., 25, 26, 27).
- Identify the hour that a time is "before" (e.g., it's before 10) and count by 5s and 1s to tell about how many minutes before the hour.
- Write the time in digital notation (4:20).
- Match the time on a digital clock to an analog clock.

These skills need to be developed over a long period of time, and children need to have clocks with movable hands. Encourage parents or caregivers to help you teach telling time.

You can begin problem solving with young children if they have clocks or watches. For example, as soon as children can tell time on the hour, you can ask questions such as, "What time will it be in 2 hours?" As the children become more familiar with the clock, you can give more challenging questions such as the ones in Figure 16-8.

Although reading digital clocks is easier than reading analog clocks, solving the problems in Figure 16-8 is more difficult with a digital clock. In learning to read a regular clock, a child learns the relationship of the minutes and hours and has a model to use in solving such problems. If all clocks were digital, teachers would need to spend less time on how to read time but more on how our time system works.

Students often have difficulty with elapsed time, partly because of the 60 minutes in an hour. For example, a common response to the amount of time from 2:48 to 4:05 is 1 hour and 57 minutes. Do you see how the students with this response subtracted? The article, "Tracking Time" (Dixon, 2008) shows how children can use a number line to calculate elapsed time.

To integrate math and history, an interdisciplinary unit on the history of time and clocks could prove to be interesting to older students. Students enjoy seeing how the measurement of time has evolved through the years. A good source (Branley, 1993) described in the Book Nook for Children, for history and for making sand clocks, sundials, or other primitive instruments.

Protractors The protractor is often a difficult instrument for students who do not understand angles. The most common protractors have two scales on them, so students first have to make a decision of which scale they are using. Young children who have been asked to compare angles, especially to the benchmark of a right angle, have less trouble with the protractor scales because they already know whether the angle is greater or less than 90°. Estimating sizes of angles (it is about half of a right angle, or 45°; it is about a five-minute turn, or $\frac{1}{3}$ of a right turn, or 30°) will also help students make sense of the protractor. The CCSSM calls for these skills to

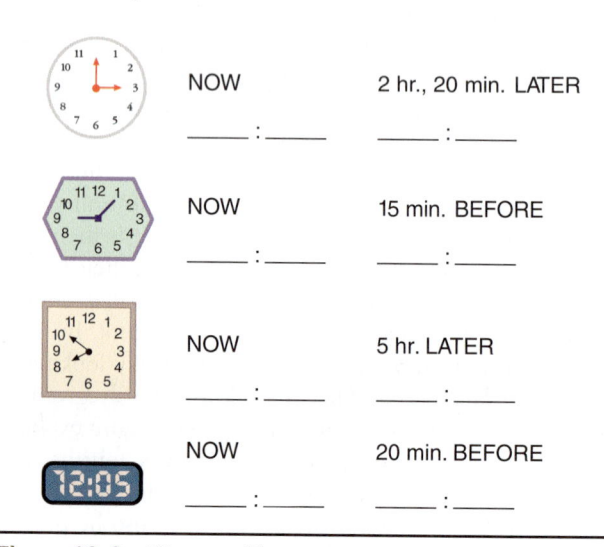

	NOW	2 hr., 20 min. LATER
	____ : ____	____ : ____
	NOW	15 min. BEFORE
	____ : ____	____ : ____
	NOW	5 hr. LATER
	____ : ____	____ : ____
12:05	NOW	20 min. BEFORE
	____ : ____	____ : ____

Figure 16-8 *Time problems.*

be developed in Grade 4, earlier than in most programs. The article by Browning and others in *Teaching Children Mathematics* (Dec 2007/Jan 2008) will give you more ideas of the struggle and success that students can have with angles.

USING FORMULAS Formulas for area, perimeter, volume, and surface area are usually introduced in the upper grades. Although formulas are necessary in many measurement situations, they should not take the place of careful development of measurement attributes and the measuring process. The skill of using formulas should be developed, but not at the expense of helping students build meaning for the formulas.

Before considering area formulas, students should be given the opportunity to compare areas of regions with and without units. Comparing areas without the aid of units by placing one region on top of the other and by cutting one region in order to make the comparison has been described in the section on learning about the attribute.

When introducing units of area, provide students with experiences in covering a region with a variety of types of units—squares, triangles, and rectangles. To find the area of a region, they should count the number of units. Covering many different shapes helps students see the need to approximate and to use smaller units. For example, this region was first covered by one size of squares and then by smaller squares in order to more adequately report its area.

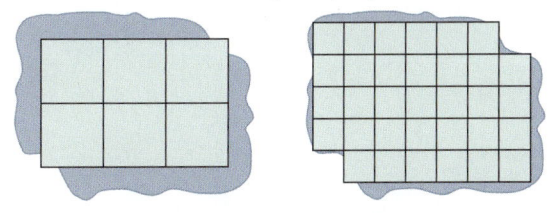

Rectangle The formula for the area of a rectangle is often the first formula children encounter. Use of the rectangle is appropriate because it can be developed easily, building on models that children may have used for multiplication. Figure 16-9 shows a sequence that can be used to develop instructional activities that lead to the formula for the area of a rectangle, $A = b \times a$ (where A is area, b is base, and a is altitude). This form of the formula generalizes to other polygons better than $A = l \times w$.

When children measure with square units, the base and altitude may not be an exact number of units. You may have them begin by estimating how many squares it would take to cover the shape. Later, you can develop the idea of using smaller units or fractional parts of the unit.

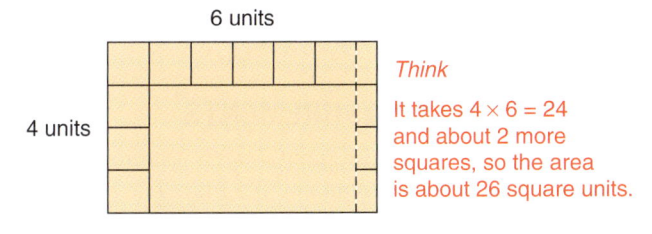

6 units
4 units

Think
It takes 4 × 6 = 24 and about 2 more squares, so the area is about 26 square units.

You should not teach the steps as listed in Figure 16-9. It is given as an outline to help you sequence experiences and assess students' understandings. Although some of the steps could be combined into a single activity, step 3 should not be

Developing the Formula for the Area of a Rectangle

Prerequisites:
- Identifies rectangle
- Compares areas directly
- Assigns a measurement by covering with units
- Models multiplication as an array

Step 1: Covering with Nonstandard Units
- Reviews covering of a rectangle with units
- Develops finding the area by multiplying the number of rows by the number in each row

Step 2: Covering with Standard Units
- Uses standard square units such as square centimeters or square inches
- Continues to find area by multiplying the number of rows by the number in each row

Step 3: Shortcut to Covering
- Develops a shortcut to covering the entire rectangle by showing that it is only necessary to see how many rows and how many in each row

Step 4: Shorter Shortcut
- Marks how many squares could fit across and down
- Continues to multiply to find the area

Step 5: Identifying Base and Altitude
- Identifies base and altitude of rectangles (begins with cut-outs of rectangles and measures their bases and altitudes)

Step 6: Formula for Area
- Measures base and altitude
- Tells the number of squares across the base and down the altitude
- Multiplies the number of rows (altitude) by the number in each row (base) to give the number of square units
- Uses the formula $A = a \times b$

Step 7: Applying Formula to Real Objects
- Practices finding the areas of regions and of real objects

Figure 16-9 *One sequence for developing the area formula for a rectangle.*

done until children have the experiences listed in step 1 or 2. In addition, problems may arise when children are doing step 7 if they have not had prior experience in covering real objects.

One difficulty that children often have with the area of a rectangle is that they may learn, by rote, that the area is the length times the width (or longer times shorter) but not develop the underlying concepts. Thus, when they are faced with finding the area of a square, they run into difficulty. (It has no side that is longer than the others!)

We have concentrated on the formula for the rectangle because it can be used to develop the formulas for other

362 Chapter 16 • Measurement

common shapes as described below. See if you can visually make the connections between the shapes.

Parallelogram Every parallelogram can easily be transformed into a rectangle, as shown here. Students do not have difficulty seeing this; however, they often use the side of the parallelogram instead of the altitude. They need experiences in measuring (and marking) the altitudes on parallelograms. Students with special needs may need to cut out copies of parallelograms and stand them up to see the altitude (measuring from the top to the "floor" much like you measure your height)

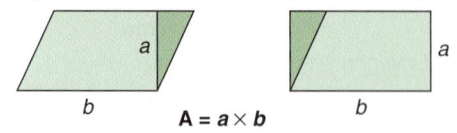

$$A = a \times b$$

Triangle Every triangle is half of a parallelogram. Before students are introduced to the formula for the area of a triangle, they should experiment with all types of triangles. Have each student cut out two congruent triangles and make a four-sided figure. After sharing their results, they should follow the argument below that the area of a triangle is half the area of the parallelogram.

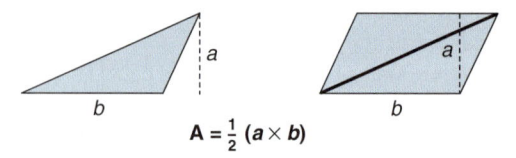

$$A = \tfrac{1}{2}(a \times b)$$

Trapezoid Two congruent trapezoids can be put together to make a parallelogram. As with the triangles, have students experiment with a variety of trapezoids. See if they can generalize to give the formula.

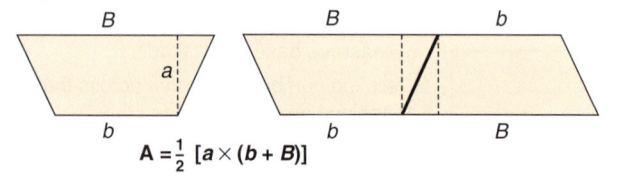

$$A = \tfrac{1}{2}[a \times (b + B)]$$

After learning how to use these formulas, children need experiences like those in In the Classroom 16-8, in which they cut the shapes into different smaller shapes for which they can use area formulas that they know. There are many different ways to subdivide the shapes. If you try this with students, have them show the different ways they subdivided the shapes.

Virtual Classroom Observation

Go to www.wiley.com/college/reys,
Access the Wiley Resource Kit
Click on the Virtual Classroom Observations Section
Module 5: Geometry: Calculating the Area of a Triangle.

1. Watch the videos: Teaching Examples #5 Posing a Challenge to Assess Understanding.
2. Think and discuss: First, find another way to calculate the area of the figure. Assess your own understanding of this problem by listening to all the ways that your colleagues found. Did all of them make sense to you?

In the Classroom 16–8

BREAKING UP IS NOT HARD TO DO

Objective: To find the area of irregular shapes using familiar shapes.

Grade Level: 4–6.

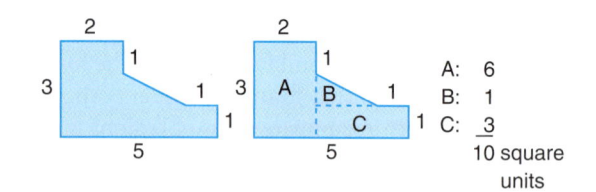

A: 6
B: 1
C: 3
10 square units

"Rectangle A is a 2 by 3 so its area is 6 square units. Rectangle C has a base of 3 and an altitude of 1, so its area is 3 square units. The base of triangle C is 2 and its altitude is 1, so its area is 1 square unit. The total area is 10 square units."

▼ Find the area of each of these shapes. The shading indicates that part is missing.

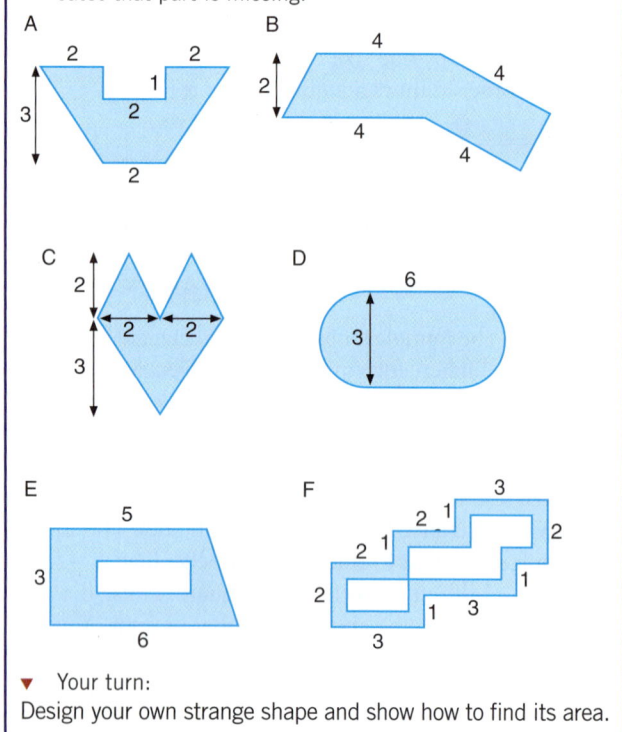

A   B

C   D

E   F

▼ Your turn:
Design your own strange shape and show how to find its area.

V. REPORTING MEASUREMENTS

This last step in the measurement process requires the students to tell both the number and the unit. The attributes that we have discussed in this chapter are continuous, in contrast to the discrete attribute of numeracy (number of objects in a set). To tell the measurement of a set of objects, we use the exact number. However, the measure of a continuous attribute is the nearest number to the actual measure using that specified unit. In essence, we have treated these attributes as being represented by discrete units. However, even young children realize that the length of a pencil is not exactly 6 inches. We encourage you to think about saying "about 6 inches" or "between 6 and 7 inches." If a more precise measurement is needed, then subdivide the original unit into smaller parts. We might report the length as $6\frac{1}{2}$ inches or another length as 4 feet 5 inches. Later, students will study precision and error of measurements.

CREATING OBJECTS GIVEN THE MEASUREMENT

We have concentrated on measuring a given object. Equally important is creating an object of a given measurement. For example, children can often measure a line segment that is 8 inches long but have trouble drawing one that is 8 inches long. More problematic is creating an area given the measurement. This is partly because there are often other restraints. For example, "create a rectangle that is 8 square inches." There are many rectangles that would have an area of 8 square inches if sides were not restricted to whole numbers. Even with this restriction, there are two different rectangles.

The following item was given on the 2003 TIMSS assessment:

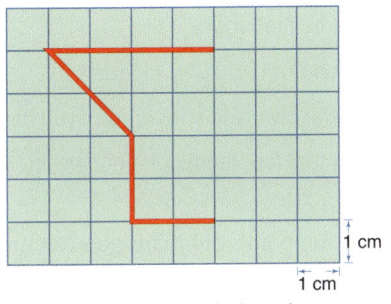

The squares in the grid above have areas of 1 square centimeter. Draw lines to complete the figure so that it has an area of 13 square centimeters.

The international average for the correct answer for students at grade 4 was 29%; for U.S. students, it was 24%. In some countries, though, over half of the students completed the figure to show 13 square centimeters. It is a good item for you to try with students, giving it and other variations—draw any figure that has an area of 13 square centimeters, draw a rectangle that has an area of 13 square centimeters, complete a figure that shows 6 squares and one half square and see if they can draw the other half (using symmetry). Which of these is an easier task? What successes did students have?

This example was shown to emphasize the difficulty that students have in reversing a task. In this case, it is not a simple reversal (perhaps drawing any figure with an area of 13 square centimeters is). It involves much more depth of knowledge because you have to bring other things to mind or ignore other conditions. In the TIMSS item, it is tempting for children to draw a symmetric figure. They have to ignore symmetry.

It was also included to remind you that children who can measure do not automatically create objects with given measurements. This should be part of your curriculum, and children love these challenges when given time and encouragement. It often allows for the creative thinkers in your class.

COMPARING MEASUREMENTS

Solving problems involving measurements often involves a comparison or an arithmetic operation. In so doing, it may be necessary to change from one unit to another (*conversion*), which relies on the *equivalence* relationship between the two units.

In this section, we examine equivalences and conversions within the customary or the metric system. As emphasized in the CCSSM document, students need to become conversant in each system and not rely on converting between systems. However, if they have benchmarks for the different units in each system and how they are related, they will make more sense of the units. For example, they should realize that a liter is a little more than a quart, a meter is a little longer than a yard, or an inch is a little more than 2 centimeters.

EQUIVALENCES

As you introduce new standard units, you should relate them to others. For example, if you are introducing the millimeter, you should relate it to a centimeter by showing that it is smaller and that it is one-tenth of a centimeter.

After using different units, children should learn certain equivalences. Some of these, such as 7 days = 1 week, 60 minutes = 1 hour, and 12 inches = 1 foot, should become known through repeated use. Children need to know that they are expected to know other equivalences. Look at your state guidelines to know which are expected of your students.

Although children are no longer required to memorize long tables of equivalences, you should expect them to know the ones that are commonly used. The task is easier with the metric system because of the standard prefixes and tens relationship between units, as indicated in Table 16-2. The table shows the relationship between metric measurements of different units of length, capacity, and mass (or weight). For example, the meter (m) is the base unit for length. A kilometer (km) is 1000 times as long as a meter. A decimeter (dm) is one-tenth of a meter.

Children need to be able to use a table of equivalences. This involves several skills, which are illustrated in the questions that follow this table:

Unit	Equivalent
1 day	24 hours
1 hour	60 minutes
1 minute	60 seconds

- How many hours in a day? (This answer only requires reading the table.)
- What part of a day is 1 hour? (This answer involves knowledge of fractional relations of the units in the table.)
- How many seconds are in an hour? (This answer involves conversion of units in the table.)

Area and volume equivalences can be difficult for many children because they are often derived from the linear equivalences. For example, knowing that 1 m = 10 dm allows you to derive 1 m² = 100 dm² and 1 m³ = 1000 dm³. In the Classroom 16-9 illustrates these relationships for area. In the Snapshot of a Lesson, the fourth graders who had measured in inches found that if they knew that 144 square inches is equivalent to a square foot, they did not have to measure again in feet.

CONVERSIONS

To change from one unit to another, children must know the equivalence or relation between the two units. By itself, however, this information is not sufficient to make conversions. Let's look at an example of a class discussion.

Mr. Bane: It seems that several of you are stumped on the assignment. Devon, please read the first exercise and let's look at it together.

Devon: Blank dm equals 5 m.

[*Mr. Bane writes on the board:* ___ dm = 5 m.]

Mr. Bane: Who can tell me what we are looking for?

George: How many decimeters there are in 5 meters.

Mr. Bane: What do you know about decimeters and meters?

Alana: A decimeter is about this big and a meter is about this big.

Mr. Bane: Could everyone see Alana's hands? Which unit is larger, the meter or the decimeter? [*meter*] Right, so will it take more or less than 5 decimeters to make 5 meters?

Karina: It'll take more. It takes 10 decimeters to make 1 meter.

[*Mr. Bane draws this on the board:*]

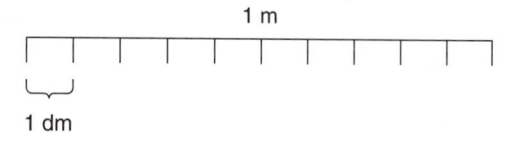

Mr. Bane: If 1 meter is 10 decimeters, then what would 5 meters be?

Devon: I see, it's 5 groups of the 10 decimeters or 50 decimeters.

Mr. Bane: Good. Let's try another: 20 cm = ___ dm

Leon: Centimeters are smaller, takes more of them to measure something; so it won't be as many as 20.

Mr. Bane: Good. We know our answer must be less than 20.

Randy: Let's draw a picture like this. Oh, there's no use drawing in all 20 marks. We know 10 centimeters makes 1 decimeter. So we want to know how many tens in 20.

Mr. Bane: Right, in this case we can just look at the picture and see that it is 2 decimeters. But what if it was 184 centimeters?

Dave: We still need to know how many tens in 184—we divide.

Jim: There are 18 decimeters.

Mr. Bane: What about the 4 left over?

Paula: Those are centimeters. We have 18 full decimeters and 4 leftover centimeters.

Mr. Bane: Let's write that down:

184 cm = 18 dm 4 cm

Ok. Try some on your own, and I'll help you if you have questions.

Sarah: Isn't 4 cm just four-tenths of a decimeter? So couldn't I write 18.4 dm?

This discussion points out many good techniques to use in developing conversions. First, Mr. Bane had the children decide whether their answer would be larger or smaller than the number given. This relies on children knowing the relative sizes of the units and understanding that the smaller the unit, the more it takes to represent the attribute. Second, Mr. Bane tried to have the children visualize the relationship between the units. Third, he related the operation to be used to their understanding of what multiplication and division mean. How would you answer Sarah?

TABLE 16-2 • Most Commonly Used Metric Units

	Kilo (k) 1000	Hecto (h) 100	Deca (da) 10	Base Unit	Deci (d) 0.1	Centi (c) 0.01	Milli (m) 0.001
Length Capacity Mass	Kilometer (km) Kilogram (kg)			Meter (m) Liter (l) Gram (g)		Centimeter (cm)	Millimeter (mm) Milliliter (ml)

In the Classroom 16–9

SOLVING CONVERSION PROBLEMS WITH SQUARE UNITS

Objective: To determine the equivalents between dm² and m².

Grade Level: 4–5.

▼ Ask students how many square decimeters (dm²) are in 1 square meter (m²). A representation of the relationship is shown below.

1 dm²

1 m²

▼ Show that each row would have 10 squares and there would be 10 rows. So 1 m² = 100 dm².

10 × 10 = 100 squares

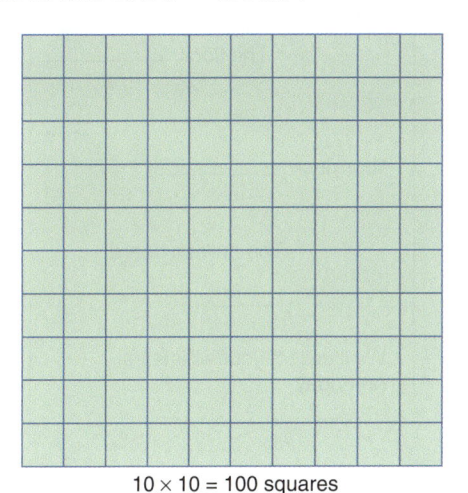

▼ Use sketches and the model above to answer the following questions.

- Conrad has a piece of cloth that is 3 m × 1 m. How many dm² is that?
- Maria has a blanket that is 15 dm × 10 dm. How many m² is it?

ESTIMATING MEASUREMENTS

Estimating is the mental process of arriving at a measurement without the aid of measuring instruments. There are many reasons to include estimating in the development of measurement. For one, it helps reinforce the size of units and the relationships among units. For another, it is a practical application—think of all the times you want to know approximately how long, how heavy, or how much something holds. Fifth-grade students were encouraged to use measurement and to make estimations of inaccessible objects. Read about their adventure at a science museum (Sedzielarz and Robinson, 2007) and think again about Galileo's statement in the opening quote of this chapter.

There are two main types of estimation. In the most common type, the attribute and object are named and the measurement is unknown. For example, about how long is your arm? In the other type, the measurement is known and the object is to be chosen. For example, what piece of furniture in your room is about 1 meter long? By keeping the two classes of estimation in mind, you can expand your repertoire of estimation activities. Identify these two types in the activities in In the Classroom 16-10.

Several common strategies can be used with either type of estimation. You can help children develop these strategies by talking through the various methods that different children use to make estimations and by presenting the following strategies.

What is the height of the door?

Using a referent: *I am about 6 feet, so the door must be about 7 feet.*

Chunking: *I know it is a yard to the knob, another yard to here, so a little more than 2 yards.*

Unitizing: *Each cement block is about 8 inches. There are 10 blocks, so about 80 inches.*

1. One strategy is to *compare to a referent*. If you know that you are 1 meter 70 centimeters tall, then you can estimate the height of a child who comes up to your waist. Or if you have to choose a board that is 2 meters long, you will have some idea of the size. The NCTM *Standards* (NCTM, 2000) call for developing common referents or benchmarks to help in estimating measurements. As suggested previously, it is helpful to have a referent for each of the units.

2. Another strategy is that of *chunking*. In this process, you break the object into subparts and estimate each part. For example, you want to know about how far you walked from your home to the library and the store and then back to your home. If you know that from your home to the library is about 1 mile, that it's

In the Classroom 16–10

IDEAS FOR ESTIMATION

Objective: To estimate measurements of objects and compare measurements with those of other objects.

Grade Level: 3–5.

About how many centimeters long?

	Estimate	Measure
Little finger	_____	_____
Nose	_____	_____
Foot	_____	_____

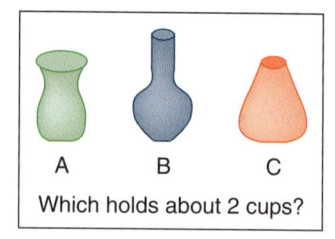

A   B   C

Which holds about 2 cups?

Make a decimeter

Draw a snake that you think is a decimeter long.

Draw a tree that you think is a decimeter high.

CONTEST FOR WEDNESDAY

About how much does the wonderful watermelon weigh?

Name	Guess
_____	_____
_____	_____

Winner gets the largest piece at Ho-Ho's picnic.

Guess which of the boxes will hold 60 sugar cubes.

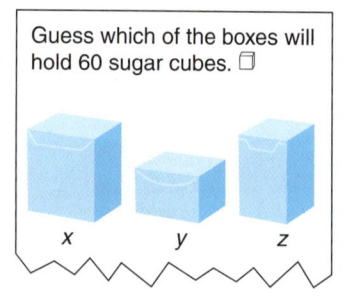

x   y   z

About how many squares?

The floor _____

Your desk _____

The bulletin board _____

We'll collect your estimates on Friday.

HUNT   HUNT

There is something in the room that weighs a kilogram.

Can you find it?

---

about that same distance from the library to the store, and twice as far from the store back to your home, you walked about 4 miles altogether.

3. A strategy related to chunking is *unitizing*. In this case, you estimate one part and see how many parts are in the whole. For example, someone asks you to cut a piece of string that is about 3 meters long. You estimate 1 meter and take three of these.

When including estimation in your program, you should try to make it a natural part of measurement activities.

1. Encourage children to see if they can tell about how long or heavy the object is before they measure it.
2. Look for ways to include estimation in other subject areas: About how far did you jump?
   What size paper do you need for your art project?
   About how long will it take you to read the book?

3. Plan estimating activities for their own sake or use brief ones as daily openers for several weeks throughout the year. In the Classroom 16-10 presents some ideas to get you started. Once you begin thinking about the things in your room, you will be able to come up with a lot of variations.

One thing to remember: Do not call an estimate right or wrong. Help children develop ways to make better estimates, but do not discourage them. Let them check their estimates by measuring. They will know whether they were close or not. You may be surprised to find out who are the good estimators in your class.

CONNECTING ATTRIBUTES

Activities involving two attributes can help children see how the attributes are related or how one attribute does not depend on the other. For example, by doubling the dimensions of a rectangle, children may see how the area is changed. By examining figures with the same area but different shape, children may see that area is independent of shape. We have included some sample activities with suggestions for other variations or extensions.

AREA AND SHAPE

In the Classroom 16-11 encourages children in grades 2 through 4 to investigate the different shapes they can make using two to four squares. Younger children enjoy this activity using square tiles. For older children, extend this activity to more squares and place the restrictions that the squares must have touching sides (not corners) and that two shapes are the same if one is a reflection or rotation (see Chapter 15) of the other. A challenging variation is to use triangular grid paper (see the Appendix) rather than the square grid paper. Make certain with all of these that the students can make the generalization: The shapes are the same area but different in shapes. In the upper grades, they should explore similar shapes with different areas.

VOLUME AND SHAPE

An activity such as the one in In the Classroom 16-2 can be done with cubes. Not only does this activity look at the generalization that shapes with the same volume may be of different shapes, it also ties the investigation to number theory (primes and composites).

PERIMETER AND AREA

Children are often confused about perimeter and area. This confusion may be caused partly by a lack of understanding of area and partly by premature introduction of the formulas. Many activities can help children see that a figure with a given perimeter may have many different areas.

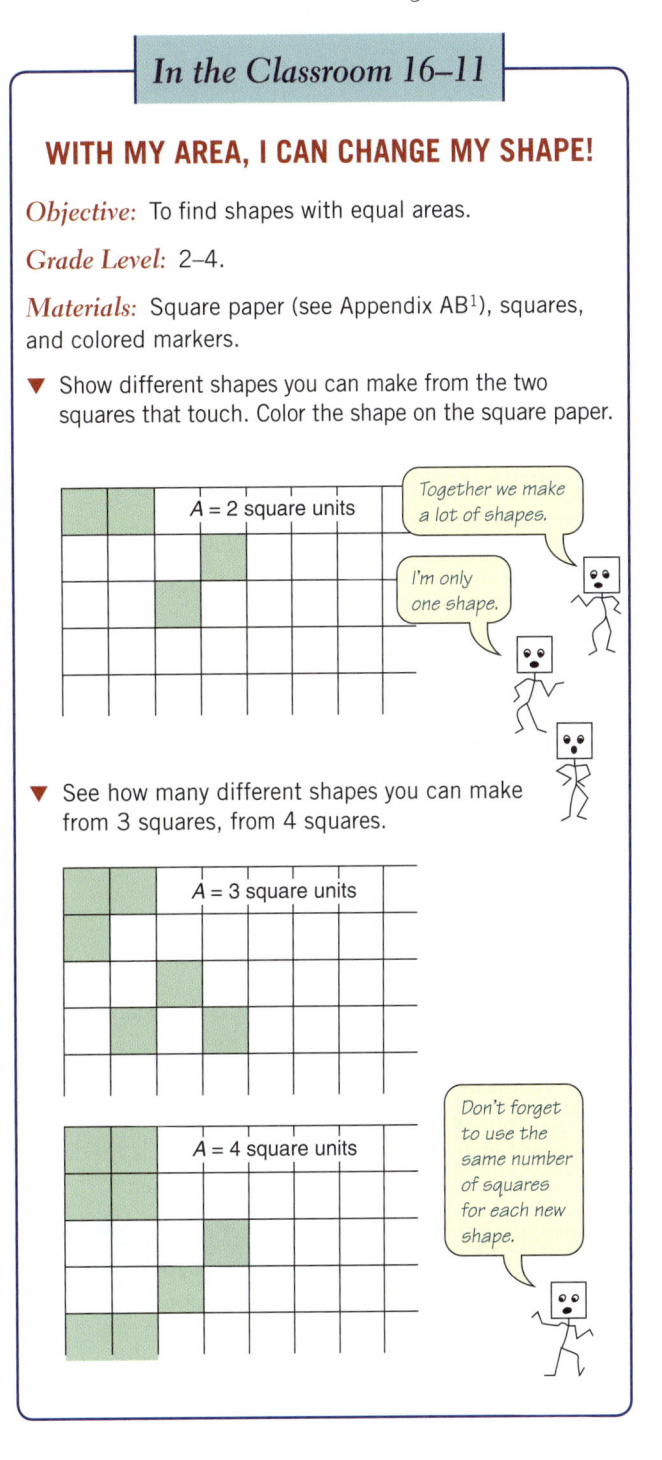

In the Classroom 16–11

WITH MY AREA, I CAN CHANGE MY SHAPE!

Objective: To find shapes with equal areas.

Grade Level: 2–4.

Materials: Square paper (see Appendix AB[1]), squares, and colored markers.

▼ Show different shapes you can make from the two squares that touch. Color the shape on the square paper.

A = 2 square units

Together we make a lot of shapes.

I'm only one shape.

▼ See how many different shapes you can make from 3 squares, from 4 squares.

A = 3 square units

A = 4 square units

Don't forget to use the same number of squares for each new shape.

In the Classroom 16-13 is a sample of this type of activity. Children also should realize that figures with the same area can have different perimeters. You can modify In the Classroom 16-11 by asking the perimeter of each of the shapes or by challenging students to take five squares and see what shape they can make that has the largest perimeter or the smallest perimeter. The electronic geoboard (see Math Links 16.4) will allow students to make a shape and then check the area and the perimeter. For example, have students explore and record (see geoboard paper in the Appendix)

In the Classroom 16–12

SAME VOLUME, DIFFERENT SHAPE

Objective: To explore the different shapes of solids that can be made with a given volume.

Grade Level: 5–8

Materials: Cubes

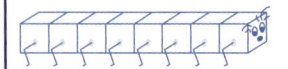

I'm 8 cubes—8 by 1 by 1.
My volume is
8 cubic units.

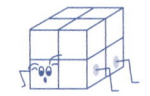

I'm 8 cubes—8 by 2 by 2.
My volume is
8 cubic units, too!

▼ See how many different rectangular solids you can make with 12 cubes. Record the dimensions and volume of each.

▼ Now use 2, 3, 4, ..., 17, 18 cubes

▼ How many different solids can you make if the number of cubes is

- prime? ____
- a product of two primes? ____
- a perfect square? ____

▼ How many solids can you make with 24 cubes? ____

In the Classroom 16–13

DO YOU KNOW HOW TO CONNECT PERIMETER AND AREA?

Objective: To explore different areas that can be made from a given perimeter.

Grade Level: 5–8.

Materials: Centimeter grid (Appendix) and pencil.

▼ How many different rectangles can you make with a perimeter of 16 cm? Find the area of each.

Here's one that I made.

I see you followed the rules and made each side a whole number.

$P = 16$ cm
$A = 12$ cm^2

▼ With three other students, try each of the following perimeters: 8 cm, 9 cm, 10 cm, 11 cm, 12 cm, and 13 cm. Make a table of the number of different rectangles you can make from each perimeter.

▼ How many different rectangles do you think you could make if the perimeter were 52 cm? What is your conjecture?

---

the areas and perimeters of various shapes, then choose an area (say, 6 squares) and see how many shapes with different perimeters they can make. What shape has the largest perimeter? What shape has the smallest perimeter?

Math Links 16.4

The virtual electronic manipulatives site has an electronic geoboard. Become skillful in handling it. It has been referenced in the Connecting Attributes (Perimeter and Area) section. Try one of the suggestions there. You can access this at http://nlvm.usu.edu/en/nav/index.html or from this book's Web site.

www.wiley.com/college/reys

VOLUME AND SURFACE AREA

Just as the area of a figure does not depend on the perimeter, the volume does not depend on the surface area. The experiment in In the Classroom 16-14 looks at the relation of lateral surface area to volume. You can vary this activity by having children fold the papers into thirds (sixths) to make prisms or make cylinders (a long, thin one and a short, fat one). In middle school, after developing the formula for a rectangular solid, students could calculate the volume of each of the tubes in In the Classroom 16-14.

CULTURAL CONNECTIONS

Measurement combined with children's literature is a wonderful way to bring the different cultures of your classroom and of the world to life. We have selected only one, food, of many topics to wet your appetite. We give a brief description of the book including the ethnic group and one idea for using it.

Everybody Cooks Rice (Dooley, N., 1990, Minneapolis: Carolrhoda) tells the story of Carrie and Andy as they sample ethnic rice dishes. Recipes from Barbados, Puerto Rico,

In the Classroom 16–14

WHAT IS THE CONNECTION BETWEEN VOLUME AND SURFACE AREA?

Objective: To explore the volume of prisms and cylinders with the same lateral area.

Grade Level: 5–8.

Materials: Construction paper (9 in. by 12 in.), tape, and dry filler.

▼ Use construction paper to make two tubes:

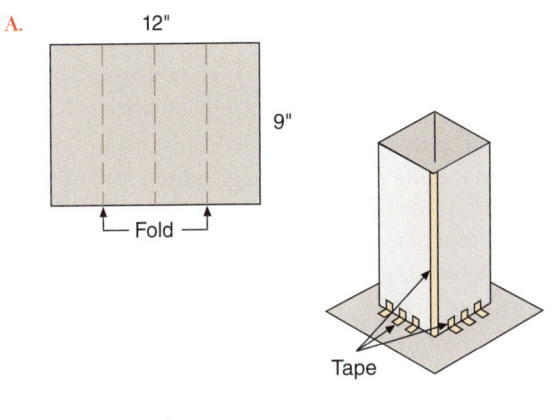

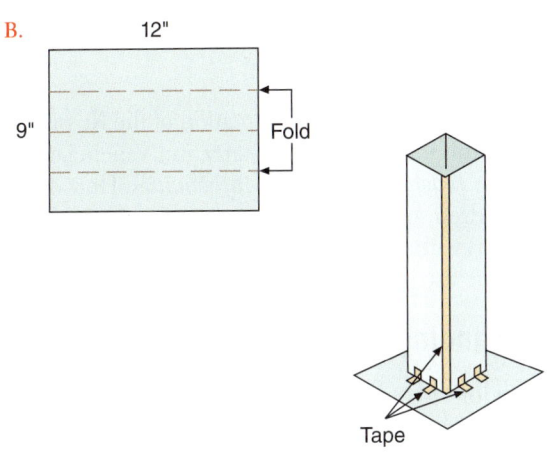

A. 12" × 9" (Fold) — Tape

B. 12" × 9" (Fold) — Tape

▼ Guess which tube holds more, or do they hold the same? _____
▼ Fill and see.
▼ Try the same with triangular prisms, hexagonal prisms, octagonal prisms, and cylinders.
▼ Do you have a conjecture about which shape, the shorter or the taller of each pair, will hold the most?

Vietnam, India, China, Italy and Haiti are provided. Different groups could illustrate the different recipes on a large sheet of paper (drawing the number of cups, etc.) and then a class discussion could follow on how they are alike and how they are different.

Bee-bim Bop! (Park, L, 2005, NY: Clarion Books) is a rhyming text that shares the tradition of creating a favorite traditional Korean meal. Collect food ads and have students shop for the food needed. You can find prices of Korean specialty food on line.

Grandma Lena's Big Ol" Turnip (Hester, D., 2005, Morton Grove, IL: Albert Whitman) is a tale of a very large turnip and the African American family that shares it with the town. While you may not have a turnip to measure, bring a pumpkin, a large zucchini, or other vegetables to weigh and measure length and distance around.

The Popcorn Book (DePaola, T., 1978, NY: Holiday House) is the story of the history of popcorn (Native American) as well as facts about popcorn. This wonderful story can encourage estimating and comparing capacities of containers. Pop corn and see which container holds the most!

The Perfect Piñata (Dominguez, K., 2002, Morton Grove, IL: Albert Whitman) is the story of Marissa and her choice of a piñata for traditional Mexican birthday. Another book to encourage estimating.

Chapter 3 gives overall suggestions for using children's literature in teaching mathematics. These types of books rarely become dated, so begin to make a collection of titles. You will find those that complement your teaching of measurement and will help you meet the goal of fourth reason, engaging children, that we gave for including measurement in your classroom.

A GLANCE AT WHERE WE'VE BEEN

Measuring is a process that may be used when determining the size of many attributes. The measurement process consists of identifying the attribute, choosing a unit, comparing the object being measured to that unit, finding the number of units, and reporting the number of units. To help children understand the attribute, we suggest comparing two objects on that attribute: perceptually, directly, and indirectly. In selecting a unit, there are nine concepts that need to be developed over time. Comparing an object to a unit entails finding the number of units that would represent the object on the attribute in question; for example, comparing a pencil to an inch and finding out how many inches the pencil is in length. Children may first assign a number by counting and later by using instruments or formulas. Other suggested ways to help children learn about measuring are estimating sizes of objects, finding equivalent measurements, and relating two attributes.

By including measuring in your classroom, you have the opportunity to show how mathematics is practical, to develop other mathematical ideas, to relate mathematics to other topics, and to make mathematics meaningful for many children.

Things to Do: From What You've Read

1. What are the five steps in the measurement process?
2. What attributes, units, and instruments are included in most elementary mathematics programs?
3. Why should you include measurement in your mathematics program? Choose one reason and give examples of how measurement fulfills that reason.
4. What nine concepts related to units need to be developed as children have experiences with measurements?
5. Describe three difficulties children have with measuring instruments.
6. Why do students need to be able to convert from one unit to another?
7. Show how you would find the areas of each of the figures in In the Classroom 16-8.
8. Give three examples of connections that can be made between different attributes.

Things to Do: Going Beyond This Book

In the Field

1. [1]*Measuring Length with Arbitrary Units.* Develop and teach the lesson described in In the Classroom 16-6. What ideas about units of measurement do children grasp?
2. [1]*Comparing Height and Circumference.* Try the activity in In the Classroom 16-4 with students. What did students learn? Modify this activity to show how to approximate the circumference of a can and try it with upper-level students.
3. [1]*What Is the Connection Between Volume and Surface Area?* Try the activities in In the Classroom 16-14 with students. Describe your experience, including whether the children could make the connections expected.

In Your Journal

4. You have a child in your class who physically cannot handle a ruler. What would you do in this situation? Give at least two solutions.
5. Describe a set of activities that you could use to introduce a centimeter, a pound, a liter, or a square inch based on those for introducing a decimeter.
6. Design an activity to have students investigate a fixed area and varying perimeters (e.g., a fixed area of four square units made with four square tiles and find the different perimeters). Include the answer to the investigation.
7. For a grade level of your choice, design five estimation activities that include both types of estimation (estimating measurement for a known object and choosing an object that fits a known measurement).

With Additional Resources

8. Read an article about the adaptation of the Russian curriculum in Hawaii (Dougherty and Venenciano, 2007) in *Teaching Children Mathematics*. How do the children solve the problem about comparing line segments without moving them? What do you think about the use of letters to describe their comparisons?

With Technology

9. Watch the video *Pencil Box Staining*, excerpted in the Snapshot of a Lesson. What strategies did the students use to find the area?

---

[1] Additional activities, suggestions, and questions are available in the field experience manual on the Student Companion site at www.wiley.com/college/reys.

Note to Instructors: You can find additional resources, learning activities, and blackline masters in this text's accompanying Instructor's Manual at www.wiley.com/college/reys.

Book Nook for Children

Branley, F. M. *Keeping Time: From the Beginning and into the 21st Century.* Boston: Houghton Mifflin, 1993.

Early timekeepers were the sun, moon, and stars. Directions for making ancient timekeepers such as the sundial, candle clock, water clock, and sand clock are given. Mechanical clocks are introduced, and the need for more accuracy is discussed. An index and bibliography of other books on time for young and more advanced readers are included.

Davies, N. *Just the Right Size: Why Big Animals are Big and Little Animals are Little.* Somerville, MA: Candlewick Press, 2009.

Examines why small animals can do things that large animals cannot do based on the volume, weight, and surface area. A fun book that makes you think about size.

Hightower, S. *Twelve Snails to One Lizard: A Tale of Mischief and Measurement.* New York: Simon & Schuster Books for Young Readers, 1997.

Animals introduce measurement facts in a delightful book for young children. A break in the dam sets the need for nonstandard and standard measurements. This book succeeds in introducing customary length measures and the relationships among such measures in a humorous way.

Murphy, S. J. *Room for Ripley.* New York: HarperCollins, 1999.

Carlos visited a pet store to buy his favorite type of fish, a guppy. His sister Ana shows Carlos how to set up a home for his guppy using different types of liquid measurements. Carlos decides to buy another guppy, which prompts him to add more water to the fish bowl using different liquid measures. There are activities and games for teachers and parents to use.

Myller, R. *How Big Is a Foot?* New York: Dell Publishing, 1991.

This is a humorous tale about nonstandard measures in which a king decides to have a bed made for the queen as a surprise for her birthday. The large king marks off the dimensions for the proposed bed with his feet. Unfortunately, when the bed is delivered, it is the wrong size. The apprentice who made the bed solves the problem from his jail cell.

Reisberg, J. *Zackary Zormer: Shape Transformer.* Watertown, MA: Charlesbridge, 2006.

This is the story of Zack Zormer's favorite day: Measurement Day. Zack has forgotten to bring something to show at school. All he has is a piece of paper and an imagination. Zack cleverly shows length, width, perimeter, and area with his paper.

CHAPTER 17

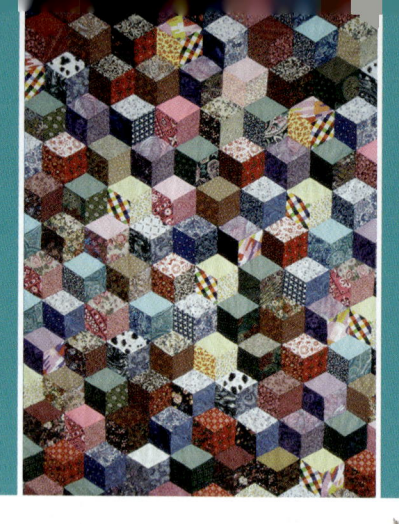

Data Analysis, Statistics, and Probability

> "FACTS ARE STUBBORN THINGS, BUT STATISTICS ARE MORE PLIABLE."
> — Mark Twain

>> SNAPSHOT OF A LESSON

KEY IDEAS

1. Show students how to calculate the measures of central tendency: the mode, median, and mean.
2. Show students how to calculate a measure of variation: the range.
3. Have students use the calculations to rank data for meaningful interpretation.
4. Help students use data to support their conclusions.

BACKGROUND

This lesson is based on the video *Looking Behind the Numbers*, from the Modeling Middle School Mathematics project. To view the video, go to http://www.mmmproject.org and click on the video matrix.

Amy Doherty, an eighth-grade teacher, is using the *MathScape* curriculum in her class and has introduced the concepts of mode, median, mean, and range in previous lessons. Students have also had a chance to work with all of the measures previous to this lesson. The students have been given the data from three basketball games to analyze. Their task is to prepare a letter to the basketball coach nominating the Most Valuable Player (MVP) based on these data.

Ms. Doherty: OK, now that you've all calculated the statistics for each of the basketball players, let's put it up here on the board. So for player B, can someone list the scores for us?

Student: The high score was 21. The median was 18. The mean was 17. The mode was 18, and the range was 8.

Ms. Doherty: Now, for player C.

Student: The high score was 25, the median was 16, the mean was 18, the mode was 14, and the range was 14.

Ms. Doherty: Now that we've calculated all of the statistical measures, we're going to use this information to compare the three basketball players and find out who is the best player. So what I'd like you to do is on the next sheet we're going to rank the players according to the different statistical measures. Which player had the best score, the highest score? Grecia?

	High Score	Median	Mean	Mode	Range
Player A	20	13	14	12	9
Player B	21	18	17	18	8
Player C	25	16	18	14	14

Grecia: Player C and he had 25.

Ms. Doherty: OK, and who was second best?

Grecia: Player B; he had 21.

Ms. Doherty: OK, third best?

Grecia: Player A; he had 20.

Ms. Doherty: OK, now what I would like you to do with your partners is to rank the rest of the measures. So rank by median

372

score, by mean score, by the mode, and by the range. Take a minute to do that with your partners.

Student: I have player C first.

Student: And the second best would be player B, with 17.

Ms. Doherty: OK, let's fill in the rest of this chart. For the median, who had the best (ranking) median, Amy?

Amy: B—18?

[*The students continue with the activity.*]

Ms. Doherty: OK, now for the mode, Gregor?

Gregor: B had 18?

Ms. Doherty: Yup.

Gregor: C had 14. And A had 12.

Ms. Doherty: OK, the last one we need to rank is the range. Jackie?

Jackie: The best was C—14; the second best was A with 9; and the third best was B—8.

Ms. Doherty: Why do we have so many hands up? Derrick?

Derrick: Because the best score is supposed to be B with 8 and A with 9 and C with 14.

Ms. Doherty: How come?

Derrick: On the range you want the lowest, because that means that they are the most consistent and they score around the same number of points a game.

FOCUS QUESTIONS

1. What different graphs can be used to represent data? How does the type of graph used relate to the data?
2. What descriptive statistics are appropriate to introduce in the elementary grades? What are some examples of ways they can be introduced?
3. What are some common misconceptions young students have about probability?

INTRODUCTION

Data analysis, statistics, and probability provide a meaningful context for promoting problem solving and critical thinking, enhancing communication, developing number sense, and applying computation. The study of these topics supports a problem-solving or investigative approach to learning and doing mathematics. As a result, statistics and probability are now highly visible topics in elementary school mathematics programs. Figure 17-1 highlights specific recommendations from the Common Core State Standards for Mathematics (CCSSM). These recommendations reflect the growing importance of analyzing data in our daily lives.

Let's look at some reasons for including the study of data analysis, statistics, and probability in elementary school.

- *Children encounter ideas of statistics and probability outside of school every day.*

Radio, television, newspapers, and the Internet bombard us with information. For example, news reports present national economic and social statistics; opinion polls; weather reports; and medical, business, and financial data.

The current demand for information-processing skills is much greater than it was 25 years ago, and technological advances will place a far greater premium on such skills in the years ahead. Many consumer and business decisions are based on market research and sales projections. If these data are to be understood and used widely, every educated person must be able to process such information effectively and efficiently. An intelligent consumer must be able to understand and use statistics and probability to make informed decisions.

The data students encounter outside of school are often presented in a graphical, statistical, or probabilistic form:

- *Graphical.*
  Which company holds the majority of the global soccer sales?

Global Soccer Sales by Company

- *Statistical.*
  The median household income in the USA is $56,000.
- *Probabilistic.*
  The probability of rain today is 0.35.

Each of these statements needs to be understood if meaningful interpretations are to be made. The context and format of the way such information is presented vary greatly, but correct interpretation of the information often requires the application of mathematics. Consider, for example, the mathematical concepts involved in weather reports (decimals, percents, and probability); public opinion polls (sampling techniques and errors of measurement); advertising claims (hypothesis testing); and monthly government reports involving

> **Develop Understanding of Statistical Variability**
> - Recognize a statistical question as one that anticipates variability in the data related to the question and accounts for it in the answers. For example, 'How old am I?' is not a statistical question, but 'How old are the students in my school?' is a statistical question because one anticipates variability in students' ages.
> - Understand that a set of data collected to answer a statistical question has a distribution which can be described by its center, spread and overall shape.
> - Recognize that a measure of center for numerical data set summarizes all of its values with a single number, while a measure of variation describes how its values vary with a single number.

Figure 17-1 *Some Statistics and Probability standards for grade 6 from CCSSM. This excerpt of the CCSSM standards was made by the authors of Helping Children Learn Mathematics. The full document (CCSSI, 2010) may be obtained at www.corestandards.org/thestandards/mathematics.*

unemployment, inflation, and energy supplies (percents, prediction, and extrapolation).

- *Data analysis, statistics, and probability provide connections to other mathematics topics or school subjects.*

The study of data provides an excellent opportunity for curriculum integration. For example, each day kindergarten children create a picture graph with their photos to show who is present and absent. In doing so, they practice basic counting skills. Third graders may estimate the number of raisins in single-serving packages and graph their estimates. After counting the actual number of raisins, a second graph may be created, displaying the actual number of raisins in each package. The children are able to compare their estimates with the actual results. Fourth graders measure the height of each student in the class, and then students find the mean, median, mode, and range so they can better understand and summarize their data. As sixth graders conduct a probability simulation with dice, they use fractions, decimals, and percents to report their results.

Statistics and probability are also easily integrated into other school subjects. For example, second graders might use graphs in reading to keep track of numbers and types of books they have read. Fifth-grade students conducting a science experiment on rolling a car down ramps of various heights may calculate the mean distance for the number of trials at each height. In social studies, graphs and charts are frequently used to display information about populations or geographic areas. In physical education class, students may graph pretest data from physical fitness tests and compare it with posttest results at the end of the year. Effective teachers find ways to make data analysis an integral part of the elementary and middle school curriculum.

- *Data analysis, statistics, and probability provide opportunities for computational activity in a meaningful context.*

Data are not merely numbers but numbers with a context. The number 12 in the absence of a context carries no information, but saying that a baby weighed 12 pounds at birth makes it easy to comment about its size! Data provide many opportunities to think, use, understand, and interpret numbers, rather than simply carrying out arithmetic operations. Using data helps further develop number sense. Working with real data requires judgment in choosing methods and interpreting results. Thus, statistics and probability are not taught in elementary school for their own sake but because they provide an effective way to develop quantitative understanding and mathematical thinking.

- *Data analysis, statistics, and probability provide opportunities for developing critical thinking skills.*

When students learn how to design and carry out experiments that utilize data analysis and probability, they develop skills to help them answer questions that often involve uncertainty, and they draw conclusions based on their interpretations of the data. As students learn how to approach situations statistically, they can face up to prejudices, think more consistently about arguments, and justify their thinking with numerical information (Burns, 2000).

This approach has applications in many areas of our lives—social and political. For example, the opening Snapshot of a Lesson requires students to make thoughtful analysis of the data in determining the MVP.

Data analysis, statistics, and probability should not be viewed or treated in isolation. Their study provides numerous opportunities to review and apply much mathematics in a variety of real-world situations. For example, whole numbers, fractions, decimals, percents, ratios, and proportion are essential ideas for understanding a wide variety of situations. Many computational skills are reviewed and polished as they are applied in graphing or doing statistics and probability.

> **Math Links 17.1**
>
> The full-text electronic version of the *Common Core State Standards Initiative* for mathematics is available at http://www.corestandards.org/
>
> *Guidelines for Assessment and Instruction in Statistics Education* is available at the ASA's Web site at www.amstat.org/education/gaise or by linking from this book's Web site.
>
> www.wiley.com/college/reys

The American Statistical Association developed the *Guidelines for Assessment and Instruction in Statistics Education* (GAISE) in 2005. The GAISE report outlines a

ASA: GUIDELINES

Process Component	Level A	Level B
I. Formulate Question	• Teachers pose questions of interest • Questions restricted to the classroom	• Students begin to pose their own questions of interest • Questions not restricted to the classroom
II. Collect Data	• Do not yet design for differences • Census of classroom • Simple experiment	• Beginning awareness of design for differences • Sample surveys: begin to use random selection • Comparative experiment: begin to use random allocation
III. Analyze Data	• Use particular properties of distributions in the context of a specific example • Display variability within a group • Compare individual to individual • Compare individual to group • Beginning awareness of group to group • Observe association between two variables.	• Learn to use particular properties of distributions as tools of analysis • Quantify variability within a group • Compare group to group in displays • Acknowledge sampling error • Some quantification of association; simple models for association.
IV. Interpret Results	• Students do not look beyond the data • No generalization beyond the classroom • No differences between two individuals with different conditions • Observe association in displays	• Students acknowledge that looking beyond the data is feasible • Acknowledge that a sample may or may not be representative of the larger population

Figure 17-2 *Excerpt of a framework for statistics education from American Statistical Association's Guidelines for Assessment and Instruction Statistics Education Report (2005).*

curricular framework for the development of statistical literacy in preK–12 students. The framework (see Figure 17-2) describes how statistical literacy is developed over two levels: level A, elementary; and level B, middle grades. The GAISE framework promotes the development of statistical literacy by providing opportunities for students to formulate questions, collect data, analyze data, and interpret results.

for which they can collect data available in their classroom. At the middle school level, students should be encouraged to ask questions that will require the collection of data outside of their classroom. Questions may come from a variety of sources on many different topics. In the Classroom 17-1 gives a few examples. These ideas may be shared with children to help them brainstorm the kind of questions they might ask.

FORMULATING QUESTIONS

Begin with a good question or problem that interests students, one for which the answer is not immediately obvious, and one that also clearly gives students a reason for collecting and analyzing data and then interpreting the results. There is a real benefit to having students identify their own questions or problems, for they will take ownership of the investigation and their motivation will be high (Bohan, Irby, and Vogel, 1995; Riskowski, Gayla & Wilson, 2010). At the elementary level, the teacher may have to help in formulating the questions to be sure students are focused on questions

COLLECTING DATA

Once a suitable question has been identified, students will need to plan how to collect the data needed to answer it. Communication skills are very important during this stage. To be successful, students must be able to develop clear survey questions and logical steps for their experiments or simulations. They must communicate with others to negotiate the details of the investigation. They must find a clear and efficient method of recording their data. Students may collect data from surveys, experiments, and simulations, all of which typically involve counting or

In the Classroom 17–1

LET'S FIND OUT

Objective: To plan and conduct a survey.

Grade Level: 4–5.

Steps:

1. Think of a question you would like to answer. Here are some ideas to get you started.

   - **Questions about ourselves:** Who can whistle a tune? How far can we throw a softball? What is our class's typical height, eye color, shoe size, number in family, amount of allowance, pets . . . ?

   - **Question about opinions or feelings:** How do you feel about fractions? Does life exist on other planets? What should be done about pollutions? What country do you want to study in social studies? What is your favorite television show, song, book, sport, color, food . . . ?

   - **Questions about the world:** Which month has the most birthdays? What is the most popular color of car in the school parking lot? Which brand of cookie has the most chocolate chips? How many paper towels do we use in one day? What is the effect of fertilizer on bean plant growth? What type of paper airplane will fly the farthest?

   My Question: _____
   _____

2. Plan the survey by answering these questions.

   a. Where or from whom will I collect the data?
   b. How will I collect the data?
   c. How much data will I collect?
   d. When will I collect the data?
   e. How will I record the data as I collect them?
   f. What else do I need to do before I start collecting data?

3. Collect your data!

---

measuring. Data may be recorded in a variety of ways, such as using tally marks or placing information in a table. Computer spreadsheets may also be used for recording data.

SURVEYS

Survey data result from collecting information. These data may range from taking a national public opinion poll or observing cars passing the window, to simply tallying the ages of students in a class. A wide array of data are available from the U.S. Census Bureau (Math Links 17.2). The actual data used depends on student interest and maturity, but survey data collected by students provides a freshness that increases student interest and sustains persistence in related problem-solving activities. Computer software is also available that allows students to design a survey and have the persons surveyed enter their responses directly into the computer. Once the data have been collected, the program can display results in a table or in several types of graphs.

Each question from In the Classroom 17–1 gives students an opportunity to collect data themselves. In order to sharpen data-collecting techniques, students may consider the following questions:

- What questions will this survey answer?
- Where should I conduct my survey?
- When should I conduct the survey?

Why is it important for students to think about these questions before conducting their survey? In planning a survey, students are required to refine and polish their questions to get whatever information they are seeking, which in itself is an important and valuable experience. A host of other idea starters are available (Bright, Frierson, Tarr, and Thomas, 2003; Lindquist, Lauquire, Gardner, and Shekaramin, 1992).

EXPERIMENTS

Experiments may be somewhat more advanced than surveys. When students conduct experiments, in addition to using observation and recording skills, they often incorporate the use of the scientific method. For example, students may design an experiment to compare flight times of different paper airplanes. They may try to determine which brand of tissue is the most absorbent. Students may play a spinner game to determine if it is fair to all players.

SIMULATIONS

Although a simulation is similar to an experiment, random number tables or devices such as coins, dice, spinners, or computer programs are also used to model real-world occurrences. Students may start with a probability question such as "If I flip a coin 20 times, how many times will it land on heads?" Then they carry out the simulation and record data as they are generated. Simulations help students gain insight and understanding of empirical or experimental probability.

Sampling is another method of data collection that students can simulate. In statistics, the whole group you are studying is called the *population*. In real-life data collection, there are times when it is impossible or impractical to collect data from a complete population. A *sample* is a subset of a population. Samples are often collected to learn more about public health issues or the public's buying habits, or to predict election results. It is important for students to realize that the use of a particular sample may make a survey biased and to discuss ways to reduce bias. For example, consider some students surveying the students in their room in order to purchase some new games to be used during recess. The population to be surveyed is Ms. Smith's fifth-grade class; however, if the students interview only the girls in the class (a sample), their results will probably not reflect the wishes of both boys and girls. Students to be interviewed could be randomly selected by drawing names out of a hat, which would more likely reflect the total population.

Researchers have developed methods of collecting samples so they may learn more about a population that would be impossible to survey completely. For example, when wildlife biologists want to count the number and types of fish in a lake, they use a mark-recapture technique. The biologist captures a number of fish, counts them, marks them, and releases them. More samples are collected, and the number of marked fish in the sample are compared in a ratio to the unmarked fish in the sample. The biologist may then use the data from the samples collected to estimate the total population of fish and the ratio of marked and unmarked fish. These techniques and how they can be simulated with elementary and middle school students using beans, crackers, or games are explained further in Morita (1999), and Quinn and Wiest (1999).

ANALYZING DATA: GRAPHICAL ORGANIZATION

After data have been collected, the first step of analysis is to organize the information so that the results may be interpreted. A *graph* is a type of diagram that may be used to visually present or organize data. At the elementary and middle school levels, students frequently encounter and create real, picture, bar, and circle graphs. As they move through school, the types of graphs they read and create increase and become more complex. Knowledge related to constructing and interpreting various types of graphs is an important part of mathematics instruction and should begin in the primary grades. Children gain competence with age and experience.

For young children, initial work with data means they begin by working with concrete objects in their environment. Collecting and counting objects, sorting them into categories, and then displaying them in an organized fashion is a good introduction. For example, children may each bring a favorite book and then collaborate to organize their collection. A *real graph*, or concrete graph, is developed as the actual books are arranged in rows. As students work with their "data," they also practice counting skills. Older students may work with larger collections of objects, and, by grouping the objects into groups of tens, place value may be reinforced. You must also find ways to help children move from the concrete, real-graph representations to more symbolic representations.

An example of an early experience is to sort real fruit. First, ask each child to choose one piece of his or her favorite fruit from a basket and position this piece of fruit on a table, as shown in Figure 17-3(a). The resulting rows of fruit represent the children's preferences in a concrete fashion as a real graph. Next, ask each child to draw the fruit he or she chose on an index card, then have the children use the cards to build a picture-bar graph, as in Figure 17-3(b). Although this graph is a less concrete means of showing the information, most children still find it a meaningful way to represent their preferences. Finally, this same information can be expressed more symbolically in the bar graph in Figure 17-3(c). Children can also be given blank paper and encouraged to organize and report their data on the page in a way that makes sense to them. In this way, children learn to communicate their information in ways that are meaningful both to themselves and others (Folkson, 1996).

Regardless of how the data are presented, pertinent questions can be asked to encourage thoughtful interpretation of the graphs. Such questions might include the following:

What do you notice about the graph?
How many children prefer apples?
What is the favorite fruit?
How many different fruits are shown?
How many children contributed to the graph?

These types of questions result naturally from the data and provide valuable opportunities for students to ask as well as answer questions.

> ### Math Links 17.2
>
> There are a number of excellent Web sites that provide opportunities for children to collect data via simulations and graph their results. Here are several:
>
> **Go to the**
>
> - U. S. Census Bureau at http//:www.census.gov. There you will find the latest data regarding population, income, employment, and economic indicators.
> - National Library of Virtual Manipulatives at htpp://www.nlvm.usu.edu. There you will find a number of tools that help to build different graphs, such as Bar charts, Pie Chart, and Histogram.
> - National Council of Teachers of Mathematics Illuminations Web site http://illuminations.nctm.org/ and go to activities to find an array of applets, such as Bar Grapher, Circle Grapher, and Histogram Tool, that allow children to create their own graphs.
>
> www.wiley.com/college/reys

QUICK AND EASY GRAPHING METHODS

For elementary school children, the process of constructing a graph helps them learn about its critical features and is a valuable activity; however, some graphs are more difficult to create than others, and construction may develop into a lengthy process. The applets found in Math Links 17.2 provide ways of collecting data and shows different ways to represent the results geographically that are consistent with the CCSSM recommendations shown in Figure 17-1.

If the main purpose of visually displaying the data is simply to learn more about the data and easily examine results, several alternatives exist to having students create

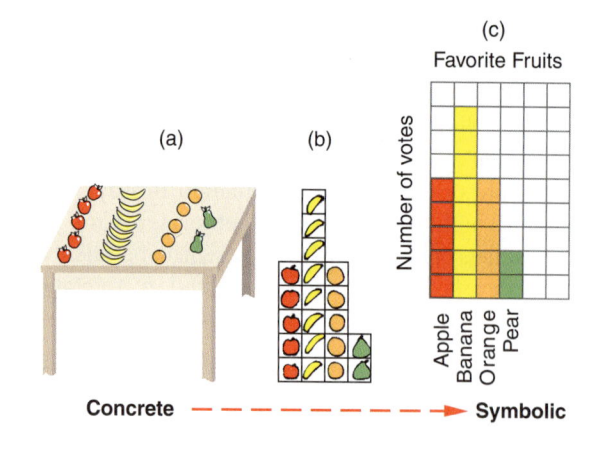

Figure 17-3 *One method for introducing graphs.*

complete publication-quality graphs by hand. By using some of the materials illustrated in Figure 17-4, data may be displayed immediately as they are being collected. These quick graphs, called "sketch graphs" or "working graphs," can be created quickly and provide a visual representation of the shape of the data. They should be clear, but they may not be neat. Sketch graphs don't require labels or titles and don't require time-consuming attention to construction. Sketch graphs may be made with concrete materials or with paper and pencil.

Spreadsheets or statistical programs such as Tinkerplots™ may be used to quickly graph data. As an added benefit, once the data have been entered, students can quickly and easily see their data displayed and printed in more than one type of graph or with a different scale. The graphing calculator is another tool that may be used for making graphs. Graphing calculators designed for middle grades allow students to create picture graphs, bar graphs, and pie graphs.

PLOTS

A *plot* is another type of graph used to visually display data. In recent years, plots have been used frequently in magazines and newspapers because they provide efficient ways of showing information as well as comparing different sets of data. Some plots, such as line plots and stem-and-leaf plots, are quick and easy to make and can be used as sketch graphs to get an initial look at the shape of the data.

LINE PLOTS A line plot may be used to quickly display numerical data with a small range. The range and distribution of the data may be clearly seen in the display. Line plots may be successfully used at all levels. Young children can more easily create a line plot than a bar graph and older students will enjoy the quick feedback they receive when sketching a line plot. When creating line plots, the data should be plotted with the same scale. This allows for a quick visual interpretation of the data when attempting to identify trends in the data.

Suppose that at a class party, some third-grade students receive raisins in single-serving packages. The children open the packs and someone suggests they count to see how many raisins they each received. The class begins counting,

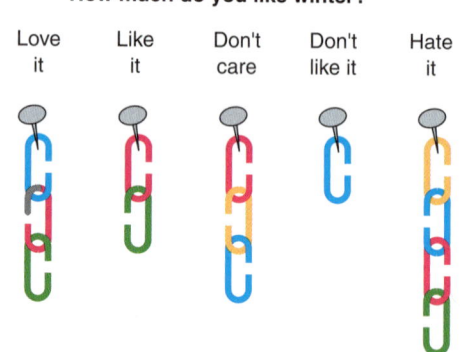

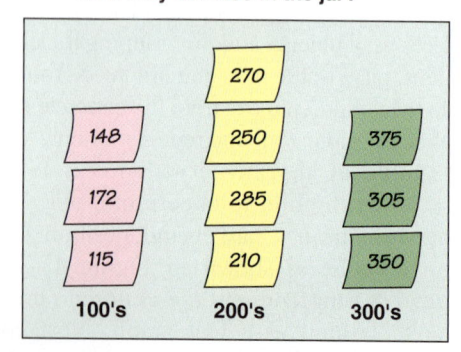

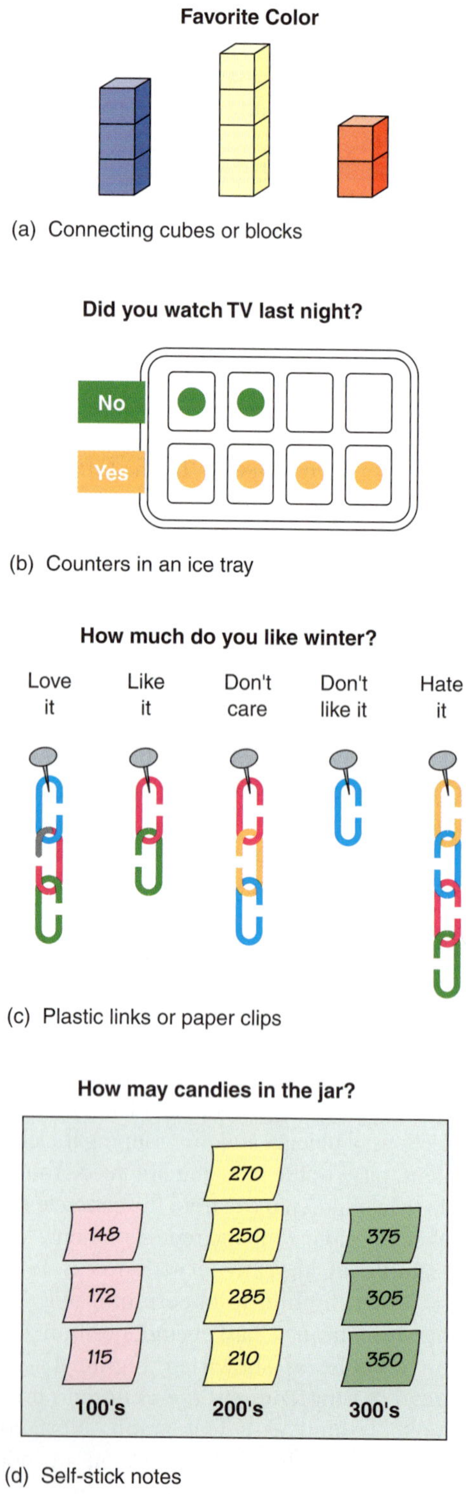

Figure 17-4 *Quick and easy graphing materials.*

and soon each child begins calling out his or her total. You write the totals on the chalkboard as they are called out:

17, 19, 21, 20, 15, 18, 22, 17, 20, 18, 17, 18, 22, 17, 20

Discussion continues as students try to draw conclusions from the data. You suggest putting the data into a line graph so it will be easier to see and understand. The children notice that the smallest value is 15 and the largest is 22, so they direct you to draw a horizontal number line beginning with 15 and ending with 22. Next, an x is placed above each number to mark its frequency in the set of data. Figure 17-5 shows the completed line plot. The students are easily able to see that 17 occurred most often, followed by 18 and 20. The student who had only 15 in the package felt cheated, and those who had 22 felt lucky. The class might talk about why there might not always be the same amount in each package.

STEM-AND-LEAF PLOTS A stem-and-leaf plot is another quick way to display data and provides a quite different representation than when the data are arranged in a line plot or bar graph. It works best with data that span several decades, since the plot is usually organized by tens. This plot is a little more abstract than the line plot, but it may be used successfully with students in the intermediate grades.

It is often useful to display data in more than one way. Consider the raisin data just displayed in a line plot and arrange it in a stem-and-leaf plot (Figure 17-6). To begin, divide each value into tens and ones. The tens become the "stem," and the ones will be the "leaves." Notice that the data fall into two decades, the tens and twenties. A vertical line is drawn with the tens values (1 and 2) on the left of the line. The ones values are placed on the right side, evenly spaced and in numerical order from lowest to highest for each "leaf." By examining the plot, it is easy to see that 17 occurred most often in this set of data. From this arrangement you can also see that more packages contained

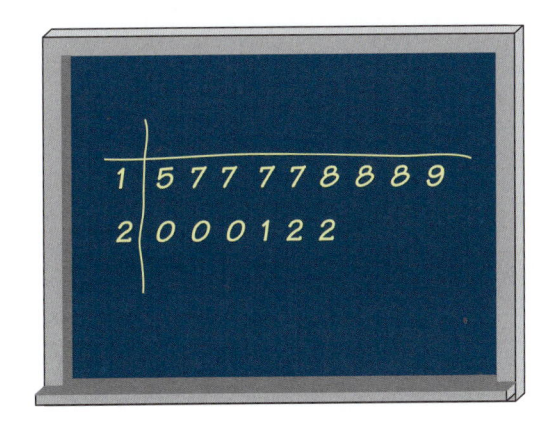

Figure 17-6 *Stem-and-leaf plot of number of raisins per package.*

candy amounts in the tens than in the twenties. Now all those students who had 20 or more feel lucky.

Stem-and-leaf plots may also be used to compare two sets of data in the same plot. Suppose you wanted to explore some questions related to the height of students in a fourth-grade class. Comparing the heights of the boys and girls will generate some interesting discussions. Questions such as "Which group is tallest?" and "Which group has the most variability?" are natural. After some conjectures have been made, it is time to have students measure their heights and begin to analyze those data. Here are some steps that lead to stem-and-leaf plots. The heights of the 27 fourth-grade students (15 girls and 12 boys) are reported in centimeters in the following table:

Boys				Girls			
118	132	135	137	122	155	114	125
120	125	147	129	155	137	136	137
133	148	153	125	134	130	133	145
				148	148	147	

Rather than use a traditional frequency distribution, the values are organized in a stem-and-leaf plot, shown in Figure 17-7. The stem represents the hundreds and tens places of the data on student height, and the leaves represent the ones place.

Thus, in the bottom row of Figure 17-7, |11| 4 means one girl has a height of 114 cm and one boy is 118 cm tall. The stem-and-leaf plot preserves the individual measures while revealing the general shape of the organized data. Thus, it presents all of the information, in this case for both groups, and provides a clear visual picture of it.

In addition to the quick graphs and plots mentioned previously, students should also be involved with graphs that require more construction time and effort.

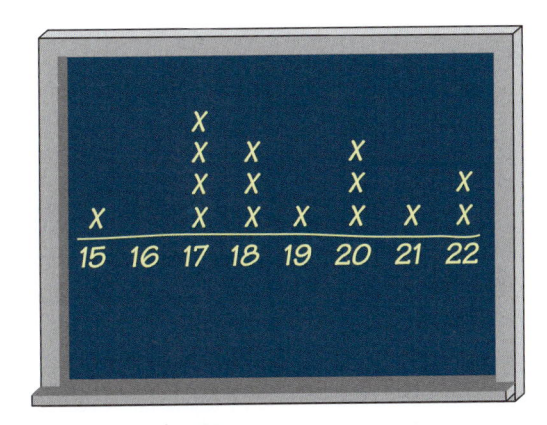

Figure 17-5 *Line plot of number of raisins per package.*

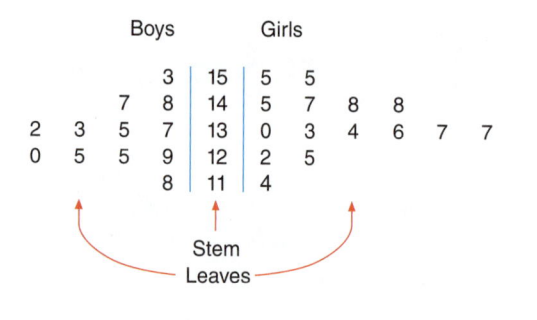

```
         Boys       Girls
            3  | 15 | 5  5
      7  8  | 14 | 5  7  8  8
 2  3  5  7 | 13 | 0  3  4  6  7  7
 0  5  5  9 | 12 | 2  5
            8  | 11 | 4
                 Stem
                Leaves
```

Figure 17-7 *Stem-and-leaf plot of heights of boys and girls.*

BOX PLOTS A *box plot* (also called a box-and-whisker plot) summarizes data and provides a visual means of showing variability—the spread of the data. The box plot shown in Figure 17-8 compares the heights of girls to boys. The median is a key reference point; we will talk more about the median later in the chapter. The lower hinge, or lower quartile, is the median of the lower half of the data; the upper hinge, or upper quartile, is the median of the upper half of the data. These are found by computing the medians of the data in the lower and upper halves, respectively. The interquartile range (IQR) is a measure of variability and is the difference between the upper and lower quartiles. In Figure 17-8, the IQR for the boys is 17 (142–125) and that for the girls is 18 (148–130). The smallest and largest heights represent the lower extreme and the upper extreme. The lines (also called "whiskers") extending from the top of the upper quartile to the maximum value and then from the bottom of the lower quartile to the minimum value provide another visual indication of variability.

The box plot shows many things. For example, it shows that the median height for the girls is greater than that for the boys. Although the groups have about the same IQR, the boys are a bit more evenly distributed throughout the box than are the girls. (Why? Because the median of the boys is closer to the middle of the box than is the median of the girls.) The box plot is derived naturally from a stem-and-leaf plot or line plot. The box plot shows many important characteristics of a group visually and, when two or more groups are shown on the same graph, it allows comparisons to be made easily. Since each section represents 25% of the data, multiplicative comparisons can be made. For example, about 75% of the girls are taller than about 50% of the boys (the median for the boys and the lower quartile for the girls are about the same height). The box plot is an appropriate display for middle school students to construct. It provides a nice connection to descriptive statistics such as median and range. In addition, many standardized test results are reported as box plots, so this is an important kind of graph for teachers to understand even if they don't teach it. Finally, box plots allow two groups with unequal amounts of data to be compared to each other since both sets are converted to fractional representations and are shown on the box plot.

PICTURE GRAPHS

In *picture graphs*, data are represented by pictures. For example, children may graph pictures of their favorite food or the pets they own. A picture can represent one object (Figure 17-9) or several (Figure 17-10). To properly interpret picture graphs, children must know how much each object represents. Research shows that students often ignore such coding information when interpreting graphs (Bright and Hoeffner, 1993; Friel, Curcio and Bright, 2001).

BAR GRAPHS AND HISTOGRAMS

Bar graphs are used mostly for discrete, or separate and distinct, data. For example, they might graph the number of children's birthdays in each month or the number of students who travel to school by bus, by car, or on foot. Figure 17-3(c) showed that values can be read from the axis. Figure 17-11 shows that other times the values are reported directly on the graph. Bar graphs are often used for quick visual comparisons of categories of data and are appropriate for all ages.

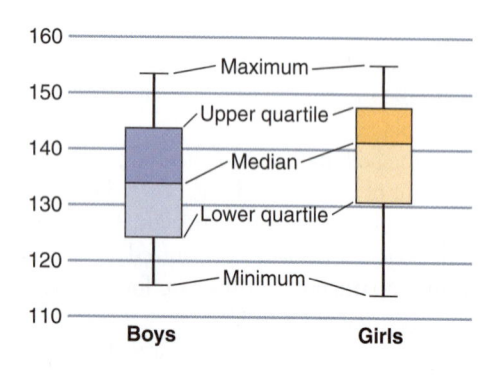

Figure 17-8 *Box plots for heights of boys and girls.*

Figure 17-9 *A picture graph in which each picture represents one object.*

Estimated school retention rates in the United States, 1982 to 2000

Each figure represents 10 persons:

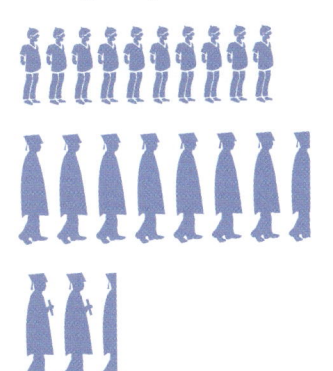

For every 100 students in the 9th grade in Fall 2006

76% graduated from high school by Fall 2010

About 25% will earn a bachelor's degrees by 2015

Figure 17-10 *A picture graph in which each picture represents several objects. (From surveys, estimates, and projections of the National Center for Education Statistics.)*

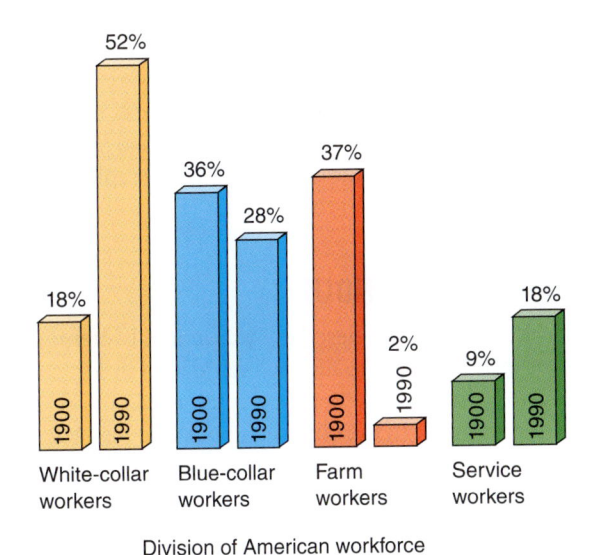

Division of American workforce

White-collar workers: 1900—18%, 1990—52%
Blue-collar workers: 1900—36%, 1990—28%
Farm workers: 1900—37%, 1990—2%
Service workers: 1900—9%, 1990—18%

Figure 17-11 *A bar graph with values shown directly on the bars.*

Although a *histogram* looks like a bar graph (Figure 17-12), there are some key differences that discriminate the two from each other. A histogram is used with continuous data, not discrete data. Therefore the data are represented with connected bars, each representing an interval. Any data that falls within that interval appears in the bar. If a student uses the Internet for 22 hours, his or her data point is located between 20 and 29.99 hours. The interval includes all of the numbers up to, but not including, the maximum value of the interval. Therefore, individual data points may not be distinguishable from reading a histogram as they would

Analyzing Data: Graphical Organization 381

Hours of Internet usage in 1 month reported by 47 students

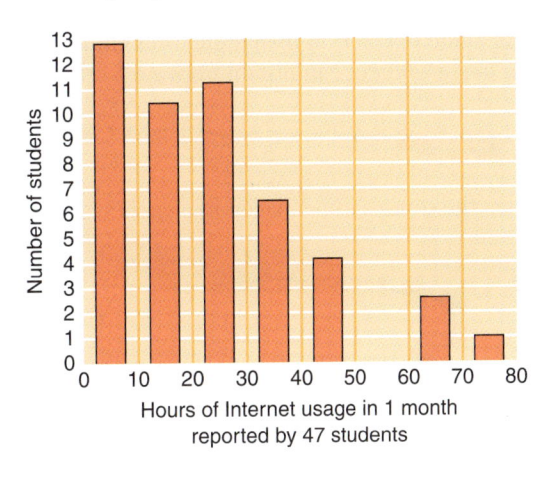

Figure 17-12 *A histogram showing students' hours of Internet usage in 1 month.*

be in a bar graph. Histograms are more appropriate for middle grades students, although many younger children draw them mistakenly when attempting to create bar graphs.

PIE GRAPHS

A *pie graph* is a circle representing the whole, with wedges reporting percents of the whole, as illustrated in Figure 17-13. The pie graph is popular because it is easy to interpret; however, it has major limitations in that it represents only a fixed moment in time, and it cannot exceed 100 percent. For example, students may graph how they spend their weekly allowance or the favorite colors of the class. While students in the upper elementary grades can read and interpret pie graphs, effective construction may require the use of fractions, percents, proportions, and measurement of angles. Pie graph construction is more appropriate for middle school students; however, there is an effective, concrete method of connecting pie graphs to bar graphs, which elementary students may successfully complete.

As illustrated with the line and stem-and-leaf plots, showing the same data in different displays can be both useful and effective. For example, the data shown in the bar graph in Figure 17-3 can be easily shown in a pie graph.

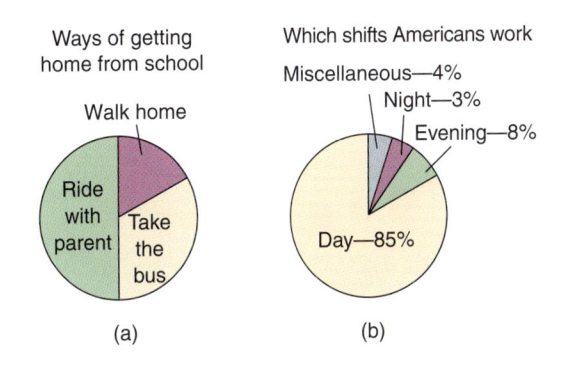

Ways of getting home from school: Walk home, Ride with parent, Take the bus
(a)

Which shifts Americans work: Miscellaneous—4%, Night—3%, Evening—8%, Day—85%
(b)

Figure 17-13 *Pie graphs.*

382 Chapter 17 • Data Analysis, Statistics, and Probability

Cubes of different colors to match the fruits could be strung together as shown here equidistant from each other:

Then the string can be placed in a circle:

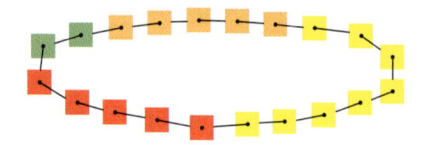

This experience helps make a connection between bar and pie graphs. It also provides a natural context for fractions and percents. For example, the pie graph suggests that one-fourth of the fruits are oranges.

This model can be extended by placing a one-meter tape, or a 100-centimeter strip of paper marked with similar units, around the circle of blocks to form concentric circles, as shown in Figure 17-14(a). Comparing the sections suggested by the different groups of colored blocks with the markers on the strip or tape will identify percents that can be easily read.

As a more concrete visualization, 20 children can be arranged in a circle and the "wedges" of a circle graph duplicated with string, as illustrated in Figure 17-14(b). Both of the models in Figure 17-14 make it easy to estimate or read the percents and conclude, for example, that more than 50% of the children chose apples or bananas.

Similar observations might have been made directly from the bar graph in Figure 17-3, but conclusions involving fractions and percents are much more obvious from the pie graph. Technology applications, such as spreadsheets, are also useful tools that allow students to create bar graphs and pie graphs of the data. The process of moving from a bar graph to a pie graph provides different perspectives for the same set of data, and research suggests that developing such multiple perspectives helps promote greater understanding (Shaughnessy, 2007).

LINE GRAPHS

Line graphs are effective for showing trends over time. In line graphs, points on a grid are used to represent continuous, or uninterrupted, data. Each axis is clearly labeled so the data shown can be interpreted properly. A wide variety of line graphs exist and are used, but three basic assumptions are inherent:

1. The data are continuous rather than discrete. This means that the data are grouped along a continuous scale and cannot be "counted."
2. Data can occur between points with continuous data. Because of this, the line can also be used to interpret values between plotted data.
3. Change is accurately represented with linear functions (i.e., by lines) rather than some other curve.

As Figure 17-15 shows, line graphs are particularly good for showing variations or changes over time, such as hours of daylight, temperatures, rainfall, and so on. For example, students could graph plant growth in a science experiment. Line graphs are also an effective visual means of comparing several sets of data, as illustrated in Figure 17-16. Constructing or interpreting line graphs requires students to examine both horizontal and vertical axes, which is good preparation for understanding a coordinate system. Line graphs are used most often at the middle school level.

GRAPHICAL ROUNDUP

Each of these graphs deserves instructional attention as students examine ways to display their data. Children need experience constructing them and interpreting information that is represented. In the children's book *The Best Vacation Ever* (Murphy, 1997), a little girl collects and displays data

(a) (b)

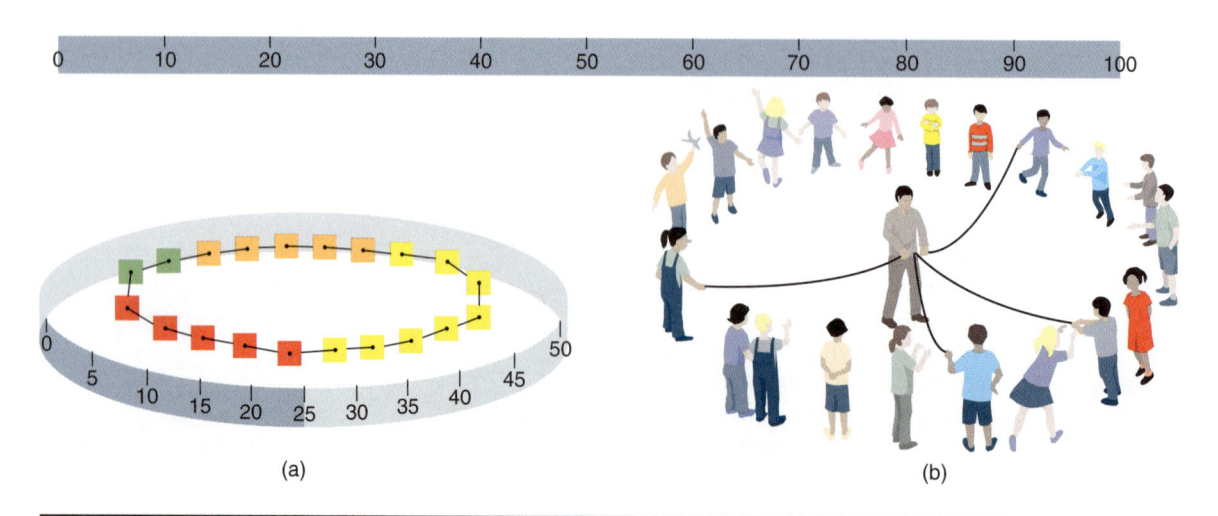

Figure 17-14 *Models for interpreting circle graph data.*

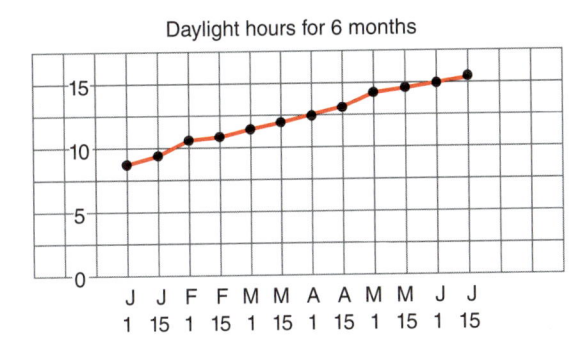

Analyzing Data: Descriptive Statistics   **383**

Figure 17-15 A *line graph of a single set of varying data.*

for her family so they can plan a vacation. In the Classroom 17-2 gives students ideas to think about when planning to display a collection of data.

The availability of graphing calculators and graphing programs allows for easy construction of a variety of graphs. This availability of different graphs via technology places a greater premium on interpreting and understanding the graphs that are so easily produced. As children become familiar with different graphs, they should recognize some characteristics associated with them. Figure 17-17 highlights specific characteristics of graphs that are encountered in elementary and middle school. The focus here is not on memorizing characteristics of these graphs, but rather on becoming aware that each type has strengths. The selection of a graph should capitalize on these strengths while recognizing any of its limitations. Figure 17-18 shows graphical complexity and a suggested progression for introducing different types of graphs that are consistent with the

---

> ### *In the Classroom 17–2*
>
> **SHARING DATA**
>
> *Objective:* To develop techniques for handling data.
>
> *Grade Level:* 4–5
>
> Getting Started
> - Collect some data:
> - Before sharing our data, we need to agree on some things:
>
> Before Graphing:
> - Shall we start with a sketch graph?
> - What kind of graph will best display our data?
> - Can we show our data in more than one way?
> - How shall we label our graph?
>
> Before Reporting Results:
> - What questions can we answer with our data?
> - Who might be interested in our results?
> - How will we report our findings?
>
> ▼ Why do these questions need to be answered before graphing and planning our report?
>
> ▼ Name two other important questions that need to be answered before beginning.
>
>   1. _____
>   2. _____

CCSSM and GAISE expectations shown in Figures 17-1 and 17-2. Again, memorizing this information is not the intent. Instead, be aware of how some graphs, due to their complexity, are better understood at different levels, which includes students' ability to read and interpret a graph versus progress for introduction of types of graphs (includes creating the graph, both reading and constructing graphs).

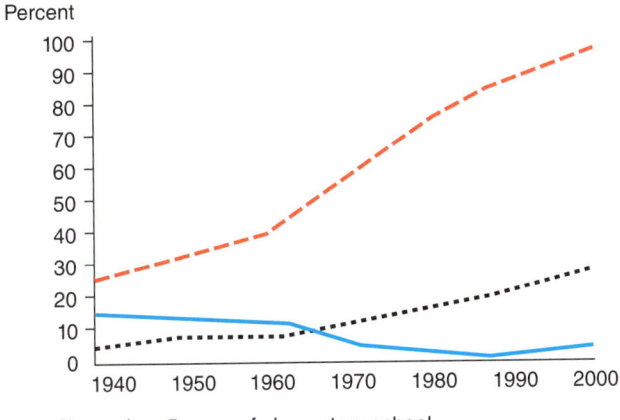

ANALYZING DATA: DESCRIPTIVE STATISTICS

Another way to analyze data is to use descriptive statistics. So much information exists today that it must often be simplified or reduced in ways other than by graphs. The organization and summarization of data is called *descriptive statistics*. Descriptive statistics are in common use. They are introduced in the primary grades through data collection and graphs, and then extended with further exploration and practice activities in the intermediate and middle school grades. Here are some familiar examples:

Figure 17-16 A *line graph comparing several sets of data.*

384　Chapter 17 • Data Analysis, Statistics, and Probability

Type of Graph	Characteristics
Real	Used by young children Actual objects are placed on graph
Line plot	Provides a quick way to examine the shape and variation or spread of data Gives a bar-graph-like representation
Stem-and-leaf plot	Efficient way to show detailed data Provides similar visual patterns as a bar graph but more detailed information Uses stem-and-leaf coding that needs to be understood Technology has facilitated its use
Picture	Frequently encountered in newspapers and reports Generally easy to use and interpret but visuals may be misleading Codes/keys which accompany graph need to be understood
Bar	Mostly used for discrete data Frequently encountered in newspapers and reports Easy to interpret Uses scales/codes that need to be understood
Histogram	Used for continuous data, data grouped in intervals Similar to bar graph, but individual data grouped in intervals Effective to display data with a large range but limited values
Pie	Frequently encountered in newspapers and reports Shows fractional parts, which are based on a whole or 100% Easy to use and interpret Difficult to construct by hand, easy with technology
Line	Frequently encountered in newspapers and reports Used for continuous data Effective to show patterns, trends, comparisons, and change over time Uses vertical and horizontal scales that need to be understood Provides good readiness for coordinate graphs
Box plot	Provides useful information about the variability of data Requires knowledge of range, median, and quartiles to interpret Can be used to compare two data sets with unequal amount of data Technology has facilitated its use

Figure 17-17 *Characteristics of graphs.*

Tables as representational or as organizing tools

Grades K–2 → Grades 3–5 → Grades 6–8

- Object graphs
- Picture graphs
- Line plots
- Bar graphs (with use of grid lines to facilitate reading frequencies; labeling of bars with numerical values)

- Bar graphs (stacked or using multiple sets of data)
- Stem plots
- Pie graphs (reading primary emphasis)

- Bar graphs (reading and constructing)
- Histograms
- Box plots
- Line graphs

Introduction and use of scale

Developing mathematics knowledge

Complexity of data

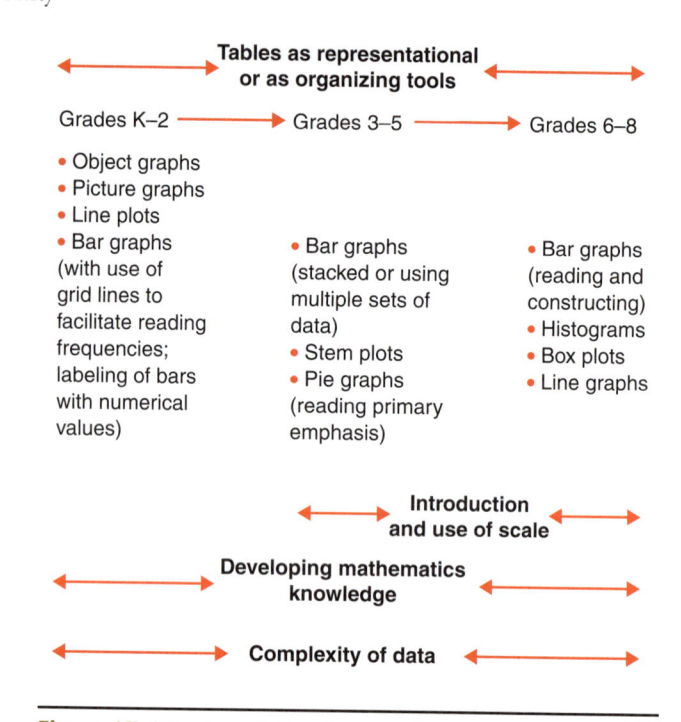

Figure 17-18 *Graph/display complexity: suggested progress for introduction of types of graphs (includes both reading and constructing graphs).*

"Most children in the fifth grade are 10 years old."

"The median family income is $56,000."

"The average temperature today was 29°F."

Each of these statements uses a number to summarize what is typical for a current situation or condition. Two of the most common types of descriptive statistics include measures of variation and measures of central tendency. For each of the descriptive statistics that follow, we provide a concrete example using cubes and an example using numbers. For some children, manipulating the physical model not only helps them understand the formula but also promotes retention.

MEASURES OF CENTRAL TENDENCY OR AVERAGES

> **Virtual Classroom Observation**
>
> Go to www.wiley.com/college/reys,
> Access the Wiley Resource Kit
> Click on the Virtual Classroom Observations Section
> Data Analysis and Probability: Measures to Center
> Teaching Examples:
> Video: Guessing versus estimating
>
> What are some strategies used as students moved from guessing to estimating?

The word *average* is a popular statistical term that many children have heard. It is used to report such things as average temperature, average family income, test averages, batting averages, and average life expectancy.

Any number that is used to describe the center or middle of a set of values is called an *average* of those values. Many different averages exist, but three—mode, median, and mean—are commonly encountered in elementary and middle school. Simply being able to state the algorithm for finding these statistics is not enough. To support the development of data sense, each of these should be developed meaningfully through concrete activities before introducing computation. Such experiences provide greater conceptual knowledge or understanding of the concept of average. Spreadsheets and software programs provide explorations that allow middle school students to investigate what happens to various descriptive statistics when different values are used (Wilson and Kraft, 1995).

MODE The *mode* is the value that occurs most frequently in a collection of data. Using concrete materials such as cubes, the mode is easily identified as the tower height that occurs most often. In Figure 17-19 it is easy to see that 9 is the mode because it occurs twice, while the other values only occur one time each. In graphical terms, this is also the largest portion—the tallest column in a bar graph, for example. In Figure 17-20(a), the most frequently occurring test score is 90 (it occurred twice), so the mode is 90. The mode is a versatile average in that it may be used with both numeric and nonnumeric data, also called categorical data. The mode is easy to find and is affected very little by extreme scores. Young children are often interested in which item occurs most. Therefore, their initial experience with average begins with mode. The children are comfortable using the mathematical term, *mode*, as the teacher introduces it to them.

Students' ages within a class provide an excellent application of mode, because within a given class, a large number of children are the "same" age. Businesses also frequently rely on the mode to select merchandise. Suppose, for example, that you own a shoe store. The modal shoe size has practical value for restocking because you want to stock the sizes most people wear. In some cases, a data set will have multiple values that occur most frequently. For example, if a class of students is asked how old they are, and there are nine students who are 10 and nine students who are 11, and no age occurs more then nine times, then the data set has two modes. In the case of two, the set is considered to be bimodal; three modes is called trimodal, and so on.

MEDIAN The *median* is another type of average that can easily be identified without the need for computation. It is used with numeric data only and is the middle value in a data set of ordered data. The median divides the data into two equal subsets. Thus the same number of values are above as below the median. The median is easy to illustrate. Look again at the children's ages in Figure 17-19. Before looking for the median, the data must be ordered. For example, these towers are arranged smallest to largest. Once that has been done, the blocks in the outside towers can be removed, one at a time, until the middle value remains. It is easy to see that the middle age is going to be between 7 and 9 years. Reference to a highway median will remind students that a median in statistics is a middle position. Notice that the median is not the middle of the range of data because data aren't always spread symmetrically over the range. One way of visually representing this is to provide children a strip of grid paper that has exactly as many boxes as data values. Have them put each ordered data point into a box and fold the strip in half. The median is the fold. In the children's age example, the median simply tells us that there are as many children in the group from ages 2–7 as there are in the group containing ages 7–9.

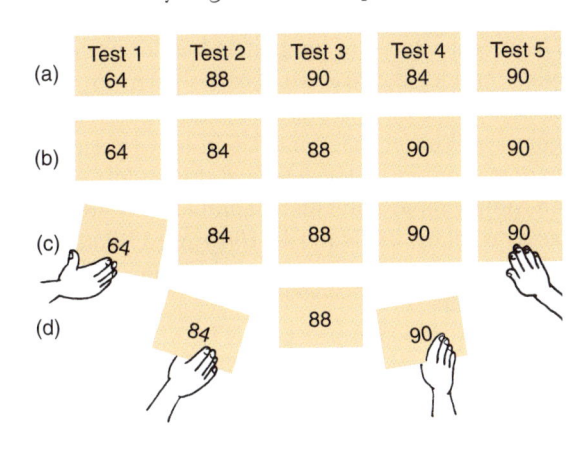

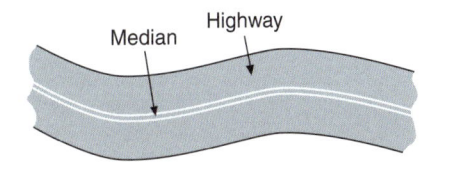

Figure 17-20 *Model for finding the median of five test scores.*

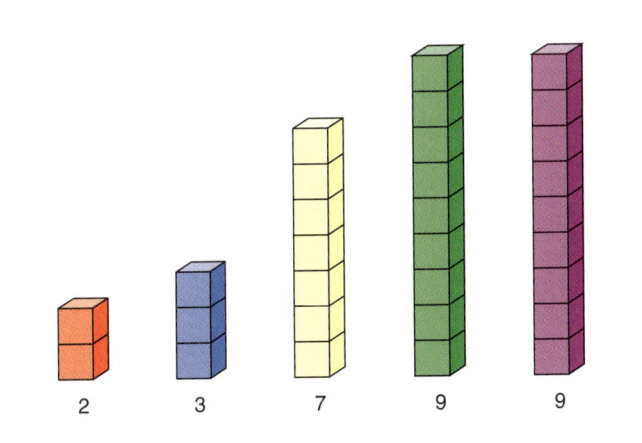

Figure 17-19 *Ages of children.*

The median, too, can be modeled with numbers. For example, consider the five test scores shown on cards in Figure 17-20(a). Ordering them from lowest to highest, as in Figure 17-20(b), provides practice in using greater than, less than, and ordering skills. To find the middle score, or median, simply remove the highest and lowest cards simultaneously, as shown in Figures 17-20(c) and 17-20(d). Continue this process until the middle card remains. This score, 88, is the median.

There are five scores in Figure 17-20. Suppose a sixth score of 17 was made. Ordering the six test scores, as shown in Figure 17-21(a), could make a new arrangement. Again, remove the highest and lowest cards simultaneously until two cards remain. In this case, as shown in Figure 17-21(b), the median is the middle point between these two scores, or 86. If you add together the two remaining cards and divide them by 2, you will obtain the median as well (84 + 88)/2 = 86. It should be noted that the GAISE document recommends that younger children first be introduced to the median using data sets with an odd number of entries so that the median is one of the entries in the data set.

MEAN The *mean* is called the arithmetic average because it is determined by adding all the values involved and dividing by the number of addends. The mean is the most difficult to compute, although it can be understood by children beginning in the upper elementary grades. It is used with numeric data only. When people talk about finding the average of a set of data, they are often referring to calculating the mean. It is important that children realize that the mean is not the only type of average.

Simply being able to state or use the algorithm does not indicate understanding of the mean. Difficulties in interpreting the mean were shown on a national assessment. Figure 17-22 shows a question that required eighth graders to determine what data would be reasonable for a given mean. Less than 40% answered all four choices correctly. These assessment results illustrate that many middle grade students are able to calculate averages when asked to do so, but the depth of their understanding of the concept of average is shallow.

Providing conceptual interpretations for the mean helps develop understanding of average. Say that four students have 9, 5, 3, and 7 trading cards. The cubes in Figure 17-23(a) show the number of cards each child has. The first interpretation involves the concept of equal distribution or sharing,

Akira read from a book on Monday, Tuesday, and Wednesday. He read an average of 10 pages per day. Circle whether each of the following is possible or not possible.

Possible	Not Possible		Monday	Pages Read Tuesday	Wednesday
A	Ⓐ	(a)	4 pages	4 pages	2 pages
Ⓑ	B	(b)	9 pages	10 pages	11 pages
Ⓒ	C	(c)	5 pages	10 pages	15 pages
D	Ⓓ	(d)	10 pages	15 pages	20 pages

Figure 17-22 *Eighth-grade national assessment question on interpreting an average (correct responses are circled).*

an idea very familiar to students. In this interpretation, the mean is identified as the number that describes the data if each piece of data was "evened out" or the same as all others. People often think of this as what is typical for the data. If children are asked to even out or share the cards so that each student has a fair or equal share of cards, this evening-out process produces a mean of six cards per student (Figure 17-23(b)). As students gain experience and become comfortable with the equal distribution interpretation, they also begin to realize that with large data sets, such as whole-class data, trying to share equally becomes cumbersome. Students often discover the add-then-divide algorithm on their own, or you can introduce it with concrete materials. The mean also could have been determined by computing:

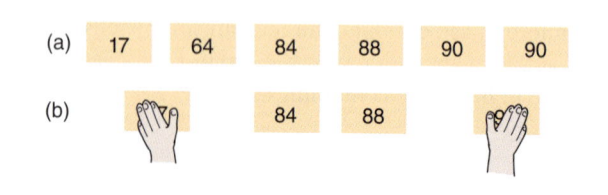

Figure 17-21 *Model for finding the median of six test scores.*

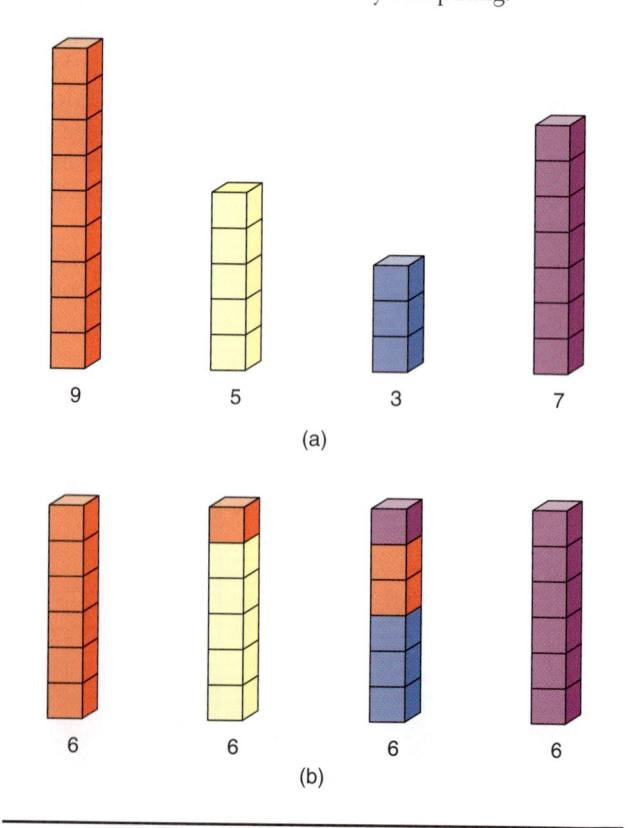

Figure 17-23 *Equal distribution or sharing model for the mean.*

Analyzing Data: Descriptive Statistics   387

would be 88, 64, and 0. The length of the tape is still 152 cm. But if we fold the tape into thirds, rather than in half, the length of the strip is a little over 50 cm. Therefore, the mean test score dropped over 25 points! This technique is appealing and enlightening.

Not all data sets divide evenly, and once students are comfortable with the concept of evening out, they can discuss what to do with remainders, which makes a natural connection to decimals. They can also discuss how some data cannot be divided and some averages that are calculated are not realistic. For example, in one data set the mean indicated that the average number of children per family was 2.5. Children will be quick to point out that you can't have five-tenths or one-half of a child.

These types of experiences help students understand some fundamental notions related to the mean; namely, that the mean must be somewhere between the values averaged and that it is the typical value or balancing point for a set of data.

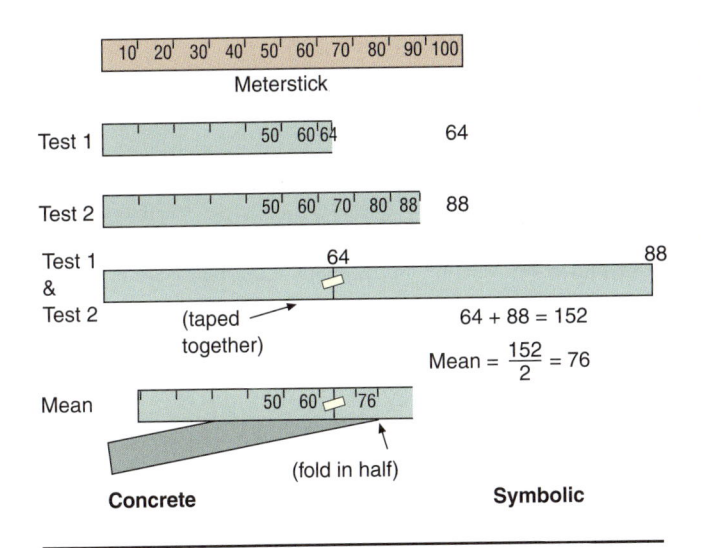

Figure 17-24 *Model for finding the mean using lengths of adding machine tape to indicate test scores.*

$$\text{mean} = \frac{9+5+3+7}{4} = \frac{24}{4} = 6$$

Figure 17-24 shows a way to model the add-then-divide algorithm for the mean. Test scores are returned to children on pieces of adding machine tape, and the length of each strip is determined by the score (e.g., a score of 88 is 88 cm long and a score of 64 is 64 cm long). Scores can be physically compared using the tapes (e.g., it is clear that the score on Test 2 was higher). To show the mean score, simply tape the two strips of paper together (add) and then fold the resulting strip in half (divide by 2). It also illustrates the effect of a test score of zero when the fold number on the tape is increased by 1. Using this example, the three test scores

CHOOSING AVERAGES   As students use averages, they must be aware of how averages are influenced by data. For example, in Figure 17-20 the mean for 5 tests is 83 and the median is 88. When the extremely low score of 17 is included in Figure 17-21, the mean drops greatly (from 83 to 72) but the median changes only slightly (from 88 to 86). One negative characteristic of the mean is that its value is affected by extreme scores.

Grasp of the concept of average is a powerful tool in estimation and problem solving. Problems such as the one in Figure 17-25 provide opportunities to apply averages and estimation in everyday situations.

Finding the mean, median, and mode for the same data can generate discussion about when certain averages should be used. You can use In the Classroom 17-3 to

Figure 17-25 *Example problem for developing averaging and estimating skills.*

In the Classroom 17–3

PEANUTS

Objective: To choose the best average.

Grade Level: 4–8.

Suppose you have opened some Nutty Bars to check the company's claim of an "average" of 8 peanuts per bar. Here is what you found after opening 10 bars.

Bar	1st	2nd	3rd	4th	5th	6th	7th	8th	9th	10th
Number of peanuts	5	8	8	8	11	7	8	6	6	6

▼ Create some different sketch graphs so you can examine the shape of the data. Which graph would the company probably use to promote their product? Why?

▼ Calculate the averages. You may use counters such as beans and grid paper to represent the peanuts if you wish.
  - What is the mean number of nuts?
  - What is the median number of nuts?
  - What is the modal number of nuts?
  - Which average did the company probably use? Why?

▼ Write at least three questions that can be answered by your graphs and statistics.

1. _____
2. _____
3. _____

---

allow children to calculate averages with concrete materials or numbers and to discuss how one might select a particular average to describe the data. For another example, look at In the Classroom 17-4. Calculating the mean, median, and mode provides practice in computational skills. More important, however, is deciding which of these averages to report. The median salary of $500,000 or the modal salary of $480,000 seems more representative than the mean salary of $2,030,000. If salary negotiations were taking place, the players might cite one average and the owners a very different average. Discussing which averages are appropriate for what purpose helps students better understand why different ones exist and are used. Once children have collected their own data, they should determine which averages to calculate and report.

Each of the averages can be modeled and developed in ways that are appealing, interesting, and meaningful. No new mathematics is required, yet learning about averages provides a vehicle for applying many mathematical concepts and skills that students are developing.

Care must be taken to ensure that statistics is viewed as more than a series of skills or techniques. For example, finding an average is an important skill that should be developed; however, the teaching of statistics must not stop with the "how to." Rather, it must raise questions such as "When is an average useful?"

For example, the ice cream survey in Figure 17-9 reports that eight students like chocolate, five like vanilla, and three like strawberry. This picture graph clearly and accurately shows student ice cream preferences. These preferences are an example of nonnumeric data, or categorical data. A mean or median of these data could be computed but would be inappropriate. In fact, a mean or median is meaningless for these data! Before any statistics are computed, challenge students to decide what questions are to be answered and discuss what statistics, if any, are needed to answer them. Nonnumeric data such as ice cream flavors are better analyzed with frequency tables or graphs. If an average is desired, the mode would be the appropriate average to use.

Additional questions might include "Why should the average be reported?" "What average is most appropriate?" "Why?" "What degree of precision is needed?" These questions are essential and must be asked regularly. This sort of discussion will support students as they develop their data sense. The teaching of statistics in elementary school must

Interpreting Results 389

In the Classroom 17–4

WHAT'S THE AVERAGE?

Objective: To calculate mean, median, and mode in deciding which average is most appropriate to use for a particular purpose.

Grade Level: 4–8.

$380,000	600,000
420,000	5,740,000
480,000	2,700,000
480,000	8,500,000
480,000	
520,000	

Here are yearly salaries of one professional basketball team!

- What is the mean salary for the team? _____
- What is the median salary? _____
- What is the mode? _____
▼ Decide on the best average to report for the team:
  _____
  - Why is your choice better than other averages?
  _____
▼ Use the Internet to find other professional team salaries. Find and report the averages.

aim higher than skill development. Students should know how to get a statistic, but they must also know what the statistics tells them.

Math Links 17.3

To view a middle school lesson where students are collecting data and determining averages, you can view *Trashketball* from Modeling Middle School Mathematics by going to http://www.mmmproject.org or linking through this book's Web site. Click on video matrix to find this video.

www.wiley.com/college/reys

MEASURES OF VARIATION

As students investigate measures of central tendency, they should be presented with multiple data sets that have the same median and mode. Figure 17-26 shows the number of points scored by two basketball players for their last 10 games. When comparing the data, it should be noted that the mean, median, and mode do not help in distinguishing between the two players. In this case, a measure of variability could distinguish their performance by illustrating a difference in the consistency of the two players. Measures of variability are used to describe how much the data are spread out. One measure of variability is the *range*. The range is a simple measure that tells the difference between the maximum and minimum values in the data set. Students may find the range by comparing the maximum value to see how much greater it is than the minimum value. The range for player 1's scores may be found by subtracting 16 from 20, resulting in a range of 4; for player 2, you would subtract 0 from 40, resulting in a range of 40. The difference in the ranges shows that player 2 has a great deal more variability in his scores.

We have already mentioned another measure of variability, the interquartile range (IQR), which is found when subtracting the first from the third quartile. (See the earlier discussion of box plots.) One more measure of variability that may be explored by middle school students is the *mean absolute deviation*. The mean absolute deviation has to do with the distance (always positive) of each data entry from the mean. The distances may be found by taking the absolute value of the difference between the mean and the values. These distances are then averaged to obtain the mean absolute deviation of the data from the mean. Figure 17-27 shows the calculation of the mean absolute deviation of scores for the basketball player data seen in Figure 17-26. The mean absolute deviation shows that player 1 was much more consistent than player 2.

The introduction of range, IQR, and mean absolute variation to measure variability are important for elementary and middle grades students to prepare them for the study of variance and standard deviation in secondary grades.

INTERPRETING RESULTS

DATA SENSE

Once data have been collected and displayed, they should be analyzed and interpreted. Just as children develop knowledge about numbers or number sense, they can also develop knowledge about statistics or *data sense*. Data sense is gradually developed as students formulate questions, collect data, construct graphs, find descriptive statistics, and interpret them in a variety of contexts. Students with data sense are able to determine how data should be interpreted. They are able to read and evaluate statistical information being presented, such as material presented by the media. Students also display data sense when they are able to use statistical language when reasoning about data. The goal is for students to develop both procedural knowledge (how to

	Points scored in last 10 games	Mean	Median	Mode
Player 1	16, 20, 20, 18, 22, 24, 20, 20, 20, 20	20	20	20
Player 2	10, 2, 20, 36, 4, 20, 38, 0, 30, 40	20	20	20

Figure 17-26 *Comparison of scores of basketball players for the last 10 games.*

Player 1		Player 2	
16	$|16 - 20| = 4$	10	$|10 - 20| = 10$
20	$|20 - 20| = 0$	2	$|2 - 20| = 18$
20	$|20 - 20| = 0$	20	$|20 - 20| = 0$
18	$|18 - 20| = 2$	36	$|36 - 20| = 16$
22	$|22 - 20| = 2$	4	$|4 - 20| = 0$
24	$|24 - 20| = 4$	20	$|20 - 20| = 0$
20	$|20 - 20| = 0$	38	$|38 - 20| = 18$
20	$|20 - 20| = 0$	0	$|0 - 20| = 20$
20	$|20 - 20| = 0$	30	$|30 - 20| = 10$
20	$|20 - 20| = 0$	40	$|40 - 20| = 20$
Sum	12	**Sum**	128
Average (mean absolute deviation)	$12 \div 10 = \mathbf{1.2}$	**Average (mean absolute deviation)**	$128 \div 10 = \mathbf{12.8}$

Figure 17-27 *Calculation of mean absolute deviation for basketball players' scores for the last 10 games.*

construct a graph or calculate a statistic) and conceptual knowledge (understanding what a graph or statistic is communicating) (Friel, 1998).

One way to begin interpreting data is through the use of questions. As in the opening Snapshot of a Lesson, students should be encouraged to examine their results and discuss questions that may be answered by the data. There are three levels of graph comprehension, which progress from lower-level to higher-level questions. Your goal should be to teach students to move beyond lower-level thinking and to ask and answer higher-level questions.

- *Reading the data.*
  The student is able to answer specific questions for which the answer is prominently displayed. For example, "Which player averaged the most points?"
- *Reading between the data.*
  The student is able to find relationships in the data, such as comparison, and is able to operate on the data. For example, "How many players had a median less than their mean?"
- *Reading beyond the data.*
  The student is able to predict or make inferences. For example, "Which player had the greatest range? The smallest range? What do these numbers tell you about the player?"

Data may also be interpreted by describing the shape of the data in the graph. Corwin and Friel (1990) suggest first having students use informal language such as "clumps," "holes", and "spread out," to describe features of the data. You may want to create a word wall, or visible word list, for students to refer to when describing data. Second, students should attempt to develop theories about why the data look the way they do. This second step encourages students to read beyond the data.

Another beneficial graphical interpretation task is to give students mystery graphs that are missing some of their labels and have students predict what the data might be. For example, groups of students may have each measured a body part such as arms, legs, or distance around heads. A graph that contains measurements from 18 to 24 inches is displayed. Students would be asked to hypothesize which body parts might be illustrated on the graph. This activity also reinforces the importance of using labels on graphs.

A variation of the mystery graph activity involves the use of a graphing calculator and a calculator-based laboratory (CBL). With a CBL motion detector, graphs may be created to show movements. Students may be given a CBL graph, and the challenge is to move so they can duplicate the graph. For example, Figure 17-28 shows the graphs of three different people walking. Time is shown on the horizontal axis, and the distance from a motion detector is

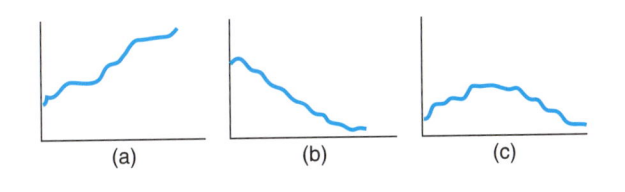

Figure 17-28 *Graphs showing distance and time.*

shown on the vertical axis. Examine parts a, b, and c, and decide which of these graphs shows movement toward the motion detector. Part c is different because, instead of showing continuous movement away or toward, it shows a person leaving and then returning. These sorts of activities encourage students to consider the axes of the graphs and think carefully about how data are placed on a graph.

MISLEADING GRAPHS

Another important component in developing data sense is the ability to critically examine graphs and correctly interpret the data presented. "Although viewing data graphically allows the reader to see the trend of the data easily, the choice of scaling along the x- and y-axes could influence their interpretation of the data" (Harper, 2004, p. 341). Sometimes even simple graphs may be misleading. For example, consider the graph shown in Figure 17-29. Eighth graders were asked to explain why this 200% graph was misleading.

Although the captions in the graph indicate that the amount of trash has doubled in two decades, the visual elements reflect a doubling of both the width and height to produce a figure whose area is four times greater and a volume that is actually eight times greater.

People may focus on the visual graph and ignore the numerical data that accompany the graph. In fact, less than 10% of eighth graders identified the critical problem associated

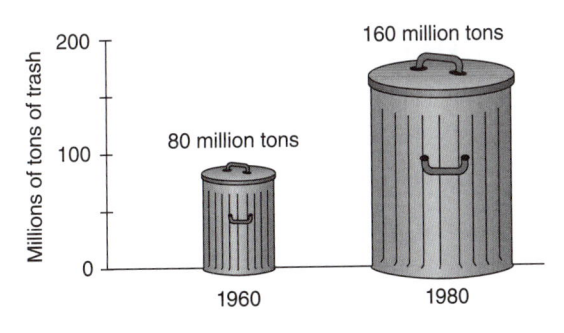

Figure 17-29 *Eighth-grade national assessment question on interpreting graphs.*

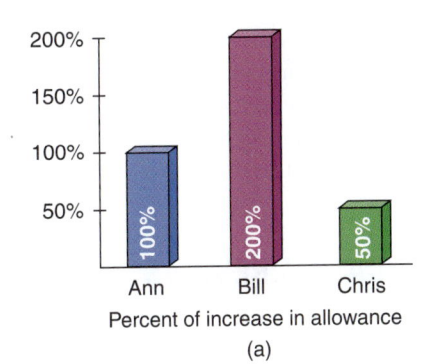

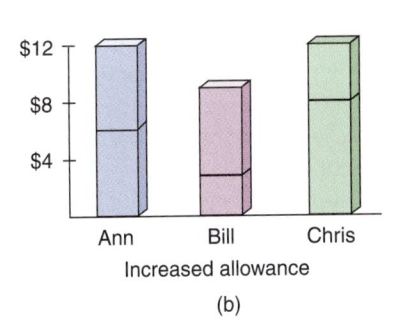

Figure 17-30 *Example of distortion of data in graphs.*

with the graph in Figure 17-30, which suggests that instructional attention needs to be given to helping children examine graphs with a careful and suspicious eye.

Graphs may also be deceptive in other ways. For example, the graph in Figure 17-30(a) reports changes in allowances for three children. It shows that Ann's allowance was doubled, Bill's tripled, and Chris's increased by one-half. Based on this information, Bill may be feeling philanthropic and Chris complaining of hard times. What is wrong with the graph? Technically, it is correct, but it doesn't tell the entire story because the original allowances were not the same. Let's look at the actual data:

	Original Allowance	Size of Increase	Amount of Increase	New Allowance
Ann	$6.00	Double	$6.00	$12.00
Bill	$3.00	Triple	$6.00	$9.00
Chris	$8.00	Half	$4.00	$12.00

As Figure 17-30(b) shows, a graph with a vertical axis labeled differently reflects the situation more accurately. These different graphs of the same data demonstrate how graphs can distort and sometimes misrepresent information. Developing a healthy skepticism of graphical displays is an important part of developing graphing skills. Harper (2004) suggests having students construct misleading graphs and then discussing the message being conveyed by the graphs.

This can help students develop a better sense of misleading or inaccurate graphs, and what to look for when analyzing a display. According to Harper, students are more likely to notice scaling features, but "they are not as keen to observe pictorial embellishments or consider the source of the graph" (p. 343).

COMMUNICATING RESULTS

Once data have been collected, analyzed, and interpreted, it is appropriate for students to communicate their findings. Just as in problem solving, students should be encouraged to look back at their results. Communication can help students clarify their ideas during this process. For example, students might complete regular journal entries throughout their project. It is also useful to have large- or small-group discussions about the data.

Research supports having students go through a writing process when analyzing data, similar to one used in language arts (Shaughnessy, 2007). First, the students are involved in planning or prewriting. Second, a rough draft is created. Third, students revise their work. Finally, the work is published and shared with others. Students could craft business letters to a company or group that would be interested in the data and mail them, similar to the task in the Snapshot of a Lesson. This is an opportunity to integrate across subject areas.

It is valuable for students to learn to communicate their results with others. Their final presentation may include oral or written communication or both. They must learn to report clearly the answers to their original questions and hypotheses and select the most appropriate way to communicate their findings in a graphic format or by using other statistical measures. The presentation may be for other classmates; however, a presentation for someone outside of the classroom can be particularly meaningful. For example, one class calculated the area of classrooms in their building and compared the average area of the fifth-grade classrooms to the sixth-grade classrooms. Their findings were presented to the principal who made room assignments for the building (Scavo and Petraroja, 1998). Another group of students collected data to find out which school lunches were the most and least popular. Once their findings were collected and displayed, they presented their findings to the director of food services, who planned the monthly menus.

PROBABILITY

In daily conversations, it is common to speak of events in terms of their chances of occurring. *Probability* is used to predict the chance of something happening. The terms *chance* and *probability* are often applied to those situations where the outcome cannot completely be determined in advance. Here are some examples of common probabilistic statements:

"The chance of rain today is 40%."
"The Cardinals are a 3-to-1 favorite to win."
"The probability of an accident on the job is less than 1 in 100."
"The patient has a 50–50 chance of recovering."
"If I study, I will probably pass the test."
"I am sure we will have a test Friday."
"We will have milk in the cafeteria today."

The first four statements are commonly heard and relate directly to probability. The last three illustrate a subtle but frequent use of probability in many everyday situations. In all of these cases, the utilitarian role of probability makes it an important basic skill. One way to increase students' awareness of the use of probability is to have them make a daily or weekly list of probability statements they have seen (in newspapers, in magazines, or on television) or heard (on radio and television).

The study of probability is intertwined with the study of statistics. "Often times, probability is addressed as a subset of concepts addressed within statistics and little connection is made between data analysis, descriptive statistics, and probability in school mathematics. Some of the most powerful and useful ways to use probability involve making sense of a statistic derived from samples" (Stohl, 2005, p. 350.) In fact, according to the NCTM *Standards* (2000), ideas from probability serve as a foundation for the collection, description, and interpretation of data. Probability will not and should not be learned from formal definitions; rather, the presentation of varied examples and activities helps illustrate and clarify important concepts. In the early grades, the treatment of probability should be informal. At all stages of instruction, though, you must use correct language to describe what is happening. This language serves as a model for children as they begin developing probability concepts and simultaneously add new probabilistic terms to their vocabulary. Let's look at appropriate ways for elementary students to experience some key concepts and terms.

PROBABILITY OF AN EVENT

Look at these statements that involve probability:

The probability of tossing a head is $\frac{1}{2}$.
The probability of rolling a four on a standard six-sided die is $\frac{1}{6}$.
The probability of having a birthday on February 30 is 0.

In these examples, tossing a head, rolling a four, and having a birthday on February 30 are *events*, or *outcomes*. Probability assigns a number (from 1 to 0) to an event. The more likely an event is to occur, the larger the number assigned to it, and so the probability is 1.0 when something is certain to happen. These numbers are commonly assigned percents as well. For example, the probability of students in an elementary class having been born is 1, or 100%. On the other hand, the probability of something impossible happening is 0, or 0%. For example, the probability of students in the class having been born on February 30 is 0. Therefore, all probabilities lie between 0 and 1, or 0% and 100%. Through the grades, students should be able to move from situations in which the probability of an event can be determined to situations in which sampling and simulations help them quantify the likelihood of an uncertain outcome. Students will also move from fractions to decimals and percents in later grades.

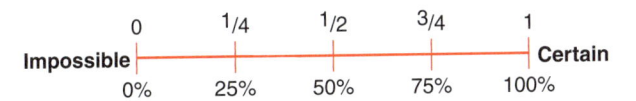

```
    0      1/4     1/2     3/4      1
Impossible |-------|-------|-------|-------| Certain
    0%     25%     50%     75%    100%
```

Long before children are ready to calculate probabilities of specific events, it is important that terms such as *certain, uncertain, impossible, likely,* and *unlikely* be introduced and discussed. Most students, even in primary grades, are familiar with the terms *impossible* and *certain* and can give meaningful examples. Although *likely* and *unlikely* are less familiar and require more careful development, using activities such as In the Classroom 17-5 with children provides a good start. As each card is sorted, an explanation or argument for placing it in the specific box should be given. This rationale is essential in refining and developing a clear understanding of these important terms.

An excellent extension at the bottom of In the Classroom 17-5 is to have students write statements to be sorted into the same categories. Each student should write several original statements and then exchange papers so that someone else classifies them. Once these general probabilistic terms become familiar, more specific probabilities can be determined.

The activities from In the Classroom 17-6 ask students to determine the likelihood of a particular event. Experiments involving blocks, spinners, dice, and coins are often used to introduce probability concepts and notation. Students should be given many opportunities to create and explore activities that ask them to answer questions about the likelihood of events, using the vocabulary of probability.

Sample space is a fundamental concept that must be established or at least understood before the probabilities of specific events can be determined. The sample space for a probability problem represents all possible outcomes.

In the Classroom 17–5

WHAT ARE THE CHANCES?

Objective: To identify the likelihood of an event.
Grade Level: 3–5.

▼ Sort these statements into the best box.

| Impossible | Unlikely | Likely | Certain |

A. The sun will rise in the west.
B. The cafeteria will serve chocolate milk.
C. A boy in our class will be 2 meters tall.
D. Everyone in this room is alive.
E. Most people in our class have brown eyes.
F. There are more right-handed people in this room than left-handed.
G. The price of gas will be higher next year.
H. It will rain today.

▼ Write statements to be sorted in the boxes.

Explain your reasons for each.

In the Classroom 17–6

WHAT'S MORE LIKELY?

Objective: Determine the likelihood of an event.
Grade Level: 4–5.

Activity One:
1. If you put four red blocks in a paper bag, can you be sure what color block you will pull out? Why or why not?
2. If you put a green block in the bag with the red blocks, can you be sure what color block you will pull out? Why or why not? Try this experiment. Record your results.
3. If you put three green blocks in a bag with seven red blocks, is one color more likely to be pulled out? Why or why not? Try this experiment. Record your results.
4. Try a blocks-in-a-bag experiment of your own. Record your results.

Activity Two:
▼ Mary flipped a coin four times. It came up heads four times. She flipped the coin a fifth time. What is she likely to get on the fifth flip of the coin?

  a. Heads
  b. Tails
  c. It is equally likely to be heads or tails.

▼ Explain your answer. How could you test your idea? Try your experiment and record your results.

394 Chapter 17 • Data Analysis, Statistics, and Probability

TABLE 17-1 • Sample Spaces of Some Events and Their Probabilities

Questions	Sample Space	Number of Successes	Probability
What is the probability of getting a head on a single toss of a coin?	H, T	1	$\frac{1}{2}$
What is the probability of getting two heads when two coins are tossed?	HH, HT TH, TT	1	$\frac{1}{4}$
What is the probability of getting a five on a single roll of a die?	1, 2, 3, 4, 5, 6	1	$\frac{1}{6}$
What is the probability of drawing or a spade from a deck of 52 playing cards?	52 cards	13	$\frac{13}{52}$ or $\frac{1}{4}$
If each letter of the alphabet is written on a piece of paper, what is the probability of drawing a vowel?	26 letters of the alphabet	5 (a, e, i, o, u)	$\frac{5}{26}$

Another suggestion to help children think about what outcomes are possible for a particular event is to have them examine some common probabilistic situations: (1) If I flip a coin, will it land heads up, on its edge, tails up, or just float in the air? (2) If I drop a glass, will it hit and break, float in the air, or hit but not break? In each case, have the students discuss whether the event could happen. Consider, for example, the situation in which a coin is tossed. Some children may realize that a coin will not land on its edge, nor will it float. Thus only two outcomes can happen, and these possible outcomes comprise the sample space. Having children create their own events and determine possible outcomes is also a beneficial experience.

Once the sample space is known, the calculation of specific probabilities usually follows naturally. When a coin is flipped, as described in Table 17-1, the probability of a head is the number of ways a head can occur divided by the total number of outcomes (head or tail). Specific probabilities rest heavily on fractions, which provide a direct and convenient means of reporting and interpreting probabilities.

Discussion of possible outcomes helps identify the sample space and clarify notions of probability. Using the example of an equally likely spinner, students can be questioned about the probability of events occurring. Questions along these lines might get the discussion started:

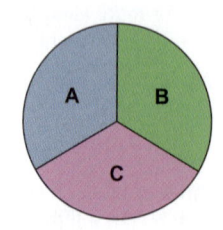

Can the spinner stop in region D?
Can the spinner stop on a line?

The first outcome is impossible. Even though the second outcome is unlikely, it can happen, and a plan of action should be specified if it does. (Maybe you spin again.)

In the Classroom 17-7 involves collecting data, graphing results, and exploring patterns. It uses several valuable ideas of probability, including sample space and probability of an event, in a natural and interesting setting. As children are involved in this process, they are developing and practicing basic facts, as was shown earlier in Chapter 9. Such an activity further illustrates how mathematical topics are interrelated and how important connections can be made.

Once you have tried In the Classroom 17-7 with children, another perspective can be obtained by examining Table 17–2, which summarizes the results when two dice are added (a) and multiplied (b). One diagonal of Table 17–2(a) shows all the ways that a sum of 7 can be obtained. Table 17–2a

TABLE 17-2 • Results of Operations with two dice

a.

+	1	2	3	4	5	6
1	2	3	4	5	6	7
2	3	4	5	6	7	8
3	4	5	6	7	8	9
4	5	6	7	8	9	10
5	6	7	8	9	10	11
6	7	8	9	10	11	12

a

b.

×	1	2	3	4	5	6
1	1	2	3	4	5	6
2	2	4	6	8	10	12
3	3	6	9	12	15	18
4	4	8	12	16	20	24
5	5	10	15	20	25	30
6	6	12	18	24	30	36

b

Randomness

the product of two even numbers is even and that the product of an odd number and an even number is also even.

RANDOMNESS

Randomness is an important concept underlying all learning in probability. When something is random, it means that it is not influenced by any factors other than chance. In the Classroom 17-8 could be used to build on In the Classroom 17-5 and provides an opportunity to discuss randomness in a specific context. Here students are encouraged to think about events based on their classmates and decide about where these events would be placed on a probability number line that shows 0 and 1. Here are some starter questions:

Why is it important that the "name will be randomly picked"?

Should the names be seen by the person doing the drawing?

Would it matter if some people wrote their names on large pieces of paper and others on small pieces of paper?

In the Classroom 17-7

ROLLING AND RECORDING

Objective: To conduct a probability experiment.
Grade Level: 4–8.
Activity One:
1. Choose a partner, and each of you make a chart like the one shown. Predict which sum of two dice will come up most often.
2. Each of you take turns rolling two dice.
3. On a turn, find the sum of the spots on the two dice, and place a tally mark in that column on your chart.
4. Continue rolling and recording until one of you has 10 tally marks in one column. Compare your prediction with the actual result.

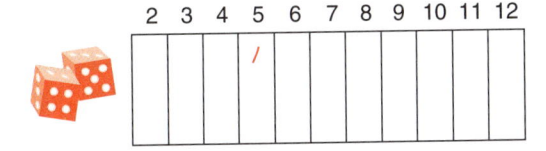

2	3	4	5	6	7	8	9	10	11	12
			/							

Putting It Together:
- Why doesn't the chart need a ones column? A thirteens column?
- In which column did you or your partner reach 10?
- Compare your results, and tell how they are similar. Different.
- Tell why you would expect more sums of 7 than 2.
- Complete the following. "I would expect about the same number of sums of 4 as …"
- Would you expect to get about the same number of even sums as odd sums? Tell why.

Putting It Together:
Suppose you multiplied the numbers on the dice instead of adding them.
- How would the values along the top of the chart change?
- How many values (i.e., different products) would be needed?
- Which values would be least likely?
- Would you expect to get about the same number of even values as odd values? Tell why?

In the Classroom 17-8

ARE YOU A WINNER?

Objective: To explore the concept of randomness.
Grade Level: 4–8.
Our class is having a drawing. Each person gets to place their name in the drawing one time. One name will be randomly picked, and that person will be the winner.

▼ Read each of the following statements.
▼ Think about the people in our class.
▼ Then check the number line below and decide about where the following statements should be placed:

A. The winner will be left-handed.
B. The winner will be a girl.
C. The winner will be someone in our class.
D. The number of letters in the first name of the winner will be less than the number of letters in their last name.
E. The winner's first name will begin with a vowel.
F. The winner will wear glasses.
G. You will be the winner.
H. The winner will be wearing socks.
I. You will not be the winner.

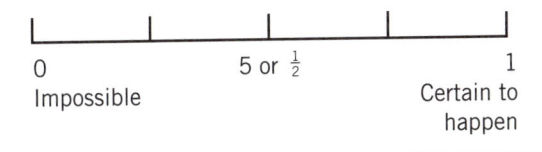

```
|--------|--------|--------|--------|
0                 5 or 1/2          1
Impossible                    Certain to
                              happen
```

---

also shows the different ways that each of the other sums can result. Is the sum of two dice more likely to be even or odd? An examination of Table 17–2a shows that the even sums will occur 18 out of 36 times, or half the time.

Is the product of two dice more likely to be even or odd? The shaded cells in Table 17–2b show that the even products are much more likely. In fact, an even product would be expected to occur 27 times out of 36, or three-fourths of the time. Analyzing and discussing why this happens helps connect probability to properties and relationships between numbers and operations. For example, it is a reminder that

If the names are seen or if people don't all write their names on the same size of paper, the drawing might not be random. When this happens, some people would have an advantage and the notions of *fair* and *unfair* become important. The term *fair* is often used in describing a situation. For example, to say "a fair coin" or "fair dice" makes it clear that no inherent biases exist that would affect randomness. A person may be asked to toss (not scoot) a die to ensure that one face is not favored. If Ping-Pong balls are drawn from a bowl, it is important that the balls be thoroughly mixed and the person doing the drawing be blindfolded to ensure both randomness and fairness. The children's book *No Fair!* (Holtzman, 1997) describes two children who are trying to determine when activities such as drawing an item from a bag are fair.

Discussions of the consequences of unfairness and absence of randomness should be a regular part of developing probability. For example, would it be fair if two dice are rolled and player A wins if the product is even and player B wins if the product is odd? Table 17-2(b) shows that A will win much more than B, so this game is not fair. On the other hand, if the game is based on the sum of two dice (Table 17-2(a)), each player has an equal chance of winning and the game is fair. Suppose you modify the sum-of-two-dice game to play with three players:

Player A wins if the sum is 1, 2, 3, or 4.

Player B wins if the sum is 5, 6, 7, or 8.

Player C wins if the sum is 9, 10, 11, or 12.

Is this game fair for each of the players? Does each player have the same chance of winning? An analysis of Table 17-2(a) suggests that Player B will win more often than either of the other players. As children explore this game, you might challenge them to tell how the game might be modified to make it fair for everyone.

INDEPENDENCE OF EVENTS

Independence of events is an important concept in probability, but one that does not develop naturally from intuition. If two events are independent, one event in no way affects the outcome of the other. Thus, if a coin is tossed, lands heads, and then is tossed again, it is still equally likely to land heads or tails. This sounds simple enough, but consider this question: Suppose four consecutive sixes have occurred on four rolls of a fair die. What is the probability of getting a six on the next roll?

A majority of middle grade students miss this question. Many students suggest that the die has a "memory" and things would "even out." Most did not conclude that the probability was unchanged, regardless of what had already happened. If an event has occurred a number of times in a row, most people falsely presume that the "law of averages" makes it unlikely that the event will occur on the next trial. The same thing happens when a coin is flipped multiple times, generating a string of the same outcomes. If the coin lands heads four times in a row, it has an equally likely chance of landing heads or tails on the fifth flip. This basic misunderstanding contradicts the notion of independence of certain events. Research suggests that experiences exploring independence should be encountered in middle school (Jones and Langrall, 2007; Tarr and Jones, 1997).

Having children collect data and discuss the results can help dispel some of this erroneous thinking. "Working with real data allows students to appreciate the difference between empirical phenomena and probabilistic models; it shows them the usefulness of these models in explaining, predicting and controlling a variety of real phenomena beyond pure games of chance" (Batanero and Sanchez, 2005, p. 262). When you use In the Classroom 17-9 with children, different students will produce different results, yet the answers to the questions will be very similar. Why? Because these events, the rolls of a die, are independent of one another.

Tossing a coin and recording the outcomes in sequence will likely generate some long runs of an occurrence, even though each outcome is independent of the others. Although the probability of a head is $\frac{1}{2}$, children might flip a coin 10 times and get eight, nine, or even ten heads in a row. Consider this record of 20 tosses of a coin:

TTTTTHTHHHHHTTHHTHHHH

In the Classroom 17–9

CAN YOU MAKE PREDICTIONS?

Objective: To identify independence of events.

Grade Level: 4–6.

▼ Roll a die six times and record the results:

1	2	3	4	5	6

- Did each face appear once? _____
- Does knowing what happened on the first roll help predict the second? _____ the third? _____

▼ Roll a die 24 times and record the results.

1	2	3	4	5	6	7	8	9	10	11	12

13	14	15	16	17	18	19	20	21	22	23	24

- Did each face appear once? _____
- The same number of times? _____
- What face appeared most? _____
- Does this mean the die is unfair? _____
- Does this record tell you what will occur on the next roll? _____

There are two sequences of four consecutive heads and one of five consecutive tails. Overall, 11 heads appeared. Such analysis and discussion helps children understand that things don't even out on each flip. As the number of flips gets very large, however, the ratio of heads to the total number of flips will get closer and closer to the theoretical expected value of $\frac{1}{2}$. This latter point is very important, but it often baffles elementary students. Using simulation programs to repeat an event many times helps children better understand the notion of the expression *in the long run*.

There are, of course, times when one event may depend on another. For example, suppose you wanted to roll two dice and obtain a sum of 8. If a 1 is shown on the first die, it is impossible to get a sum of 8. This leads toward notions of *conditional probability*.

Instruction is needed to develop the necessary techniques to solve simple probability problems. There are other more complicated situations for which probabilities are difficult to calculate. What is the probability of the New York Mets winning the World Series? Of a woman being elected president? Such questions do not lend themselves to simple solutions, but experts can approximate their probabilities. Regardless of who determines the numerical probabilities, the knowledge and interpretative skills developed in simpler probability situations can be successfully applied.

Math Links 17.4

There are a number of excellent Web sites that provide opportunities for children to simulate many trials with dice and spinners.

Go to the

- National Library of Virtual Manipulatives at http://www.nlvm.usu.edu. There you will find a number of tools that help to simulate different probability models (Spinners, Coin Tossing, Box Model, and Histogram).
- National Council of Teachers of Mathematics Illuminations Web site at http://illuminations.nctm.org/ to find an array of lessons and tools, such as an Adjustable Spinner and Random Drawing Tool, that allow children to gather data via simulations.

www.wiley.com/college/reys

MISCONCEPTIONS ABOUT PROBABILITY

Young children often hold common misconceptions about various aspects of probability. For example, they make predictions based on preference, such as their favorite color or number (Jones & Langrall, 2007). They also hold biases against certain numbers, believing, for example, that it may be hardest to throw a 6 on a die. Young children are not surprised by extremely unlikely events and do not search for underlying causes. They may expect all outcomes in an experiment to be equally likely. It is also difficult for many children to make inferences from data.

So do not be surprised when your own students reflect misconceptions in probability. The more opportunities you give them to explore a variety of probability notions through hands-on activities, the better they will be able to develop and evaluate inferences and predictions that are based on data and apply basic concepts of probability.

CULTURAL CONNECTIONS

Several previous chapters have discussed the success of the Asian nations on the Trends in International Mathematics and Science Study (TIMSS). Many of these nations have shown success in mathematics overall and high performance in computation in particular. However, an analysis by strand shows some shockingly different results, especially in data analysis and probability.

The TIMSS 2003 mathematics test was designed to test fourth and eighth graders in five content areas. The "Data" portion of the test enables reporting over four topics of data analysis and probability: (1) data collection and organization of data, (2) representing data, (3) interpretation of data, and (4) probability and uncertainty (Mullis et al., 2004). All four areas were tested in the eighth grade; however, probability was not included on the fourth grade test.

The five top scoring nations on the TIMSS 2003 eighth-grade test were Singapore, the Republic of Korea, Hong Kong, Chinese Taipei, and Japan. All five nations scored significantly higher than the international average scale score. However, four of these nations' scores in data (Japan excluded) were not only their lowest, but the scores in some cases were significantly lower than the next closest strand. The United States, on the other hand, scored the highest in data, 17 points higher than any other area in eighth grade.

Based on the reported use of the curriculum, many nations were not emphasizing data analysis and probability in their curricula. This lack of emphasis is highlighted in Mullis et al. (2004), "with 95% of students across countries having been taught the TIMSS number topics (strand) by eighth grade. This was followed by measurement (78%), geometry (69%), algebra (66%), and data (46%)" (p. 189). In eighth grade, the top five nations ranged from 6% to 59% of the data strand covered, while the United States averaged 83%. Because data are not a key component of curricula in many parts of the world, students are scoring significantly lower on questions related to data.

Why is the United States spending more time on data and scoring much higher? Jones and Langrall (2007) suggest two major factors are the growing recognition of the importance of probabilistic thinking and the increasing amount of research. As a result, many mathematics curricula materials in the United States are devoting more

Chapter 17 • Data Analysis, Statistics, and Probability

attention to data analysis, statistics, and probability, thereby increasing the students' opportunity to learn.

A GLANCE AT WHERE WE'VE BEEN

Data analysis, statistics, and probability provide an opportunity for students to develop problem-solving and critical-thinking skills as well as to make connections to other mathematical topics and school subjects. Students engage in the process of data analysis by formulating questions that are meaningful to them, collecting the data to answer the questions, and displaying the data in a way that helps them analyze and interpret the results. Teachers and students pose questions about the data together, extend students' abilities to analyze their results, and communicate their findings in a way that represents the data in a truthful manner.

In the elementary and early middle school years, students are exposed to many types of graphs: line plots, stem-and-leaf plots, picture graphs, bar graphs, histograms, pie graphs, line graphs, and box plots. Graphing calculators and other technological tools and programs allow students opportunities to experiment with different ways of displaying data, therefore developing multiple perspectives for the same set of data. This helps promote data sense.

Students in the elementary grades learn to analyze data using descriptive statistics, especially measures of variation and central tendency—mean, median, and mode. The goal is to understand fundamental notions related to these statistics and to be able to choose averages that are most appropriate and describe the data in meaningful ways.

Probabilistic thinking develops in elementary students through a variety of hands-on activities and experiments that challenge their intuitive notions of what is fair and unfair. It is important that children know the probability of an event ranges between 0 and 1, and be able to calculate the probability of events. The ability to use the language of probability is an important skill in our daily lives, and it can be explored in many ways that help build excitement for learning mathematics.

Things to Do: From What You've Read

1. Here are the results on three tests: 68, 78, 88. What are the mean and the median? Explain why the mode is of little value. What score would be needed on the next test to get an average of 81? Describe two different ways you could determine this score.

2. Arrange interlocking cubes together in lengths of 3, 6, 6, and 9.
    a. Describe how you could use the blocks to find the mean, median, and mode.
    b. Suppose you introduce another length of 10 cubes. Has the mean changed? The median? The mode?

3. Answer each of the following questions, and tell why you answered as you did.
    a. Is it possible for a set of data to have more than one mode? Give an example.
    b. Is it easier to find the median of 25 or 24 student scores?
    c. Could the mean be as large as the largest value in a set of data? Tell how.

4. Describe the differences between a bar graph, a histogram and a line plot.

5. Describe two activities that could be used to discuss probabilities of zero and one.

6. Ten cards are marked 0, 1, 2, 3, ... , 9 and placed face down. If the cards are shuffled and then one card is drawn, tell why the following statements are true:
    a. The sample space has 10 events.
    b. The probability of drawing the 6 is $\frac{1}{10}$.
    c. The probability of drawing the 3 is the same as the probability of drawing the 7.

7. Many state lotteries advertise with the slogan "If you don't play, you can't win." Does that mean if you do play, you will win? Explain.

8. Give an example where measure of variation (range, absolute deviation or interquartile range) would be useful in interpreting a set of data.

9. If a lottery consists of picking 6 numbers from the numbers 1 through 45, which group of 6 numbers would you pick? Explain your choice.
    a. 1, 2, 3, 4, 5, 6
    b. 3, 10, 17, 21, 35, 43
    c. Both are equally likely to win.

Things to Do: Going Beyond This Book

In the Field

1. [1]Use In the Classroom 17-1 with a small group of children as they plan and conduct a survey. Describe how it went.

2. Examine the graphs shown by Sakshaug (2000). Use these graphs and questions with some children and compare your results with those reported by Sakshaug and Wohlhuter (2001).

3. Use In the Classroom 17-2 with some children in grades 4–8. Which key features of various graphs and statistics did you discuss with the children? How did you guide the children to create higher-level questions?

4. [1]Select a middle school student and use In the Classroom 17-6 to guide some questions related to probability. What did you learn about the child's thinking?

5. Examine the scope-and-sequence chart for an elementary textbook series used in your school. At what level is graphing first introduced? What kinds of graphing skills are highlighted? What important statistical topics are included? At what levels are they taught? Do the same for probability.

6. What would you say to a principal who accused you of allowing your students to "play games" when they were conducting dice experiments?

With Additional Resources

7. Read the article "100 Students" (Riskowski, Olbricht & Wilson, 2010). Summarize questionnaire and the process used to collect and summarize data. Describe some implications it has for developing a lesson related to statistics?

8. Examine one of the NCTM's *Navigations Series* books for Data Analysis and Probability (preK–2, Sheffield et al., 2002; 3–5, Chapin, Komiol, MacPherson, and Rezba, 2002; 6–8 probability, Bright, Frierson et al., 2003; or 6–8, data analysis, Bright, Brewer et al., 2003) and present one of the grade-appropriate activities to a group of students. Discuss what happened.

9. Examine newspapers or magazines. Start a file of graphs students can examine. Make a list of the different kinds of graphs used. Select a few and tell why you think a particular graph (e.g., picture graph, pie graph, bar graph) was used in each situation.

10. Games are fun and can help develop a better understanding of probability. Play one of the games from *What Are My Chances?* (Shulte and Choate, 1996), or play the "Cover Up Game" from *Chance Encounters: Probability in Games and Simulation* (Brutlag, 1996). Identify some of the mathematics learned in these games. Tell how you might use these games with students.

11. Read one of the research articles related to probability or statistics listed in Jones and Langrall (2007), Shaughnessy (2007), or Friel et al. (2001). Discuss the nature of the research. Also identify an instructional idea or activity suggested by the research that you think would be effective in helping children learn a particular concept.

With Technology

12. Use technology such as a spreadsheet, graphing or survey software, or a graphing calculator to display and analyze data. Evaluate how these tools might be used with children. Better yet, try it out with children and then evaluate.

13. Check the Web for the NCTM's *Illuminations* activities. Look for an activity that can be used in helping elementary children understand probability. If you can, try this activity with some students.

---

[1]Additional activities, suggestions, and questions are provided in *Teaching Elementary Mathematics: A Resource for Field Experiences*, found on this text's accompanying Web site, at www.wiley.com/college/reys.

Note to Instructors: You can find additional resources, learning activities, and black-line masters in this text's accompanying Instructor's Manual, at www.wiley.com/college/reys.

Book Nook for Children

Arnold, C. *Charts and Graphs: Fun, Facts, and Activities.* New York: Franklin Watts, 1984.

Photographs and drawings are used to illustrate real-world applications and examples using pie, bar, and line graphs and pictographs. Illustrations are colorful and interesting and provide a multicultural focus for the book. The glossary is written in terms children can understand and is helpful in comparing various ways of presenting information.

Cushman, J. *Do You Wanna Bet? Your Chance to Find Out About Probability.* New York: Clarion Books, 1991.

Whether flipping coins to decide what television program to watch or analyzing which events are "certain," "impossible," or "maybe," Danny and Brian become involved in everyday situations, both in and out of school, that involve probability. Several important probability concepts are woven into an interesting story line.

Davies, J. *Tricking The Tallyman,* Alfred A. Knopf, New York, New York 2009

In 1790, Secretary of State, Thomas Jefferson, took on the responsibility of the first census. With the help of 650 assistants, they were to ride from every village and town in the United States. In the book, the residents of a small Vermont town try and trick the tallyman that has been sent to count their population.

Holtzman, C. *No Fair!* New York: Scholastic, 1997.

When Kristy and David cannot agree on which game to play, David brings out a bag of marbles and says they can play Kristy's game if she chooses a blue marble, but Kristy objects when she learns that only one marble was blue.

Leedy, L. *The Great Graph Contest,* New York: Holiday House, 2006.

The book will show the reader how to make and understand such things as: bar graphs, circle graphs, quantity graphs, and Venn diagrams.

Murphy, S. J. *The Best Vacation Ever.* New York: HarperCollins, 1997.

This entry in the MathStart series demonstrates that collecting data and s tabulating results can help the decision-making process. A girl with an active, overscheduled family puts math to work to coordinate a family expedition. She uses charts to plot her family's preferences.

ANNOTATED RESOURCES

Bright, G. W., Brewer, W., McClain, K., and Mooney, E. S. *Navigating Through Data Analysis in Grades 6 through 8.* Reston, VA: NCTM, 2003.

Bright, G. W., Frierson, D., Jr., Tarr, J. E., and Thomas, C. *Navigating Through Probability in Grades 6 through 8.* Reston, VA: NCTM, 2003.

Burns, M. (2000). *About Teaching Mathematics: A K-8 Resource.* Sausalito, CA: Math Solutions.

Bright, G., Brewer, W., McClain, K. & Mooney, E. (2003) *Navigating through data analysis in grades 6–8.* Reston, VA: National Council of Teachers of Mathematics

Bright, G. W., and Hoeffner, K. "Measurement, probability, statistics and graphing". In *Research Ideas for the Classroom: Middle Grades Mathematics* (ed. D. T. Owens). Reston, VA: NCTMand New York: Macmillan, 1993, pp. 78–98.

This book considers daily experiences with data and chance as it addresses the role of both in middle school mathematics, grades 5–8. Activities are presented to help naturally build data and chance abilities with students.

Chapin, S., Koziol, A., MacPherson, J., and Rezba, C. *Navigating Through Data Analysis and Probability in Grades 3 Through 5. Reston,* VA: NCTM, 2002.

Each of these books in the *Navigation* series provides a wide range of activities to help children develop fundamental concepts of data representation and probability. The accompanying CD-ROM provides grade-level-appropriate applets that engage children in a range of simulations and tools for representing data.

Friel, S. "Teaching statistics: What's average?" In *The Teaching and Learning of Algorithms in School Mathematics, 1998 Yearbook* (ed. L. J. Morrow). Reston, VA: NCTM, 1998, pp. 208–217.

This article discusses strategies for building conceptual and procedural understandings in teaching mean, median, and mode. Developing "data sense" includes a consideration of the "when and why" of the use of the representations and statistics.

Quinn, R., and Wiest, L. "Reinventing Scrabble with middle school students." *Mathematics Teaching in the Middle School,* 5 (December 1999), pp. 210–213.

Students use the Scrabble game to explore frequency distribution and other topics as they work cooperatively to collect, analyze, and interpret data.

Sakshaug, L E. "Which graph is which?" *Teaching Children Mathematics,* 6 (March 2000), pp. 454–455.

Ideas for engaging elementary children in graphing. Problem posed.

CHAPTER 18

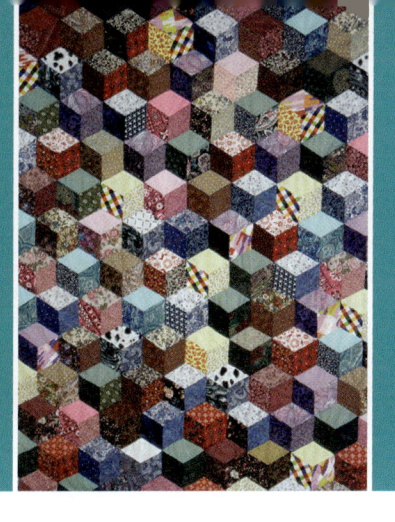

Number Theory

> "WHAT WE HAVE ONCE ENJOYED WE CAN NEVER LOSE."
> — Helen Keller

> SNAPSHOT OF A LESSON

KEY IDEA

1. Motivate the study of number relationships

BACKGROUND

This snapshot is from *The Great Stringdini*, part of the For Real segment of *Cyberchase*[1], *the Emmy award winning math mystery cartoon on* PBS KIDS GO! A PBS mathematics-based cartoon program for school-age children. To view it, go to http://pbskids.org/cyberchase/forreal/301_for_real_hi.html. It is different from those in previous chapters in that it is not a classroom scene but a suggestion for a motivational opener when studying number relationships. We will show how it relates to number theory topics in the chapter itself.

Harry, the main character, is trying out for a magic show. After a nervous wait and an unacceptable beginning with simple magic tricks such as making coins appear in an empty can and pulling a skunk from a hat, he turns to the "real thing."

Harry: These tricks aren't working. I need a volunteer from the audience. How about you?

He chooses the biggest heckler and tells him that this simple string will tell more about him than just being a person with a big mouth.

Harry: May I see your thumb.

He takes the string and puts it around the thumb.

Harry: Twice around your thumb is once around your wrist

Much to the surprise of the heckler, Harry shows him that twice around his thumb is once around his wrist.

Harry: Suppose I want to buy a shirt for you birthday. If I asked you for your collar size, you wouldn't be surprised. But twice around your wrist is your neck size.

Again, the heckler is amazed.

Harry: And three times around your head is???

Harry goes ahead and shows the heckler, but we will leave it to you to find what body length is three times your head or to watch the short clip for a solution.

Although this clip seems to be about measurement and proportion, we will tie it to number theory as we examine Fibonacci numbers.

FOCUS QUESTIONS

1. Why study number theory topics in elementary school?
2. What number theory topics are appropriate for students in elementary school?
3. How does number theory complement the teaching and learning of mathematics in elementary school?

---

[1] Cyberchase is produced by Thirteen/NNET, New York. All copyrights of the Cyberchase characters are owned by Educational Broadcasting Company.

Chapter 18 • Number Theory

INTRODUCTION

Number theory is a branch of mathematics, mainly concerned with the integers, that has been a topic of study for centuries. For many years its application to practical situations was limited, but with the power of computers it has become important to cryptography, random number generation, and coding theory. These applications are beyond most elementary students, but there are many topics that are appropriate for elementary students to explore. In exploring these topics, students can be given the opportunity to view mathematics in a way that will bring Helen Keller's quote to mind. It is a chance for many to enjoy mathematics and make it a part of their own lives.

This chapter is meant to be a culminating chapter in which you reflect on what you have learned and show your willingness to learn more mathematics. Things to Do at the end of the chapter contains questions that will challenge you with a mathematical activity or with a pedagogical question that often refers to the previous chapters in this text.

WHY STUDY NUMBER THEORY

There are four reasons we think number theory topics provide opportunities for students to look at mathematics in a different way. We briefly discuss each reason and illustrate it with an activity. You should take the time to do at least one of these activities before reading the remainder of the chapter. If there are terms that you do not understand in these activities look up the term on the site in Math Links 18.1.

Math Links 18.1

You may find definitions of number theory terms and more about those terms by asking Dr. Math. This site may be assessed at http://mathforum.org/dr.math/faq or from this book's Web site.

www.wiley.com/college/reys

Reason 1. Number theory is a prime source to show that numbers can be fascinating. From the time of the ancient civilizations, people have been fascinated with numbers and number patterns. Ancient peoples often thought numbers had mystical qualities, and the branch of mathematical called *numerology* was popular. Number theory, however, looks at the mathematical properties of numbers. Have you thought about why we have 60 minutes in an hour? What are the properties of 60 that make this appropriate? Would 40 have worked as well?

To begin your thinking about how a different arrangement of a familiar number sequence can bring out various patterns, consider the 100-chart arranged in the triangle, In the Classroom 18-1 (see Appendix for extra copies) Shows the triangle and includes a few questions about patterns that show up in this arrangement.

In the Classroom 18–1

WHAT DO YOU SEE IN ME?

Objective: To explore patterns in 100-chart.

Grade Level: 4 and above.

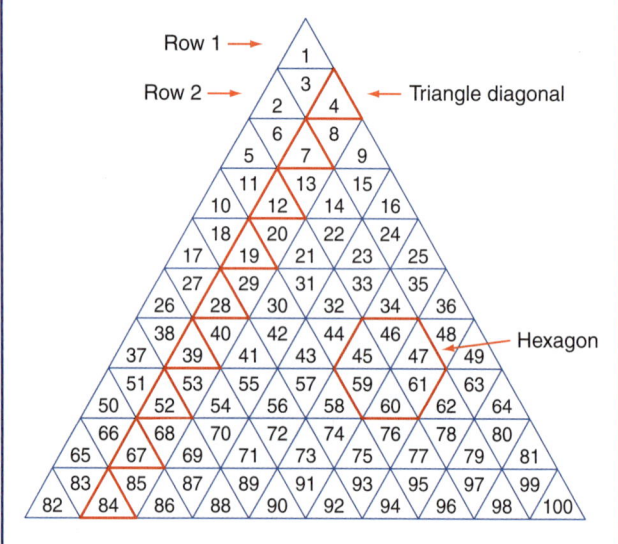

1. Where are the perfect squares?
2. Where are the odd numbers?
3. Find the sum of the upright diamonds (such as 1 and 3, 2 and 6). If you do this in order, what is the pattern of the sums?
4. Find the sum of the numbers in each row. Find a shortcut for finding the sum.
5. Choose a hexagon. What is the sum of the numbers? Find a shortcut.
6. Take any two adjacent numbers in a triangle diagonal—for example, 4 and 7. Where is the product? Does this always work?
7. Find another pattern and describe it.

A more detailed plan for this activity is available on the Student Companion site at www.wiley.com/college/reys.

Reason 2. Number theory opens the doors to many mathematical conjectures. One of the practices describe in the CCSSM (CCSSI, 2010) involves constructing viable arguments. They call for students to make conjectures and build a logical progression of statements to explore the truth of their conjecture beginning with elementary students. Students at this level can construct arguments using concrete referents such as objects, drawings, diagrams, and actions. Such arguments can make sense and be correct, even though they are not generalized or made formal.

Many conjectures in number theory are easy to state but may be difficult to prove. For example, the famous conjecture by Goldbach is easy enough for fourth and fifth graders to understand. In 1732, Goldbach conjectured that *any even number greater than 2 could be written as the sum of two primes*. For example, 48 is the sum of the two primes, 19 and 29. Some numbers, such as 10, have more than one way to express the sum. This conjecture has never been proved or disproved, so don't give up on reasonable conjectures that students make but cannot prove.

Children get a lot of practice in finding primes and seeing patterns if they test the conjecture for the first hundred even numbers greater than 2. Teachers often put up a chart (see Figure 18-1) with the even numbers and let children write the sums they find.

Watch for children who write 7 + 9 for 16 because they see a pattern. What is wrong with that sum? Does it satisfy the conjecture? No, an important part of the conjecture is that the addends must be primes.

In the Classroom 18-2 presents five conjectures with made-up names that are not true. As discussed in Chapter 14, it only takes one example to disprove a conjecture. Can you find such an example? This activity is mainly for you; however, you will have some students who need such a challenge and like working on such problems. You could use a chart like that suggested for Goldbach's conjecture to engage students in exploring these conjectures.

Through this chapter, you will find many conjectures that are well within the ability of elementary students to explore and to make convincing arguments about their validity.

Reason 3. Number theory provides an avenue to extend and practice mathematical skills. We often need ways to help differentiate instruction within a topic. For example, suppose you are beginning the year in a fifth-grade class and you need to review finding factors of whole numbers with some children. Although all children may benefit from some review, they will not need the same amount. What do you do with those students who do not need as much review? Number theory often provides a way to both review finding factors and challenge students. In this case, you might have them investigate *abundant, deficient, and perfect* numbers (see In the Classroom 18-3).

You will see other examples of ways to practice skills in this chapter as students investigate different number theory topics.

Number	Sum of 2 primes
4	2 + 2
6	3 + 3
8	3 + 5
10	5 + 5, 3 + 7
12	5 + 7
14	7 + 7
16	
18	

Figure 18-1 *Testing Goldbach's conjecture.*

In the Classroom 18–2

DO YOU BELIEVE THAT?

Objective: To disprove conjectures.

Grade Level: 5 and above.

Investigate one of these conjectures about **positive whole numbers.** See if you can find an example that disproves it. The examples show that it works for some numbers.

1. **Tinbach Conjecture:**

Every number can be expressed as the difference of two primes. Examples:

$$14 = 17 - 3 \quad 5 = 7 - 2$$
$$28 = 31 - 3 \quad 11 = 13 - 2$$

2. **Zincbach Conjecture:**

Every number can be expressed as the sum of 3 squares (0 permitted). Examples:

$$3 = 1^2 + 1^2 + 1^2 \quad 14 = 1^2 + 2^2 + 3^2$$
$$9 = 3^2 + 0^2 + 0^2$$

3. **Aluminumbach Conjecture:**

Every odd number can be expressed as the sum of 3 primes. Examples:

$$15 = 5 + 5 + 5 \quad 11 = 3 + 3 + 5$$
$$21 = 7 + 7 + 7$$

4. **Brassbach Conjecture:**

Every square number has exactly 3 divisors. Examples:

1, 2, and 4 are the 3 divisors of 4

1, 5, and 25 are the 3 divisors of 25

5. **Copperbach Conjecture:**

The product of any number of primes is odd. Examples:

$5 \times 3 \times 7 = 105$ (3 primes) $7 \times 3 \times 3 \times 5 = 315$ (4 primes)

With a minor changing in the wording, some of these conjectures are true. See if you can find which ones and make the change.

A more detailed plan for this activity is available on the Student Companion site at www.wiley.com/college/reys.

Reason 4. Number theory offers a source of recreation. Did you ever think of mathematics as a recreation? Number theory provides many puzzle-type activities that children enjoy. It is one more way to individualize your curriculum. Just as different children enjoy different sports or games, different children will find different puzzles recreational. The puzzle must be challenging yet in reach of the child.

In solving such puzzles, students are practicing skills, developing number sense, and using problem-solving strategies. For example, in the magic square (In the Classroom 18-4) children soon learn that the two largest numbers cannot go in the same row, column, or diagonal. Although

In the Classroom 18–3

AM I ABUNDANT, DEFICIENT, OR PERFECT?

Objective: To examine the sum of factors.

Grade Level: 5 or above.

How To Tell

1. Find all the factors of the number. Add all the factors except the number itself.
2. If the sum is
   - greater than the number, then the number is abundant.
   - less than the number, then the number is deficient.
   - equal to the number, then the number is perfect.

Examples

- 18 $1 + 2 + 3 + 6 + 9 = 21$
  $21 > 18$ 18 is abundant
- 35 $1 + 5 + 7 = 13$
  $13 < 35$ 13 is deficient
- 28 $1 + 2 + 4 + 7 + 14 = 28$
  $28 = 28$ 28 is perfect

Tasks

1. With your classmates, classify the first 100 numbers as abundant, deficient, or perfect.
2. Are more numbers abundant or deficient?
3. Show that 496 and 8128 are perfect.
4. Explore the conjecture: If you multiply two abundant numbers, is the product abundant?
5. Explore the conjecture: If you multiply two deficient numbers, is the product deficient?
6. Explore the conjecture: If you multiply two perfect numbers, is the product perfect?

In the Classroom 18–4

MAGIC SQUARES

Objective: To explore magic squares.

Grade Level: 1 or above.

Materials: Number tiles or paper squares (1–9) and a 3-by-3 square.

A magic square is a square in which the sum of the numbers in each row, each column, and each main diagonal is the sum.

In the magic square below, arrange the number tiles so the sum is 15.

A Magic Square

Gear Down:

Place some of the tiles in the proper places on the 3 × 3 square. The more numbers you specify, the easier the puzzle will be.

Gear Up:

Use different numbers.

- Use the first nine even numbers. The sum will be 30.
- Use the first nine multiples of 5 (including 5). The sum will be 75.
- Use 4, 7, 10, 13, 16, 19, 22, 25, and 28. The sum will be 48.
- Use decimals (0.2, 0.5, 0.8, 1.1, 1.4, 1.7, 2.0, 2.3, and 2.6). The sum will be 4.2.
- Use fractions ($\frac{1}{5}$, $\frac{3}{5}$, 1, $\frac{7}{5}$, $\frac{9}{5}$, $\frac{11}{5}$, $\frac{13}{5}$, 3, $\frac{17}{5}$). The sum will be $5\frac{1}{5}$.

---

this activity may be done with paper and pencil, we suggest using number cards. First, the cards make the activity seem more like a puzzle to many children. Second, it saves erasing wrong tries until holes appear on the paper! The disadvantage of the cards is that children have no record of their attempts, so they may repeat the same attempt over and over. This may lead to the need to record attempts. You can increase the size of the square in the puzzle to any size or use fractions or decimals. For more information about magic squares, begin with the site suggested in Math Links 18.2.

Math Links 18.2

More information and activities about Magic Squares and Stars may be found at PBS Mathline at http://www.pbs.org/teachers/mathline/concepts/historyandmathematics/activity2.shtm or from this book's Web site.

www.wiley.com/college/reys

NUMBER THEORY IN ELEMENTARY SCHOOL MATHEMATICS

ODDS AND EVENS

Classifying numbers as odd or even is one of the first number theory topics that children encounter. As children count by 2—2, 4, 6, 8, 10—they learn that there is something special about these numbers; these are the even numbers, and all the other positive whole numbers are odd. More precisely, even numbers are those that are divisible by 2 (remainder is 0 when divided by 2). See if you can use this definition to justify that 0 is even.

Children often rely on counting to justify which numbers are even—every other number beginning with 0 is even (and every other number beginning with 1 is odd). The CCSSM introduces even and odd numbers in Grade 2 and encourages children to note that an even number is the sum of the two numbers which are the same. For example, 12 is even since 12 is the sum of 6 and 6. Many programs build the idea that if the number can be grouped by (or paired in) twos, then it is even. In this case, 12 is even since it has 6 pairs of two. Later, students learn that all even numbers are divisible by 2 (no remainder when divided by 2) and the odds have a remainder of 1.

While this seems to be a rather simple idea, only 65% of students in fourth grade could classify a set of numbers involving two or three digits as even or odd, and fewer students could apply their knowledge of even and odd to a word problem about students walking in pairs. They had a better grasp of which numbers less than 50 were even or odd. (Warfield and Meier, 2007, p. 45)

In Chapter 5, there is an illustration of a student justifying the conjecture that the sum of two odd numbers is even (see Figure 5-4). It uses ideas closely associated with pairing by twos and pictorial representations.

The following conjectures are well within the reach of second and third graders. You might show them the justification by the student in Chapter 5 and have them decide whether these are true and justify them—or disprove them if they are false (only one is false). You can make the conjectures more appropriate for older students by using similar statements about products instead of sums or have children make and check their own conjectures. One of the conjectures is false.

- The sum of two odd numbers is even.
- The sum of two even numbers is even.
- The sum of three odd numbers is odd.
- The sum of any number of odd numbers is odd.
- The sum of any number of even numbers is even.
- The sum of two odd numbers and an even number is even.

Let's return to the example of a magic square (see In the Classroom 18-4) to investigate how odds and evens can help us solve the magic square. If we use the numbers 1–9, how many odd numbers are we using (and how many even numbers)? The arrangement shown below is one of the arrangements of odd (o) and even (e) numbers. Check it out. Does it have the correct number of odds and evens? Do the sums of each row, each column, and the two diagonals produce an odd number (remember, the sum must be 15 for the magic square involving 1–9)? Why can't all the odds be placed on the two diagonals? Use the conjectures about sums of three numbers to explain your answer.

e	o	e
o	o	o
e	o	e

See if using this helps you complete the magic square. Look at the row and column with odd numbers. What do you notice? What three odd numbers add to 15? [(3, 5, and 7) and (1, 5, and 9)]. Place those in the puzzle. Note that there are several ways to do this; one way is shown below. Now see if you can place 2, 4, 6, and 8. You must use some number sense. Does it make sense to place 8 in the same row as 9 or 2 in the column with 1?

e	3	e
1	5	9
e	7	e

You can find many times to use evens and odds naturally. Many of you will remember having to do all the odd-numbered exercises since the answers to the even-numbered exercises were in the back of the text. A simple game involving odds and evens is explained in In the Classroom 18-5. You may have older students sum the first 50 odd numbers (see Figure 18-6 for a hint about adding these using a staircase approach). Students could find the probability of throwing an even sum with two dice (see Table 17-2).

FACTORS AND MULTIPLES

While learning about multiplication and division, children have also begun learning about factors and multiples. The CCSSM recommends looking at finding factor pairs and determining whether or not a number is a multiple of a given number for the numbers 1–100. We discuss factors and multiples separately, then look at common factors and common multiples which the CCSSM recommends as a topic for Grade 6.

Factors A factor of a number divides that number with no remainder. For example, 3 is a factor of 15, since 15 divided by 3 is 5 (no remainder). Another way to look at factors is to look at which two numbers have a product of 15. Since $3 \times 5 = 15$, both 3 and 5 are factors of 15.

Young children can begin exploring factors by using materials. Figure 18-2 shows a story written by a third-grade student. Her teacher gave each child a number and asked them to explore its factors. Pam decided to write about it as well as to draw the groupings she had made with candies.

Using square tiles or square graph paper to explore the factors encourages children to make the connection to area. For example, have children investigate the factors of 12 by making rectangles with 12 squares.

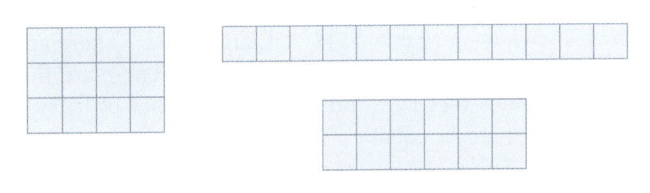

In the Classroom 18–5

MORRA: A GAME OF ODDS AND EVENS

I'm even, but you win since sum is odd.

I'm odd. You won!

Objective: To decide if a sum is even or odd

Grade Level: 1 or above.

Morra is a game from ancient times. The Greeks and the Romans had variations of it. The game described here is for two players.

Rules

One player is designated the "odd" and the other player is "even."

Each player puts out zero to five fingers on "go." (Have the pair, count "1, 2, 3, go.")

If the sum is odd, the odd player gets 1 point. If the sum is even, the even player gets 1 point.

Play until one player gets the designated target number (say 10).

Variations

1. Play with two pairs of two as teams. One team is odd and one is even. Find the sum of the fingers of all four players and assign points as to odd or even. This is the same game but allows for larger numbers and more practice in adding four numbers.

2. Play with three players. Designate one player as 0, the second player as 1, and the third player as 2. Follow the same rules of putting out zero to five fingers. Divide the sum by 3. The player whose number is the remainder receives the point.

This shows that 12 is 3 × 4, 1 × 12, and 2 × 6. Note that there are actually six rectangles (if you turn each of the three shown to show 4 × 3, 12 × 1, and 6 × 2). There are six factors of 12 in all: 1, 2, 3, 4, 6, and 12. These are the same numbers that Pam found for the number of friends that could share 12 equally.

Students in the upper grades need other methods of finding all the factors of larger numbers. For example, what are all the factors of 84? Read through this discussion that Mr. Thomas is having with a small group of bright sixth-grade students.

Mr. Thomas: How can we find all the factors of 84.

Zach: We can divide by all the numbers, 1, 2, up to 84?

Victor: No way, that's 84 division problems; it's got to be simpler.

Madeline: Well, I know 1 divides everything. So 1 is a factor—that means 84 is a factor, too.

Mr. Thomas: Let's record those two like this. You will see why.

Rileigh: 2 is easy. 2 × 42 is 84.

Victor: I'll check 3. Zach, you check 4. Madeline, you check 5, and Rileigh 6.

Mr. Thomas: Good, and I will record what you find. I have to put 2 and 42 on the chart.

Victor reports that 3 and 28 are factors, Zach says 4 × 21, Madeline says 5 won't work, and Rileigh says 6 is a factor and that means 14 is, too. Mr. Thomas records their findings, putting the pairs in colors.

Victor: Gosh, we still have a lot of work. Let's see, 7, 8, 9, 10—we have four more to try now.

Madeline: I tried 7 when 5 wouldn't work. It is 7 × 12.

The students work on 8, 9, and 10 and find they do not divide 84 evenly.

1										84
1	2	3	4	6			14	21	28	42	84
1	2	3	4	6	7	12	14	21	28	42	84

The 12 Story

I had to share 12 candies with friends. No one could get more. I drew pictures to show my shares.

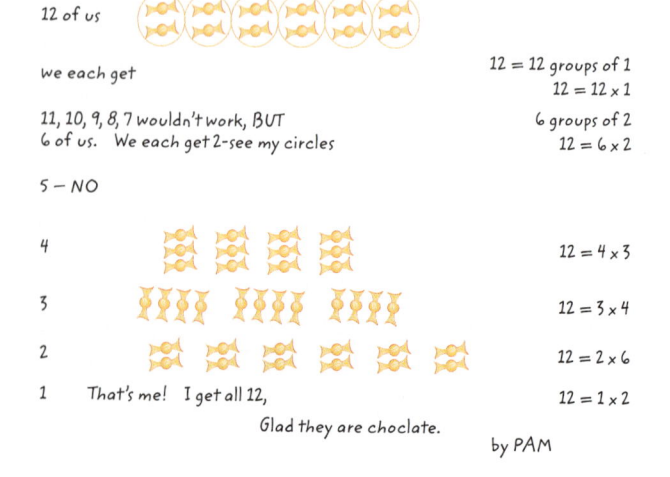

12 of us
we each get   12 = 12 groups of 1
                   12 = 12 × 1

11, 10, 9, 8, 7 wouldn't work, BUT
6 of us. We each get 2—see my circles   6 groups of 2
                                          12 = 6 × 2

5 – NO

4   12 = 4 × 3

3   12 = 3 × 4

2   12 = 2 × 6

1   That's me! I get all 12,   12 = 1 × 2

     Glad they are choclate.
                              by PAM

Figure 18-2 *The story of 12 by Pam.*

Rileigh: "When I tried 9 on the calculator, I got 9.333. Then I tried 11 and got 7 and some decimal. The other number just gets smaller—and we tried all the small ones. I think we can stop."

Zach: I see why you recorded that way; you come into sort of the middle. When the numbers almost bump, you are done.

Mr. Thomas explored with the group what Rileigh and Zach said; they seemed to understand that at some point you have tried enough numbers. Later, he will return to this in other examples to discuss "when the numbers bump."

Victor: Thank goodness. It's not as bad as I thought.

The game, Factor Me Out (see In the Classroom 18-6), is a classical game that requires finding factors of the numbers 2–36. As children discover a good strategy, they are beginning to explore primes and composite numbers.

MULTIPLES A *multiple* of a number is the product of that number and any other whole number. For example, 36 is a multiple of 4 since $4 \times 9 = 36$. In looking for a multiple of a number, we usually begin with the number and generate multiples of it. The positive multiples of 4 are

$$4, 8, 12, 16, 20, 24, 28, \ldots$$

Multiplying 4 by 1, 2, 3, 4 and so forth generates this sequence.

The concept of multiple is not difficult; a multiple of a number is just the product of that number and another. However, there is often confusion between the two words *multiple* and *factor*. Confusion also rises when we ask, "What is 36 a multiple of?" We are actually asking for a factor. (36 is a multiple of 4, of 6, and of seven other numbers.) We can also ask, "What is 4 a factor of?" In this case, we are asking for a multiple of 4. We need to be able to think in both directions: to list all the multiples of 7 and to find what 42 is a multiple of. But when you are first beginning the study of multiples and factors, keep the language to finding factors and to finding multiples. Gradually build in the other questions when students are firm in their understandings.

GREATEST COMMON FACTOR AND LEAST COMMON MULTIPLE
We have looked at factors and multiples of individual numbers; now we will look at pairs of numbers and ask, "What is the largest number that is a factor of both numbers (greatest *common factor*)?" "What is the smallest number that is a multiple of both the numbers (*least common multiple*)?"

Understanding these concepts depends mainly on knowing how to find factors and multiples and keeping straight which is which. Algorithms for finding the greatest common factor (GCF) and least common multiple (LCM) are very similar. While this has an advantage, it can confuse students who do not understand factors and multiples or why they are finding the least or the greatest.

The simplest algorithm for each is just to list the factors (or multiples) of the two numbers and identify the greatest (smallest) number that is common to both lists. At the elementary level, it provides a good source of problem

---

In the Classroom 18–6

FACTOR ME OUT

Objective: To explore factors of numbers, both prime and composite.

Grade Level: 4 and above.

▼ Choose a partner and make this chart:

●	2	3	4	5	6
7	8	9	10	11	12
13	14	15	16	17	18
19	20	21	22	23	24
25	26	27	28	29	30
31	32	33	34	35	36

▼ Rules:
- Player 1 chooses a number. He or she gets that many points. The opponent gets points equal to the sum of all the factors that are left in play.
- Make a table to record scores:

Action	Score	Action	Score
Player 1	Player 1	Player 2	Player 2
Picks 10	10	Factors 10	2+5=7

- Mark out the number and the factors; these cannot be used again.
- Repeat with Player 2 choosing the number.
- Alternate turns until no numbers are left.

The player with the most points wins.
A more detailed plan for this activity is available on the Student Companion site at www.wiley.com/college/reys.

---

solving. In the Classroom 18-7 has problems from an old textbook with corresponding simpler versions of each problem. Some students will be able to solve the more complicated version after solving the simpler one.

PRIMES AND COMPOSITES

A *prime number* is a whole number greater than 1 that has exactly two factors, 1 and itself. Any number with more than two factors is a *composite number*. For example, 13 is prime since 1 and 13 are its only factors; 10 is composite since

In the Classroom 18–7

PROBLEM OF THE DAY FROM OLDEN DAYS

Objective: To solve problems involving multiples and factors.

Grade Level: 4 and above.

The first problem in each pair was taken from an 1887 arithmetic book. The second problem uses smaller numbers and the language of today. Solve one of each pair.

Cover It

Three rooms are 168, 196, and 224 inches wide, respectively. What is the widest carpeting that can be contained exactly in each room?

Mr. Jud has to buy carpet for three rooms that are 16, 24, and 36 feet wide. He does not want to cut the carpet's width. What is the widest width he can use to carpet each room?

When Will We Meet Again?

Four men start at the same place to walk around a garden. A can go around in 9 minutes, B in 10 minutes, C in 12 minutes, and D in 15 minutes. In what time will they all meet at the starting place?

Ann, Bella, Cid, and Dan walk around and around a path. They begin at the same place at the same time. Ann can go around once in 4 minutes, Bella in 6 minutes, Cid in 8 minutes, and Dan in 9 minutes. How many minutes will it be until they are all back at the starting place at the same time?

Digging

A can dig 14 rods of ditch in a week, B can dig 18 rods, C 22 rods, and D 24 rods. What is the least number of rods that would afford an exact number of a week's work for each one of them.

Andy can dig 4 ditches in a day, Bev can dig 5 ditches in a day, Cal can dig 6 ditches in a day, and Donna can dig 7 ditches in a day. Mr. B says, "I have enough ditches to dig so that any one of you could work all day for a number of days. Of course, some of you would have to work more days than others." What is the least number of ditches Mr. B could have?

---

the first prime, you circle it and mark out all the multiples of 2. You proceed to the next prime (the next number not marked out) and circle it, then mark out all its multiples. If you continue this process, only the primes will not have dropped through the sieve.

> ### Math Links 18.3
> You can learn more about the sieve and connect to other sites from wikipedia at http://en.wikipedia.org/wiki/Sieve_of_Eratosthenes or from this book's Web site.
>
> www.wiley.com/college/reys

In the Classroom 18-9 has a variation of the Sieve of Eratosthenes in which the numbers are arranged in rows of six.

PRIME FACTORIZATION

The *fundamental theorem of arithmetic* is the following: Every composite number may be uniquely expressed as a product of primes if the order is ignored. For example, 12 may be written as the product: $2 \times 2 \times 3$. This is called the prime factorization of 12.

The most commonly used method in elementary school to find the prime factorization of a number is the factor tree. Two factor trees for 3190 are shown below. Both show that the prime factorization of 3190 is $2 \times 5 \times 11 \times 29$.

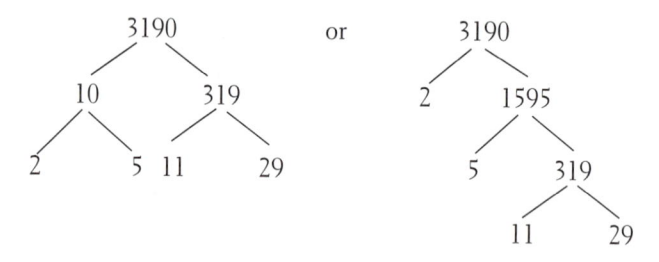

```
      3190                    3190
     /    \         or       /    \
   10     319               2    1595
   / \    / \                    /   \
  2   5  11  29                 5    319
                                     /  \
                                    11   29
```

Another method, the division method, involves repeated division by primes. All the divisors are primes. How do you know when you reach the end? What type of number is the last quotient?

$$\begin{array}{r}2\overline{)3190}\\5\overline{)1595}\\11\overline{)319}\\29\end{array}$$ The quotient of 3190 divided by 2

Do you see that the two methods are analogous if a prime factor must be used for each branch of the tree? No matter which method your students use, make certain that they write the prime factorization as $3190 = 2 \times 5 \times 11 \times 29$. Other factorizations, such as $3190 = 10 \times 319$ or $3190 = 2 \times 5 \times 319$, are true statements, but they are not the *prime* factorization.

---

it has factors of 1, 2, 5, and 10. This topic, often begun in grade 4 is rich enough to be revisited each year.

The activity about hopping on the number line (see In the Classroom 18-8) introduces one model for determining which numbers are primes. Arranging squares to make rectangles, discussed in the section on finding factors, is another model. If only two rectangles can be made with a given number of squares, then the number is prime. For example, 13 is prime because only two rectangles can be made (1×13 and 13×1).

Another way to find the primes is to use the Sieve of Eratosthenes (pronounced "air-a-toss-the-knees"). The numbers are arranged in order in rows of 10. Beginning with 2,

In the Classroom 18–8

THE HOPPERS

Objective: To find primes.

Grade Level: 4 and above.

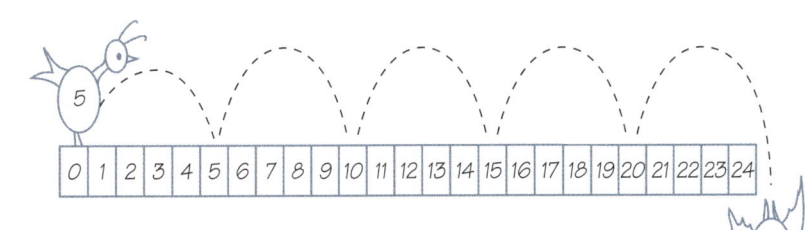

The hoppers are strange characters that can only hop a certain length. For example the 5 hopper hops 5 spaces each time. They all begin at 0.

See where each hopper will land. The 1-hopper and the 2-hopper have been done for you.

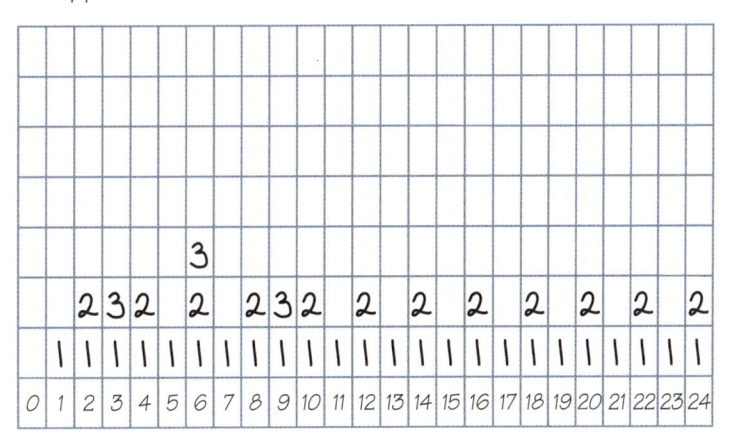

Be a 3-hopper. You will land at 3 (write a 3 above the 3) you will land at 6 (write a 3 above the 6)...Keep going. (You should have 3's above 9, 12, 15, 18, 21, 24).

Do the same for the 4-hopper, the 5-hopper, all the way to a 24-hopper.

Questions:

1. What numbers had two hopper stoppers? __2, 3__
2. What type of numbers are these? _____
3. What numbers had more than two hopper stoppers? __6__
4. What type of numbers are these? _____
5. What about 1? _____

In the Classroom 18–9

SIEVE: THE PRIMES REMAIN

Objective: To generate the prime numbers from 2 to 102.

Grade Level: 5 and above.

Directions:

1. Circle the 2 and then mark out <u>every multiple</u> of 2. We have begun this.
2. Circle the 3 and then mark out every multiple of 3. We have begun this
3. Continue with the next number that is not circled or marked out. Circle it and mark out all its multiples.
4. Continue to 102. Then, answer the questions below.

②	3	4̸	5	6̸	
7	8̸	9̸	10	11	12
13	14	15	16	17	18
19	20	21	22	23	24
25	26	27	28	29	30
31	32	33	34	35	36
37	38	39	40	41	42
43	44	45	46	47	48
49	50	51	52	53	54
55	56	57	58	59	60
61	62	63	64	65	66
67	68	69	70	71	72
73	74	75	76	77	78
79	80	81	82	83	84
85	86	87	88	89	90
91	92	93	94	95	96
97	98	99	100	101	102

Questions:

1. What type of numbers is not marked out?
2. In which columns are the primes?
3. All the numbers in these columns are not prime. Which ones are not primes?
4. Check to see that each of the primes (except 2 and 3) are of the form
   $(6 \times n) + 1$ or $(6 \times m) - 1$, for some whole number n or m. For example, $13 = (6 \times 2) + 1$ and $89 = (6 \times 15) - 1$.

A fun java script that allows you to make factor trees for single numbers and to explore using factor trees and Venn diagrams for finding the greatest common factor and the least common multiple can be found at Math Links 18.4.

Math Links 18.4

The java script may be found at http://nlvm.usu.edu/en/nav/frames_asid_202_g_3_t_1.html or from this book's Web site.

www.wiley.com/college/reys

DIVISIBILITY

We will look at two aspects of divisibility—divisibility rules and conjectures about divisibility. Any number can be divided by any other number, but we say that a number is *divisible* by another number if there is no remainder. 84 is divisible by 2, but it is not divisible by 5.

Divisibility rules (tests or procedures) were important as a check for computation when all arithmetic was done mentally or with paper and pencil. At one time students had to learn the divisibility rules for 2–12. One of the many divisibility rules for 7 is:

> Double the last digit (begin with the ones) and subtract it from the remaining digits. Repeat as many times as needed until the remainder is easily checked. If it is divisible by 7, then the original number is divisible by 7. For example, check 3941. Double the 1 and subtract 2 from 394. This leaves 392. Double the 2 and subtract it from 39, leaving 35. Since 35 is divisible by 7, the original number, 3941, is also.

Wouldn't you rather just divide? These rote rules meant little to students then as well. However, finding out why such a rule works is at the heart of mathematics.

Today, divisibility rules are an interesting sidelight in mathematics that provide opportunities to discover why a rule works or to discover a rule. Many children discover the rules for 2 and 5 by observing the patterns in the multiplication facts for 2 and 5. The rules for other numbers are not quite as evident. In the Classroom 18-10 provides a guided discovery of the rules for 3 and 9. The students should conjecture the following two rules:

- If the sum of the digits of a number is divisible by 3, then the number is divisible by 3.
- If the sum of the digits of a number is divisible by 9, then the number is divisible by 9.

It is important to discuss the results of the table with the class and to give more examples if the students do not answer the questions clearly. Good questions to raise would be: Does this divisibility test work for larger numbers? Is this a test (summing the digits) only for 3 and 9? Why does this test work?

Divisibility   411

In the Classroom 18–10

DIVISIBILITY DISCOVERY

Objective: To explore the divisibility rules for 3 and 9.

Grade Level: 4 and above.

▼ Use your 🖩 Fill in the chart and look for patterns.

Number	Divisible by 3?	Divisible by 9?	Sum of digits	Sum of divisible by 3?	Sum of divisible by 9?
456	yes	no	4 + 5 + 6 = 15	yes	no
891					
892					
514					
37					
78					
79					
1357					
1358					
1359					
1360					
1361					
1362					

▼ What do you think?
- A number is divisible by 3. Is it always divisible by 9?
- A number is divisible by 9. Is it always divisible by 3?
- What does the sum of the digits tell you?
- What did you notice about the sequence of numbers 1357, 1358, 1359, … 1362?

You can help students reason through the last question by examining a model of a meaning of division. Look at 243 by using the model for 243 shown in Figure 18-3. Thinking of division by 9 as sharing with 9 people, we can see that if the nine people shared one of the hundred blocks, 1 unit block would be left over. Similarly, 1 unit block would be left over from the other hundred blocks. If 9 people shared a 10-rod, 1 unit block would be left over from each rod. Therefore, if you divided 243 by 9, 2 unit blocks would be left over from the hundred blocks, 4 unit blocks would be left over from the 10-rods, and the 3 original unit blocks would also be left over. This makes a total of 9 unit blocks left over, which you can clearly share among 9 people. Thus, 243 is divisible by 9.

Students can investigate other divisibility tests in a similar manner. Using the view of sharing each of the hundreds, tens, and ones, do you see why 243 is not divisible by 2? By 5? More generally, since we can share the hundred block and the 10-rod with 5 people (or 2 people), the only thing left to check is the number of ones. So, we only need to consider the ones place when deciding whether or not a number is divisible by 2 or by 5.

The divisibility test for 4 can also be justified for three-digit numbers this way. Try it. Older children should also explore larger numbers and write a justification. Return to place value. What would happen if we have a number in the

Share 243 among 9 people

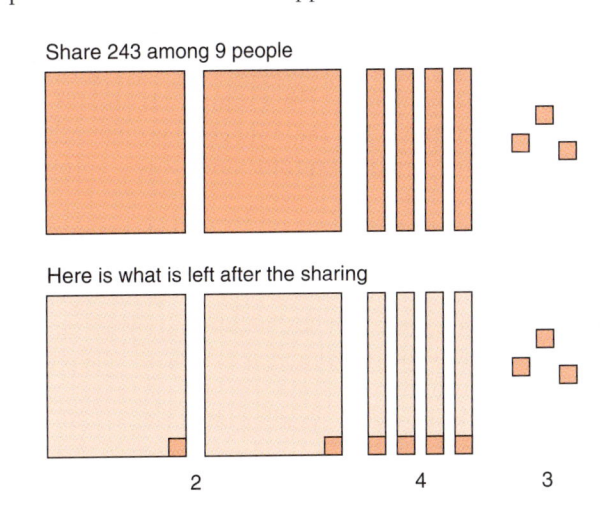

Here is what is left after the sharing

2   4   3

Figure 18-3 *Is 243 divisible by 9?*

thousands and tried to share the thousands with 9 people? How about hundred thousands?

Divisibility also provides a wonderful opportunity for students to investigate conjectures. Students can also investigate the following conjectures (two are false) about divisibility:

1. If a number is divisible by 2 and 3, it is divisible by 6.
2. If a number is divisible by 2 and 5, it is divisible by 10.
3. If a number is divisible by 12, it is divisible by 3 and 6.
4. If a number is divisible by 2 and 4, it is divisible by 8.
5. If a number is divisible by 3 and 6, it is divisible by 12.

Students (and adults) usually begin to understand the conjecture by trying numbers. For example, in the first conjecture, they may say, "I'll try 72. It is divisible by 2 and by 3—and yes, it is divisible by 6." Some students may list all the beginning multiples of 2, 3, and 6.

- Multiples of 2: 2, 4, 6, 8, 10, 12, 14, 16, 18, 20, 22, 24, 26, 28, 30 …
- Multiples of 3: 3, 6, 9, 12, 15, 18, 21, 24, 27, 30, 33, 36, 39, 42 …
- Multiples of 6: 6, 12, 18, 24, 30, 36, 42, 48 …

Notice that if a number is on the list of multiples of 2 and on the 3 list, then it is on the 6 list. Try this for the fourth conjecture. What happens?

OTHER NUMBER THEORY TOPICS

In this section, we briefly describe other number theory topics and present one of many activities associated with that topic. Many more activities can be found at Web sites for further exploration by you or by your students.

RELATIVELY PRIME PAIRS OF NUMBER

Two numbers are *relatively prime* if they have no common factors other than 1. For example, 8 and 15 are relatively prime since 1 is the only common factor, but 8 and 18 are not relatively prime since 2 is a common factor of both.

One fun exploration (Bennett, 1978) that students can begin to explore is described in In the Classroom 18-11. Although there are many conjectures about the star patterns, we focus on those that deal with relatively prime and not relatively prime pairs of numbers as well as connect the stars to geometry terms.

For this activity, use the circle paper in Appendix that has points spaced with 9, 10, 12, 24, 30, and 36. Circles with other numbers of points can be made from these by selecting only some points. For example, an 8-point circle could be made from the 24-point circle by selecting every third point.

In the Classroom 18–11

Objective: To investigate star patterns.

Grade Level: 5 and above.

A star pattern (n, s) is created from places n equally spaced points on a circle and connecting them by steps of length s. For example, a (12, 5) star is drawn by places 12 points on a circle, and then:

Connect the top point (12 o'clock) to the fifth points, clockwise (a below)

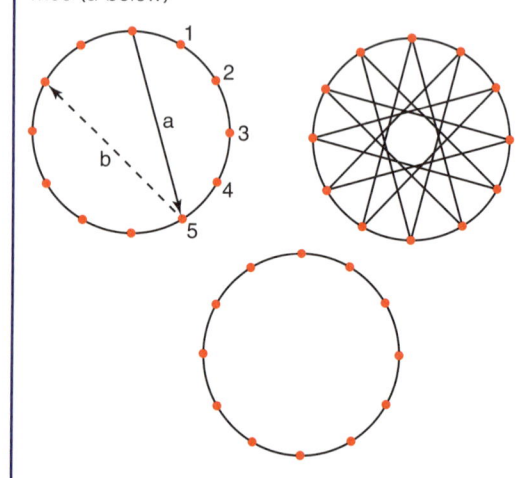

Using an empty 12-point circle like the one above, make the following stars with your classmates.

A. (12, 1)   G. (12, 7)
B. (12, 2)   H. 1(2, 8)
C. (12, 3)   I. (12, 9)
D. (12, 4)   J. (12, 10)
E. (12, 5)   K. (12, 11)
F. (12, 6)

If you get back to the starting point before all points are touched, go to the next point (clockwise) and continue.

Questions

1. Which stars are the same?
2. Which stars make one path as the (12,5) star did? (You did not get back to the starting point before having to move to the next star.) How do you describe the pair of numbers?
3. How would you describe the (12, 1), (12, 2), (12, 3), (12,4), and (12, 6) stars with geometric terms?
4. Try some other stars such as one with 18 points and various steps; see if you can tell when the stars will make only one path.

POLYGONAL NUMBERS

Polygonal or figurate numbers are numbers related to geometric shapes. You are familiar with square number such as 4, 9, and 25. However, some people only know they are called square numbers because they are 2^2, 3^2, and 5^2. They

Other Number Theory Topics 413

are also called square numbers because they can be related to squares, as shown in this illustration.

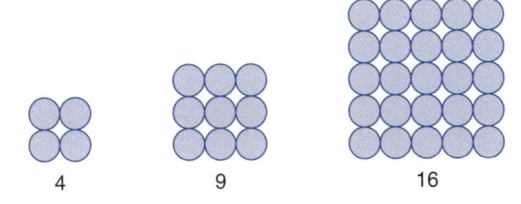

4   9   16

There are many patterns to be found in the sequence of square numbers similar to the ones that students will find in their exploration of triangular numbers in In the Classroom 18-12. Students can also explore pentagonal and hexagonal numbers.

MODULAR ARITHMETIC

In some elementary books, *modular arithmetic* is called *clock arithmetic*. It is an arithmetic based on a limited number of integers, just like the clock is based on the integers 1–12. This is arithmetic of 12 called mod 12. However, in mod 12, 0 is used instead of 12 because it is arithmetic of remainders. You will see what we mean as we explore mod 8.

In mod 8 (see the "clock" below), we use the numbers 0, 1, 2, ... 7. The numbers mean the same as they do in our number system; however, these are all the numbers we have. What do you think the sum of 6 and 7 would be? Try it on the clock and think of starting at 6 o'clock and adding 7 hours. What time would it be?

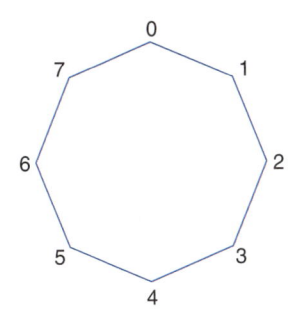

If you do enough sums on the clock, you will find a shortcut. In our number system, 6 + 7 = 13. Thirteen is once around the 8-hour clock and then 5 more. Thus, 6 + 7 = 5 mod 8. What is the remainder after subtracting 8 from 13? How many complete revolutions would you make if you had a sum of 35? How many "hours" after the last revolution?

In the Classroom 18–12

TRIANGULAR NUMBERS

Objective: To explore triangular numbers.

Grade Level: 5 and above.

A number is said to be triangular if that number of dots can be arranged in a triangle like the ones below. The sequence of numbers is found in nature and is used in solving many problems. We will use the symbol T with a subscript, such as T_3, to indicate the third triangular number.

1. Using and extending the pictures, find the first 6 triangular numbers.

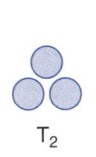

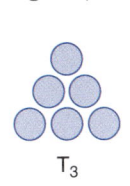

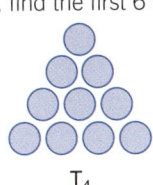

T_1 T_2 T_3 T_4 Draw T_5 Draw T_6

2. Fill in the chart below:

Triangular number	T_1	T_2	T_3	T_4	T_5	T_6	T_7	T_8	T_9
Number of dots	1			10					

Questions

1. What five numbers are you adding when you find T_5? What six numbers for T_6?
2. If you know T_4, tell a quick way to find T_5. If you know T_6, how do find T_7?
3. If T_{10} is 55, what is T_{11}?
4. What is T_{100}? Or how can you find the sum of the numbers 1–100?
5. Do you see the triangular numbers in Pascal's Triangle?

In the Classroom 18–13

MODULAR QUILT

Objective: To use modular addition or multiplication to make a quilt.

Grade Level: Grade 4 and above.

Materials: Two colored markers and graph paper shown in quilt in Figure 18-4 (see Appendix)

Instructions: These are instructions for a mod 4 quilt.

1. Color each square below with two colors. Use different colorings for each square.

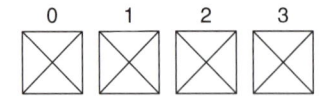

   0   1   2   3

2. Begin with only using 4 x 4 part of the graph paper. Number each of these squares across the top and down the side of the graph paper with the numbers 0-4.

3. Complete either the addition or multiplication table for modular 4.

+	0	1	2	3
0	0	1	2	3
1				
2			1	
3	4			

The blue 1 is the sum of 2 and 3, mod 4.

×	0	1	2	3
0	0	0	0	0
1	0	1	2	
2				2
3	3			1

The red 2 is the product of 2 and 3, mod 4.

4. Color in each square of the 4 by 4 square according to its sum (or product). For example, any sum that is 0 would be colored like you colored the 0 square at the top of the page.

Although there are many explorations with modular arithmetic, we have chosen one that ties it to art—making quilts based on modular 4 arithmetic. In the Classroom 18-13 requires use of an addition or multiplication table for mod 4. We have begun both tables; students would need to complete the tables before coloring their quilt.

Your students could generate many different quilts with these instructions and with other modular numbers. For more patient and persistent students, try mod 8 and use only a 8-by-8 square. Figure 18-4 shows two quilts, one based on the addition table and one on the multiplication table.

PASCAL'S TRIANGLE

Pascal's Triangle is most closely associated with probability. However, there are many number patterns in it and some surprising results, some of which are suggested in In the Classroom 18-14.

Make certain that your students see how to generate the rows before exploring the different patterns. (The number in a hexagon is the sum of the two numbers in the two hexagons immediately above it.)

An interactive Web site allowing students to look at patterns created by coloring multiples of a given number in Pascal's Triangle and a more sophisticated exploration of Pascal's Triangle involving prime factorization may be found at the sites recommended in Math Links 18.5.

Math Links 18.5

The interactive Pascal's Triangle described previously may be found at http://www.shodor.org/interactive/activities/ColoringMultiples/, and the other exploration may be accessed at http://mathforum.org/workshops/usi/pascal/petals_pascal.html. Either of these sites also can be accessed from this book's Web site.

www.wiley.com/college/reys

PYTHAGOREAN TRIPLES

A *Pythagorean triple* is a triple of numbers (a, b, c) such that $a^2 + b^2 = c^2$. You may remember the (3, 4, 5) triple from your geometry class. (*The sum of the squares of the two legs of a right triangle equals the square of the hypotenuse.*) Check to see that $3^2 + 4^2$ does equal 5^2.

Other Number Theory Topics 415

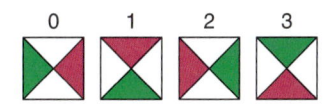

Peyton and Parker colored the four squares like this for their modular 4 quilts.

My addition quilt.

Peyton

Multiplication Mod 4 Quilt

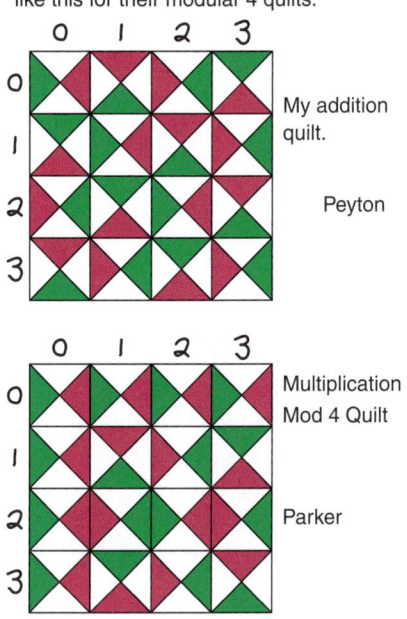

Parker

Figure 18-4 Two modular quilts.

The geometric picture of the Pythagorean relationship is shown below. Note that the square of a side gives the area of the square with that length. For example, the side that is 3 units has an area of 9 square units. The sum of the squares of the blue and yellow squares (the sides) will make the green square (the square on the hypotenuse).

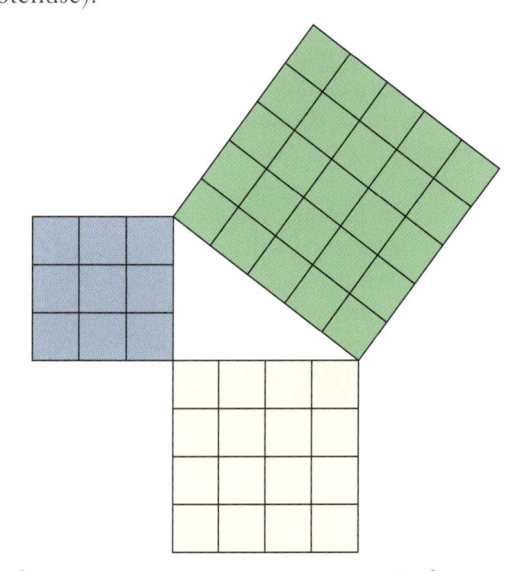

There are many ways to generate Pythagorean triples and many patterns in the triples. In the Classroom 18-15 shows one way to generate triples that is well in

---

In the Classroom 18–14

PASCAL'S TRIANGLE

Objective: To explore number patterns within Pascal's Triangle.

Grade Level: 4 and above.

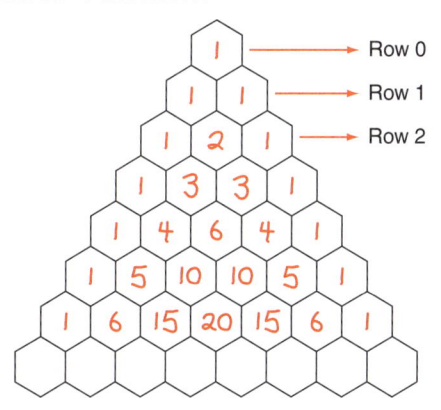

Row 0
Row 1
Row 2

Questions

1. What would be the numbers in row 7? Row 8? Row 9?
2. What patterns do you see?
3. Find the sum of the numbers in each row beginning with row 1 and ending with row 6. (Hint: A chart with each row number and its sum may be helpful.) What do you think would be the sum of row 7? Row 8? Row 20?
4. Look at the numbers in any odd row (ignore the 1's). What do you notice? Do you think that would be true in row 9?
5. Choose any number. Then, going left in that row, alternate subtracting and adding the numbers to the left. For example, choose the second 10 in row 5. Then subtract and add: $10 - 10 + 5 - 1$. The result is 4. Where is the nearest 4 in the triangle? Try more numbers. What is your conjecture?

---

reach of upper elementary students. A calculator will make the computation less tedious and allow students to have fun exploring the patterns. The activity also ties the triples to geometry and measurement, as students draw the sides of the right triangles and check to see if the hypotenuse (or long side) is the measure they found in the pattern.

FIBONACCI SEQUENCE

Some sequences of numbers are famous enough to be named. One of these is the *Fibonacci sequence of numbers*:

1 1 2 3 5 8 13 21 34 55, ...

Do you see how it is generated? The first two terms are ones; thereafter, any term is the sum of the previous two terms.

416 Chapter 18 • Number Theory

In the Classroom 18–15

PYTHAGOREAN TRIPLES

Objective: To generate and explore Pythagorean triples.

Grade Level: 6 and above.

Here is a table from which you can generate Pythagorean triples.

TRIPLE	#1	#2	#3	#4	#5	#6	#7
A	3	5	7	9			
B	4	12	24	40			
C	5	13	25	41			

1. Use a calculator and check to see if (3, 4, 5), (5, 12, 13), and (7, 24, 25) are Pythagorean triples. That is, see if $3^2 + 4^2 = 5^2$.
2. What would be the A in triple #5? _____
3. What is the relation between B and C in each triple? _____
4. The secret is to find B. Hint: Look at the three numbers of the same color. What would be B in triple #5?
5. Fill in the remaining triples in the table. Check to make certain they are Pythagorean Triples.
6. Draw a right triangle with sides 3 cm and 4 cm. Check to see that the third side is 5 cm.
7. Check one more triangle. Use millimeters.

This sequence is rich—with examples of this pattern in nature, art, and even our bodies. Many books have been written about this topic alone. We recommend Math Links 18.6 as a beginning place to take the topic in any direction you wish. All upper elementary students can find some part of this topic to investigate.

Look at the following picture of a hand. The purple shows from the tip of the finger to the first knuckle and so forth to your wrist. Have you seen these numbers before? Check your own hand. You may need to make your own scale, with 2 being the length of the tip to the first knuckle.

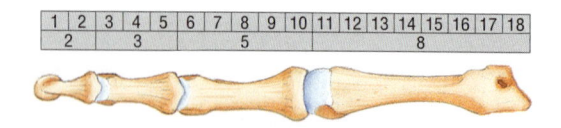

The Fibonacci sequence is also related to the Golden Ratio—another topic that is rich for you and your students to investigate. Begin at the site that looks at hands and other body parts in Math Links 18.7.

This takes us back to the opening Snapshot of a Lesson in which Harry was investigating the ratio of the lengths of thumbs to wrists, or wrists to heads. Do you remember the numbers? They involved 2 and 3. Do you think there is a relationship of 5? Do not stop now—keep looking for the wonders in mathematics all around us.

CULTURAL CONNECTIONS

The roots of number theory can be found thousands of years ago throughout the ancient world. Looking only at the names we have included in this chapter, you can find influences from many countries. For example, the game involving evens and odds (Morra) comes from the Greeks and Romans.

Pythagoras, the person given credit for the Pythagorean theorem, lived in Greece in about 500 B.C. The Pythagorean triples explored in this chapter are built on this theorem. A thousand years later a Frenchman, Fermat, claimed

Math Links 18.6

This site, which will lead to others, is full of history, nature, art, simple puzzles, and activities and may be accessed at http://www.mcs.surrey.ac.uk/Personal/R.Knott/Fibonacci/ or from this book's Web site.

www.wiley.com/college/reys

Math Links 18.7

The picture of the hand and many other relations involving the Fibonacci numbers and the Golden Ratio may be accessed from http://goldennumber.net/ or from this book's Web site.

www.wiley.com/college/reys

he had examined an extension of the Pythagorean triples. Fermat's Last Theorem stated that there was no solution involving only nonzero integers to the equation $a^n + b^n = c^n$ for any integer n greater than 2. For example, there are no integers that satisfy $a^3 + b^3 = c^3$. Try some numbers to convince yourself that this may be the case. Although Fermat called this a theorem and claimed he had a proof (which was never found), it took over 350 years before this conjecture was proven. Two British mathematicians, Wiles and Taylor, were successful in completing a proof in 1994. So don't give up if your students cannot prove their conjectures!

Fibonacci was an Italian who lived in the thirteenth Century. His famous sequence also was known early in India. The Pascal triangle is named for a Frenchman who lived in the seventeenth century. Gauss, a German whose life spanned the eighteenth and nineteenth centuries, is called the father of modular arithmetic. As a young child, he amazed his teacher when he quickly added the first 100 integers.

Benjamin Franklin was fascinated with magic squares. He found methods to generate squares of different sizes. Your students also may become fascinated with a topic in number theory. Give them that opportunity as you also learn more about numbers.

A GLANCE AT WHERE WE'VE BEEN

After examining some reasons for including number theory topics in elementary school, we have taken a quick journey through some of those topics. The suggested activities and references to others should give you lots of ideas about how to differentiate your curriculum to meet the needs of students. We have connected activities to logical reasoning, art, nature, geometry, measurement, and number ideas.

Most importantly, we have chosen activities that we know different children enjoy. We hope that the brief background information will give you enough information or inspiration to learn more so that you can engage your students in some of these topics.

Things to Do: From What You've Read

1. List the four reasons for including number theory in elementary school. Explain each in your own words.
2. Choose the reason for including number theory that resonates with you. Give one activity, other than the one suggested with it, that you think illustrates that reason. (It can be another activity in the chapter, other chapters in the book, from Math Links, or from other sources.)
3. Justify that 15 is the sum of each row of a 3 × 3 magic square using the numbers 0–9. Hint: Find the sum of all the numbers (look at Figure 5-6 in Chapter 5 for a shortcut). What is the sum of each row in a 4 × 4 magic square that uses the numbers 1–16?
4. In the discussion with Mr. Thomas about finding all the factors of a number, Zach said there was a point at which "the numbers almost bump." What did he mean? How do you know that you have come to that "bump"? What numbers would be the "bump" for 144, for 200, for 900.
5. Using the ways to assess students from Chapter 4, explain how you would assess students' understanding of odds and evens or of prime number.
6. Design a guided discovery lesson for divisibility by 8 (or 11).

Things to Do: Going Beyond the Book

In the Field

1. [2]*A Game of Odds and Evens.* Play this game (In the Classroom 18-5) with a friend and then with a young child. Do you notice any differences in their strategies?
2. [2]*Factor Me Out.* Play this game (In the Classroom 18-6) several times with a child in grades 3–5. What do you notice about the ability of the child to choose a number that will maximize his or her score for that round?
3. [2]*The Hoppers.* Do this activity (In the Classroom 18-8) with a small group of children. What questions will you ask students? What mathematical outcome do you expect?
4. Interview children about odds and evens (or prime numbers). What are the different conceptions that they hold about these numbers?

In Your Journal

5. Explain how number theory can help you differentiate instruction at a given grade level. Give specific examples of activities you could use for that grade level.

---

[2] Additional activities, suggestions, and questions are available in the field experience manual on the Student Companion site at www.wiley.com/college/reys.

418 Chapter 18 • Number Theory

6. Jasper, a sixth-grade student, completed the sieve (In the Classroom 18-9). He said, "if a number is prime, it must be one more or one less than a multiple of 6." Other than the primes 2 and 3, this is a true statement. Explain in your own words, after completing the sieve yourself, what led Jasper to this conjecture. Then, justify the conjecture. (You may want to review some of the algebraic conjectures in Chapter 14). Hint: Represent each column algebraically. For example, all of the numbers in column 2 are of the form $6n + 2$ for some integer n.

7. Write up your exploration of one of the conjectures from In the Classroom 18-2.

8. Choose your favorite number. Describe it in terms of all the number theory topics discussed in this chapter. (For example: Is it prime? What are its factors? Is it in Pascal's triangle?)

With Additional Resources

9. Examine a textbook series (grades 4–6) to see what number theory topics are included.

With Technology

10. Find a method for determining the LCM using the site suggested in Math Links 18.4. Hint: Do some small numbers first and see if you can find a relationship between the product of the two numbers and the LCM and GCF.

Note to Instructors: You can find additional resources, learning activities, and black line master sin this text's accompanying Instructor's Manual at www.wiley.com/college/reys.

Book Nook for Children

Bessinger, J., and Pless, V. *The Cryptoclub: Using Mathematics to Make or Break Secret Codes.* Wellesley, MA: A.K. Peters, 2006.

This is a challenging book, appropriate for gifted students or mathematics clubs. Boys and girls create and decode ciphers based on number theory topics such as modular arithmetic and prime numbers.

Goennel, H. *Odds and Evens.* New York: Harper Collins, 1994.

This book for young children shows 13 familiar expressions in counting, from a one-horse town to a baker's dozen. Evens and odd numbers are emphasized in the pictures associated with the numbers.

Hulme, J. N. *Wild Fibonacci: Nature's Secret Code Revealed.* Berkeley, CA: Tricycle Press, 2005.

This book shows different Fibonacci patterns in recurring sequences using numbers, plants, and animals. The author gives a brief history of Fibonacci numbers and how the reader can find these sequences in nature.

Turner, P. *Among the Odds and Evens: A Tale of Adventure.* New York: Scholastic, 1999.

The Kingdom of Wontoo is visited by two adventurers who soon learn there are two types of numbers. The Odds are eccentric, and the Evens are orderly and predictable. The reader will see how odd and even numbers are represented in a variety of pictures and will also learn about adding evens and odds. This book is a great start for algebraic reasoning for young students.

REFERENCES

Abrams, J., and Ferguson, J. "Teaching students from many nations." *Educational Leadership*, 62(4) (December 2004/January 2005), pp. 64–67.

Akers, J., Battista, M., Godrow, A., Clements, H. D, and Sarama, J. *Halves and Symmetry: Investigations in Number, Data, and Space: Grade* 2. Menlo Park, CA: Dale Seymour Publications, 1996.

Alajmi, A., and Reys, R. "Reasonable and reasonableness of answers: Kuwaiti middle school teachers' perspectives." *Educational Studies in Mathematics*, 65(1) (May 2007), pp. 77–94.

Albert, L. R., Mayotte, G., and Sohn, S. C. "Making observations interactive." *Mathematics Teaching in the Middle School*, 7(7) (March 2002), pp. 396–401.

Ameis, J. A. "Stories invite children to solve mathematical problems." *Teaching Children Mathematics*, 8(5) (January 2002), p. 260.

Amos, S. F. "Talking mathematics." *Teaching Children Mathematics*, 14(2) (September 2007), pp. 68–73.

Andrews, A., and Trafton, P. R. *Little Kids—Powerful Problem Solvers: Math Stories from a Kindergarten Classroom*. Portsmouth, NH: Heinemann, 2002.

Anthony, G. J., and Walsh, M. "Zero: A 'none' number." *Teaching Children Mathematics*, 11(1) (August 2004), pp. 38–42.

Arbaugh, F., Brown, C., Lynch, K., and McGraw, R. "Students' ability to construct responses (1992–2000): Findings from short and extended constructed-response items." In *Results and Interpretations of the 1990 Through 2000 Mathematics Assessments of the National Assessment of Educational Progress* (eds. P. Kloosterman and F. K. Lester.). Reston, VA: NCTM, 2004.

Ashlock, R. B. *Error Patterns in Computation*, 10th ed. New York: Merrill, 2010.

Badger, E. "More than testing." *Arithmetic Teacher*, 39(9) (May 1992), pp. 7–11.

Ball, D. L., and Bass, H. "Making mathematics reasonable in school." In *A Research Companion to Principles and Standards for School Mathematics* (eds. J. Kilpatrick, W. G. Martin, and D. Schifter). Reston, VA: NCTM, 2003, pp. 27–44.

Baroody, A. J., Lai, M. I., and Mix, K. S. "The development of young children's number and operation sense and its implications for early childhood education." In *Handbook of Research on the Education of Young Children* (eds. B. Spodek and O. N. Saracho). Mahwah, NJ: Erlbaum, 2006, pp. 187–221.

Barrow, A. T. "Building word problems: What does it take?" *Teaching Children Mathematics*, 17(3) (October 2010), pp. 140–148.

Bass, H. "Computational fluency, algorithms, and mathematical proficiency: One mathematician's perspective." *Teaching Children Mathematics*, 9(6) (February 2003), pp. 322–327.

Battista, M. "Learning with understanding; Principles and processes in the construction of meaning for geometric ideas." In *The Learning of Mathematics, 2007 Yearbook of the National Council of Teachers of Mathematics* (eds. G. Martin, M. Strutchens, and P. Elliot). Reston, VA: NCTM, 2007, pp. 65–80.

Battista, M. "The influence of technology on mathematics learning in elementary and middle school classrooms." In *Teaching and Learning Mathematics: Translating Research for Elementary School Teachers* (ed. D. Lambdin). Reston, VA: National Council of Teachers of Mathematics, 2010, pp. 55–60.

Battista, M. T. "Computer technologies and teaching geometry through problem solving." In *Teaching Mathematics Through Problem Solving: Prekindergarten–Grade 6* (eds. F. K. Lester and R. I. Charles). Reston, VA: NCTM, 2003, pp. 229–238.

Bay-Williams, J., and Herrera, S. "Is 'just good teaching' enough to support the learning of English language learners? Insights from sociocultural learning theory." In *The Learning of Mathematics, 2007 Yearbook of the National Council of Teachers of Mathematics* (eds. W. G. Martin, M. Strutchens, and P. C. Elliott). Reston, VA: NCTM, 2007, pp. 43–63.

Bay-Williams, J., and Karp, K. "Elementary school mathematics teachers' beliefs." In *Teaching and Learning Mathematics: Translating Research for Elementary School Teachers* (Ed. D. Lambdin), Reston, VA: National Council of Teachers of Mathematics. 2011, pp. 47–54

Bay-Williams, J., and Martinie, S. *Math and Literature: Grades 6–8*. Sausalito, CA: Math Solutions Publications, 2004.

Bennett, A. B. "Star patterns." *Arithmetic Teacher*, 26(1) (January 1978), pp. 12–14.

Blume, G.W., Galindo, E., and Walcott, C. "Performance in measurement and geometry from the viewpoint of *Principles and Standards for School Mathematics*." In *Results and Interpretations of the 2003 Mathematics Assessment of the National Assessment of Educational Progress* (eds. P. Kloosterman and F. K. Lester). Reston, VA: NCTM, 2007, pp. 95–138.

Bobis, J. "The empty number line: A useful tool or just another procedure?" *Teaching Children Mathematics*, 13(8) (April 2007), pp. 410–413.

Boerst, T. A. "Division discussions: Bridging student and teaching thinking." *Teaching Children Mathematics*, 11(3) (November, 2004), pp. 233–236.

Bohan, H., Irby, B., and Vogel, D. "Problem solving: Dealing with data in the elementary school." *Teaching Children Mathematics*, 1(5) (January 1995), pp. 256–260.

Bohl, D., Oursland, M., and Finco, K. "The legend of Paul Bunyan." *Mathematics Teaching in the Middle School*, 8(8) (April 2003), pp. 441–448.

Bresser, R. *Math and Literature: Grades 4–6*. Sausalito, CA: Math Solutions Publications, 2004.

Bright, G., Joyner, J., and Wallis, C. "Assessing proportional reasoning." *Mathematics Teaching in the Middle School*, 9(3) (November 2003), pp. 166–172.

Brodesky, A., Gross, F., McTigue, A., and Tierney, C. "Planning strategies for students with special needs: A professional development activity." *Teaching Children Mathematics*, 11(3) (October 2004), pp. 146–154.

Brown-Chidsey, R. "No more "waiting to fail" *Educational Leadership*, 65(2) (October 2007), pp. 40–46.

Brown, C. L., Cady, J. A., and Taylor, P.M. "Problem solving and the English language learner." *Mathematics Teaching in the Middle School*, (14)9, (May 2009), pp. 533–539.

Brownell, W. A. "An experiment on 'borrowing' in third-grade arithmetic." *Journal of Educational Research*, 41(3) (November 1947), pp. 161–171.

Brownell, W. A. "Psychological considerations in the learning and the teaching of arithmetic." In *The Teaching of Arithmetic, Tenth Yearbook of the National Council of Teachers of Mathematics* (ed. W. D. Reeves). Reston, VA: NCTM, 1935, pp. 1–31.

Brownell, W. A. "The revolution in Arithmetic." Reprinted From *Arithmetic Teacher* (February, 1954), pp 1–5, *Mathematics Teaching in the Middle School* 12(1) (August 2006), pp. 26-30.

Brownell, W. A., and Chazal, C. B. "The effects of premature drill in third-grade arithmetic." *Journal of Educational Research*, 29(1) (September 1935), pp. 17–28.

Browning, C., Garza-Kling, G., and Sundling, E. "What's your angle on angle?" *Teaching Children Mathematics*, 14(5) (December 2007/January 2008), pp. 283–287.

Bruner, J. *Actual Minds, Possible Worlds*. Cambridge, MA: Harvard University Press, 1986.

Brutlag, D. *Chance Encounters: Probability in Games and Simulation*. Palo Alto, CA: Creative, 1996.

Burns, M. "Introducing division through problem-solving experiences." *Arithmetic Teacher*, 38(8) (April 1991), pp. 14–18.

Burns, M. "The role of questioning." *Arithmetic Teacher*, 32(6) (February 1985), pp. 14–16.

Burns, M. *About Teaching Mathematics: A K–8 Resource*, 2nd ed. Sausalito, CA: Math Solutions Publications, 2007.

Burns, M. *Mathematics: Assessing Understanding, Grades K-6* (video). Vernon Hills, IL: ETA/Cuiseriaire, 1993.

Burns, M., and Sheffield, S. *Math and Literature: Grades 2–3*. Sausalito, CA: Math Solutions Publications, 2004b.

Burns, M., and Sheffield, S. *Math and Literature: Grades K–1*. Sausalito, CA: Math Solutions Publications, 2004a.

Burns, M., and Sheffield, S. *Math and Literature: Grades K–1*. Sausalito, CA: Math Solutions Publications, 2004a.

Burton, G., Mills, A., Lennon, C., and Parker, C. *Number Sense and Operations*. Reston, VA: NCTM, 1993.

Buschman, L. "Children who enjoy problem solving." *Teaching Children Mathematics*, 9, 2003a, pp. 539–544.

Buschman, L. (2001). "Using student interviews to guide classroom instruction: An action research project." *Teaching Children Mathematics*, 8(4), pp. 222–227.

Buschman, L. *Share and Compare: A Teacher's Story About Helping Children Become Problem Solvers in Mathematics*. Reston, VA: NCTM, 2003b.

Bush, W. S. *Mathematics Assessment: Cases and Discussion Questions for Grades K–5*. Reston, VA: NCTM, 2001.

Cady, J. "Implementing reform practices in a middle school classroom." *Mathematics Teaching in the Middle School*, 11(9) (2006), pp. 460–466.

Cai, J. "Helping elementary school students become successful mathematical problem solvers." In *Teaching and Learning Mathematics: Translating Research for Elementary School Teachers* (ed., D. V. Lambdin and F.K. Lester). Reston, VA: NCTM, 2010, pp. 9–14.

Cai, J. "What research tells us about teaching mathematics through problem solving." In *Teaching Mathematics through Problem Solving: Prekindergarten-Grade 6*, (ed., F. K. Lester and R. I. Charles). Reston, VA: NCTM, 2003, pp. 241–254.

Campbell, P. F., and Rowan, T. E. "Teacher questions 1 student language 1 diversity 1 mathematical power." In *Multicultural and Gender Equity in the Mathematics Classroom: The Gift of Diversity, 1997 Yearbook of the National Council of Teachers of Mathematics* (eds. K. Trentacosta and M. J. Kenney). Reston, VA: NCTM, 1997, pp. 60–70.

Carpenter, T. P., and Levi, L. "Building a foundation for learning algebra in the elementary grades." In *Brief, K–12 Mathematics and Science: Research & Implications*, 1(2) (Fall 2000). http://www.wcer.wisc.edu/NCISLA/publications/briefs/fall2000.pdf.

Carpenter, T. P., and Romberg, T. A. *Powerful Practices in Mathematics and Science*. Madison, WI: The Board of Regents of the University of Wisconsin System, 2004.

Carpenter, T. P., Corbitt, M. K., Kepner, H. S., Jr., Lindquist, M. M., and Reys, R. E. *Results from the Second Mathematics Assessment of the National Assessment of Educational Progress*. Reston, VA: NCTM, 1981.

Carpenter, T. P., Franke, M. L., and Levi, L. *Thinking Mathematically: Integrating Arithmetic and Algebra in the Elementary School*. Portsmouth, NH: Heinemann, 2003.

Carreher, D., and Schliemann, A. "Algebraic reasoning in elementary school classrooms." In *Teaching and Learning Mathematics: Translating Research for Elementary School Teachers* (ed., D. V. Lambdin). Reston, VA: NCTM, 2010, pp. 23–29.

Cavanagh, M., Dacey, L., Findell, C., Greenes, C., Jensen-Sheffield, L., and Small, M. *Navigating Through Number and Operations in Prekindergarten–Grade 2*. Reston, VA: NCTM, 2004.

CCSSI(2010)

CCSSO reference

Chandler, C., and Kamii, C. "Giving change when payment is made with a dime: The difficulty of tens and ones." *Journal for Research in Mathematics Education*, 40(2), (March 2009), pp. 97–118.

Chapin S. H., and O'Connor, C. "Academically productive talk: Supporting students' learning in mathematics." In *The Learning of Mathematics, 69th Yearbook of the National Council of Teachers of Mathematics* (eds. W. G. Martin and M. E. Strutchens). Reston, VA: NCTM, 2007, pp. 113–128.

Charles, R. I, Lester F. K., and O'Daffer, P. *How to Evaluate Progress in Problem Solving*. Reston, VA: NCTM, 1987.

Charles, R. I., Lester, F. K., and Lambdin, D. V. *Problem Solving Experiences: Making Sense of Mathematics. (Grades 3–8)*. Parsippany, NJ: Dale Seymour, 2005.

Chval, K., and Hicks, S. "Calculators in K-5 textbooks." *Teaching Children Mathematics*, 15(7) (March 2009), pp. 430–437.

Civil, M., Planas, N., and Quintos, B. " Immigrant parents' perspectives on their children's mathematics." *Zentralblatt fur Didaktik der Mathematik (ZDM)* 37(2) (2005), pp. 81–89. Math.arizonaedu/~civil/Civil%20et%20al_ZDM.pdf.

Clark, F. B., and Kamii, C. "Identification of multiplicative thinking in children in grades 1–5." *Journal for Research in Mathematics Education*, 27(1) (January 1996), pp. 41–51.

Clarkson, L., Robelia. B., Chahine, I., Fleming, M., and Lawrenz, F. "Rulers of different colors: Inquiry into measurement." *Teaching Children Mathematics*, 14(1) (August 2007), pp. 34–39.

Clement, L. L. "A model for understanding, using, and connecting representations." *Teaching Children Mathematics*, 11(2) (September 2004), pp. 97–102.

Clement, R. *Counting on Frank*. Milwaukee, WI: Gareth Stevens Publishing, 1991.

Clements, D. H. "Teaching and learning geometry." In *A Research Companion to Principles and Standards for School Mathematics* (eds. J. Kilpatrick, W. G. Martin, and D. Schifter). Reston, VA: NCTM, 2004, pp. 151–178.

Clements, D. H. Subitizing. What is it? Why teach it? *Teaching Children Mathematics* 5(7) (March 1999), pp. 400–405.

Clements, D. H., and Sarama, J. "Early childhood mathematics learning." In *Second Handbook of Research on Mathematics Teaching and Learning*. New York: Information Age Publishing, 2007, pp. 461–555.

Clements, D., and Battista, M. "Geometry and spatial reasoning." In *Handbook of Research on Mathematics Teaching and Learning* (ed. D. Grouws). New York: Macmillan, 1992, pp. 420–464.

Coats, G. "Middle school girls in mathematics classroom." Mathematics Teaching in the Middle School, 13(4), (November 2007), pp. 234–235.

Common Core State Standards Initiative (CCSSI) "Common Core State Standard for mathematics" In common core State Standards (College- and career-Readiness Standards and K-12 standrands in English languages Arts and Math) Washington, D.C National Governor Association Center for Best Practices and Council of Chief State School Officers (NGA Center and CCSSO), 2010.http://www.corestandars.org

Corwin, R., and Friel, S. *Used Numbers–Statistics: Prediction and Sampling*, Grades 5–6. Menlo Park, CA.: Dale Seymour, 1990.

Cramer, K., and Whitney, S. "Learning Rational Number Concepts and Skills in elementary School Classrooms." In *Teaching and Learning Mathematics: Translating Research for Elementary School Teachers* (ed. D. V. Lambdin). Reston, VA: NCTM, 2010, pp. 15–22.

Cramer, K., Post, T., and Currier, S. "Learning and teaching ratio and proportion: Research implications." In *Research Ideas for the Classroom: Middle Grades Mathematics* (ed. D. T. Owens). Reston, VA: NCTM, 1993, pp. 159–178.

Crown, W. "Using technology to enhance a problem-based approach to teaching: What will and will not work." In *Teaching Mathematics Through Problem Solving: Prekindergarten–Grade 6*. (ed. F. K. Lester). Reston, VA: NCTM, 2003, pp. 217–228.

Cuoco, A. A. (ed.). *The Roles of Representation in School Mathematics, 2001 Yearbook of the National Council of Teachers of Mathematics*. Reston, VA: NCTM, 2001.

Dienes, Z. P. *Building Up Mathematics*. London: Hutchinson Education, 1960.

Diezman, C. "On the spot assessments." *Teaching Children Mathematics*, 15(5) (December 2008), pp. 290–294.

Diezmann, C., Thornton, C., and Watters, J. "Addressing the needs of exceptional students through problem solving." In *Teaching Mathematics Through Problem Solving: Prekindergarten–Grade 6* (ed. F. K. Lester). Reston, VA: NCTM, 2003, pp. 169–182.

Dixon, J. "Tracking time." *Teaching Children Mathematics*, 15(1) (August 2008), pp. 18–24.

Dougherty, B. J., and Venenciano, L. "Measuring up for understanding." *Teaching Children Mathematics*, 13(9) (May 2007), pp. 452–456.

Driscoll, M. J. *Fostering Algebraic Thinking*. Portsmouth, NH: Heinemann, 1999.

Dubon, L. P., and Shafer, K. G. "Storyboards for meaningful patterns." *Teaching Children Mathematics*, 16(6) (February 2010), pp. 325–329.

Dwyer, J. "Ordering rectangles: Which is bigger?" In *Teaching Mathematics Through Problem Solving: Prekindergarten–Grade 6* (eds. F. K. Lester and R. I. Charles). Reston, VA: NCTM, 2003, pp. 143–147.

Earnest, D., and Balti, A. "Instructional strategies for teaching algebra in elementary school: Findings from a research-practice collaboration." *Teaching Children Mathematics* 14(9) (May 2008), pp. 519–522.

Ebeling, D., Deschenes, C., and Sprague, J. *Adapting Curriculum and Instruction in Inclusive Classrooms: Staff Development Kit*. Bloomington, IN: Institute for the Study of Developmental Disabilities, 1994.

Economopoulos, K., and Russell, S. J. *Putting Together and Taking Apart: Addition and Subtraction*. A Unit in the "Investigations in Number, Data, and Space Curriculum." White Plains, NY: Dale Seymour, 1998.

Ellemor-Collins, D. L., and Wright, R. J. "Assessing student thinking about arithmetic: Videotaped interviews." *Teaching Children Mathematics*, 15(2) (September 2008), pp. 106–111.

Ellett, K. "Making a million meaningful." *Mathematics Teaching in the Middle School*, 10(8) (April 2005), pp. 416–423.

Ellington, A. "The effects of non-CAS graphing calculators on student achievement and attitude levels in mathematics: A meta-analysis." *School Science and Mathematics*, 106(1) (January 2006), pp. 16–26.

Ellington, A. J. "A meta-analysis of the effects of calculators on students' achievement and attitude levels in precollege mathematics classes." *Journal for Research in Mathematics Education*, 34(5) (November 2003), pp. 433–463.

Ellis, M., and Yeh, C. "Creative arithmetic: Exploring alternative methods." *Teaching Children Mathematics*, 14(6) (February 2008), pp. 367–368.

Ellis, M., Yeh, C., and Stump, S. "Problem solvers: Rock, paper, scissors and solutions to the broken calculator problem." *Teaching Children Mathematics*, 14(5) (December 2007), p. 309.

Fennell, F., Ferrini-Mundy, J., Ginsburg, H., Greenes, C., Murphy, S., and Tate, W. *Silver Burdett Ginn Mathematics, Grade 4, Volume 1*. Parsippany, NJ: Silver Burdett Ginn, 1999.

Fernandes, A., Anhalt, C. O., and Civil, M. "Mathematical interviews to assess Latino students." *Teaching Children Mathematics*, 16(3) (October 2009), pp. 162–169.

Fillingim, J. G., and Barrow, A. T. "From the inside out." *Teaching Children Mathematics*, 17(2) (September 2010), pp. 80–88.

Fischbein, E., Nello, M. S., and Marino, M. S. "Factors affecting probabilistic judgments in children and adolescents." *Educational Studies in Mathematics*, 22(6) (December 1991), pp. 523–549.

Flores, A. "On my mind: The finger and the moon." *Mathematics Teaching in the Middle School*, 13(3) (October 2007), pp. 132–133.

Flores, A., Samson, J., and Yanik, H.B. "Quotient and measurement interpretations of rational numbers." *Teaching Children Mathematics*, 13(1) (August 2007), pp. 34–39.

Folkson, S. "Meaningful communication among children: Data collection." In *Communication in Mathematics, 1996 Yearbook of the National Council of Teachers of Mathematics* (ed. P. Elliott). Reston, VA: NCTM, 1996, pp. 29–34.

Franke, M. L., and Carey, D. A. "Young children's perceptions of mathematics in problem-solving environments." *Journal for Research in Mathematics Education*, 28(1) (January 1997), pp. 8–25.

Franklin, C., Kader, G., Mewborn, D., Moreno, J., Peck, R., Perry, M., and Scheaffer, R. *Guidelines for Assessment and Instruction in Statistics Education (GAISE) Report: A pre-K–12 Curriculum Framework*. Alexandria, VA: American Statistical Association, 2005.

Friel, S., Curcio, F., and Bright, G. Making sense of graphs. *Journal for Research in Mathematics Education*, 32(2) (March 2001), pp. 124–157.

Funkhouser, C. "Developing number sense and basic computational skills in students with special needs." *School Science and Mathematics*, 95(5) (May 1995), pp. 236–239.

Fuson, K. "Developing mathematical power in whole number operations." In *A Research Companion to Principles and Standards for School Mathematics* (eds. J. Kilpatrick, W. G. Martin, and D. Schifter). Reston, VA: NCTM, 2003a, pp. 68–94.

Fuson, K. C. "Toward computational fluency in multidigit multiplication and division." *Teaching Children Mathematics* 9(6) (February 2003b), pp. 300–305.

Gallenstein, N. "Creative discovery through classification." *Teaching Children Mathematics*, 11(2) (September 2004), pp. 103–108.

Gay, S., and Aichele, D. B. "Middle school students' understanding of number sense related to percent." *School Science and Mathematics*, 97(1) (January 1997), pp. 27–36.

Glanfield, F. "Amelia's mental images of numbers." Teaching Children Mathematics, 16 (8) (April 2010), p. 504.

Glanfield, F., Stenmark, J. K., and Bush, W. S. *Mathematics Assessment: A Practical Handbook for Grades K–2*. Reston, VA: NCTM, 2001.

Goldin, G. A. "Representation in school mathematics: A unifying research perspective." In *A Research Companion to Principles and Standards for School Mathematics* (eds. J. Kilpatrick, W. G. Martin, and D. Schifter). Reston, VA: NCTM, 2003, pp. 275–285.

Good, T. L., and Brophy, J. E. *Looking in Classrooms*, 10th ed. New York: Longmans, 2010.

Goodrow, A., and Kidd, K. "Counting school days, decomposing number, and determining place value." *Teaching Children Mathematics*, 15(2) (September 2008), pp. 74–79.

Gray, L., Thomas, N., and Lewis, L. *Teachers' Use of Educational Technology in U.S. Public Schools: 2009* (NCES 2010-040). Washington, DC. National Center for Education Statistics, Institute of Education Sciences, U.S. Department of Education.

Greer, B. "Multiplication and division as models of situations." In *Handbook of Research on Mathematics Teaching and Learning* (ed. Douglas A. Grouws). New York: Macmillan, 1992, pp. 276–295.

Gregg, J., and Gregg, D. "Interpreting the standard division algorithm in a 'candy factory' context". *Teaching Children Mathematics* 14(1) (August, 2007), pp. 25–31.

Haberman, M. (1991). "The pedagogy of poverty versus good teaching," *Phi Delta Kappan*, 73, pp. 290–294.

Harper, S. "Students' interpretations of misleading graphs." Mathematics Teaching in the Middle School, 9(6) (February 2004), pp. 340–343.

Hartweg, K. "Solutions to the decorative plate hanging problem." *Teaching Children Mathematics*, 10(5) (December 2004–January 2005), pp. 280–284.

Hedges, M., Huinker, D., and Steinmeyer, M. Unpacking division to build teachers' mathematical knowledge. *Teaching Children Mathematics* 11 (May 2005), pp. 478–483.

Hembree, R., and Dessart, D. "Research on calculators in mathematics education." *Calculators in Mathematics Education*. 1992 Yearbook (eds. J. T. Fey and C. R. Hirsch). Reston, VA: NCTM, 1992, pp. 23–32.

Herrell, A. L. *Fifty Strategies for Teaching English Language Learners*. Upper Saddle River, NJ: Merrill, 2000.

Hiebert, J. "Relationships between research and the NCTM standards." *Journal for Research in Mathematics Education*, 30 (January 1999), pp. 3–19.

Hiebert, J. "Signposts for teaching mathematics through problem solving." In *Teaching Mathematics Through Problem Solving: Prekindergarten–Grade 6* (eds. R. I. Charles and F. K. Lester). Reston, VA: NCTM, 2003, pp. 53–61.

Hiebert, J., and Grouws, D. "Effective teaching for the development of skill and conceptual understanding of number: What is most effective?" In *Effective Instruction Briefs* (ed. J. Reed). Reston, VA: NCTM, 2007.

Hildebrand, C., Ludeman, C. J., and Mullin, J. "Integrating mathematics with problem solving using the mathematician's chair." *Teaching Children Mathematics*, 5(7) (March 1999), pp. 434–441.

Hirigoyen, H. "Dialectal variations in the language of mathematics: A source for multicultural experiences." In *Multicultural and Gender Equity in the Mathematics Classroom: The Gift of Diversity, 1997 Yearbook of the National Council of Teachers of Mathematics* (ed. J. Trentacosta). Reston, VA: NCTM, 1997.

Holloway, J. "Closing the minority achievement gap in math." *Educational Leadership*, 61(5) (February 2004), pp. 84–86.

Hope, J. A., Leutzinger, L., Reys, B. J., and Reys, R. E. *Mental Math in the Primary Grades*. Palo Alto, CA: Dale Seymour, 1988.

Hope, J. A., Reys, B. J., and Reys, R. E. *Mental Math in the Middle Grades*. Palo Alto, CA: Dale Seymour, 1987.

Houghton Mifflin. *Mathematics Grade 6*. Boston, MA: Houghton Mifflin, 2002

Hufferd-Ackles, K. Fuson, K. C., and Sherin, M.G. "Describing levels and components of a math-talk learning community." *Journal for Research in Mathematics Education*, 35(2) (March 2004), pp. 81–116.

Huinker, D. "Calculators as learning tools for young children's explorations of numbers." *Teaching Children Mathematics*, 8(6) (February 2002), pp. 316–321.

Hutchins, P. *The Doorbell Rang*. New York: Green Willow Books, 1986.

Jacobs, V.R., and Kusiak, J. "Got tools? Exploring children's use of mathematics tools during problem solving." *Teaching Children Mathematics*, **12**(9) (2006), pp. 470–477.

Jeon, K. "Mathematics hiding in the nets for a cube." *Teaching Children Mathematics*, 15(7) (March 2009), pp. 394–399.

Jeon, K., and Bishop, J. "Problem solvers: Solutions to the nine jumping numbers problem." *Teaching Children Mathematics*, 14(6) (February 2008), pp. 269–373.

Joanning, D., Weber, W., Heidt, C., Pearce, M., and Horner, K. "The *Polar Express* to early algebraic thinking." *Teaching Children Mathematics* **16**(5) (December 2009/January 2010), pp. 300–307.

Jones, C. "Counting is for the birds" *Teaching Children Mathematics* 16 (6) (February 1010), pp. 318–320.

Jones, G. A. (ed.). *Exploring Probability in School: Challenges for Teaching and Learning*. Dordrecht, The Netherlands: Kluwer Academic, 2005.

Jones, G. A., and Langrall, C. W. (2007). "Research in probability: Responding to classroom realities". In *Second Handbook of Research on Mathematics Teaching and Learning* (ed. F. K. Lester). Charlotte, NC: Information Age Publishing, 2006, pp. 909–956.

Jones, G. A., Langrall, C. W., Thornton, C., and Mogill, A. T. "Students' probabilistic thinking in instruction." *Journal for Research in Mathematics Education*, 30(5) (November 1999), pp. 487–519.

Jordan, W. J. "Defining equity: Multiple perspectives to analyzing the performance of diverse learners." In *Review of Research in Education: What Counts as Evidence in Educational Settings? Rethinking Equity, Diversity and Reform in the 21st Century* (eds. A. Luke, J. Green, and G. Kelly). Thousand Oaks, CA: Sage. 2010, pp. 142–178.

Joslyn, R. E. "Using concrete models to teach large number concepts." *Arithmetic Teacher*, 38(March–April 2002), pp. 34–39.

Judd, W. "Instructional games with calculators." *Mathematics Teaching in the Middle School*, 12(6) (February 2007), pp. 312–314.

Juster, N. *The Phantom Tollbooth*. New York: Random House, 1961.

Kabiri, M. S., and Smith, N. L. "Turning traditional textbook problems into open-ended problems." *Mathematics Teaching in the Middle School*, 9 (3) (November 2003), pp. 186–192.

Kamii, C., and Dominick, A. "The harmful effects of algorithms in grades 1–4." In *The Teaching and Learning of Algorithms in School Mathematics: 1998 Yearbook of the National Council of Teachers of Mathematics* (eds. L. J. Morrow and M. J. Kenney). Reston, VA: NCTM, 1998, pp. 130–140.

Kamii, C., and Rummelsburg, J. "Arithmetic for first graders lacking number concepts." *Teaching Children Mathematics*, 14(7) (March 2008), pp. 389-394.

Kamii, C., Lewis, B. A., and Livingston, S. J. "Primary arithmetic: Children inventing their own procedures." *Arithmetic Teacher*, 41(4) (December 1993), pp. 200–203.

Karp, K., and Howell, P. "Building responsibility for learning in students with special needs." *Teaching Children Mathematics*, 11(3) (October 2004), pp. 118–126.

Kastberg, S. E., and Walker, V. "Insights into our understandings of large numbers." *Teaching Children Mathematics*, 14(9) (May 2008), pp. 530–536.

Kenney, P. A., and Silver, E. A. (eds.). *Results from the Sixth Mathematics Assessment of the National Assessment of Educational Progress*. Reston, VA: NCTM, 1997.

Kenney, P., Lindquist, M., and Heffernan, C. "Butterflies and caterpillars: Multiplicative and proportional reasoning in the early grades." In *Making Sense of Fractions, Ratios, and Proportions* (eds. B. Litwiller and G. Bright). Reston, VA: NCTM, 2002.

Kilpatrick, J., Swafford, B., and Findell, B. (eds.). *Adding It Up: Helping Children Learn Mathematics*. Washington, DC: National Academy Press, 2001.

King, K., and Gurian, M. "Teaching to the minds of boys." *Educational Leadership*, 64(1) (September 2006), pp. 56–61.

Kitchen, R., Cherrington, A., Gates, J., Hitchings, J., Majka, M., Merk, M., and Truybow, G. "Supporting reform through performance assessment." *Mathematics Teaching in the Middle School*, 8 (2002), pp. 24–30.

Klein, A. S., Beishuizen, M., and Treffers, A. "The empty number line in Dutch second grade." In *Lessons Learned from Research* (eds. J. Sowder and B. Schappelle). Reston, VA: NCTM, 2002, pp. 41–43.

Kloosterman, P. W., and Lester, F. K. (eds.). *Results and Interpretations of the 1990 Through 2000 Mathematics Assessment of the National Assessment of Educational Progress*. Reston, VA: NCTM, 2004.

Kloosterman, P., and Lester, F. K. (eds.). *Results and Interpretation of the 2003 Mathematics Assessment of the National Assessment of Educational Progress*. Reston, VA: NCTM, 2007.

Kloosterman, P., Warfield, J., Wearne, D., Koc, Y., Martin, G., and Strutchens, M. "Fourth-grade students' knowledge of mathematics and perceptions of learning mathematics." In *Results and Interpretations of the 1990 Through 2000 Mathematics Assessments of the Educational Progress* (eds. P. Kloosterman and F. K. Lester). Reston, VA: NCTM, 2004, pp. 71–103.

Knowles, T. *The Kids Behind the Label: An Inside Look at ADHD for Classroom Teachers*. Portsmouth, NH: Heinemann, 2006.

Knuth, E. J., Choppin, J. M., and Bieda, K. N. "Proof: Examples and beyond." *Mathematics Teaching in the Middle School*, 15(4) (November 2009), pp. 206–211.

Kostelnik, M., Soderman, A., and Whiren, A. *Developmentally Appropriate Curriculum: Best Practices in Early Childhood Education*, 3rd ed. Upper Saddle River, NJ: Pearson Education, 2004.

Kouba, V. L., and Franklin, K. "Multiplication and division: Sense making and meaning." In *Research Ideas for the Classroom: Early Childhood Mathematics* (ed. R. J. Jensen). Reston, VA: NCTM, and New York: Macmillan, 1993, pp. 103–126.

Kouba, V. L., and Franklin, K. "Research into practice: Multiplication and division: Sense making and meaning." *Teaching Children Mathematics*, 1(9) (May 1995), pp. 574–577.

Labinowicz, E. *Learning from Children: New Beginnings for Teaching Numerical Thinking*. Menlo Park, CA: Addison-Wesley, 1985.

Lambdin, D. V. *Teaching and Learning Mathematics: Translating Research for Elementary School Teachers*. Reston, VA: NCTM, 2010.

Lambdin, D. V., and Walcott, C. "Changes through the years: Connections between Psychological Learning Theories and School Mathematics Curriculum. In *The Learning of Mathematics*, 69th *Yearbook* (Eds, M. E. Strutchins and W.G. Martin), Reston, VA: NCTM, 2007, pp. 3-25.

Lamon, S. J. "Rational numbers and proportional reasoning." In *Second Handbook of Research on Mathematics Learning and Teaching*, (ed. F. K. Lester). Charlotte, NC: Information Age Publishing, (2007), pp. 629–666.

Lampert, M., and Cobb, P. "Communication and language." In *A Research Companion to Principles and Standards for School Mathematics* (eds. J. Kilpatrick, W. G. Martin, and D. Schifter). Reston, VA: NCTM, 2003, pp. 237–249.

Langrall, C., and Swafford, J. "Three balloons for two dollars: Developing proportional reasoning." *Mathematics Teaching in the Middle School*, 6(4) (December 2000), pp. 254–261.

Lappan, G. "What do we have and where do we go from here?" *Arithmetic Teacher*, **40**(9) (May 1993), pp. 524–526.

Leahy, S., Lyon, C. Thompson, M., and Wiliam, D. "Classroom assessment: Minute by minute, day by day." *Educational Leadership*, 63 (November 2005): pp. 19–24.

Leatham, K. R., Lawrence, K., and Mewborn, D. "Getting started with open-ended assessment." *Teaching Children Mathematics*, 11 (2005), pp. 413–419.

Lehrer, R. "Developing understanding in measurement." In *A Research Companion to Principles and Standards for School Mathematics* (eds. J. Kilpatrick, W. G. Martin, and D. Schifter). Reston, VA: NCTM, 2003, pp. 179–192.

Lembke, L. O., and Reys, B. J. "The development of and interaction between intuitive and school-taught ideas about percent." *Journal for Research in Mathematics Education*, 25(3) (May 1994), pp. 237–259.

Lemke, M., Sen, A., Pahlke, E., Partelow, L., Miller, D., Williams, T., Kastberg, D., and Jocelyn, L. *International Outcomes of Learning in Mathematics Literacy and Problem Solving: PISA 2003 Results from the U. S. Perspective*. Washington, DC: National Center for Education Statistics, 2004.

Lesh, R., and Landau, M. *Acquisition of Mathematical Concepts and Processes*. New York: Academic Press, 1983.

Lesh, R., and Zawojewski, J. "Problem solving and modeling." In *Second Handbook of Research on Mathematics Teaching and Learning* (ed. F. K. Lester). Greenwich, CT: Information Age Publishing, 2007.

Lesh, R., Post, T., and Behr, M. "Proportional reasoning." In *Number Concepts and Operations in the Middle Grades* (eds. J. Hiebert and M. Behr). Reston, VA: NCTM, 1998, pp. 93–118.

Lester, F. K. (ed.). *Second Handbook of Research on Mathematics Teaching and Learning*. Charlotte, NC: Information Age Publishing, 2007.

Lester, F. K., and Charles, R. I. *Teaching Mathematics Through Problem Solving: Prekindergarten–Grade 6*. Reston, VA: NCTM, 2003.

Lindquist, M. M., Brown, C.A., Carpenter, T. P., Kouba, V. L., Silver, E. A., and Swafford, J. O. *Results from the Fourth Mathematics Assessment of The National Assessment of Educational Progress*. Reston, VA: NCTM, 1988.

Lobato, J., and Ellis, A. B. *Developing Essential Understanding of Ratios, Proportions, and Proportional Reasoning* (ed. R. I. Charles). Reston, VA: NCTM, 2010.

Long, M. J., and Ben-Hur, M. "Informing learning through the clinical interview." *Arithmetic Teacher*, 38(6) (February 1991), pp. 44–46.

Losq, C. "Number concepts and special needs students: The power of ten-frame tiles." *Teaching Children Mathematics*, 11(6) (February 2005), pp. 310–314.

Lubienski, S. "What can we do about achievement disparities?" *Educational Leadership*, 65(3) (November 2007), pp. 54–59.

Lubienski, S. T., McGraw, R., and Strutchens, M. "NAEP findings regarding gender: Mathematics achievement, student affect, and learning practices." In *Results and Interpretations of the 1990 Through 2000 Mathematics Assessments of the National Assessment of Educational Progress* (eds. P. Kloosterman and F. K Lester). Reston, VA: NCTM, 2004, pp. 305–336.

Lubinski, C., and Otto, A. D. "Meaningful mathematical representations and early algebraic reasoning." *Teaching Children Mathematics*, 9(2) (October 2002), pp. 76–80.

Ma, L. *Knowing and Teaching Elementary Mathematics: Teachers' Understanding of Fundamental Mathematics in China and the United States*. Mahwah, NJ: Erlbaum, 1999.

Maclean, R. "Educational change in Asia: An overview." *Journal of Educational Change* 2 (2001), pp. 189–192.

Mann, R. L. "Balancing act: The truth behind the equals sign." *Teaching Children Mathematics*, 11(2) (September 2004), pp. 65–69.

Manouchehri, A., and Lapp, D. A. "Unveiling student understanding: The Role of Questioning in Instruction." *Mathematics Teacher*, 96(8) (November 2003), pp. 562–566.

Martin, J. F. "The goal of long division." *Teaching Children Mathematics*, 15(8) (April 2009), pp. 482–487.

Martin, W. G., and Kasmer, L. "Reasoning and sense making." *Teaching Children Mathematics*, 16(5) (December 2009/January 2010), pp. 284–291.

Martinie, S., and Bay-Williams, J. "Using literature to engage students in proportional reasoning." *Mathematics Teaching in the Middle School*, 9(3) (November 2003), pp. 142–148.

Marzano, R., Pickering, D., and Pollock, J. *Classroom Instruction That Works*. Alexandria, VA: ASCD, 2001.

Masalski, W. J., and Elliott, P. C., (eds.) *Technology-supported Mathematics Learning Environments*. Sixty-seventh Yearbook. Reston, VA: NCTM (2005).

McClain, K., and Cobb, P. "An analysis of development of sociomathematical norms in one first-grade classroom." *Journal for Research in Mathematics Education*, 32(3) (May 2001), pp. 236–266.

McGivney-Burelle, J. M. "Connecting the dots: Network problems that foster mathematical reasoning." *Teaching Children Mathematics*, 11(5) (December 2004–January 2005), pp. 272–277.

McIntosh, A., and Sparrow, L. *Beyond Written Computation*. Perth, Australia: MASTEC, Edith Cowan University, 2004.

McIntosh, A., Reys, B. J., and Reys, R. E. "Mental computation performance in Australia, Japan and the United States." *Educational Studies in Mathematics*, 29(3) (October 1995), pp. 237–258.

McIntosh, A., Reys, B. J., Reys, R. E., and Hope, J. *Number Sense: Simple Effective Number Sense Experiences: Grades 1–2, 3–4, 4–6, 6–8.* Palo Alto, CA: Dale Seymour, 1997.

Metz, M. L. "Making sense of percent." *Mathematics Teaching in the Middle School,* 9(1) (September 2003), pp. 44–45.

Mewborn, D. S., and Huberty, P. D. "Questioning your way to the standards." *Teaching Children Mathematics,* 6(4) (December 1999), pp. 226–227, 243–246.

Mewborn, D., and Cross, D. "Mathematics teachers' beliefs about mathematics and links to students' learning." In *The Learning of Mathematics, 2007 Yearbook of the National Council of Teachers of Mathematics* (eds. W. G. Martin, M. Strutchens, and P. C. Elliott). Reston, VA: NCTM, 2007, pp. 259–269.

Midgett, C., and Trafton, P. R. "Learning through problems: A powerful approach to teaching mathematics." *Teaching Children Mathematics,* 7(9) (May 2001), p. 532.

Miller, J. L., and Fey, J. T. "Proportional reasoning." *Mathematics Teaching in the Middle School,* 5(5) (January 2000), pp. 310–314.

Miller, P. C., and Endo, H. "Understanding and meeting the needs of ESL students." *Phi Delta Kappan,* 85(10) (June 2004), pp. 786–791.

Molina, M., and Ambrose, R. "The meaning of the equals sign." *Teaching Children Mathematics* 13(2) (September 2006), pp. 111–117.

Monk, S. "Representation in school mathematics: Learning to graph and graphing to learn." In *A Research Companion to Principles and Standards for School Mathematics* (eds. J. Kilpatrick, W. G. Martin, and D. Schifter). Reston, VA: NCTM, 2003, pp. 250–262.

Moomaw, S., Carr, V., Boat, M., and Barnett, D. "Preschoolers' number sense." *Teaching Children Mathematics,* 16(6) (February 2010), pp. 332–339.

Moon, C. J. "Connecting learning and teaching through assessment." *Arithmetic Teacher,* 41(1) (September 1993), pp. 13–15.

Moone, G., and de Groot, C. "Fraction Action." In *Teaching Children Mathematics,* 12(9) (Dec 2006/Jan 2007), pp. 266–271.

Morita, J. "Capture and recapture your students' interest in statistics." *Mathematics Teaching in the Middle School,* 4(6) (March 1999), pp. 412–418.

Moschkovich, J. "Language, culture, and equity in elementary school mathematics classrooms." In *Teaching and Learning Mathematics: Translating Research for Elementary School Teachers* (ed. D. Lambdin), Reston, VA: National Council of Teachers of Mathematics. 2011, pp. 67–72

Moschkovich, J. "Language, culture, and equity in elementary school mathematics classrooms." In *Teaching and Learning Mathematics: Translating Research for Elementary School Teachers* (eds., D. V. Lambdin and F. K. Lester, Jr.). Reston, VA: NCTM, 2010, pp. 67–72.

Moses, B., Bjork, E., and Goldenberg, E. P. "Beyond problem solving: Problem posing." In *Teaching and Learning Mathematics in the 1990s, 1990 Yearbook of the National Council of Teachers of Mathematics* (ed. T. J. Cooney). Reston, VA: NCTM, 1990, pp. 82–91.

Moskal, B. "An assessment model for the mathematics classroom." *Mathematics Teaching in the Middle School,* 6(3) (November 2000), pp. 192–194.

Moss, L. J., and Grover, B. W. "Not just for computation: Basic calculators can advance the process standards." *Mathematics Teaching in the Middle School,* 12(5) (January 2007), pp. 266–271.

Moyer, P. S., and Milewicz, E. "Learning to question: Categories of questioning used by preservice teachers during diagnostic mathematics interviews." *Journal of Mathematics Teacher Education,* 5(4) (December 2002), pp. 293–315.

Mullis, I. V. S., Martin, M. O., Gonzalez, E. J., and Chrostowski, S. J. *TIMSS 2003 International Mathematics Report.* Boston, MA: TIMSS and PIRLS International Study Center, 2004.

Mullis, I. V. S., Martin, M. O., Ruddock, G., O'Sullivan, C., and Preuschoff, C. *TIMSS 2011 Mathematics Assessment Framework.* Boston, MA: TIMSS and PIRLS International Study Center, 2009.

Murrey, D. "Differentiating instruction in mathematics for the English language learner." Mathematics Teacher in the Middle School 14(3) (October: 2008, pp. 146-153.

National Center for Educational Statistics. *Pursuing Excellence: A Study of U.S. Eighth-Grade Mathematics and Science Teaching, Learning, Curriculum, and Achievement in International Context.* Washington, DC, 1996.

National Council of Teachers of Mathematics. *An Agenda for Action.* Reston, VA: NCTM, 1980.

National Council of Teachers of Mathematics. *Assessment Standards for School Mathematics.* Reston, VA: NCTM, 1995.

National Council of Teachers of Mathematics. *Curriculum and Evaluation Standards for School Mathematics.* Reston, VA: NCTM, 1989.

National Council of Teachers of Mathematics. *Curriculum Focal Points for Prekindergarten through Grade 8 Mathematics.* Reston, VA: NCTM, 2006.

National Council of Teachers of Mathematics. *Navigations—Steering Through Principles and Standards: Algebra in Grades Pre-K–2 and Grades 3–5.* Reston, VA: NCTM, 2001.

National Council of Teachers of Mathematics. *Principles and Standards for School Mathematics.* Reston, VA: NCTM, 2000.

National Governors Association. *Common Core State Standards.* 2010. http://www.corestandards.org.

National Research Council. *Everybody Counts: A Report to the Nation on the Future of Mathematics Education.* Washington, DC: National Research Council, 1989.

National Research Council. *Helping Children Learn Mathematics*. Washington, DC: National Academy Press, 2002.

National Research Council. *How People Learn: Brain, Mind, Experience, and School*. Washington, D.C.: National Academy Press, 1999.

North Central Regional Educational Laboratory. *Critical Issue: Using Technology to Improve Student Achievement*. 2005.

Nugent, C. "How many blades of grass are on a football field." *Teaching Children Mathematics*, 12(6) (February 2006), pp. 282–288.

O'Donnell, B. "What effective math teachers have in common." *Teaching Children Mathematics, 16(2)* (September 2009), *pp. 118–125*.

Parks, A. N. "Can teacher questions be too open?" *Teaching Children Mathematics*, 15(7) (March 2009), pp. 424–428.

Payne, J. N., and Huinker, D. M. "Early number and numeration." In *Research Ideas for the Classroom: Early Childhood Mathematics* (ed. R. J. Jensen). Reston, VA:NCTM, and New York: Macmillan, 1993, pp. 43–70.

Payne, R. A *Framework for Understanding Poverty 4th ed*. Highlands, TX: aha! Process Inc., 2005.

Perry, J. A., and Adkins, S. L. "It's not just notation: Valuing children's representations." *Teaching Children Mathematics*, (September 2002), pp. 196–201.

Petit, M. M., and Zawojewski, J. S. "Formative assessment in elementary school mathematics classrooms." In *Teaching and Learning Mathematics: Translating Research for Elementary School Teachers* (eds., D. V. Lambdin and F.K. Lester, Jr.). Reston, VA: NCTM, 2010, pp. 73–79.

Petit, M., Laird, R., and Marsden, E. *A Focus on Fractions*. New York: Routledge, 2010.

Philipp, R. "Multicultural mathematics and alternative algorithms." *Teaching Children Mathematics* 3 (1996), pp. 128–133.

Piaget, J. *To Understand Is to Invent*. New York: Grossman, 1972.

Pitler, H., Hubbell, E. R., Kuhn, M., and Malenoski, K. *Using Technology with Classroom Instruction That Works*. Alexandria, VA: ASCD, 2007.

Polya, G. *How to Solve It*. Princeton, NJ: Princeton University Press, 1973 (1945, 1957). Worth, IL: Creative Publications.

Quinn, J., Kavanaugh, B., Boakes, N. and Caro, R. "Two thumbs way, way up." *Teaching Children Mathematics*, 15(5) (December 2008), pp. 295–303.

Rathmell, E., and Trafton, P. "Whole number computation." In *Mathematics for the Young Child* (ed. J. N. Payne). Reston, VA: NCTM, 1990, pp. 153–172.

Reeder, S. "Are we golden? Investigations with the Golden Ratio." *Mathematics Teaching in the Middle Schools*, 13(3) (October 2007), pp. 150–155.

Reinhart, S.C. "Never say anything a kid can say!" *Mathematics Teaching in the Middle School*, 5(2000), pp. 478–483.

Reys, B. J. (ed.). *The Intended Mathematics Curriculum as Represented in State-Level Curriculum Standards: Consensus or Confusion?* Charlotte, NC: Information Age Publishing, 2006.

Reys, B.J., and Barger, R. "Mental computation: Issues from the United States perspective." In *Computational Alternatives for the 21st Century: Cross Cultural Perspectives from Japan and the United States* (eds. R. E. Reys and N. Nohda). Reston, VA: NCTM, 1994, pp. 31–47.

Reys, B. J., and Reys, R. E. "Computation in the elementary curriculum: Shifting the emphasis." *Teaching Children Mathematics*, 5(4) (December 1998), pp. 236–241.

Reys, B. J., Barger, R., Bruckheimer, M., Dougherty, B., Hope, J., Lembke, L., Markovitz, A., Parnas, A., Reehm, S., Sturdevant, R., and Weber, M. *Developing Number Sense in the Middle Grades*. Reston, VA: NCTM, 1991.

Reys, R. E. "Research on computational estimation: What it tells us and some questions that need to be addressed." *Hiroshima Journal of Mathematics Education*, 1 (March 1993), pp. 16–26.

Reys, R. E., and Yang, D. C. "Relationship between computational performance and number sense among sixth and eighth grade students in Taiwan." *Journal for Research in Mathematics Education*, 29(2) (March 1998), pp. 225–237.

Reys, R. E., Reys, B. J., McIntosh, A., Emanuelsson, G., Johansson, B., and Yang, D. C. "Assessing number sense of students in Australia, Sweden, Taiwan, and the United States." *School Science and Mathematics*, 99(2) (February 1999), pp. 61–70.

Riskowski, J., Olbricht, G. & Wilson, J. "100 students" *Mathematics Teaching in the Middle School*, 15(6) (February 2010), pp. 320–327.

Roberge, M., and Cooper, L. "Map scale, proportion, and Google earth." *Mathematics Teaching in the Middle School*, 15(9) (April 2010), pp. 448–457.

Roberts, S. K. "Not all manipulatives and models are created equal." *Mathematics Teaching in the Middle School*, 13(1) (August 2007), pp. 6–9.

Romberg, T. A., and de Lange, S. *Mathematics in context*. Chicago: de Lange. Encyclopedia Britannica, 1998.

Ron, P. "My family taught me this way." In *The Teaching and Learning of Algorithms in School Mathematics, 1998 Yearbook of the National Council of Teachers of Mathematics* (eds. L. J. Morrow and M. J. Kenney). Reston, VA: NCTM, 1998, pp. 115–119.

Ross, J., McDougall, D., Hogaboam-Gray, A., and LeSage, A. "A survey measuring elementary teachers' implementation of standards-based mathematics teaching." *Journal for Research in Mathematics Education*, 34(4) (July 2003), pp. 344–363.

Rubenstein, R. "Focused strategies for middle-grades mathematics vocabulary development." *Mathematics Teaching in the Middle School*, 13(4) (November 2007), pp. 200–207.

Rubenstein, R. N. "Communication strategies to support pre-service mathematics teachers from diverse backgrounds." In *Multicultural and Gender Equity in the Mathematics Classroom: The Gift of Diversity, 1997 Yearbook of the National Council of Teachers of Mathematics* (ed. J. Trentacosta). Reston, VA: NCTM, 1997, pp. 214–221.

Russell, S. J. "Learning whole-number operations in elementary school classrooms." In *Teaching and Learning Mathematics: Translating Research for Elementary School Teachers* (eds., D. V. Lambdin and F. K. Lester). Reston, VA: NCTM, 2010, pp. 1–8.

Russell, S. J. "Mathematical reasoning in the elementary grades." In *Developing Mathematical Reasoning in Grades K–12* (ed. L. V. Stiff). Reston, VA: NCTM, 1999, pp. 1–12.

Russell, S. J. *Kindergarten Investigations in Number, Data, and Space*, Investigation 1: Exploring Patterns., Pearson Education, Inc., 2004.

Sakshaug, L. E., and Wohlhuter, K. A. "Responses to the which graph is which problem?" *Teaching Children Mathematics*, 7 (6) (February 2001), pp. 350–353, 394–400.

Sandburg, C. *Carl Sandburg Harvest Poems 1910–1960*. New York: Harcourt, Brace, and World, 1960.

Sanford, S. "Assessing measurement in the primary classroom." In *Assessment in the Mathematics Classroom, 1993 Yearbook of the National Council of Teachers of Mathematics* (ed. N. Webb). Reston, VA: NCTM, 1993.

Saxe, G., Shaughnessy, M., Shannon, A., deOsuna, J., Chinn, R., and Gearhart, M., "Learning about fractions as points on the number line." In *The Learning of Mathematics, 2007 Yearbook of the National Council of Mathematics* (eds. G. Martin, M. Strutchens, and P. Elliot). Reston, VA: NCTM, 2007, pp. 221–238.

Scavo, T. R., and Petraroja, B. "Adventures in statistics." *Teaching Children Mathematics*, 4(7) (March 1998), pp. 394–400.

Schifter, D., Russell, S., and Bastable, V. "Early algebra to reach the range of learners." *Teaching Children Mathematics* 16(4) (November 2009), pp. 230–237.

Schwerdtfeger, J., and Chan, A. "Counting collections" *Teaching Children Mathematics* 13(7) (March 2007), pp. 356–361.

Sedzielarz, M., and Robinson, C. "Measuring growth on a museum field trip: Dinosaur bones and tree cross sections." *Teaching Children Mathematics*, 13(6) (February 2007), pp. 292–298.

Sellers, P. A. "The trouble with long division." *Teaching Children Mathematics*, 16(9) (May 2010), pp. 516–520.

Shannon, B. K. "Our diets may be killing us." *Mathematics Teaching in the Middle School*, 1(6) (April–May 1995), pp. 376–382.

Sheffield, L., Cavanagh, M., Daccy, L., Findell, C., Greenes, C., & Small, M. (2002). Navigating through data analysis and probability in prekindergarten–grade 2, Reston, VA: National Council of Teachers of Mathematics.

Shifter, D., Russell, S. J., and Bastable, V. "Early algebra to reach the range of learners." *Teaching Children Mathematics*, 16(4) (November 2009), pp. 230–237.

Shigematsu, S., Iwasaki, H., and Koyama, M., "Mental computation: Evaluation, curriculum and instructional issues from the Japanese perspective." In *Computational Alternatives for the 21st Century: Cross Cultural Perspectives from Japan and the United States* (eds. R. E. Reys and N. Nohda). Reston, VA: NCTM, 1994, pp. 19–30.

Shimizu, Y. "Problem solving as a vehicle for teaching mathematics: A Japanese perspective." In *Teaching Mathematics Through Problem Solving: Prekindergarten–Grade 6* (eds. F. K. Lester and R. I. Charles). Reston, VA: NCTM, 2003, pp. 205–214.

Shockey, T., and Snyder, K. "Tessellating T-shirts." *Teaching Children Mathematics*, 14(2) (September 2007), pp. 82–87.

Shulte, A. P., and Choate, S. A. *What Are My Chances?* Palo Alto, CA: Creative, 1996.

Siebert, D., and Gaskin, N. "Creating, naming, and justifying fractions." *Teaching Children Mathematics*, 12(8) (April 2006), pp. 394–400.

Silbey, R. "What is in the daily news? Problem-solving opportunities." *Teaching Children Mathematics*, 5(7) (March 1999), pp. 390–394.

Silver, E., and Kenney, P. A. (eds.). *Results from the Seventh Mathematics Assessment of the National Assessment of Educational Progress*. Reston, VA: NCTM, 2000.

Sims, L., "Look who's talking: Differences in math talk in U.S. and Chinese classrooms." *Teaching Children Mathematics*, (15)2 (September 2008), pp.120-124.

Smith, M. and Stein, M. "Selecting and creating mathematical tasks: From research to practice." In *Growing Professionally: Readings from NCTM Publications for Grades K-8* (ed. J. Bay-Williams and K. Karp). Reston, VA: National Council of Teachers of Mathematics, 2008, pp. 147–153.

Smith, S. P. "Representation in school mathematics: Children's representations." In *A Research Companion to Principles and Standards for School Mathematics* (eds. J. Kilpatrick, W. G. Martin, and D. Schifter). Reston, VA: NCTM, 2003, pp. 263–274.

Soucie, T., Radovic, N., and Svedrec, R. "Making technology work" *Mathematics Teaching in the Middle School*, (15)8, (April 2010), pp. 467–471.

Star, J., Kenyon, M., Joiner, R., and Rittle-Johnson, B. "Comparison helps students learn to be better estimators", *Teaching Children Mathematics*, 16(9) (May 2010), pp. 557–563.

Stein, M., and Smith, M. "The role of curricular materials in elementary school mathematics classrooms." In *Teaching and Learning Mathematics: Translating Research for Elementary School Teachers* (ed. D. Lambdin). Reston, VA: National Council of Teachers of Mathematics, 2010, pp. 61–65.

Stenmark, J. K., and Bush, W. S. *Mathematics Assessment: A Practical Handbook for Grades 3–5*. Reston, VA: NCTM, 2001.

Stiggins, R. J. "Assessment crisis: The absence of assessment for learning." *Phi Delta Stiggins Kappan*, 83(10) (June 2002), pp. 758–765.

Stigler, J. W., and Hiebert, J. *The Teaching Gap: Best Ideas from the World's Teachers for Improving Education in the Classroom*. New York: The Free Press, 1999.

Stohl, H., "Probability in teacher education and development." In Exploring Probability in School: Challenges for Teaching and Learning (ed. G. A. Jones). The Netherlands: Kluwer Academic, 2005, pp. 241–266.

Strutchens, M., Lubienski, S. T., McGraw, R., and Westbrook, S. "NAEP findings regarding race and ethnicity: Students' performance, school experiences, attitudes, beliefs, and family influences." In *Results and Interpretations of the 1990 Through 2000 Mathematics Assessments of the National Assessment of Educational Progress* (eds. P. Kloosterman and F. K. Lester). Reston, VA: NCTM, 2004, pp. 169–304.

Stylianou, D., Kenney, P. A., Silver, E. A., and Alacaci, C. "Gaining insight into students' thinking through assessment tasks." *Mathematics Teaching in the Middle School*, 6(2) (October 2000), pp. 136–143.

Sutton, J., and Krueger, A. (eds.). *ED Thoughts: What We Know About Mathematics Teaching and Learning*. Aurora, CO: McRel, 2002.

Tarr, J. E., and Jones, G. A. "A framework for assessing middle school students' thinking in conditional probability and independence." *Mathematics Education Research Journal*, 9(1997), pp. 39–59.

Tayeh, C. "Solutions to the Cheerio count problem." *Teaching Children Mathematics*, 13(5) (December 2006–January 2007), pp. 258–262.

Tayeh, C. "Solutions to the what's the overlap problem." *Teaching Children Mathematics*, 13(1) (August 2006), pp. 42–44.

Tayeh, C., and Britton, B. "Make it 36." *Teaching Children Mathematics*, 11(4) (February 2005), pp. 330–331.

Telese, J. A., and Abete, J. "Diet, ratios, proportions: A healthy mix." *Mathematics Teaching in the Middle School*, 8(1) (September 2002), pp. 8–13.

Thames, M., and Ball, D.L. "What math knowledge does teaching require?" *Teaching Children Mathematics*, (17) 4 (November 2010), pp. 220–229.

Thiessen, D. (ed.) *Exploring Mathematics Through Literature*. Reston, VA: NCTM, 2004.

Thiessen, D. (ed.). *Exploring Mathematics Through Literature*. Reston, VA: NCTM, 2004.TIMSS Elementary Mathematics Curriculum Project. *Math Trailblazers* (Student Guide, Grade 2). Dubuque, IA: Kendall/Hunt, 2008.

Thompson, C., and Bush, W. "Improving middle school teachers' reasoning about proportional reasoning." *Mathematics Teaching in the Middle School*, 8(8) (April 2003), pp. 398–404.

Thompson, D., Austin, R., and Beckmann, C. "Using literature as a vehicle to explore proportional reasoning." In Making Sense of Fractions, Ratios, and Proportions, 2002 Yearbook of the National Council of Teachers of Mathematics (eds. B. Litwiller and G. Bright). Reston, VA: NCTM, pp. 130–137.

Thompson, I. "Teaching place value in the UK: Time for a reappraisal?" *Educational Review*, 51(3) (November 2000), pp. 291–299.

Thompson, T., and Sproule, S. (2005). "Calculators for students with special needs." *Teaching Children Mathematics*, 11(7) (March 2005), pp. 391–395.

Tyminski, A. M., Richardson, S. E., and Winarski, E. "Enhancing think-pair-share." *Teaching Children Mathematics*, 16(8) (April, 2010), pp. 451–455.

University of Chicago School Mathematics Project. *Everyday Mathematics, Grade 3*, 3rd ed. Wright Group/McGraw-Hill, 2007.

University of Chicago School Mathematics Project. *Everyday Mathematics: Teacher's Reference Manual (Grades K–3)*. Chicago: Everyday Learning Corporation, 2001.

Uy, F. "The Chinese numeration system and place value." *Teaching Children Mathematics*, 9(5) (January 2003), pp. 243–247.

Vacc, Nancy Nesbitt. "Questioning in the mathematics classroom." *Arithmetic Teacher*, 41 (October 1993), pp. 88–91.

Verschaffel, L., Greer, B., and De Corte, E. "Whole number concepts and operations." In *Second Handbook of Research on Mathematics Teaching and Learning* (Ed. F. K. Lester). Charlotte, NC: Information Age Publishing, 2007, pp. 557–628.

Wagener, L. "A worthwhile task to teach slope." *Mathematics Teaching in the Middle School*, 15(3) (October 2009), pp. 168–174.

Wantabe, T. "The teaching and learning of fractions: A Japanese perspective." *Teaching Children Mathematics*, 12(7) (March 2006), pp. 368–374.

Warfield, J., and Meier, S. L. "Student performance in whole-number properties and operations." In *Results and Interpretations of the 2003 Mathematics Assessment of the National Assessment of Educational Progress* (eds. P. Kloosterman and F. K. Lester). Reston, VA: NCTM, 2007, pp. 43–66.

Washburne, C. "Mental age and the arithmetic curriculum: A summary of the committee of seven-grade placement investigation to date." *Journal of Educational Research*, **23**(3) (March 1931), pp. 210–231.

Wearne, D., and Hiebert, J. "Place value and addition and subtraction." *Arithmetic Teacher*, 41(5) (January 1994), pp. 272–274.

Wearne, D., and Kouba, V. L. "Rational number properties and operations." In *Results from the Seventh*

Mathematics Assessment of the National Assessment of Educational Progress (eds. E. A. Silver and P. A. Kenny). Reston, VA: NCTM, 2000.

Weiser, E. T. "Students control their own learning: A metacognitive approach." *Teaching Children Mathematics*, 15(2) (September 2008), pp. 90–95.

Wells, P. J., and Coffey, D. C. "Are they wrong? Or did they just answer a different question?" *Teaching Children Mathematics*, 12(4) (November 2005), pp. 202–207.

Westegaard, S. K. "Using quilt blocks to construct understanding." *Mathematics Teaching in the Middle School*, 13(6) (January 2008), pp. 361–365.

Whiteford, T. "Is mathematics a universal language?" *Teaching Children Mathematics*, 16(5) (December 2009/January 2010), pp. 276–283.

Whitenack, J., Knipping, N., Noringer, S., and Underwood, G. "Facilitating children's conceptions of tens and ones: The classroom teacher's important role." In *Beyond Written Computation* (eds. A. McIntosh and L. Sparrow). Perth, Australia: Mathematics, Science and Technology Education Center, 2004, pp. 40–50.

Whitin, D. J., and Whitin, P. *New Visions for Linking Literature and Mathematics.* Urbana, IL: National Council of Teachers of English, 2004.

Whitin, D., and Whitin, P. "Why are things shaped the way they are?" *Teaching Children Mathematics*, 15(8) (April 2009), pp. 464–472.

Whitin, P. "Promoting problem-posing explorations." *Teaching Children Mathematics*, 11(2) (November 2004), pp. 180–186.

Wilson, L. D. "High-stakes testing in mathematics." In *Second Handbook of Research on Mathematics Teaching and Learning* (Ed. F. K. Lester) Greenwich, CT: Information Age Publishing, 2007, pp. 1099–1142.

Wilson, L. D., and Kenney, P.A. "Classroom and large-scale assessment." In *A Research Companion to Principles and Standards for School Mathematics* (eds. J. Kilpatrick, W. G. Martin, and D. Schifter). Reston, VA: NCTM, 2003, pp. 53–67.

Wilson, M. R., and Kraft, C. M. "Exploring mean, median and mode with a spreadsheet." *Mathematics Teaching in the Middle School*, 1(6) (September–October 1995), pp. 490–495.

Wilson, P. S., and Osborne, A. "Foundational ideas in teaching about measure." In *Teaching Mathematics in Grades K–8* (ed. T. R. Post). Toronto: Allyn and Bacon, 1988, pp. 78–110.

Wood, T., and Sellers, P. "Assessment of a problem-centered mathematics program: Third grade." *Journal for Research in Mathematics Education*, 27(3) (May 1996), pp. 337–353.

Wood, T., Williams, G., and McNeal, B. "Children's mathematical thinking in different classroom cultures." *Journal for Research in Mathematics Education*, 37 (May 2006), pp. 222–255.

Wu, Z., An, S., King, J., Ramirez, M., and Evans, S. "Second-grade professors: Using graphic organizers and the mathematician's chair enhances second graders' proficiency in solving word problems." *Teaching Children Mathematics*, 16(1) (August 2009), pp. 34–41.

Yackel, E., and Hanna, G. "Reasoning and proof". In *A Research Companion to Principles and Standards for School Mathematics* (eds. J. Kilpatrick, W. G. Martin, and D. Schifter). Reston, VA: NCTM, 2003, pp. 227–236.

Yoshikawa, S. "Computational estimation: Curriculum and instructional issues from the Japanese perspective." In *Computational Alternatives for the 21st Century: Cross Cultural Perspectives from Japan and the United States* (eds. R. E. Reys and N. Nohda). Reston, VA: NCTM, 1994, pp. 51–62.

Zaslavsky, C. "People who live in round houses." *Arithmetic Teacher*, 37(1) (September 1989), pp. 18–21.

Zaslavsky, C. "The influence of ancient Egypt on Greek and other numeration systems." *Mathematics Teaching in the Middle Grades*, 9(3) (November 2003), pp. 174–176.

APPENDIX

Masters

Full size versions of these pages are available at www.wiley.com/college/reys

- **A–1** Attribute Pieces
- **A–2** Cuisenaire Rods
- **A–3** Base 10 Blocks
- **A–4** Pattern Blocks
- **A–5** Five and Ten Frames
- **A–6** Hundred Charts
- **A–7** Variations of Hundreds Charts
- **A–8** Basic Addition and Multiplication Facts
- **A–9** 0–9 Cards
- **A–10** Place Value Chart
- **A–11** Decimal or Percent Paper
- **A–12** Fraction Strips
- **A–13A** Fraction Models and Spinners
- **A–13B** Fraction Models and Spinners
- **A–14** Rulers
- **A–15** Geoboard Template
- **A–16** Geoboard Recording Paper
- **A–17** Centimeter Dot Paper
- **A–18** Isometric Paper
- **A–19** Centimeter Grid Paper
- **A–20** Inch Grid Paper
- **A–21** Half-Inch Grid Paper
- **A–22** Quarter-Inch Grid Paper
- **A–23** Geometric Design Paper
- **A–24** Equilateral Triangle Paper
- **A–25** Tangram
- **A–26** Circle Point Paper

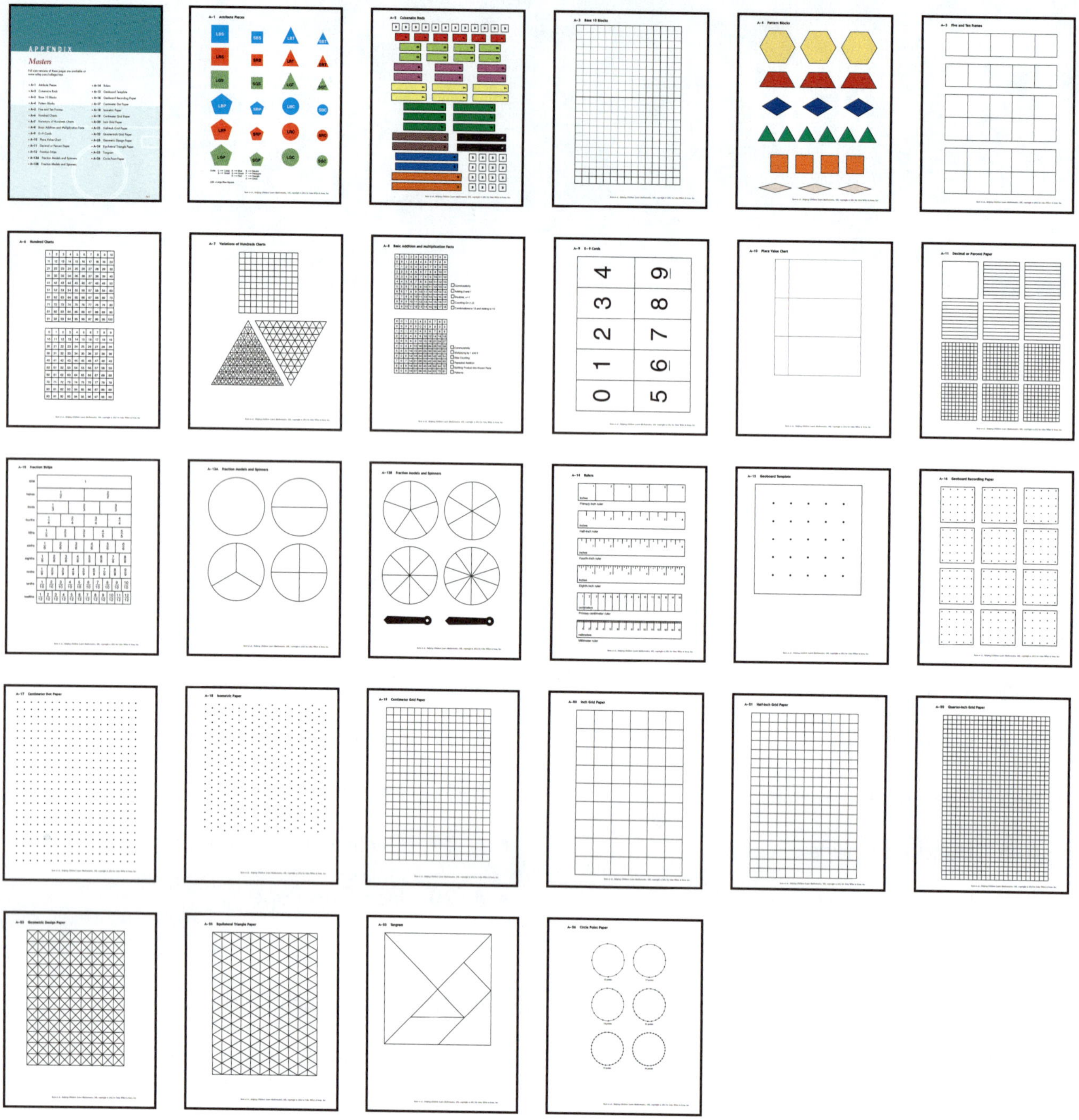
APPENDIX
Masters

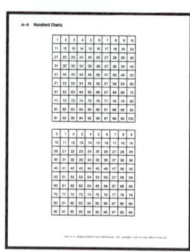

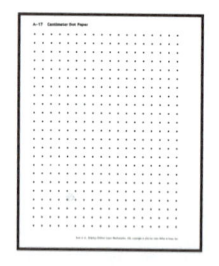

INDEX

A

Abilities of students:
   assessment of, 66–68
   in assigning small groups, 39
Abstract learning:
   moving from concrete learning to, 27–28
   of operations, 181
Abstract/relational level (geometric thought), 325
Abundant numbers, 404
Academic skills, adaptations for, 56
Accountability, 5
Achievement:
   assessment for evaluating, 65, 66
   and use of calculators, 210
Achievement gap, 5
Achievement tests, 18, 65, 66
Active involvement:
   of English-language learners, 55
   of students, 25–26
Acute triangles, 336
Adapting lessons, for students with special needs, 55–57
Adaptive reasoning, 14
Adding to 10 and beyond strategy, 194–195
Addition, 181–183
   basic facts for, 185–186
   compatible numbers for, 218
   counting on as model for, 142
   with decimals, 277
   with fractions, 268–269
   mental computation strategies for, 212–213
   models for, 182
   near doubles, 193
   and multiplication, 197
   and relationship among operations, 181, 195, 197
   repeated, 197, 198
   and subtraction, 195, 196
   thinking strategies for, 191–195, 201–202
Addition algorithms, 232–236
   higher-decade, 236
   partial-sum, 235–236
   standard, 233, 235
Additive property, 157
Adjusting estimates, 217, 218, 220
Affective attributes, 136
Africa, patterns on doors in, 344
Algebra, 3, 4, 301–302, 307, 319
Algebraic thinking, 300–320
   cultural differences in, 319–320
   functions in, 306–307, 316–317
   generalization in, 162, 313, 318–319
   justifying in, 312, 313, 318–319
   language and symbols in, 307–310
   modeling in, 310–313, 318–319
   for nonroutine problems, 302, 314–316
   patterns in, 136, 162, 302–305, 312–313
   in problems, 302
   and properties of numbers, 305, 317–318
   relations for building, 305–307, 316–318
   representing in, 310–313
   for routine problems, 302, 310–312
Algebra Standard (NCTM), 119, 310
Algorithms, 226–251
   addition, 232–236
   checking, 249
   choosing method of calculating, 228
   for computational proficiency, 249
   cultural differences in teaching, 249–251
   division, 243–249
   expectations relating to, 228–230
   mental computation vs., 214, 215
   multiplication, 240–243
   subtraction, 237–240
   teaching with understanding, 231–232
Alike-and-difference trains, 134, 135
Alternate goals, adaptation for, 57
Altitude (geometric shapes), 335, 336
American Statistical Association, 8, 374–375
Analytic rubrics, 66
Analytic scoring scale, 75
Angles, 36
   measuring, 355
   of two-dimensional shapes, 329, 334, 335
Anxiety:
   of English-language learners, 55
   math, 15–16
Approximate numbers, 171
Arabic, number symbols/names in, 174
Arbitrary units of measurement, 356
Area:
   connecting attributes for, 367, 368
   conservation of, 354
   equivalencies for, 364, 365
   formulas for, 361–362
   surface, 368, 369
   units for, 359
Area and array problems, 184
Area model (fractions), 258, 354–355
Argumentation Practice (CCSSI), 89
Arithmetic, 4
   fundamental theorem of, 408
   modular, 413–415
Arithmetic average (mean), 386–387
Arrays:
   area and array problems, 184
   and multiplication algorithms, 240, 242
Assessment(s), 61–83. See also National Assessment of Educational Progress (NAEP)
   of abilities, dispositions, and interests, 66–68
   communicating results of, 81–82
   decision making, 65
   formative, 63–66
   and goals of society, 5–6
   by interviewing, 69–70
   keeping records of, 79–81
   for learning, 63–66
   of interests of students, 66–68
   of lesson effectiveness, 56
   by observation, 68
   peer, 72, 73
   by performance tasks, 70–72
   in planning, 56, 58
   planning of, 64
   by portfolios, 74, 76
   of prior knowledge, 35–36
   by questioning, 68–69
   self-, 71–74

I-1

Assessment(s), *(Continued)*
    by standardized achievement tests, 77–79
    summative, 63
    by teacher-designed written tests, 76–78
    by work samples, 74–75
    by writings, 74, 76, 77
Assessment Principle (NCTM), 7, 63
Assessment Standards for School Mathematics (NCTM), 63
Associative property, 186
Attitude, toward mathematics, 15, 16
Attitude inventories, 72, 73
Attributes:
    in comparisons, 138
    connecting, 367–369
    identifying, 350–355
    measurements of, 349, 350
    in part-whole modeling, 257–258
    in patterns, 136
Attribute blocks, 133–135
Averages, 384–389
    choosing, 387–389
    mean, 386–387
    median, 385–386
    mode, 385

B

Balanced Assessment in Mathematics program, 66
Bar graphs, 377, 380, 381, 384
Base ten, 157
Base-ten blocks:
    addition and subtraction with, 233, 238
    electronic, 235, 239
    and place value, 158, 162, 170, 172
Basic facts, 179, 185–202
    addition, 185–186, 191–195
    building understanding of, 188–189
    and calculator use, 210
    cultural differences in learning, 201–202
    division, 186, 199–201
    expectations related to, 179, 187
    multiplication, 186, 196–199
    process for learning, 185–187
    starting point for learning, 187, 188
    strategies for remembering, 189–191
    subtraction, 186, 195–196
    thinking strategies for, 191–201
    and understanding of operations, 181
    use of language in learning, 181
    Web resources for, 180
Bean Sprouts video (Annenberg Media), 226
Behaviorism, 20–21
Beliefs, problem-solving skills and, 110–111
Benchmarks:
    fractions, 262–263
    number, 145–146
    percent, 294
Bimodal sets, 385
Binomials, foil method for, 242, 243
Box (box-and-whisker) plots, 380, 384
Breaking problems down, 126
Broken key activity, 117
Brownell, William, 4, 21, 131
Bruner, Jerome, 21–23, 27, 41
Burns, M., 42, 61, 170, 244

C

Calculators, 91
    checking computations with, 249
    computation with, 208–211
    counting and pattern recognition with, 166–167
    in early counting, 143, 145
    graphing, 378, 390, 391
    incorrect beliefs about, 210
    and learning place value, 164, 166–168
    as lesson enhancements, 44
    math skill development and use of, 7
    mental computation vs., 211
    multiplication with, 243
    problem solving with, 116–117
    regrouping with, 168
    teaching the use of, 208
Calculator-based laboratories (CBLs), 91, 390, 391
Calculator-based rangers (CBRs), 91
Calendars:
    number patterns on, 315
    numbers on, 142
CALP (Cognitive Academic Language Proficiency), 55
Capacity, 353, 359
Cardinality, 135, 140
Cardinality rule, 141
Cardinal numbers, 133, 148, 149
CBLs, *see* Calculator-based laboratories
CBRs (calculator-based rangers), 91
CCSSI (Common Core State Standards Initiative), 63, 89
CCSSM, *see Common Core State Standards for Mathematics*
Center (shapes), 329
Center for the Study of Mathematics Curriculum, 8
Central tendency, measures of, 384–389
CGI (Cognitively Guided Instruction) program, 92
Chalkboard, effective use of, 127–128
Chance, 392, 393. *See also* Probability
Children's books, as lesson enhancements, 41, 56, 58
China:
    counting system in, 150
    number symbols/names in Chinese, 151, 173
    patterns on doors in, 344

Circles, 27–28, 329
Circumference, 329, 352, 353
Classification:
    as CCSSM expectation, 132
    of geometric shapes, 336–338
    as prenumber concept, 133–135
    of solids, 327
Class records, 81
Classroom routines, managing, 125–126
Classroom working portfolios, 74
Clocks, 360
Clock arithmetic, 413
Closed sentences, 309
Cognitive Academic Language Proficiency (CALP), 55
Cognitive developmental stages, 23–25
Cognitively Guided Instruction (CGI) program, 92
Combination problems (multiplication), 184
Combinations-to-10 addition strategy, 194
Commas (in numbers), 170
Committee of Seven, 4
Common Core State Standards for Mathematics (CCSSM), 6–8, 131
    algebraic thinking, 320
    algorithms, 229–230
    area comparisons, 354, 355
    basis for, 13, 22
    benchmark scores for standardized tests, 78
    calculators, 209
    classification and number expectations, 132, 141
    communication, 95, 96
    computations, 207
    connections, 97
    data analysis, statistics, and probability, 374, 375, 377, 383
    decimals, 273, 276
    detailed information, 229
    even and odd numbers in, 405
    factors and multiples in, 405
    fractions, 255–257, 259, 268, 270
    geometric transformations, 340
    geometry, 324
    measurement, 348, 350, 363
    operations, 179
    patterns, 303
    percents, 292
    place value and ordering numbers, 156
    problem solving, 106, 307, 308
    processes of doing math, 89
    protractor skills, 360, 361
    rational counting, 142
    ratios, proportions, and percents, 284
    reasoning and proof, 92–93
    recommendations for basic fact mastery, 187
    teaching objectives from, 37
    technology recommendations in, 43

volume comparisons, 355
Common Core State Standards Initiative (CCSSI), 63, 89
Communication. *See also* Language; Talking
  about calculator use, 209
  of assessment results, 81–82
  BICS, 55
  in classification, 134, 135
  to encourage sense making, 28
  to encourage understanding, 28–29
  with parents or guardians, 41, 81, 82, 209
  in patterning, 136
  as process of doing math, 88, 95–97
  and reflection, 95
  of statistical investigation results, 392
  written, 29, 181
Communication Standard (NCTM), 88
Commutative property, 186
  in addition, 192
  in multiplication, 196–197, 271
Comparison(s), 138–139
  of fractions, 266
  in measurement, 350–355
  in multiplication, 183–184
  of objects to units of measurement, 357–362
  of ratios, 284–286
  in subtraction, 182
Compatible numbers:
  in estimation, 218, 219, 235–236
  in mental computation, 212
Compensatory strategies, 126
Composite numbers, 407, 408
Composition (in place value), 167–169
Computation(s), 205–223. *See also* Mental computation
  balancing instruction in, 208–209
  with calculators, 208–211
  checking, 249
  choosing appropriate methods for, 207
  estimation, 216–222
  instructional decisions for, 208
  recommendations for teaching of, 230–231
  speed in, 210
  written, 208, 222, 223
Computational algorithms, 228. *See also* Algorithms
Computational alternatives, 215
Computational fluency (computational proficiency):
  balance of conceptual understanding and, 230
  and calculators, 210
  cultural differences in, 222
  helping development of, 180–181
  knowledge of basic facts for, 187
  and methods of computing, 231

and understanding of operations, 181–182
Computers. *See also* Software
  for geometric shape-making, 343
  for lesson enhancement, 43–44
  problem solving with, 117
Concavity, 335
Conceptualization, 112
Conceptual knowledge:
  acquisition of, 18–19
  formal representations from, 28
Conceptual understanding, 14
Concrete learning:
  moving to abstract learning from, 27–28
  of operations, 181
Concrete models of fractions, 264
Concrete operational thinking, 22, 35
Concrete representations of algebraic problems, 310
Conditional probability, 397
Conferences, parent, 81
Congruence, 257
Congruent faces, 327
Congruent shapes, 341
Conjectures, 405
  about divisibility, 412
  in number theory, 402–403
Connections:
  to aid retention, 18
  in counting, 146–148
  between procedural and conceptual knowledge, 18–19
  as process of doing math, 88, 97–98
  showing one-to-one correspondence, 138, 139
Connectionism, 4
Connections Standard (NCTM), 88
Conservation of area, 354
Conservation of number, 137
Constructed-response tasks, 83
Construction of knowledge, 4
Constructivism, 20–23
Contexts, for math lessons, 40
Continuous quantities, 140
Control, problem-solving skills and, 111
Conversion of measurements, 364–365
Convexity, 334–335
Cookies to Share video (Annenberg Media), 12
Coordinate systems, 339, 340
Corners (two-dimensional shapes), 331–332
Council of Chief State School Officers, 6, 63
Counting, 139–148
  and basic facts, 188
  in classification, 135
  defined, 139–140
  development of, 132
  in different countries, 150–151

of fractional parts, 259–260
with hundreds charts, 162–163
in learning place value, 166–167
making connections, 146–148
number benchmarks, 145–146
and operations sense, 180
and place value concept, 158, 168
practice in, 145
principles of, 140–141
rational, 140, 142
rote, 141–142
stages of, 140–142
strategies for, 142–145
by tens, hundreds, thousands, 166–167
trouble spots in, 145
units of measurement, 357–358
Counting back, 142–144
  in division, 201
  in learning place value, 167
  as subtraction strategy, 196
Counting on, 142, 143
  as addition strategy, 193–194
  in other cultures, 201
  as subtraction strategy, 196
Criterion-referenced tests, 77, 78
Critical thinking:
  and mental computation, 214
  rewarding students for, 15
Cuban division, 250
Cube painting problem, 302
Cuisenaire rods, 136
Cultural diversity:
  in algebraic thinking, 319–320
  in computational algorithms, 249–251
  in computational proficiency, 222
  in counting and number development, 150–151
  creating connections to cultures, 56, 58
  in learning basic facts, 201–202
  and learning preferences, 53–54
  in measurement systems, 368, 369
  and NAEP test performance, 82–83
  in patterns for naming numbers, 173–174
  and Project IMPACT, 101, 102
  in socioeconomic status, 29
  in teaching fractions and decimals, 278–279
  in teaching geometry, 344
Curriculum(-a), 3–7
  adapting, 55–57
  standards-based, 41
Curriculum Focal Points for Prekindergarten through Grade 8 Mathematics (NCTM), 5, 8, 229
Curriculum guides, 45
Curriculum Principle (NCTM), 6
Cylinders, 327

D

Data analysis, 373–392
  formulating questions for, 375
  graphical organization, 377–384
  interpreting results, 389–392
  software for, 100
Data collection, 375–377
  experiments, 376
  simulations, 376–377
  surveys, 376
Data sense, 389–391
Decagons, 337
Decimals, 272–278
  cultural differences in learning, 278
  and fractions, 255, 273–275
  model and symbolization of, 293
  ordering, 275–276
  and place value, 275
  rounding, 276
  writing fractions as, 273
Decimal grid, 276
Decimeters, 358
Decomposition:
  in mental computation, 211
  in place value, 167–169
Decomposition algorithm, 237, 250
Deficient numbers, 404
Density property (fractions), 266
Describing three-dimensional objects, 325–326
Descriptive/analytic level (geometric thought), 325
Descriptive statistics, 383–389
  averages, 384–389
  measures of central tendency, 384–389
  measures of variation, 389, 390
Developmental characteristics of students, 23–25, 34–35
Developmental learning, Bruner's levels of, 22
Developmental readiness, 4
Dewey, John, 19
Diagrams:
  of geometric shapes, 342, 343
  problem solving with, 119
  Venn, 410
Diameter, 329
Dienes, Zoltan, 21–23
Digits:
  moving, 169
  in place value learning, 165–166
  position of, 158
  reading, 171
  reversing, 161
  writing blocks of, 170
Direct comparisons:
  of angles, 355
  of area, 354
  of capacity, 353
  of length, 351, 352
Direct instruction lessons, 46, 48, 50

Discrete objects, 140
Disjoint sets, 134
Dispositions of students, assessment of, 66–68
Distance, measuring, 351, 352
Distributive division algorithm, 245
Distributive property, 186, 241, 318
Diversity. *See also* Cultural diversity
  and equity, 6
  and expectations of math aptitude, 17
  and planning to meet needs of all students, 52–57
  socioeconomic, 29
  supporting, 14–19
  using manipulatives to meet needs, 42
Divisibility, 410–412
Divisibility rules, 410, 411
Division, 183–185
  basic facts for, 186
  compatible numbers for, 218, 219
  with decimals, 278
  factors, 405–408
  with fractions, 271–273
  mental computation strategies for, 213
  and multiplication, 199
  and relationship among operations, 182, 199
  thinking strategies for, 199–201
Division algorithms, 243–249
  with one-digit divisors, 244–245
  with remainders, 248–249
  with two-digit divisors, 247
Division method of prime factorization, 408
Domain, 116
Doubles:
  adding, 193
  subtracting, 196
Drawings:
  models of fractions, 261–262
  problem solving with, 119
Drills, 4
  for basic facts, 188, 189
  and development of meaning, 37
  online resources for, 189
  understanding vs., 16
Drill-and-practice software, 43
Dyslexia, 161

E

Early number development, 137–148
ECR (extended-constructed-response) items, 108
Edges (solids), 326, 327
Edge models, 329–331
Educated guesses, 121
Educational game software, 43, 189
Education-World (Web site), 98
Efficiency (of mental computation), 214
Egyptians, use of fractions by, 278
Electronic manipulatives, 342

  for algebraic concepts, 317
  base-ten blocks, 235, 239
  building graphs with, 377
  modeling numbers with, 164
  multiplication and division with, 231
  in probability models, 397
  problem solving with, 100, 112
  for ratios, fractions, and decimals, 264, 294
  visualizing basic facts with, 180
ELLs, *see* English-language learners
Embodiments, 27, 159
Enactive learning, 22
English, number names in, 150, 151, 173–174
English as a new language (ENL) learners, 101
English as a second language (ESL) learners, 101
English-language learners (ELLs):
  interviews with, 70
  Project IMPACT, 101, 102
  teaching, 54–56
Equal-additions algorithm, 250
Equal distribution model of mean, 386
Equal-groups problems, 183
Equality (in algebraic thinking), 308–309
Equations (in algebraic thinking), 309–311
Equilateral triangles, 336
Equity of educational opportunity, 53
Equity Principle (NCTM), 6, 14, 17
Equivalence:
  defined, 255
  of measurement units, 363–365
Equivalent fractions, 256
  defined, 261
  ordering, 263–267
Equivalent ratios, 289
Eratosthenes, 408, 410
Estimation, 216–222
  adjusting, 217, 218, 220
  averaging, 220–221
  background for, 216–217
  checking computation with, 249
  choosing strategies for, 221–222
  chunking, 365–366
  clustering in, 220–221
  compatible numbers in, 218, 219, 235–236
  of decimal quantities, 277
  and division algorithms, 244
  expectations for, 208
  flexible rounding in, 219, 220
  and fractions, 268
  front-end, 217, 218
  as means of computation, 208
  of measurements, 365–367
  and mental computation, 214, 216
  and multiplication algorithms, 241
  of percent, 294–295

overestimation, 221
underestimation, 221
in problem solving, 112
and proportions, 290–291
Euler's formula, 328
Europe, decimal notation in, 278
Even numbers, 404–405
Events:
  independence of, 396–397
  probability of, 392–395
  sample spaces of, 393, 394
Expectations, 16–17. *See also Common Core State Standards for Mathematics* (CCSSM)
  of equal student aptitude for math, 17
  establishing, 16–17
  grade-level, 229–230
  relating to algorithms, 228–230
Explicit equations, 312
Exploration phase (explorations), 50
Explorations, 25, 26, 46, 50, 52
Expressions (in algebraic thinking), 309–310
Extended-constructed-response (ECR) items, 108

F

Faces (solids), 326, 327
Fact families, 195
Factors, 404–408
Factor trees, 408, 410
Fairness, 396
Fermat's Last Theorem, 417
Fibonacci, Leonardo, 417
Fibonacci sequence, 305, 415–417
Field theory, 4
Figures within two-dimensional figures, 338
Five-frame, 145
Flashcards, 189
Flip cards, recording observations on, 68
Flip transformation, 340, 341
Foil method for multiplying binomials, 242, 243
Foreign languages:
  counting in, 150–151
  number names in, 173–174
Forgetting, 18
Formal axiomatic level (geometric thought), 325
Formalization, 22
Formal operational thinking, 22
Formal representations, 28
Formative assessment, 63–66
Formulas, 309, 361–362
Four-sided figures, 19
Four-stage model of problem solving, 117
Fractions, 255–272
  conceptual development of, 255–256
  controversy over study of, 255
  cultural differences in teaching, 278–279
  and decimals, 255, 273–275
  equivalent, 256, 261, 263–267
  improper, 259, 267, 269
  making sense of, 258–263
  meanings of, 256–258
  and measuring with rulers, 359
  mixed numbers, 267
  model and symbolization of, 293
  operations with, 267–273
  ordering, 263–267
  ratios vs., 286–287
  between two given fractions, 266–267, 319
  writing as decimals, 274–275
Fraction bars, 260, 264
Fraction circles, 293
France, geometry teaching in, 344
Free play, 22
Functions:
  in algebraic thinking, 306–307, 316–317
  graphing, 316
  as patterns, 316
  variables in, 309
Function machines, 306–307
Function table, 306, 308
Fundamental theorem of arithmetic, 408

G

Gagné, Robert, 20
GAISE, *see Guidelines for Assessment and Instruction in Statistics Education*
Games, basic facts practice with, 190–191. *See also specific games*
Game-based interviews, 70
Game software, 43, 189
Gathering evidence phase (assessment), 64
Gauss, Carl Friedrich, 154, 156, 417
GCF (greatest common factor), 407
Gearing down, 53
Gearing up, 53
Gender, expectations of math aptitude and, 17
Generalization(s), 22
  in algebraic thinking, 162, 313, 318–319
  in problem solving, 124
  in reasoning, 94
Generalized lesson plan, 46, 47
Geoboard, 341–342
Geometer's Sketchpad (software package), 117, 333, 341
Geometric attributes, 136
Geometric thought, 325
Geometry, 322–344
  classification schemes, 336–338
  convexity and concavity, 334–336
  cultural differences in teaching, 344
  in mathematics, 3, 4
  properties of shapes, 324, 325
  sides of geometric shapes, 331–334
  space, 338–340
  three-dimensional shapes, 325–331
  transformations, 340–342
  two-dimensional shapes, 329, 331–338
  visualization and spatial reasoning, 342–344
Geometry Standard (NCTM), 323
German, number names in, 150, 151
Germany, lesson plans in, 46
Gestalt theory, 4
Give-away-a-Flat (game), 238
Goals:
  adaptation of, 57
  behaviorist perspective on, 20–21
  of math instruction, 109
  of teaching mathematics, 2
Goldbach conjecture, 402–403
Golden Ratio, 416
Grading, scoring vs., 66
Graphs, 373, 377–384
  bar, 377, 380, 381, 384
  concrete, 377
  comparison of, 384
  for comparisons, 139
  complexity of, 384
  defined, 377
  of functions, 316
  histograms, 381
  line, 382–384
  misleading, 391–392
  picture, 377, 380, 384
  pie, 381–382
  plots, 378–380
  quick graphing methods, 377–378
  real (concrete), 377, 384
Graphing calculators, 91, 378, 390, 391
Graphing methods, 377–378
Greatest common factor (GCF), 407
Grouping, in learning place value, 157–162
Grouping students, 39
  large groups, 125
  small groups, 39, 69, 125–126
Group recognition stage (number development), 137–138
Growing patterns, 304–305
Guess and check strategy, 121, 122
Guess and test strategy, 296
Guidelines for Assessment and Instruction in Statistics Education (GAISE), 374–375, 383, 386

H

Height, circumference and, 353
Heptagons, 337
Hexagons, 337
Hidden questions, 122
Hiele, Pierre Marie van, 324
Hiele Geldof, Dina van, 324
Higher-decade addition algorithm, 236

Higher-level thinking, encouraging, 36–37
High-stakes assessments, 5–6, 63
Hindu-Arabic numeration system, 151, 157, 158
Histograms, 381, 384
Holistic rubrics, 66
How Big Is a Foot? (Myller), 357
How many sides? activity, 331, 332
Hundreds charts, 144, 162–163
Hundredths, 274
Hypotheses, 70

I

Iconic learning, 22
IDEA (Individuals with Disabilities Education Improvement Act), 53
Identity property, 186
Imprints (of geometric faces), 343
Improper fractions, 259, 267, 269
Incidental learning, 4
Indirect comparisons:
  of angles, 355
  of area, 354, 355
  of capacity, 353
  of length, 351, 352
  of weight/mass, 354
Individual instruction, 39, 126
Individualizing instruction, 53
Individuals with Disabilities Education Improvement Act (IDEA), 53
Inequality (in algebraic thinking), 308–309
Initiation phase (interview), 70
In–Out Machine activity, 307
Input, English-language learners', 55
Input adaptation, 55, 57
Instruction:
  adapting, 55–57
  direct, 46, 48, 50, 51
  goal of, 109
  individual, 39, 126
  individualizing, 53
  large-group, 125
  mental computation as part of, 215
  modifying, to meet student needs, 126
  in pairs, 126
  small-group, 39, 125–126
  value of calculators in, 210–211
  whole-class, 39
Instruct phase (direct instruction lessons), 48, 50
Intellectual stimulation, 15
Intermediate elementary students, 23–24
Internet:
  data collection Web sites, 377
  for lesson enhancement, 44
  probability simulations on, 397
  resources on, 44, 117
Interpreting evidence phase (assessment), 64

Interquartile range (IQR), 380, 389
Intersection of sets, 134, 135
Intervals, number line, 135
Interviewing, assessment by, 69–70
Inventiveness, mental computation and, 215, 216
Inventories of students' understanding, 188
Investigate phase (investigative lesson plans), 48
Investigations in Number, Data, and Space (elementary curriculum), 41, 50, 113, 114
Involvement, English-language learners', 55
IQR (interquartile range), 380, 389
Isosceles triangles, 336
Iterations, 357

J

Japan:
  computational algorithms in, 250–251
  computational proficiency in, 222
  computation strategies in, 214
  counting system in, 150
  data analysis and probability learning in, 397
  lesson plans in, 46
  number benchmarks in, 145
  number names in Japanese, 151, 173–174
  students' understanding of percent, 297
  teaching fractions in, 278, 279
  teaching problem solving in, 127–128
Justifying (in algebraic thinking), 312, 313, 318–319

K

Kindergarten students, expectations for, 179
Kites, 336
Knowledge:
  construction of, 22
  forgetting, 18
  number-related, 132
  prior, 35–36
  and problem-solving skills, 110
  procedural and conceptual, 18–19
  of teacher about math, 34
Known information, 115

L

Labeling students' methods, 127
Language. *See also* Communication; Foreign languages; Vocabulary
  of algebra, 307–310
  in classification, 134, 135
  contextualizing, 55
  for interpreting data, 390
  in learning, 28
  math as, 3, 96, 97

  for mathematical properties, 186
  in patterning, 136
  to relate symbols to operations, 181
  relating to probability, 393
Language processing adaptations, 56
Large-group instruction, 125
Lattice multiplication algorithm, 241
Launch phase:
  direct instruction lessons, 48
  explorations, 50
  investigative lessons, 48
LCM (least common multiple), 407
Learned helplessness, 17
Learning, 12–30
  abstract vs. concrete, 27–28, 181
  accountability for, 5
  different ways of, 126
  expectations for, 16–17
  focus of, 4
  improving skill retention, 17–18
  incidental, 4
  levels of, 22
  from making mistakes, 95
  mathematical proficiency model, 13–14
  of mathematics, 19–23
  and negative experiences, 15–16
  positive environment for, 14–15
  prevailing theories of, 19–20
  of procedural and conceptual knowledge, 18–19
  questions facilitating, 38
  role of language in, 28
  as social process, 22
  and socioeconomic diversity, 29
  supporting diversity of students, 14–19
  and teaching, 22
  teaching recommendations to aid, 23–29
  treating students equally, 17
Learning centers, 50. *See also* Explorations
Learning disabilities, students with, 126
Learning environment, 14–15
Learning Principle (NCTM), 6–7, 19, 20, 23
Learning stations, 50. *See also* Explorations
Least common multiple (LCM), 407
Length:
  as attribute of objects, 351–353
  comparing and finding units of, 357–358
  fractions of, 257
  units, 359
Length of duration, 355
Less is best activity, 332, 333
Lesson plans, 46–52
  analysis, 56
  competing coasters, 355
  direct instruction, 46, 48, 50, 51

daily planning, 46
explorations, 46, 50, 52
formats for, 46–48
generalized, 46, 47
investigative, 46, 48, 49
mental vs. written, 45
review-teach-practice, 46
Lesson plan books, 45
Lesson planning, behaviorist perspective on, 20–21
Letter to myself activity, 74
Line graphs, 382–384
Line plots, 378, 379, 384
Line segments, 331
Line symmetry, 332–334
"Listening to Children" (Labinowicz), 70
Literature, as lesson enhancement, 41, 56, 58
Local guidelines, 8
Locations, 338–340
Logical connectives, in classification, 134, 135
Logic blocks, 133–135
Logo, 339, 340
Look-for-a-pattern problems, 302
Looking back, 124–125
Looking for patterns strategy, 119–120

M

Magic squares, 403–405, 417
"Make-a-10" thinking strategy, 202
Manipulatives, 27–28. *See also* Electronic manipulatives
 for addition and subtraction, 182
 for algorithms, 231
 for concrete understanding, 180–181
 for fractions, 264
 for length, 351
 as lesson enhancements, 41–42
 materials for, 41–42
 for patterning, 136
 virtual, 44, 100, 112
Mark-recapture technique, 377
Mass, 353, 354
Materials for teaching math, 9–10
 algorithms, 231
 factors, 405, 406
 fractions, 258
 geometry, 342, 343
 graphing, 378
 manipulatives, 41–42
 place value, 158, 159
 selecting, 39–44
 textbooks, 40–41
Mathematics:
 curriculum for, 3–7
 defined, 2–3
 focus on, 37
 guidelines for, 7–8
 history of teaching, 10
 as language, 96, 97

learning of, 19–23
materials for study of, 9–10
research on education in, 8–9
teacher's understanding of, 34
teaching of, 10
Mathematical ideas, telling students about, 110
Mathematical learning, Dienes' levels of, 22
Mathematical memory, 94–95
Mathematical proficiency, 6–7, 13–14, 18
Mathematical properties, 185, 186
Mathematics anxiety, 15–16
"Mathematics Hiding in the Nets for a Cube" (Jeon), 329
Math Forum (Web site), 9, 66, 144, 308, 414
Math nights, 41
Mathophobia, 15
Math Trailblazers, 41, 48
Mean, 386–387
Mean absolute deviation, 389, 390
Meaning:
 as focus of learning, 4
 memorization vs., 15
Meaningful learning, 18
Measurement(s), 347–369
 choosing units, 355–357
 comparing, 363–365
 comparing objects to units, 357–362
 connecting attributes, 367–369
 creating objects based on, 363
 cultural differences in, 368, 369
 customary units, 357, 359
 estimating, 365–367
 finding number of units, 357–362
 identifying attributes, 350–355
 importance of studying, 348–350
 in learning place value, 158
 in mathematics, 3, 4
 process of, 349, 350
 reporting measurements, 362–363
Measurement problems (division), 185
Measures of central tendency, 384–389
Measures of variation, 389, 390
Measuring instruments, 358–361
Median, 380, 385–386
Memorization:
 and estimation proficiency, 216
 meaning and understanding vs., 15, 19
Memory, mathematical, 94–95
Memory adaptations, 56
Mental age, 4
Mental computation, 211–216
 algorithms vs., 214, 215
 calculators vs., 211
 as common means of computation, 208
 encouraging, 213–216
 expectations for, 208
 guidelines for developing, 214–216
 of percent, 295–296

steps in, 212–213
strategies and techniques for, 211–213
students' preferences for, 214, 215
Mental images (geometric), 343–344
Metacognition, 26, 72, 124, 125, 362–363
Meter (unit), 358
Metric units of measurement, 358, 359, 364
Mexico, patterns on doors in, 344
Minimal competency movement, 4
Minority students, expectations of, 17
Missing-addend problems, 182
Missing Digits (game), 246
Missing-number problems, 289
Mistakes, learning from, 95
Mixed numbers, 259, 267
 adding, 269
 subtracting, 269
Mode, 385
Models (modeling), 27–28
 addition, 142, 182, 235
 algebraic thinking, 310–313, 318–319
 averages, 384–387
 counting practice, 140
 division, 244–245, 273
 fractions, 257–262, 273
 geometric, 325, 342, 343
 larger numbers, 170
 as lesson enhancements, 42
 multiplication, 183, 240
 one more or one less, 146–147
 percent, 292–296
 pie graph construction, 381–382
 place value, 158–159, 165–166, 168
 representations as, 100–101
 rounding, 172
 subtraction, 143–145
 for ten, 148
 three-dimensional shapes, 328–331
 zero in, 147
Modeling Practice (CCSSI), 89, 98
Modular arithmetic, 413–415
Money:
 counting change, 145
 as nonproportional models, 158
 and ratios, 285
Morra (game), 406, 416
Motor skills, adaptations for, 56
Multiples, 405, 407, 408
Multiple-choice tasks, 83
Multiple embodiment (multiembodiment), 27
Multiple-step problems, 112
Multiplication, 183–185
 and addition, 197
 basic facts for, 186
 with decimals, 277–278
 and division, 199–201
 finger, 200
 with fractions, 269–271

I-8 Index

Multiplication, *(Continued)*
  mental computation strategies for, 213
  models for, 183
  and relationship among operations, 181, 197, 199–201
  structures for, 183–184
  thinking strategies for, 196–199
Multiplication algorithms, 240–243
  for computational proficiency, 250
  for multiplication by 10 and multiples of 10, 242
  for multiplication with large numbers, 243
  for multiplication with zeros, 242
  for one-digit multipliers, 240–241
  for two-digit multipliers, 242–243

N

National Assessment of Educational Progress (NAEP), 8, 9
  and attitudes about math, 15
  calculator use with, 244
  and computer use, 43
  cultural differences in results, 82–83
  decimals as percents, 291, 293
  figures within figures, 338
  fractions, 255, 259, 268
  functions in, 306
  location, 338–339
  measurement on, 348, 359
  open sentences, 310
  patterns, 120, 304
  recent results from, 82–83, 108
  relations between pairs of quantities, 317
National Association for the Education of Young Children (NAEYC), 8
National Center for Education Statistics (NCES), 14, 108
National Clearinghouse for English Language Acquisition (NCELA), 54, 55
National Council of Teachers of Mathematics (NCTM), 7, 8, 10
  on calculators, 44, 209
  math content distribution through grades, 2–3
  principles of, 6–7
  on reasoning and proof, 94
  on shifts in classroom environment, 15
  standards of, 5, 63
  vision of, 19
  Web site, 63, 339
National Governors Association, 6, 63
National guidelines, 7–8
National Library of Virtual Manipulatives (NLVM), 42, 100, 112, 164, 180, 231, 235, 239, 264, 294, 317, 342, 377, 397, 410
National Research Council (NRC), 19, 34, 208, 222

National Science Foundation, 41
Nation's Report Card (Web site), 108
Navigational thinking, 338–339
NCELA (National Clearinghouse for English Language Acquisition), 54, 55
NCLB, *see* No Child Left Behind Act
NCTM Illuminations (Web site), 70, 96, 112, 144, 180, 183, 231, 309, 350, 355, 377, 397
NCTM Standards (Web site), 39, 148
Negative experiences, learning and, 15–16
Nested inclusion of numbers, 142
"New math," 3
Newspapers, numbers in, 171
No Child Left Behind Act (NCLB), 5, 8, 53, 63
Nominal numbers, 149
Nonagons, 337
Nonexamples, 28
Nonproportional models, 158–159, 168, 169
Nonroutine problems, 107, 108, 302, 314–316
Norm-referenced tests, 77
Numbers:
  abundant and deficient, 404
  approximate and exact, 171
  on calendars, 142
  cardinal, 133, 148, 149
  comparing, 132
  compatible, 212, 218, 219, 235–236
  composite, 407, 408
  conservation of, 137
  exact, 171
  Fibonacci sequence of, 305, 415, 416
  Hindu-Arabic numeration system, 157, 158
  in mathematics, 3, 4
  mixed, 259, 267, 269
  Negative integers, counting back and, 143, 144
  nested inclusion of, 142
  nominal, 149
  odd and even, 404–406
  ordinal, 148–149, 151
  perfect, 404
  polygonal, 412–413
  prenumber concepts, 133–136
  prime, 402–403, 407–410, 412
  properties of, 305, 317–318
  Pythagorean triples, 414–416
  reading and writing, 168–171
  relatively prime, 412
  Roman numerals, 157
  rounding, 171–172
  square, 412–413
  symbols for, 133
  teen, 160
  triangular, 413

  whole, 131–132, 270–271
Number and Operation Standard (NCTM), 44, 180, 228
Number and Operation standards (CCSSM), 208
Number benchmarks, 145–146
Number charts, 147
Number lines, 143–144
  counting intervals on, 135
  and fractions, 262
  and fractions between fractions, 267
  two-sided, 288
Number names, 133, 139–140
  for larger numbers, 169, 170
  learning, in early childhood, 160
  patterns in, 173–174
  reading, 171
Number puzzles, 302, 314, 315
Number sense, 130–139
  approximate and exact numbers, 171
  cardinal numbers, 148, 149
  characteristics of, 131
  comparisons, 138–139
  conservation stage of, 137
  cultural differences in, 150–151
  and division algorithms, 247
  and estimation, 216
  group recognition stage of, 137–138
  helping development of, 180–181
  and mental computation, 214
  nominal numbers, 149
  one-to-one correspondence, 138, 139
  ordering, 139–140
  ordinal numbers, 148–149
  place value, 160–161
  prenumber concepts, 133–136
  stages in development of, 131–133
  Web sites for learning, 164
  writing numerals, 149–150
Number theory, 401–417
  in ancient cultures, 416–417
  divisibility, 410–412
  factors and multiples, 405–408
  Fibonacci sequence, 415, 416
  modular arithmetic, 413–415
  odds and evens, 404–406
  Pascal's Triangle, 414, 415
  polygonal numbers, 412–413
  prime factorization, 408, 410
  primes and composites, 407–410
  Pythagorean triples, 414–416
  reasons to study, 402–404
  relatively prime pairs of numbers, 412
Numerals, writing, 149–150
Numeration systems:
  in ancient cultures, 174
  in different countries, 150–151, 173–174
  Hindu-Arabic, 157, 158
Numerology, 402

O

Objects:
   attributes of, *see* Attributes
   comparing to units of measurement, 357–362
   creating, based on measurements, 363
Observation, 68, 79
Obtuse triangles, 336
Octagons, 337
Odd numbers, 404, 405
One:
   adding, 192, 193
   dividing by, 201
   multiplying by, 199
   subtracting, 196
100-chart, 402
One less, concept of, 146–147
One more, concept of, 146–147
One-more-set strategy, 197–198
One-right-answer syndrome, 222
One-to-one correspondence, 138–142
Open-ended problems, 112–113
Open-ended tasks, 40, 41
Open-ended writing, 97
Open sentences, 309–312
Operations, 177–202. *See also* Basic facts
   addition, 181–183
   algorithms based on, 231
   compatible numbers for, 218, 219
   and counting, 180
   with decimals, 276–278
   developing meanings for, 181–185
   division, 183–185
   expectations related to, 179
   with fractions, 267–273
   instructional goal for, 180–181
   mathematical properties pertaining to, 185, 186
   multiplication, 183–185
   organizing basic facts for, 187
   state standards for, 228–230
   subtraction, 181–183
   use of language in learning about, 181–182
Ordering, 139–140
   comparisons in, 139
   decimals, 275–276
   fractions, 263–267
   ordinal numbers, 148–149
Order-irrelevance rule, 140
Ordinal numbers, 148–149, 151
Output adaptation, 57

P

Pairs, working in, 126
Parallel faces, 327
Parallelograms, 329, 333, 336, 337, 362
Parallel sides, 334, 336
Parent conferences, 81

Parents or guardians, communicating with:
   about assessment results, 81, 82
   about calculator use, 209
   about new methods, 41
Partial-difference subtraction algorithm, 239–240
Partial-products multiplication algorithm, 240–241
Partial-sum addition algorithm, 235–236
Participation adaptation, 57
Participation checklist, 79
Partitioning, 255, 259
Partition problems, 185
Part-to-part ratios, 286
Part-to-whole ratios, 286
Part-whole meaning of fractions, 256–258
Part-whole problems, in subtraction, 182–183
Pascal, Blaise, 417
Pascal's Triangle, 414, 415
Patterns:
   in algebraic thinking, 162, 302–305, 312–313
   with blocks, 25
   copying, 136
   in counting, 145
   creating, 136
   extending, 136, 313
   finding next one in, 136
   functions as, 316
   in geometry, 340–341
   in learning place value, 166–167
   math as study of, 3
   for mental computation, 213, 236
   as multiplication strategy, 199
   in naming numbers, 173–174
   in number development, 146, 148
   on 100-chart, 402
   as prenumber concept, 136
   problem solving by looking for, 119–120
   ratios from, 285–286
   and subitizing, 138
   Web sites for learning, 164
Pattern blocks, 136
Peer assessments, 72, 73
Pencil-and-paper tests, 77
Pentagons, 337
Pentomino, 113, 115
Percents, 291–296
   applying, 294–296
   understanding, 292–294
Percent bars, 293, 295, 296
Perceptual comparisons:
   of angles, 355
   of area, 354
   of length, 351
   of temperature, 355
   of volume, 355

   of weight/mass, 353, 354
Perfect numbers, 404
Perfect squares, 304–305
Performance indicators, 67
Performance tasks, assessment by, 70–72
Perimeter, 352, 367, 368
Perpendicular edges, 327
Perpendicular sides, 334
Personalities, student:
   assessment of, 66–67
   in assigning small groups, 39
Phone calls, to parents, 81
Physical attributes, 136
Physical developmental stages, 24
Piaget, Jean, 20–23, 28, 137, 297
Pictorial concrete models (fractions), 264, 265, 268
Picture graphs, 377, 380, 384
Pie graphs, 381–382
Place value, 154–174
   and algorithms, 231–232, 239, 240
   beginning teaching of, 160–164
   common errors related to, 160
   composing and decomposing, 167–169
   counting and patterns, 166–167
   and decimals, 275
   development of, 157–158
   extending, 165–168
   grouping or trading, 159–160
   in Hindu-Arabic numeration system, 157
   modeling, 158–159
   patterns in naming numbers, 173–174
   pregrouped materials, 158–160
   primary interview on, 71
   reading and writing numbers, 168–171
   rounding, 171–172
Place-value grid, 275, 276
Place-value mats, 161, 162, 165, 168–170, 235
Planning, 44–59. *See also* Lesson plans
   assessment and analysis in, 56, 58
   for daily lessons, 46
   importance of, 44–45
   lesson types, 46–52
   levels of, 45–46
   to meet needs of all students, 52–57
   multicultural literature in, 56, 58
   for units, 45–46
   for the year, 45
Poincaré, Henri, 177, 179
Polishing up ideas, 128
Polya, George, 117, 118
Polygons, 337–338
Polygonal numbers, 412–413
Polyhedra, 328
Positive learning environment, 14–15
Practice:
   mental computation, 215
   necessity of, 37
Precision Practice (CCSSI), 89
Preoperational thinking, 22

Prerecognition level (geometric thought), 325
Primary elementary students:
  appropriate problems for, 91
  developmental characteristics in, 23, 24
Prime factorization, 408, 410
Prime numbers, 407–410
  even numbers great than 2 as sum of, 402–403
  prime factorization, 408, 410
  relatively prime pairs of numbers, 412
Principles and Standards for School Mathematics (NCTM), 5, 7
  algebra instruction, 301
  concept development, 255
  content and processes indicated in, 45
  curriculum implementation, 41
  on estimation and computation, 244, 249
  on fractions and decimals, 255
  on geometry, 323, 324
  high expectations and student support in, 63
  kind of teaching envisioned by, 46, 48
  on measurement, 365
  on probability, 392
  and problem-centered programs, 92
  on problem solving, 106–107
  processes of doing math, 88, 89, 101
  on representation, 99
  on role of language, 181
  and state standards, 228
  on teacher's understanding of math, 34
  teaching objectives from, 37
  teaching recommendations in, 28
Prisms, 327–329
Probability, 373–375, 392–397
  conditional, 397
  of events, 392–395
  independence of events, 396–397
  in mathematics, 3, 4
  misconceptions about, 397
  outcomes, 392
  and randomness, 395–396
  sample space, 393, 394
Problem-based lessons, *see* Investigative lessons
Problem contexts, 181
Problem solving, 5, 105–128
  by acting it out, 118–119
  with calculators, 44, 116–117
  with computers, 117
  by constructing tables, 120, 121
  defined, 89
  exercises vs., 107–108, 127
  finding problems, 114–115
  focus on methods for, 107–109
  four-stage model of, 117–118

  by guess and check, 121, 122
  having students pose problems, 115–116
  helping students with, 125–127
  importance of looking back in, 124–125
  investigation, 111–112
  in Japan, 127–128
  by looking for patterns, 119–120
  by making drawings or diagrams, 119
  modeling, 15
  open-ended problems, 112–113
  as process of doing math, 88–92
  retention of skills in, 18
  rewarding students for, 15
  self-assessment of, 72–74
  signposts in teaching with, 109–110
  by solving similar but simpler problems, 122–124
  strategies for, 117–124
  success factors for, 110–111
  types of problems, 111–113
  by working backward, 122
Problem-solving software, 43–44
Problem Solving Standard (NCTM), 88
Procedural fluency, 14
Procedural knowledge, acquisition of, 18–19
Processes of doing math, 86–102
  CCSSM Practices for, 89
  communication, 95–97
  connections, 97–98
  and ESL/ENL learners, 101, 102
  NCTM Standards for, 88, 89
  problem solving, 88–92
  reasoning and proof, 92–95
  representations, 98–101
Process standards (NCTM), 88
Productive disposition, 14
Professional organizations, 10
Professional Standards for Teaching Mathematics (NCTM):
  on communication, 76
  on equal opportunities for every child, 52, 53
  higher-level question categories in, 38
  teaching recommendations in, 22
  on worthwhile tasks, 71
Professional Teaching Standards (NCTM), 56
Proficiency:
  computational, *see* Computational fluency (computational proficiency)
  mathematical, 6–7, 13–14, 18
Progressive movement, 4
Progress monitoring, assessment for, 65
Project IMPACT, 101, 102
Proof, *see* Reasoning and proof
Properties of numbers, algebraic thinking and, 305, 317–318

Properties of Numbers Using Calculators and Hundred Boards, 148
Proportions, 287–291
Proportional models, 158–159, 168, 169
Proportional reasoning, 287–289
Protractors, 360–361
Proximal development, zone of, 23
Puzzles, number, 403, 404
Pyramids, 329
Pythagoras, 416
Pythagorean theorem, 416
Pythagorean triples, 414–416

Q

Quadrilaterals, 335–337
Quantities, continuous, 140
Questions (questioning):
  assessment by, 68–69
  to facilitate learning, 38
  hidden, 122
  for interpreting data, 390
  in interviews, 69–70
  for statistical investigations, 375
Quilts, modular arithmetic and, 414, 415
Quotient meaning of fractions, 256–257

R

Radius, 329
Randomness, 395–396
Range, 389
Ratios, 284–287
Ratio meaning of fractions, 257
Rational counting, 140, 142
Ratio tables, 285, 286, 288–291
Reading numbers, 168–171
Real graphs, 377, 384
Reasoning and proof:
  generalizations, 94
  learning from mistakes, 95
  mathematical memory, 94–95
  in problem solving, 112
  as process of doing math, 88, 92–95
  proportional, 287–289
  spatial, 342, 343
Reasoning and Proof Standard (NCTM), 88, 92
Reasoning Practice (CCSSI), 89, 92
Recordkeeping:
  of assessments, 79–81
  of problem solving techniques, 126
Rectangles, 184, 329, 336, 337
  formula for area of, 361–362
Recursive expressions, 312
Referent, comparing to, 365
Reflection, 26, 95, 340, 341
Reflectional symmetry, 332, 333
Reflections (Web site), 69
Reform texts, 41
Reformulations, 121
Region model (fractions), 257, 261
Regrouping, 238

and algorithms, 232, 233
in learning place value, 167–169
to subtract fractions, 269
Reinforcement, shaping behavior through, 20
Relationship(s), 282–297. *See also* Decimals; Fractions
for algebraic thinking, 305, 318–319
in different countries, 296–297
math as study of, 3
mathematical memory built on, 94–95
mental computation, 213
percents, 291–296
proportions, 287–291
ratios, 284–287
Relatively prime pairs of numbers, 412
Remainders, division with, 248–249
Repeated addition, 197, 198
Repeated subtraction, 185, 201
Repeating patterns, 303–304
Reporting measurements, 362–363
Representation(s), 22
in algebraic thinking, 310–313
creating and using, 99
formal, 28
of locations, 338–340
to model and interpret phenomena, 100–101
in multiplication and division, 183
of numbers, 146–148
of place value, 165
in problem solving, 111
as process of doing math, 88, 98–101
of quantities, 160–161
selecting, applying, and translating among, 99, 100
Representations Standard (NCTM), 88, 98
Resources, 7–10
cultural and international, 9
guidelines, 7–8
math problems, 113–115
other teachers, 10
professional development, 10
professional organizations, 10
Response to Intervention (RTI), 53, 65
Retention:
improving, 17–18
of isolated learnings, 21
Reversibility of thinking, 317
Review-teach-practice lessons, 46
Rhombus, 336, 337
Right prisms, 327
Right triangles, 336
Roman numerals, 157
Rotation, 340, 341
Rote learning, 19
Rounding:
decimals, 276
in estimation, 219, 220
flexible, 219, 220

in learning place value, 171–172
similarity of measuring and, 349
Routine problems, 107, 108, 302, 310–312
RTI (Response to Intervention), 53, 65
Rubrics, 66
Rulers, 358–359

S

Same number puzzle, 302
Sample space, 393, 394
Sampling, 376–377
Scales, balancing, 305
Scaled instruments, 359–360
Scalene triangles, 336
Scaling problems, 289
Scope-and-sequence charts, 35, 45
Scoring, grading vs., 66
Scoring guides, 66–68
SCR (short-constructed-response) items, 108
Self-assessments, 71–74
Self-reflective diagnosis, 16
Self-regulation adaptations, 56
Semiconcrete learning (operations), 181
Sense making, 25, 26
communication to encourage, 28
as goal of instruction, 109
Separation problems, in subtraction, 182
Sets, bimodal and trimodal, 385
Set model (fractions), 258, 262
Set operations, 134, 135
Shapes:
connecting attributes for, 367, 368
properties of, 324, 325
three-dimensional, 325–331
two-dimensional, 329, 331–338
Sharing information, problem solving and, 110
Sharing model of mean, 386
Sharing problems (division), 185
Sharing rods, 260
Short-constructed-response (SCR) items, 108
Sides (two-dimensional shapes), 331–332
Sidewalk garden problem, 302, 304, 312
Sieve of Eratosthenes, 408, 410
Similarity, 341–342
Similar rectangles, 291
Simulations, 376–377
Simulation software, 43
Singapore, 279, 397
Size adaptation, 57
Sketch graphs, 378
Skills:
behaviorism in acquisition of, 20
number-related, 132
Skinner, B. F., 20
Skip counting, 144, 145
in division, 201
for multiplication, 197

and place value, 168
by tens and hundreds, 166
Slide transformation, 340, 341
Small groups, 39, 69, 125–126
Social development, 23
Social developmental stages, 24
Social utility, 4
Society, math principle and needs of, 4–5
Sociocultural factors, problem-solving skills and, 111
Socioeconomic status (SES), 29, 54, 82
Software:
to build computational fluency, 249
data analysis, 99
descriptive statistics, 385
educational, 43–44
for factor trees and Venn diagrams, 410
for graphing, 378
for learning basic facts, 189
for location and movement activities, 339, 340
for problem-solving, 117
shape-making, 343
types of, 43–44
Solid shapes, 325
cross sections, 344
Solution methods, encouraging variety of, 128
Solution process, looking back at, 124–125
Solving similar but simpler problems strategy, 122–124
Soroban, 145
Sorting, 325–326. *See also* Classification
Spanish, number names in, 150, 173
Spatial reasoning, in geometry, 342, 343
Spatial relations, coordinate system for, 338–340
Special needs students:
adapting instruction for, 55–56
manipulatives for, 42
metacognition of, 26
problem solving help for, 125
Speed tests, understanding vs., 16
Spiral approach to teaching, 35–36
Splitting the product strategy, 197–198
Spreadsheets, 121
descriptive statistics on, 385
for graphing, 378
Sputnik, 3
Squares, 329, 336, 337
Square numbers, 412–413
Stable-order rule, 140, 141
Standards, teacher's responsibility in meeting, 37
Standard addition algorithm, 233, 235
Standards-based curricula, 41
Standards-based texts, 41
Standards for mathematical practice (CCSSI), 89
Standardized achievement tests, 77–79

Index

Standard subtraction algorithm, 237–239
Standard units of measurement, 357–359
Star patterns, 412
State guidelines, 8
State standards, 228–230
Statistics, 373–375
Statistical investigation:
  collecting data, 375–377
  communicating results, 392
  cultural diversity in teaching, 397
  data analysis, 377–392
  descriptive statistics, 383–389
  formulating questions, 375
  interpreting results, 389–392
  probability, 392–397
  randomness, 395–396
  steps in, 375–377
Stem-and-leaf plots, 379–380, 384
Stick models (three-dimensional shapes), 329–331
Strategic competence, 14
Strategies:
  for comparing fractions, 266
  compensatory, 126
  for counting, 142–145
  for estimating measurements, 365, 366
  for estimation, 217–222
  for mental computation, 211–213
  patterns in, 136
  for problem solving, 117–124
  for remembering basic facts, 189–191
  for thinking about basic facts, 191–201
Structure Practice (CCSSI), 89
Students. *See also* Special needs students
  abilities of, 39, 66–68
  active involvement of, 25–26
  assessing basic facts known by, 186, 187
  communicating assessment results to, 81
  curriculum and needs of, 4
  developmental characteristics of, 23–25, 34–35
  dispositions of, 66–68
  grouping, 39, 125–126
  interests of, 66–68
  intermediate elementary, 23–24
  kindergarten, 179
  managing needs of, 126–127
  minority, 17
  personalities of, 39, 66–67
  primary elementary, 23, 24, 91
  prior knowledge of, 35–36
  upper elementary, 91, 92
Student files, 80, 81
Subitizing, 137–138
Substitute curriculum, 57
Subtraction, 181–183
  and addition, 195, 196
  basic facts for, 186
  counting back as model of, 142–144
  with decimals, 277
  with fractions, 268–269
  models for, 182, 183
  and relationship among operations, 181, 196
  repeated, 201
  thinking strategies for, 195–196
Subtraction algorithms, 237–240
  partial-difference, 239–240
  standard, 237–239
Subtractive division algorithm, 245–247
Succession, comparisons and, 139
Summarize phase:
  direct instruction lessons, 50
  explorations, 50
  investigative lessons, 48
Support, adaptation of, 57
Surface area, connecting attributes for, 368, 369
Symbols, 27, 28
  of algebra, 307–310
  conceptual ideas represented by, 97–98
  for fractions, 260–261
  meaningfulness of, 168
  for numbers, introducing, 133
  for ratios, 285–286
  relating operations to, 181
  telling students about, 110
  understanding meaning of, 187
Symbolic learning, 22
Symbolic representation:
  of fractions, 265–266
  of place value, 161
Symbolization, 22
Symmetry, 332–334

T

Tables:
  function, 306, 308
  problem solving by constructing, 120, 121
  ratio, 285, 286, 288, 291
Taiwan, computational proficiency in, 222, 223
Talking:
  effective, encouraging, 37–38
  to encourage sense making, 28
  to relate symbols to operations, 181
Tangram puzzle, 342
Tasks:
  appropriateness of, 36–37
  open-ended, 40, 41
  performance, 70–72
Teachers:
  personal understanding of math, 34
  roles of, 33
Teacher-designed written tests, assessment by, 76–78
Teaching, 34–44
  appropriate tasks for students, 36–37
  developmental characteristics of students, 23–25, 34–35
  encouraging communication by students, 37–39
  grouping students, 39
  and learning, 22
  personal understanding of math, 34
  questions to facilitate learning, 38
  selecting materials, 39–44
  students' prior knowledge, 35–36
Teaching and Learning Mathematics (Lambdin), 8
Teaching Children Mathematics (journal), 10
Teaching Principle (NCTM), 7
"Teaching to the test," 5
Technology:
  equal access to, 44
  for lesson enhancement, 42–44
  and needs of the subject, 3–4
  for problem solving, 91
  for translating among representations, 99
Technology Principle (NCTM), 7, 42
Teen numbers, 160
Temperature:
  measuring, 355
  units for, 359
Ten:
  base of, 157
  importance of, 148
Ten-frame, 145–146, 148, 161
Tenths, 273, 274
Tests and testing. *See also* Assessment(s)
  achievement tests, 18, 65, 66
  assessment by, 76–79
  high-stakes assessments, 63
  speed tests, 16
Textbooks, 9
  addition and multiplication facts in, 189
  lessons from, 40–41
  practice material in, 37
Think addition (subtraction strategy), 195
Thinking. *See also* Mental computation
  about size of numbers, 170
  computers and skill development in, 43
  looking back at, 125
  math as way of, 3
  patterns in, 136
  Piaget's levels of, 22
  reversibility of, 317
  in use of calculators, 210
  written communication conveying, 29
Thinking strategies:
  for addition, 191–195, 201–202
  for basic facts, 188, 191–201
  for division, 199–201
  efficiency of, 188
  for multiplication, 196–199
  for subtraction, 195–196
Think multiplication (division strategy), 199, 201

Think-pair-share, 69
Thorndike, Edward L., 20
Thousand chart, 167, 168
Thousandths, 274
Three-dimensional shapes, 325–331
Three-fourths, students' understanding of, 259
Time:
   equivalencies, 364
   measuring, 355
   units for, 359
Time adaptation, 57
Time management, 125
Tinkerplots™, 99, 378
Tool, math as, 3
Tool software, 44
Tracing numerals, 149–150
Trading:
   and algorithms, 231–232
   in learning place value, 157–160, 167, 168
Transformations (geometry), 340–342
Translation, 340, 341
Trapezoids, 336, 362
   algebraic topics in, 320
   data analysis and probability in, 397
   functions in, 317
   geometry in, 344
   on lesson formats, 46
   measurement in, 348, 363
   problem solving in, 127
Triangles, 329, 331, 336, 337, 362
Triangular numbers, 413
Trimodal sets, 385
Turn transformation, 340, 341
Tutorial software, 43
Twice-as-much strategy, 198
Two-dimensional shapes, 329, 331–338
Two-sided number lines, 288–290

U

Understanding:
   communication to encourage, 28–29
   and learning, 4, 19–20
   memorization vs., 15
   speed tests and drills vs., 16

Unfairness, 396
Ungrouped materials, 58, 160
Union, 134, 135
Units, planning, 45–46
United Kingdom, number symbols/names in, 174
U.S. Census Bureau, 376, 377
Unitizing, 366
Units of measurement, 355–357
   choosing, 355–357
   comparing objects to, 357–362
   conversion of, 364–365
   counting, 357–358
   equivalences of, 363–365
   finding number of, 357–362
Upper elementary students, appropriate problems for, 91, 92
Using results phase (assessment), 64–65

V

Variables (in algebraic thinking), 309
Variation, measures of, 389, 390
Venn diagrams, 410
Verbal communication, *see* Talking
Verbal interaction (English-language learners), 55
Vertices:
   solids, 326
   two-dimensional shapes, 329
Viable arguments, 402
Virtual manipulatives, 44, 100, 112. *See also* Electronic manipulatives
Visual cues, 324
Visualization, 96. *See also* Manipulatives
   in geometry, 342–344
   of multiplication, 197
Visual level (geometric thought), 325
Visual processing adaptations, 56
Vocabulary:
   for counting, naming, and representing numbers, 160
   in developing place value, 165
   for fractions, 259
   geometric, 344
   mathematical, 38
   for two-dimensional shapes, 329–330

Volume, 355, 368, 369
   equivalencies, 364, 365
   units for, 359
Vygotsky, Lev, 23, 28

W

Web sites. *See also specific Web sites*
   for data collection, 377
   for probability simulations, 397
Weight, 353, 354, 359
Who Am I? (game), 133–134
Whole class:
   discussion with, 128
   instruction for, 39
Whole numbers, 131–132, 270–271
Working backward strategy, 122
Working-from-known-facts strategy, 198
Working graphs, 378
Work samples, assessment by, 74–75
Worksheets, computational-skill, 249
Writings:
   assessment by, 74, 76, 77
   open-ended, 97
Writing numbers, 168–171
Writing numerals, 149–150
Writing prompts, 77
Written communication:
   to convey thinking, 29
   to relate symbols to operations, 181
   of results, 392
   students' writing of problems, 116
Written computation, 208, 214, 222
Written tests, teacher-designed, 76–78

Z

Zero:
   adding, 192–193
   dividing by, 201
   as even number, 405
   in Hindu-Arabic numeration system, 157
   in Japanese counting, 151
   models showing, 147
   in multiplication, 199, 242
   subtracting, 196, 238
Zone of proximal development, 23